T0213436

PROBLEMS
AND PROOFS IN
REAL ANALYSIS
Theory of Measure and Integration

PROBLEMS
AND PROOFS IN
REAL ANALYSIS
Theory of Measure and Integration

J Yeh

University of California, Irvine, USA

 World Scientific

NEW JERSEY • LONDON • SINGAPORE • BEIJING • SHANGHAI • HONG KONG • TAIPEI • CHENNAI

Published by

World Scientific Publishing Co. Pte. Ltd.

5 Toh Tuck Link, Singapore 596224

USA office: 27 Warren Street, Suite 401-402, Hackensack, NJ 07601

UK office: 57 Shelton Street, Covent Garden, London WC2H 9HE

Library of Congress Cataloging-in-Publication Data
Yeh, J.
 Problems and proofs in real analysis : theory of measure and integration / by J. Yeh, University of California, Irvine, USA.
  pages cm
 Companion volume to: Real analysis : theory of measure and integration (3rd ed.).
 Intended as a self-study volume.
 ISBN 978-981-4578-50-9 (softcover : alk. paper)
 1. Mathematical analysis--Study and teaching. 2. Measure theory--Study and teaching.
I. Yeh, J. Real analysis. II. Title.
 QA312.Y445 2014
 515'.8--dc23

                                                         2013041974

British Library Cataloguing-in-Publication Data
A catalogue record for this book is available from the British Library.

Printed in Singapore by World Scientific Printers.

To my wife
Betty

Contents

Preface

This volume consists of proofs of the problems in the monograph Real Analysis: Theory of Measure and Integration, 3rd Edition. Alternate proofs are included when appropriate to show different approaches to the problem or different techniques in constructing a proof. We hope that this volume will be helpful to those who read Real Analysis in self-study and have no easy access to help.

J. Yeh

Corona del Mar, California
September, 2013

§1 Measure on a σ-algebra of Sets

Prob. 1.1. Given two sequences of subsets $(E_n : n \in \mathbb{N})$ and $(F_n : n \in \mathbb{N})$ of a set X.
(a) Show that

(1) $$\liminf_{n\to\infty} E_n \cup \liminf_{n\to\infty} F_n \subset \liminf_{n\to\infty}(E_n \cup F_n) \subset \liminf_{n\to\infty} E_n \cup \limsup_{n\to\infty} F_n$$
$$\subset \limsup_{n\to\infty}(E_n \cup F_n) \subset \limsup_{n\to\infty} E_n \cup \limsup_{n\to\infty} F_n.$$

(b) Show that

(2) $$\liminf_{n\to\infty} E_n \cap \liminf_{n\to\infty} F_n \subset \liminf_{n\to\infty}(E_n \cap F_n) \subset \liminf_{n\to\infty} E_n \cap \limsup_{n\to\infty} F_n$$
$$\subset \limsup_{n\to\infty}(E_n \cap F_n) \subset \limsup_{n\to\infty} E_n \cap \limsup_{n\to\infty} F_n.$$

(c) Show that if $\lim_{n\to\infty} E_n$ and $\lim_{n\to\infty} F_n$ exist, then $\lim_{n\to\infty} (E_n \cup F_n)$ and $\lim_{n\to\infty} (E_n \cap F_n)$ exist and moreover

(3) $$\lim_{n\to\infty} (E_n \cup F_n) = \lim_{n\to\infty} E_n \cup \lim_{n\to\infty} F_n,$$

(4) $$\lim_{n\to\infty} (E_n \cap F_n) = \lim_{n\to\infty} E_n \cap \lim_{n\to\infty} F_n.$$

Proof. Let $(A_n : n \in \mathbb{N})$ be a sequence of subsets of X. According to Lemma 1.7, $\liminf_{n\to\infty} A_n$ consists of every $x \in X$ such that $x \in A_n$ for all but finitely many $n \in \mathbb{N}$ and $\limsup_{n\to\infty} A_n$ consists of every $x \in X$ such that $x \in A_n$ for infinitely many $n \in \mathbb{N}$. This is the basis for the proof of the chain of inclusions.

1. Let us prove (1) by verifying the set inclusions one after another.

1.1. Let us prove $\liminf_{n\to\infty} E_n \cup \liminf_{n\to\infty} F_n \subset \liminf_{n\to\infty}(E_n \cup F_n)$.
Let $x \in \liminf_{n\to\infty} E_n \cup \liminf_{n\to\infty} F_n$. Then we have $x \in \liminf_{n\to\infty} E_n$ or $x \in \liminf_{n\to\infty} F_n$. If $x \in \liminf_{n\to\infty} E_n$ then $x \in E_n$ for all but finitely many $n \in \mathbb{N}$ and then $x \in E_n \cup F_n$ for all but finitely many $n \in \mathbb{N}$ and therefore $x \in \liminf_{n\to\infty}(E_n \cup F_n)$. Similarly if $x \in \liminf_{n\to\infty} F_n$ then $x \in \liminf_{n\to\infty}(E_n \cup F_n)$. This proves $\liminf_{n\to\infty} E_n \cup \liminf_{n\to\infty} F_n \subset \liminf_{n\to\infty}(E_n \cup F_n)$.

1.2. Let us prove $\liminf_{n\to\infty}(E_n \cup F_n) \subset \liminf_{n\to\infty} E_n \cup \limsup_{n\to\infty} F_n$.
Let $x \in \liminf_{n\to\infty}(E_n \cup F_n)$. Then $x \in E_n \cup F_n$ for all but finitely many $n \in \mathbb{N}$. Suppose $x \in F_n$ for infinitely many $n \in \mathbb{N}$. Then $x \in \limsup_{n\to\infty} F_n$. On the other hand if $x \in F_n$ for only finitely many $n \in \mathbb{N}$, then since $x \in E_n \cup F_n$ for all but finitely many $n \in \mathbb{N}$, we must have $x \in E_n$ for all but finitely many $n \in \mathbb{N}$ and then $x \in \liminf_{n\to\infty} E_n$. This shows that if $x \in \liminf_{n\to\infty}(E_n \cup F_n)$ then $x \in \limsup_{n\to\infty} F_n$ or $x \in \liminf_{n\to\infty} E_n$ so that $x \in \liminf_{n\to\infty} E_n \cup \limsup_{n\to\infty} F_n$. This proves $\liminf_{n\to\infty}(E_n \cup F_n) \subset \liminf_{n\to\infty} E_n \cup \limsup_{n\to\infty} F_n$.

1.3. Let us prove $\liminf_{n\to\infty} E_n \cup \limsup_{n\to\infty} F_n \subset \limsup_{n\to\infty}(E_n \cup F_n)$.
Since $\liminf_{n\to\infty} E_n \subset \limsup_{n\to\infty} E_n$, we have $\liminf_{n\to\infty} E_n \cup \limsup_{n\to\infty} F_n \subset \limsup_{n\to\infty} E_n \cup \limsup_{n\to\infty} F_n$.

Since $E_n \subset E_n \cup F_n$ for every $n \in \mathbb{N}$, we have $\limsup\limits_{n\to\infty} E_n \subset \limsup\limits_{n\to\infty}(E_n \cup F_n)$. Similarly $\limsup\limits_{n\to\infty} F_n \subset \limsup\limits_{n\to\infty}(E_n \cup F_n)$. Therefore $\limsup\limits_{n\to\infty} E_n \cup \limsup\limits_{n\to\infty} F_n \subset \limsup\limits_{n\to\infty}(E_n \cup F_n)$. This proves $\liminf\limits_{n\to\infty} E_n \cup \limsup\limits_{n\to\infty} F_n \subset \limsup\limits_{n\to\infty}(E_n \cup F_n)$.

1.4. Let us prove $\limsup\limits_{n\to\infty}(E_n \cup F_n) \subset \limsup\limits_{n\to\infty} E_n \cup \limsup\limits_{n\to\infty} F_n$.
Let $x \in \limsup\limits_{n\to\infty}(E_n \cup F_n)$. Then $x \in E_n \cup F_n$ for infinitely many $n \in \mathbb{N}$. If $x \in E_n$ for infinitely many $n \in \mathbb{N}$ then $x \in \limsup\limits_{n\to\infty} E_n$. On the other hand if $x \in E_n$ for only finitely many $n \in \mathbb{N}$, then since $x \in E_n \cup F_n$ for infinitely many $n \in \mathbb{N}$, we must have $x \in F_n$ for infinitely many $n \in \mathbb{N}$ and thus $x \in \limsup\limits_{n\to\infty} F_n$. This shows that if $x \in \limsup\limits_{n\to\infty}(E_n \cup F_n)$, then $x \in \limsup\limits_{n\to\infty} E_n$ or $x \in \limsup\limits_{n\to\infty} F_n$ and therefore $x \in \limsup\limits_{n\to\infty} E_n \cup \limsup\limits_{n\to\infty} F_n$. This proves $\limsup\limits_{n\to\infty}(E_n \cup F_n) \subset \limsup\limits_{n\to\infty} E_n \cup \limsup\limits_{n\to\infty} F_n$.

2. Let us prove (2) by verifying the set inclusions one after another.

2.1. Let us prove $\liminf\limits_{n\to\infty} E_n \cap \liminf\limits_{n\to\infty} F_n \subset \liminf\limits_{n\to\infty}(E_n \cap F_n)$.
Let $x \in \liminf\limits_{n\to\infty} E_n \cap \liminf\limits_{n\to\infty} F_n$. Then $x \in \liminf\limits_{n\to\infty} E_n$ and this implies that there exists $N_1 \in \mathbb{N}$ such that $x \in E_n$ for $n \geq N_1$. Similarly $x \in \liminf\limits_{n\to\infty} F_n$ implies that there exists $N_2 \in \mathbb{N}$ such that $x \in F_n$ for $n \geq N_2$. Let $N = \max\{N_1, N_2\}$. Then $x \in E_n \cap F_n$ for $n \geq N$ and this implies that $x \in \liminf\limits_{n\to\infty}(E_n \cap F_n)$. This proves $\liminf\limits_{n\to\infty} E_n \cap \liminf\limits_{n\to\infty} F_n \subset \liminf\limits_{n\to\infty}(E_n \cap F_n)$.

2.2. Let us prove $\liminf\limits_{n\to\infty}(E_n \cap F_n) \subset \liminf\limits_{n\to\infty} E_n \cap \limsup\limits_{n\to\infty} F_n$.
Let $x \in \liminf\limits_{n\to\infty}(E_n \cap F_n)$. Then $x \in E_n \cap F_n$ for all but finitely many $n \in \mathbb{N}$. Then $x \in E_n$ for all but finitely many $n \in \mathbb{N}$ so that $x \in \liminf\limits_{n\to\infty} E_n$ and similarly $x \in \liminf\limits_{n\to\infty} F_n$. Thus $x \in \liminf\limits_{n\to\infty} E_n \cap \liminf\limits_{n\to\infty} F_n \subset \liminf\limits_{n\to\infty} E_n \cap \limsup\limits_{n\to\infty} F_n$. This proves $\liminf\limits_{n\to\infty}(E_n \cap F_n) \subset \liminf\limits_{n\to\infty} E_n \cap \limsup\limits_{n\to\infty} F_n$.

2.3. Let us prove $\liminf\limits_{n\to\infty} E_n \cap \limsup\limits_{n\to\infty} F_n \subset \limsup\limits_{n\to\infty}(E_n \cap F_n)$.
Let $x \in \liminf\limits_{n\to\infty} E_n \cap \limsup\limits_{n\to\infty} F_n$. Then $x \in \liminf\limits_{n\to\infty} E_n$ so that $x \in E_n$ for all but finitely many $n \in \mathbb{N}$ and $x \in \limsup\limits_{n\to\infty} F_n$ for infinitely many $n \in \mathbb{N}$. This implies that $x \in E_n \cap F_n$ for infinitely many $n \in \mathbb{N}$ and thus $x \in \limsup\limits_{n\to\infty}(E_n \cap F_n)$. This proves $\liminf\limits_{n\to\infty} E_n \cap \limsup\limits_{n\to\infty} F_n \subset \limsup\limits_{n\to\infty}(E_n \cap F_n)$.

2.4. Let us prove $\limsup\limits_{n\to\infty}(E_n \cap F_n) \subset \limsup\limits_{n\to\infty} E_n \cap \limsup\limits_{n\to\infty} F_n$.
Let $x \in \limsup\limits_{n\to\infty}(E_n \cap F_n)$. Then $x \in E_n \cap F_n$ for infinitely many $n \in \mathbb{N}$. This implies that $x \in E_n$ for infinitely many $n \in \mathbb{N}$ so that $x \in \limsup\limits_{n\to\infty} E_n$ and similarly $x \in F_n$ for infinitely many $n \in \mathbb{N}$ so that $x \in \limsup\limits_{n\to\infty} F_n$. Thus we have $x \in \limsup\limits_{n\to\infty} E_n \cap \limsup\limits_{n\to\infty} F_n$. This proves $\limsup\limits_{n\to\infty}(E_n \cap F_n) \subset \limsup\limits_{n\to\infty} E_n \cap \limsup\limits_{n\to\infty} F_n$.

3. Let us prove (c). Let us assume that $\lim\limits_{n\to\infty} E_n$ and $\lim\limits_{n\to\infty} F_n$ exist. Then we have

(5)
$$\begin{cases} \lim\limits_{n\to\infty} E_n = \liminf\limits_{n\to\infty} E_n = \limsup\limits_{n\to\infty} E_n, \\ \lim\limits_{n\to\infty} F_n = \liminf\limits_{n\to\infty} F_n = \limsup\limits_{n\to\infty} F_n. \end{cases}$$

Substituting (5) into (1), we have

$$\lim\limits_{n\to\infty} E_n \cup \lim\limits_{n\to\infty} F_n \subset \liminf\limits_{n\to\infty}(E_n \cup F_n)$$

$$\subset \limsup\limits_{n\to\infty}(E_n \cup F_n) \subset \lim\limits_{n\to\infty} E_n \cup \lim\limits_{n\to\infty} F_n,$$

which implies

$$\liminf\limits_{n\to\infty}(E_n \cup F_n) = \limsup\limits_{n\to\infty}(E_n \cup F_n) = \lim\limits_{n\to\infty} E_n \cup \lim\limits_{n\to\infty} F_n.$$

This shows that $\lim\limits_{n\to\infty}(E_n \cup F_n)$ exists and $\lim\limits_{n\to\infty}(E_n \cup F_n) = \lim\limits_{n\to\infty} E_n \cup \lim\limits_{n\to\infty} F_n$.

Similarly by substituting (5) into (2) we show that $\lim\limits_{n\to\infty}(E_n \cap F_n)$ exists and moreover $\lim\limits_{n\to\infty}(E_n \cap F_n) = \lim\limits_{n\to\infty} E_n \cap \lim\limits_{n\to\infty} F_n.$ ∎

Prob. 1.2. (a) Let $(A_n : n \in \mathbb{N})$ be a sequence of subsets of a set X. Let $(B_n : n \in \mathbb{N})$ be a sequence obtained by dropping finitely many entries in the sequence $(A_n : n \in \mathbb{N})$. Show that $\liminf\limits_{n\to\infty} B_n = \liminf\limits_{n\to\infty} A_n$ and $\limsup\limits_{n\to\infty} B_n = \limsup\limits_{n\to\infty} A_n$. Show that $\lim\limits_{n\to\infty} B_n$ exists if and only if $\lim\limits_{n\to\infty} A_n$ exists and when they exist they are equal.

(b) Let $(A_n : n \in \mathbb{N})$ and $(B_n : n \in \mathbb{N})$ be two sequences of subsets of a set X such that $A_n = B_n$ for all but finitely many $n \in \mathbb{N}$. Show that $\liminf\limits_{n\to\infty} B_n = \liminf\limits_{n\to\infty} A_n$ and $\limsup\limits_{n\to\infty} B_n = \limsup\limits_{n\to\infty} A_n$. Show that $\lim\limits_{n\to\infty} B_n$ exists if and only if $\lim\limits_{n\to\infty} A_n$ exists and when they exist they are equal.

Proof. 1. Let us prove (a). If $x \in \liminf\limits_{n\to\infty} B_n$ then $x \in B_n$ for all but finitely many $n \in \mathbb{N}$ and hence $x \in A_n$ for all but finitely many $n \in \mathbb{N}$ and then $\liminf\limits_{n\to\infty} A_n$. This shows that $\liminf\limits_{n\to\infty} B_n \subset \liminf\limits_{n\to\infty} A_n$. By the same argument we show that $\liminf\limits_{n\to\infty} A_n \subset \liminf\limits_{n\to\infty} B_n$ and therefore we have $\liminf\limits_{n\to\infty} B_n = \liminf\limits_{n\to\infty} A_n$.

We show by the same argument as above that $\limsup\limits_{n\to\infty} B_n = \limsup\limits_{n\to\infty} A_n$.

Now $\lim\limits_{n\to\infty} B_n$ exists if and only if $\liminf\limits_{n\to\infty} B_n = \limsup\limits_{n\to\infty} B_n$ and when the equality holds then $\lim\limits_{n\to\infty} B_n = \liminf\limits_{n\to\infty} B_n = \limsup\limits_{n\to\infty} B_n$. But as we showed above, we have $\liminf\limits_{n\to\infty} B_n = \liminf\limits_{n\to\infty} A_n$ and $\limsup\limits_{n\to\infty} B_n = \limsup\limits_{n\to\infty} A_n$. Therefore $\lim\limits_{n\to\infty} B_n$ exists if and only if $\lim\limits_{n\to\infty} A_n$ exists and when they exist they are equal.

2. Let us prove (b). Now if $A_n = B_n$ for all but finitely many $n \in \mathbb{N}$ then there exists $N \in \mathbb{N}$ such that $A_n = B_n$ for $n \geq N$. Let $(A'_n : n \in \mathbb{N})$ be the sequence obtained by dropping the initial N entries in $(A_n : n \in \mathbb{N})$ and similarly let $(B'_n : n \in \mathbb{N})$ be the sequence obtained by dropping the initial N entries in $(B_n : n \in \mathbb{N})$. We have $A'_n = B'_n$ for every $n \in \mathbb{N}$. Then by (a) we have

$$\liminf_{n\to\infty} A_n = \liminf_{n\to\infty} A'_n = \liminf_{n\to\infty} B'_n = \liminf_{n\to\infty} B_n,$$

$$\limsup_{n\to\infty} A_n = \limsup_{n\to\infty} A'_n = \limsup_{n\to\infty} B'_n = \limsup_{n\to\infty} B_n.$$

Then by the same argument as in (a), $\lim\limits_{n\to\infty} B_n$ exists if and only if $\lim\limits_{n\to\infty} A_n$ exists and when they exist they are equal. ∎

Prob. 1.3. Let $(E_n : n \in \mathbb{N})$ be a disjoint sequence of subsets of a set X. Show that $\lim_{n \to \infty} E_n$ exists and $\lim_{n \to \infty} E_n = \emptyset$.

Proof. If $(E_n : n \in \mathbb{N})$ is a disjoint sequence then we have

$$\liminf_{n \to \infty} E_n = \left\{ x \in X : x \in E_n \quad \text{for all but finitely many } n \in \mathbb{N} \right\} = \emptyset,$$

$$\limsup_{n \to \infty} E_n = \left\{ x \in X : x \in E_n \quad \text{for infinitely many } n \in \mathbb{N} \right\} = \emptyset.$$

Thus we have $\liminf_{n \to \infty} E_n = \limsup_{n \to \infty} E_n = \emptyset$ and this implies that $\lim_{n \to \infty} E_n$ exists and $\lim_{n \to \infty} E_n = \emptyset$. ∎

Prob. 1.4. Let $a \in \mathbb{R}$ and let $(x_n : n \in \mathbb{N})$ be a sequence of points in $\mathbb{R}$, all distinct from a, such that $\lim_{n \to \infty} x_n = a$. Show that $\lim_{n \to \infty} \{x_n\}$ exists and $\lim_{n \to \infty} \{x_n\} = \emptyset$ and thus $\lim_{n \to \infty} \{x_n\} \neq \{a\}$.

Proof. Let $E_n = \{x_n\}$ for $n \in \mathbb{N}$ and consider the sequence $(E_n : n \in \mathbb{N})$ of sets in $\mathbb{R}$. Since $\lim_{n \to \infty} x_n = a$, we have

$$\liminf_{n \to \infty} E_n = \left\{ x \in \mathbb{R} : x \in E_n \quad \text{for all but finitely many } n \in \mathbb{N} \right\}$$

$$= \left\{ x \in \mathbb{R} : x = x_n \quad \text{for all but finitely many } n \in \mathbb{N} \right\}$$

$$= \emptyset,$$

and

$$\limsup_{n \to \infty} E_n = \left\{ x \in \mathbb{R} : x \in E_n \quad \text{for infinitely many } n \in \mathbb{N} \right\}$$

$$= \left\{ x \in \mathbb{R} : x = x_n \quad \text{for infinitely many } n \in \mathbb{N} \right\}$$

$$= \emptyset.$$

Thus we have $\liminf_{n \to \infty} E_n = \limsup_{n \to \infty} E_n = \emptyset$. This implies that $\lim_{n \to \infty} E_n$ exists and $\lim_{n \to \infty} E_n = \emptyset$. This proves that $\lim_{n \to \infty} \{x_n\}$ exists and $\lim_{n \to \infty} \{x_n\} = \emptyset$ and in particular $\lim_{n \to \infty} \{x_n\} \neq \{a\}$. ∎

Prob. 1.5. For $E \subset \mathbb{R}$ and $t \in \mathbb{R}$, let us write $E + t = \{x + t \in \mathbb{R} : x \in E\}$ and call it the translate of E by t. Let $(t_n : n \in \mathbb{N})$ be a strictly decreasing sequence in $\mathbb{R}$ such that $\lim_{n\to\infty} t_n = 0$ and let $E_n = E + t_n$ for $n \in \mathbb{N}$. Let us investigate the existence of $\lim_{n\to\infty} E_n$.

(a) Let $E = (-\infty, 0)$. Show that $\lim_{n\to\infty} E_n = (-\infty, 0]$.

(b) Let $E = \{a\}$ where $a \in \mathbb{R}$. Show that $\lim_{n\to\infty} E_n = \emptyset$.

(c) Let $E = [a, b]$ where $a, b \in \mathbb{R}$ and $a < b$. Show that $\lim_{n\to\infty} E_n = (a, b]$.

(d) Let $E = (a, b)$ where $a, b \in \mathbb{R}$ and $a < b$. Show that $\lim_{n\to\infty} E_n = (a, b]$.

(e) Let $E = Q$, the set of all rational numbers. Assume that $(t_n : n \in \mathbb{N})$ satisfies the additional condition that t_n is a rational number for all but finitely many $n \in \mathbb{N}$. Show that $\lim_{n\to\infty} E_n = E$.

(f) Let $E = Q$ as in (d) but assume that $(t_n : n \in \mathbb{N})$ satisfies the additional condition that t_n is a rational number for infinitely many $n \in \mathbb{N}$ and t_n is an irrational number for infinitely many $n \in \mathbb{N}$. Show that $\lim_{n\to\infty} E_n$ does not exist.

Proof. 1. Let us prove (a). We have $E = (-\infty, 0)$ and $E_n = E + t_n = (-\infty, t_n)$ for $n \in \mathbb{N}$. Thus $(E_n : n \in \mathbb{N})$ is a decreasing sequence of sets and this implies $\lim_{n\to\infty} E_n = \bigcap_{n\in\mathbb{N}} E_n = (-\infty, 0]$.

2. Let us prove (b). We have $E = \{a\}$ and $E_n = E + t_n = \{a + t_n\}$ for $n \in \mathbb{N}$. Now $\limsup_{n\to\infty} E_n$ consists of every $x \in \mathbb{R}$ such that $x \in E_n$ for infinitely many $n \in \mathbb{N}$. Since $\{E_n : n \in \mathbb{N}\}$ is a disjoint collection, it is impossible for any $x \in \mathbb{R}$ to be in more than one E_n. Thus $\limsup_{n\to\infty} E_n = \emptyset$. Then since $\liminf_{n\to\infty} E_n \subset \limsup_{n\to\infty} E_n$, we have $\liminf_{n\to\infty} E_n = \emptyset$ as well. Thus $\liminf_{n\to\infty} E_n = \limsup_{n\to\infty} E_n = \emptyset$ and this implies that $\lim_{n\to\infty} E_n$ exists and

$$\lim_{n\to\infty} E_n = \liminf_{n\to\infty} E_n = \limsup_{n\to\infty} E_n = \emptyset.$$

3. Let us prove (c). We have $E = [a, b]$ and $E_n = E + t_n = [a + t_n, b + t_n]$ for $n \in \mathbb{N}$. Consider $x \in (a, b]$. Now $t_n \downarrow 0$ implies $a + t_n \downarrow a$ and $b + t_n \downarrow b$. This implies that $x \in [a + t_n, b + t_n] = E_n$ for all but finitely many $n \in \mathbb{N}$ so that $x \in \liminf_{n\to\infty} E_n$. Thus

$$(1) \qquad\qquad (a, b] \subset \liminf_{n\to\infty} E_n \subset \limsup_{n\to\infty} E_n.$$

Next consider $x \in (-\infty, a]$. Since $a + t_n \downarrow a$, we have $x \notin E_n$ for any $n \in \mathbb{N}$ and thus $x \notin \limsup_{n\to\infty} E_n$. This implies that

$$(2) \qquad\qquad (-\infty, a] \cap \limsup_{n\to\infty} E_n = \emptyset.$$

Then since $\liminf_{n\to\infty} E_n \subset \limsup_{n\to\infty} E_n$, (2) implies

$$(3) \qquad\qquad (-\infty, a] \cap \liminf_{n\to\infty} E_n = \emptyset.$$

Next consider $x \in (b, \infty)$. Since $b + t_n \downarrow b$, we have $x \in [a = t_n, b + t_n] = E_n$ for only finitely many $n \in \mathbb{N}$ so that $x \notin \limsup_{n\to\infty} E_n$. This shows that

$$(4) \qquad\qquad (b, \infty) \cap \limsup_{n\to\infty} E_n = \emptyset.$$

Then since $\liminf\limits_{n\to\infty} E_n \subset \limsup\limits_{n\to\infty} E_n$, (4) implies

$$(5) \qquad\qquad (b, \infty) \cap \liminf\limits_{n\to\infty} E_n = \emptyset.$$

By (1), (2) and (4) we have $\limsup\limits_{n\to\infty} E_n = (a, b]$ and similarly by (1), (3) and (5) we have $\liminf\limits_{n\to\infty} E_n = (a, b]$. Thus we have $\liminf\limits_{n\to\infty} E_n = \limsup\limits_{n\to\infty} E_n = (a, b]$. This implies that $\lim\limits_{n\to\infty} E_n$ exists and moreover

$$\lim\limits_{n\to\infty} E_n = \liminf\limits_{n\to\infty} E_n = \limsup\limits_{n\to\infty} E_n = (a, b].$$

4. Let us prove (d). We have $E = (a, b)$ and $E_n = E + t_n = (a + t_n, b + t_n)$ for $n \in \mathbb{N}$. By the same argument as in **3**, we show that $\limsup\limits_{n\to\infty} E_n = (a, b]$ and $\liminf\limits_{n\to\infty} E_n = (a, b]$. Then we have $\lim\limits_{n\to\infty} E_n = (a, b]$. (Note that while $b \notin E$ in this case we still have $b \in \liminf\limits_{n\to\infty} E_n \subset \limsup\limits_{n\to\infty} E_n$.)

5. Let us prove (e). Since $t_n \in Q$ for all but finitely many $n \in \mathbb{N}$, there exists $N \in \mathbb{N}$ such that $t_n \in Q$ for $n \geq N$. Then for $n \geq N$ we have $E_n = E + t_n = Q + t_n = Q = E$. Then we have $\lim\limits_{n\to\infty} E_n = E$.

6. Let us prove (f). Let P be the set of all irrational numbers. Then we have $Q \cap P = \emptyset$ and $Q \cup P = \mathbb{R}$. Let us observe that

$$(6) \qquad\qquad \xi \in Q \text{ and } \eta \in P \Rightarrow \xi + \eta \in P.$$

(Suppose not. Then $\zeta := \xi + \eta \in Q$. Then $\eta = \zeta - \xi \in Q$, a contradiction to $\eta \in P$.)

To show that $\lim\limits_{n\to\infty} E_n$ does not exist, we show that $\liminf\limits_{n\to\infty} E_n \neq \limsup\limits_{n\to\infty} E_n$. We show this by showing that whereas $\limsup\limits_{n\to\infty} E_n \neq \emptyset$ we have $\liminf\limits_{n\to\infty} E_n = \emptyset$.

Our sequence $(t_n : n \in \mathbb{N})$ is such that $t_n \in Q$ for infinitely many $n \in \mathbb{N}$ and at the same time $t_n \in P$ for infinitely many $n \in \mathbb{N}$. If $t_n \in Q$ then $E_n = E + t_n = Q + t_n = Q$ and thus $Q = E_n$ for infinitely many $n \in \mathbb{N}$ and hence $Q \subset \limsup\limits_{n\to\infty} E_n$. Therefore we have

$$(7) \qquad\qquad \limsup\limits_{n\to\infty} E_n \neq \emptyset.$$

Now $\liminf\limits_{n\to\infty} E_n$ consists of every $x \in \mathbb{R}$ such that $x \in E_n$ for all but finitely many $n \in \mathbb{N}$. Let us show that no such $x \in \mathbb{R}$ exists. If $x \in \mathbb{R}$ then either $x \in Q$ or $x \in P$. Consider first the case $x \in Q$. If $t_n \in P$ then $E_n = E + t_n = Q + t_n \subset P$ by (6) and thus $x \notin E_n$. Since $t_n \in P$ for infinitely many $n \in \mathbb{N}$, we have $x \notin E_n$ for infinitely many $n \in \mathbb{N}$ and thus $x \notin \liminf\limits_{n\to\infty} E_n$. Consider next the case $x \in P$. If $t_n \in Q$ then $E_n = E + t_n = Q + t_n = Q$ so that $x \notin E_n$. Since $t_n \in Q$ for infinitely many $n \in \mathbb{N}$, we have $x \notin E_n$ for infinitely many $n \in \mathbb{N}$ and therefore $x \notin \liminf\limits_{n\to\infty} E_n$. Thus whether $x \in Q$ or $x \in P$ we have $x \notin \liminf\limits_{n\to\infty} E_n$. Therefore we have

$$(8) \qquad\qquad \liminf\limits_{n\to\infty} E_n = \emptyset.$$

Finally, (7) and (8) show that $\liminf\limits_{n\to\infty} E_n \neq \limsup\limits_{n\to\infty} E_n$ and thus $\lim\limits_{n\to\infty} E_n$ does not exist. ∎

Prob. 1.6. The characteristic function $\mathbf{1}_A$ of a subset A of a set X is a function on X defined by

$$\mathbf{1}_A(x) = \begin{cases} 1 & \text{for } x \in A, \\ 0 & \text{for } x \in A^c. \end{cases}$$

Let $(A_n : n \in \mathbb{N})$ be a sequence of subsets of X and A be a subset of X.

(a) Show that if $\lim_{n \to \infty} A_n = A$ then $\lim_{n \to \infty} \mathbf{1}_{A_n} = \mathbf{1}_A$ on X.

(b) Show that if $\lim_{n \to \infty} \mathbf{1}_{A_n} = \mathbf{1}_A$ on X then $\lim_{n \to \infty} A_n = A$.

Proof. 1. Let us prove (a). Let us assume $\lim_{n \to \infty} A_n = A$ and prove $\lim_{n \to \infty} \mathbf{1}_{A_n} = \mathbf{1}_A$ on X. Now $\lim_{n \to \infty} A_n = A$ implies $A = \liminf_{n \to \infty} A_n = \limsup_{n \to \infty} A_n$. Let $x \in X$. Then either $x \in A$ or $x \in A^c$.

Consider first the case $x \in A$. Then $\mathbf{1}_A(x) = 1$. Since $x \in A = \liminf_{n \to \infty} A_n$, we have $x \in A_n$ for all but finitely many $n \in \mathbb{N}$. Then $\mathbf{1}_{A_n}(x) = 1$ for all but finitely many $n \in \mathbb{N}$ and thus $\lim_{n \to \infty} \mathbf{1}_{A_n}(x) = 1 = \mathbf{1}_A(x)$.

Next consider the case $x \in A^c$. In this case we have $x \notin A$ and then $\mathbf{1}_A(x) = 0$. Since $x \in A^c = \left(\limsup_{n \to \infty} A_n\right)^c$, we have $x \notin \limsup_{n \to \infty} A_n$ and hence $x \in A_n$ for only finitely many $n \in \mathbb{N}$. Thus $\mathbf{1}_{A_n}(x) = 1$ for only finitely many $n \in \mathbb{N}$ and consequently $\mathbf{1}_{A_n}(x) = 0$ for all but finitely many $n \in \mathbb{N}$. Then $\lim_{n \to \infty} \mathbf{1}_{A_n}(x) = 0 = \mathbf{1}_A(x)$.

Thus for every $x \in X$, we have $\lim_{n \to \infty} \mathbf{1}_{A_n}(x) = \mathbf{1}_A(x)$.

2. Let us prove (b). Let us assume $\lim_{n \to \infty} \mathbf{1}_{A_n} = \mathbf{1}_A$ on X. To show that $\lim_{n \to \infty} A_n = A$, we show that $A = \liminf_{n \to \infty} A_n = \limsup_{n \to \infty} A_n$.

Let $x \in A$. Then $\mathbf{1}_A(x) = 1$. Since $\lim_{n \to \infty} \mathbf{1}_{A_n} = \mathbf{1}_A$ on X, we have $\lim_{n \to \infty} \mathbf{1}_{A_n}(x) = \mathbf{1}_A(x) = 1$. Since $\mathbf{1}_{A_n}(x)$ is either 1 or 0 for every $n \in \mathbb{N}$, the last convergence implies that $\mathbf{1}_{A_n}(x) = 1$ for all but finitely many $n \in \mathbb{N}$. Thus $x \in A_n$ for all but finitely many $n \in \mathbb{N}$ and therefore $x \in \liminf_{n \to \infty} A_n$. This shows that $A \subset \liminf_{n \to \infty} A_n$.

Next let $x \in \limsup_{n \to \infty} A_n$. Then $x \in A_n$ for infinitely many $n \in \mathbb{N}$ and hence $\mathbf{1}_{A_n}(x) = 1$ for infinitely many $n \in \mathbb{N}$. Since $\lim_{n \to \infty} \mathbf{1}_{A_n} = \mathbf{1}_A$ on X, we have $\lim_{n \to \infty} \mathbf{1}_{A_n}(x) = \mathbf{1}_A(x)$. Since $\mathbf{1}_{A_n}(x) = 1$ for infinitely many $n \in \mathbb{N}$, the last convergence implies that $\mathbf{1}_A(x) = 1$ and hence $x \in A$. This shows that $\limsup_{n \to \infty} A_n \subset A$.

We have shown above that $\limsup_{n \to \infty} A_n \subset A \subset \liminf_{n \to \infty} A_n$. On the other hand we have $\liminf_{n \to \infty} A_n \subset \limsup_{n \to \infty} A_n$. Therefore we have $\liminf_{n \to \infty} A_n = \limsup_{n \to \infty} A_n = A$. This implies that $\lim_{n \to \infty} A_n = A$. ∎

Prob. 1.7. Let $\mathfrak{A}$ be a σ-algebra of subsets of a set X and let Y be an arbitrary subset of X. Let $\mathfrak{B} = \{A \cap Y : A \in \mathfrak{A}\}$. Show that $\mathfrak{B}$ is a σ-algebra of subsets of Y.

Proof. To show that $\mathfrak{B}$ is a σ-algebra of subsets of Y, we verify the following conditions:
1° $Y \in \mathfrak{B}$.
2° $B \in \mathfrak{B} \Rightarrow Y \setminus B \in \mathfrak{B}$.
3° $(B_n : n \in \mathbb{N}) \subset \mathfrak{B} \Rightarrow \bigcup_{n \in \mathbb{N}} B_n \in \mathfrak{B}$.
   1. Now $\mathfrak{B} = \{A \cap Y : A \in \mathfrak{A}\}$. With $X \in \mathfrak{A}$, we have $Y = X \cap Y \in \mathfrak{B}$. This verifies 1°.
   2. Let $B \in \mathfrak{B}$. Then $B = A \cap Y$ where $A \in \mathfrak{A}$. Then we have

$$Y \setminus B = (X \cap Y) \setminus (A \cap Y) = (X \setminus A) \cap Y = A^c \cap Y.$$

Now $A \in \mathfrak{A}$ implies $A^c \mathfrak{A}$ since $\mathfrak{A}$ is a σ-algebra of subsets of X. Thus $A^c \cap Y \in \mathfrak{B}$, that is, $Y \setminus B \in \mathfrak{B}$. This verifies 2°.
   3. Let $(B_n : n \in \mathbb{N}) \subset \mathfrak{B}$. Now $B_n \in \mathfrak{B}$ implies $B_n = A_n \cap Y$ where $A_n \in \mathfrak{A}$ for $n \in \mathbb{N}$. Then we have

$$\bigcup_{n \in \mathbb{N}} B_n = \bigcup_{n \in \mathbb{N}} (A_n \cap Y) = \left(\bigcup_{n \in \mathbb{N}} A_n \right) \cap Y.$$

Now since $\mathfrak{A}$ is a σ-algebra of subsets of X, $(A_n : n \in \mathbb{N}) \subset \mathfrak{A}$ implies $A := \bigcup_{n \in \mathbb{N}} A_n \in \mathfrak{A}$. Then $\bigcup_{n \in \mathbb{N}} B_n = A \cap Y \in \mathfrak{B}$. This verifies 3°. ∎

Prob. 1.8. Let $\mathfrak{A}$ be a collection of subsets of a set X with the following properties:
1°. $X \in \mathfrak{A}$,
2°. $A, B \in \mathfrak{A} \Rightarrow A \setminus B = A \cap B^c \in \mathfrak{A}$.
Show that $\mathfrak{A}$ is an algebra of subsets of the set X.

Proof. To show that $\mathfrak{A}$ is an algebra of subsets of the set X, it suffices to verify the following conditions:
3° $A \in \mathfrak{A} \Rightarrow A^c \in \mathfrak{A}$.
4° $A, B \in \mathfrak{A} \Rightarrow A \cup B \in \mathfrak{A}$.
   1. Suppose $A \in \mathfrak{A}$. Then by 1° we have $X, A \in \mathfrak{A}$ and this implies $X \setminus A \in \mathfrak{A}$ by 2°. Thus $A^c = X \setminus A \in \mathfrak{A}$. This verifies 3°.
   2. Suppose $A, B \in \mathfrak{A}$. Now we have

$$(A \cup B)^c = A^c \cap B^c = A^c \setminus B.$$

Since $A \in \mathfrak{A}$, we have $A^c \in \mathfrak{A}$ by 3°. Then $A^c, B \in \mathfrak{A}$ and this implies $A^c \setminus B \in \mathfrak{A}$ by 2°. Thus $(A \cup B)^c \in \mathfrak{A}$ and this implies $A \cup B \in \mathfrak{A}$ by 3°. This verifies 4°. ∎

Prob. 1.9. Let $\mathfrak{A}$ be an algebra of subsets of a set X. Suppose $\mathfrak{A}$ has the property that for every increasing sequence $(A_n : n \in \mathbb{N})$ in $\mathfrak{A}$, we have $\bigcup_{n\in\mathbb{N}} A_n \in \mathfrak{A}$. Show that $\mathfrak{A}$ is a σ-algebra of subsets of the set X.

Proof. Since $\mathfrak{A}$ is an algebra of subsets of the set X, to show that $\mathfrak{A}$ is a σ-algebra of subsets of X it suffices to verify the condition:

$$(B_n : n \in \mathbb{N}) \subset \mathfrak{A} \Rightarrow \bigcup_{n\in\mathbb{N}} B_n \in \mathfrak{A}.$$

Given $(B_n : n \in \mathbb{N}) \subset \mathfrak{A}$, let us define a sequence $(A_n : n \in \mathbb{N})$ of subsets of X by setting

$$(1) \qquad\qquad A_n = \bigcup_{k=1}^{n} B_k \quad \text{for } n \in \mathbb{N}.$$

Then $(A_n : n \in \mathbb{N})$ is an increasing sequence of subsets of X and

$$(2) \qquad\qquad \bigcup_{n\in\mathbb{N}} B_n = \bigcup_{n\in\mathbb{N}} A_n.$$

Since $(B_n : n \in \mathbb{N}) \subset \mathfrak{A}$ and $\mathfrak{A}$ is an algebra of subsets of X, we have $A_n = \bigcup_{k=1}^{n} B_k \in \mathfrak{A}$ and thus $(A_n : n \in \mathbb{N})$ is an increasing sequence in $\mathfrak{A}$ and this implies that $\bigcup_{n\in\mathbb{N}} A_n \in \mathfrak{A}$ by assumption. Then by (2) we have $\bigcup_{n\in\mathbb{N}} B_n \in \mathfrak{A}$. ∎

Prob. 1.10. Let $(X, \mathfrak{A})$ be a measurable space and let $(E_n : n \in \mathbb{N})$ be an increasing sequence in $\mathfrak{A}$ such that $\bigcup_{n\in\mathbb{N}} E_n = X$.
(a) Let $\mathfrak{A}_n = \mathfrak{A} \cap E_n$, that is, $\mathfrak{A}_n = \{A \cap E_n : A \in \mathfrak{A}\}$. Show that $\mathfrak{A}_n$ is a σ-algebra of subsets of E_n for each $n \in \mathbb{N}$.
(b) Does $\bigcup_{n\in\mathbb{N}} \mathfrak{A}_n = \mathfrak{A}$ hold?

Proof. 1. $\mathfrak{A}_n$ is a σ-algebra of subsets of E_n for each $n \in \mathbb{N}$ by Prob. 1.7.
2. Clearly $\mathfrak{A}_n \subset \mathfrak{A}$ for every $n \in \mathbb{N}$ and therefore $\bigcup_{n\in\mathbb{N}} \mathfrak{A}_n \subset \mathfrak{A}$. However the equality $\bigcup_{n\in\mathbb{N}} \mathfrak{A}_n = \mathfrak{A}$ does not hold in general. For instance if $E_n \neq X$ for every $n \in \mathbb{N}$, then $X \notin \mathfrak{A}_n$ for every $n \in \mathbb{N}$ and then $X \notin \bigcup_{n\in\mathbb{N}} \mathfrak{A}_n$. But $X \in \mathfrak{A}$. Thus $\bigcup_{n\in\mathbb{N}} \mathfrak{A}_n \neq \mathfrak{A}$. ∎

Prob. 1.11. (a) Show that if $(\mathfrak{A}_n : n \in \mathbb{N})$ is an increasing sequence of algebras of subsets of a set X, then $\bigcup_{n\in\mathbb{N}} \mathfrak{A}_n$ is an algebra of subsets of X.
(b) Show that if $(\mathfrak{A}_n : n \in \mathbb{N})$ is a decreasing sequence of algebras of subsets of a set X, then $\bigcap_{n\in\mathbb{N}} \mathfrak{A}_n$ is an algebra of subsets of X.

Proof. 1. Let us prove (a). Let $(\mathfrak{A}_n : n \in \mathbb{N})$ be an increasing sequence of algebras of subsets of a set X. To show that $\bigcup_{n\in\mathbb{N}} \mathfrak{A}_n$ is an algebra of subsets of X, we verify that $\bigcup_{n\in\mathbb{N}} \mathfrak{A}_n$ satisfies the following conditions:
1° $X \in \bigcup_{n\in\mathbb{N}} \mathfrak{A}_n$.
2° $A \in \bigcup_{n\in\mathbb{N}} \mathfrak{A}_n \Rightarrow A^c \in \bigcup_{n\in\mathbb{N}} \mathfrak{A}_n$.
3° $A, B \in \bigcup_{n\in\mathbb{N}} \mathfrak{A}_n \Rightarrow A \cup B \in \bigcup_{n\in\mathbb{N}} \mathfrak{A}_n$.
    Now $X \in \mathfrak{A}_n$ for every $n \in \mathbb{N}$. Then $X \in \mathfrak{A}_n \subset \bigcup_{n\in\mathbb{N}} \mathfrak{A}_n$. This verifies 1°.
    Let $A \in \bigcup_{n\in\mathbb{N}} \mathfrak{A}_n$. Then $A \in \mathfrak{A}_{n_0}$ for some $n_0 \in \mathbb{N}$. Then since $\mathfrak{A}_{n_0}$ is an algebra of subsets of X, we have $A^c \in \mathfrak{A}_{n_0}$. Then $A^c \in \mathfrak{A}_{n_0} \subset \bigcup_{n\in\mathbb{N}} \mathfrak{A}_n$. This verifies 2°.
    Let $A, B \in \bigcup_{n\in\mathbb{N}} \mathfrak{A}_n$. Then $A \in \mathfrak{A}_{n_1}$ for some $n_1 \in \mathbb{N}$ and $B \in \mathfrak{A}_{n_2}$ for some $n_2 \in \mathbb{N}$. Let $n_0 = \max\{n_1, n_2\}$. Then since $(\mathfrak{A}_n : n \in \mathbb{N})$ is an increasing sequence, we have $\mathfrak{A}_{n_1}, \mathfrak{A}_{n_2} \subset \mathfrak{A}_{n_0}$. Then $A \in \mathfrak{A}_{n_1} \subset \mathfrak{A}_{n_0}$ and $B \in \mathfrak{A}_{n_2} \subset \mathfrak{A}_{n_0}$ and thus $A, B \in \mathfrak{A}_{n_0}$. Then since $\mathfrak{A}_{n_0}$ is an algebra, we have $A \cup B \in \mathfrak{A}_{n_0} \subset \bigcup_{n\in\mathbb{N}} \mathfrak{A}_n$. This verifies 3°. This completes the proof that $\bigcup_{n\in\mathbb{N}} \mathfrak{A}_n$ is an algebra of subsets of X.

2. Let us prove (b). Let $(\mathfrak{A}_n : n \in \mathbb{N})$ be a decreasing sequence of algebras of subsets of a set X. To show that $\bigcap_{n\in\mathbb{N}} \mathfrak{A}_n$ is an algebra of subsets of X, we verify that $\bigcap_{n\in\mathbb{N}} \mathfrak{A}_n$ satisfies the following conditions:
1° $X \in \bigcap_{n\in\mathbb{N}} \mathfrak{A}_n$.
2° $A \in \bigcap_{n\in\mathbb{N}} \mathfrak{A}_n \Rightarrow A^c \in \bigcap_{n\in\mathbb{N}} \mathfrak{A}_n$.
3° $A, B \in \bigcap_{n\in\mathbb{N}} \mathfrak{A}_n \Rightarrow A \cup B \in \bigcap_{n\in\mathbb{N}} \mathfrak{A}_n$.
    Since $\mathfrak{A}_n$ is an algebra of subsets of X, we have $X \in \mathfrak{A}_n$ for every $n \in \mathbb{N}$. Then $X \in \bigcap_{n\in\mathbb{N}} \mathfrak{A}_n$. This verifies 1°.
    Suppose $A \in \bigcap_{n\in\mathbb{N}} \mathfrak{A}_n$. Then $A \in \mathfrak{A}_n$ for every $n \in \mathbb{N}$. Then since $\mathfrak{A}_n$ is an algebra of subsets of X, we have $A^c \in \mathfrak{A}_n$ for every $n \in \mathbb{N}$. Then $A^c \in \bigcap_{n\in\mathbb{N}} \mathfrak{A}_n$. This verifies 2°.
    Suppose $A, B \in \bigcap_{n\in\mathbb{N}} \mathfrak{A}_n$. Then $A, B \in \mathfrak{A}_n$ for every $n \in \mathbb{N}$. Then since $\mathfrak{A}_n$ is an algebra of subsets of X, we have $A \cup B \in \mathfrak{A}_n$ for every $n \in \mathbb{N}$. Then $A \cup B \in \bigcap_{n\in\mathbb{N}} \mathfrak{A}_n$. This verifies 3°. This completes the proof that $\bigcap_{n\in\mathbb{N}} \mathfrak{A}_n$ is an algebra of subsets of X. ∎

Prob. 1.12. Let $(X, \mathfrak{A})$ be a measurable space. Let us call an $\mathfrak{A}$-measurable subset E of X an atom in the measurable space $(X, \mathfrak{A})$ if $E \neq \emptyset$ and $\emptyset$ and E are the only $\mathfrak{A}$-measurable subsets of E. Show that if E_1 and E_2 are two distinct atoms in $(X, \mathfrak{A})$ then they are disjoint.

Proof. Let E_1 and E_2 be two distinct atoms in $(X, \mathfrak{A})$. Then $E_1 \neq E_2$. To show that E_1 and E_2 are disjoint, that is, $E_1 \cap E_2 = \emptyset$, let us assume the contrary, that is, $E_1 \cap E_2 \neq \emptyset$. Now $E_1 \cap E_2$, being a $\mathfrak{A}$-measurable subset of an atom E_1, must be either equal to $\emptyset$ or equal to E_1. Since we are assuming $E_1 \cap E_2 \neq \emptyset$, we must have $E_1 \cap E_2 = E_1$. Similarly $E_1 \cap E_2$, being a $\mathfrak{A}$-measurable subset of an atom E_2, must be either equal to $\emptyset$ or equal to E_2. Since we are assuming $E_1 \cap E_2 \neq \emptyset$, we must have $E_1 \cap E_2 = E_2$. Then we have $E_1 = E_1 \cap E_2 = E_2$. This contradicts the assumption that $E_1 \neq E_2$. Therefore we must have $E_1 \cap E_2 = \emptyset$. ∎

Prob. 1.13. For an arbitrary collection $\mathfrak{C}$ of subsets of a set X, let $\alpha(\mathfrak{C})$ be the algebra generated by $\mathfrak{C}$, that is, the smallest algebra of subsets of X containing $\mathfrak{C}$, and let $\sigma(\mathfrak{C})$ be the σ-algebra generated by $\mathfrak{C}$. Prove the following statements:
(a) $\alpha\big(\alpha(\mathfrak{C})\big) = \alpha(\mathfrak{C})$,
(b) $\sigma\big(\sigma(\mathfrak{C})\big) = \sigma(\mathfrak{C})$,
(c) $\alpha(\mathfrak{C}) \subset \sigma(\mathfrak{C})$,
(d) if $\mathfrak{C}$ is a finite collection, then $\alpha(\mathfrak{C}) = \sigma(\mathfrak{C})$,
(e) $\sigma\big(\alpha(\mathfrak{C})\big) = \sigma(\mathfrak{C})$.
(Hint for (d): Use Prob. 1.18 below.)

Proof. (a) $\alpha\big(\alpha(\mathfrak{C})\big)$ is the smallest algebra containing $\alpha(\mathfrak{C})$. Then since $\alpha(\mathfrak{C})$ is itself an algebra we have $\alpha\big(\alpha(\mathfrak{C})\big) = \alpha(\mathfrak{C})$.

(b) $\sigma\big(\sigma(\mathfrak{C})\big)$ is the smallest σ-algebra containing $\sigma(\mathfrak{C})$. Then since $\sigma(\mathfrak{C})$ is itself a σ-algebra we have $\sigma\big(\sigma(\mathfrak{C})\big) = \sigma(\mathfrak{C})$.

(c) $\sigma(\mathfrak{C})$ is the smallest σ-algebra containing $\mathfrak{C}$ and hence an algebra containing $\mathfrak{C}$. Then since $\alpha(\mathfrak{C})$ is the smallest algebra containing $\mathfrak{C}$ we have $\alpha(\mathfrak{C}) \subset \sigma(\mathfrak{C})$.

(d) (d) is (e) of Prob. 1.18.

(e) Since $\mathfrak{C} \subset \alpha(\mathfrak{C})$, we have $\sigma(\mathfrak{C}) \subset \sigma\big(\alpha(\mathfrak{C})\big)$. On the other hand by (c) we have $\alpha(\mathfrak{C}) \subset \sigma(\mathfrak{C})$. Then $\sigma\big(\alpha(\mathfrak{C})\big) \subset \sigma\big(\sigma(\mathfrak{C})\big) = \sigma(\mathfrak{C})$ by (b). Thus we have $\sigma\big(\alpha(\mathfrak{C})\big) = \sigma(\mathfrak{C})$. ∎

Prob. 1.14. Let $(\mathfrak{A}_n : n \in \mathbb{N})$ be a monotone sequence of σ-algebras of subsets of a set X and let $\mathfrak{A} = \lim_{n \to \infty} \mathfrak{A}_n$.
(a) Show that if $(\mathfrak{A}_n : n \in \mathbb{N})$ is a decreasing sequence then $\mathfrak{A}$ is a σ-algebra.
(b) Show that if $(\mathfrak{A}_n : n \in \mathbb{N})$ is an increasing sequence then $\mathfrak{A}$ is an algebra but $\mathfrak{A}$ may not be a σ-algebra by constructing an example.

Proof. 1. Let us prove (a). Now if $(\mathfrak{A}_n : n \in \mathbb{N})$ is a decreasing sequence then we have $\mathfrak{A} = \lim_{n \to \infty} \mathfrak{A}_n = \bigcap_{n \in \mathbb{N}} \mathfrak{A}_n$. A σ-algebra is necessarily an algebra. Thus $\bigcap_{n \in \mathbb{N}} \mathfrak{A}_n$ is an algebra by Prob. 1.11. To show that $\bigcap_{n \in \mathbb{N}} \mathfrak{A}_n$ is a σ-algebra, it remains to prove

$$(A_i : i \in \mathbb{N}) \subset \bigcap_{n \in \mathbb{N}} \mathfrak{A}_n \Rightarrow \bigcup_{i \in \mathbb{N}} A_i \in \bigcap_{n \in \mathbb{N}} \mathfrak{A}_n.$$

Let $(A_i : i \in \mathbb{N}) \subset \bigcap_{n \in \mathbb{N}} \mathfrak{A}_n$. Then $(A_i : i \in \mathbb{N}) \subset \mathfrak{A}_n$ for every $n \in \mathbb{N}$. Then since $\mathfrak{A}_n$ is a σ-algebra, $(A_i : i \in \mathbb{N}) \subset \mathfrak{A}_n$ implies $\bigcup_{i \in \mathbb{N}} A_i \in \mathfrak{A}_n$. Since this holds for every $n \in \mathbb{N}$, we have $\bigcup_{i \in \mathbb{N}} A_i \in \bigcap_{n \in \mathbb{N}} \mathfrak{A}_n$. This proves that $\bigcap_{n \in \mathbb{N}} \mathfrak{A}_n$ is a σ-algebra.
    2. Let us prove (b). Assume that $(\mathfrak{A}_n : n \in \mathbb{N})$ is an increasing sequence. Then we have $\mathfrak{A} = \lim_{n \to \infty} \mathfrak{A}_n = \bigcup_{n \in \mathbb{N}} \mathfrak{A}_n$. A σ-algebra is necessarily an algebra. Thus $\bigcup_{n \in \mathbb{N}} \mathfrak{A}_n$ is an algebra by Prob. 1.11. However $\bigcup_{n \in \mathbb{N}} \mathfrak{A}_n$ need not be a σ-algebra. Let us construct two examples to show this.

Example 1. Let $X = [0, \infty)$. For each $n \in \mathbb{N}$, decompose $[0, \infty)$ into disjoint intervals

$$I_{n,k} = \left[\frac{k-1}{2^n}, \frac{k}{2^n} \right) \quad \text{for } k \in \mathbb{N},$$

and let

$$\mathfrak{I}_n = \left\{ I_{n,k} : k \in \mathbb{N} \right\}.$$

Let $\mathfrak{A}_n$ be the collection of all arbitrary unions of members of $\mathfrak{I}_n$ (including $\emptyset$ as the union of zero members of $\mathfrak{I}_n$).
    Let us show that $\mathfrak{A}_n$ is a σ-algebra of subsets of X. To start with, we have $X = \bigcup_{k \in \mathbb{N}} I_{n,k}$ and therefore $X \in \mathfrak{A}_n$. Next let $A \in \mathfrak{A}_n$. Then A is a union of members of $\mathfrak{I}_n$ which is a disjoint collection. Then since X is the union of all members of $\mathfrak{I}_n$, $A^c = X \setminus A$ is a union of members of $\mathfrak{I}_n$ and hence $A^c \in \mathfrak{A}_n$. Finally let $(A_i : i \in \mathbb{N}) \subset \mathfrak{A}_n$. Since A_i is a union of members of $\mathfrak{I}_n$ for every $i \in \mathbb{N}$, $\bigcup_{i \in \mathbb{N}} A_i$ is a union of members of $\mathfrak{I}_n$ and thus $\bigcup_{i \in \mathbb{N}} A_i \in \mathfrak{A}_n$. This completes the proof that $\mathfrak{A}_n$ is a σ-algebra of subsets of X.
    The fact that $(\mathfrak{A}_n : n \in \mathbb{N})$ is an increasing sequence (that is, $\mathfrak{A}_n \subset \mathfrak{A}_{n+1}$ for every $n \in \mathbb{N}$), is shown as follows. Let $A \in \mathfrak{A}_n$. Then A is a union of members of $\mathfrak{I}_n$. But every member of $\mathfrak{I}_n$ is the union of two members of $\mathfrak{I}_{n+1}$. Thus A is a union of members of $\mathfrak{I}_{n+1}$ and therefore $A \in \mathfrak{A}_{n+1}$. This shows that $\mathfrak{A}_n \subset \mathfrak{A}_{n+1}$.
    Now $(\mathfrak{A}_n : n \in \mathbb{N})$ is an increasing sequence of σ-algebras of subsets of X. According to Prob. 1.11 this implies that $\bigcup_{n \in \mathbb{N}} \mathfrak{A}_n$ is an algebra of subsets of X. Let us show that

$\bigcup_{n \in \mathbb{N}} \mathfrak{A}_n$ is not a σ-algebra of subsets of X. Let

$$I_1 = \left[0, \frac{1}{2}\right) \in \mathfrak{I}_1 \subset \mathfrak{A}_1,$$

$$I_2 = \left[1, \frac{1}{2^2}\right) \in \mathfrak{I}_2 \subset \mathfrak{A}_2,$$

$$I_3 = \left[1, \frac{1}{2^3}\right) \in \mathfrak{I}_3 \subset \mathfrak{A}_3,$$

$$\vdots$$

Then $(I_n : n \in \mathbb{N})$ is a sequence in $\bigcup_{n \in \mathbb{N}} \mathfrak{A}_n$. But $\bigcup_{i \in \mathbb{N}} I_n \notin \mathfrak{A}_n$ for any $n \in \mathbb{N}$ and therefore we have $\bigcup_{i \in \mathbb{N}} I_n \notin \bigcup_{n \in \mathbb{N}} \mathfrak{A}_n$. This shows that $\bigcup_{n \in \mathbb{N}} \mathfrak{A}_n$ is not a σ-algebra of subsets of X.

Example 2. Let $X = (0, 1]$. For each $n \in \mathbb{N}$, decompose $(0, 1]$ into disjoint intervals

$$I_{n,k} = \left(\frac{k-1}{2^n}, \frac{k}{2^n}\right] \quad \text{for } k = 1, \ldots, 2^n,$$

and let

$$\mathfrak{I}_n = \left\{I_{n,k} : k = 1, \ldots, 2^n\right\}.$$

Let $\mathfrak{A}_n$ be the collection of all arbitrary unions of members of $\mathfrak{I}_n$. The fact that $(\mathfrak{A}_n : n \in \mathbb{N})$ is an increasing sequence of σ-algebras of subsets of X is proved by the same arguments as in Example 1 above. Then $\bigcup_{n \in \mathbb{N}} \mathfrak{A}_n$ is an algebra of subsets of X by Prob. 1.11.

To show that $\bigcup_{n \in \mathbb{N}} \mathfrak{A}_n$ is not a σ-algebra, we show that there exists a sequence $(A_n : n \in \mathbb{N})$ in $\bigcup_{n \in \mathbb{N}} \mathfrak{A}_n$ such that $\bigcap_{n \in \mathbb{N}} A_n \notin \bigcup_{n \in \mathbb{N}} \mathfrak{A}_n$. For each $n \in \mathbb{N}$, let

$$A_n = \left(\frac{2^n - 1}{2^n}, \frac{2^n}{2^n}\right] \in \mathfrak{I}_n \subset \mathfrak{A}_n \subset \bigcup_{n \in \mathbb{N}} \mathfrak{A}_n.$$

Then we have

$$\bigcap_{n \in \mathbb{N}} A_n = \{1\} \notin \mathfrak{A}_n \quad \text{for any } n \in \mathbb{N}$$

and thus $\bigcap_{n \in \mathbb{N}} A_n \notin \bigcup_{n \in \mathbb{N}} \mathfrak{A}_n$. ∎

Prob. 1.15. Let $\mathfrak{E} = \{A_1, \ldots, A_n\}$ be a disjoint collection of nonempty subsets of a set X such that $\bigcup_{i=1}^n A_i = X$. Let $\mathfrak{F}$ be the collection of all arbitrary unions of members of $\mathfrak{E}$.
(a) Show that $\mathfrak{F} = \sigma(\mathfrak{E})$, the smallest σ-algebra of subsets of X containing $\mathfrak{E}$.
(b) Show that the cardinality of $\sigma(\mathfrak{E})$ is equal to 2^n.

Proof. 1. Let us show first that $\mathfrak{F}$ is a σ-algebra of subsets of X. For this purpose we verify the following conditions:
$1°$ $X \in \mathfrak{F}$.
$2°$ $F \in \mathfrak{F} \Rightarrow F^c \in \mathfrak{F}$.
$3°$ $(F_k : k \in \mathbb{N}) \subset \mathfrak{F} \Rightarrow \bigcup_{k\in\mathbb{N}} F_k \in \mathfrak{F}$.
 To start with, we have $X = \bigcup_{i=1}^n A_i$. Thus X is a union of members of $\mathfrak{E}$ and therefore $X \in \mathfrak{F}$. This verifies $1°$.
 Let $F \in \mathfrak{F}$. Then F is a union of members of the disjoint collection $\mathfrak{E} = \{A_1, \ldots, A_n\}$. Then since $X = \bigcup_{i=1}^n A_i$, $F^c = X \setminus F$ is a union of members of $\mathfrak{E}$ and hence $F^c \in \mathfrak{F}$. This verifies $2°$.
 Let $(F_k : k \in \mathbb{N}) \subset \mathfrak{F}$. Then F_k is a union of members of $\mathfrak{E}$ for every $k \in \mathbb{N}$ and then $\bigcup_{k\in\mathbb{N}} F_k$ is a union of members of $\mathfrak{E}$ and therefore $\bigcup_{k\in\mathbb{N}} F_k \in \mathfrak{F}$. This verifies $3°$ and completes the proof that $\mathfrak{F}$ is a σ-algebra of subsets of X.
 2. Let us show that $\mathfrak{E} \subset \mathfrak{F}$. Now $\mathfrak{F}$ is the collection of all arbitrary unions of members of the collection $\mathfrak{E} = \{A_1, \ldots, A_n\}$. Thus $A_i \in \mathfrak{F}$ for every $i = 1, \ldots, n$ and hence $\mathfrak{E} \subset \mathfrak{F}$.
 3. Let us show that $\mathfrak{F} = \sigma(\mathfrak{E})$, the smallest σ-algebra of subsets of X containing $\mathfrak{E}$. We showed above that $\mathfrak{F}$ is a σ-algebra of subsets of X containing $\mathfrak{E}$. To show that $\mathfrak{F}$ is the smallest σ-algebra of subsets of X containing $\mathfrak{E}$, let us assume the contrary. Then there exists a σ-algebra $\mathfrak{G}$ of subsets of X containing $\mathfrak{E}$ such that $\mathfrak{G} \subset \mathfrak{F}$ and $\mathfrak{G} \neq \mathfrak{F}$. Then there exists $F_0 \in \mathfrak{F}$ such that $F_0 \notin \mathfrak{G}$. Now our $F \in \mathfrak{F}$ is a union of members of the collection $\mathfrak{E}$. Since $\mathfrak{G}$ is a σ-algebra of subsets of X containing $\mathfrak{E}$, $\mathfrak{G}$ contains every finite union of members of $\mathfrak{E}$. Since $\mathfrak{E}$ is a finite collection, an arbitrary union of members of $\mathfrak{E}$ is a finite union of members of $\mathfrak{E}$. Thus $\mathfrak{G}$ contains every union of members of $\mathfrak{E}$. Then $\mathfrak{G}$ contains our F_0. This is a contradiction. Therefore $\mathfrak{F}$ must be the smallest σ-algebra of subsets of X containing $\mathfrak{E}$.
 4. Since $\sigma(\mathfrak{E}) = \mathfrak{F}$, to show that the cardinality of $\sigma(\mathfrak{E})$ is equal to 2^n, we show that the cardinality of $\mathfrak{F}$ is equal to 2^n. Now $\mathfrak{F}$ is the collection of all arbitrary unions of members of the disjoint collection $\mathfrak{E} = \{A_1, \ldots, A_n\}$ of non-empty subsets of X. To form a union of members of $\mathfrak{E}$, we either select A_1 or we do not select A_1 and then we either select A_2 or we do not select A_2 and so on until we either select A_n or we do not select A_n. Thus there are 2^n ways of forming a union of members of $\mathfrak{E}$. Any 2 different ways of forming a union result in 2 distinct unions since $\mathfrak{E}$ is a disjoint collection of non-empty sets. Thus the cardinality of $\mathfrak{F}$ is equal to 2^n. ∎

Prob. 1.16. Let $\mathfrak{E} = \{A_i : i \in \mathbb{N}\}$ be a disjoint collection of nonempty subsets of a set X such that $\bigcup_{i \in \mathbb{N}} A_i = X$. Let $\mathfrak{F}$ be the collection of all arbitrary unions of members of $\mathfrak{E}$.
(a) Show that $\mathfrak{F} = \sigma(\mathfrak{E})$, the smallest σ-algebra of subsets of X containing $\mathfrak{E}$.
(b) Show that the cardinality of $\sigma(\mathfrak{E})$ is equal to $2^{\aleph_0}$.

Proof. 1. To show that $\mathfrak{F}$ is a σ-algebra of subsets of X, we verify that $\mathfrak{F}$ satisfies the following conditions:
$1°$ $X \in \mathfrak{F}$.
$2°$ $F \in \mathfrak{F} \Rightarrow F^c \in \mathfrak{F}$.
$3°$ $(F_k : k \in \mathbb{N}) \subset \mathfrak{F} \Rightarrow \bigcup_{k \in \mathbb{N}} F_k \in \mathfrak{F}$.
   Now $X = \bigcup_{i \in \mathbb{N}} A_i$. Thus $X \in \mathfrak{F}$. This verifies $1°$.
   Let $F \in \mathfrak{F}$. Then F is a union of members of the disjoint collection $\mathfrak{E} = \{A_i : i \in \mathbb{N}\}$. Then since $X = \bigcup_{i \in \mathbb{N}} A_i$, $F^c = X \setminus F$ is a union of members of $\mathfrak{E}$ and hence $F^c \in \mathfrak{F}$. This verifies $2°$.
   Let $(F_k : k \in \mathbb{N}) \subset \mathfrak{F}$. Then F_k is a union of members of $\mathfrak{E}$ for every $k \in \mathbb{N}$ and then $\bigcup_{k \in \mathbb{N}} F_k$ is a union of members of $\mathfrak{E}$ and therefore $\bigcup_{k \in \mathbb{N}} F_k \in \mathfrak{F}$. This verifies $3°$ and completes the proof that $\mathfrak{F}$ is a σ-algebra of subsets of X.
   2. Let us show that $\mathfrak{F} = \sigma(\mathfrak{E})$. Now $\mathfrak{F}$ is the collection of all arbitrary unions of members of the collection $\mathfrak{E} = \{A_i : i \in \mathbb{N}\}$. Then A_i, as the union of one member of $\mathfrak{E}$, is in $\mathfrak{F}$ for every $i \in \mathbb{N}$. Thus $\mathfrak{E} \subset \mathfrak{F}$. Then $\mathfrak{F}$ is a σ-algebra of subsets of X containing $\mathfrak{E}$. To show that $\mathfrak{F}$ is the smallest σ-algebra of subsets of X containing $\mathfrak{E}$, assume the contrtary. Thus we assume that there exists a σ-algebra $\mathfrak{G}$ of subsets of X containing $\mathfrak{E}$ such that $\mathfrak{G} \subset \mathfrak{F}$ and $\mathfrak{G} \neq \mathfrak{F}$. Then there exists $F \in \mathfrak{F}$ such that $F \notin \mathfrak{G}$. Now $F \in \mathfrak{F}$ is a union of members of $\mathfrak{E}$. Since $\mathfrak{E}$ is a countable collection, F is a union of at most countably many members of $\mathfrak{E}$ union of members of $\mathfrak{E}$. Then since $\mathfrak{G}$ is a σ-algebra of subsets of X containing $\mathfrak{E}$, $\mathfrak{G}$ must contain every union of countably members of $\mathfrak{E}$. Thus $\mathfrak{G}$ must contain F. This is a contradiction. This proves that $\mathfrak{F} = \sigma(\mathfrak{E})$.
   3. Let us show that the cardinality of $\sigma(\mathfrak{E})$ is equal to $2^{\aleph_0}$. Since $\sigma(\mathfrak{E}) = \mathfrak{F}$, we show that the cardinality of $\mathfrak{F}$ is equal to $2^{\aleph_0}$. Now $\mathfrak{F}$ is the collection of all arbitrary unions of members of the disjoint countable collection $\mathfrak{E} = \{A_i : i \in \mathbb{N}\}$ of non-empty subsets of X. To form a union of members of $\mathfrak{E}$, we either select A_1 or we do not select A_1 and then we either select A_2 or we do not select A_2 and so on indefinitely. Thus there are $2^{\aleph_0}$ ways of forming a union of members of $\mathfrak{E}$. Any 2 different ways of forming a union result in 2 distinct unions since $\mathfrak{E}$ is a disjoint collection of non-empty sets. Thus the cardinality of $\mathfrak{F}$ is equal to $2^{\aleph_0}$. ∎

Prob. 1.17. Show that a σ-algebra of subsets of a set cannot be a countably infinite collection, that is, it is either a finite or an uncountable collection.

Proof. Let $\mathfrak{A}$ be a σ-algebra of subsets of a set X. Then $\mathfrak{A}$ may be a finite collection. For instance the collection $\{\emptyset, X\}$ is trivially a σ-algebra of subsets of X and it consists of 2 members. We show below that if a σ-algebra $\mathfrak{A}$ of subsets of a set X is an infinite collection then it is an uncountable collection.

1. We show first that if $\mathfrak{A}$ is an infinite collection then there exists a strictly decreasing sequence $(A_n : n \in \mathbb{N})$ of non-empty members of $\mathfrak{A}$, strictly decreasing in the sense that $A_n \supset A_{n+1}$ and $A_n \neq A_{n+1}$ for $n \in \mathbb{N}$.

Let $A \in \mathfrak{A}$. We say that A is divisible if there exist $A', A'' \in \mathfrak{A}$ such that $A' \neq \emptyset$, $A'' \neq \emptyset$, $A' \cap A'' = \emptyset$ and $A' \cup A'' = A$. When A is divisible, we call the collection $\{A', A''\}$ a division of A.

Observe that $A \in \mathfrak{A}$ is divisible if and only if there exists $B \in \mathfrak{A}$ such that $A \cap B \neq \emptyset$ and $A \cap B^c \neq \emptyset$. In this case, $\{A \cap B, A \cap B^c\}$ is a division of A and we say that B divides A into $\{A \cap B, A \cap B^c\}$.

Let us write $[A : A_1, \ldots, A_n]$ to indicate that $\{A_1, \ldots, A_n\}$ is a disjoint collection of non-empty members of $\mathfrak{A}$ such that $X = \bigcup_{k=1}^n A_k$.

Let us show that $X \in \mathfrak{A}$ is divisible. Since $\mathfrak{A}$ is an infinite collection, there exists $A_1 \in \mathfrak{A}$ such that $A_1 \neq \emptyset$ and $A_1 \neq X$. Let $A_2 = A_1^c$. Then we have $A_2 \neq \emptyset$, $A_1 \cap A_2 = \emptyset$ and $A_1 \cup A_2 = X$. This shows that X is divisible and $\{A_1, A_2\}$ is a division of X. Then we have $[X : A_1, A_2]$.

Let us show next that at least one in the collection $[X : A_1, A_2]$ is divisible. Since $\mathfrak{A}$ is an infinite collection, there exists $B \in \mathfrak{A}$ such that $B \neq \emptyset$, $B \neq X$, $B \neq A_1$ and $B \neq A_2$. Since $X = A_1 \cup A_2$, such $B \in \mathfrak{A}$ divides at least one of the collection $\{A_1, A_2\}$. If B divides A_1 into $\{A_{1,1}, A_{1,2}\}$ then we have $[X : A_{1,1}, A_{1,2}, A_2]$ and if B divides A_2 into $\{A_{2,1}, A_{2,2}\}$ then we have $[X : A_1, A_{2,1}, A_{2,2}]$.

Let us show that if we have $[X : A_{1,1}, A_{1,2}, A_2]$ then at least one member of the collection $[X : A_{1,1}, A_{1,2}, A_2]$ is divisible. Consider the collection:

$$\emptyset, \ X, \ A_{1,1}, \ A_{1,2}, \ A_2, \ A_{1,1} \cup A_{1,2}, \ A_{1,1} \cup A_2, \ A_{1,2} \cup A_2.$$

Since $\mathfrak{A}$ is an infinite collection there exists $B \in \mathfrak{A}$ that is not any in the finite collection above. Such $B \in \mathfrak{A}$ divides at least one of the collection $[X : A_{1,1}, A_{1,2}, A_2]$.

Similarly if we have $[X : A_1, A_{2,1}, A_{2,2}]$ then, by the same argument as above, at least one member of the collection $[X : A_1, A_{2,1}, A_{2,2}]$ is divisible.

Let us show that this process of division can be repeated indefinitely. Suppose that after dividing k times we have a collection:

$$[X : A_1, \ldots, A_{k+1}].$$

Consider the sets:

$$\emptyset, \; X, \; A_1, \ldots, A_{k+1},$$

$$\frac{(k+1)!}{2!} \text{ unions of 2 members of } \{A_1, \ldots, A_{k+1}\},$$

$$\frac{(k+1)!}{3!} \text{ unions of 3 members of } \{A_1, \ldots, A_{k+1}\},$$

$$\vdots$$

$$\frac{(k+1)!}{k!} \text{ unions of } k \text{ members of } \{A_1, \ldots, A_{k+1}\}.$$

This is a finite collection of members of $\mathfrak{A}$. Then since $\mathfrak{A}$ is an infinite collection, there exists $B \in \mathfrak{A}$ that is not any in the finite collection above. Such $B \in \mathfrak{A}$ divides at least one member of the collection $[X : A_1, \ldots, A_{k+1}]$. This shows that the process of division can be repeated indefinitely. This then implies the existence of a strictly decreasing sequence of non-empty members of $\mathfrak{A}$, $(A_{i_1}, A_{i_1,i_2}, A_{i_1,i_2,i_3}, \ldots)$, where i_n assumes the values $\{1, 0\}$ for $n \in \mathbb{N}$.

2. We showed above the existence of a strictly decreasing sequence $(A_n : n \in \mathbb{N})$ of non-empty members of $\mathfrak{A}$. For $n \in \mathbb{N}$, let $B_n = A_n \setminus A_{n+1}$. Then $\{B_n : n \in \mathbb{N}\}$ is a disjoint collection of non-empty members of $\mathfrak{A}$. Let us define

$$B_n^1 = B_n \quad \text{and} \quad B_n^0 = \emptyset \quad \text{for } n \in \mathbb{N}.$$

Let $\lambda = (\lambda_n : n \in \mathbb{N})$ where λ_n assumes the values $\{1, 0\}$ for $n \in \mathbb{N}$. Let Λ be the collection of all sequences λ. Then the cardinality of Λ is equal to $2^{\aleph_0}$. Corresponding to $\lambda \in \Lambda$ given by $\lambda = (\lambda_n : n \in \mathbb{N})$ define a subset E_λ of X by setting

$$E_\lambda = B_1^{\lambda_1} \cup B_2^{\lambda_2} \cup B_3^{\lambda_3} \cup \cdots = \bigcup_{n \in \mathbb{N}} B_n^{\lambda_n} \in \mathfrak{A}.$$

Let us define a mapping φ of Λ into $\mathfrak{A}$ by setting

$$\varphi(\lambda) = E_\lambda \in \mathfrak{A} \quad \text{for } \lambda \in \Lambda.$$

Let us show that the mapping φ is one-to-one. Suppose $\lambda', \lambda'' \in \Lambda$ and $\lambda' \neq \lambda''$. Now $\lambda' \neq \lambda''$ implies that there exists $n_0 \in \mathbb{N}$ such that $\lambda'_{n_0} \neq \lambda''_{n_0}$. There are two possible cases: Case 1 with $\lambda'_{n_0} = 1$ and $\lambda''_{n_0} = 0$ and $\lambda'_{n_0} = 0$ and $\lambda''_{n_0} = 1$. In Case 1, the n_0-th component of the union $E_{\lambda'}$ is B_{n_0} and the n_0-th component of the union $E_{\lambda''}$ is $\emptyset$. Since $\{B_n : n \in \mathbb{N}\}$ is a disjoint collection of non-empty members of $\mathfrak{A}$, this implies that $E_{\lambda'} \neq E_{\lambda''}$. In Case 2, the n_0-th component of the union $E_{\lambda'}$ is $\emptyset$ and the n_0-th component of the union $E_{\lambda''}$ is B_{n_0} and this implies that $E_{\lambda'} \neq E_{\lambda''}$. Thus we have shown that if $\lambda' \neq \lambda''$ then $E_{\lambda'} \neq E_{\lambda''}$. This proves that the mapping φ of Λ into $\mathfrak{A}$ is one-to-one. Then since the cardinality of Λ is equal to $2^{\aleph_0}$, the cardinality of $\mathfrak{A}$ is at least equal to $2^{\aleph_0}$. This shows that $\mathfrak{A}$ is an uncountable collection. ∎

Prob. 1.18. Let $\mathfrak{E} = \{E_1, \cdots, E_n\}$ be a finite collection of distinct, but not necessarily disjoint, subsets of a set X. Let $\mathfrak{D}$ be the collection of all subsets of X of the type:

$$E_1^{\lambda_1} \cap E_2^{\lambda_2} \cap \cdots \cap E_n^{\lambda_n},$$

where λ_i assumes the values $\{1, 0\}$ and $E_i^1 = E_i$ and $E_i^0 = E_i^c$ for $i = 1, \ldots, n$. Let $\mathfrak{F}$ be the collection of all arbitrary unions of members of $\mathfrak{D}$.
(a) Show that any two distinct members of $\mathfrak{D}$ are necessarily disjoint, that is, $\mathfrak{D}$ is a disjoint collection.
(b) Show that the cardinality of $\mathfrak{D}$ is at most 2^n.
(c) Show that $\mathfrak{F} = \alpha(\mathfrak{E})$.
(d) Show that the cardinality of $\alpha(\mathfrak{E})$ has at most 2^{2^n}.
(e) Show that $\sigma(\mathfrak{E}) = \alpha(\mathfrak{E})$.
Remark. For an arbitrary collection $\mathfrak{E}$ of subsets of a set X, the smallest σ-algebra of subsets of X containing $\mathfrak{E}$, $\sigma(\mathfrak{E})$, always exists according to Theorem 1.11. Prob. 1.18 presents a method of constructing $\sigma(\mathfrak{E})$ for the case that $\mathfrak{E}$ is a finite collection.

Proof. 1. Let us prove (a). Let $D', D'' \in \mathfrak{D}$. Then we have

$$D' = E_1^{\lambda_1'} \cap E_2^{\lambda_2'} \cap \cdots \cap E_n^{\lambda_n'},$$

$$D'' = E_1^{\lambda_1''} \cap E_2^{\lambda_2''} \cap \cdots \cap E_n^{\lambda_n''}.$$

Now if $E_i^{\lambda_i'} = E_i^{\lambda_i''}$ for every $i = 1, \ldots, n$, then $D' = D''$. Thus $D' \neq D''$ implies that there exists at least one i such that $E_i^{\lambda_i'} \neq E_i^{\lambda_i''}$, that is, $\lambda_i' \neq \lambda_i''$. Then we have either Case 1: $\lambda_i' = 1$ and $\lambda_i'' = 0$ so that $E_i^{\lambda_i'} = E_i$ and $E_i^{\lambda_i''} = E_i^c$ or Case 2: $\lambda_i' = 0$ and $\lambda_i'' = 1$ so that $E_i^{\lambda_i'} = E_i^c$ and $E_i^{\lambda_i''} = E_i$. In any case we have $E_i^{\lambda_i'} \cap E_i^{\lambda_i''} = \emptyset$. Then since $D' \subset E_i^{\lambda_i'}$ and $D'' \subset E_i^{\lambda_i''}$, we have $D' \cap D'' = \emptyset$. This shows that $\mathfrak{D}$ is a disjoint collection.
2. Let us prove (b). Let Λ be the collection of all n-term sequences $\lambda = (\lambda_1, \ldots, \lambda_n)$ where λ_i assumes the values $\{1, 0\}$ for $i = 1, \ldots, n$. The cardinality of the set Λ is equal to 2^n. Let us define a mapping φ of Λ into $\mathfrak{D}$ by setting

$$\varphi(\lambda) = E_1^{\lambda_1} \cap E_2^{\lambda_2} \cap \cdots \cap E_n^{\lambda_n} \in \mathfrak{D} \quad \text{for } \lambda = (\lambda_1, \ldots, \lambda_n) \in \Lambda.$$

Every $D \in \mathfrak{D}$ is given by $D = E_1^{\lambda_1} \cap E_2^{\lambda_2} \cap \cdots \cap E_n^{\lambda_n}$. With $\lambda = (\lambda_1, \ldots, \lambda_n)$, we have $\varphi(\lambda) = D$. This shows that φ maps Λ onto $\mathfrak{D}$. However the mapping φ need not be one-to-one. Consider an example in which $n = 4$ and $E_1 \subset E_2$ and $E_3 \subset E_4$. Let $\lambda' = (1, 0, 1, 1)$ and $\lambda'' = (1, 1, 1, 0)$. We have $\lambda' \neq \lambda''$. However we have

$$\varphi(\lambda') = \varphi\big((1, 0, 1, 1)\big) = E_1 \cap E_2^c \cap E_3 \cap E_4 = \emptyset \cap E_3 \cap E_4 = \emptyset,$$

$$\varphi(\lambda'') = \varphi\big((1, 1, 1, 0)\big) = E_1 \cap E_2 \cap E_3 \cap E_4^c = E_1 \cap E_2 \cap \emptyset = \emptyset,$$

so that $\varphi(\lambda') = \emptyset = \varphi(\lambda'')$. This shows that the mapping φ of Λ onto $\mathfrak{D}$ need not be one-to-one. Then the cardinality of $\mathfrak{D}$ cannot exceed that of Λ and is hence at most equal to 2^n.

3. Let us prove (c), that is, $\mathfrak{F} = \alpha(\mathfrak{C})$.

3.1. First of all let us show that $\mathfrak{F}$ is an algebra of subsets of X. Thus we are to verify:

1° $X \in \mathfrak{F}$.

2° $F \in \mathfrak{F} \Rightarrow F^c \in \mathfrak{F}$.

3° $F_1, F_2 \in \mathfrak{F} \Rightarrow F_1 \cup F_2 \in \mathfrak{F}$.

Let $x \in X$. For every $i = 1, \ldots, n$ we have $E_i^1 \cup E_i^0 = E_i \cup E_i^c = X$. Thus $x \in E_i^{\lambda_i}$ where $\lambda_i = 1$ or $\lambda_i = 0$. Since this is true for every $i = 1, \ldots, n$ we have

$$x \in E_1^{\lambda_1} \cap E_2^{\lambda_2} \cap \cdots \cap E_n^{\lambda_n} = D \in \mathfrak{D}.$$

Thus every $x \in X$ is contained in some $D \in \mathfrak{D}$. This implies that $X = \bigcup_{D \in \mathfrak{D}} D \in \mathfrak{F}$. This verifies 1°.

Let $F \in \mathfrak{F}$. Then F is a union of members of $\mathfrak{D}$. We showed above that $X = \bigcup_{D \in \mathfrak{D}} D$, that is, X is the union of all members of $\mathfrak{D}$. Now $\mathfrak{D}$ is a disjoint collection as we showed in (a). Then $F^c = X \setminus F$ is a union of members of $\mathfrak{D}$ and thus $F^c \in \mathfrak{F}$. This verifies 2°.

Let $F_1, F_2 \in \mathfrak{F}$. Then F_1 is a union of members of $\mathfrak{D}$ and so is F_2. Then $F_1 \cup F_2$ is a union of members of $\mathfrak{D}$ and therefore $F_1 \cup F_2 \in \mathfrak{F}$. This verifies 3° and completes the proof that $\mathfrak{F}$ is an algebra of subsets of X.

3.2. Let us show that $\mathfrak{C} \subset \mathfrak{F}$. Since $\mathfrak{C} = \{E_1, \cdots, E_n\}$, it suffices to show that $E_i \in \mathfrak{F}$ for every $i = 1, \ldots, n$. We showed in **2** above that $\varphi(\Lambda) = \mathfrak{D}$. For $\lambda = (\lambda_1, \ldots, \lambda_n) \in \Lambda$ having $\lambda_i = 1$, we have

$$\varphi(\lambda) = E_1^{\lambda_1} \cap \cdots \cap E_i \cap \cdots \cap E_n^{\lambda_n} \subset E_i,$$

and for $\lambda = (\lambda_1, \ldots, \lambda_n) \in \Lambda$ having $\lambda_i = 0$, we have

$$\varphi(\lambda) = E_1^{\lambda_1} \cap \cdots \cap E_i^c \cap \cdots \cap E_n^{\lambda_n} \subset E_i^c.$$

Thus we have

(1)
$$\begin{cases} \bigcup_{\lambda \in \Lambda \text{ with } \lambda_i = 1} \varphi(\lambda) \subset E_i, \\ \bigcup_{\lambda \in \Lambda \text{ with } \lambda_i = 0} \varphi(\lambda) \subset E_i^c. \end{cases}$$

Now

(2)
$$\left\{ \bigcup_{\lambda \in \Lambda \text{ with } \lambda_i = 1} \varphi(\lambda) \right\} \cup \left\{ \bigcup_{\lambda \in \Lambda \text{ with } \lambda_i = 0} \varphi(\lambda) \right\} = \bigcup_{\lambda \in \Lambda} \varphi(\lambda) = \bigcup_{D \in \mathfrak{D}} D = X.$$

On the other hand, we have

(3)
$$E_i \cup E_i^c = X \quad \text{and} \quad E_i \cap E_i^c = \emptyset.$$

Then (1), (2) and (3) imply

(4)
$$\begin{cases} \bigcup_{\lambda \in \Lambda \text{ with } \lambda_i = 1} \varphi(\lambda) = E_i, \\ \bigcup_{\lambda \in \Lambda \text{ with } \lambda_i = 0} \varphi(\lambda) = E_i^c. \end{cases}$$

Since $\varphi(\lambda)$ is a member of $\mathfrak{D}$ for every $\lambda \in \Lambda$ as we pointed out in **1**, (4) shows that E_i is a union of members of $\mathfrak{D}$ and therefore $E_i \in \mathfrak{F}$. This completes the proof that $\mathfrak{E} \subset \mathfrak{F}$.

3.3. Let us show that $\mathfrak{F} = \alpha(\mathfrak{E})$, the smallest algebra of subsets of X containing $\mathfrak{E}$. We have shown in **3.1** and **3.2** that $\mathfrak{F}$ is an algebra of subsets of X containing $\mathfrak{E}$. Let $\mathfrak{G}$ be an algebra of subsets of X containing $\mathfrak{E}$. Let us show that $\mathfrak{F} \subset \mathfrak{G}$. Since $\mathfrak{E} \subset \mathfrak{G}$, we have $E_1, \ldots, E_n \in \mathfrak{G}$. Then since $\mathfrak{G}$ is an algebra, we have $E_1^c, \ldots, E_n^c \in \mathfrak{G}$. Then again since $\mathfrak{G}$ is an algebra, we have $E_1^{\lambda_1} \cap E_2^{\lambda_2} \cap \cdots \cap E_n^{\lambda_n} \in \mathfrak{G}$. This shows that every member of $\mathfrak{D}$ is in $\mathfrak{G}$ and hence we have $\mathfrak{D} \subset \mathfrak{G}$. Then since $\mathfrak{G}$ is an algebra and $\mathfrak{D}$ is a finite collection so that an arbitrary union of members of $\mathfrak{D}$ is necessarily a union of finitely many members, the algebra $\mathfrak{G}$ contains every arbitrary union of members of $\mathfrak{D}$. Therefore we have $\mathfrak{F} \subset \mathfrak{G}$. This shows that $\mathfrak{F}$ is the smallest algebra of subsets of X containing $\mathfrak{E}$.

4. Let us prove (d). By (c), we have $\alpha(\mathfrak{E}) = \mathfrak{F}$. Thus to show that the cardinality of $\alpha(\mathfrak{E})$ is at most 2^{2^n}, we show that the cardinality of $\mathfrak{F}$ is at most 2^{2^n}. Now $\mathfrak{F}$ is the collection of all arbitrary unions of members of $\mathfrak{D} = \{D_1, \ldots, D_m\}$ where $m \leq 2^n$ by (b).

To form a union of members of $\mathfrak{D}$, for every $i = 1, \ldots, m$, we have 2 choices, that is, we either select D_i or we do not select D_i. Thus we have 2^m different choices in forming a union of members of $\mathfrak{D}$. Since $\mathfrak{D}$ is a disjoint collection, two different choices among the 2^m choices result in two distinct unions. Thus the cardinality of $\mathfrak{F}$ is equal to $2^m \leq 2^{2^n}$.

5. Let us prove (e). Now $\alpha(\mathfrak{E})$ is an algebra of subsets of X containing $\mathfrak{E}$. According to (d), $\alpha(\mathfrak{E})$ is a finite collection. Thus $\alpha(\mathfrak{E})$ is trivially a σ-algebra of subsets of X.

Let us show that $\alpha(\mathfrak{E}) = \sigma(\mathfrak{E})$, the smallest σ-algebra of subsets of X containing $\mathfrak{E}$. Let $\mathfrak{G}$ be a σ-algebra of subsets of X containing $\mathfrak{E}$. Then we have $\mathfrak{D} \subset \mathfrak{G}$. Since $\mathfrak{D}$ is a finite collection, a union of members of $\mathfrak{D}$ is necessarily a union of finitely many members of $\mathfrak{D}$. Then the σ-algebra $\mathfrak{G}$ contains every union of members of $\mathfrak{D}$. Thus we have $\mathfrak{F} \subset \mathfrak{G}$. But we have $\mathfrak{F} = \alpha(\mathfrak{E})$ by (c). Thus we have $\alpha(\mathfrak{E}) \subset \mathfrak{G}$. This shows that $\alpha(\mathfrak{E})$ is the smallest σ-algebra of subsets of X containing $\mathfrak{E}$, that is, $\alpha(\mathfrak{E}) = \sigma(\mathfrak{E})$. ∎

Prob. 1.19. Let $\mathfrak{C}$ be an arbitrary collection of subsets of a set X. Consider $\alpha(\mathfrak{C})$, the smallest algebra of subsets of X containing $\mathfrak{C}$. Show that for every $A \in \alpha(\mathfrak{C})$ there exists a finite subcollection $\mathfrak{C}_A$ of $\mathfrak{C}$ depending on A such that $A \in \alpha(\mathfrak{C}_A)$.

Proof. Let $\mathfrak{F}$ be the collection of every $A \in \alpha(\mathfrak{C})$ such that there exists a finite subcollection $\mathfrak{C}_A$ of $\mathfrak{C}$ such that $A \in \alpha(\mathfrak{C}_A)$.

    1. Let us show that $\mathfrak{F} \neq \emptyset$ and indeed we have $\mathfrak{C} \subset \mathfrak{F}$. Select $E \in \mathfrak{C}$ arbitrarily. Then $E \in \alpha(\mathfrak{C})$. Moreover with the finite subcollection $\{E\}$ of $\mathfrak{C}$, we have $E \in \alpha(\{E\})$. Thus $E \in \mathfrak{F}$. Since this holds for an arbitrary $E \in \mathfrak{C}$, we have $\mathfrak{C} \subset \mathfrak{F}$.

    2. Let us show that $\mathfrak{F}$ is an algebra of subsets of X. For this purpose we verify that $\mathfrak{F}$ satisfies the following conditions:

$1°$ $X \in \mathfrak{F}$.

$2°$ $A \in \mathfrak{F} \Rightarrow A^c \in \mathfrak{F}$.

$3°$ $A, B \in \mathfrak{F} \Rightarrow A \cup B \in \mathfrak{F}$.

    For every $E \in \mathfrak{C}$ we have $E \in \alpha(\{E\})$. Then since $\alpha(\{E\})$ is an algebra of subsets of X, we have $X \in \alpha(\{E\})$. Thus $X \in \mathfrak{F}$. This verifies $1°$.

    If $A \in \mathfrak{F}$, then $A \in \alpha(\mathfrak{C}_A)$ where $\mathfrak{C}_A$ is a finite subcollection of $\mathfrak{C}$. Then since $\alpha(\mathfrak{C}_A)$ is an algebra of subsets of X, $A \in \alpha(\mathfrak{C}_A)$ implies $A^c \in \alpha(\mathfrak{C}_A)$. Then $A^c \in \mathfrak{F}$. This verifies $2°$.

    Suppose $A, B \in \mathfrak{F}$. Then $A \in \alpha(\mathfrak{C}_A)$ where $\mathfrak{C}_A$ is a finite subcollection of $\mathfrak{C}$ and $B \in \alpha(\mathfrak{C}_B)$ where $\mathfrak{C}_B$ is a finite subcollection of $\mathfrak{C}$. Let $\mathfrak{C}_{A \cup B} = \mathfrak{C}_A \cup \mathfrak{C}_B$, a finite subcollection of $\mathfrak{C}$. Then since $\mathfrak{C}_A \subset \mathfrak{C}_{A \cup B}$ and $\mathfrak{C}_B \subset \mathfrak{C}_{A \cup B}$, we have

$$A \in \alpha(\mathfrak{C}_A) \subset \alpha(\mathfrak{C}_{A \cup B}),$$

$$B \in \alpha(\mathfrak{C}_B) \subset \alpha(\mathfrak{C}_{A \cup B}).$$

Then since $\alpha(\mathfrak{C}_{A \cup B})$ is an algebra of subsets of X, we have $A \cup B \in \alpha(\mathfrak{C}_{A \cup B})$. This shows that $A \cup B \in \mathfrak{F}$ and verifies $3°$. This completes the proof that $\mathfrak{F}$ is an algebra of subsets of X.

    3. We have shown above that $\mathfrak{F}$ is an algebra of subsets of X containing $\mathfrak{C}$. Then since $\alpha(\mathfrak{C})$ is the smallest algebra of subsets of X containing $\mathfrak{C}$, we have $\alpha(\mathfrak{C}) \subset \mathfrak{F}$. Then every $A \in \alpha(\mathfrak{C})$ is in $\mathfrak{F}$ and thus for every $A \in \alpha(\mathfrak{C})$ there exists a finite subcollection $\mathfrak{C}_A$ of $\mathfrak{C}$ such that $A \in \alpha(\mathfrak{C}_A)$. $\blacksquare$

Prob. 1.20. Let $\mathfrak{C}$ be an arbitrary collection of subsets of a set X. Consider $\sigma(\mathfrak{C})$, the smallest σ-algebra of subsets of X containing $\mathfrak{C}$. Show that for every $A \in \sigma(\mathfrak{C})$ there exists an at most countable subcollection $\mathfrak{C}_A$ of $\mathfrak{C}$ depending on A such that $A \in \sigma(\mathfrak{C}_A)$.

Proof. Let $\mathfrak{F}$ be the collection of every $A \in \sigma(\mathfrak{C})$ such that there exists an at most countable subcollection $\mathfrak{C}_A$ of $\mathfrak{C}$ such that $A \in \sigma(\mathfrak{C}_A)$.

1. Let us show that $\mathfrak{F} \neq \emptyset$ and moreover we have $\mathfrak{C} \subset \mathfrak{F}$. Pick $E \in \mathfrak{C}$ arbitrarily. Then $E \in \sigma(\mathfrak{C})$. Furthermore with the finite subcollection $\{E\}$ of $\mathfrak{C}$, we have $E \in \sigma(\{E\})$. Thus $E \in \mathfrak{F}$. Since this holds for an arbitrary $E \in \mathfrak{C}$, we have $\mathfrak{C} \subset \mathfrak{F}$.

2. Let us show that $\mathfrak{F}$ is a σ-algebra of subsets of X. Thus we are to verify that $\mathfrak{F}$ satisfies the following conditions:

1° $X \in \mathfrak{F}$.
2° $A \in \mathfrak{F} \Rightarrow A^c \in \mathfrak{F}$.
3° $(A_n : n \in \mathbb{N}) \subset \mathfrak{F} \Rightarrow \bigcup_{n \in \mathbb{N}} A_n \in \mathfrak{F}$.

For every $E \in \mathfrak{C}$ we have $E \in \sigma(\{E\})$. Then since $\sigma(\{E\})$ is a σ-algebra of subsets of X, we have $X \in \sigma(\{E\})$. Thus $X \in \mathfrak{F}$. This verifies 1°.

If $A \in \mathfrak{F}$, then $A \in \sigma(\mathfrak{C}_A)$ where $\mathfrak{C}_A$ is an at most countable subcollection of $\mathfrak{C}$. Then since $\sigma(\mathfrak{C}_A)$ is a σ-algebra of subsets of X, $A \in \sigma(\mathfrak{C}_A)$ implies $A^c \in \sigma(\mathfrak{C}_A)$. Then $A^c \in \mathfrak{F}$. This verifies 2°.

Let $(A_n : n \in \mathbb{N})$ be a sequence in $\mathfrak{F}$. Then $A_n \in \mathfrak{F}$ implies that there exists an at most countable subcollection $\mathfrak{C}_{A_n}$ of $\mathfrak{C}$ such that $A_n \in \sigma(\mathfrak{C}_{A_n})$. Let us define an at most countable subcollection of $\mathfrak{C}$ by setting

$$(1) \qquad \mathfrak{C}_{\bigcup_{n \in \mathbb{N}} A_n} = \bigcup_{n \in \mathbb{N}} \mathfrak{C}_{A_n}.$$

Then we have $\mathfrak{C}_{A_n} \subset \mathfrak{C}_{\bigcup_{n \in \mathbb{N}} A_n}$ for every $n \in \mathbb{N}$ and this implies

$$(2) \qquad A_n \in \sigma(\mathfrak{C}_{A_n}) \subset \sigma(\mathfrak{C}_{\bigcup_{n \in \mathbb{N}} A_n}) \quad \text{for } n \in \mathbb{N}.$$

Then since $\sigma(\mathfrak{C}_{\bigcup_{n \in \mathbb{N}} A_n})$ is a σ-algebra of subsets of X, (2) implies

$$(3) \qquad \bigcup_{n \in \mathbb{N}} A_n \in \sigma(\mathfrak{C}_{\bigcup_{n \in \mathbb{N}} A_n}).$$

This shows that $\bigcup_{n \in \mathbb{N}} A_n \in \mathfrak{F}$ and verifies 3°. Thus $\mathfrak{F}$ is a σ-algebra of subsets of X.

3. In **1** and **2**, we showed that $\mathfrak{F}$ is a σ-algebra of subsets of X containing $\mathfrak{C}$. Then since $\sigma(\mathfrak{C})$ is the smallest σ-algebra of subsets of X containing $\mathfrak{C}$, we have $\sigma(\mathfrak{C}) \subset \mathfrak{F}$. Then every $A \in \sigma(\mathfrak{C})$ is in $\mathfrak{F}$ and thus for every $A \in \sigma(\mathfrak{C})$ there exists an at most countable subcollection $\mathfrak{C}_A$ of $\mathfrak{C}$ such that $A \in \sigma(\mathfrak{C}_A)$. ∎

Prob. 1.21. Let μ be a measure on a σ-algebra $\mathfrak{A}$ of subsets of a set X and let $\mathfrak{A}_0$ be a sub-σ-algebra of $\mathfrak{A}$, that is, $\mathfrak{A}_0$ is a σ-algebra of subsets of X and $\mathfrak{A}_0 \subset \mathfrak{A}$. Show that the restriction of μ to $\mathfrak{A}_0$ is a measure on $\mathfrak{A}_0$.

Proof. Since μ is a measure on a σ-algebra $\mathfrak{A}$ of subsets of X, μ satisfies the defining conditions for a measure:
$1°$ $\mu(E) \in [0, \infty]$ for every $E \in \mathfrak{A}$.
$2°$ $\mu(\emptyset) = 0$.
$3°$ $(E_n : n \in \mathbb{N}) \subset \mathfrak{A}$, disjoint $\Rightarrow \mu\left(\bigcup_{n \in \mathbb{N}} E_n\right) = \sum_{n \in \mathbb{N}} \mu(E_n)$.
Then since $\mathfrak{A}_0 \subset \mathfrak{A}$, $1°$, $2°$ and $3°$ imply:
$4°$ $\mu(E) \in [0, \infty]$ for every $E \in \mathfrak{A}_0$.
$5°$ $\mu(\emptyset) = 0$.
$6°$ $(E_n : n \in \mathbb{N}) \subset \mathfrak{A}_0$, disjoint $\Rightarrow \mu\left(\bigcup_{n \in \mathbb{N}} E_n\right) = \sum_{n \in \mathbb{N}} \mu(E_n)$.
This shows that μ is a measure on the σ-algebra $\mathfrak{A}_0$. ∎

Prob. 1.22. Let $(X, \mathfrak{A}, \mu)$ be a measure space. Show that for any $E_1, E_2 \in \mathfrak{A}$ we have the equality: $\mu(E_1 \cup E_2) + \mu(E_1 \cap E_2) = \mu(E_1) + \mu(E_2)$.

Proof. Let us write

$$E_1 \cup E_2 = (E_1 \cap E_2) \cup \left(E_1 \setminus (E_1 \cap E_2)\right) \cup \left(E_2 \setminus (E_1 \cap E_2)\right),$$

where the right side is a union of 3 disjoint sets. Then we have

$$\mu(E_1 \cup E_2) = \mu(E_1 \cap E_2) + \mu\left(E_1 \setminus (E_1 \cap E_2)\right) + \mu\left(E_2 \setminus (E_1 \cap E_2)\right).$$

Then adding $\mu(E_1 \cap E_2)$ on both sides of the last equality, we have

$$\mu(E_1 \cup E_2) + \mu(E_1 \cap E_2) = \left\{\mu\left(E_1 \setminus (E_1 \cap E_2)\right) + \mu(E_1 \cap E_2)\right\}$$
$$+ \left\{\mu\left(E_1 \setminus (E_2 \cap E_2)\right) + \mu(E_1 \cap E_2)\right\}$$
$$= \mu(E_1) + \mu(E_2). \ ∎$$

Prob. 1.23. Let $(X, \mathfrak{A})$ be a measurable space. Let μ_k be a measure on the σ-algebra $\mathfrak{A}$ of subsets of X and let $\alpha_k \geq 0$ for every $k \in \mathbb{N}$. Define a set function μ on $\mathfrak{A}$ by setting $\mu = \sum_{k \in \mathbb{N}} \alpha_k \mu_k$. Show that μ is a measure on $\mathfrak{A}$.

Proof. To show that μ is a measure on $\mathfrak{A}$, we verify that μ satisfies the following conditions:
1° $\mu(E) \in [0, \infty]$ for every $E \in \mathfrak{A}$.
2° $\mu(\emptyset) = 0$.
3° $(E_n : n \in \mathbb{N}) \subset \mathfrak{A}$, disjoint $\Rightarrow \mu\left(\bigcup_{n \in \mathbb{N}} E_n\right) = \sum_{n \in \mathbb{N}} \mu(E_n)$.

Now for every $E \in \mathfrak{A}$, since $\mu_k(E) \in [0.\infty]$ and $\alpha_k \geq 0$ for every $k \in \mathbb{N}$, we have $\mu(E) = \sum_{k \in \mathbb{N}} \alpha_k \mu_k(E) \in [0, \infty]$. This verifies 1°.

Secondly we have $\mu_k(\emptyset) = 0$ for every $k \in \mathbb{N}$ and then $\mu(\emptyset) = \sum_{k \in \mathbb{N}} \alpha_k \mu_k(\emptyset) = 0$. This verifies 2°.

Let $(E_n : n \in \mathbb{N})$ be a disjoint sequence in $\mathfrak{A}$. Then we have

$$\mu\left(\bigcup_{n \in \mathbb{N}} E_n\right) = \sum_{k \in \mathbb{N}} \alpha_k \, \mu_k\left(\bigcup_{n \in \mathbb{N}} E_n\right) = \sum_{k \in \mathbb{N}} \alpha_k \left\{\sum_{n \in \mathbb{N}} \mu_k(E_n)\right\}$$

$$= \sum_{n \in \mathbb{N}} \left\{\sum_{k \in \mathbb{N}} \alpha_k \, \mu_k(E_n)\right\} = \sum_{n \in \mathbb{N}} \mu(E_n).$$

This verifies 3°. ∎

Prob. 1.24. Let $X = (0, \infty)$ and let $\mathfrak{J} = \{J_k : k \in \mathbb{N}\}$ where $J_k = (k - 1, k]$ for $k \in \mathbb{N}$. Let $\mathfrak{A}$ be the collection of all arbitrary unions of members of $\mathfrak{J}$. For every $A \in \mathfrak{A}$ let us define $\mu(A)$ to be the number of elements of $\mathfrak{J}$ that constitute A.
(a) Show that $\mathfrak{A}$ is a σ-algebra of subsets of X.
(b) Show that μ is a measure on the σ-algebra $\mathfrak{A}$.
(c) Let $(A_n : n \in \mathbb{N}) \subset \mathfrak{A}$ where $A_n = (n, \infty)$ for $n \in \mathbb{N}$. Show that for the decreasing sequence $(A_n : n \in \mathbb{N})$ we have $\lim_{n \to \infty} \mu(A_n) \neq \mu\left(\lim_{n \to \infty} A_n\right)$.

Proof. 1. To show that $\mathfrak{A}$ is a σ-algebra of subsets of X, we verify that $\mathfrak{A}$ satisfies the following conditions:
1° $X \in \mathfrak{A}$.
2° $A \in \mathfrak{A} \Rightarrow A^c \in \mathfrak{A}$.
3° $(A_n : n \in \mathbb{N}) \subset \mathfrak{A} \Rightarrow \bigcup_{n \in \mathbb{N}} A_n \in \mathfrak{A}$.
    Now $X = \bigcup_{k \in \mathbb{N}} J_k$. Thus $X \in \mathfrak{A}$. This verifies 1°.
    Suppose $A \in \mathfrak{A}$. Then A is a union of members of $\mathfrak{J}$. Since X is the union of all members of $\mathfrak{J}$ and since $\mathfrak{J}$ is a disjoint collection, $A^c = X \setminus A$ is a union of members of $\mathfrak{J}$ and therefore $A^c \in \mathfrak{A}$. This verifies 2°.
    Let $(A_n : n \in \mathbb{N}) \subset \mathfrak{A}$. Then $A_n \in \mathfrak{A}$ is a union of members of $\mathfrak{J}$ for every $n \in \mathbb{N}$ and thus $\bigcup_{n \in \mathbb{N}} A_n$ is a union of members of $\mathfrak{J}$ and therefore $\bigcup_{n \in \mathbb{N}} A_n \in \mathfrak{A}$. This verifies 3°. This completes the proof that $\mathfrak{A}$ is a σ-algebra of subsets of X.
    2. To show that μ is a measure on the σ-algebra $\mathfrak{A}$, we verify that μ satisfies the following conditions:
1° $\mu(A) \in [0, \infty]$ for every $A \in \mathfrak{A}$.
2° $\mu(\emptyset) = 0$.
3° $\{A_n : n \in \mathbb{N}\} \subset \mathfrak{A}$, disjoint $\Rightarrow \mu\left(\bigcup_{n \in \mathbb{N}} A_n\right) = \sum_{n \in \mathbb{N}} \mu(A_n)$.
    But 1°, 2° and 3° follow immediately from the definition of $\mu(A)$ as the number of members of $\mathfrak{J}$ that constitute $A \in \mathfrak{A}$.
    3. For $n \in \mathbb{N}$, let $A_n = (n, \infty)$. Then $A_n = \bigcup_{k \geq n+1} J_k$ so that $A_n \in \mathfrak{A}$ and $\mu(A_n) = \infty$. Now $(A_n; n \in \mathbb{N})$ is a decreasing sequence in $\mathfrak{A}$ and thus $\lim_{n \to \infty} A_n = \bigcap_{n \in \mathbb{N}} A_n = \emptyset$ and then $\mu\left(\lim_{n \to \infty} A_n\right) = 0$. Then we have

$$\lim_{n \to \infty} \mu(A_n) = \infty \neq 0 = \mu\left(\lim_{n \to \infty} A_n\right).$$

This proves (c). ∎

Prob. 1.25. Let $(X, \mathfrak{A}, \mu)$ be a σ-finite measure space so that there exists a sequence $(E_n : n \in \mathbb{N})$ in $\mathfrak{A}$ such that $\bigcup_{n \in \mathbb{N}} E_n = X$ and $\mu(E_n) < \infty$ for every $n \in \mathbb{N}$. Show that there exists a disjoint sequence $(F_n : n \in \mathbb{N})$ in $\mathfrak{A}$ such that $\bigcup_{n \in \mathbb{N}} F_n = X$ and $\mu(F_n) < \infty$ for every $n \in \mathbb{N}$.

Proof. Let us define a sequence $(F_n : n \in \mathbb{N})$ in $\mathfrak{A}$ by setting

$$F_1 = E_1 \quad \text{and} \quad F_n = E_n \setminus \bigcup_{j=1}^{n-1} E_j \quad \text{for } n \geq 2.$$

Then $(F_n : n \in \mathbb{N})$ is a disjoint sequence in $\mathfrak{A}$ and moreover $\bigcup_{n \in \mathbb{N}} F_n = \bigcup_{n \in \mathbb{N}} E_n$. We also have $F_n \subset E_n$ and therefore $\mu(F_n) \leq \mu(E_n) < \infty$ for every $n \in \mathbb{N}$. ∎

Prob. 1.26. Let $\mathfrak{B}_\mathbb{R}$ be the Borel σ-algebra of subsets of $\mathbb{R}$, that is, the smallest σ-algebra of subsets of $\mathbb{R}$ containing the collection of all open sets in $\mathbb{R}$. The Lebesgue measure μ_L is a measure on $\mathfrak{B}_\mathbb{R}$ with the property that for every interval I in $\mathbb{R}$, $\mu_L(I) = \ell(I)$ where $\ell(I)$ is the length of I. The Lebesgue measure μ_L on $\mathfrak{B}_\mathbb{R}$ will be constructed in §3. Here we assume its existence and pose the following problems:
(a) Construct a sequence $(E_n : n \in \mathbb{N})$ of sets in $\mathfrak{B}_\mathbb{R}$ such that $\lim_{n \to \infty} E_n$ exists but $\lim_{n \to \infty} \mu_L(E_n)$ does not exist.
(b) Construct a sequence $(E_n : n \in \mathbb{N})$ of sets in $\mathfrak{B}_\mathbb{R}$ such that $\lim_{n \to \infty} \mu_L(E_n)$ exists but $\lim_{n \to \infty} E_n$ does not exist.
(c) Construct a sequence $(E_n : n \in \mathbb{N})$ of sets in $\mathfrak{B}_\mathbb{R}$ such that both $\lim_{n \to \infty} E_n$ and $\lim_{n \to \infty} \mu_L(E_n)$ exist but $\mu_L\left(\lim_{n \to \infty} E_n\right) \neq \lim_{n \to \infty} \mu_L(E_n)$.
(d) Show that for every $x \in \mathbb{R}$, we have $\{x\} \in \mathfrak{B}_\mathbb{R}$ and $\mu_L(\{x\}) = 0$.
(e) Let Q be the set of all rational numbers in $\mathbb{R}$. Show that $Q \in \mathfrak{B}_\mathbb{R}$ and $\mu_L(Q) = 0$.
(f) Let P be the set of all irrational numbers in $\mathbb{R}$. Show that $P \in \mathfrak{B}_\mathbb{R}$ and $\mu_L(P) = \infty$.
(g) Construct an uncountable union of null sets that is not a null set.

Proof. **(a)** Let $(E_n : n \in \mathbb{N})$ be a disjoint sequence of intervals in $\mathbb{R}$ with $\mu_L(E_n) = \ell(E_n) = 1$ when n is odd and $\mu_L(E_n) = \ell(E_n) = 2$ when n is even. Then $\lim_{n \to \infty} \mu_L(E_n)$ does not exist. On the other hand, disjointness of the sequence $(E_n : n \in \mathbb{N})$ implies that $\lim_{n \to \infty} E_n$ exists and $\lim_{n \to \infty} E_n = \emptyset$ by Prob. 1.3.
    (b) Let $A, B \in \mathfrak{B}_\mathbb{R}$ be such that $A \neq \emptyset$, $B \neq \emptyset$, $A \cap B = \emptyset$ and $\mu_L(A) = \mu_L(B) = c > 0$. (For instance, let $A = (0, 1)$ and $B = (1, 2)$. Then we have $\mu_L(A) = \mu_L(B) = 1 > 0$.) Let $E_n = A$ for odd n and $E_n = B$ for even n. Then we have $\mu_L(E_n) = c$ for every $n \in \mathbb{N}$ and $\lim_{n \to \infty} \mu_L(E_n) = c$.
    On the other hand, we have

$$\liminf_{n \to \infty} E_n = \left\{ x \in \mathbb{R} : x \in E_n \quad \text{for all but finitely many } n \in \mathbb{N} \right\} = \emptyset,$$

$$\limsup_{n \to \infty} E_n = \left\{ x \in \mathbb{R} : x \in E_n \quad \text{for infinitely many } n \in \mathbb{N} \right\} = A \cup B.$$

Thus we have $\liminf\limits_{n\to\infty} E_n \neq \limsup\limits_{n\to\infty} E_n$ and this implies that $\lim\limits_{n\to\infty} E_n$ does not exist.

(c) Let $E_n = (n, \infty)$ for every $n \in \mathbb{N}$. Then $(E_n : n \in \mathbb{N})$ is a decreasing sequence of sets and this implies that $\lim\limits_{n\to\infty} E_n = \bigcap_{n\in\mathbb{N}} E_n = \emptyset$. Then we have $\mu_L\left(\lim\limits_{n\to\infty} E_n\right) = \mu_L(\emptyset) = 0$. On the other hand, we have $\mu_L(E_n) = \mu_L((n,\infty)) = \infty$ for every $n \in \mathbb{N}$ and then $\lim\limits_{n\to\infty} \mu_L(E_n) = \infty$.

(d) Let $x \in \mathbb{R}$ and consider $\{x\}$. Let $E_n = \left(x - \frac{1}{n}, x + \frac{1}{n}\right)$ for $n \in \mathbb{N}$. Then E_n is an open set in $\mathbb{R}$ and therefore $E_n \in \mathfrak{B}_{\mathbb{R}}$. Since E_n is an interval in $\mathbb{R}$, we have $\mu_L(E_n) = \ell(E_n) = \frac{2}{n}$. Now we have $\bigcap_{n\in\mathbb{N}} E_n = \{x\}$. Since $(E_n : n \in \mathbb{N}) \subset \mathfrak{B}_{\mathbb{R}}$ and $\mathfrak{B}_{\mathbb{R}}$ is a σ-algebra, we have $\bigcap_{n\in\mathbb{N}} E_n \in \mathfrak{B}_{\mathbb{R}}$. Thus we have $\{x\} \in \mathfrak{B}_{\mathbb{R}}$.

Since $(E_n : n \in \mathbb{N})$ is a decreasing sequence of sets, we have $\lim\limits_{n\to\infty} E_n = \bigcap_{n\in\mathbb{N}} E_n = \{x\}$. The sequence is contained in a set $A \in \mathfrak{B}_{\mathbb{R}}$ with $\mu_L(A) < \infty$. Indeed if we let $A = (x-1, x+1)$, then $A \in \mathfrak{B}_{\mathbb{R}}$ with $\mu_L(A) = \ell(A) = 2 < \infty$, and $E_n = \left(x - \frac{1}{n}, x + \frac{1}{n}\right) \subset (x - 1, x + 1) = A$ for every $n \in \mathbb{N}$. Then by Theorem 1.28 we have

$$\mu_L\left(\lim_{n\to\infty} E_n\right) = \lim_{n\to\infty} \mu_L(E_n) = \lim_{n\to\infty} \frac{2}{n} = 0.$$

Thus we have $\mu_L(\{x\}) = 0$.

(e) Let Q be the set of all rational numbers in $\mathbb{R}$. Since there are only countably many rational numbers, Q can be represented as $Q = \{x_n : n \in \mathbb{N}\}$. Then $Q = \bigcup_{n\in\mathbb{N}} \{x_n\}$. By (d) we have $\{x_n\} \in \mathfrak{B}_{\mathbb{R}}$ for every $n \in \mathbb{N}$. Then since $\mathfrak{B}_{\mathbb{R}}$ is a σ-algebra we have $\bigcup_{n\in\mathbb{N}} \{x_n\} \in \mathfrak{B}_{\mathbb{R}}$, that is, $Q \in \mathfrak{B}_{\mathbb{R}}$.

Now $(\{x_n\} : n \in \mathbb{N})$ is disjoint sequence of members of $\mathfrak{B}_{\mathbb{R}}$. Then by the countable additivity of μ_L on $\mathfrak{B}_{\mathbb{R}}$, we have

$$\mu_L\left(\bigcup_{n\in\mathbb{N}} \{x_n\}\right) = \sum_{n\in\mathbb{N}} \mu_L(\{x_n\}) = \sum_{n\in\mathbb{N}} 0 = 0.$$

This shows that $\mu_L(Q) = 0$.

(f) Let P be the set of all irrational numbers in $\mathbb{R}$. Then $P \cup Q = \mathbb{R}$ and $P \cap Q = \emptyset$. Thus $P = Q^c$. By (e) we have $Q \in \mathfrak{B}_{\mathbb{R}}$. Then since $\mathfrak{B}_{\mathbb{R}}$ is a σ-algebra, we have $Q^c \in \mathfrak{B}_{\mathbb{R}}$, that is, $P \in \mathfrak{B}_{\mathbb{R}}$. Then by the additivity of μ_L on $\mathfrak{B}_{\mathbb{R}}$, we have

$$\mu_L(\mathbb{R}) = \mu_L(P \cup Q) = \mu_L(P) + \mu_L(Q) = \mu_L(P).$$

Thus we have $\mu_L(P) = \mu_L(\mathbb{R}) = \infty$.

(g) The set P of all irrational numbers is an uncountable set. Thus we can write $P = \{x_\alpha : \alpha \in A\} = \bigcup_{\alpha\in A} \{x_\alpha\}$ where A is an uncountable index set. Now $\{x_\alpha\}$ is a null set for every $\alpha \in A$ according to (e). P is then an uncountable union of null sets. But $\mu_L(P) = \infty$ by (f) and thus P is not a null set. ∎

Prob. 1.27. Consider the measurable space $(\mathbb{R}, \mathfrak{P}(\mathbb{R}))$ where $\mathfrak{P}(\mathbb{R})$ is the σ-algebra of all subsets of $\mathbb{R}$. Let us define a set function μ on $\mathfrak{P}(\mathbb{R})$ by setting $\mu(E)$ for $E \in \mathfrak{P}(\mathbb{R})$ to be equal to the number of elements in E when E is a finite set and setting $\mu(E) = \infty$ when E is an infinite set.
(a) Show that μ is a measure on $\mathfrak{P}(\mathbb{R})$.
(b) Show that the measure μ is not σ-finite.

Proof. 1. Let us show that μ is a measure on the σ-algebra $\mathfrak{P}(\mathbb{R})$ of subsets of $\mathbb{R}$. From the definition of μ we have $\mu(E) \in [0, \infty]$ for every $E \in \mathfrak{P}(\mathbb{R})$ and $\mu(\emptyset) = 0$. Let us show that μ is countably additive on $\mathfrak{P}(\mathbb{R})$, that is,

$$(E_n : n \in \mathbb{N}) \subset \mathfrak{P}(\mathbb{R}), \text{ disjoint} \Rightarrow \mu\Big(\bigcup_{n \in \mathbb{N}} E_n\Big) = \sum_{n \in \mathbb{N}} \mu(E_n).$$

Now $\mu\big(\bigcup_{n \in \mathbb{N}} E_n\big)$ is equal to the number of elements in the set $\bigcup_{n \in \mathbb{N}} E_n$ and this is equal to the sum of the number of elements in E_n which is equal to $\mu(E_n)$ for $n \in \mathbb{N}$. This proves the countable additivity of μ on $\mathfrak{P}(\mathbb{R})$ and completes the proof that μ is a measure on the σ-algebra $\mathfrak{P}(\mathbb{R})$ of subsets of $\mathbb{R}$.

2. Let us show that μ is not σ-finite. Assume that μ is σ-finite. Then there exists a disjoint sequence $(E_n : n \in \mathbb{N}) \subset \mathfrak{P}(\mathbb{R})$ such that $\bigcup_{n \in \mathbb{N}} E_n = \mathbb{R}$ and $\mu(E_n) < \infty$ for every $n \in \mathbb{N}$. Now $\mu(E_n) < \infty$ implies that E_n is a finite set for every $n \in \mathbb{N}$. Then $\mathbb{R} = \bigcup_{n \in \mathbb{N}} E_n$ has at most countably many elements. This contradicts the fact that $\mathbb{R}$ is an uncountable set. Therefore μ cannot be σ-finite. ∎

Prob. 1.28. Let $(X, \mathfrak{A}, \mu)$ be a finite measure space. Let $\mathfrak{C} = \{E_\lambda : \lambda \in \Lambda\}$ be a disjoint collection of members of $\mathfrak{A}$ such that $\mu(E_\lambda) > 0$ for every $\lambda \in \Lambda$. Show that $\mathfrak{C}$ is at most a countable collection.

Proof. Since $\mu(E_\lambda) > 0$, there exists $k \in \mathbb{N}$ such that $\mu(E_\lambda) \geq \frac{1}{k}$. For each $k \in \mathbb{N}$, let $\mathfrak{C}_k$ be the subcollection of $\mathfrak{C}$ consisting of such E_λ that $\mu(E_\lambda) \geq \frac{1}{k}$. Then every member of $\mathfrak{C}$ belongs to $\mathfrak{C}_k$ for some $k \in \mathbb{N}$ and thus we have

$$\mathfrak{C} = \bigcup_{k\in\mathbb{N}} \mathfrak{C}_k.$$

To show that the collection $\mathfrak{C}$ is at most a countable collection, assume the contrary, that is, assume that $\mathfrak{C}$ is an uncountable collection. Now if $\mathfrak{C}_k$ is a finite collection for every $k \in \mathbb{N}$ then $\mathfrak{C} = \bigcup_{k\in\mathbb{N}} \mathfrak{C}_k$ is at most a countable collection, contradicting the assumption that $\mathfrak{C}$ is an uncountable collection. Thus we must have $k_0 \in \mathbb{N}$ such that $\mathfrak{C}_{k_0}$ is an infinite collection. Then we can select a sequence $(E_n : n \in \mathbb{N})$ of distinct members of $\mathfrak{C}_{k_0}$. Being a subcollection of the disjoint collection $\mathfrak{C}$, the collection $\{E_n : n \in \mathbb{N}\}$ is a disjoint collection. Now $\bigcup_{n\in\mathbb{N}} E_n \in \mathfrak{A}$ and $\bigcup_{n\in\mathbb{N}} E_n \subset X$ and thus

$$\mu\left(\bigcup_{n\in\mathbb{N}} E_n\right) \leq \mu(X) < \infty.$$

On the other hand, the disjointness of the collection $\{E_n : n \in \mathbb{N}\} \subset \mathfrak{C}_{k_0}$ implies

$$\mu\left(\bigcup_{n\in\mathbb{N}} E_n\right) = \sum_{n\in\mathbb{N}} \mu(E_n) \geq \sum_{n\in\mathbb{N}} \frac{1}{k_0} = \infty.$$

This is a contradiction. Therefore $\mathfrak{C}$ cannot be an uncountable collection and must be at most a countable collection. ∎

Prob. 1.29. Let X be a countably infinite set and let $\mathfrak{A}$ be the σ-algebra of all subsets of X. Define a set function μ on $\mathfrak{A}$ by defining for every $E \in \mathfrak{A}$

$$\mu(E) = \begin{cases} 0 & \text{if } E \text{ is a finite set,} \\ \infty & \text{otherwise.} \end{cases}$$

(a) Show that μ is additive but not countably additive on $\mathfrak{A}$.
(b) Show that X is the limit of an increasing sequence $(E_n : n \in \mathbb{N})$ in $\mathfrak{A}$ with $\mu(E_n) = 0$ for all n, but $\mu(X) = \infty$.

Proof. 1. Let us show that μ is additive on $\mathfrak{A}$, that is,

$$E_1, E_2 \in \mathfrak{A}, \ E_1 \cap E_2 = \emptyset \Rightarrow \mu(E_1 \cup E_2) = \mu(E_1) + \mu(E_2).$$

There are 3 cases to consider.

Case 1. Suppose both E_1 and E_2 are finite sets. In this case, $E_1 \cup E_2$ is also a finite set. Thus we have $\mu(E_1) = 0$, $\mu(E_2) = 0$ and $\mu(E_1 \cup E_2) = 0$ and then

$$\mu(E_1 \cup E_2) = 0 = 0 + 0 = \mu(E_1) + \mu(E_2).$$

Case 2. Suppose E_1 is a finite set and E_2 is an infinite set. In this case, $E_1 \cup E_2$ is a finite set. Thus we have $\mu(E_1) = 0$, $\mu(E_2) = \infty$ and $\mu(E_1 \cup E_2) = \infty$ and then

$$\mu(E_1 \cup E_2) = \infty = 0 + \infty = \mu(E_1) + \mu(E_2).$$

Case 1. Suppose both E_1 and E_2 are infinite sets. In this case, $E_1 \cup E_2$ is also an infinite set. Thus we have $\mu(E_1) = \infty$, $\mu(E_2) = \infty$ and $\mu(E_1 \cup E_2) = \infty$ and then

$$\mu(E_1 \cup E_2) = \infty = \infty + \infty = \mu(E_1) + \mu(E_2).$$

2. Let us show that μ is not countably additive on $\mathfrak{A}$. Now since X is a countably infinite set, X can be represented by $X = \{x_n : n \in bbn\}$. Now $\{x_n\} \in \mathfrak{A}$ and $\mu(\{x_n\}) = 0$ for every $n \in \mathbb{N}$ and $\{\{x_n\} : n \in \mathbb{N}\}$ is a disjoint countable collection. Moreover $\bigcup_{n \in \mathbb{N}} \{x_n\} = X$. Then

$$\mu\left(\bigcup_{n \in \mathbb{N}} \{x_n\}\right) = \mu(X) = \infty \neq 0 = \sum_{n \in \mathbb{N}} \mu(\{x_n\}).$$

This shows that μ is not countably additive on $\mathfrak{A}$.

3. Let us show that X is the limit of an increasing sequence $(E_n : n \in \mathbb{N})$ in $\mathfrak{A}$ with $\mu(E_n) = 0$ for all n. With the representation of X as $X = \{x_n : n \in \mathbb{N}\}$, let

$$E_n = \{x_1, \ldots, x_n\} \quad \text{for } n \in \mathbb{N}.$$

Since E_n is a finite set we have $\mu(E_n) = 0$ for every $n \in \mathbb{N}$. Now $(E_n : n \in \mathbb{N})$ is an increasing sequence in $\mathfrak{A}$ and thus $\lim_{n \to \infty} E_n = \bigcup_{n \in \mathbb{N}} E_n = X$. ∎

Prob. 1.30. Let X be an arbitrary infinite set. We say that a subset A of X is cofinite if A^c is a finite set. Let $\mathfrak{A}$ be the collection of all the finite and the cofinite subsets of the set X.
(a) Show that $\mathfrak{A}$ is an algebra of subsets of X.
(b) Show that $\mathfrak{A}$ is not a σ-algebra.

Proof. 1. To show that $\mathfrak{A}$ is an algebra of subsets of X, we show that $\mathfrak{A}$ satisfies the following conditions:
$1°$ $X \in \mathfrak{A}$.
$2°$ $A \in \mathfrak{A} \Rightarrow A^c \in \mathfrak{A}$.
$3°$ $A, B \in \mathfrak{A} \Rightarrow A \cup B \in \mathfrak{A}$.

Consider X. We have $X^c = \emptyset$, a finite set. Thus X is a cofinite set and then $X \in \mathfrak{A}$. This verifies $1°$.

Suppose $A \in \mathfrak{A}$. Then A is either a finite set or a cofinite set. If A is a finite set, then A^c is a cofinite set so that $A^c \in \mathfrak{A}$. If A is a cofinite set then A^c is a finite set so that $A^c \in \mathfrak{A}$. This shows that if $A \in \mathfrak{A}$ then $A^c \in \mathfrak{A}$, which verifies $2°$.

Suppose $A, B \in \mathfrak{A}$. Let us show that $A \cup B \in \mathfrak{A}$. There are three cases to consider.

(i). Consider the case that both A and B are finite sets. In this case, $A \cup B$ is a finite set and therefore $A \cup B \in \mathfrak{A}$.

(ii). Consider the case that one of the two sets is a finite set and the other is a cofinite set, say A is a finite set and B is a cofinite set. Now $(A \cup B)^c = A^c \cap B^c$. Since B is a cofinite set, B^c is a finite set. Then $A^c \cap B^c \subset B^c$ is a finite set and thus $A^c \cap B^c \in \mathfrak{A}$. Therefore $(A \cup B)^c \in \mathfrak{A}$. This implies $A \cup B \in \mathfrak{A}$ according to $2°$ which we verified above.

(iii) Consider the case that both A and B are cofinite sets. In this case both A^c and B^c are finite sets and then $(A \cup B)^c = A^c \cap B^c$ is a finite set. Thus $(A \cup B)^c \in \mathfrak{A}$ and this implies $A \cup B \in \mathfrak{A}$ according to $2°$. This completes the proof that $\mathfrak{A}$ is an algebra of subsets of X.

2. Let us show that $\mathfrak{A}$ is not a σ-algebra. We show this by showing that there exists a sequence $(A_n : n \in \mathbb{N}) \subset \mathfrak{A}$ such that $\bigcup_{n \in \mathbb{N}} A_n \notin \mathfrak{A}$.

Since X is an infinite set, we can select two sequences of distinct points in X, namely $(x_n : n \in \mathbb{N})$ and $(y_n : n \in \mathbb{N})$. Then $\{x_n\}$ is a finite set so that $\{x_n\} \in \mathfrak{A}$ for $n \in \mathbb{N}$. Now $\bigcup_{n \in \mathbb{N}} \{x_n\} = \{x_n; n \in \mathbb{N}\}$ which is neither finite nor cofinite since $\{x_n; n \in \mathbb{N}\}^c \supset \{y_n : n \in \mathbb{N}\}$ is not a finite set. Thus $\bigcup_{n \in \mathbb{N}} \{x_n\} \notin \mathfrak{A}$. ∎

Prob. 1.31. Let X be an arbitrary uncountable set. We say that a subset A of X is co-countable if A^c is a countable set. Let $\mathfrak{A}$ the collection of all the countable and the co-countable subsets of the set X. Show that $\mathfrak{A}$ is a σ-algebra of subsets of X.
(This offers an example that an uncountable union of members of a σ-algebra is not a member of the σ-algebra. Indeed, let A be a subset of X such that neither A nor A^c is a countable set so that $A, A^c \notin \mathfrak{A}$. Let A be given as $A = \{x_\gamma \in X : \gamma \in \Gamma\}$ where Γ in an uncountable set. Then the finite set $\{x_\gamma\} \in \mathfrak{A}$ for every $\gamma \in \Gamma$, but $\bigcup_{\gamma \in \Gamma}\{x_\gamma\} = A \notin \mathfrak{A}$.)

Proof. To show that $\mathfrak{A}$ is a σ-algebra of subsets of X, we show that $\mathfrak{A}$ satisfies the following conditions:
1° $X \in \mathfrak{A}$.
2° $A \in \mathfrak{A} \Rightarrow A^c \in \mathfrak{A}$.
3° $(A_n : n \in \mathbb{N}) \subset \mathfrak{A} \Rightarrow \bigcup_{n \in \mathbb{N}} A_n \in \mathfrak{A}$.

We have $X^c = \emptyset$, a finite and hence countable set. Thus X is a co-countable set and then $X \in \mathfrak{A}$. This verifies 1°.

Let $A \in \mathfrak{A}$. Then A is a countable set or a co-countable set. If A is a countable set then A^c is a co-countable set so that $A^c \in \mathfrak{A}$. If A is a co-countable set then A^c is a countable set so that $A^c \in \mathfrak{A}$. This verifies 2°.

Let $(A_n : n \in \mathbb{N}) \subset \mathfrak{A}$. Now $A_n \in \mathfrak{A}$ implies that A_n is a countable set or a co-countable set for $n \in \mathbb{N}$. Thus we can write

$$\bigcup_{n \in \mathbb{N}} A_n = \left(\bigcup_{n \in \mathbb{N}} B_n\right) \cup \left(\bigcup_{n \in \mathbb{N}} C_n\right)$$

where B_n is a countable set and C_n is a co-countable set for $n \in \mathbb{N}$. Now we have

$$\left(\bigcup_{n \in \mathbb{N}} A_n\right)^c = \left(\bigcup_{n \in \mathbb{N}} B_n\right)^c \cap \left(\bigcup_{n \in \mathbb{N}} C_n\right)^c,$$

and

$$\left(\bigcup_{n \in \mathbb{N}} C_n\right)^c = \bigcap_{n \in \mathbb{N}} C_n^c.$$

Since C_n is a co-countable set, C_n^c is a countable set and then $\bigcap_{n \in \mathbb{N}} C_n^c$ is a countable set, that is, $\left(\bigcup_{n \in \mathbb{N}} C_n\right)^c$ is a countable set. This implies that $\left(\bigcup_{n \in \mathbb{N}} B_n\right)^c \cap \left(\bigcup_{n \in \mathbb{N}} C_n\right)^c$ is a countable set, that is, $\left(\bigcup_{n \in \mathbb{N}} A_n\right)^c$ is a countable set. Then $\bigcup_{n \in \mathbb{N}} A_n$ is a co-countable set so that $\bigcup_{n \in \mathbb{N}} A_n \in \mathfrak{A}$. This completes the proof that $\mathfrak{A}$ is a σ-algebra of subsets of X. ∎

Prob. 1.32. Let X be an infinite set and let $\mathfrak{A}$ be the algebra of subsets of X consisting of the finite and the cofinite subsets of X (cf. Prob. 1.30). Define a set function μ on $\mathfrak{A}$ by setting for every $A \in \mathfrak{A}$:
$$\mu(A) = \begin{cases} 0 & \text{if } A \text{ is finite,} \\ 1 & \text{if } A \text{ is cofinite.} \end{cases}$$
(Note that since X is an infinite set, no subset A of X can be both finite and cofinite although it can be neither.)
(a) Show that μ is additive on the algebra $\mathfrak{A}$.
(b) Show that when X is countably infinite, μ is not countably additive on the algebra $\mathfrak{A}$.
(c) Show that when X is countably infinite, then X is the limit of an increasing sequence $(A_n : n \in \mathbb{N})$ in $\mathfrak{A}$ with $\mu(A_n) = 0$ for every $n \in \mathbb{N}$, but $\mu(X) = 1$.
(d) Show that when X is uncountable, then μ is countably additive on the algebra $\mathfrak{A}$.

Proof. (a) Let us show that μ is additive on the algebra $\mathfrak{A}$. We are to verify
$$A_1, A_2 \in \mathfrak{A}, \ A_1 \cap A_2 = \emptyset \Rightarrow \mu(A_1 \cup A_2) = \mu(A_1) + \mu(A_2).$$

There are 3 cases to consider.

Case 1. Suppose both A_1 and A_2 are finite sets. In this case $A_1 \cup A_2$ is also a finite set. Then $\mu(A_1) = 0$, $\mu(A_2) = 0$, and $\mu(A_1 \cup A_2) = 0$ and then
$$\mu(A_1 \cup A_2) = 0 = 0 + 0 = \mu(A_1) + \mu(A_2).$$

Case 2. Suppose at least one of A_1 and A_2 is a cofinite set, say A_1 is a cofinite set. Then A_1^c is a finite set. Now $A_1 \cap A_2 = \emptyset$ implies that $A_2 \subset A_1^c$. Thus A_2 is a finite set. Then we have $\mu(A_1) = 1$ and $\mu(A_2) = 0$. But $(A_1 \cup A_2)^c = A_1^c \cap A_2^c \subset A_1^c$ so that $(A_1 \cup A_2)^c$ is a finite set and then $(A_1 \cup A_2)$ is a cofinite set and thus $\mu(A_1 \cup A_2) = 1$. Then we have
$$\mu(A_1 \cup A_2) = 1 = 1 + 0 = \mu(A_1) + \mu(A_2).$$

Case 3. Let us show that it is impossible that both A_1 and A_2 are cofinite sets. Suppose both A_1 and A_2 are cofinite sets. We showed above that if A_1 is a cofinite set then A_2 is a finite set so that A_2 is both cofinite and finite. But no subset of X can be both finite and cofinite. Therefore it is impossible that both A_1 and A_2 are cofinite sets.

This completes the proof that μ is additive on the algebra $\mathfrak{A}$.

(b) Let us show that when X is a countably infinite set, μ is not countably additive on the algebra $\mathfrak{A}$, that is,
$$\{A_n : n \in \mathbb{N}\} \subset \mathfrak{A}, \ \text{disjoint}, \ \bigcup_{n \in \mathbb{N}} A_n \in \mathfrak{A} \not\Rightarrow \mu\left(\bigcup_{n \in \mathbb{N}} A_n\right) = \sum_{n \in \mathbb{N}} \mu(A_n).$$

Let us represent the countably infinite set X as $X = \{x_n : n \in \mathbb{N}\}$. Let $A_n = \{x_n\}$ for every $n \in \mathbb{N}$. Then A_n is a finite set and thus $A_n \in \mathfrak{A}$ and $\mu(A_n) = 0$ for every $n \in \mathbb{N}$. Also $\{A_n; n \in \mathbb{N}\}$ is a disjoint collection in $\mathfrak{A}$ and $\bigcup_{n \in \mathbb{N}} A_n = X \in \mathfrak{A}$. Now $X^c = \emptyset$, a finite set, and thus X is a cofinite set and then $\mu(X) = 1$. Thus we have
$$\mu\left(\bigcup_{n \in \mathbb{N}} A_n\right) = \mu(X) = 1 \neq 0 = \sum_{n \in \mathbb{N}} \mu(A_n).$$

This shows that μ is not countably additive on the algebra $\mathfrak{A}$.

(c) With a countably infinite set X represented by $X = \{x_n : n \in \mathbb{N}\}$, let $A_n = \{x_1, \ldots, x_n\}$ for every $n \in \mathbb{N}$. Then A_n is a finite set so that $A_n \in \mathfrak{A}$ and $\mu(A_n) = 0$ for every $n \in \mathbb{N}$. Now $(A_n : n \in \mathbb{N})$ is an increasing sequence in $\mathfrak{A}$ and then $\lim_{n \to \infty} A_n = \bigcup_{n \in \mathbb{N}} A_n = X$. But $X^c = \emptyset$, a finite set, implies that X is a cofinite set so that $\mu(X) = 1$.

(d) Let us show that when X is an uncountable set then μ is countably additive on the algebra $\mathfrak{A}$, that is,

$$\{A_n : n \in \mathbb{N}\} \subset \mathfrak{A}, \text{ disjoint}, \bigcup_{n \in \mathbb{N}} A_n \in \mathfrak{A} \Rightarrow \mu\left(\bigcup_{n \in \mathbb{N}} A_n\right) = \sum_{n \in \mathbb{N}} \mu(A_n).$$

Since $\mu(A_n)$ is equal to 0 or 1 for every $n \in \mathbb{N}$, the value of $\sum_{n \in \mathbb{N}} \mu(A_n)$ is equal to a nonnegative integer or ∞. There are 3 cases to consider.

Case 1. Consider the case that we have $\sum_{n \in \mathbb{N}} \mu(A_n) = 0$. In this case we have $\mu(A_n) = 0$ for every $n \in \mathbb{N}$. Now $A_n \in \mathfrak{A}$ and $\mu(A_n) = 0$ imply that A_n is a finite set. Then $\bigcup_{n \in \mathbb{N}} A_n$ is an at most countable set. Then since X is an uncountable set, $\left(\bigcup_{n \in \mathbb{N}} A_n\right)^c$ is an infinite set. Thus $\bigcup_{n \in \mathbb{N}} A_n$ is not a cofinite set. Then since $\bigcup_{n \in \mathbb{N}} A_n \in \mathfrak{A}$ and every member of $\mathfrak{A}$ is either a finite set or a cofinite set, $\bigcup_{n \in \mathbb{N}} A_n$ must be a finite set and then $\mu\left(\bigcup_{n \in \mathbb{N}} A_n\right) = 0$. Then we have

$$\mu\left(\bigcup_{n \in \mathbb{N}} A_n\right) = 0 = \sum_{n \in \mathbb{N}} \mu(A_n).$$

Case 2. Consider the case that we have $\sum_{n \in \mathbb{N}} \mu(A_n) = 1$. In this case there exists $n_0 \in \mathbb{N}$ such that $\mu(A_{n_0}) = 1$ and $\mu(A_n) = 0$ for every $n \neq n_0$. Since $A_{n_0} \in \mathfrak{A}$, $\mu(A_{n_0}) = 1$ implies that A_{n_0} is a cofinite set and thus $A_{n_0}^c$ is a finite set. Now we have

$$\left(\bigcup_{n \in \mathbb{N}} A_n\right)^c = \bigcap_{n \in \mathbb{N}} A_n^c \subset A_{n_0}^c.$$

Thus $\left(\bigcup_{n \in \mathbb{N}} A_n\right)^c$ is a finite set so that $\bigcup_{n \in \mathbb{N}} A_n$ is a cofinite set and thus $\mu\left(\bigcup_{n \in \mathbb{N}} A_n\right) = 1$. Then we have

$$\mu\left(\bigcup_{n \in \mathbb{N}} A_n\right) = 1 = \sum_{n \in \mathbb{N}} \mu(A_n).$$

Case 3. We show that it is impossible to have $\sum_{n \in \mathbb{N}} \mu(A_n) > 1$. Suppose we have $\sum_{n \in \mathbb{N}} \mu(A_n) > 1$. Then since $\mu(A_n)$ is integer-valued, we have $\sum_{n \in \mathbb{N}} \mu(A_n) \geq 2$. This implies that there exist $n_1, n_2 \in \mathbb{N}$, $n_1 \neq n_2$, such that $\mu(A_{n_1}) = 1$ and $\mu(A_{n_2}) = 1$. Then A_{n_1} and A_{n_2} are both cofinite sets. Now since $A_{n_1} \cap A_{n_2} = \emptyset$, we have $A_{n_2} \subset A_{n_1}^c$. Since A_{n_1} is a cofinite set, $A_{n_1}^c$ is a finite set. Then the last set inclusion above implies that A_{n_2} is a finite set and thus A_{n_2} is both finite and cofinite. This contradicts the fact that no subset of X can be both finite and cofinite. This shows that it is impossible to have $\sum_{n \in \mathbb{N}} \mu(A_n) > 1$.

This completes the proof that μ is countably additive on the algebra $\mathfrak{A}$. ∎

Prob. 1.33. Let X be an uncountable set and let $\mathfrak{A}$ be the σ-algebra of subsets of X consisting of the countable and the co-countable subsets of X (cf. Prob. 1.31). Define a set function μ on $\mathfrak{A}$ by setting for every $A \in \mathfrak{A}$:
$$\mu(A) = \begin{cases} 0 & \text{if } A \text{ is countable,} \\ 1 & \text{if } A \text{ is co-countable.} \end{cases}$$
(Note that since X is an uncountable set, no subset A of X can be both countable and co-countable although it can be neither.) Show that μ is countably additive on $\mathfrak{A}$.

Proof. Let us show that μ is countably additive on the σ-algebra $\mathfrak{A}$, that is,
$$\{A_n : n \in \mathbb{N}\} \subset \mathfrak{A}, \text{ disjoint } \Rightarrow \mu\left(\bigcup_{n\in\mathbb{N}} A_n\right) = \sum_{n\in\mathbb{N}} \mu(A_n).$$

There are 3 cases to consider.

Case 1. Consider the case that we have $\sum_{n\in\mathbb{N}} \mu(A_n) = 0$. In this case we have $\mu(A_n) = 0$ for every $n \in \mathbb{N}$. This implies that A_n is a countable set for every $n \in \mathbb{N}$ and then $\bigcup_{n\in\mathbb{N}} A_n$ is a countable set and thus $\mu\left(\bigcup_{n\in\mathbb{N}} A_n\right) = 0$. Then we have
$$\mu\left(\bigcup_{n\in\mathbb{N}} A_n\right) = 0 = \sum_{n\in\mathbb{N}} \mu(A_n).$$

Case 2. Consider the case that we have $\sum_{n\in\mathbb{N}} \mu(A_n) = 1$. In this case there exists $n_0 \in \mathbb{N}$ such that $\mu(A_{n_0}) = 1$ and $\mu(A_n) = 0$ for every $n \neq n_0$. Since $A_{n_0} \in \mathfrak{A}$, $\mu(A_{n_0}) = 1$ implies that A_{n_0} is a co-countable set and A_n is a countable set for every $n \neq n_0$. Now we have
$$\left(\bigcup_{n\in\mathbb{N}} A_n\right)^c = \bigcap_{n\in\mathbb{N}} A_n^c \subset A_{n_0}^c.$$
Since A_{n_0} is a co-countable set, $A_{n_0}^c$ is a countable set and then the last set inclusion above implies that $\left(\bigcup_{n\in\mathbb{N}} A_n\right)^c$ is a countable set and then $\bigcup_{n\in\mathbb{N}} A_n$ is a co-countable set and thus $\mu\left(\bigcup_{n\in\mathbb{N}} A_n\right) = 1$. Thus we have
$$\mu\left(\bigcup_{n\in\mathbb{N}} A_n\right) = 1 = \sum_{n\in\mathbb{N}} \mu(A_n).$$

Case 3. We show that it is impossible to have $\sum_{n\in\mathbb{N}} \mu(A_n) > 1$. Suppose we have $\sum_{n\in\mathbb{N}} \mu(A_n) > 1$. Then since $\mu(A_n)$ is integer-valued, we have $\sum_{n\in\mathbb{N}} \mu(A_n) \geq 2$. This implies that there exist $n_1, n_2 \in \mathbb{N}$, $n_1 \neq n_2$, such that $\mu(A_{n_1}) = 1$ and $\mu(A_{n_2}) = 1$. Then A_{n_1} and A_{n_2} are both co-countable sets. Now since $A_{n_1} \cap A_{n_2} = \emptyset$, we have $A_{n_2} \subset A_{n_1}^c$. Since A_{n_1} is a co-countable set, $A_{n_1}^c$ is a countable set. Then the last set inclusion above implies that A_{n_2} is a countable set and thus A_{n_2} is both countable and co-countable. This contradicts the fact that no subset of X can be both countable and co-countable. This shows that it is impossible to have $\sum_{n\in\mathbb{N}} \mu(A_n) > 1$.

This completes the proof that μ is countably additive on the σ-algebra $\mathfrak{A}$. ∎

Prob. 1.34. Given a measure space $(X, \mathfrak{A}, \mu)$. We say that a collection $\{A_\lambda : \lambda \in \Lambda\} \subset \mathfrak{A}$ is almost disjoint if $\lambda_1, \lambda_2 \in \Lambda$ and $\lambda_1 \neq \lambda_2$ imply $\mu(A_{\lambda_1} \cap A_{\lambda_2}) = 0$.
(a) Show that if $\{A_n : n \in \mathbb{N}\} \subset \mathfrak{A}$ is almost disjoint then $\mu\left(\bigcup_{n \in \mathbb{N}} A_n\right) = \sum_{n \in \mathbb{N}} \mu(A_n)$.
(b) Show that if $\{A_n : n \in \mathbb{N}\} \subset \mathfrak{A}$ is such that $\mu\left(\bigcup_{n \in \mathbb{N}} A_n\right) = \sum_{n \in \mathbb{N}} \mu(A_n)$ and $\mu(A_n) < \infty$ for every $n \in \mathbb{N}$ then $\{A_n : n \in \mathbb{N}\}$ is almost disjoint.
(c) Show that if we remove the condition in (b) that $\mu(A_n) < \infty$ for every $n \in \mathbb{N}$ then the condition $\mu\left(\bigcup_{n \in \mathbb{N}} A_n\right) = \sum_{n \in \mathbb{N}} \mu(A_n)$ alone does not imply that $\{A_n : n \in \mathbb{N}\}$ is almost disjoint.

Proof. 1. Let us prove (a). Suppose $\{A_n : n \in \mathbb{N}\} \subset \mathfrak{A}$ is an almost disjoint collection. Let us define a collection $\{B_n : n \in \mathbb{N}\} \subset \mathfrak{A}$ by setting

$$B_1 = A_1 \quad \text{and} \quad B_n = A_n \setminus \bigcup_{j=1}^{n-1} A_j \quad \text{for } n \geq 2.$$

Then $\{B_n : n \in \mathbb{N}\}$ is a disjoint collection in $\mathfrak{A}$ and moreover $\bigcup_{n \in \mathbb{N}} B_n = \bigcup_{n \in \mathbb{N}} A_n$. Then we have

$$(1) \qquad \mu\left(\bigcup_{n \in \mathbb{N}} A_n\right) = \mu\left(\bigcup_{n \in \mathbb{N}} B_n\right) = \sum_{n \in \mathbb{N}} \mu(B_n).$$

It remains to show that $\mu(B_n) = \mu(A_n)$ for every $n \in \mathbb{N}$. We have $\mu(B_1) = \mu(A_1)$ since $B_1 = A_1$. Thus we need to consider only $n \geq 2$. Now for any two subsets E and F of X, we have $E \setminus F = E \setminus (E \cap F)$. Then for $n \geq 2$ we have

$$(2) \qquad B_n = A_n \setminus \bigcup_{j=1}^{n-1} A_j = A_n \setminus \left(A_n \cap \bigcup_{j=1}^{n-1} A_j\right) = A_n \setminus \bigcup_{j=1}^{n-1} (A_n \cap A_j).$$

Since $\{A_n : n \in \mathbb{N}\}$ is an almost disjoint collection, we have

$$(3) \qquad \mu\left(\bigcup_{j=1}^{n-1} (A_n \cap A_j)\right) \leq \sum_{j=1}^{n-1} \mu(A_n \cap A_j) = 0.$$

Then (2) and (3) imply

$$(4) \qquad \mu(B_n) = \mu(A_n) - \mu\left(\bigcup_{j=1}^{n-1} (A_n \cap A_j)\right) = \mu(A_n).$$

With (1) and (4) we have

$$\mu\left(\bigcup_{n \in \mathbb{N}} A_n\right) = \sum_{n \in \mathbb{N}} \mu(A_n).$$

This proves (a).
  2. Let us prove (b). Thus assume that $\{A_n : n \in \mathbb{N}\} \subset \mathfrak{A}$ satisfies the conditions:

$$(5) \qquad \mu\left(\bigcup_{n \in \mathbb{N}} A_n\right) = \sum_{n \in \mathbb{N}} \mu(A_n) \quad \text{and} \quad \mu(A_n) < \infty \quad \text{for every } n \in \mathbb{N}.$$

To show that $\{A_n : n \in \mathbb{N}\}$ is an almost disjoint collection, assume the contrary. Then there exist $j, k \in \mathbb{N}$ such that $j \neq k$ and $\mu(A_j \cap A_k) > 0$. Renumber the collection $\{A_n : n \in \mathbb{N}\}$ if necessary so that $j = 1$ and $k = 2$. Then we have $\mu(A_1 \cap A_2) > 0$. Now

$$A_1 \cup A_2 = A_1 \cup (A_2 \setminus A_1) = A_1 \cup \big(A_2 \setminus (A_1 \cap A_2)\big),$$

and the disjointness of A_1 and $A_2 \setminus (A_1 \cap A_2)$ implies

(6) $$\mu(A_1 \cup A_2) = \mu(A_1) + \mu\big(A_2 \setminus (A_1 \cap A_2)\big).$$

Since $(A_1 \cap A_2) \subset A_2$ and $\mu(A_1 \cap A_2) \leq \mu(A_1) < \infty$, we have

(7) $$\mu\big(A_2 \setminus (A_1 \cap A_2)\big) = \mu(A_2) - \mu(A_1 \cap A_2).$$

Combining (6) and (7), we have

(8) $$\mu(A_1 \cup A_2) = \mu(A_1) + \mu(A_2) - \mu(A_1 \cap A_2)$$

$$< \mu(A_1) + \mu(A_2).$$

By the countable subadditivity of the measure μ on $\mathfrak{A}$ we have

$$\mu\Big(\bigcup_{n \in \mathbb{N}} A_n \Big) \leq \mu(A_1 \cup A_2) + \sum_{n \geq 3} \mu(A_n)$$

$$< \mu(A_1) + \mu(A_2) + \sum_{n \geq 3} \mu(A_n)$$

$$= \sum_{n \in \mathbb{N}} \mu(A_n).$$

This contradicts the assumption that $\mu\big(\bigcup_{n \in \mathbb{N}} A_n\big) = \sum_{n \in \mathbb{N}} \mu(A_n)$. Therefore the collection $\{A_n : n \in \mathbb{N}\}$ must be an almost disjoint collection.

3. Let us construct an example for (c). In the measure space $(\mathbb{R}, \mathfrak{M}_L, \mu_L)$, let

$$A_1 = (-\infty, 1], \quad A_2 = (0, 2], \quad \text{and} \quad A_n = (n - 1, n] \quad \text{for } n \geq 3.$$

We have $\mu(A_1) = \infty$, $\mu(A_2) = 2$ and $\mu(A_n) = 1$ for $n \geq 3$. We have $\bigcup_{n \in \mathbb{N}} A_n = \mathbb{R}$. Thus we have

$$\mu\Big(\bigcup_{n \in \mathbb{N}} A_n \Big) = \mu(\mathbb{R}) = \infty = \sum_{n \in \mathbb{N}} \mu(A_n).$$

But $\{A_n; n \in \mathbb{N}\}$ is not an almost disjoint collection since $\mu(A_1 \cap A_2) = 1 > 0$. ∎

Prob. 1.35. Let $(X, \mathfrak{A}, \mu)$ be a measure space. The symmetric difference of two subsets A and B of a set X is defined by $A \triangle B = (A \setminus B) \cup (B \setminus A)$.
(a) Show that $A = B$ if and only if $A \triangle B = \emptyset$.
(b) Show that $A \cup B = (A \cap B) \cup (A \triangle B)$.
(c) Prove the triangle inequality for the symmetric difference of sets, that is, for any three subsets A, B, C of a set X we have $A \triangle B \subset (A \triangle C) \cup (C \triangle B)$.
(d) Show that $\mu(A \triangle B) \leq \mu(A \triangle C) + \mu(C \triangle B)$ for any $A, B, C \in \mathfrak{A}$.
(e) Show that $\mu(A \cup B) = \mu(A \cap B) + \mu(A \triangle B)$ for any $A, B \in \mathfrak{A}$.
(f) Show that if $\mu(A \triangle B) = 0$ then $\mu(A) = \mu(B)$ for any $A, B \in \mathfrak{A}$.

Proof. 1. Let us prove (a). Suppose $A \triangle B = \emptyset$. Then we have $A \setminus B = \emptyset$ and $B \setminus A = \emptyset$. Now $A \setminus B$ consists of those elements of A that are not elements of B. Thus $A \setminus B = \emptyset$ implies $A \subset B$. Similarly $B \setminus A = \emptyset$ implies $B \subset A$. Thus we have $A = B$.

Conversely suppose $A = B$. Then $A \setminus B = \emptyset$ and $B \setminus A = \emptyset$ and thus we have $A \triangle B = (A \setminus B) \cup (B \setminus A) = \emptyset$.

2. Let us prove (b). Now $(A \cap B) \subset A \cup B$ and $A \triangle B = (A \setminus B) \cup (B \setminus A) \subset A \cup B$. Thus we have

$$(1) \qquad\qquad (A \cap B) \cup (A \triangle B) \subset A \cup B.$$

Let us prove the reverse inclusion. Now $A \cup B$ consists of every point that is in at least one of the two sets A and B. Such a point is either in only one of the two sets or in both of the two sets, that is, it is either in $A \triangle B$ or in $A \cap B$. Thus we have

$$(2) \qquad\qquad A \cup B \subset (A \triangle B) \cup (A \cap B).$$

By (1) and (2), we have $A \cup B = (A \cap B) \cup (A \triangle B)$.

3. Let us prove (c). By definition we have $A \triangle B = (A \setminus B) \cup (B \setminus A)$. Now

$$A = (A \setminus C) \cup (A \cap C),$$

and then

$$(3) \qquad A \setminus B = \Big((A \setminus C) \cup (A \cap C) \Big) \setminus B = \Big((A \setminus C) \setminus B \Big) \cup \Big((A \cap C) \setminus B \Big)$$

$$\subset (A \setminus C) \cup (C \setminus B) \subset (A \triangle C) \cup (C \triangle B).$$

Interchanging the roles of A and B, we have

$$(4) \qquad\qquad B \setminus A \subset (B \triangle C) \cup (C \triangle A).$$

Then by (3) and (4) we have

$$A \triangle B = (A \setminus B) \cup (B \setminus A) \subset (A \triangle C) \cup (C \triangle B).$$

4. Let us prove (d). By (c) we have $A \triangle B \subset (A \triangle C) \cup (C \triangle B)$. Then by the finite subadditivity of the measure μ on $\mathfrak{A}$ we have

$$\mu(A \triangle B) \leq \mu\Big((A \triangle C) \cup (C \triangle B) \Big) \leq \mu(A \triangle C) + \mu(C \triangle B).$$

5. Let us prove (e). By (b) we have $A \cup B = (A \cap B) \cup (A \triangle B)$. Now $A \cap B$ and $A \triangle B$ are disjoint. Then by the additivity of μ on $\mathfrak{A}$ we have

$$\mu(A \cup B) = \mu(A \cap B) + \mu(A \triangle B).$$

6. Let us prove (f). We are to prove that if $A, B \in \mathfrak{A}$ and $\mu(A \triangle B) = 0$ then we have $\mu(A) = \mu(B)$. Let us observe first of all that if $\mu(A \cap B) = \infty$ then $\mu(A) \geq \mu(A \cap B) = \infty$ and similarly $\mu(B) \geq \mu(A \cap B) = \infty$ and then $\mu(A) = \infty = \mu(B)$ and we are done. Thus it remains to consider the case that $\mu(A \cap B) < \infty$.

Now $A \triangle B = (A \setminus B) \cup (B \setminus A)$ is the union of 2 disjoint members of $\mathfrak{A}$ and thus

$$\mu(A \triangle B) = \mu(A \setminus B) + \mu(B \setminus A).$$

Then the assumption $\mu(A \triangle B) = 0$ implies

(5) $\mu(A \setminus B) = 0 \quad \text{and} \quad \mu(B \setminus A) = 0.$

Let us write

$$A \setminus B = A \setminus (A \cap B).$$

Now $A \cap B \subset A$ and moreover $\mu(A \cap B) < \infty$. This implies

(6) $\mu(A \setminus B) = \mu\Big(A \setminus (A \cap B)\Big) = \mu(A) - \mu(A \cap B).$

Combining (6) and the first equality in (5), we have

(7) $\mu(A) = \mu(A \cap B).$

Interchanging the roles of A and B in the argument above, we have

(8) $\mu(B) = \mu(B \cap A).$

With (7) and (8), we have $\mu(A) = \mu(B)$. ∎

Preamble to Prob. 1.36. Let $(X, \mathfrak{A}, \mu)$ be a finite measure space. Then a function ρ on $\mathfrak{A} \times \mathfrak{A}$ defined by $\rho(A, B) = \mu(A \triangle B)$ for $A, B \in \mathfrak{A}$ has the following properties:
1° $\rho(A, B) \in [0, \mu(X)]$,
2° $\rho(A, B) = \rho(B, A)$,
3° $\rho(A, B) \leq \rho(A, C) + \rho(C, B)$.
However ρ need not be a metric on the set $\mathfrak{A}$ since $\rho(A, B) = 0$ does not imply $A = B$.

Prob. 1.36. Let $(X, \mathfrak{A}, \mu)$ be a finite measure space. Let a relation $\sim$ among the members of $\mathfrak{A}$ be defined by writing $A \sim B$ when $\mu(A \triangle B) = 0$.
(a) Show that $\sim$ is an equivalence relation, that is,
1° $A \sim A$,
2° $A \sim B \Rightarrow B \sim A$,
3° $A \sim B, B \sim C \Rightarrow A \sim C$.
(b) Let $[A]$ be the equivalence class to which A belongs and let $[\mathfrak{A}]$ be the collection of all the equivalence classes with respect to the equivalence relation $\sim$. Define a function ρ^* on $[\mathfrak{A}] \times [\mathfrak{A}]$ by setting $\rho^*([A], [B]) = \mu(A \triangle B)$ for $[A], [B] \in [\mathfrak{A}]$.
Show that ρ^* is well defined in the sense that its definition as given above does not depend on the particular representative A and B of the equivalence classes $[A]$ and $[B]$; in other words,
$$A' \in [A], B' \in [B] \Rightarrow \mu(A' \triangle B') = \mu(A \triangle B).$$
(c) Show that ρ^* is a metric on the set $[\mathfrak{A}]$.

Proof. 1. Let us prove (a). Thus we are to verify 1°, 2° and 3° in (a).
Now for any $A \in \mathfrak{A}$, we have $A \triangle A = \emptyset$ and then $\mu(A \triangle A) = 0$. Thus $A \sim A$.
Suppose $A \sim B$, that is, $\mu(A \triangle B) = 0$. Now $B \triangle A = A \triangle B$. Then we have $\mu(B \triangle A) = \mu(A \triangle B) = 0$, that is, $B \sim A$. This verifies 2°.
Suppose $A \sim B$ and $B \sim C$. Then we have $\mu(A \triangle B) = 0$ and $\mu(B \triangle C) = 0$. By (b) of Prob. 1.35, we have

$$\mu(A \triangle C) \leq \mu(A \triangle B) + \mu(B \triangle C) = 0,$$

and thus we have $\mu(A \triangle C) = 0$, that is, $A \sim C$. This verifies 3°.
2. Let us show that ρ^* is well defined, that is,

$$A' \in [A], B' \in [B] \Rightarrow \mu(A' \triangle B') = \mu(A \triangle B).$$

Let $A' \in [A]$ and $B' \in [B]$. Then by repeated application of (b) of Prob. 1.35 we have

$$\mu(A' \triangle B') \leq \mu(A' \triangle A) + \mu(A \triangle B')$$
$$\leq \mu(A' \triangle A) + \mu(A \triangle B) + \mu(B \triangle B').$$

Since $A' \sim A$ we have $\mu(A' \triangle A) = 0$ and since $B' \sim B$ we have $\mu(B \triangle B') = 0$. Thus the last inequalities reduce to

(1) $$\mu(A' \triangle B') \leq \mu(A \triangle B).$$

Interchanging the roles of A and A' and interchanging the roles of B and B' in the argument above, we have

(2) $\mu(A \triangle B) \leq \mu(A' \triangle B')$.

By (1) and (2), we have $\mu(A' \triangle B') = \mu(A \triangle B)$. This proves that ρ^* is well defined.

3. To show that ρ^* is a metric on the set $[\mathfrak{A}]$, we verify that ρ^* satisfies the following conditions:

1° $\rho^*([A], [B]) \in [0, \infty)$ for every $[A], [B] \in [\mathfrak{A}]$.
2° $\rho^*([A], [B]) = 0 \Leftrightarrow [A] = [B]$ for every $[A], [B] \in [\mathfrak{A}]$.
3° symmetry: $\rho^*([A], [B]) = \rho^*([B], [A])$ for every $[A], [B] \in [\mathfrak{A}]$.
4° triangle inequality: $\rho^*([A], [B]) \leq \rho^*([A], [C]) + \rho^*([C], [B])$ for $[A], [B], [C] \in [\mathfrak{A}]$.

Now for $[A], [B] \in [\mathfrak{A}]$, we have

$$\rho^*([A], [B]) = \mu(A \triangle B) \leq \mu(X) < \infty.$$

Thus we have $\rho^*([A], [B]) \in [0, \infty)$ for every $[A], [B] \in [\mathfrak{A}]$. This verifies 1°.

For every $[A] \in [\mathfrak{A}]$, we have $\rho^*([A], [A]) = \mu(A \triangle A) = \mu(\emptyset) = 0$. Conversely suppose that for some $[A], [B] \in [\mathfrak{A}]$ we have $\rho^*([A], [B]) = 0$. Then $\mu(A \triangle B) = 0$, that is, $A \sim B$, that is, $[A] = [B]$. This verifies 2°.

For $[A], [B] \in [\mathfrak{A}]$, we have

$$\rho^*([A], [B]) = \mu(A \triangle B) = \mu(B \triangle A) = \rho^*([B], [A]).$$

This proves the symmetry of ρ^*.

For $[A], [B], [C] \in [\mathfrak{A}]$, we have, by (b) of Prob. 1.35,

$$\rho^*([A], [B]) = \mu(A \triangle B) \leq \mu(A \triangle C) + \mu(C \triangle B)$$

$$= \rho^*([A], [C]) + \rho^*([C], [B]).$$

This verifies the triangle inequality for ρ^*. ∎

§2 Outer Measures

Prob. 2.1. Let μ^* be an outer measure on a set X. Show that if μ^* is additive on $\mathfrak{P}(X)$, then it is countably additive on $\mathfrak{P}(X)$.

Proof. If μ^* is an outer measure on a set X then μ^* is a nonnegative extended real-valued countably subadditive set function on the σ-algebra $\mathfrak{P}(X)$ of all subsets of X. If μ^* is additive on $\mathfrak{P}(X)$ then μ^* is both additive and countably subadditive on the σ-algebra $\mathfrak{P}(X)$ and this implies that μ^* is countably additive on $\mathfrak{P}(X)$ by Proposition 1.23. ∎

Prob. 2.2. Let μ^* be an outer measure on a set X. Show that a non μ^*-measurable subset of X exists if and only if μ^* is not countably additive on $\mathfrak{P}(X)$.

Proof. According to Theorem 2.5, we have

(1) μ^* is additive on $\mathfrak{P}(X)$

         $\Leftrightarrow \mathfrak{M}(\mu^*) = \mathfrak{P}(X)$, that is, every subset of X is μ^*-measurable

   1. Suppose a non μ^*-measurable subset of X exists, that is, $\mathfrak{M}(\mu^*) \neq \mathfrak{P}(X)$. Then by (1), μ^* is not additive on $\mathfrak{P}(X)$ and certainly not countably additive on $\mathfrak{P}(X)$ (since countable additivity implies additivity).
   2. Conversely suppose μ^* is not countably additive on $\mathfrak{P}(X)$. Now as an outer measure on X, μ^* is countably subadditive on the σ-algebra $\mathfrak{P}(X)$ of subsets of X. If μ^* were additive on $\mathfrak{P}(X)$ then μ^* would be countably additive on $\mathfrak{P}(X)$ according to Proposition 1.23. Since μ^* is assumed to be not countably additive on $\mathfrak{P}(X)$, μ^* must be not additive on $\mathfrak{P}(X)$. Then by (1) we have $\mathfrak{M}(\mu^*) \neq \mathfrak{P}(X)$ and thus a non μ^*-measurable subset of X exists. ∎

Prob. 2.3. For an arbitrary set X let us define a set function μ^* on $\mathfrak{P}(X)$ by

$$\mu^*(E) = \begin{cases} \text{number of elements of } E \text{ if } E \text{ is a finite set,} \\ \infty \text{ if } E \text{ is an infinite set.} \end{cases}$$

(a) Show that μ^* is an outer measure on X.
(b) Show that μ^* is additive on $\mathfrak{P}(X)$, that is, $\mu^*(E_1 \cup E_2) = \mu^*(E_1) + \mu^*(E_2)$ for any $E_1, E_2 \in \mathfrak{P}(X)$ such that $E_1 \cap E_2 = \emptyset$.
(c) Show that μ^* is a measure on the σ-algebra $\mathfrak{P}(X)$. (This measure is called the counting measure.)
(d) Show that $\mathfrak{M}(\mu^*) = \mathfrak{P}(X)$, that is, every $E \in \mathfrak{P}(X)$ is μ^*-measurable.

Proof. 1. To show that μ^* is an outer measure on X, we verify that μ^* satisfies the following conditions:
1° nonnegative extended real-valued: $\mu^*(E) \in [0, \infty]$ for every $E \in \mathfrak{P}(X)$.
2° $\mu^*(\emptyset) = 0$.
3° monotonicity: $E_1, E_2 \in \mathfrak{P}(X), E_1 \subset E_2 \Rightarrow \mu^*(E_1) \le \mu^*(E_2)$.
4° countable subadditivity: $(E_n : n \in \mathbb{N}) \subset \mathfrak{P}(X) \Rightarrow \mu^*\left(\bigcup_{n\in\mathbb{N}} E_n\right) \le \sum_{n\in\mathbb{N}} \mu^*(E_n)$.

Now 1°, 2° and 3° are immediate from the definition of $\mu^*(E)$ as the number of elements from E. Let us prove 4°.

(i) If at least one of $(E_n : n \in \mathbb{N})$ is an infinite set, say E_{n_0} is an infinite set, then $\bigcup_{n\in\mathbb{N}} E_n$ is an infinite set and then we have $\mu^*(E_{n_0}) = \infty$ and $\mu^*\left(\bigcup_{n\in\mathbb{N}} E_n\right) = \infty$. Then

$$\mu^*\left(\bigcup_{n\in\mathbb{N}} E_n\right) = \infty = \mu^*(E_{n_0}) \le \sum_{n\in\mathbb{N}} \mu^*(E_n).$$

(ii) Suppose E_n is a finite set for every $n \in \mathbb{N}$. If only finitely many of $(E_n : n \in \mathbb{N})$ are non-empty, say only $E_1, \ldots, E_N$ are non-empty having $p_1, \ldots p_N$ elements respectively where $p_1, \ldots, p_n \in \mathbb{N}$, then $\bigcup_{n\in\mathbb{N}} E_n$ has at most $p_1 + \cdots + p_N$ elements and therefore

$$\mu^*\left(\bigcup_{n\in\mathbb{N}} E_n\right) \le p_1 + \cdots + p_N = \sum_{n=1}^{N} \mu^*(E_n) = \sum_{n\in\mathbb{N}} \mu^*(E_n).$$

If infinitely many in $(E_n : n \in \mathbb{N})$ are non-empty then $\sum_{n\in\mathbb{N}} \mu^*(E_n) = \infty$ and then $\mu^*\left(\bigcup_{n\in\mathbb{N}} E_n\right) \le \infty = \sum_{n\in\mathbb{N}} \mu^*(E_n)$. This completes the proof of 4°.
2. Let us show that μ^* is additive on $\mathfrak{P}(X)$. Let $E_1, E_2 \in \mathfrak{P}(X)$ be such that $E_1 \cap E_2 = \emptyset$. If both E_1 and E_2 are finite sets, then $\mu^*(E_1) = p_1 \in \mathbb{Z}_+$ and $\mu^*(E_2) = p_2 \in \mathbb{Z}_+$ and then $E_1 \cup E_2$ has $p_1 + p_2$ elements. Then we have

$$\mu^*(E_1 \cup E_2) = p_1 + p_2 = \mu^*(E_1) + \mu^*(E_2).$$

If at least one of E_1 and E_2 is an infinite set, say E_2 is an infinite set, then $E_1 \cup E_2$ is an infinite set and then we have $\mu*(E_2) = \infty$ and $\mu^*(E_1 \cup E_2) = \infty$. Now $\mu^*(E_1) = p \in \mathbb{Z}_+$ or $\mu^*(E_1) = \infty$. Then we have

$$\mu^*(E_1 \cup E_2) = \infty = p + \infty = \mu^*(E_1) + \mu^*(E_2).$$

This completes the proof that μ^* is additive on $\mathfrak{P}(X)$.

3. To show that the outer measure μ^* on X is a measure on the σ-algebra $\mathfrak{P}(X)$ of subsets of X, we show that μ^* is countably additive on $\mathfrak{P}(X)$. According to Proposition 1.23, if a nonnegative extended real-valued set function γ on an algebra $\mathfrak{A}$ of subsets of a set X is additive and countably subadditive on $\mathfrak{A}$ then γ is countably additive on $\mathfrak{A}$. Now μ^*, being an outer measure on X, is countably subadditive on the σ-algebra $\mathfrak{P}(X)$ of subsets of X. Moreover we showed in (b) that μ^* is additive on $\mathfrak{P}(X)$. Then by Proposition 1.23, μ^* is countably additive on $\mathfrak{P}(X)$. This completes the proof that μ^* is a measure on the σ-algebra $\mathfrak{P}(X)$ of subsets of X.

4. Let us show that $\mathfrak{M}(\mu^*) = \mathfrak{P}(X)$. According to Theorem 2.5, $\mathfrak{M}(\mu^*) = \mathfrak{P}(X)$ if and only if μ^* is additive on $\mathfrak{P}(X)$. We showed in (b) that μ^* is additive on $\mathfrak{P}(X)$. Thus we have $\mathfrak{M}(\mu^*) = \mathfrak{P}(X)$. ∎

Prob. 2.4. Let X be an infinite set and let μ be the counting measure on the σ-algebra $\mathfrak{A}$ of all subsets of X. Show that there exists a decreasing sequence $(E_n : n \in \mathbb{N})$ in $\mathfrak{A}$ such that $E_n \downarrow \emptyset$, that is, $\lim_{n\to\infty} E_n = \emptyset$, with $\lim_{n\to\infty} \mu(E_n) \neq 0$.

Proof. Since X is an infinite set we have $\{x_n : n \in \mathbb{N}\} \subset X$. Let $E_n = \{x_k : k \geq n\}$ for $n \in \mathbb{N}$. Then $(E_n : n \in \mathbb{N})$ is a decreasing sequence and $\lim_{n\to\infty} E_n = \bigcap_{n\in\mathbb{N}} E_n = \emptyset$. Since E_n is an infinite set, we have $\mu(E_n) = \infty$ for every $n \in \mathbb{N}$. Then we have $\lim_{n\to\infty} \mu(E_n) = \infty$.

§3 Lebesgue Measure on $\mathbb{R}$

Prob. 3.1. Let a decreasing sequence $(E_n : n \in \mathbb{N}) \subset \mathfrak{M}_L$ in the Lebesgue measure space $(\mathbb{R}, \mathfrak{M}_L, \mu_L)$ be given by $E_n = [n, \infty)$ for $n \in \mathbb{N}$.
(a) Find $\lim_{n \to \infty} E_n$ and $\mu_L \left(\lim_{n \to \infty} E_n \right)$.
(b) Find $\lim_{n \to \infty} \mu_L(E_n)$.

Proof. According to Lemma 1.7, $\liminf_{n \to \infty} E_n$ consists of $x \in \mathbb{R}$ that is in E_n for all but finitely many $n \in \mathbb{N}$ and $\limsup_{n \to \infty} E_n$ consists of $x \in \mathbb{R}$ that is in E_n for infinitely many $n \in \mathbb{N}$. Thus $\liminf_{n \to \infty} E_n = \emptyset$ and $\limsup_{n \to \infty} E_n = \emptyset$ also. Thus we have $\liminf_{n \to \infty} E_n = \emptyset = \limsup_{n \to \infty} E_n$ and this implies that $\lim_{n \to \infty} E_n$ exists and $\lim_{n \to \infty} E_n = \liminf_{n \to \infty} E_n = \limsup_{n \to \infty} E_n = \emptyset$. Then $\mu_L \left(\lim_{n \to \infty} E_n \right) = \mu_L(\emptyset) = 0$. On the other hand, $\mu_L(E_n) = \infty$ for every $n \in \mathbb{N}$ and then $\lim_{n \to \infty} \mu_L(E_n) = \lim_{n \to \infty} \infty = \infty$. ∎

Prob. 3.2. Consider a sequence $(E_n : n \in \mathbb{N}) \subset \mathfrak{M}_L$ defined by $E_n = [0, 1) \cup [n, n + 1)$ when n is odd and $E_n = [0, 1) \cup [n, n + 2)$ when n is even.
(a) Show that $\lim_{n \to \infty} E_n$ exists and find $\lim_{n \to \infty} E_n$.
(b) Show that $\lim_{n \to \infty} \mu_L(E_n)$ does not exist.

Proof. We have $\liminf_{n \to \infty} E_n = [0, 1)$ and $\limsup_{n \to \infty} E_n = [0, 1)$. Then we have $\liminf_{n \to \infty} E_n = \limsup_{n \to \infty} E_n$ and thus $\lim_{n \to \infty} E_n$ exists and $\lim_{n \to \infty} E_n = \liminf_{n \to \infty} E_n = \limsup_{n \to \infty} = [0, 1)$. Then we have $\mu_L \left(\lim_{n \to \infty} E_n \right) = \mu_L([0, 1)) = 1$.

On the other hand, $\mu_L(E_n) = 2$ for odd $n \in \mathbb{N}$ and $\mu_L(E_n) = 3$ for even $n \in \mathbb{N}$ so that $\lim_{n \to \infty} \mu_L(E_n)$ does not exist. ∎

Prob. 3.3. For each of the following sequences $(E_n : n \in \mathbb{N}) \subset \mathfrak{M}_L$,
(a) show that $\lim\limits_{n\to\infty} E_n$ exists and find $\lim\limits_{n\to\infty} E_n$,
(b) show that $\lim\limits_{n\to\infty} \mu_L(E_n)$ exists and $\lim\limits_{n\to\infty} \mu_L(E_n) \neq \mu_L\big(\lim\limits_{n\to\infty} E_n\big)$.

(1) $\qquad E_n = [0, 1) \cup [n, n+1)$ for $n \in \mathbb{N}$,
(2) $\qquad E_n = [0, 1) \cup [n, 2n)$ for $n \in \mathbb{N}$,
(3) $\qquad E_n = [n, n+1)$ for $n \in \mathbb{N}$.

Proof.

(1) $\qquad\qquad E_n = [0, 1) \cup [n, n+1) \quad \text{for } n \in \mathbb{N},$

$$\liminf_{n\to\infty} E_n = [0, 1) \text{ and } \limsup_{n\to\infty} E_n = [0, 1),$$

$$\lim_{n\to\infty} E_n = \liminf_{n\to\infty} E_n = \limsup_{n\to\infty} E_n = [0, 1),$$

$$\mu_L\big(\lim_{n\to\infty} E_n\big) = \mu_L\big([0, 1)\big) = 1,$$

$$\mu_L(E_n) = 2 \text{ for every } n \in \mathbb{N}, \ \lim_{n\to\infty} \mu_L(E_n) = 2.$$

(2) $\qquad\qquad E_n = [0, 1) \cup [n, 2n) \quad \text{for } n \in \mathbb{N},$

$$\liminf_{n\to\infty} E_n = [0, 1) \text{ and } \limsup_{n\to\infty} E_n = [0, 1),$$

$$\lim_{n\to\infty} E_n = \liminf_{n\to\infty} E_n = \limsup_{n\to\infty} E_n = [0, 1),$$

$$\mu_L\big(\lim_{n\to\infty} E_n\big) = \mu_L\big([0, 1)\big) = 1,$$

$$\mu_L(E_n) = 1 + n \text{ for every } n \in \mathbb{N}, \ \lim_{n\to\infty} \mu_L(E_n) = \infty.$$

(3) $\qquad\qquad E_n = [n, n+1) \quad \text{for } n \in \mathbb{N},$

$$\liminf_{n\to\infty} E_n = \emptyset \text{ and } \limsup_{n\to\infty} E_n = \emptyset,$$

$$\lim_{n\to\infty} E_n = \liminf_{n\to\infty} E_n = \limsup_{n\to\infty} E_n = \emptyset,$$

$$\mu_L\big(\lim_{n\to\infty} E_n\big) = \mu_L(\emptyset) = 0,$$

$$\mu_L(E_n) = 1 \text{ for every } n \in \mathbb{N}, \ \lim_{n\to\infty} \mu_L(E_n) = 1. \ \blacksquare$$

Prob. 3.4. Let $\mathfrak{J}$ be the collection of $\emptyset$ and all finite open intervals in $\mathbb{R}$. Define an outer measure μ^* on $\mathbb{R}$ by setting for each $E \in \mathfrak{P}(\mathbb{R})$

$$\mu^*(E) = \inf \left\{ \sum_{n \in \mathbb{N}} \ell(I_n) : (I_n : n \in \mathbb{N}) \subset \mathfrak{J}, \bigcup_{n \in \mathbb{N}} I_n \supset E \right\}.$$

Show that $\mu^* = \mu_o^*$ defined in Observation 3.2.

Proof. The set $\mathfrak{J}$ is the collection of $\emptyset$ and all finite open intervals in $\mathbb{R}$ and the set $\mathfrak{J}_o$ in Observation 3.2 is the collection of $\emptyset$ and all open intervals in $\mathbb{R}$. Thus $\mathfrak{J} \subset \mathfrak{J}_o$. For every $E \in \mathfrak{P}(\mathbb{R})$ we have

$$\mu_o^*(E) = \inf \left\{ \sum_{n \in \mathbb{N}} \ell(I_n) : (I_n : n \in \mathbb{N}) \subset \mathfrak{J}_o, \bigcup_{n \in \mathbb{N}} I_n \supset E \right\}.$$

Then $\mathfrak{J} \subset \mathfrak{J}_o$ implies

(1) $\mu_o^*(E) \leq \mu^*(E).$

Let us show that we actually have

(2) $\mu_o^*(E) = \mu^*(E).$

Now if $E \in \mathfrak{P}(\mathbb{R})$ is such that $\mu_o^*(E) = \infty$ then we have $\mu^*(E) = \infty$ by (1) and thus $\mu_o^*(E) = \mu^*(E)$ and (2) holds. It remains to consider the case $\mu_o^*(E) < \infty$. Assume that we have $\mu_o^*(E) < \infty$. If $(I_n : n \in \mathbb{N})$ is a sequence in $\mathfrak{J}_o$ such that $\bigcup_{n \in \mathbb{N}} I_n \supset E$ and if I_{n_0} is an infinite open interval for some $n_0 \in \mathbb{N}$, then $\ell(I_{n_0}) = \infty$ and then $\sum_{n \in \mathbb{N}} \ell(I_n) = \infty$ and this implies that such a sequence $(I_n : n \in \mathbb{N})$ in $\mathfrak{J}_o$ has no effect on our $\mu_o^*(E) < \infty$. Thus for $\mu_o^*(E) < \infty$ we may drop any sequence $(I_n : n \in \mathbb{N})$ in $\mathfrak{J}_o$ which includes at least one infinite open interval, that is, we may confine ourselves to sequences of finite open intervals. Therefore we have

$$\mu_o^*(E) = \inf \left\{ \sum_{n \in \mathbb{N}} \ell(I_n) : (I_n : n \in \mathbb{N}) \subset \mathfrak{J}, \bigcup_{n \in \mathbb{N}} I_n \supset E \right\} = \mu^*(E). \quad \blacksquare$$

Prob. 3.5. Let $\mathfrak{E}$ be a disjoint collection of members of $\mathfrak{M}_L$ in the Lebesgue measure space $(\mathbb{R}, \mathfrak{M}_L, \mu_L)$. Show that if $\mu_L(E) > 0$ for every $E \in \mathfrak{E}$, then the collection $\mathfrak{E}$ is at most countable.

Proof. 1. Let $\mathfrak{E} = \{E_\alpha : \alpha \in A\}$ be an arbitrary disjoint collection of $\mathfrak{M}_L$-measurable subset E_α of $\mathbb{R}$ with $\mu_L(E_\alpha) > 0$. Here A is an arbitrary indexing set. Let us show that $\mathfrak{E}$ is at most a countable collection.

Let $I_n = (-n, n) \subset \mathbb{R}$ for $n \in \mathbb{N}$. Then $(I_n : n \in \mathbb{N})$ is an increasing sequence of sets in $\mathfrak{M}_L$ and $\lim\limits_{n \to \infty} I_n = \bigcup_{n \in \mathbb{N}} I_n = \mathbb{R}$.

For an arbitrary $E_\alpha \in \mathfrak{E}$, $(E_\alpha \cap I_n : n \in \mathbb{N})$ is an increasing sequence of sets in $\mathfrak{M}_L$ and we have

$$\lim_{n \to \infty} (E_\alpha \cap I_n) = \bigcup_{n \in \mathbb{N}} (E_\alpha \cap I_n) = E_\alpha \cap \left(\bigcup_{n \in \mathbb{N}} I_n \right) = E_\alpha \cap \mathbb{R} = E_\alpha$$

and then

$$\lim_{n \to \infty} \mu_L (E_\alpha \cap I_n) = \mu_L \left(\lim_{n \to \infty} (E_\alpha \cap I_n) \right) = \mu_L(E_\alpha).$$

Since $\mu_L(E_\alpha) > 0$, there exists $n(\alpha) \in \mathbb{N}$ such that

$$(1) \qquad \mu_L (E_\alpha \cap I_{n(\alpha)}) > 0.$$

Let us define a subset F_α of E_α by setting

$$(2) \qquad F_\alpha = E_\alpha \cap I_{n(\alpha)}.$$

Let us define a collection $\mathfrak{F}$ of $\mathfrak{M}_L$-measurable subsets of $\mathbb{R}$ with positive measure μ_L by setting

$$(3) \qquad \mathfrak{F} = \{F_\alpha : \alpha \in A\}.$$

Since $F_\alpha \subset E_\alpha$ for every $\alpha \in A$ and since $\mathfrak{E} = \{E_\alpha : \alpha \in A\}$ is a disjoint collection, our $\mathfrak{F} = \{F_\alpha : \alpha \in A\}$ is a disjoint collection. Also $\mathfrak{F}$ has the same cardinality as $\mathfrak{E}$. Thus to show that $\mathfrak{E}$ is at most a countable collection, it suffices to show that $\mathfrak{F}$ is at most a countable collection.

2. Let us show that $\mathfrak{F}$ is at most a countable collection. Assume the contrary, that is, assume that $\mathfrak{F}$ is an uncountable collection. Now every $F_\alpha \in \mathfrak{F}$ as defined by (2) is contained in an open interval I_n for some $n \in \mathbb{N}$. For $n \in \mathbb{N}$, let $\mathfrak{F}_n$ be the subcollection of $\mathfrak{F}$ consisting of those members of $\mathfrak{F}$ that are contained in I_n. Then we have

$$(4) \qquad \mathfrak{F} = \bigcup_{n \in \mathbb{N}} \mathfrak{F}_n.$$

Since $\mathfrak{F}$ is an uncountable collection, there exists $n_0 \in \mathbb{N}$ such that $\mathfrak{F}_{n_0}$ is an uncountable collection. Being a subcollection of the disjoint collection $\mathfrak{F}$, $\mathfrak{F}_{n_0}$ is a disjoint collection. Now for every $F_\alpha \in \mathfrak{F}_{n_0}$ we have $\mu_L(F_\alpha) > 0$. Thus there exists $k \in \mathbb{N}$ such that $\mu_L(F_\alpha) \geq \frac{1}{k}$. Let $\mathfrak{F}_{n_0, k}$ be the subcollection of $\mathfrak{F}_{n_0}$ consisting of those members F_α of

$\mathfrak{F}_{n_0}$ with $\mu_L(F_\alpha) \geq \frac{1}{k}$. As a subcollection of the disjoint collection $\mathfrak{F}_{n_0}$, $\mathfrak{F}_{n_0,k}$ is a disjoint collection. Now we have

(5)
$$\mathfrak{F}_{n_0} = \bigcup_{k \in \mathbb{N}} \mathfrak{F}_{n_0,k}.$$

Then since $\mathfrak{F}_{n_0}$ is an uncountable collection, there exists $k_0 \in \mathbb{N}$ such that $\mathfrak{F}_{n_0,k_0}$ is an uncountable collection. Then we can select distinct members F_m of $\mathfrak{F}_{n_0,k_0}$ for $m \in \mathbb{N}$. Since $\mathfrak{F}_{n_0,k_0}$ is a disjoint collection, its subcollection $\{F_m : m \in \mathbb{N}\}$ is a disjoint collection. Then we have

(6)
$$\mu_L\Big(\bigcup_{m \in \mathbb{N}} F_m\Big) = \sum_{m \in \mathbb{N}} \mu_L(F_m) \geq \sum_{m \in \mathbb{N}} \frac{1}{k_0} = \infty.$$

On the other hand, $F_m \in \mathfrak{F}_{n_0,k_0}$ and this implies that $F_m \subset I_{n_0}$ and then $\bigcup_{m \in \mathbb{N}} F_m \subset I_{n_0}$. Then we have

(7)
$$\mu_L\Big(\bigcup_{m \in \mathbb{N}} F_m\Big) \leq \mu_L(I_{n_0}) = 2n_0.$$

This contradicts (6). Therefore $\mathfrak{F}$ cannot be an uncountable collection and $\mathfrak{F}$ is at most a countable collection. This implies that $\mathfrak{E}$ is at most a countable collection. ∎

Prob. 3.6. For $E \in \mathfrak{M}_L$ with $\mu_L(E) < \infty$, define a real-valued function φ_E on $\mathbb{R}$ by setting
$$\varphi_E(x) = \mu_L\big(E \cap (-\infty, x]\big) \quad \text{for } x \in \mathbb{R}.$$
(a) Show that φ_E is an increasing function on $\mathbb{R}$.
(b) Show that $\lim\limits_{x \to -\infty} \varphi_E(x) = 0$ and $\lim\limits_{x \to \infty} \varphi_E(x) = \mu_L(E)$.
(c) Show that φ_E satisfies the Lipschitz condition on $\mathbb{R}$, that is,
$$|\varphi_E(x') - \varphi_E(x'')| \le |x' - x''| \quad \text{for } x', x'' \in \mathbb{R}.$$
(d) Show that φ_E is uniformly continuous on $\mathbb{R}$.

Proof. 1. Let us show that φ_E is an increasing function on $\mathbb{R}$. Let $x', x'' \in \mathbb{R}$ and $x' < x''$. Then we have

(1)
$$\varphi_E(x'') - \varphi_E(x') = \mu_L\big(E \cap (-\infty, x'']\big) - \mu_L\big(E \cap (-\infty, x']\big).$$

Now $(-\infty, x''] = (-\infty, x'] \cup (x', x'']$ and this implies
$$E \cap (-\infty, x''] = \big(E \cap (-\infty, x']\big) \cup \big(E \cap (x', x'']\big),$$

and then

(2)
$$\mu_L\big(E \cap (-\infty, x'']\big) = \mu_L\big(E \cap (-\infty, x']\big) + \mu_L\big(E \cap (x', x'']\big).$$

Since $\mu_L\big(E \cap (-\infty, x']\big) \le \mu_L(E) < \infty$, $\mu_L\big(E \cap (-\infty, x']\big)$ is a nonnegative real number. Subtracting this nonnegative real number from both sides of (2) we have

(3)
$$\mu_L\big(E \cap (-\infty, x'']\big) - \mu_L\big(E \cap (-\infty, x']\big) = \mu_L\big(E \cap (x', x'']\big) \ge 0.$$

Substituting (3) in (1), we have
$$\varphi_E(x'') - \varphi_E(x') \ge 0.$$

This shows that φ_E is an increasing function on $\mathbb{R}$.

2. Since φ_E is an increasing function on $\mathbb{R}$, to show that $\lim\limits_{x \to \infty} \varphi_E(x) = \mu_L(E)$, it suffices to show that there exists an increasing sequence $(\xi_k : k \in \mathbb{N})$ in $\mathbb{R}$ such that $\lim\limits_{k \to \infty} \xi_k = \infty$ and $\lim\limits_{k \to \infty} \varphi_E(\xi_k) = \mu_L(E)$, and similarly to show that $\lim\limits_{x \to -\infty} \varphi_E(x) = 0$, it suffices to show that there exists decreasing sequence $(\eta_k : k \in \mathbb{N})$ in $\mathbb{R}$ such that $\lim\limits_{k \to \infty} \eta_k = -\infty$ and $\lim\limits_{k \to \infty} \varphi_E(\eta_k) = 0$.

For $k \in \mathbb{Z}$, let $I_k = (k-1, k]$ and $E_k = E \cap I_k$. Then $\{I_k : k \in \mathbb{Z}\}$ is a disjoint collection with $\bigcup_{k \in \mathbb{Z}} I_k = \mathbb{R}$. Thus we have
$$E = E \cap \mathbb{R} = E \cap \Big(\bigcup_{k \in \mathbb{Z}} I_k\Big) = \bigcup_{k \in \mathbb{Z}}(E \cap I_k) = \bigcup_{k \in \mathbb{Z}} E_k.$$

Let us select a sequence $(\xi_k : k \in \mathbb{N})$ in $\mathbb{R}$ by setting $\xi_k = k$ for $k \in \mathbb{N}$. Then $(\xi_k : k \in \mathbb{N})$ is an increasing sequence with $\lim\limits_{k \to \infty} \xi_k = \infty$. We have

(4)
$$\varphi_E(\xi_k) = \varphi_E(k) = \mu_L\big(E \cap (-\infty, k]\big).$$

Now $\big(E \cap (-\infty, k] : k \in \mathbb{N}\big)$ is an increasing sequence in $\mathfrak{M}_L$ and

$$\lim_{k \to \infty} \big(E \cap (-\infty, k]\big) = \bigcup_{k \in \mathbb{N}} \big(E \cap (-\infty, k]\big) = E \cap \bigcup_{k \in \mathbb{N}} (-\infty, k] = E \cap \mathbb{R} = E$$

and then

(5) $\lim_{k \to \infty} \mu_L \big(E \cap (-\infty, k]\big) = \mu_L \Big(\lim_{k \to \infty} \big(E \cap (-\infty, k]\big) \Big) = \mu_L(E).$

By (4) and (5), we have $\lim_{k \to \infty} \varphi_E(\xi_k) = \mu_L(E)$. This proves $\lim_{x \to \infty} \varphi_E(x) = \mu_L(E)$.

Next select a sequence $(\eta_k : k \in \mathbb{N})$ in $\mathbb{R}$ by letting $\eta_k = -k$ for $k \in \mathbb{N}$. Then $(\eta_k : k \in \mathbb{N})$ is a decreasing sequence with $\lim_{k \to \infty} \eta_k = -\infty$. We have

(6) $\varphi_E(\eta_k) = \varphi_E(-k) = \mu_L \big(E \cap (-\infty, -k]\big).$

Observe that $\big(E \cap (-\infty, -k] : k \in \mathbb{N}\big)$ is a decreasing sequence in $\mathfrak{M}_L$ and

$$\lim_{k \to \infty} \big(E \cap (-\infty, -k]\big) = \bigcap_{k \in \mathbb{N}} \big(E \cap (-\infty, -k]\big) = E \cap \bigcap_{k \in \mathbb{N}} (-\infty, -k] = E \cap \emptyset = \emptyset.$$

Since $E \cap (-\infty, -k] \subset E$ and $\mu_L(E) < \infty$, we have by Theorem 1.26 that

(7) $\lim_{k \to \infty} \mu_L \big(E \cap (-\infty, -k]\big) = \mu_L \Big(\lim_{k \to \infty} \big(E \cap (-\infty, -k]\big) \Big) = \mu_L(\emptyset) = 0.$

By (6) and (7), we have $\lim_{k \to \infty} \varphi_E(\eta_k) = 0$. This proves $\lim_{x \to -\infty} \varphi_E(x) = 0$.

3. To show that φ_E satisfies the Lipschitz condition, let $x', x'' \in \mathbb{R}$ and $x' < x''$. Then since φ_E is an increasing function on $\mathbb{R}$ we have

(8) $|\varphi_E(x') - \varphi_E(x'')| = \varphi_E(x'') - \varphi_E(x')$

$$= \mu_L \big(E \cap (-\infty, x'']\big) - \mu_L \big(E \cap (-\infty, x']\big)$$

$$= \mu_L \big(E \cap (x', x'']\big) \quad \text{by (3)}$$

$$\leq \mu_L \big((x', x'']\big) = |x' - x''|.$$

4. Let us show that φ_E is uniformly continuous on $\mathbb{R}$. Let $\varepsilon > 0$ be arbitrarily given. Let $\delta = \varepsilon > 0$. Then for $x', x'' \in \mathbb{R}$ such that $|x' - x''| < \delta$, we have by (8)

$$|\varphi_E(x') - \varphi_E(x'')| \leq |x' - x''| < \delta = \varepsilon.$$

This proves the uniform continuity of φ_E on $\mathbb{R}$. ∎

Prob. 3.7. Let $E \in \mathfrak{M}_L$ with $\mu_L(E) < \infty$. Show that for every $\alpha \in (0, 1)$, there exists a subset E_α of E such that $E_\alpha \in \mathfrak{M}_L$ and $\mu_L(E_\alpha) = \alpha \mu_L(E)$.
(Hint: Use the continuity of the function φ_E in Prob. 3.6.)
(Thus the Lebesgue measure space $(\mathbb{R}, \mathfrak{M}_L, \mu_L)$ not only does not have any atoms but in fact every Lebesgue-measurable set with finite measure has a measurable subset with an arbitrarily designated fraction of the measure. See Prob. 3.8 for $E \in \mathfrak{M}_L$ with $\mu_L(E) = \infty$.)

Proof. Observe that if $\mu_L(E) = 0$ then for every $\alpha \in (0, 1)$ we have $\mu_L(E) = 0 = \alpha \mu_L(E)$ and the claim is trivially true. Thus we need to consider only the case $0 < \mu_L(E) < \infty$.

Consider a real-valued function φ_E on $\mathbb{R}$ defined by

$$\varphi_E(x) = \mu_L\big(E \cap (-\infty, x]\big) \quad \text{for } x \in \mathbb{R}.$$

According to Prob. 3.6, φ_E is continuous and increasing on $\mathbb{R}$ with $\lim_{x \to -\infty} \varphi_E(x) = 0$ and $\lim_{x \to \infty} \varphi_E(x) = \mu_L(E)$. Let $\alpha \in (0, 1)$ be arbitrarily given. Then let $\alpha_1, \alpha_2 \in \mathbb{R}$ be such that

$$0 < \alpha_1 < \alpha < \alpha_2 < 1.$$

Since $\lim_{x \to -\infty} \varphi_E(x) = 0$, the inequality $\alpha_1 \mu_L(E) > 0$ implies that there exists $x_1 \in \mathbb{R}$ such that

$$0 \le \varphi_E(x_1) < \alpha_1 \mu_L(E),$$

and similarly since $\lim_{x \to \infty} \varphi_E(x) = \mu_L(E)$, the inequality $\alpha_2 \mu_L(E) < \mu_L(E)$ implies that there exists $x_2 \in \mathbb{R}$ such that

$$\alpha_2 \mu_L(E) < \varphi_E(x_2) \le \mu_L(E).$$

Then for our $\alpha \in (\alpha_1, \alpha_2)$ we have

$$\varphi_E(x_1) < \alpha \mu_L(E) < \varphi_E(x_2).$$

Then since φ_E is continuous on $\mathbb{R}$, the Intermediate Value Theorem for continuous functions implies that there exists $x_0 \in (x_1, x_2)$ such that $\varphi_E(x_0) = \alpha \mu_L(E)$, that is,

$$\mu_L\big(E \cap (-\infty, x_0]\big) = \alpha \mu_L(E).$$

Let $E_\alpha = E \cap (-\infty, x_0]$. Then E_α is a $\mathfrak{M}_L$-measurable subset of E and $\mu_L(E_\alpha) = \alpha \mu_L(E)$. ∎

Prob. 3.8. Let $E \in \mathfrak{M}_L$ with $\mu_L(E) = \infty$. Show that for every $\lambda \in [0, \infty)$, there exists a subset E_λ of E such that $E_\lambda \in \mathfrak{M}_L$ and $\mu_L(E_\lambda) = \lambda$.

Proof. Let us observe that for $\lambda = 0$, $\emptyset$ is a $\mathfrak{M}_L$-measurable subset of E with $\mu_L(\emptyset) = 0 = \lambda$. Thus we need to consider only the case $\lambda \in (0, \infty)$.

Let $E_n = E \cap (-n, n)$ for $n \in \mathbb{N}$. Observe that E_n is a $\mathfrak{M}_L$-measurable subset of E and $(-n, n)$ and moreover $\mu_L(E_n) \leq \mu_L((-n, n)) = 2n < \infty$. Now $(E_n : n \in \mathbb{N})$ is an increasing sequence in $\mathfrak{M}_L$ and we have

(1) $$\lim_{n\to\infty} E_n = \bigcup_{n\in\mathbb{N}} E_n = \bigcup_{n\in\mathbb{N}} E \cap (-n, n) = E \cap \bigcup_{n\in\mathbb{N}}(-n, n) = E \cap \mathbb{R} = E$$

and then

(2) $$\lim_{n\to\infty} \mu_L(E_n) = \mu_L\left(\lim_{n\to\infty} E_n\right) = \mu_L(E) = \infty.$$

Let $\lambda \in (0, \infty)$. Then (2) implies that there exists $n_0 \in \mathbb{N}$ such that $\lambda < \mu_L(E_{n_0}) \leq 2n_0$. Let

(3) $$\alpha = \frac{\lambda}{\mu_L(E_{n_0})} \in (0, 1).$$

Now we have $\mu_L(E_{n_0}) < \infty$ and $\alpha \in (0, 1)$. Then by Prob. 3.7 there exists a $\mathfrak{M}_L$-measurable sunset F of E_{n_0} such that

(4) $$\mu_L(F) = \alpha\mu_L(E_{n_0}) = \frac{\lambda}{\mu_L(E_{n_0})} \mu_L(E_{n_0}) = \lambda.$$

Let $E_\lambda = F$. Then we have $\mu_L(E_\lambda) = \lambda$. ∎

Prob. 3.9. Let $E \subset \mathbb{R}$ and $\mu_L^*(E) = 0$. Show that E^c is a dense subset of $\mathbb{R}$, that is, every non-empty open set O in $\mathbb{R}$ contains some points of E^c, in other words no non-empty open set O in $\mathbb{R}$ can be disjoint from E^c.

Proof. For every open interval $I \subset \mathbb{R}$ we have $\mu_L^*(I) > 0$. If $\mu_L^*(E) = 0$ then E cannot contain any open interval I since $I \subset E$ implies $\mu_L^*(E) \geq \mu_L^*(I) > 0$. Thus for every open interval I we have $E^c \cap I \neq \emptyset$. Now every non-empty open set O is a union of open intervals. This implies that $E^c \cap O \neq \emptyset$, that is, O is not disjoint from E^c. ∎

Prob. 3.10. Consider the measure space $(\mathbb{R}, \mathfrak{B}_\mathbb{R}, \mu_L)$.
(a) Show that if $E \in \mathfrak{B}_\mathbb{R}$ and $t \in \mathbb{R}$, then $E + t \in \mathfrak{B}_\mathbb{R}$ and $\mu_L(E + t) = \mu_L(E)$.
Let $\mathfrak{B}_\mathbb{R} + t := \{E + t : E \in \mathfrak{B}_\mathbb{R}\}$. Show that $\mathfrak{B}_\mathbb{R} + t = \mathfrak{B}_\mathbb{R}$ for every $t \in \mathbb{R}$.
(b) Show that if $E \in \mathfrak{B}_\mathbb{R}$ and $\alpha \in \mathbb{R}$, then $\alpha E \in \mathfrak{B}_\mathbb{R}$ and $\mu_L(\alpha E) = |\alpha| \mu_L(E)$.
Let $\alpha \mathfrak{B}_\mathbb{R} := \{\alpha E : E \in \mathfrak{B}_\mathbb{R}\}$. Show that $\alpha \mathfrak{B}_\mathbb{R} = \mathfrak{B}_\mathbb{R}$ for every $\alpha \in \mathbb{R}$ such that $\alpha \neq 0$.

Proof. (a) and (b) can be proved in the same way as Theorem 3.16 and Theorem 3.18 are proved. Here we present a different proof based on some basic results in §4.
   1. Let $E \subset \mathbb{R}$. According to Observation 4.3, $E \in \mathfrak{B}_\mathbb{R}$ if and only if $\mathbf{1}_E$ is a $\mathfrak{B}_\mathbb{R}$-measurable function on $\mathbb{R}$.
   According to Theorem 4.6, a real-valued function on $\mathbb{R}$ is $\mathfrak{B}_\mathbb{R}$-measurable if and only if it is a $\mathfrak{B}_\mathbb{R}/\mathfrak{B}_\mathbb{R}$-measurable mapping of $\mathbb{R}$ into $\mathbb{R}$.
   According to Theorem 4.27, a real-valued continuous function on $\mathbb{R}$ is a $\mathfrak{B}_\mathbb{R}$-measurable function.
   2. Let us prove (a). Let $E \in \mathfrak{B}_\mathbb{R}$ and $t \in \mathbb{R}$. To show that $E + t \in \mathfrak{B}_\mathbb{R}$ we show that $\mathbf{1}_{E+t}$ is a $\mathfrak{B}_\mathbb{R}/\mathfrak{B}_\mathbb{R}$-measurable mapping of $\mathbb{R}$ into $\mathbb{R}$. Now we have

$$(1) \qquad \mathbf{1}_{E+t}(x) = \left\{ \begin{array}{ll} 1 & \text{for } x \in E + t, \text{ that is, } x - t \in E \\ 0 & \text{for } x \in (E + t)^c, \text{ that is, } x - t \in E^c \end{array} \right\} = \mathbf{1}_E(x - t).$$

For $t \in \mathbb{R}$, let τ_t be a real-valued function on $\mathbb{R}$ defined by setting

$$(2) \qquad \qquad \qquad \tau_t(x) = x - t \quad \text{for } x \in \mathbb{R}.$$

Then τ_t is a real-valued continuous function on $\mathbb{R}$ and is thus a $\mathfrak{B}_\mathbb{R}$-measurable function on $\mathbb{R}$, that is, a $\mathfrak{B}_\mathbb{R}/\mathfrak{B}_\mathbb{R}$-measurable mapping of $\mathbb{R}$ into $\mathbb{R}$. By (1) and (2), we have

$$(3) \qquad \mathbf{1}_{E+t}(x) = \mathbf{1}_E(x - t) = \mathbf{1}_E(\tau_t(x)) = (\mathbf{1}_E \circ \tau_t)(x) \quad \text{for } x \in \mathbb{R}.$$

Now τ_t is a $\mathfrak{B}_\mathbb{R}/\mathfrak{B}_\mathbb{R}$-measurable mapping of $\mathbb{R}$ into $\mathbb{R}$ and $\mathbf{1}_E$ is a $\mathfrak{B}_\mathbb{R}/\mathfrak{B}_\mathbb{R}$-measurable mapping of $\mathbb{R}$ into $\mathbb{R}$. This implies that $\mathbf{1}_E \circ \tau_t$ is a $\mathfrak{B}_\mathbb{R}/\mathfrak{B}_\mathbb{R}$-measurable mapping of $\mathbb{R}$ into $\mathbb{R}$ according to Theorem 1.40 (Chain Rule for Measurable Mappings). Thus $\mathbf{1}_{E+t}$ is a $\mathfrak{B}_\mathbb{R}/\mathfrak{B}_\mathbb{R}$-measurable mapping of $\mathbb{R}$ into $\mathbb{R}$. This shows that $E + t \in \mathfrak{B}_\mathbb{R}$.
   The measure space $(\mathbb{R}, \mathfrak{M}_L, \mu_L)$ is a translation invariant measure space according to Theorem 3.16. Thus for every $E \in \mathfrak{M}_L$ and $t \in \mathbb{R}$ we have $\mu_L(E + t) = \mu_L(E)$. Then since $\mathfrak{B}_\mathbb{R} \subset \mathfrak{M}_L$, for every $E \in \mathfrak{B}_\mathbb{R}$ and $t \in \mathbb{R}$ we have $\mu_L(E + t) = \mu_L(E)$.
   Let $\mathfrak{B}_\mathbb{R} + t := \{B + t : B \in \mathfrak{B}_\mathbb{R}\}$. We showed above that if $E \in \mathfrak{B}_\mathbb{R}$ and $t \in \mathbb{R}$ then $E + t \in \mathfrak{B}_\mathbb{R}$. This implies that $\mathfrak{B}_\mathbb{R} + t \subset \mathfrak{B}_\mathbb{R}$. Since this holds for every $t \in \mathbb{R}$ and since for every $t \in \mathbb{R}$ we have $-t \in \mathbb{R}$, we have $\mathfrak{B}_\mathbb{R} - t \subset \mathfrak{B}_\mathbb{R}$. Then $\mathfrak{B}_\mathbb{R} = (\mathfrak{B}_\mathbb{R} - t) + t \subset \mathfrak{B}_\mathbb{R} + t$. Thus we have $\mathfrak{B}_\mathbb{R} + t \subset \mathfrak{B}_\mathbb{R} \subset \mathfrak{B}_\mathbb{R} + t$ and therefore $\mathfrak{B}_\mathbb{R} + t = \mathfrak{B}_\mathbb{R}$. This completes the proof of (a).
   3. Let us prove (b). Let $E \in \mathfrak{B}_\mathbb{R}$ and $\alpha \in \mathbb{R}$. To show that $\alpha E \in \mathfrak{B}_\mathbb{R}$ we show that $\mathbf{1}_{\alpha E}$ is a $\mathfrak{B}_\mathbb{R}/\mathfrak{B}_\mathbb{R}$-measurable mapping of $\mathbb{R}$ into $\mathbb{R}$. Now if $\alpha = 0$ then $\alpha E = \{0\} \in \mathfrak{B}_\mathbb{R}$. Thus it remains to consider the case $\alpha \neq 0$. Now we have

$$(4) \qquad \mathbf{1}_{\alpha E}(x) = \left\{ \begin{array}{ll} 1 & \text{for } x \in \alpha E, \text{ that is, } \frac{1}{\alpha}x \in E \\ 0 & \text{for } x \notin \alpha E, \text{ that is, } \frac{1}{\alpha}x \notin E \end{array} \right\} = \mathbf{1}_E\left(\frac{1}{\alpha}x\right).$$

For $\gamma \in \mathbb{R}, \gamma \neq 0$, let π_γ be a real-valued function on $\mathbb{R}$ defined by setting

$$(5) \qquad\qquad \pi_\gamma(x) = \gamma x \quad \text{for } x \in \mathbb{R}.$$

Then π_γ is a real-valued continuous function on $\mathbb{R}$ and is thus a $\mathfrak{B}_\mathbb{R}$-measurable function on $\mathbb{R}$, that is, a $\mathfrak{B}_\mathbb{R}/\mathfrak{B}_\mathbb{R}$-measurable mapping of $\mathbb{R}$ into $\mathbb{R}$. By (4) and (5), we have

$$(6) \qquad \mathbf{1}_{\alpha E}(x) = \mathbf{1}_E\big(\tfrac{1}{\alpha}x\big) = \mathbf{1}_E\big(\pi_{\frac{1}{\alpha}}(x)\big) = \big(\mathbf{1}_E \circ \pi_{\frac{1}{\alpha}}\big)(x) \quad \text{for } x \in \mathbb{R}.$$

Since $\pi_{\frac{1}{\alpha}}$ is a $\mathfrak{B}_\mathbb{R}/\mathfrak{B}_\mathbb{R}$-measurable mapping of $\mathbb{R}$ into $\mathbb{R}$ and $\mathbf{1}_E$ is a $\mathfrak{B}_\mathbb{R}/\mathfrak{B}_\mathbb{R}$-measurable mapping of $\mathbb{R}$ into $\mathbb{R}$, $\mathbf{1}_E \circ \pi_{\frac{1}{\alpha}}$ is a $\mathfrak{B}_\mathbb{R}/\mathfrak{B}_\mathbb{R}$-measurable mapping of $\mathbb{R}$ into $\mathbb{R}$ according to Theorem 1.40 (Chain Rule for Measurable Mappings). Thus $\mathbf{1}_{\alpha E}$ is a $\mathfrak{B}_\mathbb{R}/\mathfrak{B}_\mathbb{R}$-measurable mapping of $\mathbb{R}$ into $\mathbb{R}$. This shows that $\alpha E \in \mathfrak{B}_\mathbb{R}$.

Consider the measure space $(\mathbb{R}, \mathfrak{M}_L, \mu_L)$. According to Theorem 3.18 (Positive Homogeneity of the Lebesgue Measure Space), for every $E \in \mathfrak{M}_L$ and $\alpha \in \mathbb{R}$ we have $\alpha E \in \mathfrak{M}_L$ and $\mu_L(\alpha E) = |\alpha|\mu_L(E)$. Then since $\mathfrak{B}_\mathbb{R} \subset \mathfrak{M}_L$, for every $E \in \mathfrak{B}_\mathbb{R}$ and $\alpha \in \mathbb{R}$ we have $\mu_L(\alpha E) = |\alpha|\mu_L(E)$.

Let $\alpha\mathfrak{B}_\mathbb{R} := \{\alpha E : E \in \mathfrak{B}_\mathbb{R}\}$. We showed above that if $E \in \mathfrak{B}_\mathbb{R}$ and $\alpha \in \mathbb{R}$ then $\alpha E \in \mathfrak{B}_\mathbb{R}$. This implies that $\alpha\mathfrak{B}_\mathbb{R} \subset \mathfrak{B}_\mathbb{R}$. Now if $\alpha \neq 0$ then $\frac{1}{\alpha} \in \mathbb{R}$ and then $\frac{1}{\alpha}\mathfrak{B}_\mathbb{R} \subset \mathfrak{B}_\mathbb{R}$. Then we have $\mathfrak{B}_\mathbb{R} = \alpha\big(\frac{1}{\alpha}\mathfrak{B}_\mathbb{R}\big) \subset \alpha\mathfrak{B}_\mathbb{R}$. Thus we have $\alpha\mathfrak{B}_\mathbb{R} \subset \mathfrak{B}_\mathbb{R} \subset \alpha\mathfrak{B}_\mathbb{R}$ and therefore we have $\alpha\mathfrak{B}_\mathbb{R} = \mathfrak{B}_\mathbb{R}$. This completes the proof of (b). ∎

Prob. 3.11. Let E and F be two subsets of $\mathbb{R}$. Suppose $E \subset \alpha F$ for some $\alpha \in \mathbb{R}$. Show that $\mu_L^*(E) \leq |\alpha|\mu_L^*(F)$.

Proof. If $E \subset \alpha F$ then by the monotonicity of the Lebesgue outer measure μ_L^* we have $\mu_L^*(E) \leq \mu_L^*(\alpha F)$. By Lemma 3.17, we have $\mu_L^*(\alpha F) = |\alpha|\mu_L^*(F)$. Therefore we have $\mu_L^*(E) \leq |\alpha|\mu_L^*(F)$. ∎

Prob. 3.12. Let f be a real-valued function on an open interval $I = (a, b) \subset \mathbb{R}$ such that the derivative f' exists and $|f'(x)| \le \gamma$ for $x \in I$ where γ is a nonnegative real number.
(a) Show that for every $I_0 = (a_0, b_0) \subset (a, b)$ the image $f(I_0)$ is either a singleton or an interval in $\mathbb{R}$ and
$$\mu_L\big(f(I_0)\big) \le \gamma \, \mu_L(I_0).$$
(b) Show that for an arbitrary subset $E \subset I$ we have
$$\mu_L^*\big(f(E)\big) \le \gamma \, \mu_L^*(E).$$

Proof. 1. Let us prove (a). Differentiability of f on I implies continuity of f on I and in particular f is continuous on $I_0 \subset I$. Then by the Intermediate Value Theorem for continuous functions, the image of the open interval I_0, $f(I_0)$, is either a singleton or an interval in $\mathbb{R}$. If $f(I_0)$ is a singleton then $\mu_L\big(f(I_0)\big) = 0 \le \gamma \, \mu_L(I_0)$. Consider the case that $f(I_0)$ is an interval in $\mathbb{R}$. Let $x', x'' \in I_0$ and $x' < x''$. Then by the Mean Value Theorem there exists $\xi \in (x', x'')$ such that $f(x'') - f(x') = f'(\xi)(x'' - x')$ and then we have

$$|f(x'') - f(x')| = |f'(\xi)|(x'' - x') \le \gamma(x'' - x')$$

$$\le \gamma \, \ell(I_0) = \gamma \, \mu_L(I_0).$$

This shows that the distance between any two points in the interval $f(I_0)$ does not exceed $\gamma \, \mu_L(I_0)$ and thus the length of the interval $f(I_0)$ does not exceed $\gamma \, \mu_L(I_0)$. Then we have $\mu_L\big(f(I_0)\big) = \ell\big(f(I_0)\big) \le \gamma \, \mu_L(I_0)$. This completes the proof of (a).

2. Let us prove (b). Let $E \subset (a, b) = I$. Now by definition of μ_L^*, we have

(1) $\qquad \mu_L^*\big(f(E)\big) = \inf\Big\{ \sum_{k \in \mathbb{N}} \ell(J_k) : (J_k : n \in \mathbb{N}) \subset \mathfrak{I}_o, \bigcup_{k \in \mathbb{N}} J_k \supset f(E) \Big\}.$

In order to show that $\mu_L^*\big(f(E)\big) \le \gamma \, \mu_L^*(E)$, we show first that for every $\varepsilon > 0$ there exists a sequence $(J_k : k \in \mathbb{N}) \subset \mathfrak{I}_o$ such that

(2) $\qquad \bigcup_{k \in \mathbb{N}} J_k \supset f(E) \quad \text{and} \quad \sum_{k \in \mathbb{N}} \ell(J_k) \le \gamma \, \mu_L^*(E) + \varepsilon.$

Now $E \subset (a, b)$. Then by Lemma 3.21, for every $\varepsilon > 0$ there exists an open set O in $\mathbb{R}$ such that $E \subset O$ and

(3) $\qquad \mu_L^*(E) \le \mu_L^*(O) \le \mu_L^*(E) + \varepsilon.$

Let $O_0 = O \cap (a, b)$. Then O_0 is an open set in $\mathbb{R}$ and since $E \subset (a, b)$ and $E \subset O$ also we have $E \subset O_0$ and this implies by the monotonicity of the outer measure μ_L^* and by (3) that

(4) $\qquad \mu_L^*(E) \le \mu_L^*(O_0) \le \mu_L^*(O) \le \mu_L^*(E) + \varepsilon.$

Since O_0 is an open set in $\mathbb{R}$, there exists a disjoint collection of open intervals $\{(a_k, b_k) : k \in \mathbb{N}\}$ such that $O_0 = \bigcup_{k \in \mathbb{N}} (a_k, b_k)$. Then we have

$$E \subset O_0 = \bigcup_{k \in \mathbb{N}} (a_k, b_k)$$

and then

$$f(E) \subset f(O_0) = f\left(\bigcup_{k \in \mathbb{N}} (a_k, b_k)\right) = \bigcup_{k \in \mathbb{N}} f((a_k, b_k)).$$

Now $(a_k, b_k) \subset O_0 \subset (a, b)$ for every $k \in \mathbb{N}$. According to (a) this implies that $f((a_k, b_k))$ is either a singleton or an interval in $\mathbb{R}$ and moreover

$$\ell(f((a_k, b_k))) \le \gamma \, \ell((a_k, b_k)).$$

For each $k \in \mathbb{N}$, let J_k be an open interval in $\mathbb{R}$ such that

$$J_k \supset f((a_k, b_k)) \quad \text{and} \quad \ell(J_k) < \ell(f((a_k, b_k))) + \frac{\varepsilon}{2^k}.$$

For the sequence $(J_k : k \in \mathbb{N}) \subset \mathfrak{I}_o$, we have

$$(5) \qquad\qquad f(E) \subset \bigcup_{k \in \mathbb{N}} f((a_k, b_k)) \subset \bigcup_{k \in \mathbb{N}} J_k,$$

and

$$(6) \qquad \sum_{k \in \mathbb{N}} \ell(J_k) \le \sum_{k \in \mathbb{N}} \left\{ \ell(f((a_k, b_k))) + \frac{\varepsilon}{2^k} \right\} \le \gamma \sum_{k \in \mathbb{N}} \ell((a_k, b_k)) + \varepsilon$$

$$= \gamma \, \mu_L(O_0) + \varepsilon \le \gamma \left\{ \mu_L^*(E) + \varepsilon \right\} + \varepsilon$$

$$= \gamma \, \mu_L^*(E) + (\gamma + 1)\varepsilon.$$

Our sequence $(J_k : k \in \mathbb{N}) \subset \mathfrak{I}_o$ has the properties (5) and (6). This implies then

$$\mu_L^*(f(E)) = \inf \left\{ \sum_{k \in \mathbb{N}} \ell(J_k) : (J_k : n \in \mathbb{N}) \subset \mathfrak{I}_o, \bigcup_{k \in \mathbb{N}} J_k \supset f(E) \right\}$$

$$\le \gamma \, \mu_L^*(E) + (\gamma + 1)\varepsilon.$$

Since this holds for every $\varepsilon > 0$, we have $\mu_L^*(f(E)) \le \gamma \, \mu_L^*(E)$. ∎

Prob. 3.13. From the interval $[0, 1]$ remove the open middle third $\left(\frac{1}{3}, \frac{2}{3}\right)$. Remove the open middle third from each of the two remaining intervals $\left[0, \frac{1}{3}\right]$ and $\left[\frac{2}{3}, 1\right]$. This leaves us four closed intervals. Remove the open middle third from each of the four. Continue this process of removal indefinitely. The resulting set T is called the Cantor ternary set.
(a) Show that T is a Borel set and indeed T is a compact set in $\mathbb{R}$.
(b) Show that $\mu_L(T) = 0$ and thus T is a null set in the measure space $(\mathbb{R}, \mathfrak{B}_\mathbb{R}, \mu_L)$.
(c) Show that T is nowhere dense. (A set A in a topological space $(X, \mathfrak{D})$ is called nowhere dense, or non-dense, if the interior of its closure is empty, that is, $(\overline{A})^\circ = \emptyset$.)
(d) Show that T is a perfect set. (A set A in a topological space $(X, \mathfrak{D})$ is called perfect if $A = A'$ where A' is the derived set of A, that is, the set consisting of all the limit points of the set A. A set A is a closed set if and only if $A \supset A'$. A point in a set A is called an isolated point of A if it is contained in an open set which contains no other point of A. Thus a set A is perfect if and only if it is closed and has no isolated point in it.)
(e) Show that T is an uncountable set and in fact a one-to-one correspondence between T and $[0, 1]$ can be established.
(Thus the Cantor ternary set is an example of a null set in $(\mathbb{R}, \mathfrak{M}_L, \mu_L)$ which is an uncountable set.)

Proof. (a) Let T_n be the resulting set after the n-th step of removal of open subintervals, then $(T_n : n \in \mathbb{N})$ is a decreasing sequence of closed sets. Then $T = \bigcap_{n \in \mathbb{N}} T_n$ is a closed set. Then since $\mathfrak{B}_\mathbb{R}$ is the smallest σ-algebra of subsets of $\mathbb{R}$ containing all open sets in $\mathbb{R}$, the closed set T is in $\mathfrak{B}_\mathbb{R}$. Moreover T is contained in the finite interval $[0, 1]$ and thus T is a bounded closed set in $\mathbb{R}$ and this implies that T is a compact set in $\mathbb{R}$.
    (b) Observe that $\mu_L(T_1) = \frac{2}{3}$ and $\mu_L(T_{k+1}) = \frac{2}{3}\mu_L(T_k)$ for every $k \in \mathbb{N}$. Thus by mathematical induction we have $\mu_L(T_n) = \left(\frac{2}{3}\right)^n$ for every $n \in \mathbb{N}$. Then since $(T_n : n \in \mathbb{N})$ is a decreasing sequence and $\lim_{n \to \infty} T_n = T$ and since $\mu_L(T_1) < \infty$, we have

$$\mu_L(T) = \lim_{n \to \infty} \mu_L(T_n) = \lim_{n \to \infty} \left(\tfrac{2}{3}\right)^n = 0.$$

    (c) Let us show that T is nowhere dense, that is, $(\overline{T})^\circ = \emptyset$. Now since T is a closed set, we have $\overline{T} = T$ and then $(\overline{T})^\circ = T^\circ$. Consider the open set T° in $\mathbb{R}$. Now a non-empty open set in $\mathbb{R}$ is the union of countably many disjoint open intervals. Note that $\mu_L(T) = 0$ implies that T does not contain any open interval and then its subset T° does not contain any open interval and therefore the open set T° cannot be non-empty. This shows that $T^\circ = \emptyset$ and then $(\overline{T})^\circ = T^\circ = \emptyset$.
    (d) Since T is a closed set, to show that it is a perfect set it remains to show that it contains no isolated points. Thus let us show that if $x \in T$ then x is a limit point of T. Let $x \in T$. Then since $T = \bigcap_{n \in \mathbb{N}} T_n$, we have $x \in T_n$ for every $n \in \mathbb{N}$. Let $\varepsilon > 0$ and consider the open interval $(x - \varepsilon, x + \varepsilon)$. Since $x \in T_n$, we have $x \in J_{n,k}$, one of the 2^n closed intervals of length $\frac{1}{3^n}$ constituting T_n. Let $N \in \mathbb{N}$ be so large that $\frac{1}{3^N} < \varepsilon$. Then for $n \geq N$, the endpoints of the closed interval $J_{n,k}$ are contained in $(x - \varepsilon, x + \varepsilon)$. Now the endpoints of $J_{n,k}$ are points of T. Thus x is a limit point of T.
    (e) Express every number in $[0, 1]$ as a ternary number, that is, $0.a_1a_2a_3 \cdots$ where $a_i = 0, 1,$ or 2 for each i. Then T consists of those ternary numbers in which the numeral

1 never appears as a_i for any i. To establish a one-to-one correspondence between T and $[0, 1]$, express every number in $[0, 1]$ as a binary number. ∎

Prob. 3.14. Construct a collection $\{E_{n,k} : n \in \mathbb{N}, \, k \in \mathbb{N}\}$ of subsets of $\mathbb{R}$ such that

$$\bigcup_{n\in\mathbb{N}}\left(\bigcap_{k\in\mathbb{N}} E_{n,k}\right) \neq \bigcap_{k\in\mathbb{N}}\left(\bigcup_{n\in\mathbb{N}} E_{n,k}\right).$$

Proof. Let $Q = \{r_n : n \in \mathbb{N}\}$ be the set of all rational numbers in $\mathbb{R}$. For $n \in \mathbb{N}$ and $k \in \mathbb{N}$, let $I_{n,k}$ be the open interval in $\mathbb{R}$ with center at r_n and with radius equal to $\frac{1}{k}$, that is,

$$I_{n,k} = \left(r_n - \tfrac{1}{k}, r_n + \tfrac{1}{k}\right).$$

For each fixed $n \in \mathbb{N}$, we have $\bigcap_{k\in\mathbb{N}} I_{n,k} = \{r_n\}$ and then

$$(1) \qquad\qquad \bigcup_{n\in\mathbb{N}}\left(\bigcap_{k\in\mathbb{N}} I_{n,k}\right) = \bigcup_{n\in\mathbb{N}}\{r_n\} = Q.$$

On the other hand, for each fixed $k \in \mathbb{N}$, every $x \in \mathbb{R}$ is contained in $I_{n,k}$ for some $n \in \mathbb{N}$ and thus $\bigcup_{n\in\mathbb{N}} I_{n,k} = \mathbb{R}$. This implies

$$(2) \qquad\qquad \bigcap_{k\in\mathbb{N}}\left(\bigcup_{n\in\mathbb{N}} I_{n,k}\right) = \bigcap_{k\in\mathbb{N}} \mathbb{R} = \mathbb{R}.$$

(1) and (2) show that $\bigcup_{n\in\mathbb{N}}\left(\bigcap_{k\in\mathbb{N}} I_{n,k}\right) \neq \bigcap_{k\in\mathbb{N}}\left(\bigcup_{n\in\mathbb{N}} I_{n,k}\right)$. ∎

Prob. 3.15. Let Q be the set of all rational numbers in $\mathbb{R}$. For an arbitrary $\varepsilon > 0$, construct an open set O in $\mathbb{R}$ such that $O \supset Q$ and $\mu_L^*(O) \leq \varepsilon$.

Proof. The set Q of all rational numbers is a countable subset of $\mathbb{R}$. Thus $Q = \{r_n : n \in \mathbb{R}\}$ where r_n is a rational number for every $n \in \mathbb{N}$. Let $\varepsilon > 0$ be arbitrarily given. For each $n \in \mathbb{N}$, let

$$I_n = \left(r_n - \frac{\varepsilon}{2^{n+1}}, r_n + \frac{\varepsilon}{2^{n+1}}\right).$$

Then $r_n \in I_n$ for every $n \in \mathbb{N}$ and then we have $Q \subset \bigcup_{n\in\mathbb{N}} I_n$. Let $O = \bigcup_{n\in\mathbb{N}} I_n$. Then O is an open set containing Q and we have

$$\mu_L^*(O) = \mu_L^*\left(\bigcup_{n\in\mathbb{N}} I_n\right) \leq \sum_{n\in\mathbb{N}} \mu_L^*(I_n) = \sum_{n\in\mathbb{N}} \ell(I_n)$$

$$= \sum_{n\in\mathbb{N}} 2\frac{\varepsilon}{2^{n+1}} = \sum_{n\in\mathbb{N}} \frac{\varepsilon}{2^n} = \varepsilon. \quad ∎$$

Prob. 3.16. Let Q be the set of all rational numbers in $\mathbb{R}$.
(a) Show that Q is a null set in $(\mathbb{R}, \mathfrak{B}_{\mathbb{R}}, \mu_L)$.
(b) Show that Q is a F_σ-set.
(c) Show that there exists a G_δ-set G such that $G \supset Q$ and $\mu_L(G) = 0$.
(d) Show that the set of all irrational numbers in $\mathbb{R}$ is a G_δ-set.

Proof. **1.** Since the set Q of all rational numbers is a countable subset of $\mathbb{R}$, we can represent Q as $Q = \{r_n : n \in \mathbb{R}\}$ where r_n is a rational number for every $n \in \mathbb{N}$. Then we have $Q = \bigcup_{n \in \mathbb{N}} \{r_n\}$. Now the singleton $\{r_n\}$ is a null set in $(\mathbb{R}, \mathfrak{B}_{\mathbb{R}}, \mu_L)$. Then Q, as a countable union of null sets, is a null set.

2. We showed above that $Q = \bigcup_{n \in \mathbb{N}} \{r_n\}$. Now a singleton $\{r_n\}$ in $\mathbb{R}$ is a closed set in $\mathbb{R}$. Then Q is a countable union of closed set, that is, Q is an F_σ-set.

3. Let us show that there exists a G_δ-set G such that $G \supset Q$ and $\mu_L(G) = 0$. Now we have $Q = \{r_n : n \in \mathbb{R}\}$. For $n \in \mathbb{N}$ and $k \in \mathbb{N}$, let

$$I_{n,k} = \left(r_n - \frac{1}{2^{n+1}} \frac{1}{k}, r_n + \frac{1}{2^{n+1}} \frac{1}{k} \right),$$

and

$$O_k = \bigcup_{n \in \mathbb{N}} I_{n,k}.$$

Then O_k is an open set containing Q. We have

$$\mu_L(O_k) = \mu_L \left(\bigcup_{n \in \mathbb{N}} I_{n,k} \right) \le \sum_{n \in \mathbb{N}} \mu_L(I_{n,k}) = \frac{1}{k} \sum_{n \in \mathbb{N}} \frac{1}{2^n} = \frac{1}{k}.$$

Let $G = \bigcap_{k \in \mathbb{N}} O_k$. Then G is a G_δ-set containing Q and moreover we have

$$\mu_L(G) = \mu_L \left(\bigcap_{k \in \mathbb{N}} O_k \right) \le \mu_L(O_k) = \frac{1}{k} \quad \text{for every } k \in \mathbb{N}.$$

This implies that $\mu_L(G) = 0$.

4. Let P be the set of all irrational numbers in $\mathbb{R}$. Then $P = Q^c$. We showed above that Q is a F_σ-set. Thus $Q = \bigcup_{n \in \mathbb{N}} F_n$ where F_n is a closed set in $\mathbb{R}$ for $n \in \mathbb{N}$. Then we have

$$P = Q^c = \left(\bigcup_{n \in \mathbb{N}} F_n \right)^c = \bigcap_{n \in \mathbb{N}} F_n^c.$$

Since F_n is a closed set in $\mathbb{R}$, F_n^c is an open set in $\mathbb{R}$. Then P as the intersection of countably many open sets is a G_δ-set. ∎

Prob. 3.17. Let P_0 be the set of all irrational numbers in the interval $(0, 1)$. Show that for every $\varepsilon \in (0, 1)$ there exists a closed set C in $\mathbb{R}$ such that $C \subset P_0$ and $\mu_L(C) > 1 - \varepsilon$.

Proof. Let P_0 be the set of all irrational numbers in the interval $(0, 1)$ and let Q_0 be the set of all rational numbers in the interval $(0, 1)$. Then $P_0 = (0, 1) \setminus Q_0$. Now $(0, 1)$ is a $\mathfrak{M}_L$-measurable subset of $\mathbb{R}$ and the countable set Q_0 is a $\mathfrak{M}_L$-measurable subset of $\mathbb{R}$. Thus we have $P_0 = (0, 1) \setminus Q_0 \in \mathfrak{M}_L$ and then $\mu_L(P_0) = \mu_L\big((0, 1)\big) - \mu_L(Q_0) = 1$.

Let E be an arbitrary subset of $\mathbb{R}$. According to Definition 3.31, the Lebesgue inner measure $\mu_{*,L}$ on $\mathbb{R}$ is defined by

$$\mu_{*,L}(E) = \sup\big\{\mu_L(C) : C \subset E, C \in \mathfrak{C}_\mathbb{R}\big\},$$

where $\mathfrak{C}_\mathbb{R}$ is the collection of all closed sets in $\mathbb{R}$. According to Theorem 3.37, $P_0 \in \mathfrak{M}_L$ implies

$$\mu_{*,L}(P_0) = \mu_L^*(P_0) = \mu_L(P_0) = 1.$$

Let $\varepsilon > 0$ be arbitrarily given. Then $1 - \varepsilon = \mu_{*,L}(P_0) - \varepsilon$ is not an upper bound of the set $\big\{\mu_L(C) : C \subset P_0, C \in \mathfrak{C}_\mathbb{R}\big\}$ and thus there exists $C \in \mathfrak{C}_\mathbb{R}$ such that $C \subset P_0$ and $\mu_L(C) > 1 - \varepsilon$. ∎

Prob. 3.18. Consider the measure space $\big(\mathbb{R}, \mathfrak{M}_L, \mu_L\big)$.
(a) Show that every null set N in $\big(\mathbb{R}, \mathfrak{M}_L, \mu_L\big)$ is a subset of a G_δ set G which is itself a null set in $\big(\mathbb{R}, \mathfrak{M}_L, \mu_L\big)$.
(b) Show that every $E \in \mathfrak{M}_L$ can be written as $E = G \setminus N$ where G is a G_δ set and N is a null set in $\big(\mathbb{R}, \mathfrak{M}_L, \mu_L\big)$ contained in G.
(c) Show that every $E \in \mathfrak{M}_L$ can be written as $E = F \cup N$ where F is an F_σ set and N is a null set in $\big(\mathbb{R}, \mathfrak{M}_L, \mu_L\big)$ disjoint from F.

Proof. 1. Let N be a null set in $\big(\mathbb{R}, \mathfrak{M}_L, \mu_L\big)$, that is, $N \in \mathfrak{M}_L$ and $\mu_L(N) = 0$. According to Theorem 3.22, $N \in \mathfrak{M}_L$ implies that there exists a G_δ-set $G \supset N$ with $\mu_L(G \setminus N) = 0$. Now $G = (G \setminus N) \cup N$ and then $\mu_L(G) = \mu_L(G \setminus N) + \mu_L(N) = 0$. This shows that our G_δ-set G is a null set.

    2. Let $E \in \mathfrak{M}_L$. According to Theorem 3.22, there exists a G_δ-set $G \supset E$ with $\mu_L(G \setminus E) = 0$. If we let $N = G \setminus E$ then N is a null set contained in G. Then we have $E = G \setminus (G \setminus E) = G \setminus N$.

    3. Let $E \in \mathfrak{M}_L$. According to Theorem 3.22, there exists a F_σ-set $F \subset E$ with $\mu_L(E \setminus F) = 0$. Let $N = E \setminus F$. Then N is a null set disjoint from F. Then we have $E = F \cup (E \setminus F) = F \cup N$. ∎

Prob. 3.19. Let $E \subset \mathbb{R}, E \in \mathfrak{M}_L$, and $\mu_L(E) < \infty$. Show that for every $\varepsilon > 0$ there exists a compact set $C \subset E$ such that $\mu_L(E \setminus C) < \varepsilon$.

Proof. For $n \in \mathbb{N}$, let $I_n = (-n, n) \subset \mathbb{R}$. Then $(I_n : n \in \mathbb{N})$ is an increasing sequence of bounded $\mathfrak{M}_L$-measurable sets and $\mathbb{R} = \bigcup_{n\in\mathbb{N}} I_n$. We have

$$E = E \cap \mathbb{R} = E \cap \left(\bigcup_{n\in\mathbb{N}} I_n \right) = \bigcup_{n\in\mathbb{N}} (E \cap I_n)$$

$$= \bigcup_{n\in\mathbb{N}} E_n \quad \text{where } E_n = E \cap I_n.$$

Now $(E_n : n \in \mathbb{N})$ is an increasing sequence of bounded $\mathfrak{M}_L$-measurable sets so that $E = \bigcup_{n\in\mathbb{N}} E_n = \lim_{n\to\infty} E_n$ and then

$$\mu_L(E) = \mu_L\left(\lim_{n\to\infty} E_n \right) = \lim_{n\to\infty} \mu_L(E_n).$$

Since $\mu_L(E) < \infty$, for every $\varepsilon > 0$ there exists $n_0 \in \mathbb{N}$ such that

$$\mu_L(E) - \mu_L(E_{n_0}) < \tfrac{\varepsilon}{2}.$$

Since $E_{n_0} \subset E$ and $\mu_L(E_{n_0}) \leq \mu_L(E) < \infty$, we have

$$\mu_L(E \setminus E_{n_0}) = \mu_L(E) - \mu_L(E_{n_0}) < \tfrac{\varepsilon}{2}.$$

Since $E_{n_0} \in \mathfrak{M}_L$, by (iv) of Theorem 3.22 there exists a closed set $C \subset E_{n_0}$ such that

$$\mu_L(E_{n_0} \setminus C) < \tfrac{\varepsilon}{2}.$$

Since $C \subset E_{n_0} \subset E$, we have

$$\mu_L(E \setminus C) = \mu_L(E \setminus E_{n_0}) + \mu_L(E_{n_0} \setminus C) < \tfrac{\varepsilon}{2} + \tfrac{\varepsilon}{2} = \varepsilon.$$

Since $C \subset E_{n_0}$ and E_{n_0} is a bounded set, C is a bounded set. Since C is a bounded closed set in $\mathbb{R}$, C is a compact set. ∎

Prob. 3.20. Show that for every increasing sequence $(E_n : n \in \mathbb{N})$ of subsets of $\mathbb{R}$, we have $\lim_{n \to \infty} \mu_L^*(E_n) = \mu_L^*\big(\lim_{n \to \infty} E_n \big)$.

Proof. According to Theorem 2.11, if μ^* is a regular outer measure on a set X then for every increasing sequence $(E_n : n \in \mathbb{N})$ of subsets of X we have

$$\lim_{n \to \infty} \mu^*(E_n) = \mu^*\big(\lim_{n \to \infty} E_n \big).$$

According to Lemma 3.21, the Lebesgue outer measure μ_L^* on $\mathbb{R}$ is a Borel regular outer measure and hence a regular outer measure on $\mathbb{R}$. Thus for every increasing sequence $(E_n : n \in \mathbb{N})$ of subsets of $\mathbb{R}$ we have

$$\lim_{n \to \infty} \mu_L^*(E_n) = \mu_L^*\big(\lim_{n \to \infty} E_n \big). \quad \blacksquare$$

Prob. 3.21. (a) Consider the μ_L^*-measurability condition on $E \in \mathfrak{P}(\mathbb{R})$:

(1) $\qquad \mu_L^*(A) = \mu_L^*(A \cap E) + \mu_L^*(A \cap E^c)$ for every $A \in \mathfrak{P}(\mathbb{R})$.

Show that this condition is equivalent to the following condition:

(2) $\qquad \mu_L^*(I) = \mu_L^*(I \cap E) + \mu_L^*(I \cap E^c)$ for every $I \in \mathfrak{I}_o$.

(b) Let $\mathfrak{J}$ be the collection of all open intervals in $\mathbb{R}$ with rational endpoints. This is a countable subcollection of $\mathfrak{I}_o$. Show that condition (2) is equivalent to the following apparently weaker condition :

(3) $\qquad \mu_L^*(J) = \mu_L^*(J \cap E) + \mu_L^*(J \cap E^c)$ for every $J \in \mathfrak{J}$.

Proof. 1. Let us prove the equivalence of (1) and (2). Clearly (1) implies (2). It remains to show that (2) implies (1).

Assume (2). Let $A \in \mathfrak{P}(\mathbb{R})$ be arbitrarily chosen. Let $(I_n : n \in \mathbb{N})$ be an arbitrary sequence in $\mathfrak{I}_o$ such that $\bigcup_{n \in \mathbb{N}} I_n \supset A$. By (2) we have

$$\mu_L^*(I_n) = \mu_L^*(I_n \cap E) + \mu_L^*(I_n \cap E^c) \quad \text{for every } n \in \mathbb{N}.$$

Summing the equalities above over $n \in \mathbb{N}$, we have

$$\sum_{n \in \mathbb{N}} \mu_L^*(I_n) = \sum_{n \in \mathbb{N}} \mu_L^*(I_n \cap E) + \sum_{n \in \mathbb{N}} \mu_L^*(I_n \cap E^c)$$

$$\geq \mu_L^*\Big(\bigcup_{n \in \mathbb{N}}(I_n \cap E)\Big) + \mu_L^*\Big(\bigcup_{n \in \mathbb{N}}(I_n \cap E^c)\Big)$$

$$= \mu_L^*\Big(\big(\bigcup_{n \in \mathbb{N}} I_n \big) \cap E\Big) + \mu_L^*\Big(\big(\bigcup_{n \in \mathbb{N}} I_n \big) \cap E^c\Big)$$

$$\geq \mu_L^*(A \cap E) + \mu_L^*(A \cap E^c),$$

where the first inequality is by the countable subadditivity of the outer measure μ_L^* and the second inequality is by the monotonicity of the outer measure μ_L^*. Since $\mu_L^*(I_n) = \ell(I_n)$, we have

(4)
$$\sum_{n\in\mathbb{N}} \ell(I_n) \geq \mu_L^*(A\cap E) + \mu_L^*(A\cap E^c).$$

Since (4) holds for every sequence $(I_n : n\in\mathbb{N}) \subset \mathfrak{I}_o$ such that $\bigcup_{n\in\mathbb{N}} I_n \supset A$, we have

(5)
$$\mu_L^*(A) = \inf\left\{\sum_{n\in\mathbb{N}} \ell(I_n) : (I_n : n\in\mathbb{N}) \subset \mathfrak{I}_o, \bigcup_{n\in\mathbb{N}} I_n \supset A\right\}$$
$$\geq \mu_L^*(A\cap E) + \mu_L^*(A\cap E^c).$$

On the other hand we have $A = (A\cap E) \cup (A\cap E^c)$ and then by the subadditivity of the outer measure μ_L^* we have

(6)
$$\mu_L^*(A) \leq \mu_L^*(A\cap E) + \mu_L^*(A\cap E^c).$$

By (5) and (6), we have (1). This completes the proof of equivalence of (1) and (2).

   2. Let us prove the equivalence of (2) and (3). Clearly (2) implies (3). It remains to show that (3) implies (2).

   Let us assume (3). Let $I \in \mathfrak{I}_o$ be arbitrarily chosen. Now we have $I = (a, b)$ where $-\infty \leq a < b \leq \infty$. Let $(a_n : n\in\mathbb{N})$ and $(b_n : n\in\mathbb{N})$ be two sequences of rational numbers in (a, b) such that $a_n < b_n$ for every $n\in\mathbb{N}$ and $a_n \downarrow a$ and $b_n \uparrow b$ as $n\to\infty$. Let $J_n = (a_n, b_n) \in \mathfrak{J}$ for $n\in\mathbb{N}$. By assumption of (3) we have

(7)
$$\mu_L^*(J_n) = \mu_L^*(J_n\cap E) + \mu_L^*(J_n\cap E^c) \quad \text{for every } n\in\mathbb{N}.$$

Now $(J_n : n\in\mathbb{N})$, $(J_n\cap E : n\in\mathbb{N})$ and $(J_n\cap E^c : n\in\mathbb{N})$ are increasing sequences of subsets of $\mathbb{R}$ and

$$\lim_{n\to\infty} J_n = \bigcup_{n\in\mathbb{N}} J_n = I,$$

$$\lim_{n\to\infty} (J_n\cap E) = \bigcup_{n\in\mathbb{N}} (J_n\cap E) = I\cap E,$$

$$\lim_{n\to\infty} (J_n\cap E^c) = \bigcup_{n\in\mathbb{N}} (J_n\cap E^c) = I\cap E^c.$$

Then by Prob. 3.20, we have

$$\lim_{n\to\infty} \mu_L^*(J_n) = \mu_L^*\left(\lim_{n\to\infty} J_n\right) = \mu_L^*(I),$$

$$\lim_{n\to\infty} \mu_L^*(J_n\cap E) = \mu_L^*\left(\lim_{n\to\infty} (J_n\cap E)\right) = \mu_L^*(I\cap E),$$

$$\lim_{n\to\infty} \mu_L^*(J_n\cap E^c) = \mu_L^*\left(\lim_{n\to\infty} (J_n\cap E^c)\right) = \mu_L^*(I\cap E^c).$$

Then letting $n\to\infty$ in (7), we have

$$\mu_L^*(I) = \mu_L^*(I\cap E) + \mu_L^*(I\cap E^c).$$

This proves (2) and completes the proof of equivalence of (2) and (3). ∎

Prob. 3.22. Prove the following statements:

(a) Let $A \subset \mathbb{R}$. If there exists $\varepsilon_0 \in (0, 1)$ such that $\mu_L^*(A \cap I) \leq \varepsilon_0 \ell(I)$ for every open interval I then $\mu_L^*(A) = 0$.

(b) Let $E \subset \mathbb{R}$. If $\mu_L^*(E) > 0$, then for every $\varepsilon \in (0, 1)$ there exists a finite open interval I_ε such that

$$\varepsilon \ell(I_\varepsilon) < \mu_L^*(E \cap I_\varepsilon).$$

(c) Let $E \subset \mathbb{R}$. If $\mu_L^*(E) > 0$, then there exists a finite open interval I_0 and $\delta_0 > 0$ such that

$$(E \cap I_0 + h) \cap (E \cap I_0) \neq \emptyset \quad \text{for every } h \in \mathbb{R} \text{ such that } |h| \leq \delta_0.$$

(d) For $E \subset \mathbb{R}$, we define $\Delta E = \{z \in \mathbb{R} : z = x - y \text{ where } x, y \in E\}$. If $\mu_L^*(E) > 0$, then ΔE contains an open interval.

Proof. 1. Let us prove (a). Let $A \subset \mathbb{R}$. Let $\mathfrak{I}$ be the collection of all open intervals in $\mathbb{R}$. Suppose there exists $\varepsilon_0 \in (0, 1)$ such that

$$(1) \qquad\qquad \mu_L^*(A \cap I) \leq \varepsilon_0 \ell(I) \quad \text{for every } I \in \mathfrak{I}.$$

Let us show that $\mu_L^*(A) = 0$. We show this first for the particular case that $\mu_L^*(A) < \infty$ and then for the general case.

Case 1. Assume that $\mu_L^*(A) < \infty$ and A satisfies the condition (1). To show that $\mu_L^*(A) = 0$, let us assume the contrary, that is, $\mu_L^*(A) > 0$. Then $\mu_L^*(A) \in (0, \infty)$ and this implies

$$(2) \qquad\qquad (1 + \delta)\mu_L^*(A) > \mu_L^*(A) \quad \text{for every } \delta > 0.$$

Now

$$(3) \qquad\qquad \mu_L^*(A) = \inf \left\{ \sum_{n \in \mathbb{N}} \ell(I_n) : (I_n : n \in \mathbb{N}) \subset \mathfrak{I}, \bigcup_{n \in \mathbb{N}} I_n \supset A \right\}.$$

Then (2) and (3) imply that there exists $(I_n : n \in \mathbb{N}) \subset \mathfrak{I}$ such that $\bigcup_{n \in \mathbb{N}} I_n \supset A$ and

$$(4) \qquad\qquad \sum_{n \in \mathbb{N}} \ell(I_n) < (1 + \delta)\mu_L^*(A).$$

Since $A \subset \bigcup_{n \in \mathbb{N}} I_n$, we have $A = A \cap \left(\bigcup_{n \in \mathbb{N}} I_n \right) = \bigcup_{n \in \mathbb{N}} (A \cap I_n)$ and then by the countable subadditivity of the Lebesgue outer measure μ_L^* and by (1) we have

$$\mu_L^*(A) = \mu_L^* \left(\bigcup_{n \in \mathbb{N}} (A \cap I_n) \right) \leq \sum_{n \in \mathbb{N}} \mu_L^*(A \cap I_n) \leq \varepsilon_0 \sum_{n \in \mathbb{N}} \ell(I_n).$$

Dividing by $\varepsilon_0 > 0$ and applying (4), we have

$$\frac{1}{\varepsilon_0} \mu_L^*(A) \leq \sum_{n \in \mathbb{N}} \ell(I_n) < (1 + \delta)\mu_L^*(A).$$

Then dividing by $\mu_L^*(A) \in (0, \infty)$, we have

$$(5) \qquad\qquad \frac{1}{\varepsilon_0} < 1 + \delta \quad \text{for every } \delta > 0.$$

But $\varepsilon_0 \in (0, 1)$ implies $\frac{1}{\varepsilon_0} > 1$ and then (5) cannot hold. This is a contradiction. Therefore we must have $\mu_L^*(A) = 0$.

Case 2. Assume that $A \subset \mathbb{R}$ satisfies the condition (1). For every $m \in \mathbb{N}$, let us define

$$I_m = (-m, m) \quad \text{and} \quad A_m = A \cap I_m.$$

Now $\bigcup_{m \in \mathbb{N}} I_m = \mathbb{R}$ and $A = A \cap \mathbb{R} = A \cap \left(\bigcup_{m \in \mathbb{N}} I_m \right) = \bigcup_{m \in \mathbb{N}} (A \cap I_m) = \bigcup_{m \in \mathbb{N}} A_m$.
Observe also that $\mu_L^*(A_m) = \mu_L^*(A \cap I_m) \le \mu_L^*(I_m) = 2m < \infty$.
For every $I \in \mathfrak{J}$, we have $I_m \cap I \in \mathfrak{J}$. Then for every $I \in \mathfrak{J}$ we have

$$\mu_L^*(A_m \cap I) = \mu_L^*(A \cap I_m \cap I) \le \varepsilon_0 \ell(I_m \cap I) \quad \text{by (1)}$$

$$\le \varepsilon_0 \ell(I).$$

Thus A_m satisfies condition (1). We have $\mu_L^*(A_m) < \infty$ also. Then we have $\mu_L^*(A_m) = 0$ by our result in Case 1 above. Then since $\mu_L^*(A_m) = 0$ for every $m \in \mathbb{N}$, we have

$$\mu_L^*(A) = \mu_L^*\left(\bigcup_{n \in \mathbb{N}} A_m \right) \le \sum_{m \in \mathbb{N}} \mu_L^*(A_m) = 0.$$

This completes the proof of (a).

2. Let us prove (b). Let $E \subset \mathbb{R}$ and suppose $\mu_L^*(E) > 0$. By (a), $\mu_L^*(E) > 0$ implies that there does not exist $\varepsilon \in (0, 1)$ with the property that

$$\mu_L^*(E \cap I) \le \varepsilon \ell(I) \quad \text{for every } I \in \mathfrak{J}$$

and then for every $\varepsilon \in (0, 1)$ there exists $I_\varepsilon \in \mathfrak{J}$ such that

$$\mu_L^*(E \cap I_\varepsilon) > \varepsilon \ell(I_\varepsilon).$$

This implies that $\ell(I_\varepsilon) < \infty$ and thus the open interval I_ε is a finite open interval. This completes the proof of (b).

3. Let us prove (c). Let $E \subset \mathbb{R}$ and suppose $\mu_L^*(E) > 0$. According to (b), $\mu_L^*(E) > 0$ implies that for every $\varepsilon \in (0, 1)$ there exists a finite open interval I_ε such that $\varepsilon \ell(I_\varepsilon) < \mu_L^*(E \cap I_\varepsilon)$. Now let

$$(6) \qquad \varepsilon_0 \in (0, 1) \quad \text{and} \quad 2\varepsilon_0 - 1 > 0, \quad \text{that is,} \quad \varepsilon_0 \in \left(\tfrac{1}{2}, 1 \right).$$

Then there exists a finite open interval I_0 such that

$$(7) \qquad \mu_L^*(E \cap I_0) > \varepsilon_0 \ell(I_0).$$

Let

$$(8) \qquad \delta_0 = (1 - \varepsilon_0) \ell(I_0).$$

Let us show that

$$(E \cap I_0 + h) \cap (E \cap I_0) \ne \emptyset \quad \text{for every } h \in \mathbb{R} \text{ such that } |h| \le \delta_0.$$

Let $h \in \mathbb{R}$ and $|h| \leq \delta_0$. Consider first the case that $h \geq 0$. Let the finite open interval I_0 be represented as $I_0 = (a, b)$ where $a, b \in \mathbb{R}$ and $a < b$. We have

(9) $$0 \leq h \leq \delta_0 = (1 - \varepsilon_0)\ell(I_0),$$

(10) $$E \cap I_0 \subset (a, b),$$

(11) $$E \cap I_0 + h \subset (a + h, b + h).$$

Then we have

(12) $$(E \cap I_0 + h) \setminus (E \cap I_0) \subset [b, b + h).$$

Now
$$E \cap I_0 + h = \big[(E \cap I_0 + h) \cap (E \cap I_0)\big] \cup \big[(E \cap I_0 + h) \setminus (E \cap I_0)\big].$$
Then by the subadditivity of the Lebesgue outer measure μ_L^* we have

$$\mu_L^*\big(E \cap I_0 + h\big) \leq \mu_L^*\big((E \cap I_0 + h) \cap (E \cap I_0)\big) + \mu_L^*\big((E \cap I_0 + h) \setminus (E \cap I_0)\big)$$

and then

$$\mu_L^*\big((E \cap I_0 + h) \cap (E \cap I_0)\big)$$
$$\geq \mu_L^*\big(E \cap I_0 + h\big) - \mu_L^*\big((E \cap I_0 + h) \setminus (E \cap I_0)\big)$$
$$\geq \mu_L^*\big(E \cap I_0 + h\big) - \mu_L^*\big([b, b + h)\big) \quad \text{by (12)}$$
$$= \mu_L^*\big(E \cap I_0\big) - h \quad \text{by the translation invariance of } \mu_L^*$$
$$> \varepsilon_0 \ell(I_0) - (1 - \varepsilon_0)\ell(I_0) \quad \text{by (7) and (9)}$$
$$= (2\varepsilon_0 - 1)\ell(I_0) > 0 \quad \text{by (6).}$$

(Translation invariance of the Lebesgue outer measure μ_L^* follows immediately from the definition of μ_L^* by Definition 3.1.) Thus we have $\mu_L^*\big((E \cap I_0 + h) \cap (E \cap I_0)\big) > 0$ and this implies that $(E \cap I_0 + h) \cap (E \cap I_0) \neq \emptyset$. The case that $h \leq 0$ is handled by similar argument. This completes the proof of (c).

4. Let us prove (d). Let $E \subset \mathbb{R}$ and suppose $\mu_L^*(E) > 0$. According to (c), $\mu_L^*(E) > 0$ implies that there exists a finite open interval I_0 and $\delta_0 > 0$ such that

$$(E \cap I_0 + h) \cap (E \cap I_0) \neq \emptyset \quad \text{for every } h \in \mathbb{R} \text{ such that } |h| \leq \delta_0.$$

Let $h \in \mathbb{R}$ be such that $|h| \leq \delta_0$. Then we have $(E \cap I_0 + h) \cap (E \cap I_0) \neq \emptyset$. Let $x \in (E \cap I_0 + h) \cap (E \cap I_0)$. Then $x \in E \cap I_0 \subset E$ and $x \in E \cap I_0 + h$. The latter implies that there exists $y \in E \cap I_0$ such that $x = y + h$. Then we have $h = x - y$ where $x \in E$ and $y \in E$. Thus $h \in \Delta E$. We have just shown that if $h \in \mathbb{R}$ and $|h| \leq \delta_0$ then $h \in \Delta E$. This implies then that $(-\delta_0, \delta_0) \subset \Delta E$. This completes the proof of (d). ∎

Prob. 3.23. For $E \subset \mathbb{R}$, we define $E + E = \{z \in \mathbb{R} : z = x + y \text{ where } x, y \in E\}$. We say that E is symmetric with respect to $0 \in \mathbb{R}$ if $-E = E$. Show that if E is symmetric and $\mu_L^*(E) > 0$ then $E + E$ contains an open interval.

Proof. Let $E \subset \mathbb{R}$ and assume that $\mu_L^*(E) > 0$ and E is symmetric. Now symmetry of E implies that $y \in E$ if and only if $-y \in E$. Thus we have

$$E + E = \{z \in \mathbb{R} : z = x + y \text{ where } x, y \in E\}$$
$$= \{z \in \mathbb{R} : z = x - (-y) \text{ where } x, -y \in E\}$$
$$= \Delta(E).$$

According to (d) of Prob. 3.22, $\mu_L^*(E) > 0$ implies that $\Delta(E)$ contains an open interval. Thus $E + E$ contains an open interval. ∎

Prob. 3.24. For $E \subset \mathbb{R}$, we define $E + E = \{z \in \mathbb{R} : z = x + y \text{ where } x, y \in E\}$. Show that if $\mu_L^*(E) > 0$ then $E + E$ contains an open interval.

Proof. 1. Suppose $\mu_L^*(E) > 0$. Let $\varepsilon_0 \in (0, 1)$ be arbitrarily chosen and fixed. According to (b) of Prob. 3.22, there exists a finite open interval I_0 such that

$$(1) \qquad\qquad \varepsilon_0 \ell(I_0) < \mu_L^*(E \cap I_0).$$

Let us consider the special case that I_0 is symmetric with respect to $0 \in \mathbb{R}$, that is,

$$(2) \qquad\qquad I_0 = (-\gamma, \gamma) \quad \text{where } \gamma > 0.$$

We have $-I_0 = I_0$. Let us show

$$(3) \qquad (E \cap I_0 + h) \cap -(E \cap I_0) \neq \emptyset \quad \text{for any } h \in \mathbb{R} \text{ such that } |h| < \varepsilon_0 \gamma.$$

Let $h \in \mathbb{R}$ be such that $|h| < \varepsilon_0 \gamma$. Let us consider first the case that $h \geq 0$. Now we have

$$E \cap I_0 \subset (-\gamma, \gamma),$$

$$E \cap I_0 + h \subset (-\gamma + h, \gamma + h),$$

$$-(E \cap I_0) \subset -I_0 = I_0 = (-\gamma, \gamma)$$

and then

$$(E \cap I_0 + h) \setminus -(E \cap I_0) \subset [\gamma, \gamma + h).$$

Thus we have

$$(4) \qquad \mu_L^*\big[(E \cap I_0 + h) \setminus -(E \cap I_0)\big] \leq \mu_L^*\big([\gamma, \gamma + h)\big) = h.$$

Let us write

$$E \cap I_0 + h = \big[(E \cap I_0 + h) \cap -(E \cap I_0)\big] \cup \big[(E \cap I_0 + h) \setminus -(E \cap I_0)\big].$$

Then we have

$$\mu_L^*(E \cap I_0 + h) \leq \mu_L^*\big[(E \cap I_0 + h) \cap -(E \cap I_0)\big] + \mu_L^*\big[(E \cap I_0 + h) \setminus -(E \cap I_0)\big]$$

and then

$$\mu_L^*\big[(E \cap I_0 + h) \cap -(E \cap I_0)\big]$$

$$\geq \mu_L^*(E \cap I_0 + h) - \mu_L^*\big[(E \cap I_0 + h) \setminus -(E \cap I_0)\big]$$

$$\geq \mu_L^*(E \cap I_0) - h \quad \text{by the translation invariance of } \mu_L^* \text{ and by (4)}$$

$$> 2\gamma \varepsilon_0 - \gamma \varepsilon_0 = \gamma \varepsilon_0 > 0,$$

by (1) and $\ell(I_0) = 2\gamma$ and $0 \leq h < \gamma \varepsilon_0$. This shows that for $0 \leq h < \varepsilon_0 \gamma$, we have $\mu_L^*\big[(E \cap I_0 + h) \cap -(E \cap I_0)\big] > 0$ and thus $(E \cap I_0 + h) \cap -(E \cap I_0) \neq \emptyset$. By similar argument we show that $(E \cap I_0 + h) \cap -(E \cap I_0) \neq \emptyset$ for $-\varepsilon_0 \gamma < h \leq 0$. This completes the proof of (3).

Now let $h \in \mathbb{R}$ be such that $|h| < \varepsilon_0 \gamma$. Then by (3), we have $(E \cap I_0 + h) \cap -(E \cap I_0) \neq \emptyset$. Select $z \in (E \cap I_0 + h) \cap -(E \cap I_0)$ arbitrarily. Then $z \in E \cap I_0 + h$ and this implies that there exists $x \in E$ such that $z = x + h$. We also have $z \in -(E \cap I_0)$ and this implies that there exists $y \in E$ such that $z = -y$. Then we have $-y = z = x + h$ so that $-h = x + y$ where $x, y \in E$. Thus $-h \in E + E$. This shows that if $h \in \mathbb{R}$ satisfies the condition $|h| < \varepsilon_0 \gamma$ then $-h \in E + E$. This implies that $(-\varepsilon_0 \gamma, \varepsilon_0 \gamma) \subset E + E$ and thus $E + E$ contains an open interval.

2. Consider the general case where the finite open interval I_0 in (1) is given by $I_0 = (2a, 2b)$ where $a, b \in \mathbb{R}$ and $a < b$. Let φ be a translation on $\mathbb{R}$ defined by

$$\varphi(x) = x - (a + b) \quad \text{for } x \in \mathbb{R}.$$

Then we have

$$\varphi(E) = E - (a + b)$$

and

$$\varphi(I_0) = I_0 - (a + b) = \big(2a - (a + b), 2b - (a + b)\big)$$
$$= \big(-(b - a), b - a\big) = (-\gamma, \gamma) \quad \text{where } \gamma := b - a > 0.$$

Consider the set $E - (a + b)$ and the finite open interval $I_0 - (a + b)$. By the translation invariance of the Lebesgue outer measure μ_L^*, we have

$$\mu_L^*\big(E - (a + b)\big) = \mu_L^*(E) > 0$$

and

$$\varepsilon_0 \ell\big(I_0 - (a + b)\big) = \varepsilon_0 \ell(I_0) < \mu_L^*(E \cap I_0) \quad \text{by (1)}$$
$$= \mu_L^*\big([E - (a + b)] \cap [I_0 - (a + b)]\big).$$

This shows that the set $E - (a + b)$ and the finite open interval $I_0 - (a + b)$ have the relationship (1). Moreover $I_0 - (a + b) = (-\gamma, \gamma)$ is symmetric with respect to $0 \in \mathbb{R}$. This implies, according to our result in **1**, that $[E - (a + b)] + [E - (a + b)]$ contains an open interval J. Then $E + E$ contains an open interval $J + (a + b)$. ∎

Prob. 3.25. For $E, F \subset \mathbb{R}$, we define $E + F = \{z \in \mathbb{R} : z = x + y$ where $x \in E$ and $y \in F\}$ and $E - F = \{z \in \mathbb{R} : z = x - y$ where $x \in E$ and $y \in F\}$. Prove the following statements:
(a) If $\mu_L^*(E \cap F) > 0$, then $E + F$ contains an open interval.
(b) If $\mu_L^*(E \cap F) > 0$, then $E - F$ contains an open interval.

Proof. 1. Let us prove (a). If $\mu_L^*(E \cap F) > 0$ then $(E \cap F) + (E \cap F)$ contains an open interval by Prob. 3.24. Now $(E \cap F) \subset E$ and $(E \cap F) \subset F$ also. This implies that $[(E \cap F) + (E \cap F)] \subset E + F$. Thus $E + F$ contains an open interval.

2. Let us prove (b). Let us observe

$$E - E = \{z \in \mathbb{R} : z = x - y \text{ where } x, y \in E\} = \Delta E,$$

as defined in Prob. 3.22. According to Prob. 3.22, if $\mu_L^*(E) > 0$ then $E - E = \Delta E$ contains an ope interval. Thus if $\mu_L^*(E \cap F) > 0$ then $(E \cap F) - (E \cap F)$ contains an open interval. Now $(E \cap F) \subset E$ and $(E \cap F) \subset F$ also. This implies that $[(E \cap F) - (E \cap F)] \subset E - F$. Thus $E - F$ contains an open interval. ∎

Prob. 3.26. Show that there exist two null sets E and F in the Lebesgue measure space $(\mathbb{R}, \mathfrak{M}_L, \mu_L)$ such that $E + F = \mathbb{R}$.

Proof. 1. Let us show first that there exist two null sets A and B in the Lebesgue measure space $(\mathbb{R}, \mathfrak{M}_L, \mu_L)$ such that $A + B = [0, 1]$. Let

(1) $\quad A = \left\{ x \in [0, 1] : x = \dfrac{i_1}{2} + \dfrac{i_2}{2^3} + \dfrac{i_3}{2^5} + \cdots \text{ where } i_k \in \{0, 1\} \text{ for } k \in \mathbb{N} \right\},$

(2) $\quad B = \left\{ x \in [0, 1] : x = \dfrac{i_1}{2^2} + \dfrac{i_2}{2^4} + \dfrac{i_3}{2^6} + \cdots \text{ where } i_k \in \{0, 1\} \text{ for } k \in \mathbb{N} \right\}.$

Then since every $x \in [0, 1]$ has a binary expansion, we have

$$x = \sum_{n \in \mathbb{N}} \frac{i_n}{2^n} = \left\{ \frac{i_1}{2} + \frac{i_2}{2^3} + \frac{i_3}{2^5} + \cdots \right\} + \left\{ \frac{i_1}{2^2} + \frac{i_2}{2^4} + \frac{i_3}{2^6} + \cdots \right\} \in A + B.$$

Thus we have $A + B = [0, 1]$.

It remains to show that A and B are null sets in the measure space $(\mathbb{R}, \mathfrak{M}_L, \mu_L)$, that is, $\mu_L(A) = 0$ and $\mu_L(B) = 0$. Now

(3) $\quad B = \left\{ x \in [0, 1] : x = \dfrac{i_1}{2^2} + \dfrac{i_2}{2^4} + \dfrac{i_3}{2^6} + \cdots \text{ where } i_k \in \{0, 1\} \text{ for } k \in \mathbb{N} \right\}$

$\qquad = \left\{ x \in [0, 1] : x = \dfrac{i_1}{4} + \dfrac{i_2}{4^2} + \dfrac{i_3}{4^3} + \cdots \text{ where } i_k \in \{0, 1\} \text{ for } k \in \mathbb{N} \right\}.$

By (1) and then by (3), we have

(4) $\quad \dfrac{1}{2} A = \left\{ x \in [0, 1] : x = \dfrac{i_1}{2^2} + \dfrac{i_2}{2^4} + \dfrac{i_3}{2^6} + \cdots \text{ where } i_k \in \{0, 1\} \text{ for } k \in \mathbb{N} \right\}$

$\qquad = \left\{ x \in [0, 1] : x = \dfrac{i_1}{4} + \dfrac{i_2}{4^2} + \dfrac{i_3}{4^3} + \cdots \text{ where } i_k \in \{0, 1\} \text{ for } k \in \mathbb{N} \right\}$

$\qquad = B.$

Thus we have $A = 2B$ and then $\mu_L(A) = 2\mu_L(B)$. Then to show that $\mu_L(A) = 0$ and $\mu_L(B) = 0$, it suffices to show that $\mu_L(B) = 0$.

Let us show that $\mu_L(B) = 0$. We show that B is a subset of a Cantor-like set C with $\mu_L(C) = 0$. The set C is constructed as follows.

Let $C_0 = [0, 1]$. Decompose C_0 into 4 subintervals of equal length and remove the open 3rd and the open 4th subinterval and let C_1 be the resulting set. Then

$$C_1 = \left[0, \tfrac{1}{4}\right] \cup \left[\tfrac{1}{4}, \tfrac{2}{4}\right] \cup \left\{\tfrac{3}{4}\right\} \cup \left\{\tfrac{4}{4}\right\},$$

a union of 2 closed intervals of length $\tfrac{1}{4}$ and 2 singletons.

Decompose each one of the 2 closed intervals in C_1 into 4 subintervals of equal length and remove the open 3rd and the open 4th subinterval and let C_2 be the resulting set. Then

$$C_2 = \left[0, \tfrac{1}{4^2}\right] \cup \left[\tfrac{1}{4^2}, \tfrac{2}{4^2}\right] \cup \left\{\tfrac{3}{4^2}\right\} \cup \left\{\tfrac{4}{4^2}\right\}$$
$$\cup \left[\tfrac{4}{4^2}, \tfrac{5}{4^2}\right] \cup \left[\tfrac{5}{4^2}, \tfrac{6}{4^2}\right] \cup \left\{\tfrac{7}{4^2}\right\} \cup \left\{\tfrac{8}{4^2}\right\}$$
$$\cup \left\{\tfrac{3}{4}\right\} \cup \left\{\tfrac{4}{4}\right\},$$

a union of 2^2 closed intervals of length $\frac{1}{4^2}$ and $2 + 2^2$ singletons.

We iterate this process of subdivision and removal indefinitely. Then for every $n \in \mathbb{N}$, C_n is a union of 2^2 closed intervals of length $\frac{1}{4^n}$ and $\sum_{k=1}^{n} 2^k$ singletons. Note that C_n is a closed set in $\mathbb{R}$ and is thus $\mathfrak{B}_{\mathbb{R}}$-measurable and then $\mathfrak{M}_L$-measurable with $\mu_L(C_n) = \left(\frac{2}{4}\right)^n$. Now $(C_n : n \in \mathbb{N})$ is a decreasing sequence of subsets of C_0. We define

$$C = \bigcap_{n \in \mathbb{N}} C_n = \lim_{n \to \infty} C_n.$$

Then C is a closed set in $\mathbb{R}$ and moreover we have

$$\mu_L(C) = \lim_{n \to \infty} \mu_L(C_n) = \lim_{n \to \infty} \left(\tfrac{2}{4}\right)^n = 0.$$

Let us show that $B \subset C$. Let $x \in B$. Then by the description (3) of the set B, we have

$$x = \sum_{k \in \mathbb{N}} \frac{i_k}{4^k} \quad \text{where } i_k \in \{0, 1\} \text{ for } k \in \mathbb{N}.$$

For every $n \in \mathbb{N}$, let

$$x_n = \sum_{k=1}^{n} \frac{i_k}{4^k}.$$

Then we have

$$x = \lim_{n \to \infty} x_n.$$

It follows from the construction of C that $x_n \in C$ for every $n \in \mathbb{N}$. Then since C is a closed set in $\mathbb{R}$, we have $\lim_{n \to \infty} x_n \in C$, that is, $x \in C$. This shows that $B \subset C$.

Since $(\mathbb{R}, \mathfrak{M}_L, \mu_L)$ is a complete measure space, the fact that $\mu_L(C) = 0$ and the fact that $B \subset C$ imply that $B \in \mathfrak{M}_L$ and $\mu_L(B) = 0$.

This completes the proof that the sets A and B defined by (1) and (2) respectively are null sets in $(\mathbb{R}, \mathfrak{M}_L, \mu_L)$ and moreover $A + B = [0, 1]$.

2. Let us construct two null sets E and F in the measure space $(\mathbb{R}, \mathfrak{M}_L, \mu_L)$ such that $E + F = \mathbb{R}$. Let A and B be as defined by (1) and (2) respectively. For every $m \in \mathbb{Z}$, let us define

$$(5) \quad A_m = \left\{ x \in \mathbb{R} : x = m + \frac{i_1}{2} + \frac{i_2}{2^3} + \frac{i_3}{2^5} + \cdots \quad \text{where } i_k \in \{0, 1\} \text{ for } k \in \mathbb{N} \right\}.$$

Let

$$(6) \quad E = \bigcup_{m \in \mathbb{Z}} A_m \quad \text{and} \quad F = B.$$

We showed in **1** that if $x \in [0, 1]$ then $x = x' + x''$ where $x' \in A$ and $x'' \in B$ so that $x \in A + B$. Now if $y \in \mathbb{R}$ then we have $y = m + x$ where $m \in \mathbb{Z}$ and $x \in [0, 1)$. Since $x = x' + x''$ where $x' \in A$ and $x'' \in B$, we have $y = m + x' + x''$ where $m + x' \in A_m$ and $x'' \in B$. Thus $y \in A_m + B \subset E + F$. This shows that $\mathbb{R} \subset E + F$. Since $E + F \subset \mathbb{R}$, we have $E + F = \mathbb{R}$.

It remains to show that E and F are null sets in the measure space $(\mathbb{R}, \mathfrak{M}_L, \mu_L)$. Now by (6) we have $F = B$ and thus F is a null set in $(\mathbb{R}, \mathfrak{M}_L, \mu_L)$. According to (5), A_m is a translate of the null set A by m. Then by the translation invariance of the measure space $(\mathbb{R}, \mathfrak{M}_L, \mu_L)$, $A_m \in \mathfrak{M}_L$ and $\mu_L(A_m) = \mu_L(A) = 0$. Thus A_m is a null set. Then $E = \bigcup_{m \in \mathbb{Z}} A_m$, being a countable union of null sets, is a null set. ∎

Prob. 3.27. Let Q be the set of all rational numbers in $\mathbb{R}$. Show that for any $x \in \mathbb{R}$ we have $\lim_{t \to 0} \mathbf{1}_{Q+t}(x) \neq \mathbf{1}_{Q}(x)$ and indeed $\lim_{t \to 0} \mathbf{1}_{Q+t}(x)$ does not exist.

Proof. 1. Let $x_0 \in \mathbb{R}$ be arbitrarily chosen. Let us define a real-valued function φ on $\mathbb{R}$ by setting

$$\varphi(t) = \mathbf{1}_{Q+t}(x_0) \quad \text{for } t \in \mathbb{R}.$$

Let us show that $\lim_{t \to 0} \varphi(t)$ does not exist.

Assume that $\lim_{t \to 0} \varphi(t)$ exists. Now φ assumes two values only, namely 0 and 1. Thus if $\lim_{t \to 0} \varphi(t)$ exists then either $\lim_{t \to 0} \varphi(t) = 0$ or $\lim_{t \to 0} \varphi(t) = 1$.

2. Suppose $\lim_{t \to 0} \varphi(t) = 0$. Then for $\varepsilon = \frac{1}{2} > 0$ there exists $\delta > 0$ such that

$$|\varphi(t) - 0| < \frac{1}{2} \quad \text{for } t \in (-\delta, \delta),$$

that is,

$$-\frac{1}{2} < \varphi(t) < \frac{1}{2} \quad \text{for } t \in (-\delta, \delta).$$

Then since the range of φ is $\{0, 1\}$, we have

(1) $$\varphi(t) = 0 \quad \text{for } t \in (-\delta, \delta).$$

Since $\varphi(t) = \mathbf{1}_{Q+t}(x_0)$, $\varphi(t) = 0$ if and only if $x_0 \notin Q+t$, that is, $x_0 - t \notin Q$, that is, $x_0 - t$ is an irrational number. Then (1) implies that $x_0 - t$ is an irrational number for $t \in (-\delta, \delta)$, that is, $(x_0 - \delta, x_0 + \delta)$ contains only irrational numbers. This is impossible.

3. Suppose $\lim_{t \to 0} \varphi(t) = 1$. Then for $\varepsilon = \frac{1}{2} > 0$ there exists $\delta > 0$ such that

$$|\varphi(t) - 1| < \frac{1}{2} \quad \text{for } t \in (-\delta, \delta),$$

that is,

$$\frac{1}{2} < \varphi(t) < \frac{3}{2} \quad \text{for } t \in (-\delta, \delta).$$

Then since the range of φ is $\{0, 1\}$, we have

(2) $$\varphi(t) = 1 \quad \text{for } t \in (-\delta, \delta).$$

Since $\varphi(t) = \mathbf{1}_{Q+t}(x_0)$, $\varphi(t) = 1$ if and only if $x_0 \in Q + t$, that is, $x_0 - t \in Q$, that is, $x_0 - t$ is a rational number. Then (2) implies that $x_0 - t$ is a rational number for $t \in (-\delta, \delta)$, that is, $(x_0 - \delta, x_0 + \delta)$ contains only rational numbers. This is impossible.

4. This completes the proof that $\lim_{t \to 0} \varphi(t)$ cannot exist. ∎

Prob. 3.28. Let $\mathfrak{J}$ be a collection of disjoint intervals in $\mathbb{R}$. Show that $\mathfrak{J}$ is a countable collection.

Proof. There can be at most 2 infinite intervals in any collection of disjoint intervals in $\mathbb{R}$. Thus to show that every collection of disjoint intervals in $\mathbb{R}$ is a countable collection, it suffices to show that every collection of disjoint finite intervals in $\mathbb{R}$ is a countable collection.

Let $\mathfrak{J}$ be a collection of disjoint finite intervals in $\mathbb{R}$. Let us show that $\mathfrak{J}$ is a countable collection. For every $n \in \mathbb{N}$, let $\mathfrak{J}_n$ be the subcollection of $\mathfrak{J}$ consisting of those members of $\mathfrak{J}$ that are contained in the interval $(-n, n)$. Since every member of $\mathfrak{J}$ is a finite interval, it is contained in $(-n, n)$ for sufficiently large $n \in \mathbb{N}$ and thus it is contained in $\mathfrak{J}_n$ for some $n \in \mathbb{N}$. Thus we have

$$\mathfrak{J} = \bigcup_{n \in \mathbb{N}} \mathfrak{J}_n.$$

Then to show that $\mathfrak{J}$ is a countable collection, it suffices to show that $\mathfrak{J}_n$ is a countable collection for every $n \in \mathbb{N}$. To show that $\mathfrak{J}_n$ is a countable collection for every $n \in \mathbb{N}$, let us assume the contrary, that is, let us assume that $\mathfrak{J}_n$ is an uncountable collection for some $n \in \mathbb{N}$.

Now for every interval I in $\mathbb{R}$, we have $\mu_L(I) = \ell(I)$, the length of I, which is a positive real number. For $k \in \mathbb{N}$, let $\mathfrak{J}_{n,k}$ be the subcollection of $\mathfrak{J}_n$ consisting of $I \in \mathfrak{J}_n$ with $\mu_L(I) > \frac{1}{k}$. Then since $\mu_L(I) > 0$, we have $\mu_L(I) > \frac{1}{k}$ for some $k \in \mathbb{N}$. Thus we have

$$\mathfrak{J}_n = \bigcup_{k \in \mathbb{N}} \mathfrak{J}_{n,k}.$$

If $\mathfrak{J}_{n,k}$ were a finite collection for every $k \in \mathbb{N}$, $\mathfrak{J}_n$ would be a countable collection, contradicting the assumption that $\mathfrak{J}_n$ is an uncountable collection. Thus there exists $k_0 \in \mathbb{N}$ such that $\mathfrak{J}_{n,k_0}$ is an infinite collection. Then we can select $I_m \in \mathfrak{J}_{n,k_0}$ for every $m \in \mathbb{N}$. Since $I_m \in \mathfrak{J}_{n,k_0} \subset \mathfrak{J}_n$, we have $I_m \subset (-n, n)$ and then $\bigcup_{m \in \mathbb{N}} I_m \subset (-n, n)$. This implies that

$$\mu_L \left(\bigcup_{m \in \mathbb{N}} I_m \right) \le \mu_L \left((-n, n) \right) = 2n.$$

On the other hand, since $\{ I_m : m \in \mathbb{N} \}$ is a disjoint collection and $\mu_L(I_m) > \frac{1}{k_0}$, we have

$$\mu_L \left(\bigcup_{m \in \mathbb{N}} I_m \right) = \sum_{m \in \mathbb{N}} \mu_L(I_m) \ge \sum_{m \in \mathbb{N}} \frac{1}{k_0} = \infty.$$

This is a contradiction. ∎

An Alternate Proof. Let $\mathfrak{J}$ be indexed as $\mathfrak{J} = \{ I_\alpha : \alpha \in A \}$. Assume that $\mathfrak{J}$ is an uncountable collection. Then the indexing set A is an uncountable set. Let $\mathbb{Q}$ be the set of all rational numbers. For every $\alpha \in A$, I_α is an interval in $\mathbb{R}$ and thus $I_\alpha \cap \mathbb{Q} \ne \emptyset$. Pick a rational number in $I_\alpha \cap \mathbb{Q}$ arbitrarily and label it as r_α. Consider the collection of rational numbers $\{ r_\alpha : \alpha \in A \}$. Since the intervals I_α for $\alpha \in A$ are disjoint, the rational numbers r_α for $\alpha \in A$ are distinct. Thus $\{ r_\alpha : \alpha \in A \}$ is an uncountable collection of rational numbers. This contradicts the fact that $\mathbb{Q}$ is a countable set. Thus $\mathfrak{J}$ cannot be an uncountable collection. Therefore $\mathfrak{J}$ is a countable collection. ∎

§4 Measurable Functions

Prob. 4.1. Given a measurable space $(X, \mathfrak{A})$. Let f be an extended real-valued function on a set $D \in \mathfrak{A}$. Let Q be the collection of all rational numbers.
(a) Show that if $\{x \in D : f(x) < r\} \in \mathfrak{A}$ for every $r \in Q$ then f is $\mathfrak{A}$-measurable on D.
(b) What subsets of $\mathbb{R}$ other than Q have this property?
(c) Show that if f is $\mathfrak{A}$-measurable on D, then there exists a countable subcollection $\mathfrak{C}$ of $\mathfrak{A}$, depending on f, such that f is $\sigma(\mathfrak{C})$-measurable on D.

Proof. (a) To show that f is $\mathfrak{A}$-measurable on $D \in \mathfrak{A}$, we show that for every $\alpha \in \mathbb{R}$ we have $\{x \in D : f(x) < \alpha\} \in \mathfrak{A}$.

Let $\alpha \in \mathbb{R}$. Then there exists a strictly increasing sequence of rational numbers $(r_n : n \in \mathbb{N})$ such that $r_n \uparrow \alpha$. Now for any $x \in D$, if $f(x) < r_n$ for some $n \in \mathbb{N}$ then $f(x) < \alpha$. Thus

$$\{x \in D : f(x) < r_n\} \subset \{x \in D : f(x) < \alpha\}$$

and then

(1)
$$\bigcup_{n \in \mathbb{N}} \{x \in D : f(x) < r_n\} \subset \{x \in D : f(x) < \alpha\}.$$

On the other hand, if $x_0 \in \{x \in D : f(x) < \alpha\}$ then $f(x_0) < \alpha$ and this implies that there exists $n_0 \in \mathbb{N}$ such that $f(x_0) < r_{n_0} < \alpha$ and then

$$x_0 \in \{x \in D : f(x) < r_{n_0}\} \subset \bigcup_{n \in \mathbb{N}} \{x \in D : f(x) < r_n\}.$$

Since this holds for an arbitrary $x_0 \in \{x \in D : f(x) < \alpha\}$, we have

(2)
$$\{x \in D : f(x) < \alpha\} \subset \bigcup_{n \in \mathbb{N}} \{x \in D : f(x) < r_n\}.$$

By (1) and (2), we have

(3)
$$\{x \in D : f(x) < \alpha\} = \bigcup_{n \in \mathbb{N}} \{x \in D : f(x) < r_n\}.$$

By assumption we have $\{x \in D : f(x) < r_n\} \in \mathfrak{A}$ for every $n \in \mathbb{N}$. Then since $\mathfrak{A}$ is closed under countable union we have $\{x \in D : f(x) < \alpha\} \in \mathfrak{A}$.

(b) Any dense subset of $\mathbb{R}$ has this property.

(c) Suppose f is $\mathfrak{A}$-measurable on $D \in \mathfrak{A}$. Let $Q = \{r_i ; i \in \mathbb{N}\}$ be the set of all rational numbers. Since f is $\mathfrak{A}$-measurable on D, we have $\{x \in D : f(x) < \alpha\} \in \mathfrak{A}$ for every $\alpha \in \mathbb{R}$ and thus in particular we have

$$D_i := \{x \in D : f(x) < r_i\} \in \mathfrak{A} \quad \text{for } i \in \mathbb{N}.$$

Let $\mathfrak{C} = \{D_i : i \in \mathbb{N}\} \subset \mathfrak{A}$ and consider $\sigma(\mathfrak{C}) \subset \mathfrak{A}$.

To show that f is $\sigma(\mathfrak{C})$-measurable on D, we show that for every $\alpha \in \mathbb{R}$ we have $\{x \in D : f(x) < \alpha\} \in \sigma(\mathfrak{C})$. Let $\alpha \in \mathbb{R}$ and let $(r_n : n \in \mathbb{N})$ be a strictly increasing sequence of rational numbers such that $r_n \uparrow \alpha$. Then by (3) above we have

$$\{x \in D : f(x) < \alpha\} = \bigcup_{n \in \mathbb{N}} \{x \in D : f(x) < r_n\} = \bigcup_{n \in \mathbb{N}} D_n.$$

Then since $D_n \in \mathfrak{C} \subset \sigma(\mathfrak{C})$ for every $n \in \mathbb{N}$ and since $\sigma(\mathfrak{C})$ as a σ-algebra of subsets of X is closed under countable union, we have $\bigcup_{n \in \mathbb{N}} D_n \in \sigma(\mathfrak{C})$, that is, $\{x \in D : f(x) < \alpha\} \in \sigma(\mathfrak{C})$. ∎

Prob. 4.2. Let $\mathfrak{E}$ be a collection of subsets of a set X. Consider $\sigma(\mathfrak{E})$, the smallest σ-algebra of subsets of X containing $\mathfrak{E}$. Let f be an extended real-valued $\sigma(\mathfrak{E})$-measurable function on X. Show that there exists a countable subcollection $\mathfrak{C}$ of $\mathfrak{E}$ such that f is $\sigma(\mathfrak{C})$-measurable on X.

Proof. Let $Q = \{r_i : i \in \mathbb{N}\}$ be the collection of all rational numbers. Since f is $\sigma(\mathfrak{E})$-measurable function on X, we have $\{x \in X : f(x) < \alpha\} \in \sigma(\mathfrak{E})$ for every $\alpha \in \mathbb{R}$ and in particular we have

$$A_i := \big\{x \in X : f(x) < r_i\big\} \in \sigma(\mathfrak{E}) \quad \text{for every } i \in \mathbb{N}.$$

According to Prob. 1.20, for every $A \in \sigma(\mathfrak{E})$ there exists a countable subcollection $\mathfrak{E}_A$ of $\mathfrak{E}$ such that $A \in \sigma(\mathfrak{E}_A)$. Thus for each $i \in \mathbb{N}$ there exists a countable subcollection $\mathfrak{C}_i$ of $\mathfrak{E}$ such that $A_i \in \sigma(\mathfrak{C}_i)$. Let $\mathfrak{C} = \bigcup_{i \in \mathbb{N}} \mathfrak{C}_i$, a countable subcollection of $\mathfrak{E}$, and consider $\sigma(\mathfrak{C})$. Then for every $r_i \in Q$ we have

$$\big\{x \in X : f(x) < r_i\big\} = A_i \in \sigma(\mathfrak{C}_i) \subset \sigma(\mathfrak{C}).$$

This shows that f is $\sigma(\mathfrak{C})$-measurable on X according to Prob. 4.1. ∎

Prob. 4.3. Consider the measurable space $(\mathbb{R}, \mathfrak{B}_{\mathbb{R}})$.
(a) Show that if $E \subset \mathbb{R}$ is a countable set then $E \in \mathfrak{B}_{\mathbb{R}}$.
(b) Show that every extended real-valued function f defined on a countable set $E \subset \mathbb{R}$ is $\mathfrak{B}_{\mathbb{R}}$-measurable on E.

Proof. 1. $\mathfrak{B}_{\mathbb{R}}$ is the σ-algebra of subsets of $\mathbb{R}$ generated by the open sets in $\mathbb{R}$. Thus $\mathfrak{B}_{\mathbb{R}}$ contains all open sets and then all closed sets in $\mathbb{R}$. Let $x \in \mathbb{R}$. A singleton $\{x\}$ in $\mathbb{R}$ is a closed set in $\mathbb{R}$ and thus $\{x\} \in \mathfrak{B}_{\mathbb{R}}$. A countable set E in $\mathbb{R}$ is the union of countably many singletons and therefore $E \in \mathfrak{B}_{\mathbb{R}}$.

    2. Let f be an extended real-valued function defined on a countable set E in $\mathbb{R}$. By (a), we have $\mathfrak{B}_{\mathbb{R}}$. To show that f is $\mathfrak{B}_{\mathbb{R}}$-measurable on E we verify that for every $\alpha \in \mathbb{R}$ we have

$$\{x \in E : f(x) \leq \alpha\} \in \mathfrak{B}_{\mathbb{R}}.$$

Now E is a countable set and this implies that its subset $\{x \in E : f(x) \leq \alpha\}$ is a countable set and hence it is in $\mathfrak{B}_{\mathbb{R}}$ by (a). ∎

Prob. 4.4. Let f be an extended real-valued Borel measurable function on a set $D \in \mathfrak{B}_{\mathbb{R}}$. Show that if we redefine f arbitrarily on a countable subset D_0 of D the resulting function is still Borel measurable on D.

Proof. Being a countable set in $\mathbb{R}$, D_0 is a $\mathfrak{B}_{\mathbb{R}}$-measurable set in $\mathbb{R}$ by Prob. 4.3. Then $D \setminus D_0$ is a $\mathfrak{B}_{\mathbb{R}}$-measurable set in $\mathbb{R}$. Let g be an extended real-valued function on D such that $g = f$ on $D \setminus D_0$. Since f is $\mathfrak{B}_{\mathbb{R}}$-measurable on D, f is $\mathfrak{B}_{\mathbb{R}}$-measurable on $D \setminus D_0$ by Lemma 4.7. Then since $g = f$ on $D \setminus D_0$, g is $\mathfrak{B}_{\mathbb{R}}$-measurable on $D \setminus D_0$. On the other hand, since D_0 is a countable set in $\mathbb{R}$, g is $\mathfrak{B}_{\mathbb{R}}$-measurable on D_0 by Prob. 4.3. Then g is $\mathfrak{B}_{\mathbb{R}}$-measurable on $D_0 \cup (D \setminus D_0) = D$ by Lemma 4.7. ∎

Prob. 4.5. Let $f(x)$ be a real-valued increasing function on $\mathbb{R}$. Show that f is Borel measurable, and hence Lebesgue measurable also, on its domain of definition.

Proof. A subset E of $\mathbb{R}$ is an interval if and only if E has at least two elements and any point between two elements of E is an element of E. Then an interval is a set of the forms

$$(\alpha, \beta), \quad [\alpha, \beta), \quad (\alpha, \beta], \quad [\alpha, \beta],$$

$$(\alpha, \infty), \quad [\alpha, \infty), \quad (-\infty, \beta), \quad (-\infty, \beta],$$

$$(-\infty, \infty).$$

Thus every interval is a $\mathfrak{B}_{\mathbb{R}}$-measurable set.

Let f be a real-valued increasing function on $\mathbb{R}$. To show that f is $\mathfrak{B}_{\mathbb{R}}$-measurable on $\mathbb{R}$ we show that for every $\alpha \in \mathbb{R}$ we have

$$E_\alpha := \{x \in \mathbb{R} : f(x) < \alpha\} \in \mathfrak{B}_{\mathbb{R}}.$$

Consider the set E_α. There are three possibilities. The set E_α may be an empty set, or a singleton, or a set with at least two elements. If E_α is an empty set, then $E_\alpha \in \mathfrak{B}_{\mathbb{R}}$. If E_α is singleton, then $E_\alpha \in \mathfrak{B}_{\mathbb{R}}$. Consider the case that E_α has at least two elements. Let $x_1, x_2 \in E_\alpha$ and $x_1 < x_2$. Let $x_0 \in \mathbb{R}$ be such that $x_1 < x_0 < x_2$. Since $x_2 \in E_\alpha$, we have $f(x_2) < \alpha$. Since f is an increasing function on $\mathbb{R}$, we have $f(x_0) \le f(x_2) < \alpha$ and thus $x_0 \in E_\alpha$. This shows that E_α is an interval. Then $E_\alpha \in \mathfrak{B}_{\mathbb{R}}$. ∎

An Alternate Proof. Let f be a real-valued increasing function on $\mathbb{R}$. Then f has at most countably many points of discontinuity. Let E be the set of all discontinuity points of f. Then E is a countable set in $\mathbb{R}$ and then $E \in \mathfrak{B}_{\mathbb{R}}$. Now $\mathbb{R} \setminus E \in \mathfrak{B}_{\mathbb{R}}$ and $\mathbb{R} \setminus E$ is the set of all continuity points of f. Then f is continuous at every point in $\mathbb{R} \setminus E$ and thus f is continuous on $\mathbb{R} \setminus E$. This implies f is $\mathfrak{B}_{\mathbb{R}}$-measurable on $\mathbb{R} \setminus E$ by Theorem 4.27. On the other hand, since E is a countable set in $\mathbb{R}$, f is $\mathfrak{B}_{\mathbb{R}}$-measurable on E by Prob. 4.3. Now that f is $\mathfrak{B}_{\mathbb{R}}$-measurable on $\mathbb{R} \setminus E$ and $\mathfrak{B}_{\mathbb{R}}$-measurable on E, f is $\mathfrak{B}_{\mathbb{R}}$-measurable on $E \cup (\mathbb{R} \setminus E) = \mathbb{R}$ by Lemma 4.7. ∎

Prob. 4.6. Show that the following functions defined on $\mathbb{R}$ are all Borel measurable, and hence Lebesgue measurable also, on $\mathbb{R}$:
(a) $f(x) = 0$ if x is rational and $f(x) = 1$ if x is irrational.
(b) $f(x) = x$ if x is rational and $f(x) = -x$ if x is irrational.
(c) $f(x) = \sin x$ if x is rational and $f(x) = \cos x$ if x is irrational.

Proof. (a) Let $\mathbb{Q}$ be the set of all rational numbers. Then $\mathbb{Q}$ is a countable set. This implies that $\mathbb{Q} \in \mathfrak{B}_{\mathbb{R}}$ and f is $\mathfrak{B}_{\mathbb{R}}$-measurable on $\mathbb{Q}$ according to Prob. 4.3. Now $\mathbb{R} \setminus \mathbb{Q} \in \mathfrak{B}_{\mathbb{R}}$ and since f is constant on $\mathbb{R} \setminus \mathbb{Q}$, f is $\mathfrak{B}_{\mathbb{R}}$-measurable on $\mathbb{R} \setminus \mathbb{Q}$. Thus f is $\mathfrak{B}_{\mathbb{R}}$-measurable on $\mathbb{Q}$ and $\mathfrak{B}_{\mathbb{R}}$-measurable on $\mathbb{R} \setminus \mathbb{Q}$ and this implies that f is $\mathfrak{B}_{\mathbb{R}}$-measurable on $\mathbb{Q} \cup (\mathbb{R} \setminus \mathbb{Q}) = \mathbb{R}$ by Lemma 4.7.

   (b) f is $\mathfrak{B}_{\mathbb{R}}$-measurable on $\mathbb{Q}$ as in **(a)** above. We have $f(x) = -x$ for $x \in \mathbb{R} \setminus \mathbb{Q}$ and thus f is continuous on $\mathbb{R} \setminus \mathbb{Q}$ and this implies that f is $\mathfrak{B}_{\mathbb{R}}$-measurable on $\mathbb{R} \setminus \mathbb{Q}$ by Theorem 4.27. Now that f is $\mathfrak{B}_{\mathbb{R}}$-measurable on $\mathbb{Q}$ and $\mathfrak{B}_{\mathbb{R}}$-measurable on $\mathbb{R} \setminus \mathbb{Q}$, f is $\mathfrak{B}_{\mathbb{R}}$-measurable on $\mathbb{Q} \cup (\mathbb{R} \setminus \mathbb{Q}) = \mathbb{R}$ by Lemma 4.7.

   (c) f is $\mathfrak{B}_{\mathbb{R}}$-measurable on $\mathbb{R}$ by the same reason as in **(b)**. ∎

Prob. 4.7. Let f be an extended real-valued function defined on an interval I in $\mathbb{R}$. Let Q be an arbitrary countable subset of $\mathbb{R}$. Show that if f is continuous at every $x \in I \setminus Q$, then f is Borel measurable on I.
(As an example of functions of this type, let f be a real-valued function on $\mathbb{R}$ given by
$$f(x) = \begin{cases} \tan x & \text{for } x \neq (2k+1)\frac{\pi}{2} & \text{for } k \in \mathbb{Z}, \\ 0 & \text{for } x = (2k+1)\frac{\pi}{2} & \text{for } k \in \mathbb{Z}. \end{cases}$$
Note also that the countable subset Q of I may be a dense subset of I. See Prob. 4.8.)

Proof. An interval I in $\mathbb{R}$ is a member of $\mathfrak{B}_{\mathbb{R}}$. A countable subset Q of $\mathbb{R}$ is a member of $\mathfrak{B}_{\mathbb{R}}$ according to Prob. 4.3. Thus $I \cap Q \in \mathfrak{B}_{\mathbb{R}}$ and $I \setminus (I \cap Q) \in \mathfrak{B}_{\mathbb{R}}$. Now f is continuous at every $x \in I \setminus (I \cap Q)$ and thus f is continuous on $I \setminus (I \cap Q)$. Then f is $\mathfrak{B}_{\mathbb{R}}$-measurable on $I \setminus (I \cap Q)$ by Theorem 4.27. Since $I \cap Q$ is a countable set in $\mathbb{R}$, f is $\mathfrak{B}_{\mathbb{R}}$-measurable on $I \cap Q$ by Prob. 4.3. Thus f is $\mathfrak{B}_{\mathbb{R}}$-measurable on $I \setminus (I \cap Q)$ and $\mathfrak{B}_{\mathbb{R}}$-measurable on $I \cap Q$. Then f is $\mathfrak{B}_{\mathbb{R}}$-measurable on $(I \setminus (I \cap Q)) \cup (I \cap Q) = I$ by Lemma 4.7. ∎

Prob. 4.8. Let Q be the collection of all rational numbers in $(0, \infty)$. Let each $x \in Q$ be expressed as $\frac{q}{p}$ where p and q are positive integers without common factors other than 1. Define a function f on $(0, \infty)$ by setting

$$f(x) = \begin{cases} \frac{1}{p} & \text{if } x \in Q \text{ and } x = \frac{q}{p}, \\ 0 & \text{if } x \in (0, \infty) \setminus Q. \end{cases}$$

Show that f is discontinuous at every rational $x \in (0, \infty)$ and continuous at every irrational $x \in (0, \infty)$.

Proof. 1. Let us show that f is discontinuous at every rational number in $(0, \infty)$. A real-valued function f defined on $(0, \infty)$ is continuous at $x_0 \in (0, \infty)$ if and only if for every sequence $(x_n : n \in \mathbb{N})$ in $(0, \infty)$ such that $\lim_{n \to \infty} x_n = x_0$ we have $\lim_{n \to \infty} f(x_n) = f(x_0)$. Now let $x_0 \in (0, \infty)$ be a rational number. Then we have $f(x_0) = \frac{1}{p}$ where $p \in \mathbb{N}$. Let $(x_n : n \in \mathbb{N})$ be a sequence of irrational numbers in $(0, \infty)$ such that $\lim_{n \to \infty} x_n = x_0$. Then $f(x_n) = 0$ for every $n \in \mathbb{N}$ and consequently we have $\lim_{n \to \infty} f(x_n) = 0 \neq \frac{1}{p} = f(x_0)$. Thus f cannot be continuous at x_0.

    2. Let us show that f is continuous at every irrational $x_0 \in (0, \infty)$. For this we show that for every $\varepsilon > 0$ there exists $\delta > 0$ such that

(1) $$|f(x) - f(x_0)| < \varepsilon \quad \text{for } x \in (x_0 - \delta, x_0 + \delta) \cap (0, \infty).$$

Now since $f(x) \geq 0$ for every $x \in (0, \infty)$ and $f(x_0) = 0$, the condition (1) reduces to

(2) $$0 \leq f(x) < \varepsilon \quad \text{for } x \in (x_0 - \delta, x_0 + \delta) \cap (0, \infty).$$

Let $N \in \mathbb{N}$ be so large that $\frac{1}{N} < \varepsilon$ and also $J := \left(x_0 - \frac{1}{2N}, x_0 + \frac{1}{2N}\right) \subset (0, \infty)$. The open interval J has length $\frac{1}{N}$. Now if $p \in \mathbb{N}$ and $p \leq N$ then the open interval J contains at most one rational number with denominator p since the distance between two rational numbers with denominator p is at least $\frac{1}{p} \geq \frac{1}{N}$. Thus the open interval J contains at most N rational numbers with denominators $p = 1, \ldots, N$. By the finiteness of the number N we can select $\delta > 0$ so small that the interval $(x_0 - \delta, x_0 + \delta)$ does not contain any of the N rational numbers with denominators $p = 1, \ldots, N$. Now let $x \in (x_0 - \delta, x_0 + \delta)$. Then either x is an irrational number so that $f(x) = 0 < \varepsilon$ or x is a rational number with denominator $p \geq N$ so that $f(x) = \frac{1}{p} \leq \frac{1}{N} < \varepsilon$. This proves (2). Thus f is continuous at every irrational $x \in (0, \infty)$ ∎

Prob. 4.9. Let f be an extended real-valued $\mathfrak{M}_L$-measurable function on a set $D \in \mathfrak{M}_L$. Show that $\sqrt{|f|}$ is a $\mathfrak{M}_L$-measurable function on D.

Proof. To show that $\sqrt{|f|}$ is $\mathfrak{M}_L$-measurable on D, we show that for every $\alpha \in \mathbb{R}$ we have

$$\left\{ x \in D : \sqrt{|f(x)|} < \alpha \right\} \in \mathfrak{M}_L.$$

Now since $\sqrt{|f(x)|} \geq 0$ for every $x \in D$, for $\alpha < 0$ we have

$$\left\{ x \in D : \sqrt{|f(x)|} < \alpha \right\} = \emptyset \in \mathfrak{M}_L.$$

Thus consider $\alpha \geq 0$. Then we have

(1) $$\left\{ x \in D : \sqrt{|f(x)|} < \alpha \right\} = \left\{ x \in D : |f(x)| < \alpha^2 \right\}.$$

Now the $\mathfrak{M}_L$-measurability of f on D implies the $\mathfrak{M}_L$-measurability of $|f|$ on D. Thus we have

(2) $$\left\{ x \in D : |f(x)| < \alpha^2 \right\} \in \mathfrak{M}_L.$$

By (1) and (2), we have

$$\left\{ x \in D : \sqrt{|f(x)|} < \alpha \right\} \in \mathfrak{M}_L. \quad \blacksquare$$

Prob. 4.10. Let $(f_n : n \in \mathbb{N})$ and f be extended real-valued functions on a set D. Show that if $\lim_{n\to\infty} f_n(x) = f(x) \in \overline{\mathbb{R}}$ at some $x \in D$, then we have both $\lim_{n\to\infty} f_n^+(x) = f^+(x)$ and $\lim_{n\to\infty} f_n^-(x) = f^-(x)$.

Proof. Suppose $\lim_{n\to\infty} f_n(x) = f(x) \in \overline{\mathbb{R}}$.

1. Let us show that $\lim_{n\to\infty} f_n^+(x) = f^+(x)$.

1.1. Consider the case that $f(x) > 0$. In this case we have $f^+(x) = f(x)$. Now $\lim_{n\to\infty} f_n(x) = f(x) > 0$ implies that there exists $N \in \mathbb{N}$ such that $f_n(x) > 0$ for $n \geq N$. Then we have $f_n^+(x) = f_n(x)$ for $n \geq N$. Then $\lim_{n\to\infty} f_n(x) = f(x)$ can be rewritten as $\lim_{n\to\infty} f_n^+(x) = f^+(x)$.

1.2. Consider the case that $f(x) < 0$. In this case we have $f^+(x) = 0$. Now $\lim_{n\to\infty} f_n(x) = f(x) < 0$ implies that there exists $N \in \mathbb{N}$ such that $f_n(x) < 0$ for $n \geq N$. Then we have $f_n^+(x) = 0$ for $n \geq N$. Thus we have $\lim_{n\to\infty} f_n^+(x) = 0 = f^+(x)$.

1.3. Consider the case that $f(x) = 0$. In this case we have $f^+(x) = 0$. Now we have

(1) $$|f_n^+(x) - f^+(x)| = |f_n^+(x)| = f_n^+(x) \leq |f_n(x)|.$$

Recall that for a sequence of extended real numbers $(a_n : n \in \mathbb{N})$ we have $\lim_{n\to\infty} a_n = 0$ if and only if $\lim_{n\to\infty} |a_n| = 0$. Thus our assumption $\lim_{n\to\infty} f_n(x) = f(x) = 0$ implies $\lim_{n\to\infty} |f_n(x)| = 0$. Then from (1) we have

$$\lim_{n\to\infty} |f_n^+(x) - f^+(x)| \leq \lim_{n\to\infty} |f_n(x)| = 0.$$

This then implies

$$\lim_{n\to\infty} \{f_n^+(x) - f^+(x)\} = 0,$$

that is,

$$\lim_{n\to\infty} f_n^+(x) = f^+(x).$$

2. Let us show that $\lim_{n\to\infty} f_n^-(x) = f^-(x)$. We have

(2) $$f_n^- = (-f_n)^+ \quad \text{and} \quad f^- = (-f)^+.$$

Now $\lim_{n\to\infty} f_n(x) = f(x)$ implies $\lim_{n\to\infty} (-f_n)(x) = (-f)(x)$. By **1** above, this implies

(3) $$\lim_{n\to\infty} (-f_n)^+(x) = (-f)^+(x).$$

Then by (2) and (3), we have

$$\lim_{n\to\infty} f_n^-(x) = f^-(x). \quad \blacksquare$$

Prob. 4.11. Let D be a null set in the Lebesgue measure space $(\mathbb{R}, \mathfrak{M}_L, \mu_L)$. Show that every extended real-valued function f defined on D is $\mathfrak{M}_L$-measurable on D.

Proof. The measure space $(\mathbb{R}, \mathfrak{M}_L, \mu_L)$ is a complete measure space, that is, every subset of a null set in $(\mathbb{R}, \mathfrak{M}_L, \mu_L)$ is a member of $\mathfrak{M}_L$. Let f be an arbitrary extended real-valued function defined on a null set D of the measure space $(\mathbb{R}, \mathfrak{M}_L, \mu_L)$. Then for every $\alpha \in \mathbb{R}$, the set $\{x \in D : f(x) < \alpha\}$ being a subset of the null set D is a member of $\mathfrak{M}_L$. Thus f is $\mathfrak{M}_L$-measurable on D. ∎

Prob. 4.12. Consider the Lebesgue measure space $(\mathbb{R}, \mathfrak{M}_L, \mu_L)$. Let $D \in \mathfrak{M}_L$ and let f be an extended real-valued function defined on D such that f is continuous at almost every $x \in D$. Show that f is $\mathfrak{M}_L$-measurable on D.

Proof. Since f is continuous at almost every $x \in D$, there exists a null set $D_0 \subset D$ of the measure space $(\mathbb{R}, \mathfrak{M}_L, \mu_L)$ such that f is continuous at every $x \in D \setminus D_0$. Now continuity of f at every $x \in D \setminus D_0 \in \mathfrak{M}_L$ implies that f is $\mathfrak{M}_L$-measurable on $D \setminus D_0$ according to Theorem 4.27. On the other hand, since D_0 is a null set in the measure space $(\mathbb{R}, \mathfrak{M}_L, \mu_L)$, f is $\mathfrak{M}_L$-measurable on D_0 by Prob. 4.11. Thus f is $\mathfrak{M}_L$-measurable on $D \setminus D_0$ and $\mathfrak{M}_L$-measurable on D_0. Then f is $\mathfrak{M}_L$-measurable on $(D \setminus D_0) \cup D_0 = D$ by Lemma 4.7. ∎

Prob. 4.13. Consider the Lebesgue measure space $(\mathbb{R}, \mathfrak{M}_L, \mu_L)$. Let f be an extended real-valued function defined on $\mathbb{R}$. Consider the following two concepts:
(i) f is continuous a.e. on $\mathbb{R}$, that is, there exists a null set N in $(\mathbb{R}, \mathfrak{M}_L, \mu_L)$ such that f is continuous at every $x \in \mathbb{R} \setminus N$,
(ii) f is equal to a continuous function a.e. on $\mathbb{R}$, that is, there exist a continuous function g on $\mathbb{R}$ and a null set N in $(\mathbb{R}, \mathfrak{M}_L, \mu_L)$ such that $f(x) = g(x)$ for every $x \in \mathbb{R} \setminus N$.
To show that these are two different concepts:
(a) Construct a function which satisfies condition (i) but not (ii),
(b) Construct a function which satisfies condition (ii) but not (i).

Proof. (a) Let us define a real-valued function f on $\mathbb{R}$ by setting

$$f(x) = \begin{cases} \frac{1}{x} & \text{for } x \neq 0, \\ 0 & \text{for } x = 0. \end{cases}$$

Then f is continuous at every $x \in \mathbb{R} \setminus \{0\}$. Since $\{0\}$ is a null set in the measure space $(\mathbb{R}, \mathfrak{M}_L, \mu_L)$, f is continuous a.e. on $\mathbb{R}$.
Let us show that there cannot exist a continuous function g on $\mathbb{R}$ and a null set N in $(\mathbb{R}, \mathfrak{M}_L, \mu_L)$ such that $f = g$ on $\mathbb{R} \setminus N$. Assume the contrary. Then the assumption that N is a null set in $(\mathbb{R}, \mathfrak{M}_L, \mu_L)$ implies that $N^c = \mathbb{R} \setminus N$ is a dense subset of $\mathbb{R}$, that is, every non-empty open set in $\mathbb{R}$ contains points of N^c. Then we can select a strictly decreasing sequence $(x_n : n \in \mathbb{N})$ in N^c such that $x_n \downarrow 0$. Then the continuity of g at $x = 0$ implies that $\lim_{n \to \infty} g(x_n) = g(0) \in \mathbb{R}$. On the other hand, since $f = g$ on N^c and since $x_n \in N^c$, we have $g(x_n) = f(x_n) = \frac{1}{x_n}$ and this implies that

$$g(0) = \lim_{n \to \infty} g(x_n) = \lim_{n \to \infty} \frac{1}{x_n} = \infty.$$

This is a contradiction. This shows that our function f satisfies (i) but not (ii).
(b) Let Q be the set of all rational numbers. Then Q is a null set in $(\mathbb{R}, \mathfrak{M}_L, \mu_L)$. Let f be a real-valued function f on $\mathbb{R}$ defined by setting

$$f(x) = \begin{cases} 1 & \text{for } x \in Q, \\ 0 & \text{for } x \in \mathbb{R} \setminus Q. \end{cases}$$

Let g be the identically vanishing function on $\mathbb{R}$. Then g is a continuous function on $\mathbb{R}$ and moreover $f = g$ on $\mathbb{R} \setminus Q$. Thus f satisfies (ii). On the other hand f is discontinuous at every $x \in \mathbb{R}$ and thus it does not satisfy (i). ∎

Prob. 4.14. Let f and g be two real-valued functions on $\mathbb{R}$.
(a) Show that if f and g are $\mathfrak{B}_{\mathbb{R}}$-measurable on $\mathbb{R}$, then so is $g \circ f$.
(b) Show that if f is $\mathfrak{M}_{L}$-measurable on $\mathbb{R}$ and g is $\mathfrak{B}_{\mathbb{R}}$-measurable on $\mathbb{R}$, then $g \circ f$ is $\mathfrak{M}_{L}$-measurable on $\mathbb{R}$.

Proof. Let us review measurable mappings.

$1°$ Let $(X, \mathfrak{A})$ and $(Y, \mathfrak{B})$ be measurable spaces. A mapping f of X into Y is said to be $\mathfrak{A}/\mathfrak{B}$-measurable if for every $B \in \mathfrak{B}$ we have $f^{-1}(B) \in \mathfrak{A}$, that is, if $f^{-1}(\mathfrak{B}) \subset \mathfrak{A}$. (See Definition 1.38.)

$2°$ Let $(X, \mathfrak{A})$, $(Y, \mathfrak{B})$ and $(Z, \mathfrak{C})$ be measurable spaces. Let f be a $\mathfrak{A}/\mathfrak{B}$-measurable mapping of X into Y and let g be a $\mathfrak{B}/\mathfrak{C}$-measurable mapping of Y into Z. Then $g \circ f$ is a $\mathfrak{A}/\mathfrak{C}$-measurable mapping of X into Z. (See Theorem 1.40.)

$3°$ Let $(X, \mathfrak{A})$ be an arbitrary measurable space. A real-valued function f is $\mathfrak{A}$-measurable on X if and only if f is a $\mathfrak{A}/\mathfrak{B}_{\mathbb{R}}$-measurable mapping of $(X, \mathfrak{A})$ into $(\mathbb{R}, \mathfrak{B}_{\mathbb{R}})$. (See Theorem 4.6.)

1. Let us prove (a). Thus let f and g be real-valued $\mathfrak{B}_{\mathbb{R}}$-measurable functions on $\mathbb{R}$. Then by $3°$, f is a $\mathfrak{B}_{\mathbb{R}}/\mathfrak{B}_{\mathbb{R}}$-measurable mapping of $(\mathbb{R}, \mathfrak{B}_{\mathbb{R}})$ into $(\mathbb{R}, \mathfrak{B}_{\mathbb{R}})$ and similarly g is a $\mathfrak{B}_{\mathbb{R}}/\mathfrak{B}_{\mathbb{R}}$-measurable mapping of $(\mathbb{R}, \mathfrak{B}_{\mathbb{R}})$ into $(\mathbb{R}, \mathfrak{B}_{\mathbb{R}})$. Then by $2°$, $g \circ f$ is a $\mathfrak{B}_{\mathbb{R}}/\mathfrak{B}_{\mathbb{R}}$-measurable mapping of $(\mathbb{R}, \mathfrak{B}_{\mathbb{R}})$ into $(\mathbb{R}, \mathfrak{B}_{\mathbb{R}})$, that is, $g \circ f$ is a $\mathfrak{B}_{\mathbb{R}}$-measurable function on $\mathbb{R}$ by $3°$. This proves (a).

2. Let us prove (b). Let f be a real-valued $\mathfrak{M}_{L}$-measurable function on $\mathbb{R}$. Then by $3°$, f is a $\mathfrak{M}_{L}/\mathfrak{B}_{\mathbb{R}}$-measurable mapping of $(\mathbb{R}, \mathfrak{M}_{L})$ into $(\mathbb{R}, \mathfrak{B}_{\mathbb{R}})$. Let g be a real-valued $\mathfrak{B}_{\mathbb{R}}$-measurable function on $\mathbb{R}$. Then by $3°$, g is a $\mathfrak{B}_{\mathbb{R}}/\mathfrak{B}_{\mathbb{R}}$- measurable mapping of $(\mathbb{R}, \mathfrak{B}_{\mathbb{R}})$ into $(\mathbb{R}, \mathfrak{B}_{\mathbb{R}})$. Then by $2°$, $g \circ f$ is a $\mathfrak{M}_{L}/\mathfrak{B}_{\mathbb{R}}$-measurable mapping of $(\mathbb{R}, \mathfrak{M}_{L})$ into $(\mathbb{R}, \mathfrak{B}_{\mathbb{R}})$, that is, $g \circ f$ is a $\mathfrak{M}_{L}$-measurable function on $\mathbb{R}$ by $3°$. This proves (b). ∎

Prob. 4.15. Let X be a set and let $(Y, \mathfrak{G})$ be a measurable space. Let T be a mapping from X to Y. Show that $T^{-1}(\mathfrak{G})$ is a σ-algebra of subsets of the set $\mathfrak{D}(T)$, and in particular, if $\mathfrak{D}(T) = X$ then $T^{-1}(\mathfrak{G})$ is a σ-algebra of subsets of the set X.

Proof. To show that $T^{-1}(\mathfrak{G})$ is a σ-algebra of subsets of the set $\mathfrak{D}(T)$, we verify:

1° $\mathfrak{D}(T) \in T^{-1}(\mathfrak{G})$.
2° $E \in T^{-1}(\mathfrak{G}) \Rightarrow \mathfrak{D}(T) \setminus E \in T^{-1}(\mathfrak{G})$.
3° $E_n \in T^{-1}(\mathfrak{G})$ for $n \in \mathbb{N} \Rightarrow \bigcup_{n \in \mathbb{N}} E_n \in T^{-1}(\mathfrak{G})$.

    1. Now $\mathfrak{D}(T) = T^{-1}(Y)$ and $Y \in \mathfrak{G}$. Thus $\mathfrak{D}(T) \in T^{-1}(\mathfrak{G})$. This verifies 1°.
    2. Suppose $E \in T^{-1}(\mathfrak{G})$. Then $E = T^{-1}(G)$ for some $G \in \mathfrak{G}$. Since $\mathfrak{G}$ is a σ-algebra of subsets of Y, $G \in \mathfrak{G}$ implies $Y \setminus G \in \mathfrak{G}$. Then

$$T^{-1}(Y \setminus G) = T^{-1}(Y) \setminus T^{-1}(G) = \mathfrak{D}(T) \setminus E.$$

Thus we have

$$\mathfrak{D}(T) \setminus E = T^{-1}(Y \setminus G) \in T^{-1}(\mathfrak{G}).$$

This verifies 2°.
    3. Suppose $E_n \in T^{-1}(\mathfrak{G})$ for $n \in \mathbb{N}$. Then there exists $G_n \in \mathfrak{G}$ such that $E_n = T^{-1}(G_n)$ for $n \in \mathbb{N}$. Since $\mathfrak{G}$ is a σ-algebra of subsets of Y, we have $\bigcup_{n \in \mathbb{N}} G_n \in \mathfrak{G}$. Then

$$T^{-1}\left(\bigcup_{n \in \mathbb{N}} G_n \right) = \bigcup_{n \in \mathbb{N}} T^{-1}(G_n) = \bigcup_{n \in \mathbb{N}} E_n.$$

Thus we have

$$\bigcup_{n \in \mathbb{N}} E_n = T^{-1}\left(\bigcup_{n \in \mathbb{N}} G_n \right) \in T^{-1}(\mathfrak{G}).$$

This verifies 3°. ∎

Prob. 4.16. Given two topological spaces $(X, \mathfrak{O}_X)$ and $(Y, \mathfrak{O}_Y)$. Let $\mathfrak{B}_X = \sigma(\mathfrak{O}_X)$, that is, the Borel σ-algebra in X, and similarly let $\mathfrak{B}_Y = \sigma(\mathfrak{O}_Y)$. Show that if T is a homeomorphism of $(X, \mathfrak{O}_X)$ onto $(Y, \mathfrak{O}_Y)$, then $T^{-1}(\mathfrak{B}_Y) = \mathfrak{B}_X$ and $T(\mathfrak{B}_X) = \mathfrak{B}_Y$. (Hint: Show first the fact that $\mathfrak{B}_Y$ is a σ-algebra of subsets of Y implies that $T^{-1}(\mathfrak{B}_Y)$ is a σ-algebra of subsets of X.)

Proof. Let $(X, \mathfrak{O}_X)$ and $(Y, \mathfrak{O}_Y)$ be two topological spaces. A homeomorphism is a one-to-one mapping T of X onto Y such that T is continuous on X and T^{-1} is continuous on Y, that is,

(1) $$T^{-1}(\mathfrak{O}_Y) \subset \mathfrak{O}_X,$$

(2) $$\left(T^{-1}\right)^{-1}(\mathfrak{O}_X) \subset \mathfrak{O}_Y \quad \text{that is,} \quad T(\mathfrak{O}_X) \subset \mathfrak{O}_Y.$$

Let us observe that (1) and (2) imply

(3) $$T^{-1}(\mathfrak{O}_Y) = \mathfrak{O}_X,$$

(4) $$T(\mathfrak{O}_X) = \mathfrak{O}_Y.$$

Indeed from (1) we have $T\left(T^{-1}(\mathfrak{O}_Y)\right) \subset T(\mathfrak{O}_X)$ so that $\mathfrak{O}_Y \subset T(\mathfrak{O}_X) \subset \mathfrak{O}_Y$ by (2) and thus $T(\mathfrak{O}_X) = \mathfrak{O}_Y$ proving (4). Similarly from (2) we have $T^{-1}\left(T(\mathfrak{O}_X)\right) \subset T^{-1}(\mathfrak{O}_Y) \subset \mathfrak{O}_X$ by (1), that is, $\mathfrak{O}_X \subset T^{-1}(\mathfrak{O}_Y) \subset \mathfrak{O}_X$ so that $T^{-1}(\mathfrak{O}_Y) = \mathfrak{O}_X$ proving (3).

Let $\mathfrak{B}_X = \sigma(\mathfrak{O}_X)$ and $\mathfrak{B}_Y = \sigma(\mathfrak{O}_Y)$. Let us prove

(5) $$T^{-1}(\mathfrak{B}_Y) = \mathfrak{B}_X,$$

(6) $$T(\mathfrak{B}_X) = \mathfrak{B}_Y.$$

Now the fact that $\mathfrak{B}_Y$ is a σ-algebra of subsets of Y implies that $T^{-1}(\mathfrak{B}_Y)$ is a σ-algebra of subsets of X according to Prob. 4.15. Now

$$T^{-1}(\mathfrak{B}_Y) = T^{-1}\left(\sigma(\mathfrak{O}_Y)\right) = \sigma\left(T^{-1}(\mathfrak{O}_Y)\right) \quad \text{by Theorem 1.14}$$

$$= \sigma(\mathfrak{O}_X) = \mathfrak{B}_X \quad \text{by (3).}$$

This proves (5). Then by (5) we have

$$T(\mathfrak{B}_X) = T\left(T^{-1}(\mathfrak{B}_Y)\right) = \mathfrak{B}_Y.$$

This proves (6). ∎

Prob. 4.17. Let g be a real-valued continuous function on an interval I in $\mathbb{R}$. Let Q be a countable dense subset of I. (The set of all rational numbers in I is an example.) Show that
$$\inf_{y \in I} g(y) = \inf_{r \in Q} g(r),$$

$$\sup_{y \in I} g(y) = \sup_{r \in Q} g(r).$$

Proof. The two equalities are proved in the same way. Let us prove the second one here.

1. Consider the case that g is not bounded above on I. In this case we have by definition $\sup_{y \in I} g(y) = \infty$. Since g is not bounded above on I, for every $M > 0$ there exists $y_M \in I$ such that $g(y_M) \geq M$. Continuity of g at y_M implies that for every sequence $(y_k : k \in \mathbb{N})$ in I such that $\lim_{k \to \infty} y_k = y_M$ we have $\lim_{k \to \infty} g(y_k) = g(y_M)$. Since Q is dense in I, we can select a sequence $(r_k : k \in \mathbb{N})$ in Q such that $\lim_{k \to \infty} r_k = y_M$. Then we have $\lim_{k \to \infty} g(r_k) = g(y_M)$. This implies that there exists $r_M \in Q$ such that

$$g(r_M) \geq \frac{1}{2} g(y_M) \geq \frac{M}{2}.$$

Since this holds for every $M > 0$, g is not bounded above on Q and therefore we have

$$\sup_{r \in Q} g(r) = \infty = \sup_{y \in I} g(y).$$

2. Consider the case that g is bounded above on I. Here we have $\sup_{y \in I} g(y) = \beta \in \mathbb{R}$. Since $Q \subset I$, we have $\sup_{r \in Q} g(r) \leq \sup_{y \in I} g(y) = \beta$. Let us show that actually we have $\sup_{r \in Q} g(r) = \beta$. To prove this, assume the contrary, that is, assume that $\sup_{r \in Q} g(r) < \beta$. Then $\sup_{r \in Q} g(r) = \beta - \varepsilon$ where $\varepsilon > 0$. Now $\beta = \sup_{y \in I} g(y)$ implies that there exists $y_0 \in I$ such that $g(y_0) > \beta - \frac{\varepsilon}{2}$. Since Q is dense in I, we can select a sequence $(r_k : k \in \mathbb{N})$ in Q such that $\lim_{k \to \infty} r_k = y_0$. Then the continuity of g at y_0 implies that

$$\lim_{k \to \infty} g(r_k) = g(y_0) > \beta - \frac{\varepsilon}{2}.$$

This implies that there exists $k \in \mathbb{N}$ such that $g(r_k) > \beta - \frac{\varepsilon}{2}$. Thus we have

$$\sup_{r \in Q} g(r) \geq g(r_k) > \beta - \frac{\varepsilon}{2} > \beta - \varepsilon = \sup_{r \in Q} g(r),$$

that is, we have $\sup_{r \in Q} g(r) > \sup_{r \in Q} g(r)$, a contradiction. Therefore we have

$$\sup_{r \in Q} g(r) = \beta = \sup_{y \in I} g(y). \quad \blacksquare$$

Prob. 4.18. Let $D \in \mathfrak{M}_L$ and let F be a real-valued function on $D \times (c, d)$ satisfying the conditions:

$1°$. For each $y \in (c, d)$, $F(\cdot, y)$ is a $\mathfrak{M}_L$-measurable function on D.

$2°$. For each $x \in D$, $F(x, \cdot)$ is a continuous function on (c, d).

Show that φ and ψ defined by $\varphi(x) = \inf_{y \in (c,d)} F(x, y)$ and $\psi(x) = \sup_{y \in (c,d)} F(x, y)$ for $x \in D$ are $\mathfrak{M}_L$-measurable on D.

Proof. Let Q be a countable dense subset of the interval (c, d). (For example, Q is the set of all rational numbers in (c, d).) By Prob. 4.17, we have

$$\varphi(x) = \inf_{y \in (c,d)} F(x, y) = \inf_{r \in Q} F(x, r) \quad \text{for } x \in D.$$

Now for each $r \in Q$, $F(\cdot .r)$ is a $\mathfrak{M}_L$-measurable function on D. Then the countability of the set Q implies that $\inf_{r \in Q} F(\cdot, r)$ is a $\mathfrak{M}_L$-measurable function on D according to Theorem 4.22. Thus φ is a $\mathfrak{M}_L$-measurable function on D.

Similarly by Prob. 4.17 we have

$$\psi(x) = \sup_{y \in (c,d)} F(x, y) = \sup_{r \in Q} F(x, r) \quad \text{for } x \in D.$$

Then ψ is a $\mathfrak{M}_L$-measurable function on D by the same reason as for φ above. ∎

Prob. 4.19. Let F be as in Prob. 4.18. Show that if $y_0 = c$ or d, then $\liminf\limits_{y \to y_0} F(\cdot, y)$ and $\limsup\limits_{y \to y_0} F(\cdot, y)$ are $\mathfrak{M}_L$-measurable functions on D.

Proof. Let us show that $\liminf\limits_{y \to c} F(\cdot, y)$ is a $\mathfrak{M}_L$-measurable function on D. By definition we have

$$\liminf_{y \to c} F(\cdot, y) = \lim_{\delta \downarrow 0} \inf_{y \in (c, c+\delta)} F(\cdot, y).$$

Now we have

$$\inf_{y \in (c, c+\delta)} F(\cdot, y) \uparrow \quad \text{as} \quad \delta \downarrow .$$

Thus we can write

$$\lim_{\delta \downarrow 0} \inf_{y \in (c, c+\delta)} F(\cdot, y) = \lim_{n \to \infty} \inf_{y \in (c, c+\frac{1}{n})} F(\cdot, y).$$

For each $n \in \mathbb{N}$, $\inf_{y \in (c, c+\frac{1}{n})} F(\cdot, y)$ is a $\mathfrak{M}_L$-measurable function on D according to Prob. 4.18. Since this is valid for every $n \in \mathbb{N}$, $\lim\limits_{n \to \infty} \inf\limits_{y \in (c, c+\frac{1}{n})} F(\cdot, y)$ is a $\mathfrak{M}_L$-measurable function on D by Theorem 4.21. This proves that $\liminf\limits_{y \to c} F(\cdot, y)$ is a $\mathfrak{M}_L$-measurable function on D.

The $\mathfrak{M}_L$-measurability of $\limsup\limits_{y \to c} F(\cdot, y)$, $\liminf\limits_{y \to d} F(\cdot, y)$, and $\limsup\limits_{y \to d} F(\cdot, y)$ on D is proved likewise. ∎

Prob. 4.20. Let F be as in Prob. 4.18. Show that if $y_0 = c$ or d, the subset of D on which the function $\lim\limits_{y \to y_0} F(\cdot, y)$ exists is a $\mathfrak{M}_L$-measurable set and the function $\lim\limits_{y \to y_0} F(\cdot, y)$ is $\mathfrak{M}_L$-measurable on the subset.

Proof. Recall that $\lim\limits_{n \to \infty} F(\cdot, y)$ exists if and only if $\liminf\limits_{y \to y_0} F(\cdot, y) = \limsup\limits_{y \to y_0} F(\cdot, y)$. According to Prob. 4.19, $\liminf\limits_{y \to y_0} F(\cdot, y)$ and $\limsup\limits_{y \to y_0} F(\cdot, y)$ are $\mathfrak{M}_L$-measurable functions on D. This implies according to Theorem 4.16 that

$$D_0 := \left\{ x \in D : \liminf_{y \to y_0} F(x, y) = \limsup_{y \to y_0} F(x, y) \right\} \in \mathfrak{M}_L.$$

This shows that the subset D_0 of D on which $\lim\limits_{y \to y_0} F(\cdot, y)$ exists is a $\mathfrak{M}_L$-measurable set.

On the $\mathfrak{M}_L$-measurable subset D_0 of D, we have $\lim\limits_{y \to y_0} F(\cdot, y) = \liminf\limits_{y \to y_0} F(\cdot, y) = \limsup\limits_{y \to y_0} F(\cdot, y)$. Then since $\lim\limits_{y \to y_0} F(\cdot, y)$ and $\limsup\limits_{y \to y_0} F(\cdot, y)$ are $\mathfrak{M}_L$-measurable on D and in particular on D_0, $\lim\limits_{y \to y_0} F(\cdot, y)$ is $\mathfrak{M}_L$-measurable on D_0. ∎

Prob. 4.21. Let f be a real-valued function on $\mathbb{R}$. Consider the derivative of f at $x \in \mathbb{R}$ defined by

$$(Df)(x) = \lim_{h \to 0} \frac{f(x+h)-f(x)}{h} \quad \text{if the limit exists in } \overline{\mathbb{R}}.$$

(We say that f is differentiable at $x \in \mathbb{R}$ only if $(Df)(x) \in \mathbb{R}$.)
Show that if f is continuous on $\mathbb{R}$ then $\mathfrak{D}(Df) \in \mathfrak{B}_\mathbb{R}$ and Df is $\mathfrak{B}_\mathbb{R}$-measurable on $\mathfrak{D}(Df)$.

Proof. 1. Consider the right-derivative and left-derivative of f at $x \in \mathbb{R}$ given by

$$(D_r f)(x) = \lim_{h \downarrow 0} \frac{f(x+h) - f(x)}{h} \quad \text{if it exists in } \overline{\mathbb{R}},$$

$$(D_\ell f)(x) = \lim_{h \uparrow 0} \frac{f(x+h) - f(x)}{h} \quad \text{if it exists in } \overline{\mathbb{R}}.$$

Now $(Df)(x)$ exists in $\overline{\mathbb{R}}$ if and only if both $(D_r f)(x)$ and $(D_\ell f)(x)$ exist in $\overline{\mathbb{R}}$ and moreover $(D_r f)(x) = (D_\ell f)(x)$. When this is the case we have $(Df)(x) = (D_r f)(x) = (D_\ell f)(x)$.

2. Let us show that $\mathfrak{D}(D_r f) \in \mathfrak{B}_\mathbb{R}$ and $D_r f$ is an extended real-valued $\mathfrak{B}_\mathbb{R}$-measurable function on $\mathfrak{D}(D_r f)$. For this purpose let us define a function F on $\mathbb{R} \times (0, \infty)$ by setting

$$(1) \qquad\qquad F(x, h) = \frac{f(x+h) - f(x)}{h} \quad \text{for } (x, h) \in \mathbb{R} \times (0, \infty).$$

Then we have

$$(2) \qquad\qquad\qquad (D_r f)(x) = \lim_{h \downarrow 0} F(x, h).$$

Since f is a continuous function on $\mathbb{R}$, $F(\cdot, h)$ is a continuous function on $\mathbb{R}$ for each fixed $h \in (0, \infty)$ and $F(x, \cdot)$ is a continuous function on $(0, \infty)$ for each fixed $x \in \mathbb{R}$.

Let Q be a countable dense subset of $\mathbb{R}$. (Q may be the set of all rational numbers.) Now we have

$$(3) \qquad\qquad\qquad \limsup_{h \downarrow 0} F(\cdot, h) = \lim_{\delta \downarrow 0} \sup_{h \in (0,\delta)} F(\cdot, h).$$

Then since $\sup_{h \in (0,\delta)} F(\cdot, h) \downarrow$ as $\delta \downarrow$, we can write

$$(4) \qquad\qquad \lim_{\delta \downarrow 0} \sup_{h \in (0,\delta)} F(\cdot, h) = \lim_{n \to \infty} \sup_{h \in \left(0, \frac{1}{n}\right)} F(\cdot, h).$$

According to Prob. 4.17, continuity of $F(x, \cdot)$ on $(0, \infty)$ implies that

$$(5) \qquad\qquad \sup_{h \in \left(0, \frac{1}{n}\right)} F(\cdot, h) = \sup_{h \in Q \cap \left(0, \frac{1}{n}\right)} F(\cdot, h).$$

By (3), (4) and (5), we have

$$(6) \qquad\qquad \limsup_{h \downarrow 0} F(\cdot, h) = \lim_{n \to \infty} \sup_{h \in Q \cap \left(0, \frac{1}{n}\right)} F(\cdot, h).$$

For each fixed $h \in (0, \infty)$, $F(\cdot, h)$ is a continuous function on $\mathbb{R}$ and hence $F(\cdot, h)$ is a $\mathfrak{B}_{\mathbb{R}}$-measurable function on $\mathbb{R}$. For each fixed $n \in \mathbb{N}$, $\sup_{h \in Q \cap (0, \frac{1}{n})} F(\cdot, h)$ is the supremum of countably many $\mathfrak{B}_{\mathbb{R}}$-measurable functions on $\mathbb{R}$ and is therefore a $\mathfrak{B}_{\mathbb{R}}$-measurable function on $\mathbb{R}$ by Theorem 4.22. Since this holds for every $n \in \mathbb{N}$, $\lim_{n \to \infty} \sup_{h \in Q \cap (0, \frac{1}{n})} F(\cdot, h)$ is a $\mathfrak{B}_{\mathbb{R}}$-measurable function on $\mathbb{R}$ by Theorem 4.21. This shows that $\limsup_{h \downarrow 0} F(\cdot, h)$ is a $\mathfrak{B}_{\mathbb{R}}$-measurable function on $\mathbb{R}$.

We show by similar argument that $\liminf_{h \downarrow 0} F(\cdot, h)$ is a $\mathfrak{B}_{\mathbb{R}}$-measurable function on $\mathbb{R}$. Now the $\mathfrak{B}_{\mathbb{R}}$-measurability of both $\limsup_{h \downarrow 0} F(\cdot, h)$ and $\liminf_{h \downarrow 0} F(\cdot, h)$ implies according to Theorem 4.16 that

$$\left\{ x \in \mathbb{R} : \limsup_{h \downarrow 0} F(x, h) = \liminf_{h \downarrow 0} F(x, h) \right\} \in \mathfrak{B}_{\mathbb{R}}.$$

Then since

$$\mathfrak{D}(D_r f) = \left\{ x \in \mathbb{R} : \limsup_{h \downarrow 0} F(x, h) = \liminf_{h \downarrow 0} F(x, h) \right\},$$

we have $\mathfrak{D}(D_r f) \in \mathfrak{B}_{\mathbb{R}}$.

For $x \in \mathfrak{D}(D_r f)$ we have $(D_r f)(x) = \limsup_{h \downarrow 0} F(x, h) = \liminf_{h \downarrow 0} F(x, h)$. Then since $\limsup_{h \downarrow 0} F(\cdot, h)$ is $\mathfrak{B}_{\mathbb{R}}$-measurable on $\mathbb{R}$ and in particular on $\mathfrak{D}(D_r f)$, $D_r f$ is $\mathfrak{B}_{\mathbb{R}}$-measurable on $\mathfrak{D}(D_r f)$. This completes the proof that $\mathfrak{D}(D_r f) \in \mathfrak{B}_{\mathbb{R}}$ and $D_r f$ is a $\mathfrak{B}_{\mathbb{R}}$-measurable function on $\mathfrak{D}(D_r f)$.

3. We show that $\mathfrak{D}(D_\ell f) \in \mathfrak{B}_{\mathbb{R}}$ and $D_\ell f$ is $\mathfrak{B}_{\mathbb{R}}$-measurable on $\mathfrak{D}(D_\ell f)$ by the same argument as above. For this start with a function G on $\mathbb{R} \times (-\infty, 0)$ defined by setting

$$G(x, h) = \frac{f(x + h) - f(x)}{h} \quad \text{for } (x, h) \in \mathbb{R} \times (-\infty, 0).$$

Then we have

$$(D_\ell f)(x) = \lim_{h \uparrow 0} G(x, h).$$

Then we show that $\mathfrak{D}(D_\ell f) \in \mathfrak{B}_{\mathbb{R}}$ and $D_\ell f$ is $\mathfrak{B}_{\mathbb{R}}$-measurable on $\mathfrak{D}(D_\ell f)$ by the same line of argument as in **2.**

4. Now $\mathfrak{D}(D_r f) \in \mathfrak{B}_{\mathbb{R}}$ and $\mathfrak{D}(D_\ell f) \in \mathfrak{B}_{\mathbb{R}}$ imply that

$$\mathfrak{D}(D_r f) \cap \mathfrak{D}(D_\ell f) \in \mathfrak{B}_{\mathbb{R}}.$$

Since $D_r f$ is a $\mathfrak{B}_{\mathbb{R}}$-measurable function on $\mathfrak{D}(D_r f)$ and $D_\ell f$ is $\mathfrak{B}_{\mathbb{R}}$-measurable on $\mathfrak{D}(D_\ell f)$, both $D_r f$ and $D_\ell f$ are $\mathfrak{B}_{\mathbb{R}}$-measurable on $\mathfrak{D}(D_r f) \cap \mathfrak{D}(D_\ell f)$. Then by Theorem 4.16, we have

$$\left\{ x \in \mathfrak{D}(D_r f) \cap \mathfrak{D}(D_\ell f) : (D_r f)(x) = (D_\ell f)(x) \right\} \in \mathfrak{B}_{\mathbb{R}}.$$

Then we have

$$\mathfrak{D}(Df) = \left\{ x \in \mathfrak{D}(D_r f) \cap \mathfrak{D}(D_\ell f) : (D_r f)(x) = (D_\ell f)(x) \right\} \in \mathfrak{B}_{\mathbb{R}}.$$

Since $Df = D_r f = D_\ell f$ on $\mathfrak{D}(Df)$ and $D_r f$ and $D_\ell f$ are $\mathfrak{B}_{\mathbb{R}}$-measurable on $\mathfrak{D}(Df)$, Df is $\mathfrak{B}_{\mathbb{R}}$-measurable on $\mathfrak{D}(Df)$. $\blacksquare$

Prob. 4.22. Let F be a real-valued continuous function on $[a, b] \times [c, d]$. Show that the function f defined on $[a, b]$ by $f(x) = \int_c^d F(x, y)\, dy$ for $x \in [a, b]$ is a $\mathfrak{B}_\mathbb{R}$-measurable function on $[a, b]$.

Proof. For each fixed $x \in [a.b]$, $F(x, \cdot)$ is a real-valued continuous function on $[c, d]$ and thus the Riemann integral $f(x) = \int_c^d F(x, y)\, dy$ exists as a real number. Now that $\int_c^d F(x, y)\, dy$ exists, it can be calculated as follows.

For each $n \in \mathbb{N}$, let us partition the interval $[c.d]$ into 2^n subintervals with length $\frac{d-c}{2^n}$ with partition points $c = y_0 < y_1 < \cdots < y_{2^n} = d$. Then we have

$$\int_c^d F(x, y)\, dy = \lim_{n \to \infty} \sum_{k=1}^{2^n} F(x, y_k)(y_k - y_{k-1}).$$

Since $F(\cdot, y_k)$ is a real-valued continuous function on $[a, b]$, $F(\cdot, y_k)$ is a $\mathfrak{B}_\mathbb{R}$-measurable function on $[a.b]$. Then $F(\cdot, y_k)(y_k - y_{k-1})$ is a $\mathfrak{B}_\mathbb{R}$-measurable function on $[a, b]$. This implies that $\sum_{k=1}^{2^n} F(x, y_k)(y_k - y_{k-1})$ is a $\mathfrak{B}_\mathbb{R}$-measurable function on $[a, b]$ for each $n \in bbn$. This implies that $\lim_{n \to \infty} \sum_{k=1}^{2^n} F(x, y_k)(y_k - y_{k-1})$ is a $\mathfrak{B}_\mathbb{R}$-measurable function on $[a, b]$. This shows that f is a $\mathfrak{B}_\mathbb{R}$-measurable function on $[a, b]$. ∎

Prob. 4.23. Let $(X, \mathfrak{A}, \mu)$ be a measure space and let f be an extended real-valued $\mathfrak{A}$-measurable function on X. For $\lambda \in \mathbb{R}$, let $E_\lambda = \{x \in X : f(x) > \lambda\}$ and define a function φ on $\mathbb{R}$ by setting $\varphi(\lambda) = \mu(E_\lambda)$ for $\lambda \in \mathbb{R}$.
(a) Show that φ is a nonnegative extended real-valued decreasing function on $\mathbb{R}$.
(b) Show that if $\mu(X) = M > 0$ then φ is bounded above by M on $\mathbb{R}$.
(c) Show that if $\varphi(\lambda_0) < \infty$ for some $\lambda_0 \in \mathbb{R}$, then φ is right-continuous on $[\lambda_0, \infty)$.
(d) Show that if $\varphi(\lambda_0) < \infty$ for some $\lambda_0 \in \mathbb{R}$, there may be $s \in (s_0, \infty)$ at which φ is not left-continuous.

Proof. 1. Clearly φ is a nonnegative extended real-valued function on $\mathbb{R}$. To show that φ is decreasing on $\mathbb{R}$, let $\lambda', \lambda'' \in \mathbb{R}$ and $\lambda' < \lambda''$. Then $E_{\lambda'} \supset E_{\lambda''}$ so that $\mu(E_{\lambda'}) \geq \mu(E_{\lambda''})$, that is, $\varphi(\lambda') \geq \varphi(\lambda'')$.

2. If $\mu(X) = M > 0$ then since $E_\lambda \subset X$ we have $\varphi(\lambda) = \mu(E_\lambda) \leq \mu(X) = M$.

3. Suppose $\varphi(\lambda_0) < \infty$ for some $\lambda_0 \in \mathbb{R}$. Then since φ is decreasing on $\mathbb{R}$, φ is real-valued on $[\lambda_0, \infty)$. Let us show that φ is right-continuous on $[\lambda_0, \infty)$, that is, φ is right-continuous at every $\lambda \in [\lambda_0, \infty)$. Since φ is decreasing on $\mathbb{R}$, it suffices to show that for every decreasing sequence $(\lambda_n : n \in \mathbb{N})$ in $[\lambda_0, \infty)$ such that $\lim_{n \to \infty} \lambda_n = \lambda$ we have $\lim_{n \to \infty} \varphi(\lambda_n) = \varphi(\lambda)$.

Now if $(\lambda_n : n \in \mathbb{N})$ is a decreasing sequence then $(E_{\lambda_n} : n \in \mathbb{N})$ is an increasing sequence in $\mathfrak{A}$ and hence $\lim_{n \to \infty} E_{\lambda_n} = \bigcup_{n \in \mathbb{N}} E_{\lambda_n}$. Then $\lambda < \lambda_n$ implies $E_\lambda \supset E_{\lambda_n}$ for every $n \in \mathbb{N}$ and thus we have $E_\lambda \supset \bigcup_{n \in \mathbb{N}} E_{\lambda_n}$. On the other hand, if $x \in E_\lambda$ then $f(x) > \lambda$. Since $\lambda_n \downarrow \lambda$, we have $f(x) > \lambda_n$ for sufficiently large $n \in \mathbb{N}$. For such $n \in \mathbb{N}$, we have $f(x) > \lambda_n$ and thus $x \in E_{\lambda_n}$. This shows that every $x \in E_\lambda$ is contained in some

E_{λ_n} so that $E_\lambda \subset \bigcup_{n\in\mathbb{N}} E_{\lambda_n}$. Therefore we have

$$E_\lambda = \bigcup_{n\in\mathbb{N}} E_{\lambda_n}.$$

Then we have

$$\mu(E_\lambda) = \mu\Big(\bigcup_{n\in\mathbb{N}} E_{\lambda_n}\Big) = \mu\Big(\lim_{n\to\infty} E_{\lambda_n}\Big) = \lim_{n\to\infty} \mu(E_{\lambda_n}),$$

that is, $\varphi(\lambda) = \lim_{n\to\infty} \varphi(\lambda_n)$.

4. Suppose $\varphi(\lambda_0) < \infty$ for some $\lambda_0 \in \mathbb{R}$. Let us show that there may be $s \in (s_0, \infty)$ at which φ is not left-continuous by constructing an example.

Consider a measure space $(\mathbb{R}, \mathfrak{A}, \mu)$ where $\mathfrak{A} = \{\emptyset, (-\infty, 0), [0, \infty).\mathbb{R}\}$ and $\mu(\emptyset) = 0$, $\mu\big((-\infty, 0)\big) = 1$, $\mu\big([0, \infty)\big) = 1$, and $\mu(\mathbb{R}) = 2$. Let f be a real-valued function on $\mathbb{R}$ defined by setting

$$f(x) = \begin{cases} 1 & \text{for } x \in (-\infty, 0), \\ 2 & \text{for } x \in [0, \infty). \end{cases}$$

Then f is a $\mathfrak{A}$-measurable function on $\mathbb{R}$. For $\lambda \in \mathbb{R}$, let $E_\lambda = \{x \in X : f(x) > \lambda\}$ and define $\varphi(\lambda) = \mu(E_\lambda)$ for $\lambda \in \mathbb{R}$. Then we have

$$\begin{cases} \lambda \in (-\infty, 1) & \Rightarrow \; E_\lambda = \mathbb{R} & \Rightarrow \; \mu(E_\lambda) = 2 \Rightarrow \; \varphi(\lambda) = 2, \\ \lambda \in [1, 2) & \Rightarrow \; E_\lambda = [0, \infty) & \Rightarrow \; \mu(E_\lambda) = 1 \Rightarrow \; \varphi(\lambda) = 1, \\ \lambda \in [2, \infty) & \Rightarrow \; E_\lambda = \emptyset & \Rightarrow \; \mu(E_\lambda) = 0 \Rightarrow \; \varphi(\lambda) = 0. \end{cases}$$

Therefore we have

$$\varphi(\lambda) = \begin{cases} 2 & \text{for } \lambda \in (-\infty, 1), \\ 1 & \text{for } \lambda \in [1, 2), \\ 0 & \text{for } \lambda \in [2, \infty). \end{cases}$$

This shows that φ is not left-continuous at $\lambda = 1$ and at $\lambda = 2$. ∎

Prob. 4.24. Let $(X, \mathfrak{A}, \mu)$ be a measure space and let f be an extended real-valued $\mathfrak{A}$-measurable function on X. For $\lambda \in \mathbb{R}$, let $F_\lambda = \{x \in X : f(x) \geq \lambda\}$ and define a function ψ on $\mathbb{R}$ by setting $\psi(\lambda) = \mu(F_\lambda)$ for $\lambda \in \mathbb{R}$.
(a) Show that ψ is a nonnegative extended real-valued decreasing function on $\mathbb{R}$.
(b) Show that if $\mu(X) = M > 0$ then ψ is bounded above by M on $\mathbb{R}$.
(c) Show that if $\psi(\lambda_0) < \infty$ for some $\lambda_0 \in \mathbb{R}$, then ψ is left-continuous on (λ_0, ∞).
(d) Show that if $\psi(\lambda_0) < \infty$ for some $\lambda_0 \in \mathbb{R}$, there may be $s \in [s_0, \infty)$ at which ψ is not right-continuous.

Proof. 1. Clearly ψ is a nonnegative extended real-valued function on $\mathbb{R}$. To show that ψ is decreasing on $\mathbb{R}$, let $\lambda', \lambda'' \in \mathbb{R}$ and $\lambda' < \lambda''$. Then $F_{\lambda'} \supset F_{\lambda''}$ so that $\mu(F_{\lambda'}) \geq \mu(F_{\lambda''})$, that is, $\psi(\lambda') \geq \psi(\lambda'')$.
   2. If $\mu(X) = M > 0$ then since $F_\lambda \subset X$ we have $\psi(\lambda) = \mu(F_\lambda) \leq \mu(X) = M$.
   3. If $\psi(\lambda_0) < \infty$ for some $\lambda_0 \in \mathbb{R}$ then since ψ is decreasing on $\mathbb{R}$, ψ is real-valued on $[\lambda_0, \infty)$. Let us show that ψ is left-continuous at every $\lambda \in (\lambda_0, \infty)$. Since ψ is decreasing on $\mathbb{R}$, it suffices to show that for every increasing sequence $(\lambda_n : n \in \mathbb{N})$ in (λ_0, ∞) such that $\lim_{n\to\infty} \lambda_n = \lambda$ we have $\lim_{n\to\infty} \psi(\lambda_n) = \psi(\lambda)$. Now if $(\lambda_n : n \in \mathbb{N})$ is an increasing sequence then $(F_{\lambda_n} : n \in \mathbb{N})$ is a decreasing sequence in $\mathfrak{A}$ and then we have $\lim_{n\to\infty} F_{\lambda_n} = \bigcap_{n\in\mathbb{N}} F_{\lambda_n}$. Since $\lambda_n < \lambda$ implies $F_{\lambda_n} \supset F_\lambda$ for every $n \in \mathbb{N}$, we have $\bigcap_{n\in\mathbb{N}} F_{\lambda_n} \supset F_\lambda$. On the other hand, if $x \in \bigcap_{n\in\mathbb{N}} F_{\lambda_n}$ then $x \in F_{\lambda_n}$ for every $n \in \mathbb{N}$ and hence $f(x) \geq \lambda_n$ for every $n \in \mathbb{N}$. Then since $\lambda_n \uparrow \lambda$, we have $f(x) \geq \lambda$ and thus $x \in F_\lambda$. This shows $\bigcap_{n\in\mathbb{N}} F_{\lambda_n} \subset F_\lambda$ and therefore we have

$$F_\lambda = \bigcap_{n\in\mathbb{N}} F_{\lambda_n}.$$

Now $(\lambda_n : n \in \mathbb{N})$ is an increasing sequence in (λ_0, ∞), we have $\lambda_n > \lambda_0$ and then $F_{\lambda_n} \subset F_{\lambda_0}$ for every $n \in \mathbb{N}$. We have $\mu(F_{\lambda_0}) = \psi(\lambda_0) < \infty$. Thus the decreasing sequence $(F_{\lambda_n} : n \in \mathbb{N})$ in $\mathfrak{A}$ is contained in the set F_{λ_0} with finite measure μ. Thus we have by Theorem 1.26

$$\mu\left(\lim_{n\to\infty} F_{\lambda_n}\right) = \lim_{n\to\infty} \mu(F_{\lambda_n}).$$

Then we have

$$\mu(F_\lambda) = \mu\left(\bigcap_{n\in\mathbb{N}} F_{\lambda_n}\right) = \mu\left(\lim_{n\to\infty} F_{\lambda_n}\right) = \lim_{n\to\infty} \mu(F_{\lambda_n}),$$

that is, $\psi(\lambda) = \lim_{n\to\infty} \psi(\lambda_n)$. This proves the left-continuity of ψ at $\lambda \in (\lambda_0, \infty)$.
   4. Suppose $\psi(\lambda_0) < \infty$ for some $\lambda_0 \in \mathbb{R}$. Let us show that there may be $s \in [s_0, \infty)$ at which ψ is not right-continuous by constructing an example.
   Consider a measure space $(\mathbb{R}, \mathfrak{A}, \mu)$ where $\mathfrak{A} = \{\emptyset, (-\infty, 0], (0, \infty). \mathbb{R}\}$ and $\mu(\emptyset) = 0$, $\mu\big((-\infty, 0]\big) = 1$, $\mu\big((0, \infty)\big) = 1$, and $\mu(\mathbb{R}) = 2$. Let f be a real-valued function on $\mathbb{R}$ defined by setting

$$f(x) = \begin{cases} 1 & \text{for } x \in (-\infty, 0], \\ 2 & \text{for } x \in (0, \infty). \end{cases}$$

Then f is a $\mathfrak{A}$-measurable function on $\mathbb{R}$. For $\lambda \in \mathbb{R}$, let $F_\lambda = \{x \in X : f(x) \geq \lambda\}$ and define $\psi(\lambda) = \mu(F_\lambda)$ for $\lambda \in \mathbb{R}$. Then we have

$$\begin{cases} \lambda \in (-\infty, 1] \;\Rightarrow\; F_\lambda = \mathbb{R} & \Rightarrow\; \mu(F_\lambda) = 2 \Rightarrow\; \psi(\lambda) = 2, \\ \lambda \in (1, 2] & \Rightarrow\; F_\lambda = (0, \infty) \;\Rightarrow\; \mu(F_\lambda) = 1 \Rightarrow\; \psi(\lambda) = 1, \\ \lambda \in (2, \infty) & \Rightarrow\; F_\lambda = \emptyset & \Rightarrow\; \mu(F_\lambda) = 0 \Rightarrow\; \psi(\lambda) = 0. \end{cases}$$

Therefore we have

$$\psi(\lambda) = \begin{cases} 2 & \text{for } \lambda \in (-\infty, 1], \\ 1 & \text{for } \lambda \in (1, 2], \\ 0 & \text{for } \lambda \in (2, \infty). \end{cases}$$

This shows that ψ is not right-continuous at $\lambda = 1$ and at $\lambda = 2$. ∎

Prob. 4.25. Let $(X, \mathfrak{A}, \mu)$ be a measure space. Let $(f_n : n \in \mathbb{N})$ be an increasing sequence of extended real-valued $\mathfrak{A}$-measurable functions on X and let $f = \lim_{n\to\infty} f_n$ on X. For $\lambda \in \mathbb{R}$, let $E_n = \{X : f_n > \lambda\}$ for $n \in \mathbb{N}$ and let $E = \{X : f > \lambda\}$. Show that $\lim_{n\to\infty} E_n = E$ and $\lim_{n\to\infty} \mu(E_n) = \mu(E)$.

Proof. 1. Since $(f_n : n \in \mathbb{N})$ is an increasing sequence of extended real-valued $\mathfrak{A}$-measurable functions on X, $(E_n : n \in \mathbb{N})$ is an increasing sequence of sets in $\mathfrak{A}$ and thus we have $\lim_{n\to\infty} E_n = \bigcup_{n\in\mathbb{N}} E_n$.

2. Let us show that $\lim_{n\to\infty} E_n = E$. Let $x \in \lim_{n\to\infty} E_n = \bigcup_{n\in\mathbb{N}} E_n$. Then $x \in E_n$ for some $n \in \mathbb{N}$ so that $f_n(x) > \lambda$. Then since $f_n(x) \uparrow f(x)$, we have $f(x) > \lambda$. Thus $x \in E$. This shows that $\lim_{n\to\infty} E_n \subset E$. Conversely if $x \in E$, then $f(x) > \lambda$. Since $f_n(x) \uparrow f(x)$ there exists $n \in \mathbb{N}$ such that $f_n(x) > \lambda$. Then $x \in E_n \subset \bigcup_{n\in\mathbb{N}} E_n = \lim_{n\to\infty} E_n$. This shows that $E \subset \lim_{n\to\infty} E_n$. This completes the proof that $\lim_{n\to\infty} E_n = E$.

3. Since $(E_n : n \in \mathbb{N})$ is an increasing sequence of sets in $\mathfrak{A}$, we have

$$\lim_{n\to\infty} \mu(E_n) = \mu\left(\lim_{n\to\infty} E_n \right) = \mu(E). \quad \blacksquare$$

Prob. 4.26. Let $(X, \mathfrak{A}, \mu)$ be a measure space. Let $(f_n : n \in \mathbb{N})$ be an increasing sequence of extended real-valued $\mathfrak{A}$-measurable functions on X and let $f = \lim_{n\to\infty} f_n$ on X. For $\lambda \in \mathbb{R}$, let $F_n = \{X : f_n \geq \lambda\}$ for $n \in \mathbb{N}$ and let $F = \{X : f \geq \lambda\}$. Construct $(f_n : n \in \mathbb{N})$ such that $\lim_{n\to\infty} F_n \neq F$ and $\lim_{n\to\infty} \mu(F_n) < \mu(F)$.

Proof. Let our measure space $(X, \mathfrak{A}, \mu)$ be such that $X \neq \emptyset$ and $\mu(X) > 0$. Let $\lambda \in \mathbb{R}$ and define function f and a sequence of functions $(f_n : n \in \mathbb{N})$ on X by setting

$$f(x) = \lambda \qquad \text{for } x \in X,$$
$$f_n(x) = \lambda - \tfrac{1}{n} \quad \text{for } x \in X, \text{ for } n \in \mathbb{N}.$$

Then $(f_n : n \in \mathbb{N})$ is an increasing sequence of real-valued $\mathfrak{A}$-measurable functions on X and $f = \lim_{n\to\infty} f_n$ on X.

Now we have $F_n = \{X : f_n \geq \lambda\} = \emptyset$ for $n \in \mathbb{N}$ and $F = \{X : f \geq \lambda\} = X$. Thus we have $\lim_{n\to\infty} F_n = \emptyset \neq X = F$ and $\lim_{n\to\infty} \mu(F_n) = \lim_{n\to\infty} \mu(\emptyset) = 0 < \mu(X) = \mu(F)$. $\quad \blacksquare$

Prob. 4.27. Let us call a sequence $(a_n : n \in \mathbb{N})$ of extended real numbers eventually monotone if there exists $n_0 \in \mathbb{N}$ such that $(a_n : n > n_0)$ is monotone, that is, either increasing or decreasing.

Let $(X, \mathfrak{A}, \mu)$ be a measure space. Let $(f_n : n \in \mathbb{N})$ be a sequence of extended real-valued $\cdot$ $\mathfrak{A}$-measurable functions on X such that $(f_n(x) : n \in \mathbb{N})$ is eventually monotone for every $x \in X$. Let $f = \lim_{n \to \infty} f_n$ on X. For $\alpha \in \mathbb{R}$ let $E_n = \{X : f_n > \alpha\}$ for $n \in \mathbb{N}$ and let $E = \{X : f > \alpha\}$.

(a) Show that $\lim_{n \to \infty} E_n = E$.

(b) Show that if $(X, \mathfrak{A}, \mu)$ is a finite measure space then $\lim_{n \to \infty} \mu(E_n) = \mu(E)$.

Proof. Observe that since for every $x \in X$ the sequence $(f_n(x) : n \in \mathbb{N})$ is an eventually monotone sequence $\lim_{n \to \infty} f_n(x)$ exists in $\overline{\mathbb{R}}$.

1. Let us consider $\liminf_{n \to \infty} E_n$ and $\limsup_{n \to \infty} E_n$ for the sequence $(E_n : n \in \mathbb{N})$ in $\mathfrak{A}$. Now $x \in \liminf_{n \to \infty} E_n$ if and only if $x \in E_n$ for all but finitely many $n \in \mathbb{N}$ and $x \in \lim_{n \to \infty} E_n$ for infinitely many $n \in \mathbb{N}$. Since the sequence $(f_n(x) : n \in \mathbb{N})$ is eventually monotone, $f_n(x) > \alpha$ for infinitely many $n \in \mathbb{R}$ if and only if $f_n(x) > \alpha$ for all but finitely many $n \in \mathbb{R}$. Thus $\liminf_{n \to \infty} E_n = \limsup_{n \to \infty} E_n$ and therefore $\lim_{n \to \infty} E_n$ exists and we have

$$\lim_{n \to \infty} E_n = \liminf_{n \to \infty} E_n = \limsup_{n \to \infty} E_n.$$

Let us show that $E = \lim_{n \to \infty} E_n$. Now since $(f_n(x) : n \in \mathbb{N})$ is eventually monotone, we have

$$\lim_{n \to \infty} f_n(x) > \alpha \Leftrightarrow f_n(x) > \alpha \quad \text{for all but finitely many } n \in \mathbb{N}.$$

This implies that $E = \liminf_{n \to \infty} E_n = \lim_{n \to \infty} E_n$.

2. If $(X, \mathfrak{A}, \mu)$ is a finite measure space then according to Theorem 1.28, $\lim_{n \to \infty} E_n = E$ implies

$$\mu(E) = \mu\left(\lim_{n \to \infty} E_n\right) = \lim_{n \to \infty} \mu(E_n).$$

This completes the proof. ∎

Prob. 4.28. Let $(f_n : n \in \mathbb{N})$ and f be extended real-valued functions defined on a set D such that $\lim_{n\to\infty} f_n(x) = f(x)$ for every $x \in D$.

(a) For $\alpha \in \mathbb{R}$, let $E = \{x \in D : f(x) > \alpha\}$ and $E_n = \{x \in E : f_n(x) > \alpha\}$ for $n \in \mathbb{N}$. Show that $\lim_{n\to\infty} E_n$ exists and moreover $\lim_{n\to\infty} E_n = E$.

(b) For $\alpha \in \mathbb{R}$, let $E = \{x \in D : f(x) < \alpha\}$ and $E_n = \{x \in E : f_n(x) < \alpha\}$ for $n \in \mathbb{N}$. Show that $\lim_{n\to\infty} E_n$ exists and moreover $\lim_{n\to\infty} E_n = E$.

Proof. 1. Let us prove (a). Let us show that $\lim_{n\to\infty} E_n$ exists, that is, $\liminf_{n\to\infty} E_n = \limsup_{n\to\infty} E_n$. Now since we have $\liminf_{n\to\infty} E_n \subset \limsup_{n\to\infty} E_n$, it remains to show that $\limsup_{n\to\infty} E_n \subset \liminf_{n\to\infty} E_n$. For this purpose let $x \in \limsup_{n\to\infty} E_n$. Then $x \in E_n$ for infinitely many $n \in \mathbb{N}$ and thus $f_n(x) > \alpha$ for infinitely many $n \in \mathbb{N}$. Now $x \in E_n \subset E$ implies that $f(x) > \alpha$ and thus $\lim_{n\to\infty} f_n(x) = f(x) > \alpha$. This implies that $f_n(x) > \alpha$ for all but finitely many $n \in \mathbb{N}$ and thus $x \in E_n$ for all but finitely many $n \in \mathbb{N}$. This shows that $x \in \liminf_{n\to\infty} E_n$ and then $\limsup_{n\to\infty} E_n \subset \liminf_{n\to\infty} E_n$.

We proved above that $\lim_{n\to\infty} E_n$ exists. Let us show that $\lim_{n\to\infty} E_n = E$. Now we have $\lim_{n\to\infty} E_n = \liminf_{n\to\infty} E_n = \limsup_{n\to\infty} E_n$. If $x \in \lim_{n\to\infty} E_n$, then $x \in \limsup_{n\to\infty} E_n$ so that $x \in E_n$ for infinitely many $n \in \mathbb{N}$. But $E_n \subset E$. Thus $x \in E$. This shows that $\lim_{n\to\infty} E_n \subset E$.

Conversely, let $x \in E$. Then we have $f(x) > \alpha$. Since $f(x) = \lim_{n\to\infty} f_n(x)$, we have $\lim_{n\to\infty} f_n(x) > \alpha$. This implies that we have $f_n(x) > \alpha$ for all but finitely many $n \in \mathbb{N}$ and hence $x \in E_n$ for all but finitely many $n \in \mathbb{N}$. Thus $x \in \liminf_{n\to\infty} E_n = \lim_{n\to\infty} E_n$. This shows that $E \subset \lim_{n\to\infty} E_n$. This proves $\lim_{n\to\infty} E_n = E$.

2. Let us show that (a) implies (b). Observe that

$$E = \left\{x \in D : f(x) < \alpha\right\} = \left\{x \in D : (-f)(x) > -\alpha\right\} := G,$$

and for $n \in \mathbb{N}$

$$E_n = \left\{x \in F : f_n(x) < \alpha\right\} = \left\{x \in G : (-f_n)(x) > -\alpha\right\} := G_n.$$

Now $\lim_{n\to\infty} f_n(x) = f(x)$ implies $\lim_{n\to\infty} (-f_n)(x) = (-f)(x)$ for $x \in D$. Thus (a) implies that $\lim_{n\to\infty} G_n$ exists and $\lim_{n\to\infty} G_n = G$, that is, $\lim_{n\to\infty} E_n$ exists and $\lim_{n\to\infty} E_n = E$. ∎

Prob. 4.29. Let $(X, \mathfrak{A}, \mu)$ be a measure space. Let $(f_n : n \in \mathbb{N})$ and f be extended real-valued $\mathfrak{A}$-measurable functions on a set $D \in \mathfrak{A}$ such that $\lim\limits_{n \to \infty} f_n = f$ on D. Then for every $\alpha \in \mathbb{R}$ we have

(1) $\qquad \mu\{D : f > \alpha\} \leq \liminf\limits_{n \to \infty} \mu\{D : f_n > \alpha\}.$

(2) $\qquad \mu\{D : f < \alpha\} \leq \liminf\limits_{n \to \infty} \mu\{D : f_n < \alpha\}.$

Proof. 1. Let us prove (1). Let $E = \{D : f > \alpha\}$ and $E_n = \{E : f_n > \alpha\}$ for $n \in \mathbb{N}$. Then we have $E = \lim\limits_{n \to \infty} E_n$ by Prob. 4.28. Then by Theorem 1.28 we have

$$\mu(E) = \mu\big(\lim_{n \to \infty} E_n\big) \leq \liminf_{n \to \infty} \mu(E_n).$$

Thus we have

$$\mu\{D : f > \alpha\} \leq \liminf_{n \to \infty} \mu\{E : f_n > \alpha\} \leq \liminf_{n \to \infty} \mu\{D : f_n > \alpha\},$$

since $E \subset D$. This proves (1).

2. Let us show that (2) follows from (1). Observe that $\lim\limits_{n \to \infty} f_n(x) = f(x)$ implies that $\lim\limits_{n \to \infty} (-f_n)(x) = (-f)(x)$ for $x \in D$. Then by (1) we have for every $-\alpha \in \mathbb{R}$

$$\mu\{D : -f > -\alpha\} \leq \liminf_{n \to \infty} \mu\{D : -f_n > -\alpha\}.$$

But $\{D : -f > -\alpha\} = \{D : f < \alpha\}$ and $\{D : -f_n > -\alpha\} = \{D : f_n < \alpha\}$. Therefore we have

$$\mu\{D : f < \alpha\} \leq \liminf_{n \to \infty} \mu\{D : f_n < \alpha\}.$$

This proves (2). ∎

§5 Completion of Measure Space

Prob. 5.1. A measure space constructed by means of an outer measure is always a complete measure space. (See Theorem 2.9.) Use this fact to construct a complete extension of an arbitrary measure space.

Proof. 1. Let $(X, \mathfrak{A}, \mu)$ be an arbitrary measure space. Since $\mathfrak{A}$ is a σ-algebra of subsets of X, we have $\emptyset, X \in \mathfrak{A}$. This makes $\mathfrak{A}$ a covering class for X in the sense of Definition 2.20. Since μ is a measure on $\mathfrak{A}$, μ is a set function on $\mathfrak{A}$ satisfying the condition $\mu(A) \in [0, \infty]$ for every $A \in \mathfrak{A}$ and the condition $\mu(\emptyset) = 0$. Let us define a set function ν^* on $\mathfrak{P}(X)$ by setting for every $E \in \mathfrak{P}(X)$

$$(1) \qquad \nu^*(E) = \inf \left\{ \sum_{n \in \mathbb{N}} \mu(A_n) : (A_n : n \in \mathbb{N}) \subset \mathfrak{A}, \bigcup_{n \in \mathbb{N}} A_n \supset E \right\}.$$

According to Theorem 2.21, ν^* is an outer measure on X. According to Theorem 2.8, if we let $\mathfrak{F}$ be the collection of all $E \in \mathfrak{P}(X)$ that satisfy the ν^* measurability condition

$$(2) \qquad \nu^*(B) = \nu^*(B \cap E) + \nu^*(B \cap E^c) \quad \text{for every } B \in \mathfrak{P}(X),$$

then $\mathfrak{F}$ is a σ-algebra of subsets of X. According to Theorem 2.9, if we write ν for the restriction of ν^* to $\mathfrak{F} \subset \mathfrak{P}(X)$ then $(X, \mathfrak{F}, \nu)$ is a complete measure space.

To show that $(X, \mathfrak{F}, \nu)$ is a complete extension of $(X, \mathfrak{A}, \mu)$, we verify that $\mathfrak{A} \subset \mathfrak{F}$ and $\nu = \mu$ on $\mathfrak{A}$.

2. To show that $\mathfrak{A} \subset \mathfrak{F}$, we show that every $A \in \mathfrak{A}$ satisfies the ν^* measurability condition

$$(3) \qquad \nu^*(B) = \nu^*(B \cap A) + \nu^*(B \cap A^c) \quad \text{for every } B \in \mathfrak{P}(X).$$

Let $B \in \mathfrak{P}(X)$. By the definition of $\nu^*(B)$ by (1) as an infimum, for every $\varepsilon > 0$ there exists a sequence $(A_n : n \in \mathbb{N})$ in $\mathfrak{A}$ such that $\bigcup_{n \in \mathbb{N}} A_n \supset B$ and

$$(4) \qquad \sum_{n \in \mathbb{N}} \mu(A_n) \leq \nu^*(B) + \varepsilon.$$

By the additivity of the measure μ on the σ-algebra $\mathfrak{A}$, we have

$$(5) \qquad \mu(A_n) = \mu(A_n \cap A) = \mu(A_n \cap A^c).$$

Substituting (5) in (4) we have

$$(6) \qquad \sum_{n \in \mathbb{N}} \mu(A_n \cap A) + \sum_{n \in \mathbb{N}} \mu(A_n \cap A^c) \leq \nu^*(B) + \varepsilon.$$

Since $B \subset \bigcup_{n \in \mathbb{N}} A_n$, we have

$$B \cap A \subset \left(\bigcup_{n \in \mathbb{N}} A_n \right) \cap A = \bigcup_{n \in \mathbb{N}} (A_n \cap A),$$

$$B \cap A^c \subset \left(\bigcup_{n \in \mathbb{N}} A_n \right) \cap A^c = \bigcup_{n \in \mathbb{N}} (A_n \cap A^c).$$

Thus $(A, A_n : n \in \mathbb{N})$ is a sequence in $\mathfrak{A}$ that covers $B \cap A$ and $(A^c, A_n : n \in \mathbb{N})$ is a sequence in $\mathfrak{A}$ that covers $B \cap A^c$. Then the definition of ν^* by (1) implies

$$\nu^*(B \cap A) \leq \sum_{n \in \mathbb{N}} \mu(A_n \cap A),$$

$$\nu^*(B \cap A^c) \leq \sum_{n \in \mathbb{N}} \mu(A_n \cap A^c).$$

Substituting these in (6), we have

$$\nu^*(B \cap A) + \nu^*(B \cap A^c) \leq \nu^*(B) + \varepsilon.$$

Since this holds for every $\varepsilon > 0$, we have

$$\nu^*(B \cap A) + \nu^*(B \cap A^c) \leq \nu^*(B).$$

The reverse inequality holds by the subadditivity of the outer measure ν^*. This proves (3).

3. Let us prove that $\nu^* = \mu$ on $\mathfrak{A}$. Let $A \in \mathfrak{A}$. Then by the definition of ν^* by (1), we have

$$(7) \qquad \nu^*(A) = \inf \Big\{ \sum_{n \in \mathbb{N}} \mu(A_n) : (A_n : n \in \mathbb{N}) \subset \mathfrak{A}, \bigcup_{n \in \mathbb{N}} A_n \supset A \Big\}.$$

Now if $\bigcup_{n \in \mathbb{N}} A_n \supset A$, then by the monotonicity and countable subadditivity of the measure μ on $\mathfrak{A}$ we have

$$\mu(A) \leq \mu\Big(\bigcup_{n \in \mathbb{N}} A_n\Big) \leq \sum_{n \in \mathbb{N}} \mu(A_n).$$

This implies

$$\mu(A) \leq \inf \Big\{ \sum_{n \in \mathbb{N}} \mu(A_n) : (A_n : n \in \mathbb{N}) \subset \mathfrak{A}, \bigcup_{n \in \mathbb{N}} A_n \supset A \Big\}.$$

On the other hand, if we select the sequence $(A_n : n \in \mathbb{N}) = (A, \emptyset, \emptyset, \emptyset, \ldots)$ then we have $\sum_{n \in \mathbb{N}} \mu(A_n) = \mu(A)$. Thus

$$\mu(A) = \inf \Big\{ \sum_{n \in \mathbb{N}} \mu(A_n) : (A_n : n \in \mathbb{N}) \subset \mathfrak{A}, \bigcup_{n \in \mathbb{N}} A_n \supset A \Big\} = \nu^*(A).$$

This shows that $\nu^* = \mu$ on $\mathfrak{A}$. ∎

Prob. 5.2. Let $\{(X, \mathfrak{F}_\alpha, \nu_\alpha) : \alpha \in A\}$ be the collection of all complete extensions of a measure space $(X, \mathfrak{A}, \mu)$. Construct the completion of $(X, \mathfrak{A}, \mu)$ from this collection.

Proof. Let $(X, \mathfrak{A}, \mu)$ be an arbitrary measure space. A complete extension of $(X, \mathfrak{A}, \mu)$ is a complete measure space $(X, \mathfrak{F}, \nu)$ such that $\mathfrak{F} \supset \mathfrak{A}$ and $\nu = \mu$ on $\mathfrak{A}$. A completion of $(X, \mathfrak{A}, \mu)$ is a complete extension $(X, \mathfrak{F}_0, \nu_0)$ of $(X, \mathfrak{A}, \mu)$ such that for any complete extension $(X, \mathfrak{F}, \nu)$ of $(X, \mathfrak{A}, \mu)$ we have $\mathfrak{F} \supset \mathfrak{F}_0$ and $\nu = \nu_0$ on $\mathfrak{F}_0$. Theorem 5.4 shows that a completion of $(X, \mathfrak{A}, \mu)$ exists. Let us construct a completion of $(X, \mathfrak{A}, \mu)$ on the basis of the fact that it exists.

Let $\{(X, \mathfrak{F}_\alpha, \nu_\alpha) : \alpha \in A\}$ be the collection of all complete extensions of a measure space $(X, \mathfrak{A}, \mu)$. Since a completion $(X, \mathfrak{F}_0, \nu_0)$ of $(X, \mathfrak{A}, \mu)$ is a particular case of complete extensions of $(X, \mathfrak{A}, \mu)$, $(X, \mathfrak{F}_0, \nu_0)$ is in the collection $\{(X, \mathfrak{F}_\alpha, \nu_\alpha) : \alpha \in A\}$. Then since $\mathfrak{A}_\alpha \supset \mathfrak{F}_0$ for every $\alpha \in A$, we have

$$\mathfrak{F}_0 = \bigcap_{\alpha \in A} \mathfrak{F}_\alpha.$$

Since $\nu_\alpha = \nu_0$ on $\mathfrak{F}_0$ for every $\alpha \in A$, pick $\alpha_1 \in A$ arbitrarily and then we have $\nu_{\alpha_1} = \nu_0$ on $\mathfrak{F}_0$. Define

$$\nu_0 = \nu_{\alpha_1} \quad \text{on } \mathfrak{F}_0.$$

Then $(X, \mathfrak{F}_0, \nu_0)$ is a completion of $(X, \mathfrak{A}, \mu)$. ∎

Prob. 5.3. Let $(X, \overline{\mathfrak{A}}, \mu)$ be the completion of a measure space $(X, \mathfrak{A}, \mu)$. Show that every subset of a null set in $(X, \mathfrak{A}, \mu)$ is $\overline{\mathfrak{A}}$-measurable.

Proof. According to Theorem 5.4, the σ-algebra $\overline{\mathfrak{A}}$ is constructed by setting

$$\overline{\mathfrak{A}} = \sigma(\mathfrak{A} \cup \overline{\mathfrak{N}}),$$

where $\mathfrak{N}$ is the collection of all null sets in $(X, \mathfrak{A}, \mu)$ and $\overline{\mathfrak{N}}$ is the collection of all subsets of members of $\mathfrak{N}$. Then every subset of a null set in $(X, \mathfrak{A}, \mu)$ is certainly in $\overline{\mathfrak{A}}$. ∎

§6 Convergence a.e. and Convergence in Measure

Assumptions. For problems in this section, unless otherwise stated, it is assumed that $(f_n : n \in \mathbb{N})$ is a sequence of extended real-valued $\mathfrak{A}$-measurable functions on a set $D \in \mathfrak{A}$ in a measure space $(X, \mathfrak{A}, \mu)$ and f is a real-valued $\mathfrak{A}$-measurable function on D. This is to ensure that $f_n - f$ is defined everywhere on D.

Prob. 6.1. Let $(X, \mathfrak{A}, \mu)$ be a measure space and let $D \in \mathfrak{A}$. Let $(f_n : n \in \mathbb{N})$ and f be extended real-valued $\mathfrak{A}$-measurable functions on D and let f be a real-valued on D. For an arbitrary $\varepsilon > 0$ let $D_{\varepsilon,n} = \{D : |f_n - f| \geq \varepsilon\}$ for $n \in \mathbb{N}$. Then $\lim_{n \to \infty} f_n = f$ a.e. on D if and only if $\mu\big(\limsup_{n \to \infty} D_{\varepsilon,n}\big) = 0$ for every $\varepsilon > 0$.

Proof. For brevity let us write $E_\varepsilon = \limsup_{n \to \infty} D_{\varepsilon,n}$. Recall that E_ε consists of every $x \in D$ such that $x \in D_{\varepsilon,n}$ for infinitely many $n \in \mathbb{N}$ according to Lemma 1.7.

1. Suppose $\lim_{n \to \infty} f_n = f$ a.e. on D. Let $\varepsilon > 0$. To show that $\mu(E_\varepsilon) = 0$, assume the contrary, that is, $\mu(E_\varepsilon) > 0$. Now if $x \in E_\varepsilon$ then $x \in D_{\varepsilon,n}$ for infinitely many $n \in \mathbb{N}$, that is, $|f_n(x) - f(x)| \geq \varepsilon$ for infinitely many $n \in \mathbb{N}$ and thus $\lim_{n \to \infty} f_n(x) = f(x)$ does not hold. Then $\mu(E_\varepsilon) > 0$ contradicts the assumption that $\lim_{n \to \infty} f_n = f$ a.e. on D.

2. Conversely suppose $\mu(E_\varepsilon) = 0$ for every $\varepsilon > 0$. Let us show that $\lim_{n \to \infty} f_n = f$ a.e. on D. Now for every $m \in \mathbb{N}$ we have $\mu(E_{\frac{1}{m}}) = 0$ by our assumption. Consider $\bigcup_{m \in \mathbb{N}} E_{\frac{1}{m}}$. We have $\mu\big(\bigcup_{m \in \mathbb{N}} E_{\frac{1}{m}}\big) \leq \sum_{m \in \mathbb{N}} \mu(E_{\frac{1}{m}}) = 0$, Consider $D \setminus \bigcup_{m \in \mathbb{N}} E_{\frac{1}{m}}$. Let $x \in D \setminus \bigcup_{m \in \mathbb{N}} E_{\frac{1}{m}}$. Then $x \notin E_{\frac{1}{m}}$ for every $m \in \mathbb{N}$. Thus for every $mn \in \mathbb{N}$, $x \in D_{\frac{1}{m},n}$ for at most finitely many $n \in \mathbb{N}$ that is, $|f_n(x) - f(x)| < \frac{1}{m}$ for all but finitely many $n \in \mathbb{N}$. Thus we have $\lim_{n \to \infty} f_n(x) = f(x)$ for $x \in D \setminus \bigcup_{m \in \mathbb{N}} E_{\frac{1}{m}}$. Since $\bigcup_{m \in \mathbb{N}} E_{\frac{1}{m}}$ is a null set in $(X, \mathfrak{A}, \mu)$, we have $\lim_{n \to \infty} f_n = f$ a.e. on D. ∎

Prob. 6.2. Let $(X, \mathfrak{A}, \mu)$ be a measure space. Let $(f_n : n \in \mathbb{N})$ be a sequence of extended real-valued $\mathfrak{A}$-measurable functions on X and let f be an extended real-valued $\mathfrak{A}$-measurable function which is finite a.e. on X. Suppose $\lim_{n\to\infty} f_n = f$ a.e. on X. Let $\alpha \in [0, \mu(X))$ be arbitrarily chosen. Show that for every $\varepsilon > 0$ there exists $N \in \mathbb{N}$ such that $\mu\{X : |f_n - f| < \varepsilon\} \geq \alpha$ for $n \geq N$.

Proof. Let $\varepsilon > 0$ and let us write for brevity $F_{\varepsilon,n} = \{X : |f_n - f| < \varepsilon\}$ for $n \in \mathbb{N}$. We are to show that for every $\alpha \in [0, \mu(X))$ there exists $N \in \mathbb{N}$ such that $\mu(F_{\varepsilon,n}) \geq \alpha$ for $n \geq N$.

Now by Lemma 1.7 we have

$$\liminf_{n\to\infty} F_{\varepsilon,n} = \{x \in X : x \in F_{\varepsilon,n} \quad \text{for all but finitely many } n \in \mathbb{N}\},$$

and then

$$\left(\liminf_{n\to\infty} F_{\varepsilon,n}\right)^c = \{x \in X : x \notin F_{\varepsilon,n} \quad \text{for infinitely many } n \in \mathbb{N}\}$$

$$= \{x \in X : |f_n(x) - f(x)| \geq \varepsilon \quad \text{for infinitely many } n \in \mathbb{N}\}$$

$$\subset \{x \in X : \lim_{n\to\infty} f_n(x) \neq f(x) \quad \text{or} \quad \lim_{n\to\infty} f_n(x) \text{ does not exist}\}.$$

Now since $\lim_{n\to\infty} f_n = f$ a.e. on X, the measure of the last set is equal to 0 and thus we have $\mu\left(\left(\liminf_{n\to\infty} F_{\varepsilon,n}\right)^c\right) = 0$ and consequently

$$\mu\left(\liminf_{n\to\infty} F_{\varepsilon,n}\right) = \mu(X).$$

Now $\liminf_{n\to\infty} F_{\varepsilon,n} = \lim_{n\to\infty} \bigcap_{k\geq n} F_{\varepsilon,k}$ and $\left(\bigcap_{k\geq n} F_{\varepsilon,k} : n \in \mathbb{N}\right)$ is an increasing sequence. Thus we have

$$\mu(X) = \mu\left(\liminf_{n\to\infty} F_{\varepsilon,n}\right) = \mu\left(\lim_{n\to\infty} \bigcap_{k\geq n} F_{\varepsilon,k}\right) = \lim_{n\to\infty} \mu\left(\bigcap_{k\geq n} F_{\varepsilon,k}\right).$$

Then for $\alpha \in [0, \mu(X))$ there exists $N \in \mathbb{N}$ such that $\mu\left(\bigcap_{k\geq n} F_{\varepsilon,k}\right) \geq \alpha$ for $n \geq N$ and consequently $\mu(F_{\varepsilon,n}) \geq \mu\left(\bigcap_{k\geq n} F_{\varepsilon,k}\right) \geq \alpha$ for $n \geq N$. ∎

Prob. 6.3. Let $(f_n : n \in \mathbb{N})$ be a sequence of real-valued continuous functions on $\mathbb{R}$. Let E be the set of all $x \in \mathbb{R}$ such that the sequence $(f_n(x) : n \in \mathbb{N})$ of real numbers converges. Show that E is an $F_{\sigma\delta}$-set, that is, the intersection of countably many F_σ-sets. (Hint: Use the expression $E = \bigcap_{m\in\mathbb{N}} \bigcup_{N\in\mathbb{N}} \bigcap_{p\in\mathbb{N}} \{x \in \mathbb{R} : |f_{N+p}(x) - f_N(x)| \le \frac{1}{m}\}$.)

Proof. For $x \in \mathbb{R}$, the sequence of real numbers $(f_n(x) : n \in \mathbb{N})$ converges if and only if it is a Cauchy sequence, that is, for every $\varepsilon > 0$ there exists $N \in \mathbb{N}$ such that

$$|f_n(x) - f'_n(x)| < \varepsilon \quad \text{for } n, n' \ge N ,$$

or equivalently for every $m \in \mathbb{N}$ there exists $N \in \mathbb{N}$ such that

$$|f_{N+p}(x) - f_N(x)| < \tfrac{1}{m} \quad \text{for every } p \in \mathbb{N} .$$

Thus the set E of all $x \in \mathbb{R}$ such that the sequence $(f_n(x) : n \in \mathbb{N})$ of real numbers converges can be expressed as

$$E = \bigcap_{m\in\mathbb{N}} \bigcup_{N\in\mathbb{N}} \bigcap_{p\in\mathbb{N}} \{x \in \mathbb{R} : |f_{N+p}(x) - f_N(x)| \le \tfrac{1}{m}\}.$$

Now

$$\{x \in \mathbb{R} : |f_{N+p}(x) - f_N(x)| \le \tfrac{1}{m}\} = |f_{N+p} - f_N|^{-1}([0, \tfrac{1}{m}]).$$

Since f_N and f_{N+p} are real-valued continuous functions on $\mathbb{R}$, $f_{N+p} - f_N$ is a real-valued continuous function on $\mathbb{R}$ and then $|f_{N+p} - f_N|$ is a real-valued continuous function on $\mathbb{R}$. Continuity of $|f_{N+p} - f_N|$ implies that the preimage $|f_{N+p} - f_N|^{-1}([0, \tfrac{1}{m}])$ of the closed set $[0, \tfrac{1}{m}]$ in $\mathbb{R}$ is a closed set in $\mathbb{R}$. Let us write $C_{N,p}$ for this closed set. Then

$$E = \bigcap_{m\in\mathbb{N}} \bigcup_{N\in\mathbb{N}} \bigcap_{p\in\mathbb{N}} C_{N,p} .$$

Now $\bigcap_{p\in\mathbb{N}} C_{N,p}$, an intersection of closed sets, is a closed set. Then $\bigcup_{N\in\mathbb{N}} \bigcap_{p\in\mathbb{N}} C_{N,p}$, a countable union of closed sets, is an F_σ-set. Then $\bigcap_{m\in\mathbb{N}} \bigcup_{N\in\mathbb{N}} \bigcap_{p\in\mathbb{N}} C_{N,p}$, a countable intersection of F_σ-sets, is an $F_{\sigma\delta}$-set. ∎

Prob. 6.4. Conditions $1°$ and $2°$ in Theorem 6.7 are sufficient for $(f_n : n \in \mathbb{N})$ to converge to f a.e. on D, but they are not necessary. To show this, consider the following example: In $(\mathbb{R}, \mathfrak{M}_L, \mu_L)$, let $D = (0, 1)$ and let $f_n(x) = 1$ for $x \in (0, \frac{1}{n})$ and $f_n(x) = 0$ for $x \in [\frac{1}{n}, 1)$ for $n \in \mathbb{N}$ and let $f(x) = 0$ for $x \in (0, 1)$. Then $(f_n : n \in \mathbb{N})$ converges to f everywhere on D. Show that for any sequence of positive numbers $(\varepsilon_n : n \in \mathbb{N})$ such that $\lim_{n\to\infty} \varepsilon_n = 0$, we have $\sum_{n\in\mathbb{N}} \mu_L\{x \in (0, 1) : |f_n(x) - f(x)| \ge \varepsilon_n\} = \infty$.

Proof. Since $\lim_{n\to\infty} \varepsilon_n = 0$ there exists $N \in \mathbb{N}$ such that $\varepsilon_n < 1$ for $n \ge N$. Then for $n \ge N$ we have

$$\{x \in (0, 1) : |f_n(x) - f(x)| \ge \varepsilon_n\} = \{x \in (0, 1) : f_n(x) \ge \varepsilon_n\} = (0, \tfrac{1}{n}).$$

Thus we have

$$\mu_L\{x \in (0, 1) : |f_n(x) - f(x)| \ge \varepsilon_n\} = \mu_L(0, \tfrac{1}{n}) = \tfrac{1}{n} \quad \text{for } n \ge N.$$

Then we have

$$\sum_{n\in\mathbb{N}} \mu_L\{x \in (0, 1) : |f_n(x) - f(x)| \ge \varepsilon_n\} \ge \varepsilon_n\} \ge \sum_{n\ge N} \tfrac{1}{n} = \infty. \quad \blacksquare$$

Prob. 6.5. To show that Egoroff's Theorem holds only on sets with finite measures, we have the following example:

Consider $(\mathbb{R}, \mathfrak{M}_L, \mu_L)$. Let $D = [0, \infty)$ with $\mu_L(D) = \infty$. Let $f_n(x) = \frac{x}{n}$ for $x \in D$ for $n \in \mathbb{N}$ and let $f(x) = 0$ for $x \in D$. Then the sequence $(f_n : n \in \mathbb{N})$ converges to f everywhere on D. To show that $(f_n : n \in \mathbb{N})$ does not converge to f almost uniformly on D, we show that for some $\eta > 0$ there does not exist a $\mathfrak{M}_L$-measurable subset E of D with $\mu_L(E) < \eta$ such that $(f_n : n \in \mathbb{N})$ converges to f uniformly on $D \setminus E$. Show that actually for any $\mathfrak{M}_L$-measurable subset E of D with $\mu_L(E) < \infty$, $(f_n : n \in \mathbb{N})$ does not converge to f uniformly on $D \setminus E$.

Proof. Let $E \subset D$, $E \in \mathfrak{M}_L$ and $\mu_L(E) < \infty$. Let us show that $(f_n : n \in \mathbb{N})$ does not converge to f uniformly on $D \setminus E$. Assume the contrary, that is, assume that $(f_n : n \in \mathbb{N})$ converges to f uniformly on $D \setminus E$. Then for every $\varepsilon > 0$ there exists $N \in \mathbb{N}$ such that

$$|f_n(x) - f(x)| < \varepsilon \quad \text{for all } x \in D \setminus E \text{ and } n \geq N,$$

that is,

(1) $$\frac{x}{n} < \varepsilon \quad \text{for all } x \in D \setminus E \text{ and } n \geq N.$$

Now since $\mu_L(D) = \infty$ and $\mu_L(E) < \infty$ we have $\mu_L(D \setminus E) = \infty$ and this implies that $D \setminus E$ is an unbounded subset of $D = [0, \infty)$. Then there exists $x_0 \in D \setminus E$ such that $\frac{x_0}{N} > \varepsilon$. This contradicts (1). This shows that $(f_n : n \in \mathbb{N})$ does not converge to f uniformly on $D \setminus E$. ∎

Prob. 6.6. By definition, $f_n \xrightarrow{\mu} f$ on D if for every $\varepsilon > 0$

$$\lim_{n \to \infty} \mu\{x \in D : |f_n(x) - f(x)| \geq \varepsilon\} = 0.$$

Show that this condition is equivalent to the condition that for every $\varepsilon > 0$

$$\lim_{n \to \infty} \mu\{x \in D : |f_n(x) - f(x)| > \varepsilon\} = 0.$$

Proof. Consider the two conditions:

(1) $\qquad \lim_{n \to \infty} \mu\{x \in D : |f_n(x) - f(x)| \geq \varepsilon\} = 0 \quad$ for every $\varepsilon > 0$.

(2) $\qquad \lim_{n \to \infty} \mu\{x \in D : |f_n(x) - f(x)| > \varepsilon\} = 0 \quad$ for every $\varepsilon > 0$.

Now for every $\varepsilon > 0$ we have $\{D : |f_n - f| > \varepsilon\} \subset \{D : |f_n - f| \geq \varepsilon\}$ and then $\mu\{D : |f_n - f| > \varepsilon\} \leq \mu\{D : |f_n - f| \geq \varepsilon\}$. Thus if we assume (1) then we have

$$\lim_{n \to \infty} \mu\{D : |f_n - f| > \varepsilon\} \leq \lim_{n \to \infty} \mu\{D : |f_n - f| \geq \varepsilon\} = 0,$$

that is, (2) holds.

On the other hand for every $\varepsilon > 0$ we have $\{D : |f_n - f| \geq \varepsilon\} \subset \{D : |f_n - f| > \frac{\varepsilon}{2}\}$ and then $\mu\{D : |f_n - f| \geq \varepsilon\} \leq \mu\{D : |f_n - f| > \frac{\varepsilon}{2}\}$. Thus if we assume (2) then we have

$$\lim_{n \to \infty} \mu\{D : |f_n - f| \geq \varepsilon\} \leq \lim_{n \to \infty} \mu\{D : |f_n - f| > \frac{\varepsilon}{2}\} = 0,$$

that is, (1) holds. This completes the proof of the equivalence of the two conditions (1) and (2). ∎

Prob. 6.7. Let $(X, \mathfrak{A}, \mu)$ be a measure space and let $D \in \mathfrak{A}$. Let $(f_n : n \in \mathbb{N})$ be a sequence of extended real-valued $\mathfrak{A}$-measurable functions on D and let f be an extended real-valued $\mathfrak{A}$-measurable function on D that is real-valued a.e. on D. Consider the condition:

1° $\quad\quad \lim_{n \to \infty} \mu\{x \in D : |f_n(x) - f(x)| > 0\} = 0.$

(a) Show that condition 1° implies that $f_n \xrightarrow{\mu} f$ on D.
(b) Show by example that the converse of (a) is false.
(c) Show that condition 1° implies that for a.e. $x \in D$ we have $f_n(x) = f(x)$ for infinitely many $n \in \mathbb{N}$.

Proof. 1. For every $\varepsilon > 0$ we have $\{D : |f_n - f| \geq \varepsilon\} \subset \{D : |f_n - f| > 0\}$ and then $\mu\{D : |f_n - f| \geq \varepsilon\} \leq \mu\{D : |f_n - f| > 0\}$. Thus if we assume condition 1° then

$$\lim_{n \to \infty} \mu\{D : |f_n - f| \geq \varepsilon\} \leq \lim_{n \to \infty} \mu\{D : |f_n - f| > 0\} = 0,$$

that is, the sequence $(f_n : n \in \mathbb{N})$ converges to f in measure on D.

2. Let $D \in \mathfrak{A}$ with $\mu(D) > 0$. Let $f_n = \frac{1}{n}$ on D for $n \in \mathbb{N}$ and let $f = 0$ on D. For an arbitrary $\varepsilon > 0$, consider the set $\{D : |f_n - f| \geq \varepsilon\} = \{D : \frac{1}{n} \geq \varepsilon\}$ for $n \in \mathbb{N}$. Let $N \in \mathbb{N}$ be so large that $\frac{1}{N} < \varepsilon$. Then for $n \geq N$ we have $\{D : |f_n - f| \geq \varepsilon\} = \{D : \frac{1}{n} \geq \varepsilon\} = \emptyset$ and then $\mu\{D : |f_n - f| \geq \varepsilon\} = \mu(\emptyset) = 0$. Therefore we have $\lim_{n \to \infty} \mu\{D : |f_n - f| \geq \varepsilon\} = 0$. This shows that the sequence $(f_n : n \in \mathbb{N})$ converges to f in measure on D.

On the other hand we have $\{D : |f_n - f| > 0\} = \{D : \frac{1}{n} > 0\} = D$ for $n \in \mathbb{N}$ and thus $\mu\{D : |f_n - f| > 0\} = \mu(D)$ for $n \in \mathbb{N}$. Then $\lim_{n \to \infty} \mu\{D : |f_n - f| > 0\} = \mu(D) > 0$ and condition 1° is not satisfied.

3. Assume that the sequence $(f_n : n \in \mathbb{N})$ satisfies condition 1°. For $n \in \mathbb{N}$, let $A_n = \{D : |f_n - f| > 0\}$. Then by 1° we have $\lim_{n \to \infty} \mu(A_n) = 0$. Then by Theorem 1.28, we have

$$\mu\left(\liminf_{n \to \infty} A_n\right) \leq \liminf_{n \to \infty} \mu(A_n) = \lim_{n \to \infty} \mu(A_n) = 0.$$

We have

$$\liminf_{n \to \infty} A_n = \{x \in D : x \in A_n \text{ for all but finitely many } n \in \mathbb{N}\}$$

$$= \{x \in D : x \notin A_n \text{ for at most finitely many } n \in \mathbb{N}\},$$

and then

$$D \setminus \liminf_{n \to \infty} A_n = \{x \in D : x \notin A_n \text{ for infinitely many } n \in \mathbb{N}\}.$$

Now

$$x \notin A_n \Leftrightarrow |f_n(x) - f(x)| = 0 \Leftrightarrow f_n(x) = f(x).$$

Thus we have

$$D \setminus \liminf_{n \to \infty} A_n = \{x \in D : f_n(x) = f(x) \text{ for infinitely many } n \in \mathbb{N}\}.$$

Then since $\mu\left(\liminf_{n \to \infty} A_n\right) = 0$, we have $f_n(x) = f(x)$ for infinitely many $n \in \mathbb{N}$ for a.e. $x \in D$. ∎

Prob. 6.8. Let $(a_n : n \in \mathbb{N})$ be a sequence of extended real numbers and let a be a real number. Show that if every subsequence $(a_{n_k} : k \in \mathbb{N})$ of the sequence $(a_n : n \in \mathbb{N})$ has a subsequence $(a_{n_{k_\ell}} : \ell \in \mathbb{N})$ converging to a then the sequence $(a_n : n \in \mathbb{N})$ converges to a.

Proof. Suppose the statement that the sequence $(a_n : n \in \mathbb{N})$ converges to a is false. Then either the sequence does not converge or the sequence converges to a real number distinct from a. Then in any case there exists $\varepsilon_0 > 0$ such that $a_n \notin (a - \varepsilon_0, a + \varepsilon_0)$ for infinitely many $n \in \mathbb{N}$. Then there exists a subsequence $(a_{n_k} : k \in \mathbb{N})$ of the sequence $(a_n : n \in \mathbb{N})$ such that $a_{n_k} \notin (a - \varepsilon_0, a + \varepsilon_0)$ for every $k \in \mathbb{N}$. Then no subsequence $(a_{n_{k_\ell}} : \ell \in \mathbb{N})$ of this subsequence $(a_{n_k} : k \in \mathbb{N})$ can converge to a. This contradicts the assumption that every subsequence $(a_{n_k} : k \in \mathbb{N})$ of the sequence $(a_n : n \in \mathbb{N})$ has a subsequence $(a_{n_{k_\ell}} : \ell \in \mathbb{N})$ converging to a. Therefore the sequence $(a_n : n \in \mathbb{N})$ must converge to a.

Prob. 6.9. Let $(X, \mathfrak{A}, \mu)$ be a measure space and let $D \in \mathfrak{A}$. Let $(f_n : n \in \mathbb{N})$ be a sequence of extended real-valued $\mathfrak{A}$-measurable functions on D and let f be an extended real-valued $\mathfrak{A}$-measurable function on D that is real-valued a.e. on D.
(a) Show that if every subsequence $(f_{n_k} : k \in \mathbb{N})$ of the sequence $(f_n : n \in \mathbb{N})$ has a subsequence $(f_{n_{k_\ell}} : \ell \in \mathbb{N})$ which converges to f in measure on D, then the sequence $(f_n : n \in \mathbb{N})$ itself converges to f in measure on D.
(b) Show that if $\mu(D) < \infty$ and if every subsequence $(f_{n_k} : k \in \mathbb{N})$ of the sequence $(f_n : n \in \mathbb{N})$ has a subsequence $(f_{n_{k_\ell}} : \ell \in \mathbb{N})$ which converges to f a.e. on D, then the sequence $(f_n : n \in \mathbb{N})$ itself converges to f in measure on D.

Proof. 1. Let us prove (a). Let $\varepsilon > 0$ and define a sequence of real numbers $(a_n : n \in \mathbb{N})$ by setting

$$a_n = \mu\{D : |f_n - f| \geq \varepsilon\} \quad \text{for } n \in \mathbb{N}.$$

Suppose every subsequence $(f_{n_k} : k \in \mathbb{N})$ of the sequence $(f_n : n \in \mathbb{N})$ has a subsequence $(f_{n_{k_\ell}} : \ell \in \mathbb{N})$ which converges to f in measure on D. Then every subsequence $(a_{n_k} : k \in \mathbb{N})$ of the sequence $(a_n : n \in \mathbb{N})$ has a subsequence $(a_{n_{k_\ell}} : \ell \in \mathbb{N})$ converging to 0. According to Prob. 6.8 this implies that the sequence $(a_n : n \in \mathbb{N})$ converges to 0, that is,

$$\lim_{n \to \infty} \mu\{D : |f_n - f| \geq \varepsilon\} = 0.$$

This shows that sequence $(f_n : n \in \mathbb{N})$ converges to f in measure on D.
   2. Let us prove (b). Let $\mu(D) < \infty$. Suppose every subsequence $(f_{n_k} : k \in \mathbb{N})$ of the sequence $(f_n : n \in \mathbb{N})$ has a subsequence $(f_{n_{k_\ell}} : \ell \in \mathbb{N})$ which converges to f a.e. on D. Since $\mu(D) < \infty$, convergence of $(f_{n_{k_\ell}} : \ell \in \mathbb{N})$ to f a.e. on D implies convergence of $(f_{n_{k_\ell}} : \ell \in \mathbb{N})$ to f in measure D according to Theorem 6.22 (H. Lebesgue). Thus every subsequence $(f_{n_k} : k \in \mathbb{N})$ of the sequence $(f_n : n \in \mathbb{N})$ has a subsequence $(f_{n_{k_\ell}} : \ell \in \mathbb{N})$ which converges to f in measure on D. According to (a), this implies that the sequence $(f_n : n \in \mathbb{N})$ converges to f in measure on D. ∎

Prob. 6.10. Show that if $f_n \overset{\mu}{\to} f$ on D and if $\mu(D) < \infty$, then for every real-valued continuous function F on $\mathbb{R}$ we have $F \circ f_n \overset{\mu}{\to} F \circ f$ on D. (Assume that f_n is real-valued for every $n \in \mathbb{N}$ to make the composition $(F \circ f_n)(x) = F(f_n(x))$ possible at every $x \in D$.)

Proof. Consider the sequence $(F \circ f_n : n \in \mathbb{N})$ of real-valued $\mathfrak{A}$-measurable functions and the real-valued $\mathfrak{A}$-measurable function $F \circ f$ on $D \in \mathfrak{A}$ with $\mu(D) < \infty$.

Let us show that the sequence $(F \circ f_n : n \in \mathbb{N})$ converges to $F \circ f$ in measure on D. According to (b) of Prob. 6.9, it suffices to show that every subsequence $(F \circ f_{n_k} : k \in \mathbb{N})$ of the sequence $(F \circ f_n : n \in \mathbb{N})$ has a subsequence $(F \circ f_{n_{k_\ell}} : \ell \in \mathbb{N})$ which converges to $F \circ f$ a.e. on D.

Take an arbitrary subsequence $(f_{n_k} : k \in \mathbb{N})$ of the sequence $(f_n : n \in \mathbb{N})$. Since the sequence $(f_n : n \in \mathbb{N})$ converges to f in measure on D, the subsequence $(f_{n_k} : k \in \mathbb{N})$ converges to f in measure on D. Then by Theorem 6.24 (F. Riesz) there exists a subsequence $(f_{n_{k_\ell}} : \ell \in \mathbb{N})$ that converges to f a.e. on D. Then continuity of F on $\mathbb{R}$ implies that $(F \circ f_{n_{k_\ell}} : \ell \in \mathbb{N})$ converges to $F \circ f$ a.e. on D. ∎

Prob. 6.11. The following are two modifications of Prob. 6.10:

(a) Show that if we remove the condition $\mu(D) < \infty$, we still have $F \circ f_n \xrightarrow{\mu} F \circ f$ on D if we assume that F is uniformly continuous on $\mathbb{R}$.

(b) Show that if we remove the condition $\mu(D) < \infty$, we still have $F \circ f_n \xrightarrow{\mu} F \circ f$ on D if we assume that the sequence $(f_n : n \in \mathbb{N})$ and f are uniformly bounded on D.

Proof. 1. Let us prove (a). Assume that F is a real-valued uniformly continuous function on $\mathbb{R}$. Then for every $\varepsilon > 0$ there exists $\delta > 0$ such that

(1) $\qquad |F(y') - F(y'')| < \varepsilon$ whenever $y', y'' \in \mathbb{R}$ and $|y' - y''| < \delta$.

Then $|f_n(x) - f(x)| < \delta$ implies $\left|F(f_n(x)) - F(f(x))\right| < \varepsilon$ and thus we have

(2) $\qquad \left\{x \in D : |f_n(x) - f(x)| < \delta\right\} \subset \left\{x \in D : \left|F(f_n(x)) - F(f(x))\right| < \varepsilon\right\}$

and then

(3) $\qquad \left\{x \in D : |f_n(x) - f(x)| \geq \delta\right\} \supset \left\{x \in D : \left|F(f_n(x)) - F(f(x))\right| \geq \varepsilon\right\}$

and therefore

(4) $\qquad \mu\left\{D : |F \circ f_n - F \circ f| \geq \varepsilon\right\} \leq \mu\left\{D : |f_n - f| \geq \delta\right\}.$

Since $(f_n : n \in \mathbb{N})$ converges to f in measure on D, we have

$$\lim_{n \to \infty} \mu\left\{D : |f_n - f| \geq \delta\right\} = 0.$$

Then

(5) $\qquad \lim_{n \to \infty} \mu\left\{D : |F \circ f_n - F \circ f| \geq \varepsilon\right\} \leq \lim_{n \to \infty} \mu\left\{D : |f_n - f| \geq \delta\right\} = 0.$

This shows that the sequence $(F \circ f_n : n \in \mathbb{N})$ converges to $F \circ f$ in measure on D.

2. Let us prove (b). Suppose the sequence $(f_n : n \in \mathbb{N})$ and f are uniformly bounded on D, that is, there exists $M > 0$ such that $|f_n(x)| \leq M$ for all $x \in D$ and $n \in \mathbb{N}$ and $|f(x)| \leq M$ for all $x \in D$. Now continuity of F on $\mathbb{R}$ implies uniform continuity of F on the compact set $[-M, M] \subset \mathbb{R}$. Then for every $\varepsilon > 0$ there exists $\delta > 0$ such that

$$|F(y') - F(y'')| < \varepsilon \quad \text{whenever } y', y'' \in [-M, M] \text{ and } |y' - y''| < \delta.$$

Then since $f_n(x) \in [-M, M]$ for all $x \in D$ and $n \in \mathbb{N}$ and $f(x) \in [-M, M]$ for all $x \in D$, $|f_n(x) - f(x)| < \delta$ implies $\left|F(f_n(x)) - F(f(x))\right| < \varepsilon$ and thus we have

$$\left\{x \in D : |f_n(x) - f(x)| < \delta\right\} \subset \left\{x \in D : \left|F(f_n(x)) - F(f(x))\right| < \varepsilon\right\}.$$

This is precisely the expression (3) above. Then (4) and (5) follow. This shows that the sequence $(F \circ f_n : n \in \mathbb{N})$ converges to $F \circ f$ in measure on D. ∎

Prob. 6.12. Let $(X, \mathfrak{A}, \mu)$ be a measure space and let $D \in \mathfrak{A}$. Let $(f_n : n \in \mathbb{N})$ be a sequence of real-valued $\mathfrak{A}$-measurable functions on D and let f be a real-valued $\mathfrak{A}$-measurable function on D. Show that if $f_n \overset{\mu}{\to} f$ on D, then for every $c \in \mathbb{R}$ we have $cf_n \overset{\mu}{\to} cf$ on D.

Proof. When $c = 0$, the claim is trivially true. Thus consider the case $c \neq 0$. Then for every $\varepsilon > 0$ we have

$$\{D : |cf_n - cf| \geq \varepsilon\} = \{D : |f_n - f| \geq \tfrac{\varepsilon}{|c|}\}$$

and then

$$\mu\{D : |cf_n - cf| \geq \varepsilon\} = \mu\{D : |f_n - f| \geq \tfrac{\varepsilon}{|c|}\}.$$

Since the sequence $(f_n : n \in \mathbb{N})$ converges to f in measure on D, we have

$$\lim_{n \to \infty} \mu\{D : |f_n - f| \geq \tfrac{\varepsilon}{|c|}\} = 0.$$

Then we have

$$\lim_{n \to \infty} \mu\{D : |cf_n - cf| \geq \varepsilon\} = \lim_{n \to \infty} \mu\{D : |f_n - f| \geq \tfrac{\varepsilon}{|c|}\} = 0.$$

This shows that the sequence $(cf_n : n \in \mathbb{N})$ converges to cf in measure on D. ∎

Prob. 6.13. Let $(X, \mathfrak{A}, \mu)$ be a measure space and let $D \in \mathfrak{A}$. Let $(f_n : n \in \mathbb{N})$ be a sequence of real-valued $\mathfrak{A}$-measurable functions on D and let f be a real-valued $\mathfrak{A}$-measurable function on D. Similarly let $(g_n : n \in \mathbb{N})$ be a sequence of real-valued $\mathfrak{A}$-measurable functions on D and let g be a real-valued $\mathfrak{A}$-measurable function on D. Show that if $f_n \xrightarrow{\mu} f$ on D and $g_n \xrightarrow{\mu} g$ on D, then $f_n + g_n \xrightarrow{\mu} f + g$ on D.

Proof. We are to show that for every $\varepsilon > 0$ we have

(1) $$\lim_{n \to \infty} \mu\{D : |(f_n + g_n) - (f + g)| \geq \varepsilon\} = 0.$$

Now

$$|(f_n + g_n) - (f + g)| = |(f_n - f) + (g_n - g)| \leq |f_n - f| + |g_n - g|$$

and then

(2) $$\{D : |(f_n + g_n) - (f + g)| \geq \varepsilon\} \subset \{D : |f_n - f| + |g_n - g| \geq \varepsilon\}.$$

Then we have

(3) $$\{D : |f_n - f| + |g_n - g| \geq \varepsilon\} \subset \{D : |f_n - f| \geq \tfrac{\varepsilon}{2}\} \cup \{D : |g_n - g| \geq \tfrac{\varepsilon}{2}\},$$

as can be verified by a contradiction argument. Combining (2) and (3), we have

$$\{D : |(f_n + g_n) - (f + g)| \geq \varepsilon\} \subset \{D : |f_n - f| \geq \tfrac{\varepsilon}{2}\} \cup \{D : |g_n - g| \geq \tfrac{\varepsilon}{2}\},$$

and then

$$\mu\{D : |(f_n + g_n) - (f + g)| \geq \varepsilon\} \leq \mu\{D : |f_n - f| \geq \tfrac{\varepsilon}{2}\} + \mu\{D : |g_n - g| \geq \tfrac{\varepsilon}{2}\}.$$

Thus we have

$$\lim_{n \to \infty} \mu\{D : |(f_n + g_n) - (f + g)| \geq \varepsilon\} \leq \lim_{n \to \infty} \mu\{D : |f_n - f| \geq \tfrac{\varepsilon}{2}\}$$
$$+ \lim_{n \to \infty} \mu\{D : |g_n - g| \geq \tfrac{\varepsilon}{2}\} = 0.$$

This proves (1). ∎

Prob. 6.14. Let $(f_n : n \in \mathbb{N})$, f, $(g_n : n \in \mathbb{N})$, and g be as in Prob. 6.13. Assume that $\mu(D) < \infty$. Show that if $f_n \xrightarrow{\mu} f$ on D and $g_n \xrightarrow{\mu} g$ on D, then $f_n g_n \xrightarrow{\mu} fg$ on D.

Proof. Let us show that the sequence $(f_n g_n : n \in \mathbb{N})$ converges to fg in measure on D. According to (b) of Prob. 6.9, it suffices to show that every subsequence $(f_{n_k} g_{n_k} : k \in \mathbb{N})$ of the sequence $(f_n g_n : n \in \mathbb{N})$ has a subsequence $(f_{n_{k_\ell}} g_{n_{k_\ell}} : \ell \in \mathbb{N})$ which converges to fg a.e. on D,

   Let $(f_{n_k} g_{n_k} : k \in \mathbb{N})$ be a subsequence of the sequence $(f_n g_n : n \in \mathbb{N})$. Since the sequence $(f_n : n \in \mathbb{N})$ converges to f in measure on D, the subsequence $(f_{n_k} : k \in \mathbb{N})$ converges to f in measure on D. Then by Theorem 6.24 (F. Riesz), there exists a subsequence $(f_{n_{k_\ell}} : \ell \in \mathbb{N})$ which converges to f a.e. on D. Now since the sequence $(g_n : n \in \mathbb{N})$ converges to g in measure on D, the subsequence $(g_{n_{k_\ell}} : \ell \in \mathbb{N})$ converges to g in measure on D. Then by Theorem 6.24 (F. Riesz), there exists a subsequence $(g_{n_{k_{\ell_m}}} : m \in \mathbb{N})$ which converges to g a.e. on D. Then since $(f_{n_{k_\ell}} : \ell \in \mathbb{N})$ converges to f a.e. on D, $(f_{n_{k_{\ell_m}}} : m \in \mathbb{N})$ converges to f a.e. on D. Therefore $(f_{n_{k_{\ell_m}}} g_{n_{k_{\ell_m}}} : m \in \mathbb{N})$ converges to fg a.e. on D. Note that $(f_{n_{k_{\ell_m}}} g_{n_{k_{\ell_m}}} : m \in \mathbb{N})$ is a subsequence of the arbitrary subsequence $(f_{n_k} g_{n_k} : k \in \mathbb{N})$ of the sequence $(f_n g_n : n \in \mathbb{N})$. ∎

Prob. 6.15. Returning to Prob. 1.32, show that $([\mathfrak{A}], \rho^*)$ is a complete metric space, that is, if a sequence $\{[A_n] : n \in \mathbb{N}\}$ in $[\mathfrak{A}]$ is a Cauchy sequence with respect to the metric ρ^* then there exists $[A] \in [\mathfrak{A}]$ to which the sequence converges with respect to the metric ρ^*.

Proof. Let $\{[A_n] : n \in \mathbb{N}\}$ be a Cauchy sequence in $[\mathfrak{A}]$ with respect to the metric ρ^*, that is, for every $\eta > 0$ there exists $N \in \mathbb{N}$ such that

$$(1) \qquad\qquad \rho^*\big([A_m], [A_n]\big) < \eta \quad \text{for } m, n \geq N.$$

Our goal is to show that there exists a set $A \in \mathfrak{A}$ such that

$$(2) \qquad\qquad \lim_{n \to \infty} \rho^*\big([A_n], [A]\big) = 0.$$

This is done in the following steps.

1. According to Observation 4.3, if $E \in \mathfrak{A}$ then the characteristic function $\mathbf{1}_E$ is an $\mathfrak{A}$-measurable function on X. Let us show first that if a sequence $\{[A_n] : n \in \mathbb{N}\}$ in $[\mathfrak{A}]$ is a Cauchy sequence with respect to the metric ρ^* then $(\mathbf{1}_{A_n} : n \in \mathbb{N})$ is a Cauchy sequence with respect to convergence in measure on X.

Let $E, F \in \mathfrak{A}$ and consider $\mathbf{1}_E$ and $\mathbf{1}_F$. Since 0 and 1 are the only values the functions $\mathbf{1}_E$ and $\mathbf{1}_F$ can assume, 0 and 1 are the only values the function $|\mathbf{1}_E - \mathbf{1}_F|$ can assume and moreover $|\mathbf{1}_E - \mathbf{1}_F|$ assumes the value 1 on the set $E \triangle F$. Thus we have

$$(3) \qquad\qquad \big\{X : |\mathbf{1}_E - \mathbf{1}_F| = 1\big\} = E \triangle F$$

and then by the definition of ρ^* we have

$$(4) \qquad\qquad \mu\big\{X : |\mathbf{1}_E - \mathbf{1}_F| = 1\big\} = \mu(E \triangle F) = \rho^*\big([E], [F]\big).$$

Then for every $\varepsilon > 0$, we have

$$(5) \qquad\qquad \mu\big\{X : |\mathbf{1}_E - \mathbf{1}_F| \geq \varepsilon\big\} \leq \mu\big\{X : |\mathbf{1}_E - \mathbf{1}_F| > 0\big\}$$
$$= \mu\big\{X : |\mathbf{1}_E - \mathbf{1}_F| = 1\big\}$$
$$= \mu(E \triangle F) = \rho^*\big([E], [F]\big),$$

where the first equality is from the fact that 0 and 1 are the only values the function $|\mathbf{1}_E - \mathbf{1}_F|$ can assume.

Now let $\{[A_n] : n \in \mathbb{N}\}$ be a Cauchy sequence in $[\mathfrak{A}]$ with respect to the metric ρ^*. Then for every $\varepsilon > 0$, we have by (4) and (1)

$$\mu\big\{X : |\mathbf{1}_{A_m} - \mathbf{1}_{A_n}| \geq \varepsilon\big\} \leq \rho^*\big([A_m], [A_n]\big) < \eta \quad \text{for } m, n \geq N.$$

This shows that $(\mathbf{1}_{A_n} : n \in \mathbb{N})$ is a Cauchy sequence with respect to convergence in measure on X.

2. According to Theorem 6.27, if a sequence $(f_n : n \in \mathbb{N})$ of $\mathfrak{A}$-measurable functions on a set $D \in \mathfrak{A}$ is a Cauchy sequence with respect to convergence in measure then there exists an $\mathfrak{A}$-measurable function f on D such that $(f_n : n \in \mathbb{N})$ converges to f in measure on D.

Then since $(\mathbf{1}_{A_n} : n \in \mathbb{N})$ is a Cauchy sequence with respect to convergence in measure on X, there exists $\mathfrak{A}$-measurable function f on X such that $(\mathbf{1}_{A_n} : n \in \mathbb{N})$ converges to f in measure on X. Let us determine this function f.

According to Theorem 6.24 (F. Riesz), convergence of the sequence $(\mathbf{1}_{A_n} : n \in \mathbb{N})$ to f in measure on X implies the existence of a subsequence $(\mathbf{1}_{A_{n_k}} : n \in \mathbb{N})$ converging to f a.e. on X. Thus there exists a null set E_0 in the measure space $(X, \mathfrak{A}, \mu)$ such that

$$\lim_{k \to \infty} \mathbf{1}_{A_{n_k}}(x) = f(x) \quad \text{for } x \in X \setminus E_0.$$

Then since 0 and 1 are the only values the function $\mathbf{1}_{A_{n_k}}$ can assume, we have

$$\lim_{k \to \infty} \mathbf{1}_{A_{n_k}}(x) = 0, \text{ or } 1 \quad \text{for } x \in X \setminus E_0.$$

Thus we have

$$f(x) = 0, \text{ or } 1 \quad \text{for } x \in X \setminus E_0.$$

If we let $A = \{X \setminus E_0 : f = 1\}$, then $f = \mathbf{1}_A$ on $X \setminus E_0$. Note that $\mathfrak{A}$-measurability of the function f on X implies the set A defined above is in $\mathfrak{A}$. Now the sequence $(\mathbf{1}_{A_n} : n \in \mathbb{N})$ converges to $f = \mathbf{1}_A$ in measure on $X \setminus E_0$. Since E_0 is a null set, this implies that $(\mathbf{1}_{A_n} : n \in \mathbb{N})$ converges to $f = \mathbf{1}_A$ in measure on X. Then for every $\varepsilon > 0$, we have

(6) $$\lim_{n \to \infty} \mu\{X : |\mathbf{1}_{A_n} - \mathbf{1}_A| \geq \varepsilon\} = 0.$$

With $\varepsilon_0 \in (0, 1]$, we have $\{X : |\mathbf{1}_{A_n} - \mathbf{1}_A| = 1\} \subset \{X : |\mathbf{1}_{A_n} - \mathbf{1}_A| \geq \varepsilon_0\}$ and thus

(7) $$\mu\{X : |\mathbf{1}_{A_n} - \mathbf{1}_A| = 1\} \leq \mu\{X : |\mathbf{1}_{A_n} - \mathbf{1}_A| \geq \varepsilon_0\}.$$

Then by (4) and (7), we have

$$\rho^*([A_n], [A]) = \mu\{X : |\mathbf{1}_{A_n} - \mathbf{1}_A| = 1\} \leq \mu\{X : |\mathbf{1}_{A_n} - \mathbf{1}_A| \geq \varepsilon_0\}.$$

Then by (6), we have

$$\lim_{n \to \infty} \rho^*([A_n], [A]) \leq \lim_{n \to \infty} \mu\{X : |\mathbf{1}_{A_n} - \mathbf{1}_A| \geq \varepsilon_0\} = 0.$$

This proves (2) and completes the proof. ∎

Prob. 6.16. A real valued function f on $\mathbb{R}$ is said to be left-continuous at $x \in \mathbb{R}$ if $\lim_{y \uparrow x} f(y) = f(x)$. Similarly f is said to be right-continuous at $x \in \mathbb{R}$ if $\lim_{y \downarrow x} f(y) = f(x)$.

(a) Show that if f is left-continuous everywhere on $\mathbb{R}$, then f is $\mathfrak{B}_{\mathbb{R}}$-measurable on $\mathbb{R}$.

(b) Show that if f is right-continuous everywhere on $\mathbb{R}$, then f is $\mathfrak{B}_{\mathbb{R}}$-measurable on $\mathbb{R}$.

Proof. 1. Let f be left-continuous everywhere on $\mathbb{R}$. For each $n \in \mathbb{N}$, let us decompose $\mathbb{R}$ into intervals $\left(\frac{k-1}{2^n}, \frac{k}{2^n}\right]$ for $k \in \mathbb{Z}$. Let us define a left-continuous step function f_n on $\mathbb{R}$ by setting

$$f_n(x) = f\left(\frac{k-1}{2^n}\right) \quad \text{for } x \in \left(\frac{k-1}{2^n}, \frac{k}{2^n}\right].$$

For every $\alpha \in \mathbb{R}$, the set $\{\mathbb{R} : f_n \leq \alpha\}$ is the union of at most countably many intervals in $\mathbb{R}$ and such a set is a $\mathfrak{B}_{\mathbb{R}}$-measurable set. Thus f_n is a $\mathfrak{B}_{\mathbb{R}}$-measurable function on $\mathbb{R}$.

Let us show that $\lim_{n \to \infty} f_n(x) = f(x)$ for every $x \in \mathbb{R}$. Let $x \in \mathbb{R}$. Then for every $n \in \mathbb{N}$ there exists $k_n \in \mathbb{Z}$ such that $x \in \left(\frac{k_n - 1}{2^n}, \frac{k_n}{2^n}\right]$. Then by the definition of f_n we have

$$f_n(x) = f\left(\frac{k_n - 1}{2^n}\right) \quad \text{for } n \in \mathbb{N}.$$

Now we have

$$\frac{k_n - 1}{2^n} < x \quad \text{and} \quad x - \frac{k_n - 1}{2^n} \leq \frac{1}{2^n} \quad \text{for every } n \in \mathbb{N}.$$

Thus $\frac{k_n - 1}{2^n} \uparrow x$ as $n \to \infty$ and then by the left-continuity of f at x we have

$$f(x) = \lim_{n \to \infty} f\left(\frac{k_n - 1}{2^n}\right) = \lim_{n \to \infty} f_n(x).$$

This shows that $\lim_{n \to \infty} f_n = f$ on $\mathbb{R}$. Then the $\mathfrak{B}_{\mathbb{R}}$-measurability of f_n on $\mathbb{R}$ for every $n \in \mathbb{N}$ implies the $\mathfrak{B}_{\mathbb{R}}$-measurability of f on $\mathbb{R}$ by Theorem 4.22.

2. The $\mathfrak{B}_{\mathbb{R}}$-measurability of a right-continuous function f on $\mathbb{R}$ can be proved by the same argument as in **1** but using a sequence of right-continuous step functions. Here we present an alternate proof utilizing the result above that a left-continuous function is $\mathfrak{B}_{\mathbb{R}}$-measurable and the fact that if $f(x)$ is a right-continuous function of $x \in \mathbb{R}$ then $f(-x)$ is a left-continuous function of $x \in \mathbb{R}$.

Let I be the identity mapping of a linear space V into V. A linear mapping T of V into V is called invertible if there exists a linear mapping U of V into V satisfying the condition

$$UT = TU = I.$$

U is called an inverse mapping of T. If an inverse mapping of T exists then it is unique. This can be shown as follows. Suppose U and W are inverse mappings of T. Then we have

$$UT = TU = I$$

$$WT = TW = I.$$

This implies that $TU = I = TW$ and then $UTU = UTW$, that is, $IU = IW$, that is, $U = W$. This proves the uniqueness of the inverse mapping of T if it exists. Let us write T^{-1} for the inverse mapping of T.

Now let ϑ be a linear mapping of $\mathbb{R}$ into $\mathbb{R}$ defined by setting

(1) $$\vartheta(x) = -x \quad \text{for } x \in \mathbb{R}.$$

Then $(\vartheta \circ \vartheta)(x) = -(-x) = x$ for every $x \in \mathbb{R}$. Thus we have

(2) $$\vartheta \circ \vartheta = I.$$

By our observations above, (2) implies that ϑ is invertible and moreover we have

(3) $$\vartheta^{-1} = \vartheta.$$

Let us write

(4) $$u = \vartheta(x) \quad \text{for } x \in \mathbb{R}.$$

Let $x_0 \in \mathbb{R}$ and $u_0 = \vartheta(x_0)$. Then we have $\vartheta(u_0) = (\vartheta \circ \vartheta)(x_0) = I(x_0) = x_0$ by (2). Thus we have

$$(f \circ \vartheta)(u_0) = f(x_0) = \lim_{x \downarrow x_0} f(x) \quad \text{by the right-continuity of } f$$

$$= \lim_{x \downarrow x_0} (f \circ \vartheta \circ \vartheta)(x) \quad \text{by (2)}$$

$$= \lim_{u \uparrow u_0} (f \circ \vartheta)(u) \quad \text{by (4)}.$$

This shows that $f \circ \vartheta$ is left-continuous on $\mathbb{R}$ and therefore $f \circ \vartheta$ is a $\mathfrak{B}_\mathbb{R}/\mathfrak{B}_\mathbb{R}$-measurable mapping of $(\mathbb{R}, \mathfrak{B}_\mathbb{R})$ into $(\mathbb{R}, \mathfrak{B}_\mathbb{R})$. Now ϑ is a $\mathfrak{B}_\mathbb{R}/\mathfrak{B}_\mathbb{R}$-measurable mapping of $(\mathbb{R}, \mathfrak{B}_\mathbb{R})$ into $(\mathbb{R}, \mathfrak{B}_\mathbb{R})$. Then the composite mapping $(f \circ \vartheta) \circ \vartheta$ is a $\mathfrak{B}_\mathbb{R}/\mathfrak{B}_\mathbb{R}$-measurable mapping of $(\mathbb{R}, \mathfrak{B}_\mathbb{R})$ into $(\mathbb{R}, \mathfrak{B}_\mathbb{R})$ by Theorem 1.40 (Chain Rule for Measurable Mappings). Now $(f \circ \vartheta) \circ \vartheta = f \circ (\vartheta \circ \vartheta) = f \circ I = f$ by (2). Thus f is a $\mathfrak{B}_\mathbb{R}/\mathfrak{B}_\mathbb{R}$-measurable mapping of $(\mathbb{R}, \mathfrak{B}_\mathbb{R})$ into $(\mathbb{R}, \mathfrak{B}_\mathbb{R})$. ∎

Prob. 6.17. Let f be an extended real-valued function on $\mathbb{R}$.
(a) Suppose $\lim_{y\uparrow x} f(y)$ exists in $\overline{\mathbb{R}}$ for every $x \in \mathbb{R}$ and define an extended real-valued function $\Lambda_\ell f$ on $\mathbb{R}$ by setting $(\Lambda_\ell f)(x) = \lim_{y\uparrow x} f(y)$ for $x \in \mathbb{R}$. Let us call $\Lambda_\ell f$ the left limit of f. Show that $\Lambda_\ell f$ is $\mathfrak{B}_{\mathbb{R}}$-measurable on $\mathbb{R}$.
(b) Suppose $\lim_{y\downarrow x} f(y)$ exists in $\overline{\mathbb{R}}$ for every $x \in \mathbb{R}$ and define an extended real-valued function $\Lambda_r f$ on $\mathbb{R}$ by setting $(\Lambda_r f)(x) = \lim_{y\downarrow x} f(y)$ for $x \in \mathbb{R}$. Let us call $\Lambda_r f$ the right limit of F. Show that $\Lambda_r f$ is $\mathfrak{B}_{\mathbb{R}}$-measurable on $\mathbb{R}$.

Proof. Let us prove (a). For each $n \in \mathbb{N}$, let us decompose $\mathbb{R}$ into intervals $\left(\frac{k-1}{2^n}, \frac{k}{2^n}\right]$ for $k \in \mathbb{Z}$. Let us define an extended real-valued step function f_n on $\mathbb{R}$ by setting

$$f_n(x) = f\left(\frac{k-1}{2^n}\right) \in \overline{\mathbb{R}} \quad \text{for } x \in \left(\frac{k-1}{2^n}, \frac{k}{2^n}\right].$$

Then for every $\alpha \in \mathbb{R}$, the set $\{\mathbb{R} : f_n \le \alpha\}$ is the union of at most countably many intervals in $\mathbb{R}$ and such a set is a $\mathfrak{B}_{\mathbb{R}}$-measurable set. Thus f_n is a $\mathfrak{B}_{\mathbb{R}}$-measurable function on $\mathbb{R}$.

Let us show that $\lim_{n\to\infty} f_n(x) = (\Lambda_\ell f)(x)$ for every $x \in \mathbb{R}$. Let $x \in \mathbb{R}$. Then for every $n \in \mathbb{N}$ there exists $k_n \in \mathbb{Z}$ such that $x \in \left(\frac{k_n-1}{2^n}, \frac{k_n}{2^n}\right]$. Then by the definition of f_n we have

$$f_n(x) = f\left(\frac{k_n-1}{2^n}\right) \in \overline{\mathbb{R}} \quad \text{for } n \in \mathbb{N}.$$

Now we have

$$\frac{k_n-1}{2^n} < x \quad \text{and} \quad x - \frac{k_n-1}{2^n} \le \frac{1}{2^n} \quad \text{for every } n \in \mathbb{N}.$$

Thus $\frac{k_n-1}{2^n} \uparrow x$ as $n \to \infty$ and then we have

$$\lim_{n\to\infty} f_n(x) = \lim_{y\uparrow x} f(y) = (\Lambda_\ell f)(x).$$

This shows that $\lim_{n\to\infty} f_n = \Lambda_\ell f$ on $\mathbb{R}$. Then the $\mathfrak{B}_{\mathbb{R}}$-measurability of f_n on $\mathbb{R}$ for every $n \in \mathbb{N}$ implies the $\mathfrak{B}_{\mathbb{R}}$-measurability of $\Lambda_\ell f$ on $\mathbb{R}$ by Theorem 4.22.
2. (b) is proved likewise. ∎

Prob. 6.18. Let f be an extended real-valued function on $\mathbb{R}$.
(a) Suppose $\lim_{y\uparrow x} f(y)$ exists in $\overline{\mathbb{R}}$ for every $x \in \mathbb{R}$ and define an extended real-valued function $\Lambda_\ell f$ on $\mathbb{R}$ by setting $(\Lambda_\ell f)(x) = \lim_{y\uparrow x} f(y)$ for $x \in \mathbb{R}$. Show that the set $\{R : \Lambda_\ell f \neq f\}$ is a countable set.
(b) Suppose $\lim_{y\downarrow x} f(y)$ exists in $\overline{\mathbb{R}}$ for every $x \in \mathbb{R}$ and define an extended real-valued function $\Lambda_r f$ on $\mathbb{R}$ by setting $(\Lambda_r f)(x) = \lim_{y\downarrow x} f(y)$ for $x \in \mathbb{R}$. Show that the set $\{R : \Lambda_r f \neq f\}$ is a countable set.

Proof. 1. Let us prove (a). Let

$$(1) \qquad E = \big\{\mathbb{R} : \Lambda_\ell f \neq f\big\} = \Big\{x \in \mathbb{R} : \lim_{y\uparrow x} f(y) \neq f(x)\Big\}.$$

Let us define a function J on E by setting

$$(2) \qquad J(x) = \Big|\lim_{y\uparrow x} f(y) - f(x)\Big| \in (0, \infty] \quad \text{for } x \in E.$$

For $n \in \mathbb{N}$, let us define

$$(3) \qquad E_n = \Big\{x \in E : J(x) > \frac{1}{n}\Big\}.$$

Then we have $E = \bigcup_{n\in\mathbb{N}} E_n$. To show that E is a countable set, it suffices to show that E_n is a countable set for every $n \in \mathbb{N}$.
2. Let A be an arbitrary non-empty subset of $\mathbb{R}$. We say that $a \in A$ is a left-isolated point in A if there exists $\varepsilon > 0$ such that $(a - \varepsilon, a) \cap A = \emptyset$. Suppose $a, b \in A$ and $a < b$. If both a and b are left-isolated points in A then there exist $\varepsilon > 0$ and $\eta > 0$ such that $(a - \varepsilon, a) \cap A = \emptyset$ and $(b - \eta, b) \cap A = \emptyset$. Now we have $(a - \varepsilon, a) \cap (b - \eta, b) = \emptyset$ for otherwise we would have $(a - \varepsilon, a) \cap (b - \eta, b) \neq \emptyset$ and then $a \in (b - \eta, b)$ contradicting the assumption that b is a left-isolated point in A. Now any collection of disjoint intervals in $\mathbb{R}$ is a countable collection. (See Prob. 3.28.) This implies that any non-empty subset of $\mathbb{R}$ has at most countably many left-isolated points in it.
3. Let us show that for every $n \in \mathbb{N}$ the set E_n defined by (3) is a countable set. Assume that E_n is an uncountable set. We showed above that any non-empty subset of $\mathbb{R}$ has at most countably many left-isolated points in it. Thus the uncountable set E_n has uncountably many points that are not left-isolated points.
Select $x_0 \in E_n$ such that x_0 is not a left-isolated point in E_n and there exist uncountably many points of E_n less than x_0. Then there exists a sequence $(x_k : k \in \mathbb{N}) \subset E_n$ such that $x_k \uparrow x_0$. Since $x_k \in E_n$, we have

$$(4) \qquad \Big|\lim_{y\uparrow x_k} f(y) - f(x_k)\Big| > \frac{1}{n}.$$

For each $k \in \mathbb{N}$, let $y_k \in \mathbb{R}$ be such that $x_{k-1} < y_k < x_k$ and

$$(5) \qquad |f(y_k) - f(x_k)| > \frac{1}{n}.$$

Consider the sequence $y_1 < x_1 < y_2 < x_2 < y_3 < x_3 < \cdots$. This is an increasing sequence converging to x_0. But the sequence $f(y_1), f(x_1), f(y_2). f(x_2), f(y_3), f(x_3), \ldots$ does not tend to any limit on account of (5). This contradicts the assumption that $\lim_{y\uparrow x} f(y)$ exists in $\overline{\mathbb{R}}$ for every $x \in \mathbb{R}$. Thus E_n must be a countable set for every $n \in \mathbb{N}$. Then $E = \bigcup_{n\in\mathbb{N}} E_n$ is a countable set. This completes the proof of (a).

4. We prove (b) by the same argument as for (a). ∎

Prob. 6.19. Let f be an extended real-valued function on $\mathbb{R}$.
(a) Assume $\lim_{y\uparrow x} f(y)$ exists in $\overline{\mathbb{R}}$ for every $x \in \mathbb{R}$. Show that f is $\mathfrak{B}_{\mathbb{R}}$-measurable on $\mathbb{R}$.
(b) Assume $\lim_{y\downarrow x} f(y)$ exists in $\overline{\mathbb{R}}$ for every $x \in \mathbb{R}$. Show that f is $\mathfrak{B}_{\mathbb{R}}$-measurable on $\mathbb{R}$.

Proof. 1. In order to prove (a), we make the following observation on $\mathfrak{B}_{\mathbb{R}}$, the Borel σ-algebra of subsets of $\mathbb{R}$. For every $x \in \mathbb{R}$ the singleton $\{x\}$ is a closed set in $\mathbb{R}$ and thus $\{x\} \in \mathfrak{B}_{\mathbb{R}}$. A countable set $E \subset \mathbb{R}$ is a union of countably many singletons and thus $E \in \mathfrak{B}_{\mathbb{R}}$. Then every extended real-valued function g defined on a countable set E is $\mathfrak{B}_{\mathbb{R}}$-measurable on E since for every $\alpha \in \mathbb{R}$ the set $\{E : g \leq \alpha\}$, as a subset of the countable set E, is a countable set and hence a $\mathfrak{B}_{\mathbb{R}}$-measurable set.

2. Consider the extended real-valued $\Lambda_\ell f$ on $\mathbb{R}$ defined in Prob. 6.17 by setting

$$\Lambda_\ell f(x) = \lim_{y\uparrow x} f(y) \quad \text{for } x \in \mathbb{R}.$$

By Prob. 6.17, $\Lambda_\ell f$ is $\mathfrak{B}_{\mathbb{R}}$-measurable on $\mathbb{R}$. Let

$$E = \{\mathbb{R} : \Lambda_\ell f \neq f\}.$$

According to Prob. 6.18, E is a countable set and then by our observation above we have $E \in \mathfrak{B}_{\mathbb{R}}$. As we observed above every extended real-valued function g defined on a countable set E is $\mathfrak{B}_{\mathbb{R}}$-measurable on E. Thus our function f restricted to E is $\mathfrak{B}_{\mathbb{R}}$-measurable on E. Consider the set

$$E^c = \{\mathbb{R} : \Lambda_\ell f = f\}.$$

Since $E \in \mathfrak{B}_{\mathbb{R}}$, we have $E^c \in \mathfrak{B}_{\mathbb{R}}$ and we also have $E \cup E^c = \mathbb{R}$. Since $\Lambda_\ell f$ is $\mathfrak{B}_{\mathbb{R}}$-measurable on $\mathbb{R}$, its restriction to E^c is $\mathfrak{B}_{\mathbb{R}}$-measurable on E^c. Then since $f = \Lambda_\ell f$ on E^c, the restriction of f to E^c is $\mathfrak{B}_{\mathbb{R}}$-measurable on E^c. Thus we have shown that the restriction of f to E is $\mathfrak{B}_{\mathbb{R}}$-measurable on E and the restriction of f to E^c is $\mathfrak{B}_{\mathbb{R}}$-measurable on E^c. This implies that f is $\mathfrak{B}_{\mathbb{R}}$-measurable on $E \cup E^c = \mathbb{R}$. This completes the proof of (a).

3. We prove (b) by the same argument as for (a). ∎

Prolog to Prob. 6.20 and Prob. 6.21. Let $(X, \mathfrak{A}, \mu)$ be a measure space and let $(E_n : n \in \mathbb{N})$ be a sequence of real-valued $\mathfrak{A}$-measurable functions on a set $D \in \mathfrak{A}$. We seek a sequence $(a_n : n \in \mathbb{N})$ of positive real numbers such that the sequence of functions $(a_n f_n : n \in \mathbb{N})$ converges to 0. Prob. 6.20 is a preliminary step to Prob. 6.21.

Prob. 6.20. Let $(X, \mathfrak{A}, \mu)$ be a measure space and let $D \in \mathfrak{A}$ with $\mu(D) < \infty$. Let $(f_n : n \in \mathbb{N})$ be an arbitrary sequence of real-valued $\mathfrak{A}$-measurable functions on D. Show that for every $\varepsilon > 0$ there exist $E \subset D, E \in \mathfrak{A}$ with $\mu(E) < \varepsilon$ and a sequence $(c_n : n \in \mathbb{N})$ of positive real numbers in $(0, 1]$ such that

$$\lim_{n \to \infty} c_n |f_n| = 0 \quad \text{on } D \setminus E.$$

Proof. 1. Let us show first that if f is a real-valued $\mathfrak{A}$-measurable function on $D \in \mathfrak{A}$ with $\mu(D) < \infty$ then for every $\varepsilon > 0$ there exists $b > 0$ such that

(1) $\mu\{x \in D : |f(x)| > b\} < \varepsilon.$

For every $k \in \mathbb{N}$, let

$$D_k = \{D : |f| \le k\}.$$

Then since f is a real-valued function on D we have $D = \bigcup_{k \in \mathbb{N}} D_k$. Since $(D_k : k \in \mathbb{N})$ is an increasing sequence in $\mathfrak{A}$ we have

$$\mu(D) = \mu\Big(\bigcup_{k \in \mathbb{N}} D_k\Big) = \mu\Big(\lim_{n \to \infty} D_k\Big) = \lim_{k \to \infty} \mu(D_k).$$

Then since $\mu(D) < \infty$ there exists $k_0 \in \mathbb{N}$ such that $\mu(D) - \mu(D_{k_0}) < \varepsilon$. This implies

$$\mu(D \setminus D_{k_0}) = \mu(D) - \mu(D_{k_0}) < \varepsilon.$$

Now $D \setminus D_{k_0} = \{D : |f| > k_0\}$. Thus we have $\mu\{D : |f| > k_0\} < \varepsilon$. This proves (1) with $b = k_0$.

2. Let $(f_n : n \in \mathbb{N})$ be a sequence of real-valued $\mathfrak{A}$-measurable functions on $D \in \mathfrak{A}$ with $\mu(D) < \infty$. Let $\varepsilon > 0$ be arbitrarily given. Then for every $n \in \mathbb{N}$, according to (1) there exists $b_n > 0$ such that

$$\mu\{D : |f_n| > b_n\} < \frac{\varepsilon}{2^n}.$$

Let

$$E_n = \{D : |f_n| > b_n\}$$

and

$$c_n = \frac{1}{n}\frac{1}{b_n}.$$

Then we have

(2) $c_n |f_n| = \frac{1}{n}\frac{1}{b_n}|f_n| \le \frac{1}{n}\frac{1}{b_n}b_n = \frac{1}{n} \quad \text{on } D \setminus E_n.$

Let $E = \bigcup_{n\in\mathbb{N}} E_n$. Then $\mu(E) \leq \sum_{n\in\mathbb{N}} \mu(E_n) < \sum_{n\in\mathbb{N}} \frac{\varepsilon}{2^n} = \varepsilon$. We also have

$$D \setminus E = D \cap E^c = D \cap \left(\bigcup_{n\in\mathbb{N}} E_n \right)^c = D \cap \left(\bigcap_{n\in\mathbb{N}} E_n^c \right)$$

$$= \bigcap_{n\in\mathbb{N}} \left(D \cap E_n^c \right) = \bigcap_{n\in\mathbb{N}} D \setminus E_n.$$

Thus if $x \in D \setminus E$ then $x \in D \setminus E_n$ for every $n \in \mathbb{N}$ and then by (2) we have $c_n|f_n(x)| \leq \frac{1}{n}$ for every $n \in \mathbb{N}$. This shows that

(3) $$\lim_{n\to\infty} |c_n f_n(x)| \leq \lim_{n\to\infty} \frac{1}{n} = 0 \quad \text{for } x \in D \setminus E.$$

Let $c_n' = c_n \wedge 1 \in (0, 1]$ for $n \in \mathbb{N}$. Then $0 < c_n' \leq c_n$ for $n \in \mathbb{N}$ and then by (3) we have

$$\lim_{n\to\infty} c_n'|f_n(x)| \leq \lim_{n\to\infty} c_n|f_n(x)| = 0 \quad \text{for } x \in D \setminus E.$$

This completes the proof. ∎

Prob. 6.21. Let $(X, \mathfrak{A}, \mu)$ be a measure space and let $D \in \mathfrak{A}$ with $\mu(D) < \infty$. Let $(f_n : n \in \mathbb{N})$ be an arbitrary sequence of real-valued $\mathfrak{A}$-measurable functions on D. Show that there exists a sequence of positive real numbers $(a_n : n \in \mathbb{N})$ such that

$$\lim_{n \to \infty} a_n f_n(x) = 0 \quad \text{for a.e. } x \in D.$$

Proof. 1. Let $\varepsilon > 0$ be arbitrarily given. By Prob. 6.20, there exist an $\mathfrak{A}$-measurable subset E_1 of D with $\mu(E_1) < \varepsilon$ and a sequence $(c_{1,n} : n \in \mathbb{N})$ of positive real numbers in $(0, 1]$ such that

(1) $$\lim_{n \to \infty} c_{1,n} |f_n| = 0 \quad \text{on } D \setminus E_1.$$

For the sequence $(c_{1,n} f_n : n \in \mathbb{N})$ of real-valued $\mathfrak{A}$-measurable functions on D, consider its restriction to $E_1 \subset D$. By Prob. 6.20, there exist an $\mathfrak{A}$-measurable subset E_2 of E_1 with $\mu(E_1) < \frac{\varepsilon}{2}$ and a sequence $(c_{2,n} : n \in \mathbb{N})$ of positive real numbers in $(0, 1]$ such that

$$\lim_{n \to \infty} c_{1,n} c_{2,n} |f_n| = 0 \quad \text{on } E_1 \setminus E_2.$$

Since $c_{2,n} \in (0, 1]$ we have $c_{1,n} c_{2,n} \le c_{1,n}$. Then (1) implies that $\lim_{n \to \infty} c_{1,n} c_{2,n} |f_n| = 0$ on $D \setminus E_1$. This implies then

(2) $$\lim_{n \to \infty} c_{1,n} c_{2,n} |f_n| = 0 \quad \text{on } D \setminus E_2.$$

For the sequence $(c_{1,n} c_{2,n} f_n : n \in \mathbb{N})$ of real-valued $\mathfrak{A}$-measurable functions on D, consider its restriction to $E_2 \subset D$. By Prob. 6.20, there exist an $\mathfrak{A}$-measurable subset E_3 of E_2 with $\mu(E_3) < \frac{\varepsilon}{3}$ and a sequence $(c_{3,n} : n \in \mathbb{N})$ of positive real numbers in $(0, 1]$ such that

$$\lim_{n \to \infty} c_{1,n} c_{2,n} c_{3,n} |f_n| = 0 \quad \text{on } E_2 \setminus E_3.$$

Since $c_{3,n} \in (0, 1]$ we have $c_{1,n} c_{2,n} c_{3,n} \le c_{1,n} c_{2,n}$. Then (2) implies that we have

$$\lim_{n \to \infty} c_{1,n} c_{2,n} c_{3,n} |f_n| = 0 \quad \text{on } D \setminus E_2.$$

This implies then

(3) $$\lim_{n \to \infty} c_{1,n} c_{2,n} c_{3,n} |f_n| = 0 \quad \text{on } D \setminus E_3.$$

Thus iterating, for every $k \in \mathbb{N}$ there exist an $\mathfrak{A}$-measurable subset E_k of E_{k-1} with $\mu(E_k) < \frac{\varepsilon}{k}$ and a sequence $(c_{k,n} : n \in \mathbb{N})$ of positive real numbers in $(0, 1]$ such that

(4) $$\lim_{n \to \infty} c_{1,n} c_{2,n} c_{3,n} \cdots c_{k,n} |f_n| = 0 \quad \text{on } D \setminus E_k.$$

Consider the decreasing sequence $(E_n : n \in \mathbb{N})$ in $\mathfrak{A}$. Let $E = \bigcap_{n \in \mathbb{N}} E_k = \lim_{k \to \infty} E_k$. Then since $E_k \subset D$ and $\mu(D) < \infty$, we have

(5) $$\mu(E) = \mu\left(\lim_{k \to \infty} E_k \right) = \lim_{k \to \infty} \mu(E_k) \le \lim_{k \to \infty} \frac{\varepsilon}{k} = 0.$$

This shows that the subset E of D is a null set in $(X, \mathfrak{A}, \mu)$. Observe also that

$$(6) \qquad D \setminus E = D \cap E^c = D \cap \left(\bigcap_{k \in \mathbb{N}} E_k \right)^c = D \cap \left(\bigcup_{k \in \mathbb{N}} E_k^c \right)$$

$$= \bigcup_{k \in \mathbb{N}} (D \cap E_k^c) = \bigcup_{k \in \mathbb{N}} (D \setminus E_k).$$

2. Consider the array of real-valued $\mathfrak{A}$-measurable functions on D

$$(7) \qquad \begin{bmatrix} F_{1,1} & F_{1,2} & F_{1,3} & \cdots \\ F_{2,1} & F_{2,2} & F_{2,3} & \cdots \\ F_{3,1} & F_{3,2} & F_{3,3} & \cdots \\ \vdots & \vdots & \vdots & \end{bmatrix}$$

$$= \begin{bmatrix} c_{1,1}|f_1| & c_{1,2}|f_2| & c_{1,3}|f_3| & \cdots \\ c_{1,1}c_{2,1}|f_1| & c_{1,2}c_{2,2}|f_2| & c_{1,3}c_{2,3}|f_3| & \cdots \\ c_{1,1}c_{2,1}c_{3,1}|f_1| & c_{1,2}c_{2,2}c_{3,2}|f_2| & c_{1,3}c_{2,3}c_{3,3}|f_3| & \cdots \\ \vdots & \vdots & \vdots & \end{bmatrix}.$$

Let us observe that since $c_{i,j} \in (0, 1]$ for $i, j \in \mathbb{N}$ each column in the array (7) is a decreasing sequence of functions on D. Observe also that (4) implies that for every $k \in \mathbb{N}$ the k-th row in the array (7), that is, the sequence $(F_{k,n} : n \in \mathbb{N})$ converges to 0 on $D \setminus E_k$.

3. Let us show that the diagonal sequence $(F_{n,n} : n \in \mathbb{N})$ in the array (7) converges to 0 on $D \setminus E$. Let $x \in D \setminus E$. Then by (6) we have $x \in D \setminus E_{k_0}$ for some $k_0 \in \mathbb{N}$. Then since $(F_{k_0,n} : n \in \mathbb{N})$ converges to 0 on $D \setminus E_{k_0}$, we have

$$(8) \qquad \lim_{n \to \infty} F_{k_0,n}(x) = 0 \quad \text{for our } x \in D \setminus E.$$

Now since each column in the array (7) is a decreasing sequence of functions, we have

$$0 \leq F_{n,n} \leq F_{k_0,n} \quad \text{for } n \geq k_0.$$

Thus (8) implies

$$(9) \qquad \lim_{n \to \infty} F_{n,n}(x) = 0 \quad \text{for } x \in D \setminus E.$$

From (7) we have for every $n \in \mathbb{N}$

$$F_{n,n} = c_{1,n}c_{2,n}c_{3,n} \cdots c_{n,n}|f_n|.$$

Let us define

$$a_n = c_{1,n}c_{2,n}c_{3,n} \cdots c_{n,n}.$$

Then $F_{n,n} = a_n|f_n|$ for every $n \in \mathbb{N}$ and (9) is transcribed as

$$(10) \qquad \lim_{n \to \infty} a_n|f_n(x)| = 0 \quad \text{for } x \in D \setminus E.$$

Since for every sequence of real numbers $(\gamma_n : n \in \mathbb{N})$ we have $\lim\limits_{n \to \infty} \gamma_n = 0$ if and only if we have $\lim\limits_{n \to \infty} |\gamma_n| = 0$, (10) is equivalent to

(11) $$\lim_{n \to \infty} a_n f_n(x) = 0 \quad \text{for } x \in D \setminus E.$$

Then since E is a null set in $(X, \mathfrak{A}, \mu)$ contained in D, we have

$$\lim_{n \to \infty} a_n f_n(x) = 0 \quad \text{for a.e. } x \in D.$$

This completes the proof. ∎

§7 Integration of Bounded Functions on Sets of Finite Measure

Prob. 7.1. Let $(X, \mathfrak{A}, \mu)$ be a measure space. Let f be a real-valued $\mathfrak{A}$-measurable function on a set $D \in \mathfrak{A}$ with $\mu(D) < \infty$. Show that for every $\varepsilon > 0$ there exists a set $E \in \mathfrak{A}$ such that $E \subset D$, $\mu(E) < \varepsilon$ and f is bounded on $D \setminus E$.

Proof. Let $D_k = \{D : |f| \leq k\}$ for $k \in \mathbb{N}$. Then $(D_k : k \in \mathbb{N})$ is an increasing sequence in $\mathfrak{A}$ and $D = \bigcup_{k \in \mathbb{N}} D_k$. Thus we have

$$\mu(D) = \mu\Big(\bigcup_{k \in \mathbb{N}} D_k\Big) = \lim_{k \to \infty} \mu(D_k).$$

Since $\mu(D) < \infty$, for every $\varepsilon > 0$ there exists $k_0 \in \mathbb{N}$ such that

$$\mu(D) - \mu(D_{k_0}) < \varepsilon.$$

Let $E = D \setminus D_{k_0}$. Then $\mu(E) = \mu(D) - \mu(D_{k_0}) < \varepsilon$. Now we have $D \setminus E = D_{k_0}$. On D_{k_0} we have $|f| \leq k_0$. Thus f is bounded on $D \setminus E$. ∎

Prob. 7.2. Let f be an extended real-valued $\mathfrak{A}$-measurable function on a set $D \in \mathfrak{A}$ in a measure space $(X, \mathfrak{A}, \mu)$. For $M_1, M_2 \in \mathbb{R}$, $M_1 < M_2$, let the truncation of f at M_1 and M_2 be defined by

$$g(x) = \begin{cases} M_1 & \text{if } f(x) < M_1, \\ f(x) & \text{if } f(x) \in [M_1, M_2], \\ M_2 & \text{if } f(x) > M_2. \end{cases}$$

Show that g is $\mathfrak{A}$-measurable on D.

Proof. Let us define three subsets of D by setting

$$D_1 = \{x \in D : f(x) < M_1\},$$
$$D_2 = \{x \in D : f(x) \in [M_1, M_2]\},$$
$$D_3 = \{x \in D : f(x) > M_2\}.$$

Then $\{D_i : i = 1, 2, 3\}$ is a disjoint collection and $D = \bigcup_{i=1}^{3}$. Moreover the $\mathfrak{A}$-measurability of f on D implies that $D_i \in \mathfrak{A}$ for $i = 1, 2, 3$.

Let us define functions g_1, g_2 and g_3 on D_1, D_2 and D_3 respectively by setting

$$g_1(x) = M_1 \quad \text{for } x \in D_1,$$
$$g_2(x) = f(x) \quad \text{for } x \in D_2,$$
$$g_3(x) = M_2 \quad \text{for } x \in D_3.$$

Then g_1, g_2 and g_3 are $\mathfrak{A}$-measurable on D_1, D_2 and D_3 respectively. This implies that g is $\mathfrak{A}$-measurable on D. ∎

Prob. 7.3. Let f be a bounded real-valued $\mathfrak{A}$-measurable function on a set $D \in \mathfrak{A}$ in a measure space $(X, \mathfrak{A}, \mu)$.
(a) Show that there exists an increasing sequence of simple functions on D, $(\varphi_n : n \in \mathbb{N})$, such that $\varphi_n \uparrow f$ uniformly on D.
(b) Show that there exists a decreasing sequence of simple functions on D, $(\psi_n : n \in \mathbb{N})$, such that $\psi_n \downarrow f$ uniformly on D.

Proof. Let $M > 0$ be a bound of f on D. Then we have $|f(x)| \le M$ for $x \in D$. The interval $[-M, M]$ contains the range of the function f. For every $n \in \mathbb{N}$, let us decompose the interval $[-M, M]$ with length $2M$ into $2 \cdot 2^n$ disjoint intervals of length $\frac{M}{2^n}$ by setting

$$I_{n,k} = \left[\frac{k}{2^n}M, \frac{k+1}{2^n}M\right) \quad \text{for } k = -2^n, -2^n + 1, -2^n + 2, \ldots, 2^n - 2,$$
$$I_{n,2^n-1} = \left[\frac{2^n-1}{2^n}M, M\right].$$

Let
$$D_{n,k} = f^{-1}(I_{n,k}) \quad \text{for } k = -2^n, -2^n + 1, -2^n + 2, \ldots, 2^n - 2, 2^n - 1.$$

Then $\{D_{n,k} : k = -2^n, -2^n + 1, -2^n + 2, \ldots, 2^n - 2, 2^n - 1\}$ is a disjoint collection in $\mathfrak{A}$ and

$$\bigcup_{k=-2^n}^{2^n-1} D_{n,k} = D.$$

Let us define two simple functions on D, φ_n and ψ_n, by setting

$$\varphi_n(x) = \frac{k}{2^n} \quad \text{for } x \in D_{n,k}, \text{ for } k = -2^n, \ldots, 2^n - 1,$$
$$\psi_n(x) = \frac{k+1}{2^n} \quad \text{for } x \in D_{n,k}, \text{ for } k = -2^n, \ldots, 2^n - 1.$$

Then we have $\varphi_n(x) \uparrow$ and $\psi_n(x) \downarrow$ as $n \to \infty$ for every $x \in D$ and moreover

$$\varphi_n(x) \le f(x) \le \psi_n(x) \quad \text{for every } x \in D,$$

and

$$f(x) - \varphi_n(x) \le \frac{1}{2^n}M \quad \text{for every } x \in D,$$
$$\psi_n(x) - f(x) \le \frac{1}{2^n}M \quad \text{for every } x \in D.$$

This shows that $\varphi_n \uparrow f$ and $\psi_n \downarrow f$ uniformly on D. ∎

Prob. 7.4. Let f be a real-valued $\mathfrak{A}$-measurable function on a set $D \in \mathfrak{A}$ in a measure space $(X, \mathfrak{A}, \mu)$. Suppose f is not bounded on D.
(a) Show that it is not possible that for an arbitrary $\varepsilon > 0$ there exists a simple function φ on D such that $|\varphi(x) - f(x)| < \varepsilon$ for all $x \in D$.
(b) Show that there cannot be any sequence of simple function which converges to f uniformly on D.

Proof. 1. Let us prove (a) by contradiction argument. Thus assume that for an arbitrary $\varepsilon > 0$ there exists a simple function φ on D such that

(1) $$|\varphi(x) - f(x)| < \varepsilon \quad \text{for all } x \in D.$$

Let the simple function φ be represented by

(2) $$\varphi(x) = \sum_{k=1}^{n} a_k \mathbf{1}_{D_k}(x) \quad \text{for } x \in D,$$

where $\{D_k : k = 1, \ldots, n\}$ is a disjoint collection of $\mathfrak{A}$-measurable subsets of D with $\bigcup_{k=1}^{n} D_k = D$ and $\{a_k : k = 1, \ldots, n\} \subset \mathbb{R}$. Then (1) and (2) imply

$$|a_k - f(x)| < \varepsilon \quad \text{for all } x \in D_k \text{ for } k = 1, \ldots, n$$

and then

$$\big||a_k| - |f(x)|\big| \leq |a_k - f(x)| < \varepsilon \quad \text{for all } x \in D_k \text{ for } k = 1, \ldots, n$$

and thus

(3) $$|f(x)| \leq |a_k| + \varepsilon \quad \text{for all } x \in D_k \text{ for } k = 1, \ldots, n.$$

Let $M = \max\{|a_k| : k = 1, \ldots, n\}$. Then (3) implies

$$|f(x)| \leq M + \varepsilon \quad \text{for all } x \in D.$$

This contradicts the assumption that f is not bounded on D.
   2. We prove (b) by contradiction argument. Assume that there exists a sequence $(\varphi_n : n \in \mathbb{N})$ of simple functions on D that converges to f uniformly on D. Then for every $\varepsilon > 0$ there exists $N \in \mathbb{N}$ such that

$$|\varphi_n(x) - f(x)| < \varepsilon \quad \text{for all } x \in D \text{ for } n \geq N.$$

This contradicts (a). ■

Prob. 7.5. Given a measure space $(X, \mathfrak{A}, \mu)$. Let us call a function φ an elementary function if it satisfies the following conditions:

1° $\mathfrak{D}(\varphi) \in \mathfrak{A}$,

2° φ is $\mathfrak{A}$-measurable on $\mathfrak{D}(\varphi)$,

3° φ assumes only countably many real values, that is, $\mathfrak{R}(\varphi)$ is a countable subset of $\mathbb{R}$.

Let f be a real-valued $\mathfrak{A}$-measurable function on a set $D \in \mathfrak{A}$.

(a) Show that there exists an increasing sequence of elementary functions on D, $(\varphi_n : n \in \mathbb{N})$, such that $\varphi_n \uparrow f$ uniformly on D.

(b) Show that there exists a decreasing sequence of elementary functions on D, $(\psi_n : n \in \mathbb{N})$, such that $\psi_n \downarrow f$ uniformly on D.

Proof. $\mathbb{R} = (-\infty, \infty)$ contains the range of f. For each $n \in \mathbb{N}$, let us decompose $\mathbb{R}$ into countably many disjoint intervals of length $\frac{1}{2^n}$ by setting

$$I_{n,k} = \left[\tfrac{k}{2^n}, \tfrac{k+1}{2^n} \right) \quad \text{for } k \in \mathbb{Z}.$$

Let

$$D_{n,k} = f^{-1}(I_{n,k}) \quad \text{for } k \in \mathbb{Z}.$$

Then $\{ D_{n,k} : k \in \mathbb{Z} \}$ is a countable disjoint collection in $\mathfrak{A}$ and $\bigcup_{k \in \mathbb{Z}} D_{n,k} = D$.

Let us define two elementary functions on D, φ_n and ψ_n, by setting

$$\varphi_n(x) = \tfrac{k}{2^n} \quad \text{for } x \in D_{n,k}, \text{ for } k \in \mathbb{Z},$$

$$\psi_n(x) = \tfrac{k+1}{2^n} \quad \text{for } x \in D_{n,k}, \text{ for } k \in \mathbb{Z}.$$

Then we have $\varphi_n(x) \uparrow$ and $\psi_n(x) \downarrow$ as $n \to \infty$ for every $x \in D$ and moreover

$$\varphi_n(x) \leq f(x) \leq \psi_n(x) \quad \text{for every } x \in D,$$

and

$$f(x) - \varphi_n(x) \leq \tfrac{1}{2^n} \quad \text{for every } x \in D,$$

$$\psi_n(x) - f(x) \leq \tfrac{1}{2^n} \quad \text{for every } x \in D.$$

This shows that $\varphi_n \uparrow f$ and $\psi_n \downarrow f$ uniformly on D. ∎

Prob. 7.6. Given a measure space $(X, \mathfrak{A}, \mu)$. Let φ be a nonnegative elementary function on a set $D \in \mathfrak{A}$ with $\mu(D) < \infty$. Let $\{\alpha_i : i \in \mathbb{N}\}$ be the set of nonnegative real values assumed by φ and let $D_i = \{x \in D : \varphi(x) = \alpha_i\}$ for $i \in \mathbb{N}$. We call the expression

$$\varphi = \sum_{i \in \mathbb{N}} \alpha_i \mathbf{1}_{D_i}$$

the canonical representation of φ. We define

$$\int_D \varphi \, d\mu = \sum_{i \in \mathbb{N}} \alpha_i \mu(D_i) \in [0, \infty].$$

Let f be a nonnegative real-valued function on D. Let Φ be the collection of all nonnegative elementary functions φ such that $0 \le \varphi \le f$ on D and let Ψ be the collection of all nonnegative elementary functions φ such that $0 \le f \le \psi$ on D.
(a) Show that $\sup_{\varphi \in \Phi} \int_D \varphi \, d\mu \le \inf_{\psi \in \Psi} \int_D \psi \, d\mu$.
(b) Show that if f is $\mathfrak{A}$-measurable on D then $\sup_{\varphi \in \Phi} \int_D \varphi \, d\mu = \inf_{\psi \in \Psi} \int_D \psi \, d\mu$.
(Then for a not necessarily bounded nonnegative real-valued $\mathfrak{A}$-measurable function f on $D \in \mathfrak{A}$ with $\mu(D) < \infty$, we define $\int_D f \, d\mu = \sup_{\varphi \in \Phi} \int_D \varphi \, d\mu = \inf_{\psi \in \Psi} \int_D \psi \, d\mu$.)
(c) Show that if $\sup_{\varphi \in \Phi} \int_D \varphi \, d\mu = \inf_{\psi \in \Psi} \int_D \psi \, d\mu$ then f is $\mathfrak{A}$-measurable on D, provided that $(X, \mathfrak{A}, \mu)$ is a complete measure space.

Proof. 1. Let us prove (a). Let an arbitrary $\varphi \in \Phi$ and an arbitrary $\psi \in \Psi$ be represented by

(1) $$\varphi = \sum_{i \in \mathbb{N}} \alpha_i \mathbf{1}_{D_i} \quad \text{and} \quad \psi = \sum_{j \in \mathbb{N}} \beta_j \mathbf{1}_{E_j}.$$

Consider $D_i \cap E_j$ for $i \in \mathbb{N}$ and $j \in \mathbb{N}$. Since $\{D_i : i \in \mathbb{N}\}$ is a disjoint collection in $\mathfrak{A}$ with $\bigcup_{i \in \mathbb{N}} D_i = D$ and similarly $\{E_j : j \in \mathbb{N}\}$ is a disjoint collection in $\mathfrak{A}$ with $\bigcup_{j \in \mathbb{N}} E_j = D$, $\{D_i \cap E_j : i \in \mathbb{N}, j \in \mathbb{N}\}$ is a disjoint collection in $\mathfrak{A}$ and $\bigcup_{i \in \mathbb{N}} \bigcup_{j \in \mathbb{N}} D_i \cap E_j = D$. Then

(2) $$\int_D \varphi \, d\mu = \sum_{i \in \mathbb{N}} \sum_{j \in \mathbb{N}} \int_{D_i \cap E_j} \varphi \, d\mu \quad \text{and} \quad \int_D \psi \, d\mu = \sum_{i \in \mathbb{N}} \sum_{j \in \mathbb{N}} \int_{D_i \cap E_j} \psi \, d\mu.$$

Since $\varphi \le f \le \psi$ on D and in particular on $D_i \cap E_j$, we have $\int_{D_i \cap E_j} \varphi \, d\mu \le \int_{D_i \cap E_j} \psi \, d\mu$ for $i \in \mathbb{N}$ and $j \in \mathbb{N}$. Then by (2) we have

(3) $$\int_D \varphi \, d\mu \le \int_D \psi \, d\mu.$$

The inequality (3) holds for an arbitrary $\varphi \in \Phi$ and an arbitrary $\psi \in \Psi$. Then for an arbitrary $\psi_0 \in \Psi$ we have

$$\int_D \varphi \, d\mu \le \int_D \psi_0 \, d\mu \quad \text{for every } \varphi \in \Phi,$$

and this implies

$$\inf_{\varphi \in \Phi} \sup \int_D \varphi \, d\mu \leq \int_D \psi_0 \, d\mu.$$

Since this holds for an arbitrary $\psi_0 \in \Psi$, we have

$$\sup_{\varphi \in \Phi} \int_D \varphi \, d\mu \leq \inf_{\psi \in \Psi} \int_D \psi \, d\mu.$$

2. Let us prove (b). Assume that the nonnegative real-valued function f on D is $\mathfrak{A}$-measurable on D. Now $[0, \infty)$ contains the range of f. For each $n \in \mathbb{N}$ let us decompose $[0, \infty)$ into countably many disjoint subintervals given by

$$I_{n,k} = \left[\frac{k-1}{2^n}, \frac{k}{2^n} \right) \quad \text{for } k \in \mathbb{N}.$$

Let

$$D_{n,k} = \left\{ x \in D : f(x) \in I_{n,k} \right\} \quad \text{for } k \in \mathbb{N}.$$

Since f is $\mathfrak{A}$-measurable on D, we have $D_{n,k} \in \mathfrak{A}$ for $k \in \mathbb{N}$. Then $\left\{ D_{n,k} : k \in \mathbb{N} \right\}$ is a disjoint collection in $\mathfrak{A}$ and $\bigcup_{k \in \mathbb{N}} D_{n,k} = D$. Let us define nonnegative elementary functions φ_n and ψ_n on D by setting

$$\varphi_n = \sum_{k \in \mathbb{N}} \frac{k-1}{2^n} \mathbf{1}_{D_{n,k}} \quad \text{and} \quad \psi_n = \sum_{k \in \mathbb{N}} \frac{k}{2^n} \mathbf{1}_{D_{n,k}}.$$

Then we have $0 \leq \varphi_n \leq f \leq \psi_n$ so that $\varphi_n \in \Phi$ and $\psi_n \in \Psi$. Now by (a) we have

$$0 \leq \inf_{\psi \in \Psi} \int_D \psi \, d\mu - \sup_{\varphi \in \Phi} \int_D \varphi \, d\mu \leq \int_D \psi_n \, d\mu - \int_D \varphi_n \, d\mu$$

$$= \int_D \{\psi_n - \varphi_n\} \, d\mu = \sum_{k \in \mathbb{N}} \left\{ \frac{k}{2^n} - \frac{k-1}{2^n} \right\} \mu(D_{n,k}) = \frac{1}{2^n} \sum_{k \in \mathbb{N}} \mu(D_{n,k})$$

$$= \frac{1}{2^n} \sum_{k \in \mathbb{N}} \mu(D).$$

Letting $n \to \infty$, we have $0 \leq \inf_{\psi \in \Psi} \int_D \psi \, d\mu - \sup_{\varphi \in \Phi} \int_D \varphi \, d\mu \leq 0$ and thus we have

$$\inf_{\psi \in \Psi} \int_D \psi \, d\mu = \sup_{\varphi \in \Phi} \int_D \varphi \, d\mu.$$

This completes the proof of (b).

3. (c) is proved by the same argument as in the proof of (b) in Theorem 7.9. ∎

Prob. 7.7. Let f be a bounded nonnegative real-valued $\mathfrak{A}$-measurable function on a set $D \in \mathfrak{A}$ with $\mu(D) < \infty$. Show that if $\int_D f \, d\mu = 0$ then $f = 0$ a.e. on D.

Proof. By definition, we say that $f = 0$ a.e. on D if there exists a subset E of D such that E is a null set in the measure space $(X, \mathfrak{A}, \mu)$ and $f = 0$ on $D \setminus E$.

Thus if $f = 0$ a.e. on D then $\{D : f \neq 0\} \subset E$ and we have $\mu\{D : f \neq 0\} \leq \mu(E) = 0$. Conversely if $\mu\{D : f \neq 0\} = 0$, then the subset $\{D : f \neq 0\}$ of D is a null set in the measure space $(X, \mathfrak{A}, \mu)$ and $f = 0$ on $D \setminus \{D : f \neq 0\}$ so that $f = 0$ a.e. on D. Thus we have shown that

(1) $f = 0$ a.e. on $D \Leftrightarrow \mu\{D : f \neq 0\} = 0$.

Let us show that if $\int_D f \, d\mu = 0$ then $f = 0$ a.e. on D. Suppose the statement that $f = 0$ a.e. on D is false. Then according to (1) the statement that $\mu\{D : f \neq 0\} = 0$ is false and consequently we have

(2) $\mu\{D : f \neq 0\} > 0$.

Now if $\mu(D) = 0$ to start with then $\mu\{D : f \neq 0\} > 0$ is impossible. Thus in this case, $\int_D f \, d\mu = 0$ implies that $f = 0$ a.e. on D. Consider the case $\mu(D) > 0$. Since f is nonnegative on D, we have $\{D : f \neq 0\} = \{D : f > 0\}$. Thus

(3) $\alpha := \mu\{D : f > 0\} = \mu\{D : f \neq 0\} > 0$.

Now for $n \in \mathbb{N}$, let

$$D_n = \left\{ D : f \geq \tfrac{1}{n} \right\}.$$

Then $(D_n : n \in \mathbb{N})$ is an increasing sequence in $\mathfrak{A}$ and

$$\lim_{n \to \infty} D_n = \bigcup_{n \in \mathbb{N}} D_n = \{D : f > 0\}$$

and

$$\lim_{n \to \infty} \mu(D_n) = \mu\left(\lim_{n \to \infty} D_n \right) = \mu\left(\bigcup_{n \in \mathbb{N}} D_n \right) = \mu\{D : f > 0\} = \alpha.$$

This implies that there exists $n_0 \in \mathbb{N}$ such that $\mu(D_{n_0}) > \frac{\alpha}{2}$. Then we have

$$\int_D f \, d\mu \geq \int_{D_{n_0}} f \, d\mu \geq \frac{1}{n_0} \mu(D_{n_0}) > \frac{1}{n_0} \frac{\alpha}{2} > 0.$$

This contradicts the assumption that $\int_D f \, d\mu = 0$. Therefore the statement that $f = 0$ a.e. on D must be valid. ∎

Prob. 7.8. Given a measure space $(X, \mathfrak{A}, \mu)$. Let f be a bounded real-valued $\mathfrak{A}$-measurable function on a set $D \in \mathfrak{A}$ with $\mu(D) < \infty$. Suppose $|f(x)| \leq M$ for $x \in D$ for some constant $M > 0$.
(a) Show that if $\int_D f \, d\mu = M\mu(D)$, then $f = M$ a.e. on D.
(b) Show that if $f < M$ a.e. on D and if $\mu(D) > 0$, then $\int_D f \, d\mu < M\mu(D)$.

Proof. 1. Let us prove (a). Let us define a function on D by setting $g = M - f$ on D. Since $-M \leq f \leq M$ on D, we have $0 \leq g \leq 2M$ on D. Thus g is a bounded nonnegative real-valued $\mathfrak{A}$-measurable function on D. If $\int_D f \, d\mu = M\mu(D)$, then we have

$$\int_D g \, d\mu = \int_D \{M - f\} d\mu = M\mu(D) - \int_D f \, d\mu = M\mu(D) - M\mu(D) = 0.$$

Then by Prob. 7.7, we have $g = 0$ a.e. on D, that is, $f = M$ a.e. on D.
   2. Let us prove (b). Suppose $f < M$ a.e. on D. Now since $f \leq M$ on D, we have

(1)
$$\int_D f \, d\mu \leq M\mu(D).$$

Thus if $\int_D f \, d\mu < M\mu(D)$ is not valid then we have

(2)
$$\int_D f \, d\mu = M\mu(D).$$

Then we have

(3)
$$\int_D \{M - f\} d\mu = M\mu(D) - \int_D f \, d\mu = m\mu(D) - M\mu(D) = 0.$$

According to Prob. 7.7, (3) implies that $M - f = 0$ a.e. on D, that is, $f = M$ a.e. on D. Thus we have

$$\begin{cases} f < M \quad \text{a.e. on } D, \\ f = M \quad \text{a.e. on } D. \end{cases}$$

This implies that there exists a subset D_1 of D such that D_1 is a null set in the measure space $(X, \mathfrak{A}, \mu)$ and $f < M$ on $D \setminus D_1$ and there exists a subset D_2 of D such that D_2 is a null set in the measure space $(X, \mathfrak{A}, \mu)$ and $f = M$ on $D \setminus D_2$. Then the subset $D_1 \cup D_2$ of D is a null set in $(X, \mathfrak{A}, \mu)$ and

$$f < M \quad \text{and} \quad f = M \quad \text{on } D \setminus (D_1 \cup D_2).$$

Then since $D_1 \cup D_2$ is a null set in $(X, \mathfrak{A}, \mu)$ and $\mu(D) > 0$, we have $\mu(D \setminus (D_1 \cup D_2)) > 0$. Thus $D \setminus (D_1 \cup D_2)$ is a non-empty set and therefore there exists $x \in D \setminus (D_1 \cup D_2)$. Then we have $f(x) < M$ and $f(x) = M$. This is a contradiction. Therefore we must have $\int_D f \, d\mu < M\mu(D)$. ∎

Prob. 7.9. Let f be a bounded real-valued $\mathfrak{A}$-measurable function on a set $D \in \mathfrak{A}$ with $\mu(D) < \infty$ satisfying $f(x) \in (-M, M]$ for $x \in D$ for some $M > 0$.
For each $n \in \mathbb{N}$, let
$$D_{n,k} = \left\{ x \in D : f(x) \in \left((k-1)\tfrac{M}{n}, k\tfrac{M}{n} \right] \right\} \text{ for } k = -n+1, \cdots, n.$$
Define simple functions φ_n and ψ_n on D by setting
$$\varphi_n = \sum_{k=-n+1}^{n} (k-1)\tfrac{M}{n} \cdot \mathbf{1}_{D_{n,k}}$$
and
$$\psi_n = \sum_{k=-n+1}^{n} k\tfrac{M}{n} \cdot \mathbf{1}_{D_{n,k}}.$$
Show that $\lim_{n\to\infty} \int_D \varphi_n \, d\mu = \int_D f \, d\mu$ and $\lim_{n\to\infty} \int_D \psi_n \, d\mu = \int_D f \, d\mu$.

Proof. Observe that

$$(1) \qquad \int_D \varphi_n \, d\mu \le \sup_{\varphi \le f} \int_D \varphi \, d\mu = \int_D f \, d\mu \le \inf_{f \le \psi} \int_D \psi \, d\mu \le \int_D \psi_n \, d\mu$$

and

$$(2) \qquad \int_D \psi_n \, d\mu - \int_D \varphi_n \, d\mu = \int_D \{\psi_n - \varphi_n\} \, d\mu = \frac{M}{n}\mu(D).$$

Then we have

$$0 \le \lim_{n\to\infty} \left\{ \int_D f \, d\mu - \int_D \varphi_n \, d\mu \right\}$$
$$\le \lim_{n\to\infty} \left\{ \int_D \psi_n \, d\mu - \int_D \varphi_n \, d\mu \right\} \quad \text{by (1)}$$
$$= \lim_{n\to\infty} \frac{M}{n}\mu(D) = 0 \quad \text{by (2)}.$$

This implies that

$$(3) \qquad \int_D f \, d\mu = \lim_{n\to\infty} \int_D \varphi_n \, d\mu.$$

Similarly we have

$$0 \le \lim_{n\to\infty} \left\{ \int_D \psi_n \, d\mu - \int_D f \, d\mu \right\}$$
$$\le \lim_{n\to\infty} \left\{ \int_D \psi_n \, d\mu - \int_D \varphi_n \, d\mu \right\} \quad \text{by (1)}$$
$$= \lim_{n\to\infty} \frac{M}{n}\mu(D) = 0 \quad \text{by (2)}.$$

This implies that

$$(4) \qquad \int_D f \, d\mu = \lim_{n\to\infty} \int_D \psi_n \, d\mu. \quad \blacksquare$$

Prob. 7.10. Let f be a bounded real-valued $\mathfrak{A}$-measurable function on a set $D \in \mathfrak{A}$ with $\mu(D) < \infty$ satisfying $f(x) \in (-M, M]$ for $x \in D$ for some $M > 0$.
Let $\mathcal{P} = (y_0, \ldots, y_p)$ be a partition of $[-M, M]$, that is, $-M = y_0 < \cdots < y_p = M$.
Let $I_k = [y_{k-1}, y_k]$ for $k = 1, \ldots, p$ and let $|\mathcal{P}| = \max_{k=1,\ldots,p} \ell(I_k)$.
Let $E_k = \{x \in D : f(x) \in (y_{k-1}, y_k]\}$ for $k = 1, \ldots, p$.
Let $\varphi(\mathcal{P})$ and $\psi(\mathcal{P})$ be simple functions on D defined by $\varphi(\mathcal{P}) = \sum_{k=1}^{p} y_{k-1} \cdot \mathbf{1}_{E_k}$ and $\psi(\mathcal{P}) = \sum_{k=1}^{p} y_k \cdot \mathbf{1}_{E_k}$.
Let $(\mathcal{P}_n : n \in \mathbb{N})$ be a sequence of partitions such that $\lim_{n \to \infty} |\mathcal{P}_n| = 0$.
Show that $\lim_{n \to \infty} \int_D \varphi(\mathcal{P}_n)\, d\mu = \int_D f\, d\mu$ and $\lim_{n \to \infty} \int_D \psi(\mathcal{P}_n)\, d\mu = \int_D f\, d\mu$.

Proof. Observe that

(1)
$$\int_D \varphi_n\, d\mu \le \sup_{\varphi \le f} \int_D \varphi\, d\mu = \int_D f\, d\mu \le \inf_{f \le \psi} \int_D \psi\, d\mu \le \int_D \psi_n\, d\mu$$

and

(2)
$$\int_D \psi_n\, d\mu - \int_D \varphi_n\, d\mu = \int_D \{\psi_n - \varphi_n\}\, d\mu = \frac{M}{n} \mu(D).$$

Then we have

$$0 \le \lim_{n \to \infty} \left\{ \int_D f\, d\mu - \int_D \varphi_n\, d\mu \right\}$$
$$\le \lim_{n \to \infty} \left\{ \int_D \psi_n\, d\mu - \int_D \varphi_n\, d\mu \right\} \quad \text{by (1)}$$
$$= \lim_{n \to \infty} \frac{M}{n} \mu(D) = 0 \quad \text{by (2).}$$

This implies that

(3)
$$\int_D f\, d\mu = \lim_{n \to \infty} \int_D \varphi_n\, d\mu.$$

Similarly we have

$$0 \le \lim_{n \to \infty} \left\{ \int_D \psi_n\, d\mu - \int_D f\, d\mu \right\}$$
$$\le \lim_{n \to \infty} \left\{ \int_D \psi_n\, d\mu - \int_D \varphi_n\, d\mu \right\} \quad \text{by (1)}$$
$$= \lim_{n \to \infty} \frac{M}{n} \mu(D) = 0 \quad \text{by (2).}$$

This implies that

(4)
$$\int_D f\, d\mu = \lim_{n \to \infty} \int_D \psi_n\, d\mu. \quad \blacksquare$$

Prob. 7.11. Consider a sequence of functions $(f_n : n \in \mathbb{N})$ defined on $[0, 1]$ by setting
$f_n(x) = \dfrac{nx}{1 + n^2 x^2}$ for $x \in [0, 1]$.

(a) Show that $(f_n : n \in \mathbb{N})$ is a uniformly bounded sequence on $[0, 1]$ and evaluate

$$\lim_{n \to \infty} \int_{[0,1]} \frac{nx}{1 + n^2 x^2} \, \mu_L(dx).$$

(b) Show that $(f_n : n \in \mathbb{N})$ does not converge uniformly on $[0, 1]$.
(Recall that for the Riemann integral, uniform convergence of $(f_n : n \in \mathbb{N})$ on $[0, 1]$ implies
$\lim_{n \to \infty} \int_0^1 f_n(x) \, dx = \int_0^1 \lim_{n \to \infty} f_n(x) \, dx$.)

Proof. 1. For each $n \in \mathbb{N}$, f_n is a nonnegative real-valued continuous function on $[0, 1]$
and thus a $\mathfrak{B}_{\mathbb{R}}$-measurable function on $[0, 1]$. Moreover we have

$$(1) \qquad \lim_{n \to \infty} f_n(x) = \lim_{n \to \infty} \frac{nx}{1 + n^2 x^2} = \lim_{n \to \infty} \frac{x}{\frac{1}{n} + n x^2} = 0 \quad \text{for } x \in [0, 1].$$

Let us show that $(f_n : n \in \mathbb{N})$ is a uniformly bounded sequence on $[0, 1]$. Observe that

$$f_n'(x) = \frac{n(1 - n^2 x^2)}{1 + n^2 x^2} \begin{cases} > 0 & \text{for } x < \frac{1}{n}, \\ = 0 & \text{for } x = \frac{1}{n}, \\ < 0 & \text{for } x > \frac{1}{n}. \end{cases}$$

It follows from this that f_n has a maximal value at $x = \frac{1}{n}$ and moreover we have

$$(2) \qquad \max_{[0,1]} f_n = f_n\left(\frac{1}{n}\right) = \frac{n\frac{1}{n}}{1 + n^2 \frac{1}{n^2}} = \frac{1}{2}.$$

Thus we have

$$(3) \qquad |f_n(x)| = f_n(x) \le \max_{[0,1]} f_n = \frac{1}{2} \quad \text{for } x \in [0, 1] \text{ and } n \in \mathbb{N}.$$

This shows that $(f_n : n \in \mathbb{N})$ is a uniformly bounded sequence on $[0, 1]$. According to the
Bounded Convergence Theorem (Theorem 7.16), (1) and (3) imply

$$\lim_{n \to \infty} \int_{[0,1]} \frac{nx}{1 + n^2 x^2} \, \mu_L(dx) = \int_{[0,1]} \lim_{n \to \infty} \frac{nx}{1 + n^2 x^2} \, \mu_L(dx)$$

$$= \int_{[0,1]} 0 \, \mu_L(dx) = 0.$$

(Comment. Uniform boundedness of the sequence $(f_n : n \in \mathbb{N})$ on $[0, 1]$ can be proved
in several different ways. Here are some:
(i) For every $n \in \mathbb{N}$ and $x \in [0, 1]$, we have $(1 - nx)^2 \ge 0$ and thus $1 - 2nx + n^2 x^2 \ge 0$,
that is, $2nx \le 1 + n^2 x^2$. This implies $0 \le \dfrac{nx}{1 + n^2 x^2} \le \frac{1}{2}$ for $x \in [0, 1]$ and $n \in \mathbb{N}$.

(ii) For every $n \in \mathbb{N}$ and $x \in [0, 1]$, we claim that $0 \leq \dfrac{nx}{1 + n^2x^2} \leq 1$. Assume the contrary, that is, assume that $\dfrac{nx}{1 + n^2x^2} > 1$. Then we have $nx > 1 + n^2x^2$ and then

$$0 > 1 - nx + n^2x^2 = \left(nx - \tfrac{1}{2}\right)^2 + \tfrac{3}{4},$$

a contradiction.

(iii) Let us show that actually $0 \leq \dfrac{nx}{1 + n^2x^2} \leq 1$ for all $x \in [0, \infty)$ and $n \in \mathbb{N}$. We show more generally that we have

(4)
$$0 \leq \frac{\alpha}{1 + \alpha^2} \leq 1 \quad \text{for } \alpha \in [0, \infty).$$

Let $\alpha \in [0, \infty)$. If $\alpha \geq 1$ then

$$0 \leq \frac{\alpha}{1 + \alpha^2} < \frac{\alpha}{\alpha^2} = \frac{1}{\alpha} \leq 1,$$

which proves (4). On the other hand, if $\alpha \in [0, 1]$ then

$$0 \leq \frac{\alpha}{1 + \alpha^2} \leq \alpha \leq 1,$$

which proves (4). This completes the proof of (4).)

2. According to (1), we have $\lim\limits_{n \to \infty} f_n(x) = 0$ for $x \in [0, 1]$. Let us show that this convergence is not uniform on $[0, 1]$. Suppose $(f_n : n \in \mathbb{N})$ converges to 0 uniformly on $[0, 1]$. Then for every $\varepsilon > 0$, there exists $N \in \mathbb{N}$ such that

$$|f_n(x) - 0| < \varepsilon \quad \text{for all } x \in [0, 1] \text{ and } n \geq N.$$

But this is impossible since according to (2) we have $f_n\left(\tfrac{1}{n}\right) = \tfrac{1}{2}$ for every $n \in \mathbb{N}$. ∎

Prob. 7.12. Given a measure space $(X, \mathfrak{A}, \mu)$. Let $(f_n : n \in \mathbb{N})$ be a sequence of extended real-valued $\mathfrak{A}$-measurable functions on a set $D \in \mathfrak{A}$ with $\mu(D) < \infty$ and let f be an extended real-valued $\mathfrak{A}$-measurable function on D that is real-valued a.e. on D. Consider the following two conditions on $(f_n : n \in \mathbb{N})$ and f:

1° $(f_n : n \in \mathbb{N})$ converges to f in measure on D.

2° $\lim\limits_{n\to\infty} \int_D \frac{|f_n - f|}{1 + |f_n - f|} \, d\mu = 0$.

Show that 1° and 2° are equivalent, that is, 1° $\Leftrightarrow$ 2°.

Proof. For an arbitrary $\varepsilon > 0$, let us define

$$A_{\varepsilon,n} = \left\{ x \in D : |f_n(x) - f(x)| \geq \varepsilon \right\} \quad \text{for } n \in \mathbb{N}.$$

By the definition of convergence in measure, that is, Definition 6.14, $(f_n : n \in \mathbb{N})$ converges to f in measure on D if and only if

(1) $$\lim_{n\to\infty} \mu(A_{\varepsilon,n}) = 0 \quad \text{for every } \varepsilon > 0.$$

1. Let us prove 1° $\Rightarrow$ 2°. Thus assume 1°. Note that on $D \setminus A_{\varepsilon,n}$ we have $|f_n - f| < \varepsilon$ and thus $\frac{|f_n - f|}{1 + |f_n - f|} < \varepsilon$. Then we have

$$\int_D \frac{|f_n - f|}{1 + |f_n - f|} \, d\mu = \int_{A_{\varepsilon,n}} \frac{|f_n - f|}{1 + |f_n - f|} \, d\mu + \int_{D \setminus A_{\varepsilon,n}} \frac{|f_n - f|}{1 + |f_n - f|} \, d\mu$$

$$\leq \int_{A_{\varepsilon,n}} d\mu + \int_{D \setminus A_{\varepsilon,n}} \varepsilon \, d\mu = \mu(A_{\varepsilon,n}) + \varepsilon \mu(D \setminus A_{\varepsilon,n})$$

$$\leq \mu(A_{\varepsilon,n}) + \varepsilon \mu(D).$$

Then by (1) we have

$$\limsup_{n\to\infty} \int_D \frac{|f_n - f|}{1 + |f_n - f|} \, d\mu \leq \limsup_{n\to\infty} \mu(A_{\varepsilon,n}) + \varepsilon \mu(D) = \varepsilon \mu(D).$$

Since this holds for every $\varepsilon > 0$, we have

$$\limsup_{n\to\infty} \int_D \frac{|f_n - f|}{1 + |f_n - f|} \, d\mu = 0.$$

On the other hand by the nonnegativity of our integral we have

$$\liminf_{n\to\infty} \int_D \frac{|f_n - f|}{1 + |f_n - f|} \, d\mu \geq 0.$$

Therefore we have

$$\liminf_{n\to\infty} \int_D \frac{|f_n - f|}{1 + |f_n - f|} \, d\mu = \limsup_{n\to\infty} \int_D \frac{|f_n - f|}{1 + |f_n - f|} \, d\mu = 0,$$

and thus

$$\lim_{n\to\infty} \int_D \frac{|f_n - f|}{1 + |f_n - f|} \, d\mu = 0.$$

This completes the proof of $1° \Rightarrow 2°$.

2. Let us prove $2° \Rightarrow 1°$. Let us assume $2°$. To show that $1°$ holds we show that (1) holds. Now since $A_{\varepsilon,n} \subset D$ we have

(2)
$$\int_D \frac{|f_n - f|}{1 + |f_n - f|} \, d\mu \geq \int_{A_{\varepsilon,n}} \frac{|f_n - f|}{1 + |f_n - f|} \, d\mu.$$

Observe that for any $a, b \in \mathbb{R}$ such that $0 \leq a \leq b$ we have

$$\frac{a}{1 + a} \leq \frac{b}{1 + b}.$$

On $A_{\varepsilon,n}$ we have $|f_n - f| \geq \varepsilon$. Thus we have

$$\frac{\varepsilon}{1 + \varepsilon} \leq \frac{|f_n - f|}{1 + |f_n - f|} \quad \text{on } A_{\varepsilon,n}.$$

Then

(3)
$$\int_{A_{\varepsilon,n}} \frac{|f_n - f|}{1 + |f_n - f|} \, d\mu \geq \int_{A_{\varepsilon,n}} \frac{\varepsilon}{1 + \varepsilon} \, d\mu = \frac{\varepsilon}{1 + \varepsilon} \mu(A_{\varepsilon,n}).$$

Combining (2) and (3), we have

$$\int_D \frac{|f_n - f|}{1 + |f_n - f|} \, d\mu \geq \frac{\varepsilon}{1 + \varepsilon} \mu(A_{\varepsilon,n}) \geq 0.$$

Then $2°$ implies that $\lim_{n \to \infty} \frac{\varepsilon}{1+\varepsilon} \mu(A_{\varepsilon,n}) = 0$ and hence $\lim_{n \to \infty} \mu(A_{\varepsilon,n}) = 0$. This completes the proof of $2° \Rightarrow 1°$. ∎

An Alternate Proof for $1° \Rightarrow 2°$. The implication $1° \Rightarrow 2°$ can be proved by applying the Bounded Convergence Theorem. Let us assume $1°$. To prove $2°$, let us define a sequence of real numbers $(\gamma_n : n \in \mathbb{N})$ by setting

$$\gamma_n = \int_D \frac{|f_n - f|}{1 + |f_n - f|} \, d\mu \quad \text{for } n \in \mathbb{N}.$$

We are to prove that $\lim_{n \to \infty} \gamma_n = 0$.

According to Prob. 6.8, for an arbitrary sequence of extended real numbers $(a_n : n \in \mathbb{N})$ if there exists $a \in \mathbb{R}$ such that every subsequence $(a_{n_k} : k \in \mathbb{N})$ of $(a_n : n \in \mathbb{N})$ has a subsequence $(a_{n_{k_\ell}} : \ell \in \mathbb{N})$ such that $\lim_{\ell \to \infty} a_{n_{k_\ell}} = a \in \mathbb{R}$, then $\lim_{n \to \infty} a_n = a$. Therefore to show that $\lim_{n \to \infty} \gamma_n = 0$, it suffices to show that every subsequence $(\gamma_{n_k} : k \in \mathbb{N})$ of $(\gamma_n : n \in \mathbb{N})$ has a subsequence $(\gamma_{n_{k_\ell}} : \ell \in \mathbb{N})$ such that $\lim_{\ell \to \infty} \gamma_{n_{k_\ell}} = 0$.

Let $(\gamma_{n_k} : k \in \mathbb{N})$ be an arbitrary subsequence of the sequence $(\gamma_n : n \in \mathbb{N})$. Consider the corresponding subsequence $(f_{n_k} : k \in \mathbb{N})$ of the sequence $(f_n : n \in \mathbb{N})$. By assumption of $1°$, the sequence $(f_n : n \in \mathbb{N})$ converges to f in measure on D and this implies that the subsequence $(f_{n_k} : k \in \mathbb{N})$ converges to f in measure on D. According to Theorem 6.24

(F. Riesz) this implies the existence of a subsequence $(f_{n_{k_\ell}} : \ell \in \mathbb{N})$ that converges to f a.e. on D. This convergence implies

$$\lim_{\ell \to \infty} \frac{|f_{n_{k_\ell}} - f|}{1 + |f_{n_{k_\ell}} - f|} = 0 \quad \text{a.e. on } D.$$

Note that we have

$$\frac{|f_{n_{k_\ell}} - f|}{1 + |f_{n_{k_\ell}} - f|} \leq 1 \quad \text{a.e. on } D \text{ for } \ell \in \mathbb{N}.$$

Then the Bounded Convergence Theorem (Theorem 7.16) is applicable and we have

$$\lim_{\ell \to \infty} \int_D \frac{|f_{n_{k_\ell}} - f|}{1 + |f_{n_{k_\ell}} - f|} \, d\mu = \int_D 0 \, d\mu = 0,$$

that is, we have $\lim_{\ell \to \infty} \gamma_{n_{k_\ell}} = 0$. Thus we have shown that every subsequence $(\gamma_{n_k} : k \in \mathbb{N})$ of the sequence $(\gamma_n : n \in \mathbb{N})$ has a subsequence $(\gamma_{n_{k_\ell}} : \ell \in \mathbb{N})$ converging to 0. Therefore we have $\lim_{n \to \infty} \gamma_n = 0$. ∎

Prob. 7.13. Let $(X, \mathfrak{A}, \mu)$ be a finite measure space, that is, $\mu(X) < \infty$. Consider extended real-valued $\mathfrak{A}$-measurable functions that are real-valued a.e. on X. Let us declare two such functions equal if they are equal a.e. on X. Let Φ be the collection of all such functions. Let

$$\rho(f, g) = \int_X \frac{|f - g|}{1 + |f - g|} \, d\mu \quad \text{for } f, g \in \Phi.$$

(a) Show that ρ is a metric on Φ.
(b) Show that Φ is complete with respect to the metric ρ.
(c) Let $(f_n : n \in \mathbb{N})$ be a sequence in Φ and let $f \in \Phi$. Show that $\lim_{n \to \infty} \rho(f_n, f) = 0$ if and only if $(f_n : n \in \mathbb{N})$ converges to f in measure on D.

Proof. 1. Let us show that ρ is a metric on the set Φ. Observe that for $f, g \in \Phi$ we have

$$0 \le \frac{|f - g|}{1 + |f - g|} \le 1 \text{ and } \int_X \frac{|f - g|}{1 + |f - g|} \, d\mu \le \mu(X) < \infty$$

so that we have

$$\rho(f, g) \in [0, \mu(X)] \quad \text{for } f, g \in \Phi.$$

Clearly $\rho(f, g) = 0$ when $f = g$. Conversely assume that $\rho(f, g) = 0$, that is, assume that $\int_X \frac{|f-g|}{1+|f-g|} \, d\mu = 0$. Then the nonnegativity of the integrand implies that the integrand $\frac{|f-g|}{1+|f-g|} = 0$, a.e. on X. Then since the denominator $1 + |f - g| > 0$, we have the numerator $|f - g| = 0$, a.e. on X and thus $f = g$. This shows that $\rho(f, g) = 0$ implies that $f = g$.

Clearly $\rho(f, g) = \rho(g, f)$ for any $f, g \in \Phi$. It remains to verify the triangle inequality for ρ. For this we require the following estimates. The first one is:

(1)
$$0 \le \alpha \le \beta \Rightarrow \frac{\alpha}{1 + \alpha} \le \frac{\beta}{1 + \beta}.$$

This is proved by observing that $0 \le \alpha \le \beta$ implies $\alpha + \alpha\beta \le \beta + \alpha\beta$. The next one is:

(2)
$$a, b, c \ge 0 \text{ and } a \le b + c \Rightarrow \frac{a}{1 + a} \le \frac{b}{1 + b} + \frac{c}{1 + c}.$$

This is proved as follows. According to (1), $a \le b + c$ implies

$$\frac{a}{1 + a} \le \frac{b + c}{1 + (b + c)} = \frac{b}{1 + (b + c)} + \frac{c}{1 + (b + c)} \le \frac{b}{1 + b} + \frac{c}{1 + c}.$$

To prove the triangle inequality for ρ, let $f, g, h \in \Phi$. Then we have

$$|f - g| \le |f - h| + |h - g|$$

and according to (2) this implies

$$\frac{|f - g|}{1 + |f - g|} \le \frac{|f - h|}{1 + |f - h|} + \frac{|h - g|}{1 + |h - g|}.$$

Then integrating over X, we have

$$\int_X \frac{|f - g|}{1 + |f - g|} \, d\mu \le \int_X \frac{|f - h|}{1 + |f - h|} \, d\mu + \int_X \frac{|h - g|}{1 + |h - g|} \, d\mu,$$

that is,

$$\rho(f, g) \le \rho(f, h) + \rho(h, g).$$

This completes the proof that ρ is a metric on the set Φ.

2. Let us show that the set Φ is complete with respect to the metric ρ, that is, for every Cauchy sequence $(f_n : n \in \mathbb{N})$ with respect to the metric ρ in Φ there exists $f \in \Phi$ such that $\lim_{n \to \infty} \rho(f_n, f) = 0$.

Now a sequence $(f_n : n \in \mathbb{N})$ in Φ is a Cauchy sequence with respect to the metric ρ if for every $\varepsilon > 0$ there exists $N \in \mathbb{N}$ such that $\rho(f_m, f_n) < \varepsilon$ whenever $m, n \ge N$, that is, for every $\varepsilon > 0$ there exists $N \in \mathbb{N}$ such that

$$(3) \qquad \rho(f_m, f_n) = \int_X \frac{|f_m - f_n|}{1 + |f_m - f_n|} \, d\mu < \varepsilon \quad \text{for } m, n \ge N.$$

Let us show that if $(f_n : n \in \mathbb{N})$ is a Cauchy sequence with respect to the metric ρ then $(f_n : n \in \mathbb{N})$ is a Cauchy sequence with respect to convergence in measure on X. Assume that $(f_n : n \in \mathbb{N})$ is a Cauchy sequence with respect to the metric ρ. Let $\eta > 0$ be arbitrarily given. Then we have

$$(4) \qquad \rho(f_m, f_n) = \int_X \frac{|f_m - f_n|}{1 + |f_m - f_n|} \, d\mu \ge \int_{\{X : |f_m - f_n| \ge \eta\}} \frac{|f_m - f_n|}{1 + |f_m - f_n|} \, d\mu.$$

On the set $\{X : |f_m - f_n| \ge \eta\}$, we have $|f_m - f| \ge \eta > 0$. Then by (1) we have

$$\frac{\eta}{1 + \eta} \le \frac{|f_m - f_n|}{1 + |f_m - f_n|} \quad \text{on } \{X : |f_m - f_n| \ge \eta\}.$$

Then we have

$$(5) \qquad \int_{\{X : |f_m - f_n| \ge \eta\}} \frac{|f_m - f_n|}{1 + |f_m - f_n|} \, d\mu \ge \int_{\{X : |f_m - f_n| \ge \eta\}} \frac{\eta}{1 + \eta} \, d\mu$$

$$= \frac{\eta}{1 + \eta} \mu\{X : |f_m - f_n| \ge \eta\}.$$

Combining (4) and (5), we have

$$(6) \qquad \mu\{X : |f_m - f_n| \ge \eta\} \le \frac{1 + \eta}{\eta} \rho(f_m, f_n).$$

Since $(f_n : n \in \mathbb{N})$ is a Cauchy sequence with respect to the metric ρ, for an arbitrary $\varepsilon > 0$ there exists $N \in \mathbb{N}$ such that

$$(7) \qquad \rho(f_m, f_n) < \frac{\eta}{1 + \eta} \varepsilon \quad \text{for } m, n \ge N.$$

Then for arbitrary $\eta > 0$ and arbitrary $\varepsilon > 0$ there exists $N \in \mathbb{N}$ such that

$$(8) \quad \mu\{X : |f_m - f_n| \ge \eta\} \le \frac{1 + \eta}{\eta} \rho(f_m, f_n) < \frac{1 + \eta}{\eta} \frac{\eta}{1 + \eta} \varepsilon = \varepsilon \quad \text{for } m, n \ge N.$$

This completes the proof that if $(f_n : n \in \mathbb{N})$ is a Cauchy sequence with respect to the metric ρ then $(f_n : n \in \mathbb{N})$ is a Cauchy sequence with respect to convergence in measure on X.

According to Theorem 6.27, if a sequence $(f_n : n \in \mathbb{N})$ in Φ is a Cauchy sequence with respect to convergence in measure on X then there exists $f \in \Phi$ such that $(f_n : n \in \mathbb{N})$ converges to f in measure on X. According to Prob. 7.12, if a sequence $(f_n : n \in \mathbb{N})$ in Φ converges to some $f \Phi$ in measure on X then we have $\lim\limits_{n \to \infty} \int_X \frac{|f_n - f|}{1 + |f_n - f|} d\mu = 0$, that is, $\lim\limits_{n \to \infty} \rho(f_n, f) = 0$. This completes the proof that the set Φ is complete with respect to the metric ρ.

3. Let $(f_n : n \in \mathbb{N})$ be a sequence in Φ and let $f \in \Phi$. By the definition of ρ we have

$$\rho(f_n, f) = \int_X \frac{|f_n - f|}{1 + |f_n - f|} d\mu.$$

Then we have

$$\lim_{n \to \infty} \rho(f_n, f) = 0 \Leftrightarrow \lim_{n \to \infty} \int_X \frac{|f_n - f|}{1 + |f_n - f|} d\mu = 0$$

$$\Leftrightarrow (f_n : n \in \mathbb{N}) \text{ converges to } f \text{ in measure on } D$$

where the second equivalence relation is by Prob. 7.12. ∎

Prob. 7.14. Given a measure space $(X, \mathfrak{A}, \mu)$. Let $(f_n : n \in \mathbb{N})$ be a uniformly bounded sequence of real-valued $\mathfrak{A}$-measurable functions and let f be a bounded real-valued $\mathfrak{A}$-measurable function on $D \in \mathfrak{A}$ with $\mu(D) < \infty$.
Show that if $f_n \overset{\mu}{\to} f$ on D then $\lim_{n\to\infty} \int_D |f_n - f| \, d\mu = 0$ and $\lim_{n\to\infty} \int_D f_n \, d\mu = \int_D f \, d\mu$.
This is the Bounded Convergence Theorem under Convergence in Measure. Construct a direct proof without using the subsequence argument as in Corollary 7.17.

Proof. Let $\varepsilon > 0$ be arbitrarily given. Define a sequence of subsets of D, $(E_{\varepsilon,n} : n \in \mathbb{N})$, by setting

$$E_{\varepsilon,n} = \left\{ x \in D : |f_n(x) - f(x)| \geq \varepsilon \right\}.$$

Since $(f_n : n \in \mathbb{N})$ converges to f in measure on D, we have

$$\lim_{n\to\infty} \mu(E_{\varepsilon,n}) = 0.$$

Let $M > 0$ be a uniform bound of the sequence $(f_n : n \in \mathbb{N})$ on D, that is, $|f_n(x)| \leq M$ for $x \in D$ and $n \in \mathbb{N}$. Then we have

$$\int_D |f_n - f| \, d\mu = \int_{D \setminus E_{\varepsilon,n}} |f_n - f| \, d\mu + \int_{E_{\varepsilon,n}} |f_n - f| \, d\mu$$

$$\leq \varepsilon \mu(D \setminus E_{\varepsilon,n}) + 2M \mu(E_{\varepsilon,n})$$

$$\leq \varepsilon \mu(D) + 2M \mu(E_{\varepsilon,n}),$$

and then, since $\lim_{n\to\infty} \mu(E_{\varepsilon,n}) = 0$ implies $\limsup_{n\to\infty} \mu(E_{\varepsilon,n}) = 0$, we have

$$\limsup_{n\to\infty} \int_D |f_n - f| \, d\mu \leq \varepsilon \mu(D) + 2M \limsup_{n\to\infty} \mu(E_{\varepsilon,n}) = \varepsilon \mu(D).$$

Since the last estimate holds for every $\varepsilon > 0$, we have

$$\limsup_{n\to\infty} \int_D |f_n - f| \, d\mu = 0.$$

Then we have

$$0 \leq \liminf_{n\to\infty} \int_D |f_n - f| \, d\mu \leq \limsup_{n\to\infty} \int_D |f_n - f| \, d\mu = 0$$

and thus

$$\liminf_{n\to\infty} \int_D |f_n - f| \, d\mu = \limsup_{n\to\infty} \int_D |f_n - f| \, d\mu = 0,$$

and then

(1) $$\lim_{n\to\infty} \int_D |f_n - f| \, d\mu = 0.$$

Next we have

$$\lim_{n\to\infty}\left|\int_D f_n\,d\mu - \int_D f\,d\mu\right| = \lim_{n\to\infty}\left|\int_D \{f_n - f\}\,d\mu\right|$$

$$\leq \lim_{n\to\infty}\int_D |f_n - f|\,d\mu$$

$$= 0 \quad \text{by (1).}$$

This implies

$$\lim_{n\to\infty}\left\{\int_D f_n\,d\mu - \int_D f\,d\mu\right\} = 0,$$

and thus we have

(2) $$\lim_{n\to\infty}\int_D f_n\,d\mu = \int_D f\,d\mu. \quad \blacksquare$$

Prob. 7.15. Let Q be the collection of all rational numbers in $(0, \infty)$. Let each $x \in Q$ be expressed as $\frac{q}{p}$ where p and q are positive integers without common factors other than 1. Define a function f on $[0, 1]$ by setting

$$f(x) = \begin{cases} \frac{1}{p} & \text{if } x \in [0, 1] \cap Q \text{ and } x = \frac{q}{p}, \\ 0 & \text{if } x \in [0, 1] \setminus Q. \end{cases}$$

Show that f is Riemann integrable on $[0, 1]$ and evaluate $\int_0^1 f(x)\,dx$.

Proof. Let $D_1 = [0, 1] \cap Q$ and $D_2 = [0, 1] \setminus Q$. Then according to Prob. 4.8, f is continuous at every $x \in D_2$ and f is discontinuous at every $x \in D_1 \setminus \{0\}$. Thus $D_1 \setminus \{0\}$ is the set of points of discontinuity of f. As a subset of Q, $D_1 \setminus \{0\}$ is a countable set and then $\mu(D_1 \setminus \{0\}) = 0$. According to Theorem 7.28, this implies the Riemann integrability of f on $[0, 1]$. According to Theorem 7.27, the Riemann integrability of f on $[0, 1]$ implies

$$\int_0^1 f(x)\,dx = \int_{[0,1]} f\,d\mu_L.$$

Then since $\mu_L(D_1) = 0$ and $f = 0$ on D_2, we have

$$\int_{[0,1]} f\,d\mu_L = \int_{D_1} f\,d\mu_L + \int_{D_2} f\,d\mu_L = 0 + 0 = 0.$$

Thus we have $\int_0^1 f(x)\,dx = 0$. $\quad \blacksquare$

Prob. 7.16. Let f be an extended real-valued function on $\mathbb{R}$. Define two extended real-valued functions φ and ψ on $\mathbb{R}$ by letting $\varphi(x) = \liminf_{y \to x} f(y)$ and $\psi(x) = \limsup_{y \to x} f(y)$ for $x \in \mathbb{R}$. Show that φ and ψ are $\mathfrak{B}_\mathbb{R}$-measurable on $\mathbb{R}$. (Note that f itself is not assumed to be $\mathfrak{B}_\mathbb{R}$-measurable on $\mathbb{R}$.)

Proof. 1. For $x \in \mathbb{R}$ and $\eta > 0$, let us write

$$U(x, \eta) = (x - \eta, x + \eta),$$

$$U_0(x, \eta) = (x - \eta, x + \eta) \setminus \{x\}.$$

By definition we have

(1) $$\varphi(x) = \liminf_{y \to x} f(y) = \lim_{\eta \downarrow 0} \inf_{U_0(x, \eta)} f,$$

(2) $$\psi(x) = \limsup_{y \to x} f(y) = \lim_{\eta \downarrow 0} \sup_{U_0(x, \eta)} f.$$

For each $n \in \mathbb{N}$, let us decompose $\mathbb{R}$ into countably many disjoint subintervals defined by

(3) $$I_{n,k} = \left(\frac{k - 1}{2^n}, \frac{k}{2^n} \right] \quad \text{for } k \in \mathbb{Z}.$$

Let us define a step function φ_n on $\mathbb{R}$ by setting

(4) $$\varphi_n = \sum_{k \in \mathbb{Z}} \inf_{I_{n,k}} f \cdot \mathbf{1}_{I_{n,k}}.$$

Then φ_n is $\mathfrak{B}_\mathbb{R}$-measurable on $\mathbb{R}$, $\varphi_n \uparrow$ as $n \to \infty$ and $\lim_{n \to \infty} \varphi_n$ is $\mathfrak{B}_\mathbb{R}$-measurable on $\mathbb{R}$.

   2. Let us call $x \in \mathbb{R}$ a strict minimal point of f if there exists $\delta > 0$ such that $f(x) < f(y)$ for all $y \in U_0(x, \delta)$. Let S be the collection of all strict minimal points of f. Let us show that S is a countable set. For each $k \in \mathbb{N}$, let S_k be the collection of all $x \in \mathbb{R}$ such that

$$f(x) < f(y) \quad \text{for all } y \in U_0\left(x, \tfrac{1}{k}\right).$$

Then every $x \in S_k$ is a strict minimal point of f so that $S_k \subset S$. On the other hand for every $\delta > 0$ in the definition of a strict minimal point of f above, we have $\frac{1}{k} < \delta$ for large enough $k \in \mathbb{N}$ and thus every $x \in S$ is in S_k for some $k \in \mathbb{N}$ and therefore $S \subset \bigcup_{k \in \mathbb{N}} S_k$. This establishes $S = \bigcup_{k \in \mathbb{N}} S_k$.

   To show that S is a countable set, since $S = \bigcup_{k \in \mathbb{N}} S_k$ it suffices to show that S_k is a countable set for every $k \in \mathbb{N}$. Let us show that S_k is a countable set. Corresponding to every $x \in S_k$ consider the set $U_0\left(x, \frac{1}{3k}\right)$. We claim that $\left\{ U_0\left(x, \frac{1}{3k}\right) : x \in S_k \right\}$ is a disjoint collection. Suppose it is not a disjoint collection. Then there exist $x', x'' \in S_k$ such that $U_0\left(x', \frac{1}{3k}\right) \cap U_0\left(x'', \frac{1}{3k}\right) \neq \emptyset$. This implies that $x' \in U_0\left(x'', \frac{1}{k}\right)$ so that $f(x'') < f(x')$ and similarly $x'' \in U_0\left(x', \frac{1}{k}\right)$ so that $f(x') < f(x'')$. This is a contradiction. Therefore $\left\{ U_0\left(x, \frac{1}{3k}\right) : x \in S_k \right\}$ is a disjoint collection. Now $\mu_L\left(U_0\left(x, \frac{1}{3k}\right)\right) = \frac{2}{3k} > 0$. Thus $\left\{ U_0\left(x, \frac{1}{3k}\right) : x \in S_k \right\}$ is a disjoint collection of $\mathfrak{M}_L$-measurable subsets of $\mathbb{R}$ with positive

μ_L measure. According to Prob. 3.5 such a collection is an at most countable set. Thus S_k is a countable set. This completes the proof that S is a countable set.

3. Let E be the collection of the endpoints of the intervals $I_{n,k}$ where $k \in \mathbb{Z}$ and $n \in \mathbb{N}$. Then E is a countable set. Since S is also a countable set, $E \cup S$ is a countable set. Now a singleton in $\mathbb{R}$ is a $\mathfrak{B}_{\mathbb{R}}$-measurable set. Then $E \cup S$, being a countable union of $\mathfrak{B}_{\mathbb{R}}$-measurable sets, is a $\mathfrak{B}_{\mathbb{R}}$-measurable set. Let us show

$$\lim_{n\to\infty} \varphi_n(x) = \varphi(x) \quad \text{for } x \in (E \cup S)^c.$$

Let $x \in (E \cup S)^c$. Then x is not an endpoint of the interval $I_{n,k}$ for any $k \in \mathbb{Z}$ and $n \in \mathbb{N}$. Then for every $n \in \mathbb{N}$, there exists $k(n) \in \mathbb{Z}$ such that $x \in I^\circ_{n,k(n)}$, the interior of the interval $I_{n,k(n)}$. Let $\eta_n > 0$ be so small that

$$U(x, \eta_n) \subset I^\circ_{n,k(n)}.$$

Then for every $n \in \mathbb{N}$ we have

$$(5) \qquad \varphi(x) = \liminf_{y\to x} f(y) = \lim_{\eta\downarrow 0} \inf_{U_0(x,\eta)} f,$$

$$\geq \inf_{U_0(x,\eta_n)} f \quad \text{since } \inf_{U_0(x,\eta)} f \uparrow \text{ as } \eta \downarrow 0$$

$$\geq \inf_{U(x,\eta_n)} f \geq \inf_{I^\circ_{n,k(n)}} f$$

$$\geq \inf_{I_{n,k(n)}} f = \varphi_n(x).$$

By (1), for every $\varepsilon > 0$ there exists $\delta > 0$ such that

$$(6) \qquad \varphi(x) - \varepsilon < \inf_{U_0(x,\delta)} f \leq \varphi(x).$$

Since $\lim_{n\to\infty} \frac{1}{2^n} = 0$ and since $x \in I^\circ_{n,k(n)}$ for every $n \in \mathbb{N}$, there exists $N \in \mathbb{N}$ such that

$$I_{n,k(n)} \subset U(x, \delta) \quad \text{for } n \geq N.$$

This implies

$$(7) \qquad \inf_{I_{n,k(n)}} f \geq \inf_{U(x,\delta)} f = \inf_{U_0(x,\delta)} f \quad \text{since } x \notin S.$$

By (5), (4), (7) and (6), we have

$$\varphi(x) \geq \varphi_n(x) = \inf_{I_{n,k(n)}} f \geq \inf_{U_0(x,\delta)} f > \varphi(x) - \varepsilon \quad \text{for } n \geq N.$$

This proves

$$(8) \qquad \lim_{n\to\infty} \varphi_n(x) = \varphi(x) \quad \text{for } x \in (E \cup S)^c.$$

Since φ_n is $\mathfrak{B}_\mathbb{R}$-measurable on $(E \cup S)^c$ for every $n \in \mathbb{N}$, (8) implies that φ is $\mathfrak{B}_\mathbb{R}$-measurable on $(E \cup S)^c$. The $\mathfrak{B}_\mathbb{R}$-measurability of φ on $E \cup S$ is verified as follows. Let $\alpha \in \mathbb{R}$ and consider the set $\{x \in E \cup S : \varphi(x) \le \alpha\}$, a subset of the countable set $E \cup S$ and hence a countable set in $\mathbb{R}$. Now every countable set in $\mathbb{R}$ is in $\mathfrak{B}_\mathbb{R}$. Thus $\{x \in E \cup S : \varphi(x) \le \alpha\} \in \mathfrak{B}_\mathbb{R}$. This shows that φ is $\mathfrak{B}_\mathbb{R}$-measurable on $E \cup S$. The $\mathfrak{B}_\mathbb{R}$-measurability of φ on $(E \cup S)^c$ and the $\mathfrak{B}_\mathbb{R}$-measurability of φ on $E \cup S$ imply the $\mathfrak{B}_\mathbb{R}$-measurability of φ on $\mathbb{R}$.

   4. The $\mathfrak{B}_\mathbb{R}$-measurability of ψ on $\mathbb{R}$ follows from that of φ. Observe that for an arbitrary extended real-valued function f on $\mathbb{R}$, we have

(9) $$\limsup_{y \to x} f(y) = -\liminf_{y \to x}(-f)(y).$$

We showed in **3** that for every extended real-valued function f defined on $\mathbb{R}$ the function $\liminf_{y \to x} f(y)$ for $x \in \mathbb{R}$ is a $\mathfrak{B}_\mathbb{R}$-measurable function on $\mathbb{R}$. Thus for the extended real-valued function $-f$ on $\mathbb{R}$, the function $\liminf_{y \to x}(-f)(y)$ for $x \in \mathbb{R}$ is a $\mathfrak{B}_\mathbb{R}$-measurable function on $\mathbb{R}$. Then (9) implies that the function $\limsup_{y \to x} f(y)$ for $x \in \mathbb{R}$ is a $\mathfrak{B}_\mathbb{R}$-measurable function on $\mathbb{R}$. ∎

Prob. 7.17. Let f be an extended real-valued function on $\mathbb{R}$. Define a subset of $\mathbb{R}$ by

$$D = \left\{ x \in \mathbb{R} : \lim_{y \to x} f(y) \text{ exists in } \overline{\mathbb{R}} \right\}$$

and define an extended real-valued function Λ on D by setting

$$\Lambda(x) = \lim_{y \to x} f(y) \quad \text{for } x \in D.$$

Show that $D \in \mathfrak{B}_\mathbb{R}$ and Λ is $\mathfrak{B}_\mathbb{R}$-measurable on D.

Proof. 1. For $x \in \mathbb{R}$, $\lim_{y \to x} f(y)$ exists in $\overline{\mathbb{R}}$ if and only if $\liminf_{y \to x} f(y) = \limsup_{y \to x} f(y)$ and when the equality holds then we have $\lim_{y \to x} f(y) = \liminf_{y \to x} f(y) = \limsup_{y \to x} f(y)$. Thus

$$D = \left\{ x \in \mathbb{R} : \liminf_{y \to x} f(y) = \limsup_{y \to x} f(y) \right\}.$$

Now according to Prob. 7.16, $\liminf_{y \to x} f(y)$ and $\limsup_{y \to x} f(y)$ are $\mathfrak{B}_\mathbb{R}$-measurable functions on $\mathbb{R}$. This implies that $D \in \mathfrak{B}_\mathbb{R}$.

   2. We have $\Lambda(x) = \liminf_{y \to x} f(y)$ for $x \in D$. Now $\liminf_{y \to x} f(y)$ for $x \in \mathbb{R}$ is $\mathfrak{B}_\mathbb{R}$-measurable on $\mathbb{R}$ and hence $\mathfrak{B}_\mathbb{R}$-measurable on $D \in \mathfrak{B}_\mathbb{R}$. Thus Λ is $\mathfrak{B}_\mathbb{R}$-measurable on D. ∎

Prob. 7.18. Let (X, ρ) be a metric space with a metric topology by the metric ρ. For $x \in X$ and $r > 0$ the open ball with center x and radius r is defined by setting

$$B(x, r) = \{y \in X : \rho(x, y) < r\}.$$

Let D be a subset of X and let f be a real-valued function defined on D. We define the oscillation of f at $x \in D$ by setting

$$\omega(x) = \lim_{\delta \downarrow 0} \sup_{B(x,\delta) \cap D} f - \lim_{\delta \downarrow 0} \inf_{B(x,\delta) \cap D} f.$$

(a) Show that $\omega(x) \in [0, \infty]$ for every $x \in D$.
(b) Show that f is continuous at $x_0 \in D$ if and only if $\omega(x_0) = 0$.
(c) For every $\varepsilon > 0$, let $D_\varepsilon = \{x \in D : \omega(x) \geq \varepsilon\}$. Show that D_ε is a closed subset of D in the relative topology of D, that is, $D_\varepsilon = F \cap D$ where F is a closed set in X.
(d) Let A be the subset of D consisting of all discontinuity points of f and let B be the subset of D consisting of all continuity points of f. Show that $A = F \cap D$ where F is an F_σ-set in X and $B = G \cap D$ where G is a G_δ-set in X.

Proof. 1. To prove (a), observe that $\inf_{B(x,\delta)} \in [-\infty, f(x)]$ so that $\lim_{\delta \downarrow 0} \inf_{B(x,\delta)} \in [-\infty, f(x)]$ and similarly $\sup_{B(x,\delta)} \in [f(x), \infty]$ so that $\lim_{\delta \downarrow 0} \sup_{B(x,\delta)} \in [f(x), \infty]$. This implies

$$\lim_{\delta \downarrow 0} \sup_{B(x,\delta) \cap D} f - \lim_{\delta \downarrow 0} \inf_{B(x,\delta) \cap D} f \in [0, \infty].$$

2. Let us prove (b).
2.1. Suppose f is continuous at $x_0 \in D$. Then for every $\varepsilon > 0$ there exists $\eta > 0$ such that

$$f(x_0) - \varepsilon < f(x) < f(x_0) + \varepsilon \quad \text{for } x \in B(x_0, \eta) \cap D$$

and hence

$$f(x_0) - \varepsilon \leq \inf_{B(x_0,\eta) \cap D} f \leq \sup_{B(x_0,\eta) \cap D} f \leq f(x_0) + \varepsilon$$

and then for $\delta \in (0, \eta]$ we have

$$f(x_0) - \varepsilon \leq \inf_{B(x_0,\eta) \cap D} f \leq \inf_{B(x_0,\delta) \cap D} f$$

$$\leq \sup_{B(x_0,\delta) \cap D} f \leq \sup_{B(x_0,\eta) \cap D} f$$

$$\leq f(x_0) + \varepsilon.$$

Thus we have

$$\sup_{B(x_0,\delta) \cap D} f - \inf_{B(x_0,\delta) \cap D} f \leq 2\varepsilon \quad \text{for } \delta \in (0, \eta]$$

and then

$$0 \leq \omega(x_0) = \lim_{\delta \downarrow 0} \left\{ \sup_{B(x_0,\delta) \cap D} f - \inf_{B(x_0,\delta) \cap D} f \right\} \leq 2\varepsilon.$$

Since this holds for an arbitrary $\varepsilon > 0$, we have $\omega(x_0) = 0$.

2.2. Conversely suppose that for $x_0 \in D$ we have

$$0 = \omega(x_0) = \lim_{\delta \downarrow 0} \left\{ \sup_{B(x_0,\delta) \cap D} f - \inf_{B(x_0,\delta) \cap D} f \right\}.$$

Then for every $\varepsilon > 0$ there exists $\delta > 0$ such that

$$\sup_{B(x_0,\delta) \cap D} f - \inf_{B(x_0,\delta) \cap D} f < \varepsilon.$$

Then for every $x', x'' \in B(x_0, \delta) \cap D$ we have

$$f(x') - f(x'') \leq \sup_{B(x_0,\delta) \cap D} f - \inf_{B(x_0,\delta) \cap D} f < \varepsilon$$

and

$$f(x'') - f(x') \leq \sup_{B(x_0,\delta) \cap D} f - \inf_{B(x_0,\delta) \cap D} f < \varepsilon$$

so that

$$|f(x') - f(x'')| < \varepsilon \quad \text{for every } x', x'' \in B(x_0, \delta) \cap D$$

and in particular

$$|f(x) - f(x_0)| < \varepsilon \quad \text{for every } x \in B(x_0, \delta) \cap D.$$

This shows that f is continuous at x_0.

3. Let us prove (c). To show that D_ε is a closed subset of D in the relative topology of D, we show that D_ε contains every limit point of D_ε in the relative topology of D. Thus let $x_0 \in D$ be a limit point of D_ε in the relative topology of D. Then for every $\delta > 0$ the set $B(x_0, \delta) \cap D$ contains some point $x_1 \in D_\varepsilon$ such that $x_1 \neq x_0$. Then since $x_1 \in D_\varepsilon$, we have

$$\lim_{\eta \downarrow 0} \left\{ \sup_{B(x_1,\eta) \cap D} f - \inf_{B(x_1,\eta) \cap D} f \right\} = \omega(x_1) \geq \varepsilon$$

and then

$$\sup_{B(x_1,\eta) \cap D} f - \inf_{B(x_1,\eta) \cap D} f \geq \varepsilon \quad \text{for every } \eta > 0.$$

In particular this inequality holds for $\eta_1 > 0$ so chosen that $B(x_1, \eta_1) \subset B(x_0, \delta)$. Then

$$\sup_{B(x_1,\eta_1) \cap D} f - \inf_{B(x_1,\eta_1) \cap D} f \geq \varepsilon.$$

Then for the set $B(x_0, \delta)$, which contains $B(x_1, \eta_1)$, we have

$$\sup_{B(x_0,\delta) \cap D} f - \inf_{B(x_0,\delta) \cap D} f \geq \varepsilon.$$

Since this holds for every $\delta > 0$, we have

$$\omega(x_0) = \lim_{\delta \downarrow 0} \left\{ \sup_{B(x_0,\delta) \cap D} f - \inf_{B(x_0,\delta) \cap D} f \right\} \geq \varepsilon.$$

This shows that $x_0 \in D_\varepsilon$. Thus D_ε contains every limit point of D_ε in the relative topology of D and thus D_ε is a closed subset of D in the relative topology of D.

4. Let us prove (d). Recall that $\omega(x) \in [0, \infty]$ for $x \in D$ and $\omega(x) = 0$ if and only if f is continuous at $x \in D$. Thus f is discontinuous at $x \in D$ if and only if $\omega(x) > 0$. Recall also the set $D_\varepsilon = \{x \in D : \omega(x) \geq \varepsilon\}$ for $\varepsilon > 0$. Then we have

$$A = \left\{x \in D : \omega(x) > 0\right\} = \bigcup_{n \in \mathbb{N}} \left\{x \in D : \omega(x) \geq \frac{1}{n}\right\} = \bigcup_{n \in \mathbb{N}} D_{\frac{1}{n}}.$$

According to (b), $D_{\frac{1}{n}}$ is a closed subset of D in the relative topology of D and thus we have $D_{\frac{1}{n}} = F_n \cap D$ where F_n is a closed set in X. Let $F = \bigcup_{n \in \mathbb{N}} F_n$, an F_σ-set in X. Then

$$A = \bigcup_{n \in \mathbb{N}} D_{\frac{1}{n}} = \bigcup_{n \in \mathbb{N}} (F_n \cap D) = \left(\bigcup_{n \in \mathbb{N}} F_n\right) \cap D = F \cap D.$$

Then

$$B = D \setminus A = D \cap A^c = D \cap (F \cap D)^c = D \cap (F^c \cup D) = D \cap F^c = G \cap D$$

where $G := F^c$, a G_δ-set set in X. ∎

Prob. 7.19. Let $\mathbb{Q}$ be the set of rational numbers in $\mathbb{R}$ and let $\mathbb{P}$ be the set of irrational numbers in $\mathbb{R}$.
(a) Show that $\mathbb{Q}$ is an F_σ-set and $\mathbb{P}$ is a G_δ-set.
(b) Show that $\mathbb{Q}$ cannot be a G_δ-set and $\mathbb{P}$ cannot be an F_σ-set.

Proof. 1. Let us show that $\mathbb{Q}$ is an F_σ-set. Now $\mathbb{Q}$ is a countable set. Let us enumerate the elements of $\mathbb{Q}$ as $\mathbb{Q} = \{r_i : i \in \mathbb{N}\}$. Since $\{r_i\}$ is a closed set in $\mathbb{R}$, $\mathbb{Q} = \bigcup_{i \in \mathbb{N}} \{r_i\}$ is an F_σ-set.
2. The complement of an F_σ-set in a topological space is a G_δ-set. Then since $\mathbb{P}$ is the complement of the F_σ-set $\mathbb{Q}$ in $\mathbb{R}$, $\mathbb{P}$ is a G_δ-set.
3. Let us show that $\mathbb{Q}$ cannot be a G_δ-set. Assume the contrary, that is, assume that $\mathbb{Q}$ is a G_δ-set. Then $\mathbb{Q} = \bigcap_{i \in \mathbb{N}} G_i$ where G_i is an open set in $\mathbb{R}$. Now every non-empty open set in $\mathbb{R}$ is the union of countably many disjoint open intervals. Since $\mathbb{Q}$ is dense in $\mathbb{R}$ (that is, every open interval in $\mathbb{R}$ contains some element of $\mathbb{Q}$), the open set G_i containing $\mathbb{Q}$ must be of the form $\mathbb{R} \setminus P_i$ where P_i consists of countably many irrational numbers. Then

$$\mathbb{Q} = \bigcap_{i \in \mathbb{N}} G_i = \bigcap_{i \in \mathbb{N}} (\mathbb{R} \setminus P_i) = \bigcap_{i \in \mathbb{N}} P_i^c = \left(\bigcup_{i \in \mathbb{N}} P_i\right)^c.$$

Now $\bigcup_{n \in \mathbb{N}} P_i$ consists of countably many irrational numbers. Then since $\mathbb{P}$ is an uncountable set, $\left(\bigcup_{i \in \mathbb{N}} P_i\right)^c$ contains some irrational numbers and thus $\left(\bigcup_{i \in \mathbb{N}} P_i\right)^c \neq \mathbb{Q}$. This contradicts the last equality. Therefore $\mathbb{Q}$ cannot be a G_δ-set.
4. Let us show that $\mathbb{P}$ cannot be an F_σ-set. Suppose $\mathbb{P}$ is an F_σ-set. Then $\mathbb{Q} = \mathbb{P}^c$ is a G_δ-set. This contradicts **3**. ∎

Prob. 7.20. (a) Construct a real-valued function f on $\mathbb{R}$ such that f is discontinuous at every rational $x \in \mathbb{R}$ and continuous at every irrational $x \in \mathbb{R}$.
(b) Show that it is impossible for a real-valued function f on $\mathbb{R}$ to be continuous at every rational $x \in \mathbb{R}$ and discontinuous at every irrational $x \in \mathbb{R}$.

Proof. 1. Let $\mathbb{Q}$ be the set of all rational numbers in $\mathbb{R}$. Then $\mathbb{Q}^c = \mathbb{R} \setminus \mathbb{Q}$ is the set of all irrational numbers in $\mathbb{R}$. Let every non-zero rational number be represented as $\pm \frac{q}{p}$ where $p, q \in \mathbb{N}$ and p and q have no common factors other than $1 \in \mathbb{N}$. ($1 \in \mathbb{N}$ is a factor of every $x \in \mathbb{R}$.) In particular every $n \in \mathbb{N}$ is represented as $\frac{n}{1}$.

For $p \in \mathbb{N}$, let $\mathbb{Q}_p$ be the set of all non-zero rational numbers represented by $\pm \frac{q}{p}$. Then $\{\mathbb{Q}_p : p \in \mathbb{N}\}$ is a collection of disjoint sets and moreover

$$\mathbb{Q} = \{0\} \cup \left(\bigcup_{p \in \mathbb{N}} \mathbb{Q}_p \right).$$

Let us define a real-valued function f on $\mathbb{R}$ by setting

$$f(x) = \begin{cases} 1 & \text{if } x = 0, \\ \frac{1}{p} & \text{if } x \in \mathbb{Q}_p \text{ for } p \in \mathbb{N}, \\ 0 & \text{if } x \in \mathbb{Q}^c. \end{cases}$$

Then we have

$$f(x) > 0 \quad \text{if } x \in \mathbb{Q},$$
$$f(x) = 0 \quad \text{if } x \in \mathbb{Q}^c.$$

2. Let us show that f is discontinuous at every $x_0 \in \mathbb{Q}$. Assume the contrary, that is, assume that f is continuous at some $x_0 \in \mathbb{Q}$. Now $x_0 \in \mathbb{Q}$ implies that $f(x_0) = \alpha > 0$. Then continuity at x_0 implies that for $\frac{\alpha}{2} > 0$ there exists $\delta > 0$ such that

$$|f(x) - f(x_0)| < \frac{\alpha}{2} \quad \text{for } x \in (x_0 - \delta, x_0 + \delta),$$

that is, recalling $f(x_0) = \alpha$, we have

$$\frac{\alpha}{2} < f(x) < \frac{3}{2}\alpha \quad \text{for } x \in (x_0 - \delta, x_0 + \delta).$$

This is impossible since for $x \in (x_0 - \delta, x_0 + \delta)$ such that $x \in \mathbb{Q}^c$ we have $f(x) = 0$.

3. Let us show that f is continuous at every $x_0 \in \mathbb{Q}^c$. Thus we are to show that if $x_0 \in \mathbb{Q}^c$ then for every $\varepsilon > 0$ there exists $\delta > 0$ such that

$$|f(x) - f(x_0)| < \varepsilon \quad \text{for } x \in (x_0 - \delta, x_0 + \delta),$$

that is, recalling $f(x_0) = 0$ and $f(x) \geq 0$, we have

$$f(x) < \varepsilon \quad \text{for } x \in (x_0 - \delta, x_0 + \delta).$$

If $x \in \mathbb{Q}^c$, then $f(x) = 0 < \varepsilon$. If $x \in \mathbb{Q}_p$ for some $p \in \mathbb{N}$, then $f(x) = \frac{1}{p}$. Now there exists $N \in \mathbb{N}$ such that $\frac{1}{N} < \varepsilon$. Then for $x \in \mathbb{Q}_p$ where $p > N$, we have

$$f(x) = \frac{1}{p} < \frac{1}{N} < \varepsilon.$$

It remains to consider $x \in \mathbb{Q}_p$ where $p = 1, \ldots, N$. Now for $p = 1, \ldots, N$, if $x', x'' \in \mathbb{Q}_p$ then $|x' - x''| \geq \frac{1}{p} \geq \frac{1}{N}$. Then the open interval $\left(x_0 - \frac{1}{2N}, x_0 - \frac{1}{2N}\right)$ with length $\frac{1}{N}$ can contain at most one element of $\mathbb{Q}_p$ for $p = 1, \ldots, N$. Since the number of elements of $\mathbb{Q}_p$ for $p = 1, \ldots, N$ contained in $\left(x_0 - \frac{1}{2N}, x_0 - \frac{1}{2N}\right)$ is at most N, we can select $\delta \in \left(0, \frac{1}{2N}\right]$ so small that $(x_0 - \delta, x_0 + \delta)$ does not contain any element of $\mathbb{Q}_p$ for $p = 1, \ldots, N$. Finally take a smaller $\delta > 0$ if necessary so that $(x_0 - \delta, x_0 + \delta)$ does not contain $0 \in \mathbb{Q}$. With such $\delta > 0$ we have $f(x) < \varepsilon$ for every $x \in (x_0 - \delta, x_0 + \delta)$. This proves the continuity of f at every $x_0 \in \mathbb{Q}^c$. ∎

4. Suppose there exists a real-valued function f on $\mathbb{R}$ such that f is continuous at every rational $x \in \mathbb{R}$ and discontinuous at every irrational $x \in \mathbb{R}$. Let $\mathbb{Q}$ be the set of the rational numbers in $\mathbb{R}$. Then $\mathbb{Q}$ is the set of continuity points of our function f. According to Prob. 7.18, the set of continuity points of a real-valued function on $\mathbb{R}$ is a G_δ-set in $\mathbb{R}$. But according to Prob. 7.19, $\mathbb{Q}$ cannot be a G_δ-set in $\mathbb{R}$. This is a contradiction. ∎

§8   Integration of Nonnegative Functions

Prob. 8.1. Let $(X, \mathfrak{A}, \mu)$ be an arbitrary measure space and let $D \in \mathfrak{A}$ with $\mu(D) > 0$. Let f be a nonnegative extended real-valued $\mathfrak{A}$-measurable function on D. Show that if $f > 0$ on D then $\int_D f \, d\mu > 0$.

Proof. For $n \in \mathbb{N}$, let $E_n = \left\{ x \in D : f(x) > \frac{1}{n} \right\}$. Since f is $\mathfrak{A}$-measurable on D, we have $E_n \in \mathfrak{A}$. Since $\frac{1}{n} \downarrow$ as $n \uparrow$, $(E_n : n \in \mathbb{N})$ is an increasing sequence in $\mathfrak{A}$. Let $E = \bigcup_{n \in \mathbb{N}} E_n$. Then $E \in \mathfrak{A}$ and $\mu(E) = \lim_{n \to \infty} \mu(E_n)$.

Now since $E_n \subset D$ for every $n \in \mathbb{N}$, we have $E = \bigcup_{n \in \mathbb{N}} E_n \subset D$. On the other hand if $x \in D$ then $f(x) > 0$ so that $f(x) > \frac{1}{n}$ for sufficiently large $n \in \mathbb{N}$ and then $x \in E_n$. Thus $D \subset \bigcup_{n \in \mathbb{N}} E_n = E$. This shows that $E = D$. Then we have $\lim_{n \to \infty} \mu(E_n) = \mu(E) = \mu(D) > 0$. Thus there exists $n_0 \in \mathbb{N}$ such that $\mu(E_{n_0}) > 0$. Let us define a function φ on D by setting

$$\varphi(x) = \frac{1}{n_0} \mathbf{1}_{E_{n_0}}(x) \quad \text{for } x \in D.$$

Then $\varphi \leq f$ on D. Thus we have

$$\int_D f \, d\mu \geq \int_D \varphi \, d\mu = \frac{1}{n_0} \mu(E_{n_0}) > 0. \quad \blacksquare$$

Prob. 8.2. Let $(X, \mathfrak{A}, \mu)$ be an arbitrary measure space. Show that if there exists a non-negative extended real-valued $\mathfrak{A}$-measurable function f on X such that $f > 0$ on X and $\int_X f \, d\mu < \infty$ then $(X, \mathfrak{A}, \mu)$ is a σ-finite measure space.

Proof. To prove the statement we prove the contra-positive of the statement. Thus we are to prove that if $(X, \mathfrak{A}, \mu)$ is not σ-finite then there does not exist a nonnegative extended real-valued $\mathfrak{A}$-measurable function f on X such that $f > 0$ on X and $\int_X f \, d\mu < \infty$.

Suppose $(X, \mathfrak{A}, \mu)$ is not σ-finite. Let f be a nonnegative extended real-valued $\mathfrak{A}$-measurable function on X such that $f > 0$ on X. Let us show that $\int_X f \, d\mu = \infty$.

Let $I_n = \left[\frac{1}{n}, \frac{1}{n-1}\right)$ for $n \in \mathbb{N}$ and $I_0 = [1, \infty]$. Then $\{I_n : n \in \mathbb{Z}_+\}$ is a disjoint collection and $\bigcup_{n \in \mathbb{Z}_+} I_n = (0, \infty]$. Since f is $\mathfrak{A}$-measurable on X, we have $E_n := f^{-1}(I_n) \in \mathfrak{A}$ for every $n \in \mathbb{Z}_+$. Since $f > 0$ on X, for every $x \in X$ we have $f(x) \in I_n$ for some $n \in \mathbb{Z}_+$ and then $x \in f^{-1}(I_n) = E_n$. This shows that $X = \bigcup_{n \in \mathbb{Z}_+} E_n$. Let us define a function g on X by setting

$$g(x) = \begin{cases} \frac{1}{n} & \text{for } x \in E_n \text{ where } n \in \mathbb{N}, \\ 1 & \text{for } x \in E_0. \end{cases}$$

Then g is nonnegative real-valued $\mathfrak{A}$-measurable function on X and $g \leq f$ on X and

$$\int_X f \, d\mu \geq \int_X g \, d\mu.$$

Now since $(X, \mathfrak{A}, \mu)$ is not σ-finite and since $\{E_n : n \in \mathbb{Z}_+\}$ is a countable disjoint collection in $\mathfrak{A}$ such that $\bigcup_{n \in \mathbb{Z}_+} E_n = X$, there must exist $n_0 \in \mathbb{Z}_+$ such that $\mu(E_{n_0}) = \infty$. If $n_0 = 0$ then we have

$$\int_X g \, d\mu \geq \int_{E_{n_0}} g \, d\mu = 1 \cdot \mu(E_{n_0}) = \infty,$$

and if $n_0 \in \mathbb{N}$ then

$$\int_X g \, d\mu \geq \int_{E_{n_0}} g \, d\mu = \frac{1}{n_0} \cdot \mu(E_{n_0}) = \infty.$$

Thus we have $\int_X g \, d\mu = \infty$ and then $\int_X f \, d\mu \geq \int_X g \, d\mu = \infty$. ∎

Prob. 8.3. Let $(X, \mathfrak{A}, \mu)$ be a measure space and let $\mathfrak{E}$ be a collection of disjoint members of $\mathfrak{A}$ with positive μ measure. Show that if $(X, \mathfrak{A}, \mu)$ is a σ-finite measure space then the collection $\mathfrak{E}$ is at most countable.

Proof. If $(X, \mathfrak{A}, \mu)$ is a σ-finite measure space then there exists a countable disjoint collection $\mathfrak{D} = \{D_n : n \in \mathbb{N}\} \subset \mathfrak{A}$ such that $X = \bigcup_{n \in \mathbb{N}} D_n$ and $\mu(D_n) < \infty$ for every $n \in \mathbb{N}$.

For $n \in \mathbb{N}$, let

$$\mathfrak{F}_n = \{E \cap D_n : E \in \mathfrak{E} \text{ and } \mu(E \cap D_n) > 0\}.$$

Since $\mathfrak{E}$ is a collection of disjoint subsets of X, $\mathfrak{F}_n$ is a collection of disjoint subsets of X. Let

$$\mathfrak{F} = \bigcup_{n \in \mathbb{N}} \mathfrak{F}_n.$$

Since $\mathfrak{D}$ is a collection of disjoint subsets of X, $\mathfrak{F}$ is a collection of disjoint subsets of X. Let us define a mapping of $\mathfrak{E}$ into $\mathfrak{F}$ as follows. For every $E \in \mathfrak{E}$, since $X = \bigcup_{n \in \mathbb{N}} D_n$, we have $E = \bigcup_{n \in \mathbb{N}} (E \cap D_n)$. Since $\mathfrak{D}$ is a collection of disjoint $\mathfrak{A}$-measurable subsets of X, $\{E \cap D_n : n \in \mathbb{N}\}$ is a collection of disjoint $\mathfrak{A}$-measurable subsets of X and thus we have

$$\mu(E) = \sum_{n \in \mathbb{N}} \mu(E \cap D_n).$$

Since $\mu(E) > 0$, there exists $n_0 \in \mathbb{N}$ such that $\mu(E \cap D_{n_0}) > 0$. Let us define

$$\varphi(E) = E \cap D_{n_0} \in \mathfrak{F}_{n_0}.$$

Then φ is a mapping of $\mathfrak{E}$ into $\mathfrak{F} = \bigcup_{n \in \mathbb{N}} \mathfrak{F}_n$.

If $E_1, E_2 \in \mathfrak{E}$ and $E_1 \neq E_2$ then $E_1 \cap E_2 = \emptyset$ since $\mathfrak{E}$ is a collection of disjoint subsets of X. Now $E_1 \cap E_2 = \emptyset$ implies that $(E_1 \cap D_{n_1}) \cap (E_2 \cap D_{n_2}) = \emptyset$ for any $n_1, n_2 \in \mathbb{N}$. This shows that if $E_1 \neq E_2$ then $\varphi(E_1) \neq \varphi(E_2)$. The existence of such mapping of $\mathfrak{E}$ into $\mathfrak{F}$ implies that the cardinality of $\mathfrak{E}$ does not exceed that of $\mathfrak{F}$. Therefore to show that $\mathfrak{E}$ is at most countable, we show that $\mathfrak{F}$ is at most countable. Since $\mathfrak{F} = \bigcup_{n \in \mathbb{N}} \mathfrak{F}_n$, to show that $\mathfrak{F}$ is at most countable it suffices to show that $\mathfrak{F}_n$ is at most countable for every $n \in \mathbb{N}$.

Let $n_0 \in \mathbb{N}$ be arbitrarily chosen. To show that $\mathfrak{F}_{n_0}$ is at most countable, let us assume the contrary. Now $\mathfrak{F}_{n_0}$ is a collection of disjoint $\mathfrak{A}$-measurable subsets of D_{n_0} with positive μ measure. Thus if $F \in \mathfrak{F}_{n_0}$ then $\mu(F) > 0$ and there exists $k \in \mathbb{N}$ such that $\mu(F) > \frac{1}{k}$. For $k \in \mathbb{N}$, let $\mathfrak{F}_{n_0,k}$ be the subcollection of $\mathfrak{F}_{n_0}$ consisting of all $F \in \mathfrak{F}_{n_0}$ with $\mu(F) > \frac{1}{k}$. We have $\mathfrak{F}_{n_0} = \bigcup_{k \in \mathbb{N}} \mathfrak{F}_{n_0,k}$. Since $\mathfrak{F}_{n_0}$ is an uncountable collection there exists $k_0 \in \mathbb{N}$ such that $\mathfrak{F}_{n_0,k_0}$ is an uncountable collection. Then we can select a sequence $(F_m : m \in \mathbb{N})$ of distinct members of $\mathfrak{F}_{n_0,k_0}$. We have $\mu(F_m) > \frac{1}{k_0}$ for every $m \in \mathbb{N}$. Since $\mathfrak{F}$ is a collection of disjoint subsets of X, distinctness of F_m for $m \in \mathbb{N}$ implies the disjointness of F_m for $m \in \mathbb{N}$. Then we have

$$\mu\left(\bigcup_{m \in \mathbb{N}} F_m\right) = \sum_{m \in \mathbb{N}} \mu(F_m) \geq \sum_{m \in \mathbb{N}} \frac{1}{k_0} = \infty.$$

But we have $\bigcup_{m \in \mathbb{N}} F_m \subset D_{n_0}$ and this implies

$$\mu(D_{n_0}) \geq \mu\left(\bigcup_{m \in \mathbb{N}} F_m\right) = \infty.$$

This contradicts the fact that $\mu(D_{n_0}) < \infty$. Thus $\mathfrak{F}_{n_0}$ cannot be uncountable and hence $\mathfrak{F}_{n_0}$ is at most countable. ∎

Prob. 8.4. Given a measure space $(X, \mathfrak{A}, \mu)$. Let f be a nonnegative extended real-valued $\mathfrak{A}$-measurable function on a set $D \in \mathfrak{A}$ with $\mu(D) < \infty$. Suppose $f > 0$ a.e. on D.
(a) Show that for every $\delta > 0$ there exists $\eta > 0$ such that for every $\mathfrak{A}$-measurable subset E of D with $\mu(E) \geq \delta$, we have $\int_E f \, d\mu \geq \eta$, (that is, if $E \in \mathfrak{A}$, $E \subset D$, and $\mu(E)$ is bounded below by a positive real number then $\int_E f \, d\mu$ is bounded below by a positive real number).
(b) Show that (a) does not hold without the assumption $\mu(D) < \infty$.

Proof. 1. Let us prove (a). Let $D_+ = \{x \in D : f(x) > 0\}$. Since $f > 0$, a.e. on D, we have $\mu(D_+) = \mu(D)$. For $n \in \mathbb{N}$, let $D_n = \{x \in D : f(x) \geq \frac{1}{n}\}$. Then $(D_n : n \in \mathbb{N})$ is an increasing sequence in $\mathfrak{A}$ and $\lim\limits_{n \to \infty} D_n = \bigcup\limits_{n \in \mathbb{N}} D_n = D_+$ so that we have

(1) $$\lim_{n \to \infty} \mu(D_n) = \mu\left(\lim_{n \to \infty} D_n\right) = \mu(D_+) = \mu(D) < \infty.$$

Let $\delta > 0$ be arbitrarily given. Then (1) implies that there exists $n_0 \in \mathbb{N}$ such that

$$\mu(D_{n_0}) \geq \mu(D) - \frac{\delta}{2}, \quad \text{that is,} \quad \mu(D) - \mu(D_{n_0}) \leq \frac{\delta}{2}.$$

Then since $\mu(D_{n_0}) < \infty$, we have

(2) $$\mu(D \setminus D_{n_0}) = \mu(D) - \mu(D_{n_0}) \leq \frac{\delta}{2}.$$

Let $E \in \mathfrak{A}$, $E \subset D$, and suppose $\mu(E) \geq \delta$. Since $D = (D \setminus D_{n_0}) \cup D_{n_0}$, we have

$$E = E \cap D = \{E \cap (D \setminus D_{n_0})\} \cup \{E \cap D_{n_0}\}.$$

Then we have

$$\delta \leq \mu(E) = \mu\big(E \cap (D \setminus D_{n_0})\big) + \mu\big(E \cap D_{n_0}\big)$$
$$\leq \mu\big(D \setminus D_{n_0}\big) + \mu\big(E \cap D_{n_0}\big)$$
$$\leq \frac{\delta}{2} + \mu\big(E \cap D_{n_0}\big),$$

and then

(3) $$\mu\big(E \cap D_{n_0}\big) \geq \frac{\delta}{2}.$$

Then we have

(4) $$\int_E f \, d\mu \geq \int_{E \cap D_{n_0}} f \, d\mu \geq \frac{1}{n_0} \mu\big(E \cap D_{n_0}\big) \geq \frac{1}{n_0} \frac{\delta}{2}.$$

Let

(5) $$\eta = \frac{1}{n_0} \frac{\delta}{2} > 0.$$

Then (4) shows that

$$\int_E f \, d\mu \geq \eta \quad \text{for } E \in \mathfrak{A}, E \subset D, \text{ with } \mu(E) \geq \delta.$$

This completes the proof of (a).

2. Let us show that (a) is not valid without the assumption that $\mu(D) < \infty$. Consider the measure space $(\mathbb{R}, \mathfrak{M}_L, \mu_L)$. Let $D = (0, \infty)$. Then $\mu_L(D) = \infty$. Let $f(x) = \frac{1}{x}$ for $x \in (0, \infty)$. Then f is a nonnegative real-valued $\mathfrak{M}_L$-measurable function on D and $f > 0$ on D. Let $\delta > 0$ be arbitrarily given. For $n \in \mathbb{N}$, let $E_n = (n, n + \delta)$. Then $\mu_L(E_n) = \delta$ for every $n \in \mathbb{N}$. On the other hand we have $\int_{E_n} f \, d\mu_L \leq \frac{1}{n}\delta$ so that $\lim_{n \to \infty} \int_{E_n} f \, d\mu_L = 0$ and thus there does not exist $\eta > 0$ such that $\int_{E_n} f \, d\mu_L \geq \eta$ for all $n \in \mathbb{N}$. ∎

Prob. 8.5. Given a measure space $(X, \mathfrak{A}, \mu)$ with $\mu(X) < \infty$. Let f be a nonnegative extended real-valued $\mathfrak{A}$-measurable function on X such that $f > 0$, μ-a.e. on X. Let $(E_n : n \in \mathbb{N})$ be a sequence in $\mathfrak{A}$ such that $\lim_{n \to \infty} \int_{E_n} f \, d\mu = 0$. Prove the following:

(a) $\lim_{n \to \infty} \mu(E_n) = 0$.

(b) $\lim_{n \to \infty} \mu(E_n) = 0$ does not hold without the assumption that $f > 0$, μ-a.e. on X.

(c) $\lim_{n \to \infty} \mu(E_n) = 0$ does not hold without the assumption that $\mu(X) < \infty$.

Proof. 1. Let us prove (a) by contradiction argument. Now $\lim_{n \to \infty} \mu(E_n) = 0$ if and only if for every $\varepsilon > 0$ there exists $N \in \mathbb{N}$ such that $\mu(E_n) < \varepsilon$ for $n \geq N$. Suppose we do not have $\lim_{n \to \infty} \mu(E_n) = 0$. Then there exists $\varepsilon_0 > 0$ such that for every $N \in \mathbb{N}$ there exists $n \geq N$ such that $\mu(E_n) \geq \varepsilon_0$. Thus we can select $n_1 < n_2 < n_3 < \cdots$ such that $\mu(E_{n_k}) \geq \varepsilon_0$ for every $k \in \mathbb{N}$. Then $\{\mu(E_{n_k}) : k \in \mathbb{N}\}$ is bounded below by $\varepsilon_0 > 0$. According to Prob. 8.4, this implies that $\{\int_{E_{n_k}} f \, d\mu : k \in \mathbb{N}\}$ is bounded below by a positive real number $\eta > 0$. Thus we have

$$\int_{E_{n_k}} f \, d\mu \geq \eta > 0 \quad \text{for } k \in \mathbb{N}.$$

This contradicts the assumption that $\lim_{n \to \infty} \int_{E_n} f \, d\mu = 0$. This proves (a).

2. To prove (b), consider the measure space $([0, 1), \mathfrak{M}_L, \mu_L)$. Let f be defined by

$$f(x) = \begin{cases} 1 & \text{for } x \in [0, \tfrac{1}{2}), \\ 0 & \text{for } x \in [\tfrac{1}{2}, 1). \end{cases}$$

Let $E_n = [\tfrac{1}{2}, 1)$ for every $n \in \mathbb{N}$. Then $\int_{E_n} f \, d\mu_L = 0$ for every $n \in \mathbb{N}$ and thus $\lim_{n \to \infty} \int_{E_n} f \, d\mu_L = 0$ but $\mu_L(E_n) = \tfrac{1}{2}$ for every $n \in \mathbb{N}$ and thus $\lim_{n \to \infty} \mu_L(E_n) = \tfrac{1}{2} \neq 0$.

3. To prove (c), consider the measure space $([0, \infty), \mathfrak{M}_L, \mu_L)$. In this case we have $\mu_L([0, \infty)) = \infty$. Let $(a_n : n \in \mathbb{Z}_+)$ be a sequence in $[0, \infty)$ defined by

$$a_0 = 0, \ a_1 = a_0 + 1, \ a_2 = a_1 + 2, \ a_3 = a_2 + 3, \ \ldots, \ a_n = a_{n-1} + n, \ \ldots.$$

Let $E_n = [a_{n-1}, a_n)$ for $n \in \mathbb{N}$. Then $\{E_n : n \in \mathbb{N}\}$ is a disjoint collection in $\mathfrak{A}$, $\bigcup_{n \in \mathbb{N}} E_n = [0, \infty)$, and $\mu_L(E_n) = n$ for every $n \in \mathbb{N}$. Let f be a positive valued function on $[0, \infty)$ defined by

$$f(x) = \frac{1}{n^2} \quad \text{for } x \in E_n.$$

Then we have $\int_{E_n} f \, d\mu_L = \frac{1}{n^2} n = \frac{1}{n}$ for every $n \in \mathbb{N}$ so that $\lim_{n \to \infty} \int_{E_n} f \, d\mu_L = 0$. On the other hand, since $\mu_L(E_n) = n$ for every $n \in \mathbb{N}$, we have $\lim_{n \to \infty} \mu_L(E_n) = \infty$. ∎

Prob. 8.6. Given a measure space $(X, \mathfrak{A}, \mu)$. Let f and g be two nonnegative extended real-valued $\mathfrak{A}$-measurable function on a set $D \in \mathfrak{A}$ such that $f \leq g$ a.e. on D.
(a) Show that if $\int_D f\, d\mu = \int_D g\, d\mu < \infty$, then $f = g$ a.e. on D.
(b) Show by constructing a counterexample that if the conditions $\int_D f\, d\mu < \infty$ and $\int_D g\, d\mu < \infty$ in (a) are removed then the conclusion is not valid.

Proof. 1. Let us prove (a). Let $D_0 = \{x \in X : f(x) \neq g(x)\}$. Since f and g are $\mathfrak{A}$-measurable on D, we have $D_0 \in \mathfrak{A}$. To prove that $f = g$ a.e. on D, we prove that $\mu(D_0) = 0$. We prove this by contradiction argument. Thus assume that $\mu(D_0) \neq 0$. Then $\mu(D_0) > 0$. Let

$$D_1 = \{x \in D : f(x) < g(x)\},$$
$$D_2 = \{x \in D : f(x) > g(x)\}.$$

Then $D_1, D_2 \in \mathfrak{A}$, $D_1 \cap D_2 = \emptyset$, and $D_1 \cup D_2 = D_0$ so that $\mu(D_0) = \mu(D_1) + \mu(D_2)$. Since $f \leq g$ a.e. on D, we have $\mu(D_2) = 0$. Thus we have

$$\mu(D_1) = \mu(D_0) > 0.$$

Now

$$D_1 = \{x \in D : f(x) < g(x)\} = \{x \in D : g(x) - f(x) > 0\}$$
$$= \bigcup_{k \in \mathbb{N}} E_k \quad \text{where} \quad E_k = \{x \in D : g(x) - f(x) > \frac{1}{k}\}.$$

Then we have

$$0 < \mu(D_1) \leq \sum_{k \in \mathbb{N}} \mu(E_k).$$

This implies that there exists $k_0 \in \mathbb{N}$ such that $\mu(E_{k_0}) > 0$. On E_{k_0} we have $g > f + \frac{1}{k_0}$. Thus we have

(1) $$\int_{E_{k_0}} g\, d\mu \geq \int_{E_{k_0}} \left\{ f + \frac{1}{k_0} \right\} d\mu = \int_{E_{k_0}} f\, d\mu + \frac{1}{k_0}\mu(E_{k_0}).$$

Since $g \geq f$ a.e. on D, we have

(2) $$\int_{D \setminus E_{k_0}} g\, d\mu \geq \int_{D \setminus E_{k_0}} f\, d\mu.$$

Adding (1) and (2) side by side, we have

(3) $$\int_D g\, d\mu \geq \int_D f\, d\mu + \frac{1}{k_0}\mu(E_{k_0}).$$

Since $\int_D f\, d\mu = \int_D g\, d\mu < \infty$, subtracting this finite number from both sides of (3) we have

$$0 \geq \frac{1}{k_0}\mu(E_{k_0}).$$

This is a contradiction. Therefore we must have $\mu(D_0) = 0$.
2. To prove (b) consider the measure space $(\mathbb{R}, \mathfrak{M}_L, \mu_L)$. Let $f = 1$ on $\mathbb{R}$ and $g = 2$ on $\mathbb{R}$. Then $f \leq g$ on $\mathbb{R}$ and $\int_\mathbb{R} f\, d\mu_L = \int_\mathbb{R} g\, d\mu_L = \infty$. But we do not have $f = g$ a.e. on $\mathbb{R}$. For another example, let $f(x) = e^x$ and $g(x) = 1 + e^x$ for $x \in \mathbb{R}$. Then $f \leq g$ on $\mathbb{R}$ and $\int_\mathbb{R} f\, d\mu_L = \int_\mathbb{R} g\, d\mu_L = \infty$. But we do not have $f = g$ a.e. on $\mathbb{R}$. ∎

Prob. 8.7. Given a measure space $(X, \mathfrak{A}, \mu)$. Let f and g be two nonnegative extended real-valued $\mathfrak{A}$-measurable function on X.
(a) Show that if $\int_X f \, d\mu = \int_X g \, d\mu < \infty$ and $\int_E f \, d\mu = \int_E g \, d\mu$ for every $E \in \mathfrak{A}$, then $f = g$ a.e. on X.
(b) Show by constructing a counterexample that if the conditions $\int_X f \, d\mu < \infty$ and $\int_X g \, d\mu < \infty$ in (a) are removed then the conclusion is not valid.

Proof. 1. Let us prove (a). Let $D_0 = \{X : f \neq g\}$. Then $\mathfrak{A}$-measurability of f and g implies that $D_0 \in \mathfrak{A}$ and $f = g$ a.e. on X if and only if $\mu(D_0) = 0$.

Suppose that the statement $f = g$ a.e. on X is false. Then we have $\mu(D_0) > 0$. Let $D_1 = \{X : f < g\}$ and $D_2 = \{X : f > g\}$. $\mathfrak{A}$-measurability of f and g implies that $D_1, D_2 \in \mathfrak{A}$ and moreover we have $D_1 \cap D_2 = \emptyset$ and $D_1 \cup D_2 = D_0$ so that $\mu(D_0) = \mu(D_1) + \mu(D_2)$. Our assumption $\mu(D_0) > 0$ implies that at least one of $\mu(D_1)$ and $\mu(D_2)$ is positive. We show below that this leads to a contradiction.

Suppose $\mu(D_1) > 0$. Now we have

$$D_1 = \{X : f < g\} = \{X : g - f > 0\} = \bigcup_{k \in \mathbb{N}} E_k$$

where $E_k = \{X : g - f > \frac{1}{k}\}$.

Then we have

$$0 < \mu(D_1) \leq \sum_{k \in \mathbb{N}} \mu(E_k).$$

This implies that there exists $k_0 \in \mathbb{N}$ such that $\mu(E_{k_0}) > 0$. On E_{k_0} we have $g > f + \frac{1}{k_0}$. Thus we have

$$(1) \qquad \int_{E_{k_0}} g \, d\mu \geq \int_{E_{k_0}} \left\{ f + \frac{1}{k_0} \right\} d\mu = \int_{E_{k_0}} f \, d\mu + \frac{1}{k_0} \mu(E_{k_0}).$$

Since $g > f$ on D_1, we have

$$(2) \qquad \int_{D_1 \setminus E_{k_0}} g \, d\mu \geq \int_{D_1 \setminus E_{k_0}} f \, d\mu.$$

Adding (1) and (2) side by side, we have

$$(3) \qquad \int_{D_1} g \, d\mu \geq \int_{D_1} f \, d\mu + \frac{1}{k_0} \mu(E_{k_0}).$$

Now our assumption $\int_X f \, d\mu = \int_X g \, d\mu < \infty$ implies that $\int_{D_1} f \, d\mu < \infty$ and $\int_{D_1} g \, d\mu < \infty$. Our assumption $\int_E f \, d\mu = \int_E g \, d\mu$ for every $E \in \mathfrak{A}$ implies

$$(4) \qquad \int_{D_1} f \, d\mu = \int_{D_1} g \, d\mu < \infty.$$

Subtracting this real number from both sides of (3), we have

$$(5) \qquad 0 \geq \frac{1}{k_0} \mu(E_{k_0}).$$

This is a contradiction. We show similarly that the assumption $\mu(D_2) > 0$ leads to a contradiction. Therefore we must have $\mu(D_0) = 0$, that is, $f = g$ a.e. on X.

2. Let us prove (b). Let X be an arbitrary non-empty set and let $\mathfrak{A}$ be an arbitrary σ-algebra of subsets of X. Let us define a set function μ on $\mathfrak{A}$ by setting $\mu(\emptyset) = 0$ and $\mu(E) = \infty$ for every $E \in \mathfrak{A}$ such that $E \neq \emptyset$. Then μ is a measure on the σ-algebra $\mathfrak{A}$ and we have a measure space $(X, \mathfrak{A}, \mu)$. Let us define two real-valued $\mathfrak{A}$-measurable functions f and g by setting $f = 1$ on X and $g = 2$ on X. Then we have

$$\int_X f\, d\mu = \int_X g\, d\mu = \infty,$$

$$\int_E f\, d\mu = \int_E g\, d\mu = \infty \quad \text{for } E \in \mathfrak{A} \text{ such that } E \neq \emptyset,$$

$$\int_\emptyset f\, d\mu = \int_\emptyset g\, d\mu = \infty.$$

Observe that not only the statement $f = g$ a.e. on X fails to hold but furthermore we have $f \neq g$ everywhere on X. ∎

Prob. 8.8. Given a measure space $(X, \mathfrak{A}, \mu)$. Let f be a real-valued $\mathfrak{A}$-measurable function on a set $D \in \mathfrak{A}$ with $\mu(D) \in (0, \infty)$ such that $f(x) \in [0, 1)$ for every $x \in D$.
(a) Show that $\int_D f\, d\mu < \mu(D)$.
(b) Show by counterexample that without the condition $\mu(D) < \infty$, the conclusion in (a) is not valid.

Proof. 1. Let us prove (a). Since $f < 1$ on D, we have $\int_D f\, d\mu \leq \int_D 1\, d\mu = \mu(D)$. To show that $\int_D f\, d\mu < \mu(D)$, let us assume the contrary, that is,

$$\int_D f\, d\mu = \mu(D) = \int_D 1\, d\mu.$$

Then we have

$$\int_D \{1 - f\}\, d\mu = \int_D 1\, d\mu - \int_D f\, d\mu = 0.$$

Since $1 - f \geq 0$ on D, $\int_D \{1 - f\}\, d\mu = 0$ implies that $1 - f = 0$ a.e. on D according to (b) of Lemma 8.2. Thus $f = 1$ a.e. on D. This contradicts the assumption that $f < 1$ on D.

2. To prove (b), consider the measure space $(\mathbb{R}, \mathfrak{M}_L, \mu_L)$. Let $f = \frac{1}{2}$ on $\mathbb{R}$. Then $f < 1$ on $\mathbb{R}$ and

$$\int_\mathbb{R} f\, d\mu_L = \int_\mathbb{R} \frac{1}{2}\, d\mu_L = \infty = \mu_L(\mathbb{R}),$$

that is, we do not have $\int_\mathbb{R} f\, d\mu_L < \mu_L(\mathbb{R})$. ∎

Prob. 8.9. Given a measure space $(X, \mathfrak{A}, \mu)$. Let f be a nonnegative extended real-valued $\mathfrak{A}$-measurable function on a set $D \in \mathfrak{A}$. Let Φ be the collection of all nonnegative simple functions φ on D such that $0 \leq \varphi \leq f$ on D and let Θ be the collection of all μ-integrable nonnegative simple functions ϑ on D such that $0 \leq \vartheta \leq f$ on D. Recall that by our definition of the μ-integral of f, we have $\int_D f \, d\mu = \sup_{\varphi \in \Phi} \int_D \varphi \, d\mu$. Note that since $\Theta \subset \Phi$, we have $\sup_{\vartheta \in \Theta} \int_D \vartheta \, d\mu \leq \sup_{\varphi \in \Phi} \int_D \varphi \, d\mu$. Show that if D is a σ-finite set with respect to the measure μ, then $\sup_{\vartheta \in \Theta} \int_D \vartheta \, d\mu = \sup_{\varphi \in \Phi} \int_D \varphi \, d\mu$.

Proof. Observe that if $\mu(D) < \infty$ then every simple function on D is μ-integrable so that $\Theta = \Phi$ and then $\sup_{\vartheta \in \Theta} \int_D \vartheta \, d\mu = \sup_{\varphi \in \Phi} \int_D \varphi \, d\mu$ is trivially true.

Consider the general case that D is a σ-finite set with respect to μ. If $\Theta = \Phi$ then $\sup_{\vartheta \in \Theta} \int_D \vartheta \, d\mu = \sup_{\varphi \in \Phi} \int_D \varphi \, d\mu$ is trivially true. Thus consider the case that $\Theta \neq \Phi$. In this case there exists $\varphi_0 \in \Phi$ such that $\varphi_0 \notin \Theta$, that is, there exists $\varphi_0 \in \Phi$ such that φ_0 is not μ-integrable on D, that is, $\int_D \varphi_0 \, d\mu = \infty$. Thus in this case we have

$$\sup_{\varphi \in \Phi} \int_D \varphi \, d\mu = \infty.$$

Now since D is a σ-finite set there exists an increasing sequence $(A_k; k \in \mathbb{N})$ in $\mathfrak{A}$ such that $\lim_{k \to \infty} A_k = \bigcup_{k \in \mathbb{N}} A_k = D$ and $\mu(A_k) < \infty$ for every $k \in \mathbb{N}$. With our $\varphi_0 \in \Phi$ with $\int_D \varphi_0 \, d\mu = \infty$, let us define

$$\vartheta_k = \varphi_0 \cdot \mathbf{1}_{A_k} \quad \text{for } k \in \mathbb{N}.$$

Then ϑ_k is μ-integrable on D so that $\vartheta_k \in \Theta$. Now $\vartheta_k \uparrow \varphi_0$ as $k \to \infty$ so that by the Monotone Convergence Theorem (Theorem 8.5), we have

$$\lim_{k \to \infty} \int_D \vartheta_k \, d\mu = \int_D \varphi_0 \, d\mu = \infty$$

and thus

$$\sup_{\vartheta \in \Theta} \int_D \vartheta \, d\mu \geq \lim_{k \to \infty} \int_D \vartheta_k \, d\mu = \infty,$$

and therefore

$$\sup_{\vartheta \in \Theta} \int_D \vartheta \, d\mu = \infty = \sup_{\varphi \in \Phi} \int_D \varphi \, d\mu. \quad \blacksquare$$

Prob. 8.10. Consider the Lebesgue measure space $\left(\mathbb{R}, \mathfrak{M}_L, \mu_L\right)$.

(a) Construct nonnegative real-valued $\mathfrak{M}_L$-measurable functions $(f_n : n \in \mathbb{N})$ and f on $\mathbb{R}$ such that $(f_n : n \in \mathbb{N})$ converges to f uniformly on $\mathbb{R}$ but $\lim_{n\to\infty} \int_{\mathbb{R}} f_n \, d\mu_L \neq \int_{\mathbb{R}} f \, d\mu_L$.

(b) Construct nonnegative real-valued $\mathfrak{M}_L$-measurable functions $(f_n : n \in \mathbb{N})$ and f on $\mathbb{R}$ such that $f_n \downarrow f$ as $n \to \infty$ on $\mathbb{R}$ but $\lim_{n\to\infty} \int_{\mathbb{R}} f_n \, d\mu_L \neq \int_{\mathbb{R}} f \, d\mu_L$.

Proof. The following example serves both (a) and (b). Let $f_n = \frac{1}{n}$ on $\mathbb{R}$ for $n \in \mathbb{N}$ and let $f = 0$ on $\mathbb{R}$. Then $(f_n : n \in \mathbb{N})$ converges to f uniformly on $\mathbb{R}$ and also $f_n \downarrow f$ as $n \to \infty$ on $\mathbb{R}$. But we have $\int_{\mathbb{R}} f_n \, d\mu_L = \infty$ for every $n \in \mathbb{N}$ and $\int_{\mathbb{R}} f \, d\mu_L = 0$ so that $\lim_{n\to\infty} \int_{\mathbb{R}} f_n \, d\mu_L \neq \int_{\mathbb{R}} f \, d\mu_L$. ∎

Prob. 8.11. Let $(X, \mathfrak{A}, \mu)$ be a measure space and let f be a nonnegative extended real-valued $\mathfrak{A}$-measurable function on X that is finite μ-a.e. on X. Define a sequence of functions on X, $(f_n : n \in \mathbb{N})$, by setting

$$f_n(x) = \begin{cases} f(x) & \text{if } f(x) \leq n, \\ 0 & \text{if } f(x) > n. \end{cases}$$

Show that $\lim_{n\to\infty} \int_X f_n \, d\mu = \int_X f \, d\mu$.

Proof. As defined above, f_n is a nonnegative real-valued function on X and since f is $\mathfrak{A}$-measurable on X, f_n is $\mathfrak{A}$-measurable on X. Moreover $(f_n : n \in \mathbb{N})$ is an increasing sequence of functions on X. Let $E = \{x \in X : f(x) = \infty\}$. Since f is finite μ-a.e. on X, we have $\mu(E) = 0$. For $x \notin E$, $f_n(x) \uparrow f(x)$ as $n \to \infty$. Thus $f_n \uparrow f$, μ-a.e. on X. Then by the Monotone Convergence Theorem (Theorem 8.5), we have $\lim_{n\to\infty} \int_X f_n \, d\mu = \int_X f \, d\mu$. ∎

Prob. 8.12. (a) Let $\{c_{n,i} : n \in \mathbb{N}, i \in \mathbb{N}\}$ be an array of nonnegative extended real numbers, that is, $c_{n,i} \in [0, \infty]$ for $n \in \mathbb{N}$ and $i \in \mathbb{N}$. Show that

$$\liminf_{n \to \infty} \sum_{i \in \mathbb{N}} c_{n,i} \geq \sum_{i \in \mathbb{N}} \liminf_{n \to \infty} c_{n,i}.$$

(b) Show that if $\left(c_{n,i} : n \in \mathbb{N}\right)$ is an increasing sequence for each $i \in \mathbb{N}$ then

$$\lim_{n \to \infty} \sum_{i \in \mathbb{N}} c_{n,i} = \sum_{i \in \mathbb{N}} \lim_{n \to \infty} c_{n,i}.$$

Proof. We prove (a) and (b) by interpreting the series of nonnegative extended real numbers $\sum_{i \in \mathbb{N}} c_{n,i}$ as the integral of a nonnegative step function on the Lebesgue measure space $\left([0, \infty), \mathfrak{M}_L, \mu_L\right)$ and then applying Fatou's Lemma for (a) and the Monotone Convergence Theorem for (b). This is done as follows.

1. Let us prove (a). Consider the measure space $\left([0, \infty), \mathfrak{M}_L, \mu_L\right)$. For each $i \in \mathbb{N}$, let $D_i = [i - 1, i)$. Then $D_i \in \mathfrak{M}_L$ with $\mu_L(D_i) = 1$ for $i \in \mathbb{N}$, and $\{D_i : i \in \mathbb{N}\}$ is a disjoint collection in $\mathfrak{M}_L$ with $\bigcup_{i \in \mathbb{N}} D_i = [0, \infty)$. For each $n \in \mathbb{N}$, let us define a nonnegative extended real-valued function f_n on $[0, \infty)$ by setting

$$(1) \qquad\qquad f_n(x) = c_{n,i} \quad \text{for } x \in D_i \text{ for } i \in \mathbb{N}.$$

Then $(f_n : n \in \mathbb{N})$ is a sequence of nonnegative extended real-valued $\mathfrak{M}_L$-measurable functions on $[0, \infty)$ so that by Fatou's Lemma (Theorem 8.13) we have

$$(2) \qquad\qquad \liminf_{n \to \infty} \int_{[0,\infty)} f_n \, d\mu_L \geq \int_{[0,\infty)} \liminf_{n \to \infty} f_n \, d\mu_L.$$

Now

$$(3) \qquad\qquad \int_{[0,\infty)} f_n \, d\mu_L = \sum_{i \in \mathbb{N}} \int_{D_i} f_n \, d\mu_L = \sum_{i \in \mathbb{N}} c_{n,i} \, \mu_L(D_i) = \sum_{i \in \mathbb{N}} c_{n,i}.$$

On the other hand, by (1) we have $\liminf_{n \to \infty} f_n = \liminf_{n \to \infty} c_{n,i}$ on D_i for $i \in \mathbb{N}$ and then

$$(4) \qquad\qquad \int_{[0,\infty)} \liminf_{n \to \infty} f_n \, d\mu_L = \sum_{i \in \mathbb{N}} \liminf_{n \to \infty} c_{n,i} \, \mu_L(D_i) = \sum_{i \in \mathbb{N}} \liminf_{n \to \infty} c_{n,i}.$$

Substituting (3) and (4) in (2), we complete the proof of (a).

2. To prove (b), let us assume that $(c_{n,i} : n \in \mathbb{N})$ is an increasing sequence for each $i \in \mathbb{N}$. Then with f_n as defined by (1), $(f_n : n \in \mathbb{N})$ is an increasing sequence of nonnegative extended real-valued $\mathfrak{M}_L$-measurable functions on $[0, \infty)$ so that by the Monotone Convergence Theorem (Theorem 8.5) we have

$$(5) \qquad\qquad \lim_{n \to \infty} \int_{[0,\infty)} f_n \, d\mu_L = \int_{[0,\infty)} \lim_{n \to \infty} f_n \, d\mu_L.$$

Now

$$(6) \qquad\qquad \int_{[0,\infty)} \lim_{n \to \infty} f_n \, d\mu_L = \sum_{i \in \mathbb{N}} \lim_{n \to \infty} c_{n,i} \, \mu_L(D_i) = \sum_{i \in \mathbb{N}} \lim_{n \to \infty} c_{n,i}.$$

Substituting (3) and (6) in (5), we complete the proof of (b). ∎

Prob. 8.13. Given a measure space $(X, \mathfrak{A}, \mu)$. Let f be a nonnegative extended real-valued $\mathfrak{A}$-measurable function on a set $D \in \mathfrak{A}$ with $\mu(D) < \infty$. Let $D_n = \{x \in D : f(x) \geq n\}$ for $n \in \mathbb{Z}_+$. Show that $\int_D f \, d\mu < \infty$ if and only if $\sum_{n \in \mathbb{Z}_+} \mu(D_n) < \infty$.

Proof. 1. Observe that $(D_n : n \in \mathbb{Z}_+)$ is a decreasing sequence in $\mathfrak{A}$ and $D_0 = D$. Let $E_n = D_n \setminus D_{n+1}$ for $n \in \mathbb{Z}_+$. Then $(E_n : n \in \mathbb{Z}_+)$ is a disjoint sequence in $\mathfrak{A}$ and $\bigcup_{n \in \mathbb{Z}_+} E_n = D_0 = D$. We also have

$$E_n = \{x \in D : f(x) \in [n, n+1)\} \quad \text{for } n \in \mathbb{Z}_+.$$

Now we have

$$(1) \qquad \int_D f \, d\mu = \sum_{n \in \mathbb{Z}_+} \int_{E_n} f \, d\mu$$

and

$$(2) \qquad \sum_{n \in \mathbb{Z}_+} n \, \mu(E_n) \leq \sum_{n \in \mathbb{Z}_+} \int_{E_n} f \, d\mu \leq \sum_{n \in \mathbb{Z}_+} (n+1)\mu(E_n).$$

Since $\mu(D_n) \leq \mu(D) < \infty$ for every $n \in \mathbb{Z}_+$, we have

$$\mu(E_n) = \mu(D_n \setminus D_{n+1}) = \mu(D_n) - \mu(D_{n+1}) \quad \text{for every } n \in \mathbb{Z}_+.$$

Substituting this in (2), we have from (1) and (2)

$$(3) \qquad \sum_{n \in \mathbb{Z}_+} n \{\mu(D_n) - \mu(D_{n+1})\} \leq \int_D f \, d\mu \leq \sum_{n \in \mathbb{Z}_+} (n+1)\{\mu(D_n) - \mu(D_{n+1})\}.$$

For brevity, let $c_n = \mu(D_n)$ for $n \in \mathbb{Z}_+$. Then (3) becomes

$$(4) \qquad \sum_{n \in \mathbb{Z}_+} n \left(c_n - c_{n+1}\right) \leq \int_D f \, d\mu \leq \sum_{n \in \mathbb{Z}_+} (n+1)\left(c_n - c_{n+1}\right).$$

Now

$$\sum_{n \in \mathbb{Z}_+} n \left(c_n - c_{n+1}\right) = 0\left(c_0 - c_1\right) + 1\left(c_1 - c_2\right) + 2\left(c_2 - c_3\right) + \cdots$$

$$= c_1 + c_2 + c_3 + \cdots = \sum_{n \in \mathbb{N}} c_n$$

and

$$\sum_{n \in \mathbb{Z}_+} (n+1)\left(c_n - c_{n+1}\right) = 1\left(c_0 - c_1\right) + 2\left(c_1 - c_2\right) + 3\left(c_2 - c_3\right) + \cdots$$

$$= c_0 + c_1 + c_2 + + \cdots = \sum_{n \in \mathbb{Z}_+} c_n.$$

Substituting these in (4), we have

(5)
$$\sum_{n\in\mathbb{N}} c_n \le \int_D f \, d\mu \le \sum_{n\in\mathbb{Z}_+} c_n.$$

2. Now suppose $\sum_{n\in\mathbb{Z}_+} \mu(D_n) < \infty$, that is, $\sum_{n\in\mathbb{Z}_+} c_n < \infty$. Then $\int_D f \, d\mu < \infty$ by (5).

3. Conversely if $\int_D f \, d\mu < \infty$, then we have $\sum_{n\in\mathbb{N}} c_n < \infty$ by (5), that is, we have $\sum_{n\in\mathbb{N}} \mu(D_n) < \infty$. Then since $\mu(D_0) = \mu(D) < \infty$, we have $\sum_{n\in\mathbb{Z}_+} \mu(D_n) < \infty$. ∎

Prob. 8.14. Given a measure space $(X, \mathfrak{A}, \mu)$. Let f be a bounded nonnegative real-valued $\mathfrak{A}$-measurable function on a set $D \in \mathfrak{A}$. Let $D_n = \{x \in D : f(x) \ge n\}$ for $n \in \mathbb{N}$. Show that if $\int_D f \, d\mu < \infty$ then $\sum_{n\in\mathbb{N}} \mu(D_n) < \infty$.

Proof. Observe that $(D_n : n \in \mathbb{Z}_+)$ is a decreasing sequence $\mathfrak{A}$-measurable subsets of D. Since f is a bounded nonnegative real-valued function on D, there exists $N \in \mathbb{N}$ such that $f(x) \in [0, N]$ for every $x \in D$. This implies that

$$D_{N+1} = \big\{x \in D : f(x) \ge N + 1\big\} = \emptyset.$$

Then since $(D_n : n \in \mathbb{Z}_+)$ is a decreasing sequence we have $D_n = \emptyset$ for $n \ge N + 1$ and then $\mu(D_n) = 0$ for $n > N$. If $\int_D f \, d\mu < \infty$ then we have

$$n \, \mu(D_n) \le \int_{D_n} f \, d\mu \le \int_D f \, d\mu < \infty \quad \text{for every } n \in \mathbb{N}.$$

Thus $\mu(D_n) < \infty$ for every $n \in \mathbb{N}$. Then we have

$$\sum_{n\in\mathbb{N}} \mu(D_n) = \sum_{n=1}^{N} \mu(D_n) < \infty. \ \blacksquare$$

Prob. 8.15. Let $(X, \mathfrak{A}, \mu)$ be a measure space with $\mu(X) < \infty$. Let f be a nonnegative extended real-valued $\mathfrak{A}$-measurable function on X that is finite μ-a.e. on X. Show that f in μ-integrable on X if and only if

$$\sum_{n \in \mathbb{Z}_+} 2^n \mu\{x \in X : f(x) > 2^n\} < \infty.$$

Proof. 1. For $n \in \mathbb{Z}_+$, let

$$(1) \qquad D_n = \{x \in X : f(x) \in (2^n, \infty)\} \quad \text{for } n \in \mathbb{Z}_+.$$

We have in particular

$$D_0 = \{x \in X : f(x) \in (1, \infty)\}.$$

Since $2^n \uparrow$ as $n \uparrow$, we have $D_n \downarrow$ as $n \uparrow$ and thus $(D_n : n \in \mathbb{Z}_+)$ is a decreasing sequence in $\mathfrak{A}$. Let

$$(2) \qquad \begin{cases} D_\alpha = \{x \in X : f(x) \in [0, 1]\}, \\ D_\omega = \{x \in X : f(x) = \infty\}. \end{cases}$$

Then we have

$$(3) \qquad X = D_\alpha \cup D_0 \cup D_\omega.$$

Now $\{x \in X : f(x) > 2^n\} = D_n \cup D_\omega$ for $n \in \mathbb{Z}_+$. Since $\mu(D_\omega) = 0$, we have

$$\mu\{x \in X : f(x) > 2^n\} = \mu(D_n) + \mu(D_\omega) = \mu(D_n).$$

Thus our problem is reduced to that of proving

$$(4) \qquad \int_X f \, d\mu < \infty \Leftrightarrow \sum_{n \in \mathbb{Z}_+} 2^n \mu(D_n) < \infty.$$

To prove (4), let

$$(5) \qquad E_n = D_n \setminus D_{n+1} = \{x \in X : f(x) \in (2^n, 2^{n+1}]\} \quad \text{for } n \in \mathbb{Z}_+.$$

Then $(E_n : n \in \mathbb{Z}_+)$ is a disjoint sequence in $\mathfrak{A}$ and $\bigcup_{n \in \mathbb{Z}_+} E_n = D_0$ so that

$$(6) \qquad X = D_\alpha \cup \left(\bigcup_{n \in \mathbb{Z}_+} E_n \right) \cup D_\omega.$$

Since $\mu(X) < \infty$, we have $\mu(D_n) < \infty$ for every $n \in \mathbb{Z}_+$ and this implies

$$(7) \qquad \mu(E_n) = \mu(D_n) - \mu(D_{n+1}) \quad \text{for } n \in \mathbb{Z}_+.$$

2. Assume that $\sum_{n\in\mathbb{Z}_+} 2^n \mu(D_n) < \infty$. Let us show that $\int_X f\,d\mu < \infty$. Now

$$\int_X f\,d\mu = \int_{D_\alpha\left(\bigcup_{n\in\mathbb{Z}_+} E_n\right)\cup D_\omega} f\,d\mu = \int_{D_\alpha} f\,d\mu + \sum_{n\in\mathbb{Z}_+}\int_{E_n} f\,d\mu$$

$$\leq 1 \cdot \mu(D_\alpha) + \sum_{n\in\mathbb{Z}_+} 2^{n+1}\mu(E_n) \leq \mu(D_\alpha) + \sum_{n\in\mathbb{Z}_+} 2^{n+1}\mu(D_n)$$

$$= \mu(D_\alpha) + 2\sum_{n\in\mathbb{Z}_+} 2^n \mu(D_n) < \infty.$$

3. Conversely let us assume $\int_X f\,d\mu < \infty$. Let us show that $\sum_{n\in\mathbb{Z}_+} 2^n \mu(D_n) < \infty$. Now by (3) and (5), we have

$$(8) \qquad \sum_{n\in\mathbb{Z}_+} 2^n \mu(E_n) \leq \sum_{n\in\mathbb{Z}_+}\int_{E_n} f\,d\mu = \int_{\bigcup_{n\in\mathbb{Z}_+} E_n} f\,d\mu \leq \int_X f\,d\mu < \infty.$$

On the other hand by (7) we have

$$(9) \qquad \sum_{n\in\mathbb{Z}_+} 2^n \mu(E_n) = \sum_{n\in\mathbb{Z}_+} 2^n \big\{\mu(D_n) - \mu(D_{n+1})\big\}$$

$$= \big\{\mu(D_0) - \mu(D_1)\big\} + 2\big\{\mu(D_1) - \mu(D_2)\big\} + 2^2\big\{\mu(D_2) - \mu(D_3)\big\} + \cdots$$

$$= \mu(D_0) + \sum_{n\in\mathbb{N}} 2^{n-1}\mu(D_n) = \mu(D_0) + \frac{1}{2}\sum_{n\in\mathbb{N}} 2^n \mu(D_n).$$

Substituting (9) in (8), we have

$$\mu(D_0) + \frac{1}{2}\sum_{n\in\mathbb{N}} 2^n \mu(D_n) \leq \int_X f\,d\mu < \infty,$$

and then

$$\sum_{n\in\mathbb{Z}_+} 2^n \mu(D_n) < \infty.$$

This completes the proof. ∎

Prob. 8.16. Let $(X, \mathfrak{A}, \mu)$ be an arbitrary measure space. Let f be a bounded nonnegative real-valued $\mathfrak{A}$-measurable function on X. Show that f is μ-integrable on X if and only if

$$\sum_{n\in\mathbb{Z}_+} \frac{1}{2^n}\mu\Big\{x \in X : f(x) > \frac{1}{2^n}\Big\} < \infty.$$

Proof. 1. Since f is a bounded nonnegative real-valued function on X, there exists $M > 0$ such that $f(x) \in [0, M]$ for every $x \in X$. Let

(1)
$$\begin{cases} D_\alpha = \big\{x \in X : f(x) = 0\big\}, \\ D_n = \big\{x \in X : f(x) \in \big(\frac{1}{2^n}, M\big]\big\} \quad \text{for } n \in \mathbb{Z}_+. \end{cases}$$

Then $D_\alpha \in \mathfrak{A}$ and $D_n \in \mathfrak{A}$ for $n \in \mathbb{Z}_+$. Since $\frac{1}{2^n} \downarrow$ as $n \uparrow$, we have $D_n \uparrow$ as $n \uparrow$. Then since $f \geq 0$ on X, we have

(2)
$$X = D_\alpha \cup \bigcup_{n\in\mathbb{Z}_+} D_n.$$

Since $f(x) \in [0, M]$ for every $x \in X$, we have $\big\{x \in X : f(x) > \frac{1}{2^n}\big\} = D_n$ for $n \in \mathbb{Z}_+$. Thus we are to show that

(3)
$$\int_X f \, d\mu < \infty \Leftrightarrow \sum_{n\in\mathbb{Z}_+} \frac{1}{2^n}\mu(D_n) < \infty.$$

Let

(4)
$$\begin{cases} E_0 = D_0 = \big\{x \in X : f(x) \in (1, M]\big\}, \\ E_n = D_n \setminus D_{n-1} = \big\{x \in X : f(x) \in \big(\frac{1}{2^n}, \frac{1}{2^{n-1}}\big]\big\} \quad \text{for } n \in \mathbb{N}. \end{cases}$$

Then $\{E_n : n \in \mathbb{Z}_+\}$ is a disjoint collection in $\mathfrak{A}$ and $\bigcup_{n\in\mathbb{Z}_+} E_n = \bigcup_{n\in\mathbb{Z}_+} D_n$ and thus

(5)
$$X = D_\alpha \cup \bigcup_{n\in\mathbb{Z}_+} E_n.$$

2. Suppose $\sum_{n\in\mathbb{Z}_+} \frac{1}{2^n}\mu(D_n) < \infty$. Then we have

$$\int_X f \, d\mu = \int_{D_\alpha\cup\bigcup_{n\in\mathbb{Z}_+} E_n} f \, d\mu = \int_{\bigcup_{n\in\mathbb{Z}_+} E_n} f \, d\mu = \sum_{n\in\mathbb{Z}_+} \int_{E_n} f \, d\mu$$

$$\leq M\mu(E_0) + \sum_{n\in\mathbb{N}} \frac{1}{2^{n-1}}\mu(E_n) \leq M\mu(D_0) + \sum_{n\in\mathbb{N}} \frac{1}{2^{n-1}}\mu(D_n)$$

$$\leq M\mu(D_0) + 2\sum_{n\in\mathbb{N}} \frac{1}{2^n}\mu(D_n) < \infty.$$

3. Conversely assume that $\int_X f \, d\mu < \infty$. Since $f = 0$ on D_α, we have

$$\int_X f \, d\mu = \int_{D_\alpha \cup \bigcup_{n \in \mathbb{Z}_+} D_n} f \, d\mu = \int_{\bigcup_{n \in \mathbb{Z}_+} D_n} f \, d\mu.$$

For each $n \in \mathbb{Z}_+$, since $f > \frac{1}{2^n}$ on D_n by (1), we have

$$\frac{1}{2^n} \mu(D_n) \leq \int_{D_n} f \, d\mu \leq \int_{\bigcup_{n \in \mathbb{Z}_+} D_n} f \, d\mu < \infty$$

and thus we have

(6) $\qquad\qquad\qquad\qquad \mu(D_n) < \infty \quad \text{for } n \in \mathbb{Z}_+.$

Then since $D_n \supset D_{n-1}$ and $E_n = D_n \setminus D_{n-1}$ for $n \in \mathbb{N}$, we have

(7) $\qquad\qquad\qquad\qquad \mu(E_n) = \mu(D_n) - \mu(D_{n-1}) \quad \text{for } n \in \mathbb{N}.$

On E_n we have $f > \frac{1}{2^n}$ for $n \in \mathbb{Z}_+$ by (4). Thus we have

(8) $\qquad \displaystyle\sum_{n \in \mathbb{Z}_+} \frac{1}{2^n} \mu(E_n) \leq \sum_{n \in \mathbb{Z}_+} \int_{E_n} f \, d\mu = \int_{\bigcup_{n \in \mathbb{Z}_+} E_n} f \, d\mu = \int_X f \, d\mu < \infty.$

Now

$$\sum_{n \in \mathbb{Z}_+} \frac{1}{2^n} \mu(E_n)$$

$$= 1 \cdot \mu(E_0) + \sum_{n \in \mathbb{N}} \frac{1}{2^n} \mu(E_n) = \mu(D_0) + \sum_{n \in \mathbb{N}} \frac{1}{2^n} \{\mu(D_n) - \mu(D_{n-1})\}$$

$$= \mu(D_0) + \frac{1}{2} \{\mu(D_1) - \mu(D_0)\} + \frac{1}{2^2} \{\mu(D_2) - \mu(D_1)\} + \frac{1}{2^3} \{\mu(D_3) - \mu(D_2)\} + \cdots$$

$$= \frac{1}{2} \mu(D_0) + \frac{1}{2^2} \mu(D_1) + \frac{1}{2^3} \mu(D_2) + \cdots$$

$$= \frac{1}{2} \sum_{n \in \mathbb{Z}_+} \frac{1}{2^n} \mu(D_n).$$

Substituting this in (8), we have $\sum_{n \in \mathbb{Z}_+} \frac{1}{2^n} \mu(D_n) < \infty$. This completes the proof. ∎

Prob. 8.17. Given a measure space $(X, \mathfrak{A}, \mu)$. Let $(f_n : n \in \mathbb{N})$ and f be extended real-valued $\mathfrak{A}$-measurable functions on a set $D \in \mathfrak{A}$ and assume that f is real-valued a.e. on D. Suppose there exists a sequence of positive numbers $(\varepsilon_n : n \in \mathbb{N})$ such that
1° $\sum_{n \in \mathbb{N}} \varepsilon_n < \infty$,
2° $\int_D |f_n - f|^p \, d\mu < \varepsilon_n$ for every $n \in \mathbb{N}$ for some fixed $p \in (0, \infty)$.
Show that the sequence $(f_n : n \in \mathbb{N})$ converges to f a.e. on D. (Note that no μ-integrability of f_n, f, $|f_n|^p$, and $|f|^p$ on D is assumed.)

Proof. By 1° and 2°, we have

(1) $$\sum_{n \in \mathbb{N}} \int_D |f_n - f|^p \, d\mu \le \sum_{n \in \mathbb{N}} \varepsilon_n < \infty.$$

On the other hand by Proposition 8.7 we have

(2) $$\sum_{n \in \mathbb{N}} \int_D |f_n - f|^p \, d\mu = \int_D \sum_{n \in \mathbb{N}} |f_n - f|^p \, d\mu.$$

Then by (1) and (2), we have

(3) $$\int_D \sum_{n \in \mathbb{N}} |f_n - f|^p \, d\mu < \infty.$$

This implies according to Lemma 8.2 that $\sum_{n \in \mathbb{N}} |f_n(x) - f(x)|^p < \infty$ for a.e. $x \in D$. Now the convergence of a series implies the convergence of the terms to 0. Thus we have

(4) $$\lim_{n \to \infty} |f_n(x) - f(x)|^p = 0 \quad \text{for a.e. } x \in D.$$

Let us observe that for a sequence $(a_n : n \in \mathbb{N})$ of nonnegative real numbers and for $p \in (0, \infty)$ we have

(5) $$\lim_{n \to \infty} a_n = 0 \Leftrightarrow \lim_{n \to \infty} a_n^p = 0.$$

Now (4) and (5) imply that $\lim_{n \to \infty} |f_n(x) - f(x)| = 0$, that is, $\lim_{n \to \infty} f_n(x) = f(x)$ for a.e. $x \in D$. ∎

(Proof of (5). Suppose $\lim_{n \to \infty} a_n = 0$. To show $\lim_{n \to \infty} a_n^p = 0$, let $\varepsilon > 0$ be arbitrarily given. Then $\varepsilon^{1/p} > 0$. Since $\lim_{n \to \infty} a_n = 0$, there exists $N \in \mathbb{N}$ such that $a_n < \varepsilon^{1/p}$ for $n \ge N$. Then $a_n^p < \varepsilon$ for $n \ge N$. This shows that $\lim_{n \to \infty} a_n^p = 0$. Conversely suppose $\lim_{n \to \infty} a_n^p = 0$. To show $\lim_{n \to \infty} a_n = 0$, let $\varepsilon > 0$ be arbitrarily given. Then $\varepsilon^p > 0$. Since $\lim_{n \to \infty} a_n^p = 0$, there exists $N \in \mathbb{N}$ such that $a_n^p < \varepsilon^p$ for $n \ge N$. Then a_n, ε for $n \ge N$. This shows that $\lim_{n \to \infty} a_n = 0$.)

Prob. 8.18. Given a measure space $(X, \mathfrak{A}, \mu)$. Let $(f_n : n \in \mathbb{N})$ and f be extended real-valued $\mathfrak{A}$-measurable functions on a set $D \in \mathfrak{A}$ and assume that f is real-valued a.e. on D. Suppose there exists a sequence of positive numbers $(\varepsilon_n : n \in \mathbb{N})$ such that
1° $\quad \lim\limits_{n \to \infty} \varepsilon_n = 0$,
2° $\quad \int_D |f_n - f|^p \, d\mu < \varepsilon_n$ for every $n \in \mathbb{N}$ for some fixed $p \in (0, \infty)$.
Show that the sequence $(f_n : n \in \mathbb{N})$ has a subsequence $(f_{n_k} : k \in \mathbb{N})$ which converges to f a.e. on D.

Proof. Condition 1° implies that we can select a subsequence $(\varepsilon_{n_k} : k \in \mathbb{N})$ such that $\sum_{k \in \mathbb{N}} \varepsilon_{n_k} < \infty$. (For example, let $n_1 \in \mathbb{N}$ be so large that $\varepsilon_{n_1} < \frac{1}{2}$. Then let $n_2 > n_1$ be so large that $\varepsilon_{n_2} < \frac{1}{2^2}$. Then let $n_3 > n_2$ be so large that $\varepsilon_{n_3} < \frac{1}{2^3}$ and so on. Then we have $\sum_{k \in \mathbb{N}} \varepsilon_{n_k} \leq \sum_{k \in \mathbb{N}} \frac{1}{2^k} = 1$.) Then we have
3° $\quad \sum_{k \in \mathbb{N}} \varepsilon_{n_k} < \infty$.
Also from 2° we have
4° $\quad \int_D |f_{n_k} - f|^p \, d\mu < \varepsilon_{n_k}$ for every $k \in \mathbb{N}$ for some fixed $p \in (0, \infty)$.
Then by Prob. 8.17, the sequence $(f_{n_k} : k \in \mathbb{N})$ converges to f a.e. on D. ∎

Prob. 8.19. Given a measure space $(X, \mathfrak{A}, \mu)$. Let $(f_n : n \in \mathbb{N})$ and f be extended real-valued $\mathfrak{A}$-measurable functions on a set $D \in \mathfrak{A}$ and assume that f is real-valued a.e. on D. Suppose $\lim\limits_{n \to \infty} \int_D |f_n - f|^p \, d\mu = 0$ for some fixed $p \in (0, \infty)$. Show that the sequence $(f_n : n \in \mathbb{N})$ converges to f on D in measure.

Proof. Let $\varepsilon > 0$ be arbitrarily given. Then we have

$$\varepsilon^p \mu\{D : |f_n - f| \geq \varepsilon\} \leq \int_{\{D : |f_n - f| \geq \varepsilon\}} |f_n - f|^p \, d\mu$$

$$\leq \int_D |f_n - f|^p \, d\mu.$$

Then

$$\lim_{n \to \infty} \varepsilon^p \mu\{D : |f_n - f| \geq \varepsilon\} \leq \lim_{n \to \infty} \int_D |f_n - f|^p \, d\mu = 0$$

and thus

$$\lim_{n \to \infty} \mu\{D : |f_n - f| \geq \varepsilon\} = 0.$$

This shows that the sequence $(f_n : n \in \mathbb{N})$ converges to f on D in measure. ∎

Prob. 8.20. Let $(X, \mathfrak{A}, \mu)$ be a measure space and let f be an extended real-valued $\mathfrak{A}$-measurable function on X such that $\int_X |f|^p \, d\mu < \infty$ for some $p \in (0, \infty)$. Show that $\lim_{\lambda \to \infty} \lambda^p \mu\{X : |f| \geq \lambda\} = 0$.

Proof. Since $\int_X |f|^p \, d\mu < \infty$, we have $|f|^p < \infty$ and then $|f| < \infty$ μ-a.e. on X. Then we have $\mu(E_\omega) = 0$ where $E_\omega = \{X : |f| = \infty\}$. For $n \in \mathbb{N}$, let $E_n = \{X : |f| \in [n-1, n)\}$. Then $\{E_n : n \in \mathbb{N}\}$ is a disjoint collection in $\mathfrak{A}$ and moreover $X = \left(\bigcup_{n \in \mathbb{N}} E_n\right) \cup E_\omega$. Then since $\mu(E_\omega) = 0$, we have

$$\int_X |f|^p \, d\mu = \sum_{n \in \mathbb{N}} \int_{E_n} |f|^p \, d\mu.$$

Since $\int_X |f|^p \, d\mu < \infty$, the series of nonnegative real numbers on the right of the last equality converges. The convergence implies

$$\lim_{N \to \infty} \sum_{n > N} \int_{E_n} |f|^p \, d\mu = 0, \text{ that is, } \lim_{N \to \infty} \int_{\bigcup_{n \geq N} E_n} |f|^p \, d\mu = 0.$$

Now $\bigcup_{n > N} E_n = E_{N+1} \cup E_{N+2} \cup \cdots = \{X : |f| \in [N, \infty)\}$. Thus we have

(1) $$\lim_{N \to \infty} \int_{\{X:|f|\in[N,\infty)\}} |f|^p \, d\mu = 0.$$

Now

(2) $$\int_{\{X:|f|\in[N,\infty)\}} |f|^p \, d\mu \geq N^p \mu\{X : |f| \in [N, \infty)\}.$$

Then (1) and (2) imply

(3) $$\lim_{N \to \infty} N^p \mu\{X : |f| \in [N, \infty)\} = 0.$$

For $\lambda \geq 1$, let $N \in \mathbb{N}$ be such that $N \leq \lambda < N + 1$. Then we have

(4) $$\lambda^p \mu\{X : |f| \in [\lambda, \infty)\} \leq (N+1)^p \mu\{X : |f| \in [N, \infty)\}$$
$$= \left(1 + \frac{1}{N}\right)^p N^p \mu\{X : |f| \in [N, \infty)\}.$$

If we let $\lambda \to \infty$ then $N \to \infty$ and thus by the fact that $\lim_{N \to \infty} \left(1 + 1/N\right)^p = 1$ and by (3) we have

(5) $$\lim_{\lambda \to \infty} \lambda^p \mu\{X : |f| \in [\lambda, \infty)\} = 0.$$

Since $\mu\{X : |f| = \infty\} = 0$, we have $\mu\{X : |f| \geq \lambda\} = \mu\{X : |f| \in [\lambda, \infty)\}$. Then (5) is rewritten as

$$\lim_{\lambda \to \infty} \lambda^p \mu\{X : |f| \geq \lambda\} = 0. \quad \blacksquare$$

Alternate Proof for Prob. 8.20. Since $\int_X |f|^p \, d\mu < \infty$, we have $|f|^p < \infty$ and then $|f| < \infty$ μ-a.e. on X. Then we have $\mu(D_\omega) = 0$ where $D_\omega = \{X : |f| = \infty\}$. For $n \in \mathbb{N}$, let $D_n = \{X : |f| > n\}$. Then $(D_n : n \in \mathbb{N})$ is a decreasing sequence in $\mathfrak{A}$ and $\bigcap_{n \in \mathbb{N}} D_n = D_\omega$.

Let us define a sequence $(g_n : n \in \mathbb{N})$ of functions on X by setting

(1) $\qquad g_n(x) = \begin{cases} 0 & \text{for } x \in D_n, \\ |f(x)| & \text{for } x \in D_n^c = \{x \in X : |f(x)| \le n\}. \end{cases}$

Then g_n is nonnegative real-valued on X and $\mathfrak{A}$-measurability of f on X implies that g_n is $\mathfrak{A}$-measurable on X. For every $x \in X$, $(g_n(x) : n \in \mathbb{N})$ is an increasing sequence of nonnegative real numbers. Indeed for $x \in D_\omega^c$, we have $(g_n(x) : n \in \mathbb{N}) = (0, \cdots, 0, |f(x)|, |f(x)|, \cdots)$ or $(g_n(x) : n \in \mathbb{N}) = (|f(x)|, |f(x)|, \cdots)$ and for $x \in D_\omega$, we have $(g_n(x) : n \in \mathbb{N}) = (0, 0, \cdots)$. Thus we have $g_n(x) \uparrow |f(x)|$ as $n \to \infty$ for $x \in D_\omega^c$, that is, we have $g_n \uparrow |f|$ as $n \to \infty$, μ-a.e. on X and then $g_n^p \uparrow |f|^p$ as $n \to \infty$, μ-a.e. on X. Then by the Monotone Convergence Theorem (Theorem 8.5), we have

(2) $\qquad \lim_{n \to \infty} \int_X g_n^p \, d\mu = \int_X |f|^p \, d\mu < \infty.$

This implies then

(3) $\qquad \lim_{n \to \infty} \int_X \{|f|^p - g_n^p\} \, d\mu = \int_X |f|^p \, d\mu - \lim_{n \to \infty} \int_X g_n^p \, d\mu$

$\qquad\qquad\qquad\qquad = \int_X |f|^p \, d\mu - \int_X |f|^p \, d\mu$

$\qquad\qquad\qquad\qquad = 0.$

From the definition of g_n by (1), we have

$$|f|^p - g_n^p = \begin{cases} |f|^p & \text{on } D_n, \\ 0 & \text{on } D_n^c. \end{cases}$$

Thus

$$\int_X \{|f|^p - g_n^p\} \, d\mu = \int_{D_n} |f|^p \, d\mu.$$

Substituting this in (3), we have

(4) $\qquad \lim_{n \to \infty} \int_{D_n} |f|^p \, d\mu = 0.$

Now

$$\int_{D_n} |f|^p \, d\mu \ge n^p \mu(D_n).$$

Then by (4) we have

$$\lim_{n \to \infty} n^p \mu(D_n) \le \lim_{n \to \infty} \int_{D_n} |f|^p \, d\mu = 0,$$

that is,

(5) $$\lim_{n\to\infty} n^p \mu\{X : |f| > n\} = 0.$$

Then

$$
\begin{aligned}
(6) \quad \lim_{n\to\infty} (n+1)^p \mu\{X : |f| > n\} &= \lim_{n\to\infty} \left(\frac{n+1}{n}\right)^p n^p \mu\{X : |f| > n\} \\
&= \lim_{n\to\infty} \left(1 + \frac{1}{n}\right)^p n^p \mu\{X : |f| > n\} \\
&= \lim_{n\to\infty} \left(1 + \frac{1}{n}\right)^p \cdot \lim_{n\to\infty} n^p \mu\{X : |f| > n\} \\
&= 1 \cdot 0 = 0 \quad \text{by (5).}
\end{aligned}
$$

If $\lambda \in (1, \infty)$ then there exists $n \in \mathbb{N}$ such that $n < \lambda \leq n + 1$. For such $n \in \mathbb{N}$, we have $\mu\{X : |f| \geq \lambda\} \leq \mu\{X : |f| > n\}$ and $\lambda^p \leq (n+1)^p$ and then

$$\lambda^p \mu\{X : |f| \geq \lambda\} \leq (n+1)^p \mu\{X : |f| > n\}.$$

Then

$$\lim_{\lambda\to\infty} \lambda^p \mu\{X : |f| \geq \lambda\} \leq \lim_{n\to\infty} (n+1)^p \mu\{X : |f| > n\} = 0,$$

by (6). This completes the proof. ∎

Prob. 8.21. Let $(X, \mathfrak{A}, \mu)$ be a σ-finite measure space. Let f be an extended real-valued $\mathfrak{A}$-measurable function on X. Show that for every $p \in (0, \infty)$ we have

$$\int_X |f|^p \, d\mu = \int_{[0,\infty)} p\lambda^{p-1} \mu\{X : |f| > \lambda\} \, \mu_L(d\lambda).$$

Proof. 1. Consider first the particular case that f is a μ-integrable simple function on $(X, \mathfrak{A}, \mu)$. Then we have

$$f = \sum_{i=1}^{k} c_i \mathbf{1}_{E_i},$$

where $\{E_i : i = 1, \ldots, k\}$ is a disjoint collection in $\mathfrak{A}$ and $0 < c_1 < \cdots, c_k < \infty$. Now μ-integrability of f on X implies that $\mu(E_i) < \infty$ for $i = 1, \ldots, k$. For brevity, let us write $a_i = \mu(E_i)$ for $i = 1, \ldots, k$.

For every $\lambda \in \mathbb{R}$, since f is $\mathfrak{A}$-measurable we have $\{x \in X : |f(x)| > \lambda\} \in \mathfrak{A}$ and thus $\mu\{x \in X : |f(x)| > \lambda\}$ is defined. Let us define a nonnegative extended real-valued function g on $[0, \infty)$ by

$$g(\lambda) := \mu\{x \in X : |f(x)| > \lambda\} \quad \text{for } \lambda \in [0, \infty).$$

Now the set $\{x \in X : |f(x)| > \lambda\} \downarrow$ as $\lambda \uparrow$. Thus g is a decreasing function on $[0, \infty)$ and this implies that g is $\mathfrak{B}_\mathbb{R}$-measurable on $[0, \infty)$ and thus $\int_{[0,\infty)} p\lambda^{p-1} g(\lambda) \, \mu_L(d\lambda)$ exists in $[0, \infty]$, that is, we have

(1) $$\int_{[0,\infty)} p\lambda^{p-1} \mu\{X : |f| > \lambda\} \, \mu_L(d\lambda) \in [0, \infty].$$

Note that the last integral on $[0, \infty)$ is equal to the integral of the same integrand on $(0, \infty)$ since $\mu_L(\{0\}) = 0$. Expressing $(0, \infty)$ as a disjoint union

$$(0, \infty) = (c_0, c_1] \cup (c_1, c_2] \cup \cdots \cup (c_{k-1}, c_k] \cup (c_k, \infty)$$

where $c_0 := 0$, we have

(2) $$\int_{(0,\infty)} p\lambda^{p-1} \mu\{X : |f| > \lambda\} \, \mu_L(d\lambda) = \sum_{i=1}^{k} \int_{(c_{i-1}, c_i]} p\lambda^{p-1} \mu\{X : |f| > \lambda\} \, \mu_L(d\lambda)$$

$$+ \int_{(c_k, \infty)} p\lambda^{p-1} \mu\{X : |f| > \lambda\} \, \mu_L(d\lambda).$$

Now for $\lambda \in (c_{i-1}, c_i]$ we have $\{X : |f| > \lambda\} = E_i \cup \cdots \cup E_k$ and then

$$\mu\{X : |f| > \lambda\} = \mu(E_i \cup \cdots \cup E_k) = \sum_{j=i}^{k} \mu(E_j) = \sum_{j=i}^{k} a_j.$$

Thus for $i = 1, \ldots, k$, we have

(3)
$$\int_{(c_{i-1}, c_i]} p\lambda^{p-1} \mu\{X : |f| > \lambda\} \, \mu_L(d\lambda) = \left\{ \sum_{j=i}^{k} a_j \right\} \int_{c_{i-1}}^{c_i} p\lambda^{p-1} \, d\lambda$$

$$= \left\{ \sum_{j=i}^{k} a_j \right\} \{c_i^p - c_{i-1}^p\}.$$

On the other hand we have

(4)
$$\int_{(c_k, \infty)} p\lambda^{p-1} \mu\{X : |f| > \lambda\} \, \mu_L(d\lambda) = \int_{(c_k, \infty)} p\lambda^{p-1} \mu(\emptyset) \, \mu_L(d\lambda) = 0.$$

Substituting (3) and (4) in the right side of (2), we have

$$\int_{(0, \infty)} p\lambda^{p-1} \mu\{X : |f| > \lambda\} \, \mu_L(d\lambda)$$

$$= \left\{ \sum_{j=1}^{k} a_j \right\} \{c_1^p - c_0^p\} + \left\{ \sum_{j=2}^{k} a_j \right\} \{c_2^p - c_1^p\} + \left\{ \sum_{j=3}^{k} a_j \right\} \{c_3^p - c_2^p\} + \cdots$$

$$\cdots + \left\{ \sum_{j=k-1}^{k} a_j \right\} \{c_{k-1}^p - c_{k-2}^p\} + \left\{ \sum_{j=k}^{k} a_j \right\} \{c_k^p - c_{k-1}^p\}$$

$$= a_1 c_1^p + a_2 c_2^p + \cdots + a_k c_k^p = c_1^p \mu(E_1) + c_2^p \mu(E_2) + \cdots + a_k c_k^p \mu(E_k)$$

$$= \int_X |f|^p \, d\mu.$$

2. Now consider the general case that f is an extended real-valued $\mathfrak{A}$-measurable function on X. Then $|f|$ is a nonnegative extended real-valued $\mathfrak{A}$-measurable function on X. Since $(X, \mathfrak{A}, \mu)$ is a σ-finite measure space, according to Lemma 8.6 there exists an increasing sequence $(\varphi_n : n \in \mathbb{N})$ of μ-integrable nonnegative simple functions on $(X, \mathfrak{A}, \mu)$ such that $\varphi_n \uparrow |f|$ on X. By the Monotone Convergence Theorem (Theorem 8.5), we have

(5)
$$\lim_{n \to \infty} \int_X \varphi_n^p \, d\mu = \int_X |f|^p \, d\mu.$$

For every $n \in \mathbb{N}$, φ_n is a μ-integrable simple function on $(X, \mathfrak{A}, \mu)$ and thus by our result in **1**, we have

(6)
$$\int_X \varphi_n^p \, d\mu = \int_{[0, \infty)} p\lambda^{p-1} \mu\{X : \varphi_n > \lambda\} \, \mu_L(d\lambda).$$

For each fixed $\lambda \in [0, \infty)$, since $(\varphi_n : n \in \mathbb{N})$ is an increasing sequence of functions on X the set $\{X : \varphi_n > \lambda\} \uparrow$ as $n \uparrow$ and then $\mu\{X : \varphi_n > \lambda\} \uparrow$ as $n \uparrow$. Moreover since $\varphi_n \uparrow |f|$ on X, we have according to Prob. 4.25,

$$\lim_{n \to \infty} \mu\{X : \varphi_n > \lambda\} = \mu\{X : |f| > \lambda\}.$$

Thus applying the Monotone Convergence Theorem (Theorem 8.5) to (6), we have

$$(7) \qquad \lim_{n \to \infty} \int_X \varphi_n^p \, d\mu = \lim_{n \to \infty} \int_{[0,\infty)} p\lambda^{p-1} \mu\{X : \varphi_n > \lambda\} \, \mu_L(d\lambda)$$

$$= \int_{[0,\infty)} p\lambda^{p-1} \mu\{X : |f| > \lambda\} \, \mu_L(d\lambda).$$

Combining (5) and (7), we have

$$\int_X |f|^p \, d\mu = \int_{[0,\infty)} p\lambda^{p-1} \mu\{X : |f| > \lambda\} \, \mu_L(d\lambda). \quad \blacksquare$$

Prob. 8.22. Let $(X, \mathfrak{A}, \mu)$ be an arbitrary measure space and let f be a nonnegative extended real-valued $\mathfrak{A}$-measurable function on X. Consider $\left\{ t \in [0, \infty] : \mu\left(f^{-1}(\{t\})\right) > 0 \right\}$, a subset of the range of the function f. Show that if f is μ-integrable on X then the set is a countable set.

Proof. Assume that f is μ-integrable on X. Let us show the countability of the set $\left\{ t \in [0, \infty] : \mu\left(f^{-1}(\{t\})\right) > 0 \right\}$. Since 0 is just one point in $[0, \infty]$, to show the countability of the set it suffices to show that the set $A := \left\{ t \in (0, \infty] : \mu\left(f^{-1}(\{t\})\right) > 0 \right\}$ is countable.

Let us define a collection of subsets of $(0, \infty]$ by setting

(1) $$I_1 = [1, \infty] \text{ and } I_k = \left[\tfrac{1}{k}, \tfrac{1}{k-1}\right) \text{ for } k \geq 2.$$

Then $\{I_k : k \in \mathbb{N}\}$ is a disjoint collection and $\bigcup_{k \in \mathbb{N}} I_k = (0, \infty]$. Let us define

$$A_k = A \cap I_k \quad \text{for } k \in \mathbb{N}.$$

Then since $A \subset (0, \infty]$ and $\bigcup_{k \in \mathbb{N}} I_k = (0, \infty]$, $\left\{ A_k : k \in \mathbb{N} \right\}$ is a disjoint collection and $\bigcup_{k \in \mathbb{N}} A_k = A$. To show that A is a countable set, assume the contrary. Then there exists $k_0 \in \mathbb{N}$ such that A_{k_0} is an uncountable set. Let us observe that

(2) $$t \in A_{k_0} = A \cap I_{k_0} \Rightarrow t \geq \tfrac{1}{k_0} \text{ and } \mu\left(f^{-1}(\{t\})\right) > 0.$$

Let us define a collection of subsets of $(0, \infty]$ by setting

(3) $$J_1 = [1, \infty] \text{ and } J_\ell = \left[\tfrac{1}{\ell}, \tfrac{1}{\ell-1}\right) \text{ for } \ell \geq 2.$$

Then $\left\{ J_\ell : \ell \in \mathbb{N} \right\}$ is a disjoint collection and $\bigcup_{\ell \in \mathbb{N}} J_\ell = (0, \infty]$. Let us define

(4) $$A_{k_0, \ell} = \left\{ t \in A_{k_0} : \mu\left(f^{-1}(\{t\})\right) \in J_\ell \right\} \quad \text{for } \ell \in \mathbb{N}.$$

Then $\left\{ A_{k_0, \ell} : \ell \in \mathbb{N} \right\}$ is a disjoint collection and $\bigcup_{\ell \in \mathbb{N}} A_{k_0, \ell} = A_{k_0}$. Since A_{k_0} is an uncountable set, there exists $\ell_0 \in \mathbb{N}$ such that A_{k_0, ℓ_0} is an uncountable set. Let $\left\{ t_n : n \in \mathbb{N} \right\}$ be a selection of distinct points in A_{k_0, ℓ_0}. Now $t_n \in A_{k_0, \ell} \subset A_{k_0}$ and then by (2) we have

(5) $$t_n \geq \tfrac{1}{k_0}.$$

Also $t_n \in A_{k_0, \ell}$ implies that $\mu\left(f^{-1}(\{t_n\})\right) \in J_{\ell_0}$ so that by (3) we have

(6) $$\mu\left(f^{-1}(\{t_n\})\right) \geq \tfrac{1}{\ell_0}.$$

Now $\left\{ f^{-1}(\{t_n\}) : n \in \mathbb{N} \right\}$ is a disjoint collection in $\mathfrak{A}$. Thus we have

$$\int_X f \, d\mu \geq \int_{\bigcup_{n \in \mathbb{N}} f^{-1}(\{t_n\})} f \, d\mu = \sum_{n \in \mathbb{N}} \int_{f^{-1}(\{t_n\})} f \, d\mu$$

$$= \sum_{n \in \mathbb{N}} t_n \, \mu\left(f^{-1}(\{t_n\})\right) \geq \sum_{n \in \mathbb{N}} \frac{1}{k_0} \frac{1}{\ell_0} = \infty.$$

This contradicts the μ-integrability of f on X. Therefore A must be a countable set. ∎

Prob. 8.23. Let $(X, \mathfrak{A}, \mu)$ be an arbitrary measure space and let f be a μ-integrable non-negative extended real-valued $\mathfrak{A}$-measurable function on X. Let us define two nonnegative extended real-valued functions g and h on $[0, \infty)$ by

$$g(t) = \mu\{x \in X : f(x) > t\} \quad \text{for } t \in [0, \infty),$$

$$h(t) = \mu\{x \in X : f(x) \geq t\} \quad \text{for } t \in [0, \infty).$$

Show that
(a) $g \leq h$ on $[0, \infty)$.
(b) g and h are decreasing on $[0, \infty)$.
(c) $g(t), h(t) < \infty$ for $t \in (0, \infty)$.
(d) $\lim_{t \to \infty} g(t) = \lim_{t \to \infty} h(t) = 0$.
(e) $g(t) = h(t)$ for all but countably many $t \in [0, \infty)$.

Proof. 1. (a) and (b) are immediate from the definitions of g and h.

2. To prove (c), since $g \leq h$ it suffices to show that $h(t) < \infty$ for $t \in (0, \infty)$. Assume the contrary. Then there exists $t_0 \in (0, \infty)$ such that $h(t_0) = \infty$, that is,

$$\mu\{x \in X : f(x) \geq t_0\} = h(t_0) = \infty.$$

Then since f is nonnegative on X we have

$$\int_X f \, d\mu \geq \int_{\{x \in X : f(x) \geq t_0\}} f \, d\mu \geq t_0 \, \mu\{x \in X : f(x) \geq t_0\} = \infty.$$

This contradicts the μ-integrability of f on X. Thus we have $h(t) < \infty$ for $t \in (0, \infty)$.

3. To prove (d), let us show that $\lim_{t \to \infty} h(t) = 0$. Now since h is a nonnegative decreasing function on $[0, \infty)$, $\lim_{t \to \infty} h(t)$ exists and is nonnegative. Suppose $\lim_{t \to \infty} h(t) \neq 0$. Then $\lim_{t \to \infty} h(t) = \varepsilon > 0$. Then we have $h(k) \geq \varepsilon$ for every $k \in \mathbb{N}$. Then by the nonnegativity of f on X we have for every $k \in \mathbb{N}$

$$\int_X f \, d\mu \geq \int_{\{x \in X : f(x) \geq k\}} f \, d\mu \geq k \, \mu\{x \in X : f(x) \geq k\} = k \, h(k) \geq k \, \varepsilon.$$

Since this holds for every $k \in \mathbb{N}$ we have $\int_X f \, d\mu = \infty$. This contradicts the μ-integrability of f on X. Therefore we must have $\lim_{t \to \infty} h(t) = 0$. Then since $0 \leq g(t) \leq h(t)$ for $t \in [0, \infty)$, we have $\lim_{t \to \infty} g(t) = 0$.

4. Let us show that $g(t) = h(t)$ for all except possibly countably many $t \in [0, \infty)$. Suppose $g(t_0) \neq h(t_0)$ for some $t_0 \in [0, \infty)$. Then since $g(t) \leq h(t)$ for every $t \in [0, \infty)$, we have $g(t_0) < h(t_0)$. Now we have from the definitions of g and h

$$g(t_0) = \mu\{x \in X : f(x) > t_0\} = \mu\big(f^{-1}\big((t_0, \infty]\big)\big)$$

and

$$h(t_0) = \mu\{x \in X : f(x) \ge t_0\} = \mu\big(f^{-1}([t_0, \infty])\big)$$
$$= \mu\big(f^{-1}(\{t_0\}) \cup f^{-1}((t_0, \infty])\big)$$
$$= \mu\big(f^{-1}(\{t_0\})\big) + \mu\big(f^{-1}((t_0, \infty])\big)$$
$$= \mu\big(f^{-1}(\{t_0\})\big) + g(t_0).$$

Thus $g(t_0) < h(t_0)$ implies $\mu\big(f^{-1}(\{t_0\})\big) > 0$. Now according to Prob. 8.22, the set $\big\{t \in [0, \infty] : \mu\big(f^{-1}(\{t\})\big) > 0\big\}$ is a countable set. Thus $g(t) = h(t)$ for all except possibly countably many $t \in [0, \infty)$. ∎

Prob. 8.24. Let $(X, \mathfrak{A}, \mu)$ be a measure space and let f be a μ-integrable nonnegative extended real-valued $\mathfrak{A}$-measurable function on X.

(a) Define a function g on $[0, \infty)$ by $g(t) = \mu\{x \in X : f(x) > t\}$ for $t \in [0, \infty)$. Prove

$$(1) \qquad \int_X f \, d\mu = \int_{[0,\infty)} g(t) \, \mu_L(dt) = \int_{[0,\infty)} \mu\{x \in X : f(x) > t\} \, \mu_L(dt).$$

(b) Define a function h on $[0, \infty)$ by $h(t) = \mu\{x \in X : f(x) \geq t\}$ for $t \in [0, \infty)$. Prove

$$(2) \qquad \int_X f \, d\mu = \int_{[0,\infty)} h(t) \, \mu_L(dt) = \int_{[0,\infty)} \mu\{x \in X : f(x) \geq t\} \, \mu_L(dt).$$

Proof. 1. g is a nonnegative extended real-valued decreasing function on $[0, \infty)$ and is thus $\mathfrak{B}_{\mathbb{R}}$-measurable. Since g is nonnegative, the integral $\int_{[0,\infty)} g(t) \, \mu_L(dt)$ exists.

To prove (1), let us consider first the case that f is a μ-integrable nonnegative simple function on $(X, \mathfrak{A}, \mu)$ given by

$$f = \sum_{i=1}^{k} a_i \mathbf{1}_{E_i},$$

where $\{E_i : i = 1, \ldots, k\}$ is a disjoint collection in $\mathfrak{A}$ and $0 < a_1 < \cdots < a_k$. Note that the μ-integrability of f on X implies that $\mu(E_i) < \infty$ for $i = 1, \ldots, k$. Then we have

$$g(t) = \mu\{x \in X : f(x) > t\} = \begin{cases} \sum_{i=1}^{k} \mu(E_i) & \text{for } t \in [0, a_1), \\ \sum_{i=2}^{k} \mu(E_i) & \text{for } t \in [a_1, a_2), \\ \sum_{i=3}^{k} \mu(E_i) & \text{for } t \in [a_2, a_3), \\ \vdots \\ \sum_{i=k-1}^{k} \mu(E_i) & \text{for } t \in [a_{k-2}, a_{k-1}), \\ \mu(E_k) & \text{for } t \in [a_{k-1}, a_k), \\ 0 & \text{for } t \in [a_k, \infty), \end{cases}$$

and

$$\int_{[0,\infty)} g(t) \, \mu_L(dt)$$
$$= \left\{ \sum_{i=1}^{k} \mu(E_i) \right\} a_1 + \left\{ \sum_{i=2}^{k} \mu(E_i) \right\} (a_2 - a_1) + \left\{ \sum_{i=3}^{k} \mu(E_i) \right\} (a_3 - a_2)$$
$$+ \cdots + \left\{ \sum_{i=k-1}^{k} \mu(E_i) \right\} (a_{k-1} - a_{k-2}) + \mu(E_k)(a_k - a_{k-1})$$
$$= \mu(E_1)a_1 + \mu(E_2)a_2 + \cdots + \mu(E_k)a_k = \int_X f \, d\mu.$$

This shows that (1) holds if f is a μ-integrable simple function on $(X, \mathfrak{A}, \mu)$.

Let f be a μ-integrable nonnegative extended real-valued function on X. According to Lemma 8.6 there exists an increasing sequence $(f_n : n \in \mathbb{N})$ of nonnegative simple functions on $(X, \mathfrak{A}, \mu)$ such that $f_n \uparrow f$ on X. μ-integrability of f on X implies that of f_n for every $n \in \mathbb{N}$. By our result above we have for every $n \in \mathbb{N}$

$$(3) \qquad \int_X f_n \, d\mu = \int_{[0,\infty)} g(t) \, \mu_L(dt) = \int_{[0,\infty)} \mu\{x \in X : f_n(x) > t\} \, \mu_L(dt).$$

By Theorem 8.5 (Monotone Convergence Theorem) we have

$$(4) \qquad \lim_{n \to \infty} \int_X f_n \, d\mu = \int_X f \, d\mu.$$

With $t \in [0, \infty)$ fixed, consider the sets:

$$E_{t,n} = \{x \in X : f_n(x) > t\} \quad \text{for } n \in \mathbb{N},$$

$$E_t = \{x \in X : f(x) > t\}.$$

Since $f_n \leq f$ on X we have $E_{t,n} \subset E_t$. Since $f_n \uparrow$ on X, $(E_{t,n} : n \in \mathbb{N})$ is an increasing sequence of sets and thus $\lim_{n \to \infty} E_{t,n} = \bigcup_{n \in \mathbb{N}} E_{t,n} \subset E_t$. Let $x \in E_t$ be arbitrarily chosen. Since $f_n(x) \uparrow f(x)$ there exists $n_0 \in \mathbb{N}$ such that $f_n(x) > t$. Then $x \in E_{t,n_0}$. Thus if $x \in E_t$ then $x \in E_{t,n}$ for some $n \in \mathbb{N}$. Therefore $E_t \subset \bigcup_{n \in \mathbb{N}} E_{t,n}$. This shows that $E_t = \bigcup_{n \in \mathbb{N}} E_{t,n} = \lim_{n \to \infty} E_{t,n}$. Therefore we have $\mu(E_{t,n}) \uparrow \mu(E_t)$, that is,

$$\mu\{x \in X : f_n(x) > t\} \uparrow \mu\{x \in X : f(x) > t\}.$$

Since this holds at every $t \in [0, \infty)$, we have by Theorem 8.5 (Monotone Convergence Theorem)

$$(5) \quad \lim_{n \to \infty} \int_{[0,\infty)} \mu\{x \in X : f_n(x) > t\} \, \mu_L(dt) = \int_{[0,\infty)} \mu\{x \in X : f(x) > t\} \, \mu_L(dt).$$

Letting $n \to \infty$ in (3), we have by (4) and (5)

$$\int_X f \, d\mu = \int_{[0,\infty)} \mu\{x \in X : f(x) > t\} \, \mu_L(dt).$$

This proves (1).

2. To prove (2), note that according to (e) of Prob. 8.23, $g = h$ except possibly at countably many $t \in [0, \infty)$. Then we have

$$\int_{[0,\infty)} h(t) \, \mu_L(dt) = \int_{[0,\infty)} g(t) \, \mu_L(dt) = \int_X f \, d\mu.$$

This proves (2). ∎

§9   Integration of Measurable Functions

Prob. 9.1. Let $(X, \mathfrak{A}, \mu)$ be a measure space and let $E_1, E_2 \in \mathfrak{A}$ such that $E_1 \cap E_2 = \emptyset$. Let f be an extended real-valued $\mathfrak{A}$-measurable function on $E_1 \cup E_2$. Show that if f is μ-integrable on E_1 and on E_2 then f is μ-integrable on $E_1 \cup E_2$ and moreover

$$\int_{E_1 \cup E_2} f \, d\mu = \int_{E_1} f \, d\mu + \int_{E_2} f \, d\mu.$$

Proof. f is μ-integrable if and only if both f^+ and f^- are μ-integrable. To show that f is μ-integrable on $E_1 \cup E_2$, we show that both f^+ and f^- are μ-integrable on $E_1 \cup E_2$. Since f^+ is nonnegative, $\int_{E_1 \cup E_2} f^+ \, d\mu$ exists and moreover by Proposition 8.11 we have

$$(1) \qquad \int_{E_1 \cup E_2} f^+ \, d\mu = \int_{E_1} f^+ \, d\mu + \int_{E_2} f^+ \, d\mu < \infty.$$

Similarly we have

$$(2) \qquad \int_{E_1 \cup E_2} f^- \, d\mu = \int_{E_1} f^- \, d\mu + \int_{E_2} f^- \, d\mu < \infty.$$

Thus both f^+ and f^- are μ-integrable on $E_1 \cup E_2$ and therefore f is μ-integrable on $E_1 \cup E_2$. Furthermore by (1) and (2) we have

$$
\begin{aligned}
(3) \quad \int_{E_1 \cup E_2} f \, d\mu &= \int_{E_1 \cup E_2} f^+ \, d\mu - \int_{E_1 \cup E_2} f^- \, d\mu \\
&= \left\{ \int_{E_1} f^+ \, d\mu + \int_{E_2} f^+ \, d\mu \right\} - \left\{ \int_{E_1} f^- \, d\mu + \int_{E_2} f^- \, d\mu \right\} \\
&= \left\{ \int_{E_1} f^+ \, d\mu - \int_{E_1} f^- \, d\mu \right\} + \left\{ \int_{E_2} f^+ \, d\mu - \int_{E_2} f^- \, d\mu \right\} \\
&= \int_{E_1} f \, d\mu + \int_{E_2} f \, d\mu. \quad \blacksquare
\end{aligned}
$$

Prob. 9.2. Let $(X, \mathfrak{A}, \mu)$ be a measure space.

(a) Let $\{E_n : n \in \mathbb{N}\}$ be a disjoint collection in $\mathfrak{A}$. Let f be an extended real-valued $\mathfrak{A}$-measurable function defined on $\bigcup_{n\in\mathbb{N}} E_n$. If f is μ-integrable on E_n for every $n \in \mathbb{N}$, does $\int_{\bigcup_{n\in\mathbb{N}} E_n} f \, d\mu$ exist?

(b) Let $(F_n : n \in \mathbb{N})$ be an increasing sequence in $\mathfrak{A}$. Let f be an extended real-valued $\mathfrak{A}$-measurable function defined on $\bigcup_{n\in\mathbb{N}} F_n$. Suppose f is μ-integrable on F_n for every $n \in \mathbb{N}$ and moreover $\lim_{n\to\infty} \int_{F_n} f \, d\mu$ exists in $\mathbb{R}$. Does $\int_{\bigcup_{n\in\mathbb{N}} F_n} f \, d\mu$ exist?

Proof. The integrals $\int_{\bigcup_{n\in\mathbb{N}} E_n} f \, d\mu$ and $\int_{\bigcup_{n\in\mathbb{N}} F_n} f \, d\mu$ may not exist.

Example 1. Consider the measure space $(\mathbb{R}, \mathfrak{M}_L, \mu_L)$. Let $E_n = [(n-1)2\pi, n \cdot 2\pi)$ for $n \in \mathbb{N}$. Then $\{E_n : n \in \mathbb{N}\}$ is a disjoint collection in $\mathfrak{M}_L$ and $\bigcup_{n\in\mathbb{N}} E_n = [0, \infty)$. Let $f(x) = \sin x$ for $x \in [0, \infty)$. For each $n \in \mathbb{N}$, we have

$$\int_{E_n} f \, d\mu_L = \int_{[(n-1)2\pi, n \cdot 2\pi)} \sin x \, \mu_L(dx) = 0.$$

Thus f is μ_L-integrable on each E_n. But we have $\int_{[0,\infty)} f^+ \, d\mu_L = \infty$ and $\int_{[0,\infty)} f^- \, d\mu_L = \infty$ so that $\int_{[0,\infty)} f \, d\mu_L$ does not exist.

Let $F_n = [0, n \cdot 2\pi)$ for $n \in \mathbb{N}$. Then $(F_n : n \in \mathbb{N})$ is an increasing sequence in $\mathfrak{M}_L$ and $\bigcup_{n\in\mathbb{N}} F_n = [0, \infty)$. We have

$$\int_{F_n} f \, d\mu_L = \int_{[0,n\cdot 2\pi)} \sin x \, \mu_L(dx) = 0$$

so that $\lim_{n\to\infty} \int_{F_n} f \, d\mu = 0 \in \mathbb{R}$. But $\int_{\bigcup_{n\in\mathbb{N}} F_n} f \, d\mu_L = \int_{[0,\infty)} \sin x \, \mu_L(dx)$ does not exist.

Example 2. In the measure space $(\mathbb{R}, \mathfrak{M}_L, \mu_L)$, let $E_n = [n-1, n)$ for $n \in \mathbb{N}$. Then we have a disjoint collection $\{E_n : n \in \mathbb{N}\}$ in $\mathfrak{M}_L$ and $\bigcup_{n\in\mathbb{N}} E_n = [0, \infty)$. Let f be a real-valued function on $[0, \infty)$ defined by

$$f(x) = (-1)^n \quad \text{for } x \in E_n \text{ and } n \in \mathbb{N}.$$

For each $n \in \mathbb{N}$, we have

$$\int_{E_n} f \, d\mu_L = (-1)^n \mu_L(E_n) = (-1)^n.$$

Thus f in μ_L-integrable on E_n for every $n \in \mathbb{N}$. On the other hand, we have

$$\int_{\bigcup_{n\in\mathbb{N}} E_n} f^+ \, d\mu_L = \int_{E_2} 1 \, d\mu_L + \int_{E_4} 1 \, d\mu_L + \int_{E_6} 1 \, d\mu_L + \cdots = \infty$$

and

$$\int_{\bigcup_{n\in\mathbb{N}} E_n} f^- \, d\mu_L = \int_{E_1} 1 \, d\mu_L + \int_{E_3} 1 \, d\mu_L + \int_{E_5} 1 \, d\mu_L + \cdots = \infty.$$

Thus $\int_{\bigcup_{n\in\mathbb{N}} E_n} f \, d\mu_L = \int_{[0,\infty)} f \, d\mu_L$ does not exist.

Let $F_n = [0, 2n)$ for $n \in \mathbb{N}$. Then $(F_n : n \in \mathbb{N})$ is an increasing sequence in $\mathfrak{M}_L$ and $\bigcup_{n \in \mathbb{N}} F_n = [0, \infty)$. We have $\int_{F_n} f \, d\mu_L = 0$ for every $n \in \mathbb{N}$ and thus $\lim_{n \to \infty} \int_{F_n} f \, d\mu_L = 0 \in \mathbb{R}$. However $\int_{\bigcup_{n \in \mathbb{N}} F_n} f \, d\mu_L = \int_{[0, \infty)} f \, d\mu_L$ does not exist as we showed above.

Example 3. In the measure space $(\mathbb{R}, \mathfrak{M}_L, \mu_L)$, let f be a real-valued function on $\mathbb{R}$ defined by

$$f(x) = \begin{cases} -1 & \text{for } x \in (-\infty, 0) \\ 1 & \text{for } x \in [0, \infty). \end{cases}$$

Let $E_1 = (-1, 1)$ and $E_n = (-n, -n + 1] \cup [n - 1, n)$ for $n \geq 2$. Then $\{E_n : n \in \mathbb{N}\}$ is a disjoint sequence in $\mathfrak{M}_L$ and $\bigcup_{n \in \mathbb{N}} E_n = \mathbb{R}$. We have $\int_{E_n} f \, d\mu_L = 0$ and thus f is μ_L-integrable on E_n for every $n \in \mathbb{N}$. On the other hand we have

$$\int_{\mathbb{R}} f^+ \, d\mu_L = \int_{[0, \infty)} 1 \, \mu_L(dx) = \infty$$

and

$$\int_{\mathbb{R}} f^- \, d\mu_L = \int_{(-\infty, 0)} 1 \, \mu_L(dx) = \infty$$

so that $\int_{\bigcup_{n \in \mathbb{N}} E_n} f \, d\mu_L = \int_{\mathbb{R}} f \, d\mu_L$ does not exist.

Let $F_n = (-n, n)$ for $n \in \mathbb{N}$. Then $(F_n : n \in \mathbb{N})$ is an increasing sequence in $\mathfrak{M}_L$ and $\bigcup_{n \in \mathbb{N}} F_n = \mathbb{R}$. We have $\int_{F_n} f \, d\mu_L = 0$ for every $n \in \mathbb{N}$ and then $\lim_{n \to \infty} \int_{F_n} f \, d\mu_L = 0 \in \mathbb{R}$. However $\int_{\bigcup_{n \in \mathbb{N}} F_n} f \, d\mu_L = \int_{\mathbb{R}} f \, d\mu_L$ does not exist as we showed above. ∎

Prob. 9.3. Let $(X, \mathfrak{A}, \mu)$ be a measure space and let f be a μ-integrable extended real-valued $\mathfrak{A}$-measurable function on X.
(a) Let $E = \{x \in X : f(x) \neq 0\}$. Show that E is a σ-finite set.
(b) Show that for every $\varepsilon > 0$ there exists $A \in \mathfrak{A}$ such that $\mu(A) < \infty$ and $\int_{X \setminus A} |f| \, d\mu < \varepsilon$.

Proof. 1. Note that

$$E = \{x \in X : f(x) \neq 0\} = \{x \in X : |f(x)| > 0\}.$$

For $n \in \mathbb{N}$, let

$$E_n = \{x \in X : |f(x)| \geq \tfrac{1}{n}\}.$$

Then $E = \bigcup_{n \in \mathbb{N}} E_n$. Now we have

$$\frac{1}{n} \leq \int_{E_n} |f| \, d\mu \leq \int_X |f| \, d\mu < \infty.$$

Thus $\mu(E_n) < \infty$ for every $n \in \mathbb{N}$. this shows that E is a σ-finite set.

2. Since E is a σ-finite set, there exists a disjoint sequence $(F_n : n \in \mathbb{N})$ in $\mathfrak{A}$ such that $\bigcup_{n \in \mathbb{N}} F_n = E$ and $\mu(F_n) < \infty$ for $n \in \mathbb{N}$. Then we have

$$\sum_{n \in \mathbb{N}} \int_{F_n} |f| \, d\mu = \int_{\bigcup_{n \in \mathbb{N}} F_n} |f| \, d\mu = \int_E |f| \, d\mu = \int_X |f| \, d\mu < \infty.$$

The convergence of the series $\sum_{n \in \mathbb{N}} \int_{F_n} |f| \, d\mu$ to the sum $\int_X |f| \, d\mu < \infty$ implies that for every $\varepsilon > 0$ there exists $N \in \mathbb{N}$ such that

$$\int_X |f| \, d\mu - \sum_{n=1}^{N} \int_{F_n} |f| \, d\mu < \varepsilon.$$

Let $A = \bigcup_{n=1}^{N} F_n$. Then $\mu(A) = \sum_{n=1}^{N} \mu(F_n) < \infty$ and moreover

$$\int_{X \setminus A} |f| \, d\mu = \int_X |f| \, d\mu - \int_A |f| \, d\mu = \int_X |f| \, d\mu - \sum_{n=1}^{N} \int_{F_n} |f| \, d\mu < \varepsilon. \quad \blacksquare$$

Prob. 9.4. Let $(X, \mathfrak{A}, \mu)$ be a measure space and let f and g be extended real-valued $\mathfrak{A}$-measurable and μ-integrable functions on X. Show that if $\int_E f \, d\mu = \int_E g \, d\mu$ for every $E \in \mathfrak{A}$ then $f = g$ a.e. on X.

Proof. Let $N_1 = \{x \in X : |f(x)| = \infty\} \in \mathfrak{A}$ and $N_2 = \{x \in X : |g(x)| = \infty\} \in \mathfrak{A}$. Since f and g are μ-integrable on X, we have $\mu(N_1) = 0$ and $\mu(N_2) = 0$. Let $N = N_1 \cup N_2$. Then we have $\mu(N) = 0$. Let $X_0 = X \setminus N \in \mathfrak{A}$. On X_0 both f and g are finite so that $g - f$ and $f - g$ are defined and finite.

Let us show that $f = g$ a.e. on X_0. Assume the contrary. Then we have

$$\mu\{x \in X_0 : f(x) \neq g(x)\} > 0$$

and this implies that at least one of the two inequalities

(1) $$\mu\{x \in X_0 : f(x) < g(x)\} > 0$$

(2) $$\mu\{x \in X_0 : f(x) > g(x)\} > 0$$

must hold. Suppose (1) holds. Let

$$A = \{x \in X_0 : f(x) < g(x)\} = \{x \in X_0 : g(x) - f(x) \in (0, \infty)\}.$$

By (1) we have $\mu(A) > 0$. Now

$$A = \bigcup_{k \in \mathbb{N}} A_k \quad \text{where} \quad A_k = \{x \in X_0 : g(x) - f(x) \in [\tfrac{1}{k}, \infty)\}.$$

Since $A_k \uparrow A$, we have $\lim_{k \to \infty} \mu(A_k) = \mu(A) > 0$. This implies that there exists $k_0 \in \mathbb{N}$ such that $\mu(A_{k_0}) > 0$. Then we have

(3) $$\int_{A_{k_0}} (g - f) \, d\mu \geq \frac{1}{k_0} \mu(A_{k_0}) > 0.$$

Now μ-integrability of f and g on X implies μ-integrability of f and g on A_{k_0}. Thus we have

(4) $$\int_{A_{k_0}} (g - f) \, d\mu = \int_{A_{k_0}} g \, d\mu - \int_{A_{k_0}} f \, d\mu.$$

By (3) and (4) we have

$$\int_{A_{k_0}} g \, d\mu - \int_{A_{k_0}} f \, d\mu > 0.$$

This contradicts the assumption that $\int_E f \, d\mu = \int_E g \, d\mu$ for every $E \in \mathfrak{A}$. Similarly (2) leads to a contradiction. Therefore we must have $f = g$ a.e. on X_0. Then since $X = X_0 \cup N$ where $X_0 \cap N = \emptyset$ and $\mu(N) = 0$, we have $f = g$ a.e. on X. ∎

Prob. 9.5. Let $(X, \mathfrak{A}, \mu)$ be a σ-finite measure space and let f and g be extended real-valued $\mathfrak{A}$-measurable functions on X. Show that if $\int_E f \, d\mu = \int_E g \, d\mu$ for every $E \in \mathfrak{A}$ then $f = g$ a.e. on X.
(Note that with the σ-finiteness of $(X, \mathfrak{A}, \mu)$, the μ-integrability of f and g is no longer assumed.)

Proof. 1. Let us assume that

(1) $$\int_E f \, d\mu = \int_E g \, d\mu \quad \text{for every } E \in \mathfrak{A}.$$

Regarding the positivity and negativity of the functions f and g, let us decompose X into $\mathfrak{A}$-measurable subsets as follows:

$$X^+ = \{x \in X : f(x) \geq 0 \text{ and } g(x) \geq 0\}$$

$$X^- = \{x \in X : f(x) < 0 \text{ and } g(x) < 0\}$$

$$A = \{x \in X : f(x) \geq 0 \text{ and } g(x) < 0\}$$

$$B = \{x \in X : f(x) < 0 \text{ and } g(x) \geq 0\}.$$

Observe that $\{A, B, X^+, X^-\}$ is a disjoint collection in $\mathfrak{A}$ and $X = X^+ \cup X^- \cup A \cup B$.

Let us show that the assumption (1) implies that A and B are null sets in the measure space $(X, \mathfrak{A}, \mu)$, that is, $\mu(A) = 0$ and $\mu(B) = 0$. Suppose $\mu(A) > 0$. Then since

$$A = \bigcup_{n \in \mathbb{N}} A_n \quad \text{where} \quad A_n = \{x \in X : f(x) \geq 0 \text{ and } g(x) < -\tfrac{1}{n}\}$$

and then

$$\sum_{n \in \mathbb{N}} \mu(A_n) \geq \mu\Big(\bigcup_{n \in \mathbb{N}} A_n\Big) = \mu(A) > 0,$$

there exists $n_0 \in \mathbb{N}$ such that $\mu(A_{n_0}) > 0$. Then we have

$$\int_{A_{n_0}} f \, d\mu \geq 0 \text{ and } \int_{A_{n_0}} g \, d\mu \leq -\frac{1}{n_0} \mu(A_{n_0}) < 0$$

and thus we have

$$\int_{A_{n_0}} f \, d\mu \neq \int_{A_{n_0}} g \, d\mu.$$

This contradicts the assumption (1). This proves that $\mu(A) = 0$. That $\mu(B) = 0$ is proved by similar argument.

2. By definition we say that $f = g$ a.e. on a set $E \in \mathfrak{A}$ if there exists a null set N in the measure space $(X, \mathfrak{A}, \mu)$ contained in E such that $f = g$ on $E \setminus N$.

Let us show that $f = g$ a.e. on $X^+ \cup X^-$. It suffices to show that $f = g$ a.e. on X^+ and $f = g$ a.e. on X^-. Then let us show that $f = g$ a.e. on X^+.

Since $(X, \mathfrak{A}, \mu)$ is a σ-finite measure space, there exists a disjoint collection $\{X_n^+ : n \in \mathbb{N}\}$ in $\mathfrak{A}$ such that $X^+ = \bigcup_{n \in \mathbb{N}} X_n^+$ and $\mu(X_n^+) < \infty$ for $n \in \mathbb{N}$. Suppose the statement that $f = g$ a.e. on X^+ is false. Then there exists $n_0 \in \mathbb{N}$ such that the statement that $f = g$ a.e. on $X_{n_0}^+$ is false. Then at least one of the following two sets:

$$C = \left\{ x \in X_{n_0}^+ : f(x) < g(x) \right\}$$

$$D = \left\{ x \in X_{n_0}^+ : f(x) > g(x) \right\}$$

has a positive mu-measure. Suppose $\mu(C) > 0$. Let us define

$$C_{\ell,m} = \left\{ x \in X_{n_0}^+ : \frac{1}{m} < g(x) - f(x) \text{ and } f(x) < \ell \right\} \quad \text{for } \ell, m \in \mathbb{N}.$$

Then we have

$$C = \bigcup_{\ell,m \in \mathbb{N}} C_{\ell,m}$$

and this implies

$$\sum_{\ell,m \in \mathbb{N}} \mu(C_{\ell,m}) \geq \mu\left(\bigcup_{\ell,m \in \mathbb{N}} C_{\ell,m} \right) = \mu(C) > 0.$$

Thus there exist $\ell_0, m_0 \in \mathbb{N}$ such that $\mu(C_{\ell_0,m_0}) > 0$. Then we have

$$(2) \qquad \int_{C_{\ell_0,m_0}} (g - f) \, d\mu \geq \frac{1}{m_0} \mu(C_{\ell_0,m_0}) > 0.$$

Now $f(x) \in [0, \ell_0]$ for $x \in C_{\ell_0,m_0}$. Thus f is μ-integrable on C_{ℓ_0,m_0} and then the difference $\int_{C_{\ell_0,m_0}} g \, d\mu - \int_{C_{\ell_0,m_0}} f \, d\mu$ exists and consequently by Corollary 9.15 we have

$$(3) \qquad \int_{C_{\ell_0,m_0}} (g - f) \, d\mu = \int_{C_{\ell_0,m_0}} g \, d\mu - \int_{C_{\ell_0,m_0}} f \, d\mu.$$

Combining (2) and (3), we have

$$\int_{C_{\ell_0,m_0}} g \, d\mu - \int_{C_{\ell_0,m_0}} f \, d\mu > 0 \text{ and thus } \int_{C_{\ell_0,m_0}} f \, d\mu \neq \int_{C_{\ell_0,m_0}} g \, d\mu.$$

This contradicts the assumption (1). Thus we must have $f = g$ a.e. on X^+. We show similarly that $f = g$ a.e. on X^-. This completes the proof that $f = g$ a.e. on $X^+ \cup X^-$.

3. To show that $f = g$ a.e. on X, we show that there exists a null set N in the measure space $(X, \mathfrak{A}, \mu)$ such that $f = g$ on $X \setminus N$. We showed in **2** that $f = g$ a.e. on $X^+ \cup X^-$. Thus there exists a null set N_0 in the measure space $(X, \mathfrak{A}, \mu)$ contained in $X^+ \cup X^-$ such that $f = g$ on $(X^+ \cup X^-) \setminus N_0$. We showed in **1** that A and B are null sets in the measure space $(X, \mathfrak{A}, \mu)$. Let us define $N = N_0 \cup A \cup B$. Then N is a null set in the measure space $(X, \mathfrak{A}, \mu)$ and we have

$$X \setminus N = (X^+ \cup X^- \cup A \cup B) \setminus (N_0 \cup A \cup B) = (X^+ \cup X^-) \setminus N_0.$$

But we have $f = g$ on $(X^+ \cup X^-) \setminus N_0$. Thus we have $f = g$ on $X \setminus N$. This completes the proof that $f = g$ a.e. on X. ∎

Prob. 9.6. In Prob. 9.5, instead of the σ-finiteness of the measure space $(X, \mathfrak{A}, \mu)$ let us assume the weaker condition that for every $E \in \mathfrak{A}$ with $\mu(E) > 0$ there exists $E_0 \in \mathfrak{A}$ such that $E_0 \subset E$ and $\mu(E_0) \in (0, \infty)$.
Show that if $\int_E f d\mu = \int_E g d\mu$ for every $E \in \mathfrak{A}$ then $f = g$ a.e. on X.

Proof. The proof is the same as the proof of Prob. 9.5 except for part 2 where we showed that $f = g$ a.e. on X^+ applying the σ-finiteness of the measure space $(X, \mathfrak{A}, \mu)$. We modify the proof of the statement that $f = g$ a.e. on X^+ as follows.

Suppose the statement that $f = g$ a.e. on X^+ is false. Then at least one of the following two sets:

$$C = \left\{ x \in X_{n_0}^+ : f(x) < g(x) \right\}$$

$$D = \left\{ x \in X_{n_0}^+ : f(x) > g(x) \right\}$$

has a positive mu-measure. Suppose $\mu(C) > 0$. By our assumption on $(X, \mathfrak{A}, \mu)$ there exists $C_0 \in eua$ such that $C_0 \subset C$ and $\mu(C_0) \in (0, \infty)$. Let us define

$$C_{\ell,m} = \left\{ x \in C_0 : \frac{1}{m} < g(x) - f(x) \text{ and } f(x) < \ell \right\} \quad \text{for } \ell, m \in \mathbb{N}.$$

Then we have

$$C_0 = \bigcup_{\ell,m \in \mathbb{N}} C_{\ell,m}$$

and this implies

$$\sum_{\ell,m \in \mathbb{N}} \mu(C_{\ell,m}) \geq \mu\left(\bigcup_{\ell,m \in \mathbb{N}} C_{\ell,m} \right) = \mu(C_0) > 0.$$

Thus there exist $\ell_0, m_0 \in \mathbb{N}$ such that $\mu(C_{\ell_0,m_0}) > 0$. Then we have

(1) $$\int_{C_{\ell_0,m_0}} (g - f) d\mu \geq \frac{1}{m_0} \mu(C_{\ell_0,m_0}) > 0.$$

Now $f(x) \in [0, \ell_0]$ for $x \in C_{\ell_0,m_0}$. Thus f is μ-integrable on C_{ℓ_0,m_0} and then the difference $\int_{C_{\ell_0,m_0}} g d\mu - \int_{C_{\ell_0,m_0}} f d\mu$ exists and consequently by Corollary 9.15 we have

(2) $$\int_{C_{\ell_0,m_0}} (g - f) d\mu = \int_{C_{\ell_0,m_0}} g d\mu - \int_{C_{\ell_0,m_0}} f d\mu.$$

Combining (1) and (2), we have

$$\int_{C_{\ell_0,m_0}} g d\mu - \int_{C_{\ell_0,m_0}} f d\mu > 0 \text{ and thus } \int_{C_{\ell_0,m_0}} f d\mu \neq \int_{C_{\ell_0,m_0}} g d\mu.$$

This contradicts the assumption that $\int_E f d\mu = \int_E g d\mu$ for every $E \in \mathfrak{A}$. Thus we must have $f = g$ a.e. on X^+. ∎

Prob. 9.7. Show that there exist a measure space $(X, \mathfrak{A}, \mu)$ and extended real-valued $\mathfrak{A}$-measurable functions f and g on X such that the assumption that $\int_E f d\mu = \int_E g d\mu$ for every $E \in \mathfrak{A}$ does not imply that $f = g$ a.e. on X.

Proof. Let $(X, \mathfrak{A}, \mu)$ be a measure space such that $\mathfrak{A} = \{\emptyset, X\}$ and $\mu(\emptyset) = 0$ and $\mu(X) = \infty$. Let f and g be real-valued $\mathfrak{A}$-measurable functions on X defined by setting $f(x) = 1$ for every $x \in X$ and $g(x) = 2$ for every $x \in X$. Then we have

$$\int_\emptyset f d\mu = 0 = \int_\emptyset g d\mu \text{ and } \int_X f d\mu = \infty = \int_X g d\mu$$

so that we have $\int_E f d\mu = \int_E g d\mu$ for every $E \in \mathfrak{A}$. Yet we have $f(x) \neq g(x)$ for every $x \in X$ and certainly the statement that $f = g$ a.e. on X is false. ∎

Prob. 9.8. Let $(X, \mathfrak{A}, \mu)$ be a measure space. Show that if there exists an extended real-valued $\mathfrak{A}$-measurable function f on X such that $f > 0$ μ-a.e. on X and f is μ-integrable on X, then $(X, \mathfrak{A}, \mu)$ is a σ-finite measure space. (In other words, if $(X, \mathfrak{A}, \mu)$ is not σ-finite, then there does not exist an extended real-valued $\mathfrak{A}$-measurable function f on X such that $f > 0$ μ-a.e. on X and f is μ-integrable on X.)

Proof. Suppose there exists an extended real-valued valued $\mathfrak{A}$-measurable function f on X such that $f > 0$ μ-a.e. on X and f is μ-integrable on X. Let

$$E_0 = \{x \in X : f(x) = 0\},$$

$$E_1 = \{x \in X : f(x) \in [1, \infty]\},$$

$$E_k = \{x \in X : f(x) \in [\tfrac{1}{k}, \tfrac{1}{k-1})\} \text{ for } k \geq 2.$$

Then $\{E_k : k \in \mathbb{Z}_+\}$ is a disjoint collection in $\mathfrak{A}$ and $\bigcup_{k \in \mathbb{Z}_+} E_k = X$. Now we have $\mu(E_0) = 0$ and for $k \geq 1$ we have

$$\frac{1}{k} \leq \int_{E_k} f d\mu \leq \int_X f d\mu < \infty$$

and this implies that $\mu(E_k) < \infty$. Thus we have $\mu(E_k) < \infty$ for every $k \in \mathbb{Z}_+$. This shows that $(X, \mathfrak{A}, \mu)$ is a σ-finite measure space. ∎

Prob. 9.9. Given a measure space $(X, \mathfrak{A}, \mu)$. Let f be an extended real-valued $\mathfrak{A}$-measurable and μ-integrable function on X. Let $(E_n : n \in \mathbb{N})$ be a sequence in $\mathfrak{A}$ such that $\lim_{n\to\infty} \mu(E_n) = 0$. Show that $\lim_{n\to\infty} \int_{E_n} f \, d\mu = 0$.

Proof. Suppose f is μ-integrable function on X. Then according to Theorem 9.26, for every $\varepsilon > 0$ there exists $\delta > 0$ such that

$$\left| \int_E f \, d\mu \right| \le \int_E |f| \, d\mu \quad \text{for every } E \in \mathfrak{A} \text{ with } \mu(E) < \delta.$$

If $(E_n : n \in \mathbb{N})$ be a sequence in $\mathfrak{A}$ such that $\lim_{n\to\infty} \mu(E_n) = 0$ then there exists $N \in \mathbb{N}$ such that $\mu(E_n) < \delta$ for $n \ge N$. Then we have

$$\left| \int_{E_n} f \, d\mu \right| < \varepsilon \quad \text{for } n \ge N.$$

This shows that $\lim_{n\to\infty} \int_{E_n} f \, d\mu = 0$. ∎

Prob. 9.10. Given a measure space $(X, \mathfrak{A}, \mu)$. Let f be an extended real-valued $\mathfrak{A}$-measurable and μ-integrable function on X. Let $E_n = \{x \in X : |f(x)| \geq n\}$ for $n \in \mathbb{N}$. Show that $\lim_{n \to \infty} \mu(E_n) = 0$.

Proof. Clearly $(E_n : n \in \mathbb{N})$ is a decreasing sequence in $\mathfrak{A}$. Thus $\lim_{n \to \infty} E_n$ exists in $\mathfrak{A}$ and moreover we have

(1) $$\lim_{n \to \infty} E_n = \bigcap_{n \in \mathbb{N}} E_n = \{x \in X : |f(x)| = \infty\}.$$

We claim that

(2) $$\mu\Big(\bigcap_{n \in \mathbb{N}} E_n\Big) = 0$$

for otherwise we would have $\mu\big(\bigcap_{n \in \mathbb{N}} E_n\big) > 0$ and then by the second equality in (1)

$$\int_X |f| \, d\mu \geq \int_{\bigcap_{n \in \mathbb{N}} E_n} |f| \, d\mu = \infty \cdot \mu\Big(\bigcap_{n \in \mathbb{N}} E_n\Big) = \infty,$$

contradicting the μ-integrability of f on X.
    We claim also that

(3) $$\mu(E_1) < \infty$$

for otherwise we would have $\mu(E_1) = \infty$ and then

$$\int_X |f| \, d\mu \geq \int_{E_1} |f| \, d\mu = 1 \cdot \mu(E_1) = \infty,$$

contradicting the μ-integrability of f on X.
    Now $(E_n : n \in \mathbb{N})$ is a decreasing sequence in $\mathfrak{A}$ contained in the set E_1 having $\mu(E_1) < \infty$. Thus by Theorem 1.26 and by (1) and (2) we have

$$\lim_{n \to \infty} \mu(E_n) = \mu\Big(\lim_{n \to \infty} E_n\Big) = \mu\Big(\bigcap_{n \in \mathbb{N}} E_n\Big) = 0.$$

This completes the proof. ∎

Prob. 9.11. Given a measure space $(X, \mathfrak{A}, \mu)$ and an extended real-valued $\mathfrak{A}$-measurable function f on X. Suppose that for some increasing sequence $(A_n : n \in \mathbb{N})$ in $\mathfrak{A}$ with $\bigcup_{n \in \mathbb{N}} A_n = X$, we have $\lim_{n \to \infty} \int_{A_n} |f| \, d\mu < \infty$. Show that f is μ-integrable on X.

Proof. Since $A_n \uparrow X$ we have $\mathbf{1}_{A_n} \uparrow \mathbf{1}_X$ and then $\mathbf{1}_{A_n} \cdot |f| \uparrow |f|$. Then by Theorem 8.5 (Monotone Convergence Theorem) we have

$$\int_X |f| \, d\mu = \int_X \lim_{n \to \infty} \mathbf{1}_{A_n} \cdot |f| \, d\mu = \lim_{n \to \infty} \int_X \mathbf{1}_{A_n} \cdot |f| \, d\mu = \lim_{n \to \infty} \int_{A_n} |f| \, d\mu < \infty.$$

This shows that f is μ-integrable on X. ∎

Prob. 9.12. Given a measure space $(X, \mathfrak{A}, \mu)$. Let $(A_n : n \in \mathbb{N})$ be an increasing sequence in $\mathfrak{A}$ with $\bigcup_{n \in \mathbb{N}} A_n = X$. Let f be an extended real-valued $\mathfrak{A}$-measurable function on X such that $\int_D f \, d\mu$ exists. Show that $\lim_{n \to \infty} \int_{A_n} f \, d\mu = \int_D f \, d\mu$.

Proof. Let $A_0 = \emptyset$ and let $B_n = A_n \setminus A_{n-1}$ for $n \in \mathbb{N}$. Then $(B_n : n \in \mathbb{N})$ is a disjoint sequence in $\mathfrak{A}$ and $\bigcup_{n \in \mathbb{N}} B_n = \bigcup_{n \in \mathbb{N}} A_n = X$. Note that $\bigcup_{k=1}^n B_k = A_n$ for every $n \in \mathbb{N}$. Then we have

$$\int_X f \, d\mu = \int_{\bigcup_{n \in \mathbb{N}} B_n} f \, d\mu = \sum_{n \in \mathbb{N}} \int_{B_n} f \, d\mu = \lim_{n \to \infty} \sum_{k=1}^n \int_{B_k} f \, d\mu$$

$$= \lim_{n \to \infty} \int_{\bigcup_{k=1}^n B_k} f \, d\mu = \lim_{n \to \infty} \int_{A_n} f \, d\mu.$$

This completes the proof. ∎

Prob. 9.13. Given a measure space $(X, \mathfrak{A}, \mu)$ and an extended real-valued $\mathfrak{A}$-measurable function f on X. The existence of an increasing sequence $(A_n : n \in \mathbb{N})$ in $\mathfrak{A}$ with $\bigcup_{n \in \mathbb{N}} A_n = X$ such that $\lim_{n \to \infty} \int_{A_n} f \, d\mu$ exists does not imply the existence of $\int_X f \, d\mu$ even when the limit is finite. Show this by constructing an example.

Proof. Consider the measure space $(\mathbb{R}, \mathfrak{M}_L, \mu_L)$.

Let $f(x) = \sin x$ for $x \in \mathbb{R}$. Then $\int_{\mathbb{R}} f^+ \, d\mu_L = \infty$ and $\int_{\mathbb{R}} f^- \, d\mu_L = \infty$ also so that $\int_{\mathbb{R}} f^+ \, d\mu_L - \int_{\mathbb{R}} f^- \, d\mu_L$ does not exist and thus $\int_{\mathbb{R}} f \, d\mu_L$ does not exist.

Let $A_n = [-n\pi, n\pi]$ for $n \in \mathbb{N}$. Then $(A_n : n \in \mathbb{N})$ is an increasing sequence in $\mathfrak{M}_L$ and $\bigcup_{n \in \mathbb{N}} A_n = \mathbb{R}$. Now for every $n \in \mathbb{N}$ we have

$$\int_{A_n} f \, d\mu_L = \int_{[-n\pi, n\pi]} \sin x \, \mu_L(dx) = 0.$$

Thus we have $\lim_{n \to \infty} \int_{A_n} f \, d\mu_L = 0.$ ∎

Prob. 9.14. Construct a real-valued continuous and μ_L-integrable function f on $[0, \infty)$ for which $\lim_{x \to \infty} f(x)$ does not exist.

Proof. For each $n \in \mathbb{N}$ define an interval I_n in $\mathbb{R}$ by setting

$$I_n = \left[n - \frac{1}{n2^n}, n + \frac{1}{n2^n} \right].$$

Let f be a real-valued function on $[0, \infty)$ defined by setting

$$f(x) = \begin{cases} 0 & \text{for } x \in [0, \infty) \setminus \bigcup_{n \in \mathbb{N}} I_n, \\ n & \text{for } x = n, \\ \text{linear} & \text{for } x \in \left[n - \frac{1}{n2^n}, n \right], \\ \text{linear} & \text{for } x \in \left[n, n + \frac{1}{n2^n}, n \right]. \end{cases}$$

Then f is continuous on $[0, \infty)$ and $\lim_{x \to \infty} f(x)$ does not exist. On the other hand we have

$$\int_{[0,\infty)} f \, d\mu_L = \sum_{n \in \mathbb{N}} \int_{I_n} f \, d\mu_L = \sum_{n \in \mathbb{N}} n \cdot \frac{1}{n2^n} = \sum_{n \in \mathbb{N}} \frac{1}{2^n} = 1 < \infty.$$

This shows that f is μ_L-integrable on $[0, \infty)$. ∎

Prob. 9.15. Let f be an extended real-valued $\mathfrak{M}_L$-measurable and μ_L integrable function on the interval $[0, \infty)$. Show that if f is uniformly continuous on $[0, \infty)$ then $\lim\limits_{x \to \infty} f(x) = 0$.

Proof. To show that $\lim\limits_{x \to \infty} f(x) = 0$, we show that for every $\varepsilon > 0$ there exists $M > 0$ such that $|f(x)| < \varepsilon$ for $x \geq$ ÃŸ M.

Let $\varepsilon > 0$ be arbitrarily given. Then by the uniform continuity of f on $[0, \infty)$ there exists $\delta > 0$ such that

(1) $\qquad |f(x') - f(x")| < \dfrac{\varepsilon}{2}$ for any $x', x" \in [0, \infty)$ such that $|x' - x"| < \delta$.

Let us decompose $[0, \infty)$ into a sequence $(I_n : n \in \mathbb{N})$ of disjoint intervals of length δ given by $I_n = [(n-1)\delta, n\delta)$ for $n \in \mathbb{N}$. We have $\bigcup_{n \in \mathbb{N}} I_n = [0, \infty)$. Thus to show that there exists $M > 0$ such that $|f(x)| < \varepsilon$ for $x \geq M$, it suffices to show that there exists $N \in \mathbb{N}$ such that $|f(x)| < \varepsilon$ for $x \geq N\delta$.

Now since $[0, \infty) = \bigcup_{n \in \mathbb{N}} I_n$, we have

(2) $$\int_{[0,\infty)} |f| \, d\mu_L = \sum_{n \in \mathbb{N}} \int_{I_n} |f| \, d\mu_L.$$

Since f is μ_L-integrable on $[0, \infty)$, the integral on the left side of (2) is finite and then the series on the right side of (2) is convergent. This implies that the terms in the series converges to 0, that is,

(3) $$\lim_{n \to \infty} \int_{I_n} |f| \, d\mu_L = 0.$$

Now (3) implies that there exists $N \in \mathbb{N}$ such that

(4) $$\int_{I_n} |f| \, d\mu_L < \frac{\varepsilon}{2}\delta \quad \text{for } n \geq N.$$

Consider $n \geq N$. If $|f(x)| \geq \frac{\varepsilon}{2}$ for every $x \in I_n$ then we have $\int_{I_n} |f| \, d\mu_L \geq \frac{\varepsilon}{2}\ell(I_n) = \frac{\varepsilon}{2}\delta$, contradicting (3). Thus there exists $\xi \in I_n$ such that $|f(\xi)| < \frac{\varepsilon}{2}$. Then for every $x \in I_n$ we have $|x - \xi| < \delta$ and thus by (1) we have $|f(x) - f(\xi)| < \frac{\varepsilon}{2}$. Thus for every $x \in I_n$ when $n \geq N$ we have

$$|f(x)| = |f(x) - f(\xi) + f(\xi)| \leq |f(x) - f(\xi)| + |f(\xi)| < \frac{\varepsilon}{2} + \frac{\varepsilon}{2} = \varepsilon.$$

This shows that $|f(x)| < \varepsilon$ for $x \geq N\delta$. ∎

Prob. 9.16. Construct extended real-valued $\mathfrak{M}_L$-measurable and μ_L-integrable functions $(f_n : n \in \mathbb{N})$ and f on a set $D \in \mathfrak{M}_L$ such that
$$\lim_{n \to \infty} f_n = f \quad \text{a.e. on } D,$$
$$\lim_{n \to \infty} \int_D f_n \, d\mu_L = \int_D f \, d\mu_L,$$
but for some $E \subset D$, $E \in \mathfrak{M}_L$, we have
$$\lim_{n \to \infty} \int_E f_n \, d\mu_L \neq \int_E f \, d\mu_L.$$

Proof. Let $D = [-1, 1]$. We define a sequence of real-valued functions $(f_n : n \in \mathbb{N})$ on D as follows. For each $n \in \mathbb{N}$, decompose $D = [-1, 1]$ as

$$\left[-1, 1\right] = \left[-1, -1 + \tfrac{1}{n}\right] \cup \left[-1 + \tfrac{1}{n}, 1 - \tfrac{1}{n}\right] \cup \left[1 - \tfrac{1}{n}, 1\right].$$

Then define f_n on D by setting

$$f_n(x) = \begin{cases} 0 & \text{for } x \in \left[-1 + \tfrac{1}{n}, 1 - \tfrac{1}{n}\right], \\ -n & \text{for } x = -1, \\ n & \text{for } x = 1, \\ \text{linear} & \text{for } x \in \left[-1, -1 + \tfrac{1}{n}\right], \\ \text{linear} & \text{for } x \in \left[1 - \tfrac{1}{n}, 1\right]. \end{cases}$$

We define $f = 0$ on D.

For f_n and f defined above, we have $\lim_{n \to \infty} f_n(x) = 0 = f(x)$ for $x \in (-1, 1)$ and thus

$$\lim_{n \to \infty} f_n = f \quad \text{a.e. on } D.$$

We have $\int_D f_n \, d\mu_L = 0$ for every $n \in \mathbb{N}$ and $\int_D f \, d\mu_L = 0$. Thus we have

$$\lim_{n \to \infty} \int_D f_n \, d\mu_L = 0 = \int_D f \, d\mu_L.$$

Let $E = [0, 1]$. Then we have $\int_E f_n \, d\mu_L = \tfrac{1}{2}\tfrac{1}{n}n = \tfrac{1}{2}$ and $\int_E f \, d\mu_L = 0$. Thus we have

$$\lim_{n \to \infty} \int_E f_n \, d\mu_L = \frac{1}{2} \neq 0 = \int_E f \, d\mu_L.$$

This completes the proof. ∎

Prob. 9.17. Let $(X, \mathfrak{A}, \mu)$ be a measure space and let $(f_n : n \in \mathbb{N})$, f and g be extended real-valued $\mathfrak{A}$-measurable and μ-integrable functions on a set $D \in \mathfrak{A}$. Suppose that

1° $\lim\limits_{n \to \infty} f_n = f$ a.e. on D,

2° $\lim\limits_{n \to \infty} \int_D f_n \, d\mu = \int_D \, d\mu$,

3° Either $f_n \geq g$ on D for all $n \in \mathbb{N}$ or $f_n \leq g$ on D for all $n \in \mathbb{N}$.

Show that for every $E \subset D$ such that $E \in \mathfrak{A}$ we have $\lim\limits_{n \to \infty} \int_E f_n \, d\mu = \int_E f \, d\mu$.
(Note that as a particular case of condition 3°, we have the condition that either $f_n \geq 0$ on D for all $n \in \mathbb{N}$ or $f_n \leq 0$ on D for all $n \in \mathbb{N}$.)

Proof. 1. Consider the case that $f_n \geq g$ on D for all $n \in \mathbb{N}$. Then by 1° and by (a) of Theorem 9.19 (Generalized Fatou's Lemma), we have

(1) $$\int_E f \, d\mu = \int_E \liminf_{n \to \infty} f_n \, d\mu \leq \liminf_{n \to \infty} \int_E f_n \, d\mu.$$

Let us show that we also have

(2) $$\limsup_{n \to \infty} \int_E f_n \, d\mu \leq \int_E f \, d\mu.$$

To prove (2), assume the contrary, that is,

(3) $$\int_E f \, d\mu < \limsup_{n \to \infty} \int_E f_n \, d\mu.$$

Now (1) holds for an arbitrary $\mathfrak{A}$-measurable subset E of D. Thus (1) holds for $D \setminus E$ and then we have

(4) $$\int_{D \setminus E} f \, d\mu \leq \liminf_{n \to \infty} \int_{D \setminus E} f_n \, d\mu.$$

Then we have by (3) and (4)

(5) $$\int_D f \, d\mu = \int_E f \, d\mu + \int_{D \setminus E} f \, d\mu < \limsup_{n \to \infty} \int_E f_n \, d\mu + \liminf_{n \to \infty} \int_{D \setminus E} f_n \, d\mu.$$

For two sequences of real numbers $(a_n : n \in \mathbb{N})$ and $(b_n : n \in \mathbb{N})$, we have

$$\limsup_{n \to \infty} a_n = \limsup_{n \to \infty}(a_n + b_n - b_n) \leq \limsup_{n \to \infty}(a_n + b_n) + \limsup_{n \to \infty}(-b_n)$$

$$= \limsup_{n \to \infty}(a_n + b_n) - \liminf_{n \to \infty} b_n$$

and then

$$\limsup_{n \to \infty} a_n + \liminf_{n \to \infty} b_n \leq \limsup_{n \to \infty}(a_n + b_n).$$

Applying this inequality to the right side of (5), we have

$$\int_D f \, d\mu < \limsup_{n \to \infty} \left\{ \int_E f_n \, d\mu + \int_{D \setminus E} f_n \, d\mu \right\}$$

$$= \limsup_{n \to \infty} \int_D f_n \, d\mu = \int_D f \, d\mu \quad \text{by 2°.}$$

This is a contradiction. This proves (2). Then by (1) and (2) we have

$$\int_E f \, d\mu \leq \liminf_{n \to \infty} \int_E f_n \, d\mu \leq \limsup_{n \to \infty} \int_E f_n \, d\mu \leq \int_E f \, d\mu.$$

This shows that

$$\liminf_{n \to \infty} \int_E f_n \, d\mu = \limsup_{n \to \infty} \int_E f_n \, d\mu = \int_E f \, d\mu.$$

This shows that $\lim_{n \to \infty} \int_E f_n \, d\mu$ exists and moreover we have $\lim_{n \to \infty} \int_E f_n \, d\mu = \int_E f \, d\mu$.

2. The case that $f_n \leq$ on D for all $n \in \mathbb{N}$ is proved similarly but by applying (b) of Theorem 9.19. ∎

Prob. 9.18. Given a measure space $(X, \mathfrak{A}, \mu)$. Let $(f_n : n \in \mathbb{N})$ be an increasing sequence of extended real-valued $\mathfrak{A}$-measurable functions on a set $D \in \mathfrak{A}$ and let $f = \lim_{n \to \infty} f_n$.
Suppose
1° $\int_D |f_1| \, d\mu < \infty$, that is, f_1 is μ-integrable on D.
2° $\sup_{n \in \mathbb{N}} \int_D f_n \, d\mu < \infty$.
Show that $\lim_{n \to \infty} \int_D f_n \, d\mu = \int_D f \, d\mu$ and $\int_D |f| \, d\mu < \infty$, that is, f is μ-integrable on D.

Proof. By 1°, f_1 is μ-integrable on D. Since $(f_n : n \in \mathbb{N})$ is an increasing sequence, we have $f_n \geq f_1$ on D for $n \in \mathbb{N}$. Thus by (a) of Theorem 9.17 (Generalized Monotone Convergence Theorem), $\int_D f_n \, d\mu$ for $n \in \mathbb{N}$ and $\int_D f \, d\mu$ all exist in $\overline{\mathbb{R}}$ and moreover

(1) $$\lim_{n \to \infty} \int_D f_n \, d\mu = \int_D f \, d\mu.$$

Then by (1) and 2° we have

(2) $$\int_D f \, d\mu = \lim_{n \to \infty} \int_D f_n \, d\mu \leq \sup_{n \in \mathbb{N}} \int_D f_n \, d\mu < \infty.$$

On the other hand, since $f = \lim_{n \to \infty} f_n \geq f_1$ and f_1 is μ-integrable on D, we have

(3) $$\int_D f \, d\mu \geq \int_D f_1 \, d\mu > -\infty.$$

By (2) and (3), we have $-\infty < \int_D f \, d\mu < \infty$, that is, f is μ-integrable on D. ∎

Prob. 9.19. Given a measure space $(X, \mathfrak{A}, \mu)$. Let f be an extended real-valued $\mathfrak{A}$-measurable function on a set $D \in \mathfrak{A}$. Assume that f is μ semi-integrable on D, that is, $\int_D f \, d\mu$ exists in $\overline{\mathbb{R}}$. Show that there exists a sequence of simple functions $(\varphi_n : n \in \mathbb{N})$ on D such that

1° $|\varphi_n| \le |f|$ on D for every $n \in \mathbb{N}$ and $\lim_{n \to \infty} \varphi_n = f$ a.e. on D,

2° $\lim_{n \to \infty} \int_D \varphi_n \, d\mu = \int_D f \, d\mu$.

Proof. Let us write $f = f^+ - f^-$. The existence of $\int_D f \, d\mu$ in $\overline{\mathbb{R}}$ implies that at least one of the two integrals $\int_D f^+ \, d\mu$ and $\int_D f^- \, d\mu$ is finite and moreover

(1) $$\int_D f \, d\mu = \int_D f^+ \, d\mu - \int_D f^- \, d\mu.$$

Note that if $\int_D f^+ \, d\mu < \infty$ then $f^+ < \infty$ a.e. on D and similarly if $\int_D f^- \, d\mu < \infty$ then $f^- < \infty$ a.e. on D. Thus at least one of the two functions f^+ and f^- is finite a.e. on D.

Now f^+ and f^- are nonnegative extended real-valued $\mathfrak{A}$-measurable functions on D. Then by Lemma 8.6 there exist sequences of simple functions on D, $(\varphi_n' : n \in \mathbb{N})$ and $(\varphi_n'' : n \in \mathbb{N})$, such that

(2) $$\varphi_n' \uparrow f^+ \quad \text{and} \quad \lim_{n \to \infty} \int_D \varphi_n' \, d\mu = \int_D f^+ \, d\mu$$

(3) $$\varphi_n'' \uparrow f^- \quad \text{and} \quad \lim_{n \to \infty} \int_D \varphi_n'' \, d\mu = \int_D f^- \, d\mu.$$

Let $\varphi_n = \varphi_n' - \varphi_n''$ for $n \in \mathbb{N}$. Then $(\varphi_n : n \in \mathbb{N})$ is a sequence of simple functions on D. For every $n \in \mathbb{N}$, we have

$$|\varphi_n| = |\varphi_n' - \varphi_n''| \le |\varphi_n'| + |\varphi_n''| \le f^+ + f^- = |f|.$$

We also have

$$\lim_{n \to \infty} \varphi_n = \lim_{n \to \infty} (\varphi_n' - \varphi_n'') = f^+ - f^- = f \text{ a.e. on } D,$$

where the second equality is by the fact that $\lim_{n \to \infty} \varphi_n' = f^+$ by (2), $\lim_{n \to \infty} \varphi_n'' = f^-$ by (3) and at least one of the two functions f^+ and f^- is finite a.e. on D. This proves 1°.

To prove 2°, let us substitute (2) and (3) in (1). Then we have

$$\int_D f \, d\mu = \int_D f^+ \, d\mu - \int_D f^- \, d\mu = \lim_{n \to \infty} \int_D \varphi_n' \, d\mu - \lim_{n \to \infty} \int_D \varphi_n'' \, d\mu$$

$$= \lim_{n \to \infty} \left\{ \int_D \varphi_n' \, d\mu - \int_D \varphi_n'' \, d\mu \right\} = \lim_{n \to \infty} \int_D (\varphi_n' - \varphi_n'') \, d\mu$$

$$= \lim_{n \to \infty} \int_D \varphi_n \, d\mu.$$

This completes the proof. ∎

Prob. 9.20. Given a measure space $(X, \mathfrak{A}, \mu)$. Let $(f_n, n \in \mathbb{N})$ and f be extended real-valued $\mathfrak{A}$-measurable functions on a set $D \in \mathfrak{A}$ such that $\lim\limits_{n \to \infty} f_n = f$ a.e. on D. Suppose there exists a nonnegative extended real-valued $\mathfrak{A}$-measurable function g on D such that

$1°$ $|f_n| \leq g$ on D for every $n \in \mathbb{N}$,

$2°$ g^p is μ-integrable on D for some $p \in (0, \infty)$.

Show that

(1) $|f|^p$ is μ-integrable on D.

(2) $\lim\limits_{n \to \infty} \int_D |f_n|^p \, d\mu = \int_D |f|^p \, d\mu$.

(3) $\lim\limits_{n \to \infty} \int_D |f_n - f|^p \, d\mu = 0$.

Proof. 1. According to Theorem 9.20 (Dominated Convergence Theorem), if $(\varphi_n : n \in \mathbb{N})$ and φ are extended real-valued $\mathfrak{A}$-measurable functions on $D \in \mathfrak{A}$ such that $\lim\limits_{n \to \infty} \varphi_n = \varphi$ a.e. on D and if there exists a nonnegative extended real-valued $\mathfrak{A}$-measurable and μ-integrable function γ on D such that $|\varphi_n| \leq \gamma$ on D for every $n \in \mathbb{N}$, then

(i) φ is μ-integrable on D.

(ii) $\lim\limits_{n \to \infty} \int_D \varphi_n \, d\mu = \int_D \varphi \, d\mu$.

Consider the sequence $(|f_n|^p : n \in \mathbb{N})$. Since $\lim\limits_{n \to \infty} f_n = f$ a.e. on D, we have $\lim\limits_{n \to \infty} |f_n|^p = |f|^p$ a.e. on D. Since $|f_n| \leq g$ on D, we have $|f_n|^p \leq g^p$ on D for every $n \in \mathbb{N}$. Then by (i) and (ii), identifying $|f_n|^p$, $|f|^p$ and g^p with φ_n, φ and γ respectively we obtain

(1) $|f|^p$ is μ-integrable on D.

(2) $\lim\limits_{n \to \infty} \int_D |f_n|^p \, d\mu = \int_D |f|^p \, d\mu$.

2. We showed above that $|f|^p$ is μ-integrable on D. This implies that $|f|^p$ is finite a.e. on D and then f is finite a.e. on D. Then $f_n - f$ is defined a.e. on D for each $n \in \mathbb{N}$. Thus there exists a null set D_0 in the measure space $(X, \mathfrak{A}, \mu)$ contained in D such that the sequence $(|f_n - f|^p : n \in \mathbb{N})$ is defined on $D \setminus D_0$. Now we have

$$\lim_{n \to \infty} |f_n - f|^p = \left| \lim_{n \to \infty} f_n - f \right|^p = |f - f| = 0 \quad \text{a.e. on } D.$$

We also have

$$|f_n - f|^p \leq \left(|f_n| + |f| \right)^p \leq (g + g)^p \leq 2^p g^p.$$

Observe that since g^p is μ-integrable, $2^p g^p$ is μ-integrable. Then by (ii), identifying $|f_n - f|^p$ and $2^p g^p$ with φ_n and γ respectively, we have

$$\lim_{n \to \infty} \int_D |f_n - f|^p \, d\mu = \int_D \lim_{n \to \infty} |f_n - f|^p \, d\mu = \int_D 0 \, d\mu = 0.$$

This proves (3). ∎

Prob. 9.21. Assume the hypothesis in Theorem 9.20 (Lebesgue Dominated Convergence Theorem). Then $g + f_n \geq 0$ and $g - f_n \geq 0$ on D for every $n \in \mathbb{N}$ so that $(g + f_n : n \in \mathbb{N})$ and $(g - f_n : n \in \mathbb{N})$ are two sequences of nonnegative extended real-valued $\mathfrak{A}$-measurable functions on D and $\lim_{n \to \infty} \{g + f_n\} = g + f$ and $\lim_{n \to \infty} \{g - f_n\} = g - f$ a.e. on D. Apply Fatou's Lemma for nonnegative functions (Theorem 8.13) to construct an alternate proof to Theorem 9.20.

Proof. Applying Theorem 8.13 to the sequence of nonnegative extended real-valued $\mathfrak{A}$-measurable functions $(g + f_n : n \in \mathbb{N})$, we have

$$\int_D \{g + f\} \, d\mu \leq \liminf_{n \to \infty} \int_D \{g + f_n\} \, d\mu = \int_D g \, d\mu + \liminf_{n \to \infty} \int_D f_n \, d\mu,$$

and then subtracting $\int_D g \, d\mu$ from both sides we have

$$(1) \qquad\qquad \int_D f \, d\mu \leq \liminf_{n \to \infty} \int_D f_n \, d\mu.$$

Similarly applying Theorem 8.13 to the sequence of nonnegative extended real-valued $\mathfrak{A}$-measurable functions $(g - f_n : n \in \mathbb{N})$, we have

$$\int_D \{g - f\} \, d\mu \leq \liminf_{n \to \infty} \int_D \{g - f_n\} \, d\mu = \int_D g \, d\mu + \liminf_{n \to \infty} \left\{ - \int_D f_n \, d\mu \right\}$$

$$= \int_D g \, d\mu - \limsup_{n \to \infty} \int_D f_n \, d\mu,$$

and then subtracting $\int_D g \, d\mu$ from both sides and multiplying by -1 we have

$$(2) \qquad\qquad \int_D f \, d\mu \geq \limsup_{n \to \infty} \int_D f_n \, d\mu.$$

By (1) and (2) we have

$$(3) \qquad\qquad \limsup_{n \to \infty} \int_D f_n \, d\mu \leq \int_D f \, d\mu \leq \liminf_{n \to \infty} \int_D f_n \, d\mu.$$

The inequality $\limsup_{n \to \infty} \int_D f_n \, d\mu \geq \liminf_{n \to \infty} \int_D f_n \, d\mu$ and (3) imply that

$$\liminf_{n \to \infty} \int_D f_n \, d\mu = \limsup_{n \to \infty} \int_D f_n \, d\mu = \int_D f \, d\mu.$$

Thus we have

$$\lim_{n \to \infty} \int_D f_n \, d\mu = \int_D f \, d\mu. \quad \blacksquare$$

Prob. 9.22. (An Extension of the Dominated Convergence Theorem) Prove the following:
Given a measure space $(X, \mathfrak{A}, \mu)$. Let $(f_n : n \in \mathbb{N})$ and f be extended real-valued $\mathfrak{A}$-measurable functions on a set $D \in \mathfrak{A}$ and let $(g_n : n \in \mathbb{N})$ and g be nonnegative extended real-valued $\mathfrak{A}$-measurable functions on D. Suppose
1° $\lim_{n \to \infty} f_n = f$ and $\lim_{n \to \infty} g_n = g$ a.e. on D,
2° $(g_n : n \in \mathbb{N})$ and g are all μ-integrable on D and $\lim_{n \to \infty} \int_D g_n \, d\mu = \int_D g \, d\mu$,
3° $|f_n| \le g_n$ on D for every $n \in \mathbb{N}$.
Then f is μ-integrable on D and $\lim_{n \to \infty} \int_D f_n \, d\mu = \int_D f \, d\mu$.

Proof. Let us prove the μ-integrability of f on D. We have

$$\int_D |f| \, d\mu = \int_D \left| \lim_{n \to \infty} f_n \right| d\mu = \int_D \lim_{n \to \infty} |f_n| \, d\mu \quad \text{by } 1°$$

$$\le \liminf_{n \to \infty} \int_D |f_n| \, d\mu \quad \text{by Theorem 8.13 (Fatou)}$$

$$\le \liminf_{n \to \infty} \int_D g_n \, d\mu \quad \text{by } 3°$$

$$= \lim_{n \to \infty} \int_D g_n \, d\mu = \int_D g \, d\mu < \infty \quad \text{by } 2°.$$

This proves the μ-integrability of f on D.

Now 3° and 1° imply that $(g_n + f_n : n \in \mathbb{N})$ and $(g_n - f_n : n \in \mathbb{N})$ are sequences of nonnegative extended real-valued $\mathfrak{A}$-measurable functions on D and $\lim_{n \to \infty} \{g_n + f_n\} = g + f$ and $\lim_{n \to \infty} \{g_n - f_n\} = g - f$. Thus by Theorem 8.13 (Fatou's Lemma) we have

(1) $\quad \int_D \{g + f\} \, d\mu \le \liminf_{n \to \infty} \int_D \{g_n + f_n\} \, d\mu = \int_D g \, d\mu + \liminf_{n \to \infty} \int_D f_n \, d\mu,$

and

(2) $\quad \int_D \{g - f\} \, d\mu \le \liminf_{n \to \infty} \int_D \{g_n - f_n\} \, d\mu = \int_D g \, d\mu + \liminf_{n \to \infty} \left\{ -\int_D f_n \, d\mu \right\}$

$$= \int_D g \, d\mu - \limsup_{n \to \infty} \int_D f_n \, d\mu.$$

Now (1) implies

(3) $\quad \int_D f \, d\mu \le \liminf_{n \to \infty} \int_D f_n \, d\mu,$

and (2) implies

(4) $\quad \int_D f \, d\mu \ge \limsup_{n \to \infty} \int_D f_n \, d\mu.$

Then (3) and (4) imply that

$$\liminf_{n \to \infty} \int_D f_n \, d\mu = \limsup_{n \to \infty} \int_D f_n \, d\mu = \int_D f \, d\mu$$

and consequently $\lim_{n \to \infty} \int_D f_n \, d\mu$ exists and $\lim_{n \to \infty} \int_D f_n \, d\mu = \int_D f \, d\mu$. ∎

Prob. 9.23. Given a measure space $(X, \mathfrak{A}, \mu)$. Let $(f_n : n \in \mathbb{N})$ and f be extended real-valued $\mathfrak{A}$-measurable functions on a set $D \in \mathfrak{A}$. Suppose
1° $\lim_{n \to \infty} f_n = f$ a.e. on D,
2° $(f_n : n \in \mathbb{N})$ and f are all μ-integrable on D.
(a) Show that if $\lim_{n \to \infty} \int_D |f_n| \, d\mu = \int_D |f| \, d\mu$, then $\lim_{n \to \infty} \int_D f_n \, d\mu = \int_D f \, d\mu$.
(b) Show that the converse of (a) is false by constructing a counterexample.

Proof. 1. We prove (a) by applying Prob. 9.22 (Extension of Dominated Convergence Theorem). Assume that we have $\lim_{n \to \infty} \int_D |f_n| \, d\mu = \int_D |f| \, d\mu$.

Let $g_n = |f_n|$ for $n \in \mathbb{N}$ and let $g = |f|$. Since $\lim_{n \to \infty} f_n = f$ a.e. on D implies $\lim_{n \to \infty} |f_n| = |f|$ a.e. on D, we have $\lim_{n \to \infty} g_n = g$ a.e. on D. Also $(g_n : n \in \mathbb{N})$ and g are all μ-integrable on D and $\lim_{n \to \infty} \int_D g_n \, d\mu = \int_D g \, d\mu$. Thus conditions 1° and 2° in Prob. 9.22 are satisfied.

Next consider the sequences $(f_n^+ : n \in \mathbb{N})$ and $(f_n^- : n \in \mathbb{N})$. We have $\lim_{n \to \infty} f_n^+ = f^+$ and $\lim_{n \to \infty} f_n^- = f^=$ a.e. on D. We also have $f_n^+ \le |f_n| = g_n$ and $f_n^- \le |f_n| = g_n$ on D for every $n \in \mathbb{N}$. Thus conditions 1° and 3° in Prob. 9.22 are satisfied. Therefore by the conclusion of Prob. 9.22 we have

$$\lim_{n \to \infty} \int_D f_n^+ \, d\mu = \int_D f^+ \, d\mu \quad \text{and} \quad \lim_{n \to \infty} \int_D f_n^- \, d\mu = \int_D f^- \, d\mu.$$

Then we have

$$\lim_{n \to \infty} \int_D f_n \, d\mu = \lim_{n \to \infty} \left\{ \int_D f_n^+ \, d\mu - \int_D f_n^- \, d\mu \right\}$$

$$= \lim_{n \to \infty} \int_D f_n^+ \, d\mu - \lim_{n \to \infty} \int_D f_n^- \, d\mu$$

$$= \int_D f^+ \, d\mu - \int_D f^- \, d\mu$$

$$= \int_D f \, d\mu.$$

This proves (a).

2. Let us construct an example to show that the converse of (a) is false.
Consider the measure space $(\mathbb{R}, \mathfrak{M}_L, \mu_L)$. Let $f = 0$ on $\mathbb{R}$ and let $(f_n : n \in \mathbb{N})$ be a sequence of real-valued $\mathfrak{M}_L$-measurable functions on $\mathbb{R}$ defined by setting

$$f_n(x) = \begin{cases} n \sin nx & \text{for } x \in \left[-\frac{1}{n}\frac{\pi}{2}, \frac{1}{n}\frac{\pi}{2} \right], \\ 0 & \text{for } x \in \left[-\frac{1}{n}\frac{\pi}{2}, \frac{1}{n}\frac{\pi}{2} \right]^c. \end{cases}$$

We have $\lim_{n \to \infty} f_n(x) = 0 = f(x)$ for every $x \in \mathbb{R}$. We have $\int_{\mathbb{R}} f_n \, d\mu_L = 0 = \int_{\mathbb{R}} f \, d\mu_L$ for every $n \in \mathbb{N}$. Thus we have $\lim_{n \to \infty} \int_{\mathbb{R}} f_n \, d\mu_L = \int_{\mathbb{R}} f \, d\mu_L$.
On the other hand we have $\int_{\mathbb{R}} |f| \, d\mu_L = 0$ and for every $n \in \mathbb{N}$ we have

$$\int_{\mathbb{R}} |f_n| \, d\mu_L = 2 \int_{\left[0, \frac{1}{n}\frac{\pi}{2}\right]} n \sin nx \, dx = 2n \left[\frac{\cos nx}{n} \right]_{\frac{1}{n}\frac{\pi}{2}}^0 = 2.$$

Thus we have $\lim\limits_{n\to\infty} \int_{\mathbb{R}} |f_n| \, d\mu_L = 2 \neq 0 = \int_{\mathbb{R}} |f| \, d\mu_L$.

3. Here is another example to show that the converse of (a) is false.

Consider the measure space $(\mathbb{R}, \mathfrak{M}_L, \mu_L)$. Let $f = 0$ on $\mathbb{R}$ and let $(f_n : n \in \mathbb{N})$ be a sequence of real-valued $\mathfrak{M}_L$-measurable functions on $\mathbb{R}$ defined by setting

$$
f_n(x) = \begin{cases} \frac{1}{n} & \text{for } x \in [0, n), \\ -\frac{1}{n} & \text{for } x \in (-n, 0), \\ 0 & \text{for } x \in (-n, n)^c. \end{cases}
$$

We have $\lim\limits_{n\to\infty} f_n(x) = 0 = f(x)$ for every $x \in \mathbb{R}$. We have $\int_{\mathbb{R}} f_n \, d\mu_L = 0 = \int_{\mathbb{R}} f \, d\mu_L$ for every $n \in \mathbb{N}$. Thus we have $\lim\limits_{n\to\infty} \int_{\mathbb{R}} f_n \, d\mu_L = \int_{\mathbb{R}} f \, d\mu_L$.

On the other hand we have $\int_{\mathbb{R}} |f_n| \, d\mu_L = 2$ for every $n \in \mathbb{N}$ and then we have $\lim\limits_{n\to\infty} \int_{\mathbb{R}} |f_n| \, d\mu_L = 2 \neq 0 = \int_{\mathbb{R}} f_n \, d\mu_L$. ∎

Prob. 9.24. Given a measure space $(X, \mathfrak{A}, \mu)$. Let $(f_n : n \in \mathbb{N})$ and f be extended real-valued $\mathfrak{A}$-measurable functions on a set $D \in \mathfrak{A}$. Suppose

1° $\lim\limits_{n \to \infty} f_n = f$ a.e. on D,

2° $(f_n : n \in \mathbb{N})$ and f are all μ-integrable on D.

3° $\lim\limits_{n \to \infty} \int_D |f_n| \, d\mu = \int_D |f| \, d\mu$.

(a) Show that for every $\mathfrak{A}$-measurable subset E of D, we have $\lim\limits_{n \to \infty} \int_E |f_n| \, d\mu = \int_E |f| \, d\mu$
and $\lim\limits_{n \to \infty} \int_E f_n \, d\mu = \int_E f \, d\mu$.

(b) Show by constructing a counterexample that if condition 3° is replaced by the condition that $\lim\limits_{n \to \infty} \int_D f_n \, d\mu = \int_D f \, d\mu$, then (a) does not hold.

Proof. 1. We prove (a) by applying Prob. 9.22 (Extension of Dominated Convergence Theorem). Assume 1°, 2° and 3°.

Let $g_n = |f_n|$ for $n \in \mathbb{N}$ and let $g = |f|$. Since $\lim\limits_{n \to \infty} f_n = f$ a.e. on D implies $\lim\limits_{n \to \infty} |f_n| = |f|$ a.e. on D, we have $\lim\limits_{n \to \infty} g_n = g$ a.e. on D. Also $(g_n : n \in \mathbb{N})$ and g are all μ-integrable on D and $\lim\limits_{n \to \infty} \int_D g_n \, d\mu = \int_D g \, d\mu$. Thus conditions 1° and 2° in Prob. 9.22 are satisfied.

Let $E \subset D$ and $E \in \mathfrak{A}$. For any extended real-valued $\mathfrak{A}$-measurable function h on D let us define an extended real-valued $\mathfrak{A}$-measurable function $\widetilde{h}$ on D by setting

$$\widetilde{h}(x) = \begin{cases} |h(x)| & \text{for } x \in E, \\ 0 & \text{for } x \in D \setminus E. \end{cases}$$

Consider $(\widetilde{f}_n : n \in \mathbb{N})$ and $\widetilde{f}$. Condition 1° implies that $\lim\limits_{n \to \infty} \widetilde{f}_n = \widetilde{f}$ a.e. on D. We also have $|\widetilde{f}_n| \le |f_n| = g_n$ for $n \in \mathbb{N}$. Thus the sequence $(\widetilde{f}_n : n \in \mathbb{N})$ and $\widetilde{f}$ satisfy conditions 1° and 3° of Prob. 9.22. Thus by the conclusion of Prob. 9.22 we have

(1) $$\lim_{n \to \infty} \int_D \widetilde{f}_n \, d\mu = \int_D \widetilde{f} \, d\mu.$$

Now we have

$$\int_D \widetilde{f}_n \, d\mu = \int_E \widetilde{f}_n \, d\mu + \int_{D \setminus E} \widetilde{f}_n \, d\mu, = \int_E |f_n| \, d\mu + \int_{D \setminus E} 0 \, d\mu = \int_E |f_n| \, d\mu$$

and similarly

$$\int_D \widetilde{f} \, d\mu = \int_E |f| \, d\mu.$$

Substituting the last two equalities in (1) we obtain

(2) $$\lim_{n \to \infty} \int_E |f_n| \, d\mu = \int_E |f| \, d\mu.$$

Then according to (a) of Prob. 9.23, (2) implies

(3) $$\lim_{n \to \infty} \int_E f_n \, d\mu = \int_E f \, d\mu.$$

This completes the proof of (a).

2. Let us show that if condition 3° is replaced by the condition that $\lim_{n\to\infty} \int_D f_n \, d\mu = \int_D f \, d\mu$, then (a) does not hold.

Consider the measure space $(\mathbb{R}, \mathfrak{M}_L, \mu_L)$. Let $f = 0$ on $\mathbb{R}$ and let $(f_n : n \in \mathbb{N})$ be a sequence of real-valued $\mathfrak{M}_L$-measurable functions on $\mathbb{R}$ defined by setting

$$f_n(x) = \begin{cases} \frac{1}{n} & \text{for } x \in [0, n), \\ -\frac{1}{n} & \text{for } x \in (-n, 0), \\ 0 & \text{for } x \in (-n, n)^c. \end{cases}$$

We have $\lim_{n\to\infty} f_n(x) = 0 = f(x)$ for every $x \in \mathbb{R}$. We have $\int_{\mathbb{R}} f_n \, d\mu_L = 0 = \int_{\mathbb{R}} f \, d\mu_L$ for every $n \in \mathbb{N}$. Thus we have $\lim_{n\to\infty} \int_{\mathbb{R}} f_n \, d\mu_L = \int_{\mathbb{R}} f \, d\mu_L$.

However we have $\int_{[0,\infty)} |f_n| \, d\mu_L = 1$ for every $n \in \mathbb{N}$ and $\int_{[0,\infty)} |f| \, d\mu_L = 0$. Thus we have $\lim_{n\to\infty} \int_{[0,\infty)} |f_n| \, d\mu_L = 1 \neq 0 = \int_{[0,\infty)} |f| \, d\mu_L = 0$. ∎

Prob. 9.25. Let us replace condition 3° in Prob. 9.24 with

4° $\lim_{n\to\infty} \int_D f_n \, d\mu = \int_D f \, d\mu$.

5° There exists a μ-integrable extended real-valued $\mathfrak{A}$-measurable function g on D such that $f_n \geq g$ (or $f_n \leq g$) on D for $n \in \mathbb{N}$.

Show that for every $\mathfrak{A}$-measurable subset E of D, we have $\lim_{n\to\infty} \int_E f_n \, d\mu = \int_E f \, d\mu$.

Proof. 1. Consider the case that $f_n \geq g$ on D for $n \in \mathbb{N}$. Consider $(f_n - g : n \in \mathbb{N})$, a sequence of nonnegative extended real-valued $\mathfrak{A}$-measurable functions on D.

By condition 1° of Prob. 9.24, we have $\lim_{n\to\infty} (f_n - g) = f - g$ a.e. on D.

Since $(f_n : n \in \mathbb{N})$, f and g are all μ-integrable on D, $(f_n - g : n \in \mathbb{N})$ and $f - g$ are all μ-integrable on D.

Condition 4° implies that $\lim_{n\to\infty} \int_D (f_n - g) \, d\mu = \int_D (f - g) \, d\mu$. Then since $f_n - g$ and $f - g$ are nonnegative, we have $\lim_{n\to\infty} \int_D |f_n - g| \, d\mu = \int_D |f - g| \, d\mu$.

Thus we have shown that $(f_n - g : n \in \mathbb{N})$ and $f - g$ satisfy conditions 1°, 2° and 3° of Prob. 9.24. Thus by (a) of Prob. 9.24, for every $E \in \mathfrak{A}$ such that $E \subset D$, we have

(1) $$\lim_{n\to\infty} \int_E |f_n - g| \, d\mu = \int_E |f - g| \, d\mu,$$

and then by the nonnegativity of $f_n - g$ and $f - g$ we have

(2) $$\lim_{n\to\infty} \int_E (f_n - g) \, d\mu = \int_E (f - g) \, d\mu.$$

Then adding $\int_E g \, d\mu$ to both sides of (2), we have

(3) $$\lim_{n\to\infty} \int_E f_n \, d\mu = \int_E f \, d\mu.$$

2. The case that $f_n \leq g$ on D for $n \in \mathbb{N}$ is treated similarly by considering $(g - f_n : n \in \mathbb{N})$ and $g - f$. ∎

Prob. 9.26. Given a measure space $(X, \mathfrak{A}, \mu)$. Let $(f_n : n \in \mathbb{N})$ and f be μ-integrable extended real-valued $\mathfrak{A}$-measurable functions on a set $D \in \mathfrak{A}$.
(a) Show that if $\lim_{n\to\infty} \int_D |f_n - f| \, d\mu = 0$, then
(1) $(f_n : n \in \mathbb{N})$ converges to f on D in measure,
(2) $\lim_{n\to\infty} \int_D |f_n| \, d\mu = \int_D |f| \, d\mu$.
(b) Show that if
$1°$ $\lim_{n\to\infty} f_n = f$ a.e. on D,
$2°$ $\lim_{n\to\infty} \int_D |f_n| \, d\mu = \int_D |f| \, d\mu$,
then $\lim_{n\to\infty} \int_D |f_n - f| \, d\mu = 0$.
(c) Show by constructing a counterexample that if f is only μ semi-integrable on D, then (b) does not hold.

Proof. 1. Let us prove (a). Thus assume that we have

$$(0) \qquad\qquad\qquad\qquad \lim_{n\to\infty} \int_D |f_n - f| \, d\mu = 0.$$

Let us prove (1). For an arbitrary $\varepsilon > 0$, let us define

$$A_n = \left\{ x \in D : |f_n(x) - f(x)| \geq \varepsilon \right\} \quad \text{for } n \in \mathbb{N}.$$

We have

$$(3) \qquad\qquad\qquad \int_D |f_n - f| \, d\mu \geq \int_{A_n} |f_n - f| \, d\mu \geq \varepsilon \mu(A_n).$$

Then (0) and (3) imply that for every $\eta > 0$ there exists $N \in \mathbb{N}$ such that

$$\varepsilon \mu(A_n) \leq \int_D |f_n - f| \, d\mu < \eta \quad \text{for } n \geq N.$$

This shows that $\lim_{n\to\infty} \varepsilon \mu(A_n) = 0$ and then $\lim_{n\to\infty} \mu(A_n) = 0$. This proves (1).
Let us prove (2). Observe that we have

$$\int_D |f_n| \, d\mu - \int_D |f| \, d\mu = \int_D \left\{ |f_n| - |f| \right\} d\mu,$$

and then

$$\left| \int_D |f_n| \, d\mu - \int_D |f| \, d\mu \right| = \left| \int_D \left\{ |f_n| - |f| \right\} d\mu \right| \leq \int_D \left| |f_n| - |f| \right| d\mu$$
$$\leq \int_D |f_n - f| \, d\mu.$$

Then (0) implies that

$$\lim_{n\to\infty} \left| \int_D |f_n| \, d\mu - \int_D |f| \, d\mu \right| = 0,$$

which is equivalent to

$$\lim_{n\to\infty} \left\{ \int_D |f_n| \, d\mu - \int_D |f| \, d\mu \right\} = 0, \text{ that is, } \lim_{n\to\infty} \int_D |f_n| \, d\mu = \int_D |f| \, d\mu.$$

This proves (2).

2. Let us prove (b). Assume 1° and 2°. Consider the sequence $(|f_n - f| : n \in \mathbb{N})$ and the identically vanishing function 0 on D. Let us define a sequence $(g_n : n \in \mathbb{N})$ of functions and a function g on D by setting

$$g_n = |f_n| + |f| \qquad \text{for } n \in \mathbb{N},$$
$$g = |f| + |f| = 2|f|.$$

Let us verify that $(|f_n - f| : n \in \mathbb{N})$, 0, $(g_n : n \in \mathbb{N})$, and g satisfy conditions 1°, 2°, and 3° of Prob. 9.22.

Now 1° of (b) implies that $\lim_{n\to\infty} |f_n - f| = 0$ a.e. on D.

Since $(f_n : n \in \mathbb{N})$ and f are all μ-integrable on D, $(g_n : n \in \mathbb{N})$ and g are all μ-integrable on D. Condition 2° of (b) implies that $\lim_{n\to\infty} \int_D g_n \, d\mu = \int_D g \, d\mu$.

For every $n \in \mathbb{N}$, we have $|f_n - f| \leq |f_n| + |f| = g_n$ on D.

This verifies that $(|f_n - f| : n \in \mathbb{N})$, 0, $(g_n : n \in \mathbb{N})$, and g satisfy conditions 1°, 2°, and 3° of Prob. 9.22. Thus by the conclusion of Prob. 9.22 we have $\lim_{n\to\infty} \int_D |f_n - f| \, d\mu = 0$.

This completes the proof of (b).

3. Let us construct an example to show that if f is only μ semi-integrable on D then (b) does not hold.

Consider the measure space $(\mathbb{R}, \mathfrak{M}_L, \mu_L)$. Let $D = [0, \infty)$. Let f be a function on D defined by setting $f(x) = x$ for $x \in [0, \infty)$. Let $(f_n : n \in \mathbb{N})$ be a sequence of functions on D defined by setting

$$f_n(x) = \begin{cases} x & \text{for } x \in [0, n), \\ 0 & \text{for } x \in [n, \infty). \end{cases}$$

Then f_n is μ_L-integrable on $[0, \infty)$ with $\int_{[0,\infty)} f_n \, d\mu_L = \frac{1}{2}n^2$. On the other hand, we have $\int_{[0,\infty)} f \, d\mu_L = \infty$ and thus f is μ_L semi-integrable but not μ_L-integrable on $[0, \infty)$.

We have $\lim_{n\to\infty} f_n = f$ on $[0, \infty)$ and

$$\lim_{n\to\infty} \int_{[0,\infty)} |f_n| \, d\mu_L = \lim_{n\to\infty} \frac{1}{2}n^2 = \infty = \int_{[0,\infty)} |f| \, d\mu_L.$$

This shows that conditions 1° and 2° in (b) are satisfied. But we have

$$\int_{[0,\infty)} |f_n - f| \, d\mu_L = \int_{[n,\infty)} x \, d\mu_L = \infty$$

for every $n \in \mathbb{N}$ and then

$$\lim_{n\to\infty} \int_{[0,\infty)} |f_n - f| \, d\mu_L = \infty \neq 0.$$

This completes the proof. ∎

Prob. 9.27. (Dominated Convergence and Convergence in Measure) Given a measure space $(X, \mathfrak{A}, \mu)$. Let $(f_n : n \in \mathbb{N})$ and f be extended real-valued $\mathfrak{A}$-measurable functions on a set $D \in \mathfrak{A}$ and assume that f is real-valued a.e. on D. According to Theorem 6.22 (Lebesgue), if $(f_n : n \in \mathbb{N})$ converges to f a.e. on D and if $\mu(D) < \infty$, then $(f_n : n \in \mathbb{N})$ converges to f on D in measure. Now consider the case $\mu(D) \in [0, \infty]$:
Let $(f_n : n \in \mathbb{N})$ and f be extended real-valued $\mathfrak{A}$-measurable functions on a set $D \in \mathfrak{A}$ and assume that f is real-valued a.e. on D. Suppose
1° $(f_n : n \in \mathbb{N})$ converges to f a.e. on D.
2° There exists a μ-integrable nonnegative extended real-valued $\mathfrak{A}$-measurable function g on D such that $|f_n| \leq g$ on D for every $n \in \mathbb{N}$.
Show that $(f_n : n \in \mathbb{N})$ converges to f on D in measure.

Proof. According to Theorem 9.20 (Dominated Convergence Theorem), conditions 1° and 2° imply that

$$(1) \qquad \lim_{n \to \infty} \int_D |f_n - f|\, d\mu = 0.$$

Let us show that (1) implies that $(f_n : n \in \mathbb{N})$ converges to f on D in measure. Let $\varepsilon > 0$ and let

$$A_n = \{x \in D : |f_n(x) - f(x)| \geq \varepsilon\} \quad \text{for } n \in \mathbb{N}.$$

We have

$$(2) \qquad \int_D |f_n - f|\, d\mu \geq \int_{A_n} |f_n - f|\, d\mu \geq \varepsilon \mu(A_n).$$

By (1), for every $\eta > 0$ there exists $N \in \mathbb{N}$ such that

$$(3) \qquad \int_D |f_n - f|\, d\mu < \eta \quad \text{for } n \geq N.$$

Then by (2) and (3), for every $\eta > 0$ there exists $N \in \mathbb{N}$ such that

$$\varepsilon \mu(A_n) \leq \int_D |f_n - f|\, d\mu < \eta \quad \text{for } n \geq N.$$

This shows that $\lim_{n \to \infty} \varepsilon \mu(A_n) = 0$, which implies that $\lim_{n \to \infty} \mu(A_n) = 0$. This proves that $(f_n : n \in \mathbb{N})$ converges to f on D in measure. ∎

Prob. 9.28. Given a measure space $(X, \mathfrak{A}, \mu)$. Prove the following statements (a) and (b);
(a) There exists a μ-integrable extended real-valued $\mathfrak{A}$-measurable function f on X with
condition $\mu\{x \in X : f(x) = 0\} = 0$ if and only if $(X, \mathfrak{A}, \mu)$ is a σ-finite measure space.
(b) If f is a μ-integrable extended real-valued $\mathfrak{A}$-measurable function on X, then the set
$\{x \in X : f(x) \neq 0\}$ is a σ-finite set, that is, the union of countably many members of $\mathfrak{A}$,
each with a finite measure.

Proof. 1. Let us prove (a). Suppose there exists a μ-integrable extended real-valued
$\mathfrak{A}$-measurable function f on X with condition $\mu\{x \in X : f(x) = 0\} = 0$. Let us define

$$E_0 = \left\{x \in X : |f(x)| = 0\right\}.$$

$$E_k = \left\{x \in X : |f(x)| \in \left(\tfrac{1}{k+1}, \tfrac{1}{k}\right]\right\} \quad \text{for } k \in \mathbb{N}.$$

$$E_\infty = \left\{x \in X : |f(x)| \in \left(1, \infty\right]\right\}.$$

Then $\{E_0, E_k, k \in \mathbb{N}, E_\infty\}$ is a countable disjoint collection in $\mathfrak{A}$ and moreover we have

$$X = E_0 \cup \left(\bigcup_{k \in \mathbb{N}} E_k\right) \cup E_\infty.$$

To show that $(X, \mathfrak{A}, \mu)$ is a σ-finite measure space, it remains to verify that $\mu(E_0) < \infty$,
$\mu(E_k) < \infty$ for $k \in \mathbb{N}$ and $\mu(E_\infty) < \infty$.
Now the condition $\mu\{x \in X : f(x) = 0\} = 0$ implies that $\mu(E_0) = 0$. For any $k \in \mathbb{N}$,
we have

$$\infty > \int_X |f| \, d\mu \geq \int_{E_k} |f| \, d\mu \geq \frac{1}{k+1} \mu(E_k)$$

and this implies that $\mu(E_k) < \infty$. Finally we have

$$\infty > \int_X |f| \, d\mu \geq \int_{E_\infty} |f| \, d\mu \geq 1 \cdot \mu(E_\infty)$$

and this implies that $\mu(E_\infty) < \infty$. This completes the proof that $(X, \mathfrak{A}, \mu)$ is a σ-finite
measure space.

Conversely suppose $(X, \mathfrak{A}, \mu)$ is a σ-finite measure space. Then there exists a countable
disjoint collection $\{A_n : n \in \mathbb{N}\}$ in $\mathfrak{A}$ such that $\bigcup_{n \in \mathbb{N}} A_n = X$ and $0 < \mu(A_n) < \infty$ for
$n \in \mathbb{N}$. Let us define a function f on X by setting

$$f(x) = \frac{1}{2^n} \frac{1}{\mu(A_n)} \quad \text{for } x \in A_n.$$

Then f satisfies the condition $\mu\{x \in X : f(x) = 0\} = 0$ and moreover we have

$$\int_X f \, d\mu = \sum_{n \in \mathbb{N}} \frac{1}{2^n} \frac{1}{\mu(A_n)} \mu(A_n) = 1 < \infty,$$

that is, f is μ-integrable on X. This shows the existence of a μ-integrable extended real-
valued $\mathfrak{A}$-measurable function f on X with condition $\mu\{x \in X : f(x) = 0\} = 0$.

2. Let us prove (b). Let f be a μ-integrable extended real-valued $\mathfrak{A}$-measurable function on X. For each $k \in \mathbb{N}$, let $E_k = \{x \in X : |f(x)| > \frac{1}{k}\}$. Then we have

$$\{x \in X : f(x) \neq 0\} = \bigcup_{k \in \mathbb{N}} E_k.$$

Since f is μ-integrable on X, we have

$$\infty > \int_X |f| \, d\mu \geq \int_{E_k} |f| \, d\mu \geq \frac{1}{k}\mu(E_k).$$

This implies that $\mu(E_k) < \infty$. This shows that the set $\{x \in X : f(x) \neq 0\}$ is the union of countably many members of $\mathfrak{A}$, each with a finite measure. ∎

Prob. 9.29. Given a measure space $(X, \mathfrak{A}, \mu)$. Let $(f_n : n \in \mathbb{N})$ be a sequence of extended real-valued $\mathfrak{A}$-measurable functions on a set $D \in \mathfrak{A}$. Let us investigate convergence of the series $\sum_{n\in\mathbb{N}} f_n$ and the μ-integrability of the sum on D. Prove the following proposition:

Proposition. *If*

1° $\sum_{n\in\mathbb{N}} \int_D |f_n| \, d\mu < \infty$,

then

2° $\sum_{n\in\mathbb{N}} |f_n|$ *converges a.e. on D and hence $\sum_{n\in\mathbb{N}} f_n$ converges a.e. on D,*

3° $\sum_{n\in\mathbb{N}} f_n$ *is μ-integrable on D and $\int_D \{\sum_{n\in\mathbb{N}} f_n\} \, d\mu = \sum_{n\in\mathbb{N}} \int_D f_n \, d\mu$.*

Proof. 1. Observe first that condition 1° implies that $\int_D |f_n| \, d\mu < \infty$ and thus f_n is μ-integrable on D for every $n \in \mathbb{N}$. Condition 1° also implies

$$(1) \qquad \infty > \sum_{n\in\mathbb{N}} \int_D |f_n| \, d\mu = \lim_{n\to\infty} \sum_{k=1}^n \int_D |f_k| \, d\mu = \lim_{n\to\infty} \int_D \sum_{k=1}^n |f_k| \, d\mu.$$

Now $\left(\sum_{k=1}^n |f_k| : n \in \mathbb{N}\right)$ is an increasing sequence of nonnegative extended real-valued $\mathfrak{A}$-measurable functions on D. Then by Theorem 8.5 (Monotone Convergence Theorem) we have

$$(2) \qquad \lim_{n\to\infty} \int_D \sum_{k=1}^n |f_k| \, d\mu = \int_D \lim_{n\to\infty} \sum_{k=1}^n |f_k| \, d\mu = \int_D \sum_{n\in\mathbb{N}} |f_n| \, d\mu.$$

By (1) and (2), we have

$$(3) \qquad \int_D \sum_{n\in\mathbb{N}} |f_n| \, d\mu < \infty.$$

This shows that $\sum_{n\in\mathbb{N}} |f_n|$ is μ-integrable on D. This then implies that $\sum_{n\in\mathbb{N}} |f_n| < \infty$ a.e. on D. Then since absolute convergence of a series implies convergence of the series, the series $\sum_{n\in\mathbb{N}} f_n$ converges a.e. on D. This completes the proof of 2°.

2. Let us prove the μ-integrability of $\sum_{n\in\mathbb{N}} f_n$ on D. Observe that $\sum_{n\in\mathbb{N}} f_n = \lim_{n\to\infty} \sum_{k=1}^n f_k$. Let $g_n = \sum_{k=1}^n f_k$ for $n \in \mathbb{N}$. Then we have

$$\lim_{n\to\infty} g_n = \lim_{n\to\infty} \sum_{k=1}^n f_k = \sum_{n\in\mathbb{N}} f_n.$$

We have

$$|g_n| = \left| \sum_{k=1}^n f_k \right| \le \sum_{k=1}^n |f_k| \le \sum_{n\in\mathbb{N}} |f_n|.$$

According to (3), $\sum_{n\in\mathbb{N}} |f_n|$ is μ-integrable on D. Thus by Theorem 9.20 (Dominated Convergence Theorem), $\lim_{n\to\infty} g_n$ is μ-integrable on D and moreover we have

$$\int_D \lim_{n\to\infty} g_n \, d\mu = \lim_{n\to\infty} \int_D g_n \, d\mu.$$

In other words, $\sum_{n \in \mathbb{N}} f_n$ is μ-integrable on D and moreover we have

$$\int_D \sum_{n \in \mathbb{N}} f_n \, d\mu = \lim_{n \to \infty} \int_D \sum_{k=1}^{n} f_k \, d\mu = \lim_{n \to \infty} \sum_{k=1}^{n} \int_D f_k \, d\mu = \sum_{n \in \mathbb{N}} \int_D f_n \, d\mu.$$

This completes the proof of 3°. ∎

Prob. 9.30. With $0 < a < b$, let $(f_n : n \in \mathbb{N})$ be a sequence of functions defined by setting
$$f_n(x) = ae^{-nax} - be^{-nbx} \text{ for } x \in [0, \infty).$$

(a) Show that $\int_{[0,\infty)} f_n \, d\mu_L = 0$ for every $n \in \mathbb{N}$ so that $\sum_{n \in \mathbb{N}} \{ \int_{[0,\infty)} f_n \, d\mu_L \} = 0$.

(b) Compute $\int_{[0,\infty)} |f_n| \, d\mu_L$ for every $n \in \mathbb{N}$.

(c) Show that $\sum_{n \in \mathbb{N}} \int_{[0,\infty)} |f_n| \, d\mu_L = \infty$.

(d) Compute $\sum_{n \in \mathbb{N}} f_n$.

(e) Show that $\sum_{n \in \mathbb{N}} f_n$ does not exist.

(Note that property (c) negates assumption 1° of Prob. 9.29 and renders Prob. 9.29 not applicable here.)

Proof. (a) Let us show that $\int_{[0,\infty)} f_n \, d\mu_L = 0$ for every $n \in \mathbb{N}$. Let us define

(1)
$$\begin{cases} g_n(x) = ae^{-nax} & \text{for } x \in [0, \infty). \\ h_n(x) = be^{-nbx} & \text{for } x \in [0, \infty). \end{cases}$$

To show that g_n is μ_L-integrable on $[0, \infty)$, let us define $(g_n)_k$ for $k \in \mathbb{N}$ by setting

(2)
$$(g_n)_k(x) = \begin{cases} g_n(x) & \text{for } x \in [0, k], \\ 0 & \text{for } x \in (k, \infty). \end{cases}$$

Then $((g_n)_k : k \in \mathbb{N})$ is an increasing sequence of nonnegative real-valued $\mathfrak{M}_L$-measurable functions such that $\lim_{k \to \infty} (g_n)_k = g_n$ on $[0, \infty)$. Thus by Theorem 8.5 (Monotone Convergence Theorem) we have

(3)
$$\int_{[0,\infty)} g_n \, d\mu_L = \lim_{k \to \infty} \int_{[0,\infty)} (g_n)_k \, d\mu_L.$$

Now we have

$$\int_{[0,\infty)} (g_n)_k \, d\mu_L = \int_{[0,k]} ae^{-nax} \, \mu_L(dx) = \int_0^k ae^{-nax} \, dx$$
$$= \left[\frac{1}{n} e^{-nax} \right]_k^0 = \frac{1}{n} \left\{ 1 - e^{-nak} \right\}.$$

Substituting this in (3) we have

$$\int_{[0,\infty)} g_n \, d\mu_L = \lim_{k \to \infty} \frac{1}{n} \left\{ 1 - e^{-nak} \right\} = \frac{1}{n}.$$

Similarly we have $\int_{[0,\infty)} g_n \, d\mu_L = \frac{1}{n}$. Thus both g_n and h_n are μ_L-integrable on $[0, \infty)$. This implies that $f_n = g_n - h_n$ is μ_L-integrable on $[0, \infty)$ and moreover we have

$$\int_D f_n \, d\mu_L = \int_D g_n \, d\mu_L - \int_D h_n \, d\mu_L = \frac{1}{n} - \frac{1}{n} = 0.$$

(b) Let us evaluate $\int_{[0,\infty)} |f_n| \, d\mu_L$ for every $n \in \mathbb{N}$. We have $f_n = g_n - h_n$ as defined by (1). Now g_n and h_n are strictly decreasing continuous nonnegative real-valued functions on

$[0, \infty)$. We have $g_n(0) = a < b = h_n(0)$. However h_n descend faster than g_n and thus the graphs of g_n and h_n have one point of intersection. The x-coordinate of this point is found by solving the equation $g_n(x) = h_n(x)$ for $x \in [0, \infty)$. Starting with $ae^{-nax} = be^{-nbx}$, we have

$$\frac{e^{-nax}}{e^{-nbx}} = \frac{b}{a}; \quad e^{n(b-a)x} = \frac{b}{a}; \quad x = \ln\frac{b}{a} \cdot \frac{1}{n(b-a)}.$$

Let us write

$$\xi_n = \ln\frac{b}{a} \cdot \frac{1}{n(b-a)}.$$

Then we have

$$\begin{cases} h_n(x) \geq g_n(x) & \text{for } x \in [0, \xi_n], \\ h_n(x) \leq g_n(x) & \text{for } x \in [\xi_n, \infty). \end{cases}$$

This implies then

$$|f_n| = \begin{cases} h_n - g_n & \text{on } [0, \xi_n], \\ g_n - h_n & \text{on } [\xi_n, \infty). \end{cases}$$

Then we have

(4) $$\int_{[0,\infty)} |f_n| \, d\mu_L = \int_{[0,\xi_n]} \{h_n - g_n\} \, d\mu_L + \int_{[\xi_n,\infty)} \{g_n - h_n\} \, d\mu_L.$$

For the first integral on the right side of (4), we have

(5) $$\int_{[0,\xi_n]} \{h_n - g_n\} \, d\mu_L = \int_{[0,\xi_n]} \{be^{-nbx} - ae^{-nax}\} \, \mu_L(dx)$$

$$= \int_0^{\xi_n} \{be^{-nbx} - ae^{-nax}\} \, dx = \frac{1}{n}\left[e^{-nbx} - e^{-nax}\right]_{\xi_n}^0$$

$$= -\frac{1}{n}\{e^{-nb\xi_n} - e^{-na\xi_n}\} = \frac{1}{n}\{e^{-na\xi_n} - e^{-nb\xi_n}\}.$$

To evaluate the second integral on the right side of (4), we define a sequence of functions $\big((g_n - h_n)_k : k \in \mathbb{N}\big)$ by setting

$$(g_n - h_n)_k(x) = \begin{cases} (g_n - h_n)(x) & \text{for } x \in [\xi_n, \xi_n + k], \\ 0 & \text{for } x \in (\xi_n + k, \infty). \end{cases}$$

Then $\big((g_n - h_n)_k : k \in \mathbb{N}\big)$ is an increasing sequence of nonnegative real-valued $\mathfrak{M}_L$-measurable functions on $[\xi_n, \infty)$ and $\lim_{k\to\infty} (g_n - h_n)_k = g_n - h_n$. Thus by Theorem 8.5 (Monotone Convergence Theorem) we have

$$\int_{[\xi_n,\infty)} \{g_n - h_n\} \, d\mu_L = \lim_{k\to\infty} \int_{[\xi_n,\infty)} (g_n - h_n)_k \, d\mu_L.$$

Now

$$\int_{[\xi_n,\infty)} (g_n - h_n)_k \, d\mu_L = \int_{[\xi_n,\xi_n+k]} \left\{ ae^{-nax} - be^{-nbx} \right\} \mu_L(dx)$$

$$= \int_{\xi_n}^{\xi_n+k} \left\{ ae^{-nax} - be^{-nbx} \right\} dx = \frac{1}{n} \left[e^{-nax} - e^{-nbx} \right]_{\xi_n+k}^{\xi_n}$$

$$= \frac{1}{n} \left\{ \left\{ e^{-na\xi_n} - e^{-nb\xi_n} \right\} - \left\{ e^{-na(\xi_n+k)} - e^{-nb(\xi_n+k)} \right\} \right\},$$

and then

$$\lim_{k\to\infty} \int_{[\xi_n,\infty)} (g_n - h_n)_k \, d\mu_L = \frac{1}{n} \left\{ e^{-na\xi_n} - e^{-nb\xi_n} \right\},$$

and thus

(6) $$\int_{[\xi_n,\infty)} \left\{ g_n - h_n \right\} d\mu_L = \frac{1}{n} \left\{ e^{-na\xi_n} - e^{-nb\xi_n} \right\}.$$

Observe that (5) and (6) show that the two integrals on the right side of (4) are actually equal. Now substituting (5) and (6) in (4), we have

(7) $$\int_{[0,\infty)} |f_n| \, d\mu_L = \frac{2}{n} \left\{ e^{-na\xi_n} - e^{-nb\xi_n} \right\} = \frac{2}{n} \left\{ e^{-\frac{a \ln \frac{b}{a}}{b-a}} - e^{-\frac{b \ln \frac{b}{a}}{b-a}} \right\}.$$

(c) By (7) we have

$$\sum_{n\in\mathbb{N}} \int_{[0,\infty)} |f_n| \, d\mu_L = 2 \left\{ e^{-\frac{a \ln \frac{b}{a}}{b-a}} - e^{-\frac{b \ln \frac{b}{a}}{b-a}} \right\} \sum_{n\in\mathbb{N}} \frac{1}{n} = \infty.$$

(d) Let us compute $\sum_{n\in\mathbb{N}} f_n$. Now we have

$$\sum_{n\in\mathbb{N}} f_n(x) = \sum_{n\in\mathbb{N}} \left\{ ae^{-nax} - be^{-nbx} \right\} \quad \text{for } x \in [0,\infty).$$

For $x \in [0,\infty)$, $\sum_{n\in\mathbb{N}} ae^{-nax}$ is a geometric series with ratio $e^{-ax} \in (0,1)$ and is thus a convergent geometric series with sum

$$\sum_{n\in\mathbb{N}} ae^{-nax} = \frac{ae^{-ax}}{1 - e^{-ax}}.$$

Similarly for $\sum_{n\in\mathbb{N}} be^{-nbx}$ and thus we have

(8) $$\sum_{n\in\mathbb{N}} f_n(x) = \frac{ae^{-ax}}{1 - e^{-ax}} - \frac{be^{-bx}}{1 - e^{-bx}} \quad \text{for } x \in [0,\infty).$$

(e) Let us show that $\int_{[0,\infty)} \left\{ \sum_{n\in\mathbb{N}} f_n \right\} d\mu_L$ does not exist. Let us write

$$\left\{ \sum_{n\in\mathbb{N}} f_n \right\} = \left\{ \sum_{n\in\mathbb{N}} f_n \right\}^+ - \left\{ \sum_{n\in\mathbb{N}} f_n \right\}^-$$

where $\{\sum_{n\in\mathbb{N}} f_n\}^+$ and $\{\sum_{n\in\mathbb{N}} f_n\}^-$ are the positive part and the negative part of $\{\sum_{n\in\mathbb{N}} f_n\}$ as in Definition 9.1. By Observation 9.2, the integral $\int_{[0,\infty)}\{\sum_{n\in\mathbb{N}} f_n\}\,d\mu_L$ exists if and only if at least one of the two integrals $\int_{[0,\infty)}\{\sum_{n\in\mathbb{N}} f_n\}^+\,d\mu_L$ and $\int_{[0,\infty)}\{\sum_{n\in\mathbb{N}} f_n\}^-\,d\mu_L$ is finite. Therefore to show that $\int_{[0,\infty)}\{\sum_{n\in\mathbb{N}} f_n\}\,d\mu_L$ does not exist we show that

(9)
$$\int_{[0,\infty)}\left\{\sum_{n\in\mathbb{N}} f_n\right\}^+ d\mu_L = \infty \quad \text{and} \quad \int_{[0,\infty)}\left\{\sum_{n\in\mathbb{N}} f_n\right\}^- d\mu_L = \infty.$$

Now by (8) we have

$$\begin{cases} \left\{\displaystyle\sum_{n\in\mathbb{N}} f_n\right\}^+ (x) = \dfrac{ae^{-ax}}{1-e^{-ax}} & \text{for } x \in [0,\infty). \\[4mm] \left\{\displaystyle\sum_{n\in\mathbb{N}} f_n\right\}^- (x) = \dfrac{be^{-bx}}{1-e^{-bx}} & \text{for } x \in [0,\infty). \end{cases}$$

Thus it remains to show

(10)
$$\int_{[0,\infty)} \frac{ae^{-ax}}{1-e^{-ax}}\,\mu_L(dx) = \infty \quad \text{and} \quad \int_{[0,\infty)} \frac{be^{-bx}}{1-e^{-bx}}\,\mu_L(dx) = \infty.$$

Let us prove the former. Applying Theorem 8.5 (Monotone Convergence Theorem), we have

$$\int_{[0,\infty)} \frac{ae^{-ax}}{1-e^{-ax}}\,\mu_L(dx) = \lim_{k\to\infty} \int_{[0,k]} \frac{ae^{-ax}}{1-e^{-ax}}\,\mu_L(dx)$$
$$= \lim_{k\to\infty} \int_0^k \frac{ae^{-ax}}{1-e^{-ax}}\,dx = \lim_{k\to\infty} \left[\ln\left(1-e^{-ax}\right)\right]_0^k.$$

Now

$$\left[\ln\left(1-e^{-ax}\right)\right]_0^k = \ln\left(1-e^{-ax}\right) - \ln 0 = \ln\left(1-e^{-ax}\right) + \infty$$

and

$$\lim_{k\to\infty}\left[\ln\left(1-e^{-ax}\right)\right]_0^k = \lim_{k\to\infty}\ln\left(1-e^{-ax}\right) + \infty = \infty.$$

This proves the first equality in (10). The second equality is proved by the same argument. ∎

Prob. 9.31. If f and g are two μ-integrable extended real-valued $\mathfrak{A}$-measurable functions on set $D \in \mathfrak{A}$ in a measure space $(X, \mathfrak{A}, \mu)$, then the product fg may not be μ-integrable, and in fact it may not even be μ semi-integrable, on D. To show this, construct the following examples.

(a) Construct a μ-integrable function f such that f^2 is μ semi-integrable but not μ-integrable.

(b) Construct two μ-integrable functions f and g such that fg is not μ semi-integrable.

Proof. **(a)** Consider the measure space $(\mathbb{R}, \mathfrak{M}_L, \mu_L)$. Let $\{I_n : n \in \mathbb{N}\}$ be a disjoint collection of intervals in $\mathbb{R}$ with $\mu_L(I_n) = \frac{1}{n^3}$ for $n \in \mathbb{N}$. Let $D = \bigcup_{n \in \mathbb{N}} I_n \in \mathfrak{M}_L$ and let f be a function on D defined by setting

$$f(x) = n \quad \text{for } x \in I_n \text{ for } n \in \mathbb{N}.$$

Then we have

$$\int_D f \, d\mu_L = \sum_{n \in \mathbb{N}} n \mu_L(I_n) = \sum_{n \in \mathbb{N}} n \frac{1}{n^3} = \sum_{n \in \mathbb{N}} \frac{1}{n^2} < \infty.$$

This shows that f is μ_L-integrable on D. On the other hand we have

$$\int_D f^2 \, d\mu_L = \sum_{n \in \mathbb{N}} n^2 \mu_L(I_n) = \sum_{n \in \mathbb{N}} n^2 \frac{1}{n^3} = \sum_{n \in \mathbb{N}} \frac{1}{n} = \infty.$$

This shows that f^2 is μ_L semi-integrable but not μ_L-integrable on D.

(b) Consider the measure space $(\mathbb{R}, \mathfrak{M}_L, \mu_L)$. Let $\{I_n, J_n : n \in \mathbb{N}\}$ be a disjoint collection of intervals in $\mathbb{R}$ with $\mu_L(I_n) = \mu_L(J_n) = \frac{1}{n^3}$ for $n \in \mathbb{N}$. Let $D = \left(\bigcup_{n \in \mathbb{N}} I_n\right) \cup \left(\bigcup_{n \in \mathbb{N}} J_n\right)$. Define functions f and g on D by setting

$$f(x) = \begin{cases} n & \text{for } x \in I_n \text{ for } n \in \mathbb{N}, \\ n & \text{for } x \in I_n \text{ for } n \in \mathbb{N}, \end{cases}$$

and

$$g(x) = \begin{cases} n & \text{for } x \in I_n \text{ for } n \in \mathbb{N}, \\ -n & \text{for } x \in I_n \text{ for } n \in \mathbb{N}. \end{cases}$$

Then we have

$$\int_D f \, d\mu_L = \sum_{n \in \mathbb{N}} n \mu_L(I_n) + \sum_{n \in \mathbb{N}} n \mu_L(J_n) = 2 \sum_{n \in \mathbb{N}} n \frac{1}{n^3} = 2 \sum_{n \in \mathbb{N}} \frac{1}{n^2} < \infty.$$

This shows that f is μ_L-integrable on D.

According to Observation 9.2, g is μ_L-integrable if and only if $|g|$ is μ_L-integrable. But $|g| = f$ and f is μ_L-integrable. Thus g is μ_L-integrable.

Let us show that fg is not μ_L semi-integrable on D, that is, $\int_D fg \, d\mu_L$ does not exist. To show that $\int_D fg \, d\mu_L$ does not exist, we show that we have both $\int_D (fg)^+ \, d\mu_L = \infty$ and $\int_D (fg)^- \, d\mu_L = \infty$. Now we have

$$\int_D (fg)^+ \, d\mu_L = \int_{\bigcup_{n \in \mathbb{N}} I_n} |fg| \, d\mu_L = \sum_{n \in \mathbb{N}} n^2 \mu_L(I_n) = \sum_{n \in \mathbb{N}} n^2 \frac{1}{n^3} = \sum_{n \in \mathbb{N}} \frac{1}{n} = \infty,$$

and

$$\int_D (fg)^- d\mu_L = \int_{\bigcup_{n\in\mathbb{N}} J_n} |fg| d\mu_L = \sum_{n\in\mathbb{N}} n^2 \mu_L(J_n) = \sum_{n\in\mathbb{N}} n^2 \frac{1}{n^3} = \sum_{n\in\mathbb{N}} \frac{1}{n} = \infty. \quad \blacksquare$$

Prob. 9.32. Let f and g be two extended real-valued $\mathfrak{A}$-measurable functions on set $D \in \mathfrak{A}$ in a measure space $(X, \mathfrak{A}, \mu)$. Show that if f is bounded on D and g is μ-integrable on D then fg is μ-integrable on D.

Proof. If f is bounded on D then there exists $M > 0$ such that $|f| \le M$ on D. Then $|fg| \le M|g|$ so that

$$\int_D |fg| d\mu \le \int_D M|g| d\mu = M \int_D |g| d\mu < \infty$$

since $\int_D |g| d\mu < \infty$. This proves the μ-integrability of $|fg|$, which is equivalent to the μ-integrability of fg. $\quad \blacksquare$

Prob. 9.33. Prove the following proposition:

Proposition. Given a measure space $(X, \mathfrak{A}, \mu)$. Let $D \in \mathfrak{A}$ and $\mu(D) < \infty$. Suppose f is an extended real-valued $\mathfrak{A}$-measurable function on a set $D \in \mathfrak{A}$ such that $\int_D fg \, d\mu$ exists in $\overline{\mathbb{R}}$ for every μ-integrable extended real-valued $\mathfrak{A}$-measurable function g on D. Then f is bounded a.e. on D, that is, there exists $M > 0$ such that $\mu\{x \in D : |f(x)| \geq M\} = 0$, and moreover f is μ-integrable on D.

Proof. (Observe that if f is bounded a.e. on D then the condition $\mu(D) < \infty$ implies that f is μ-integrable on D.)

We prove the Proposition by contradiction argument. Thus we show that if f is not bounded a.e. on D then there exists a μ-integrable extended real-valued $\mathfrak{A}$-measurable function g on D such that $\int_D fg \, d\mu$ does not exist in $\overline{\mathbb{R}}$.

Suppose f is not bounded a.e. on D. Then f is either not bounded above a.e. on D or not bounded below a.e. on D or both.

1. Consider the case that f is not bounded above a.e. on D. Then f is certainly not bounded above by 1 a.e. on D. Let

$$D_n = \left\{ x \in D : f(x) \in [n, n+1) \right\} \quad \text{for } n \in \mathbb{N}.$$

$$b_n = \mu(D_n) \in [0, \infty).$$

We claim that $b_n > 0$ for infinitely many $n \in \mathbb{N}$. To prove this, assume the contrary, that is, assume that there exists $N \in \mathbb{N}$ such that $b_n = 0$ for $n \geq N$. Now we have

$$\left\{ x \in D : f(x) \geq N \right\} = \bigcup_{n \geq N} D_n$$

and

$$\mu\left\{ x \in D : f(x) \geq N \right\} = \sum_{n \geq N} \mu(D_n) = \sum_{n \geq N} b_n = 0.$$

This shows that f is bounded above a.e. on D, contradicting our assumption. Therefore we have $b_n > 0$ for infinitely many $n \in \mathbb{N}$.

Now $(b_n : n \in \mathbb{N})$ is a sequence of nonnegative numbers and $b_n > 0$ for infinitely many $n \in \mathbb{N}$. Thus we can select a subsequence $(b_{n_k} : k \in \mathbb{N})$ such that $b_{n_k} > 0$ for every $k \in \mathbb{N}$. Select a positive number $a_{n_k} > 0$ such that $a_{n_k} b_{n_k} = \frac{1}{k^2}$ for every $k \in \mathbb{N}$. Then define a function g on D by setting

$$g(x) = \begin{cases} (-1)^k a_{n_k} & \text{for } x \in D_{n_k} \text{ for } k \in \mathbb{N}, \\ 0 & \text{for } x \in D \setminus \bigcup_{k \in \mathbb{N}} D_{n_k}. \end{cases}$$

Then we have

$$\int_D |g| \, d\mu = \sum_{k \in \mathbb{N}} |(-1)^k| a_{n_k} \mu(D_{n_k}) = \sum_{k \in \mathbb{N}} a_{n_k} b_{n_k} = \sum_{k \in \mathbb{N}} \frac{1}{k^2} < \infty.$$

This shows that g is μ-integrable on D.

Next we show that $\int_D fg \, d\mu$ does not exist in $\overline{\mathbb{R}}$. We show this by showing that $\int_D (fg)^+ d\mu = \infty$ and $\int_D (fg)^- d\mu = \infty$. Now we have

$$\int_D (fg)^+ d\mu = \int_{\bigcup_{k \in \mathbb{N}} D_{n_k}} (fg)^+ d\mu = \sum_{k \in \mathbb{N}} \int_{D_{n_k}} (fg)^+ d\mu$$

$$\geq \sum_{\text{even } k} n_k a_{n_k} \mu(D_{n_k}) = \sum_{\text{even } k} a_{n_k} b_{n_k}$$

$$\geq \sum_{\text{even } k} k \frac{1}{k^2} = \sum_{\text{even } k} \frac{1}{k} = \infty,$$

and similarly

$$\int_D (fg)^- d\mu \geq \sum_{\text{odd } k} \frac{1}{k} = \infty.$$

This completes the proof that if f is not bounded above a.e. on D then there exists a μ-integrable extended real-valued $\mathfrak{A}$-measurable function g on D such that $\int_D fg \, d\mu$ does not exist in $\overline{\mathbb{R}}$.

2. Consider the case that f is not bounded below a.e. on D. In this case, $-f$ is not bounded above a.e. on D. Then by our result in **1**, there exists a μ-integrable extended real-valued $\mathfrak{A}$-measurable function g on D such that $\int_D -fg \, d\mu$ does not exist in $\overline{\mathbb{R}}$, that is, $\int_D fg \, d\mu$ does not exist in $\overline{\mathbb{R}}$. ∎

Prob. 9.34. Prove the following proposition:

Proposition. Given a measure space $(X, \mathfrak{A}, \mu)$. Let $D \in \mathfrak{A}$ and $\mu(D) < \infty$. Let f be a bounded real-valued $\mathfrak{A}$-measurable function on D and let $m = \inf_{x \in D} f(x)$ and $M = \sup_{x \in D} f(x)$. Let g be a μ-integrable extended real-valued $\mathfrak{A}$-measurable function on D. Then we have:

(a) $f g$ and $f |g|$ are μ-integrable on D.

(b) If $\int_D |g| \, d\mu = 0$, then $\int_D f |g| \, d\mu = 0$.

(c) If $\int_D |g| \, d\mu > 0$, then there exists $c \in [m, M]$ such that $\int_D f |g| \, d\mu = c \int_D |g| \, d\mu$.

Proof. 1. Let us prove (a). According to Observation 9.2, a function h is μ-integrable if and only if the function $|h|$ is μ-integrable. Thus to show that $f g$ and $f |g|$ are μ-integrable on D, we show that $|f g|$ and $|f |g||$ are μ-integrable on D,

Since f is bounded on D, there exists $B > 0$ such that $|f| \leq B$ on D. Then we have

$$\int_D |f g| \, d\mu = \int_D |f| \, |g| \, d\mu \leq \int_D B |g| \, d\mu = B \int_D |g| \, d\mu < \infty.$$

This shows that $|f g|$ is μ-integrable on D. Similarly we have

$$\int_D |f |g|| \, d\mu = \int_D |f| \, |g| \, d\mu \leq \int_D B |g| \, d\mu = B \int_D |g| \, d\mu < \infty.$$

This shows that $|f |g||$ is μ-integrable on D.

2. Let us prove (b). Suppose $\int_D |g| \, d\mu = 0$. Then $|g| = 0$ a.e. on D. Since f is bounded on D, we have $f |g| = 0$ a.e. on D. This implies that $\int_D f |g| \, d\mu = 0$.

3. Let us prove (c). Suppose $\int_D |g| \, d\mu > 0$. Now we have $m \leq f \leq M$ on D and this implies that $m |g| \leq f |g| \leq M |g|$ on D. Then we have

$$\int_D m |g| \, d\mu \leq \int_D f |g| \, d\mu \leq \int_D M |g| \, d\mu,$$

that is,

$$m \int_D |g| \, d\mu \leq \int_D f |g| \, d\mu \leq M \int_D |g| \, d\mu.$$

Multiplying by $\left(\int_D |g| \, d\mu \right)^{-1} > 0$, we have

$$m \leq \left(\int_D |g| \, d\mu \right)^{-1} \cdot \int_D f |g| \, d\mu \leq M.$$

Then let

$$c = \left(\int_D |g| \, d\mu \right)^{-1} \cdot \int_D f |g| \, d\mu \in [m, M].$$

Then we have

$$\int_D f |g| \, d\mu = c \int_D |g| \, d\mu.$$

This completes the proof. ∎

Prob. 9.35. Let f be an extended real-valued $\mathfrak{M}_L$-measurable function on $[0, \infty)$ such that
1° f is μ_L-integrable on every finite subinterval of $[0, \infty)$,
2° $\lim_{x \to \infty} f(x) = c \in \mathbb{R}$.
Let $a > 0$. Show that $\lim_{a \to \infty} \frac{1}{a} \int_{[0,a]} f \, d\mu_L = c$.
(Observe that if f is a bounded real-valued $\mathfrak{M}_L$-measurable function on $[0, \infty)$ then condition 1° is satisfied.)

Proof. Since $\lim_{x \to \infty} f(x) = c$, for every $\varepsilon > 0$ there exists $M > 0$ such that $|f(x) - c| < \varepsilon$ for $x \geq M$, that is,

$$(1) \qquad\qquad c - \varepsilon < f(x) < c + \varepsilon \quad \text{for } x \geq M.$$

Let $a > 0$. Assume that $a \geq M$. Then we have

$$(2) \qquad\qquad \int_{[0,a]} f \, d\mu_L = \int_{[0,M]} f \, d\mu_L + \int_{[M,a]} f \, d\mu_L.$$

For brevity, let us write

$$(3) \qquad\qquad K = \int_{[0,M]} f \, d\mu_L \in \mathbb{R}.$$

Observe that (1) implies

$$(4) \qquad\qquad (c - \varepsilon)(a - M) \leq \int_{[M,a]} f \, d\mu_L \leq (c + \varepsilon)(a - M).$$

Substituting (3) and (4) in (2), we have

$$(5) \qquad\qquad K + (c - \varepsilon)(a - M) \leq \int_{[0,a]} f \, d\mu_L \leq K + (c + \varepsilon)(a - M).$$

Dividing (5) by our $a \geq M > 0$, we have

$$\frac{K}{a} + (c - \varepsilon)\left(1 - \frac{M}{a}\right) \leq \frac{1}{a} \int_{[0,a]} f \, d\mu_L \leq \frac{K}{a} + (c + \varepsilon)\left(1 + \frac{M}{a}\right).$$

Then letting $a \to \infty$, we have

$$c - \varepsilon \leq \liminf_{a \to \infty} \frac{1}{a} \int_{[0,a]} f \, d\mu_L \leq \limsup_{a \to \infty} \frac{1}{a} \int_{[0,a]} f \, d\mu_L \leq c + \varepsilon.$$

Since this holds for every $\varepsilon > 0$, we have

$$c \leq \liminf_{a \to \infty} \frac{1}{a} \int_{[0,a]} f \, d\mu_L \leq \limsup_{a \to \infty} \frac{1}{a} \int_{[0,a]} f \, d\mu_L \leq c.$$

This shows that

$$\lim_{a \to \infty} \frac{1}{a} \int_{[0,a]} f \, d\mu_L = c. \quad \blacksquare$$

Prob. 9.36. Let f be a μ_L-integrable nonnegative extended real-valued $\mathfrak{M}_L$-measurable function on $\mathbb{R}$. With $c \in \mathbb{R}$ fixed, define a function $g(x) = \sum_{n \in \mathbb{N}} f(x + nc)$ for $x \in \mathbb{R}$. Show that if the function g is μ_L-integrable on $\mathbb{R}$ then $f = 0$ a.e. on $\mathbb{R}$.

Proof. Since f is $\mathfrak{M}_L$-measurable on $\mathbb{R}$, $f(\cdot + nc)$ is $\mathfrak{M}_L$-measurable on $\mathbb{R}$ by Theorem 9.31 (Translation Invariance of the Lebesgue Integral on $\mathbb{R}$). Since f is nonnegative on $\mathbb{R}$, $g(x) = \sum_{n \in \mathbb{N}} f(x + nc)$ exists in $[0, \infty]$ for every $x \in \mathbb{R}$ and g is $\mathfrak{M}_L$-measurable on $\mathbb{R}$.

Assume that g is μ_L-integrable on $\mathbb{R}$. Let us show that this implies that $f = 0$ a.e. on $\mathbb{R}$. Assume the contrary. Then there exists $E \in \mathfrak{M}_L$ such that $\mu_L(E) > 0$ and $f > 0$ on E. This implies that $\int_E f \, d\mu_L > 0$ and then $\int_{\mathbb{R}} f \, d\mu_L \geq \int_E f \, d\mu_L > 0$. Let $a = \int_{\mathbb{R}} f \, d\mu_L > 0$. Then we have

$$\int_{\mathbb{R}} g(x) \, \mu_L(dx) = \int_{\mathbb{R}} \sum_{n \in \mathbb{N}} f(x + nc) \, \mu_L(dx) = \sum_{n \in \mathbb{N}} \int_{\mathbb{R}} f(x + nc) \, \mu_L(dx)$$

$$= \sum_{n \in \mathbb{N}} \int_{\mathbb{R}} f(x) \, \mu_L(dx) \quad \text{by Theorem 9.31}$$

$$= \sum_{n \in \mathbb{N}} a = \infty.$$

This contradicts the assumption that g is μ_L-integrable on $\mathbb{R}$. Therefore we must have $f = 0$ a.e. on $\mathbb{R}$. ∎

Prob. 9.37. Let f be a real-valued uniformly continuous function on $[0, \infty)$. Show that if f is Lebesgue integrable on $[0, \infty)$ then $\lim_{x \to \infty} f(x) = 0$.

Proof. We show that if we do not have $\lim_{x \to \infty} f(x) = 0$ then f cannot be Lebesgue integrable on $[0, \infty)$. By definition, $\lim_{x \to \infty} f(x) = 0$ if and only if for every $\varepsilon > 0$ there exists $M > 0$ such that $|f(x)| < \varepsilon$ for $x > M$.

Thus if we do not have $\lim_{x \to \infty} f(x) = 0$ then there exists $\varepsilon_0 > 0$ such that for any $M > 0$ there exists $x > M$ such that $|f(x)| \geq \varepsilon_0$. Then there exists $x_1 > 1$ such that $|f(x_1)| \geq \varepsilon_0$ and then there exists $x_2 > x_1 + 1$ such that $|f(x_2)| \geq \varepsilon_0$ and then there exists $x_3 > x_2 + 1$ such that $|f(x_3)| \geq \varepsilon_0$ and so on indefinitely. Thus we have a sequence $(x_n : n \in \mathbb{N})$ in $[0, \infty)$ such that

(1) $x_1 > 1; \quad x_{n+1} > x_n + 1; \quad |f(x_n)| \geq \varepsilon_0 \quad \text{for } n \in \mathbb{N}.$

Now since f is uniformly continuous on $[0, \infty)$, for $\frac{\varepsilon_0}{2} > 0$ there exists $\delta > 0$ such that

$$|f(x') - f(x'')| < \frac{\varepsilon_0}{2} \quad \text{whenever } x', x'' \in [0, \infty) \text{ and } |x' - x''| < \delta.$$

Then we have

(2) $|f(x) - f(x_n)| < \frac{\varepsilon_0}{2} \quad \text{for } x \in (x_n - \delta, x_n + \delta).$

Let us select $\delta \in \left(0, \frac{1}{2}\right)$ so that $\{(x_n - \delta, x_n + \delta) : n \in \mathbb{N}\}$ is a disjoint collection of intervals.

Now since $|f(x_n)| \geq \varepsilon_0$ for $n \in \mathbb{N}$, $f(x_n)$ is either positive or negative. If $f(x_n) > 0$ then $f(x_n) \geq \varepsilon_0$ and then (2) implies

$$f(x) > \frac{\varepsilon_0}{2} \quad \text{for } x \in (x_n - \delta, x_n + \delta).$$

If $f(x_n) < 0$ then $f(x_n) \leq -\varepsilon_0$ and then (2) implies

$$f(x) < -\frac{\varepsilon_0}{2} \quad \text{for } x \in (x_n - \delta, x_n + \delta).$$

Now consider the case that $f(x_n) > 0$ for infinitely many $n \in \mathbb{N}$ and the case that $f(x_n) < 0$ for infinitely many $n \in \mathbb{N}$. These are the only possible cases. If $f(x_n) > 0$ for infinitely many $n \in \mathbb{N}$ then there exists a subsequence (n_k) of (n) such that $f(x_{n_k}) > 0$ for every $k \in \mathbb{N}$ and thus

$$f(x) > \frac{\varepsilon_0}{2} \quad \text{for } x \in (x_{n_k} - \delta, x_{n_k} + \delta).$$

Then we have

$$\int_{[0,\infty)} f^+ \, d\mu_L \geq \sum_{k \in \mathbb{N}} \int_{(x_{n_k} - \delta, x_{n_k} + \delta)} f \, d\mu_L \geq \sum_{k \in \mathbb{N}} \frac{\varepsilon_0}{2} 2\delta = \sum_{k \in \mathbb{N}} \varepsilon_0 \delta = \infty.$$

This contradicts the Lebesgue integrability of f on $[0, \infty)$.

On the other hand if $f(x_n) < 0$ for infinitely many $n \in \mathbb{N}$ then there exists a subsequence (n_k) of (n) such that $f(x_{n_k}) > 0$ for every $k \in \mathbb{N}$ and thus

$$f(x) < -\frac{\varepsilon_0}{2} \quad \text{for } x \in (x_{n_k} - \delta, x_{n_k} + \delta).$$

Then we have

$$\int_{[0,\infty)} f^- \, d\mu_L \geq \sum_{k \in \mathbb{N}} \int_{(x_{n_k} - \delta, x_{n_k} + \delta)} |f| \, d\mu_L \geq \sum_{k \in \mathbb{N}} \frac{\varepsilon_0}{2} 2\delta = \sum_{k \in \mathbb{N}} \varepsilon_0 \delta = \infty.$$

This contradicts the Lebesgue integrability of f on $[0, \infty)$. ∎

Prob. 9.38. Given a measure space $(X, \mathfrak{A}, \mu)$. Let $A, B \in \mathfrak{A}$ and let f be an extended real-valued $\mathfrak{A}$-measurable function that is μ-integrable on A and on B. Show that

(a) $\int_A f \, d\mu + \int_B f \, d\mu = \int_{A \setminus B} f \, d\mu + \int_{B \setminus A} f \, d\mu + 2 \int_{A \cap B} f \, d\mu.$

(b) $\int_A f \, d\mu - \int_B f \, d\mu = \int_{A \setminus B} f \, d\mu - \int_{B \setminus A} f \, d\mu.$

Proof. Observe that for any two sets A and B can be expressed as two disjoint unions:

$$A = (A \setminus B) \cup (A \cap B) \quad \text{and} \quad B = (B \setminus A) \cup (B \cap A).$$

Then we have

$$\int_A f \, d\mu_L + \int_B f \, d\mu_L = \left\{ \int_{A \setminus B} f \, d\mu_L + \int_{A \cap B} f \, d\mu_L \right\} + \left\{ \int_{B \setminus A} f \, d\mu_L + \int_{A \cap B} f \, d\mu_L \right\}$$

$$= \int_{A \setminus B} f \, d\mu_L + \int_{B \setminus A} f \, d\mu_L + 2 \int_{A \cap B} f \, d\mu_L,$$

and

$$\int_A f \, d\mu_L - \int_B f \, d\mu_L = \left\{ \int_{A \setminus B} f \, d\mu_L + \int_{A \cap B} f \, d\mu_L \right\} - \left\{ \int_{B \setminus A} f \, d\mu_L + \int_{A \cap B} f \, d\mu_L \right\}$$

$$= \int_{A \setminus B} f \, d\mu_L - \int_{B \setminus A} f \, d\mu_L.$$

This completes the proof. ∎

Prob. 9.39. Consider $(\mathbb{R}, \mathfrak{M}_L, \mu_L)$. Let f be an extended real-valued $\mathfrak{M}_L$-measurable function on $\mathbb{R}$. We say that f is locally μ_L-integrable on $\mathbb{R}$ if f is μ_L-integrable on every bounded set $E \in \mathfrak{M}_L$.

Let f be a locally μ_L-integrable function on $\mathbb{R}$. For $x \in \mathbb{R}$ and $r > 0$, let us define $B_r(x) = \{y \in \mathbb{R} : |y - x| < r\} = (x - r, x + r)$. Define a real-valued function $\widetilde{f}_r$ on $\mathbb{R}$ by setting

$$\widetilde{f}_r(x) = \int_{B_r(x)} f(y)\,\mu_L(dy) \text{ for } x \in \mathbb{R}.$$

(a) Show that if f is locally μ_L-integrable on $\mathbb{R}$ then $\widetilde{f}_r$ is a continuous real-valued function.
(b) Show that if f is μ_L-integrable on $\mathbb{R}$ then $\widetilde{f}_r$ is uniformly continuous on $\mathbb{R}$.

Proof. (a) Assume that f is locally μ_L-integrable on $\mathbb{R}$. To show that $\widetilde{f}_r$ is continuous at every $x_0 \in \mathbb{R}$, we show that for every $\varepsilon > 0$ there exists $\delta > 0$ such that

$$(1) \qquad |\widetilde{f}_r(x) - \widetilde{f}_r(x_0)| < \varepsilon \quad \text{for } x \in \mathbb{R} \text{ such that } |x - x_0| < \delta.$$

Now for any $x \in \mathbb{R}$ we have

$$(2) \qquad \widetilde{f}_r(x) - \widetilde{f}_r(x_0) = \int_{B_r(x)} f(y)\,\mu_L(dy) - \int_{B_r(x_0)} f(y)\,\mu_L(dy).$$

By (b) of Prob. 9.38, we have

$$\int_{B_r(x)} f(y)\,\mu_L)(dy) - \int_{B_r(x_0)} f(y)\,\mu_L(y) = \int_{B_r(x)\setminus B_r(x_0)} f(y)\,\mu_L(dy)$$
$$- \int_{B_r(x_0)\setminus B_r(x)} f(y)\,\mu_L(dy).$$

Substituting this equality in (2), we have

$$\widetilde{f}_r(x) - \widetilde{f}_r(x_0) = \int_{B_r(x)\setminus B_r(x_0)} f(y)\,\mu_L(dy) - \int_{B_r(x_0)\setminus B_r(x)} f(y)\,\mu_L(dy),$$

and then

$$(3) \quad |\widetilde{f}_r(x) - \widetilde{f}_r(x_0)| = \left| \int_{B_r(x)\setminus B_r(x_0)} f(y)\,\mu_L(dy) - \int_{B_r(x_0)\setminus B_r(x)} f(y)\,\mu_L(dy) \right|$$
$$\leq \left| \int_{B_r(x)\setminus B_r(x_0)} f(y)\,\mu_L(dy) \right| + \left| \int_{B_r(x_0)\setminus B_r(x)} f(y)\,\mu_L(dy) \right|$$
$$\leq \int_{B_r(x)\setminus B_r(x_0)} |f(y)|\,\mu_L(dy) + \int_{B_r(x_0)\setminus B_r(x)} |f(y)|\,\mu_L(dy).$$

We may and do confine ourselves to $x \in \mathbb{R}$ such that $|x - x_0| \leq 1$ for instance. Then let $M > 0$ be so large that

$$B_r(x), B_r(x_0) \subset [-M.M] \quad \text{for } x \in \mathbb{R} \text{ such that } |x - x_0| \leq 1.$$

Now $[-M, M]$ is a bounded $\mathfrak{M}_L$-measurable set in $\mathbb{R}$. Then the local μ_L-integrability of f implies that f is μ_L-integrable on $[-M, M]$. By Theorem 9.26, μ_L-integrability of f on $[-M, M]$ implies that for every $\varepsilon > 0$ there exists $\delta > 0$ such that $\int_E |f| \, d\mu_L < \varepsilon$ for every $\mathfrak{M}_L$-measurable set $E \subset [-M, M]$ such that $\mu_L(E) < \delta$.

Now $B_r(x) \setminus B_r(x_0) = (x - r, x + r) \setminus (x_0 - r, x_0 + r)$. Thus if $|x - x_0| < \delta$ then $\mu_L\big(B_r(x) \setminus B_r(x_0)\big) < \delta$ and consequently we have

$$(4) \qquad \int_{B_r(x) \setminus B_r(x_0)} |f(y)| \, \mu_L(dy) < \varepsilon.$$

Similarly $|x - x_0| < \delta$ implies

$$(5) \qquad \int_{B_r(x_0) \setminus B_r(x)} |f(y)| \, \mu_L(dy).$$

Thus by (3), (4) and (5), we have

$$|\widetilde{f}_r(x) - \widetilde{f}_r(x_0)| < 2\varepsilon \quad \text{for } x \in \mathbb{R} \text{ such that } |x - x_0| < \delta.$$

This proves (1).

(b) Let us assume that f is μ_L-integrable on $\mathbb{R}$. Then by Theorem 9.26, for every $\varepsilon > 0$ there exists $\delta > 0$ such that $\int_E |f| \, d\mu_L < \varepsilon$ for every $\mathfrak{M}_L$-measurable set E in $\mathbb{R}$ such that $\mu_L(E) < \delta$. Observe that $\delta > 0$ here is universal in $\mathbb{R}$ whereas $\delta > 0$ in case (a) depended on the bounded set $[-M, M] \subset \mathbb{R}$. From this follows the uniform continuity of $\widetilde{f}_r$ on $\mathbb{R}$ by adjusting the proof of case (a) above. ∎

Prob. 9.40. Consider the measure space $(\mathbb{R}, \mathfrak{M}_L, \mu_L)$. Let f be a μ_L-integrable extended real-valued $\mathfrak{M}_L$-measurable function on $\mathbb{R}$. Show that for every $\varepsilon > 0$ there exists a continuous real-valued function g on $\mathbb{R}$ which vanishes outside a finite closed interval in $\mathbb{R}$ and satisfies the condition that $\int_{\mathbb{R}} |f - g| \, d\mu_L < \varepsilon$.

Proof. For every $n \in \mathbb{N}$ define a function f_n on $\mathbb{R}$ by setting

(1)
$$
f_n(x) = \begin{cases} f(x) & \text{for } x \in \mathbb{R} \text{ such that } |x| \leq n \text{ and } |f(x)| \leq n, \\ 0 & \text{otherwise.} \end{cases}
$$

We have $\lim_{n \to \infty} f_n = f$ on $\mathbb{R}$. This implies that $\lim_{n \to \infty} |f_n - f| = 0$ on $\mathbb{R}$. We also have $|f_n - f| \leq |f_n| + |f| \leq 2|f|$, which is μ_L-integrable on $\mathbb{R}$. Thus by Theorem 9.20 (Dominated Convergence Theorem), we have

(2)
$$
\lim_{n \to \infty} \int_{\mathbb{R}} |f_n - f| \, d\mu_L = \int_{\mathbb{R}} \lim_{n \to \infty} |f_n - f| \, d\mu_L = 0.
$$

Let $\varepsilon > 0$ be arbitrarily given. Then (2) implies that there exists $N \in \mathbb{N}$ such that

(3)
$$
\int_{\mathbb{R}} |f_N - f| \, d\mu_L < \frac{\varepsilon}{2}.
$$

Now f_N is a real-valued $\mathfrak{M}_L$-measurable function on $\mathbb{R}$ such that $|f_N| \leq N$ on $\mathbb{R}$. Then according to Theorem 6.36 there exists a continuous real-valued function g_0 on $[-N, N] \subset \mathbb{R}$ such that

$$
\begin{cases} |g_0| \leq N & \text{on } [-N, N], \\ |f_N - g_0| < \frac{\varepsilon}{16N} & \text{on } [-N, N] \setminus E, \end{cases}
$$

where E is a $\mathfrak{M}_L$-measurable subset of $[-N, N]$ with $\mu_L(E) < \frac{\varepsilon}{16N}$. Then we have

(4)
$$
\int_{[-N,N]} |f_N - g_0| \, d\mu_L = \int_{[-N,N]\setminus E} |f_N - g_0| \, d\mu_L + \int_E |f_N - g_0| \, d\mu_L
$$
$$
< \frac{\varepsilon}{16N} \cdot 2N + 2N \cdot \frac{\varepsilon}{16N} = \frac{\varepsilon}{4}.
$$

To extend our continuous real-valued function g_0 defined on $[-N, N]$ to a continuous real-valued function g on $\mathbb{R}$ which vanishes outside a finite closed interval, define two intervals $I_1 = [-N - \frac{\varepsilon}{8N}, -N]$ and $I_2 = [N, N + \frac{\varepsilon}{8N}]$ and define a function g on $\mathbb{R}$ by setting

$$
g(x) = \begin{cases} g_0(x) & \text{for } x \in [-N, N], \\ 0 & \text{for } x \in \big([-N, N] \cup I_1 \cup I_2\big)^c, \\ \text{linear} & \text{on } I_1 \text{ and on } I_2. \end{cases}
$$

We have

(5)
$$
\int_{I_1} |g| \, d\mu_L \leq N \cdot \frac{\varepsilon}{8N} = \frac{\varepsilon}{8} \quad \text{and} \quad \int_{I_2} |g| \, d\mu_L \leq \frac{\varepsilon}{8}.
$$

Then we have

(6)
$$\int_{\mathbb{R}} |f_N - g|\, d\mu_L = \int_{[-N,N]\cup I_1 \cup I_2} |f_N - g|\, d\mu_L$$

$$= \int_{[-N,N]} |f_N - g_0|\, d\mu_L + \int_{I_1} |g|\, d\mu_L + \int_{I_2} |g|\, d\mu_L$$

$$< \frac{\varepsilon}{4} + \frac{\varepsilon}{8} + \frac{\varepsilon}{8} = \frac{\varepsilon}{2} \quad \text{by (4) and (5).}$$

Then we have

$$\int_{\mathbb{R}} |f - g|\, d\mu_L \leq \int_{\mathbb{R}} |f - f_N|\, d\mu_L + \int_{\mathbb{R}} |f_N - g|\, d\mu_L$$

$$< \frac{\varepsilon}{2} + \frac{\varepsilon}{2} = \varepsilon \quad \text{by (3) and (6).}$$

This completes the proof. ∎

Prob. 9.41. Prove the following theorem:

Theorem. (Continuity in the Mean) Let f be a μ_L-integrable extended real-valued $\mathfrak{M}_L$-measurable function on $\mathbb{R}$ in the measure space $(\mathbb{R}, \mathfrak{M}_L, \mu_L)$. Let $h \in \mathbb{R}$. Then we have

$$\lim_{h \to 0} \int_{\mathbb{R}} |f(x+h) - f(x)| \mu_L(dx) = 0.$$

(Note that we may not have $\lim_{h \to 0} f(x+h) = f(x)$ as f may not be continuous at any $x \in \mathbb{R}$.)

Proof. Let f be a μ_L-integrable extended real-valued $\mathfrak{M}_L$-measurable function on $\mathbb{R}$. According to Prob. 9.38, for every $\varepsilon > 0$ there exists a continuous real-valued function g on $\mathbb{R}$ which vanishes outside some finite closed interval $[-B, B]$ and satisfies the condition:

$$(1) \qquad \int_{\mathbb{R}} |f - g| \, d\mu_L < \varepsilon.$$

Let $h \in \mathbb{R}$ and $h \neq 0$. We have

$$(2) \qquad \int_{\mathbb{R}} |f(x+h) - f(x)| \, \mu_L(dx) \leq \int_{\mathbb{R}} |f(x+h) - g(x+h)| \, \mu_L(dx)$$
$$+ \int_{\mathbb{R}} |g(x+h) - g(x)| \, \mu_L(dx)$$
$$+ \int_{\mathbb{R}} |g(x) - f(x)| \, \mu_L(dx).$$

By Theorem 9.31 (Translation Invariance of the Lebesgue Integral on $\mathbb{R}$) and by (1), we have

$$\int_{\mathbb{R}} |f(x+h) - g(x+h)| \, \mu_L(dx) = \int_{\mathbb{R}} |g(x) - f(x)| \, \mu_L(dx) < \varepsilon.$$

Substituting this in (2), we have

$$(3) \qquad \int_{\mathbb{R}} |f(x+h) - f(x)| \, \mu_L(dx) \leq \int_{\mathbb{R}} |g(x+h) - g(x)| \, \mu_L(dx) + 2\varepsilon.$$

Consider the integral on the right of (3). The function g vanishes on $[-B, B]^c$. This implies that g vanishes on $[-B - 1, B + 1]^c$. Then for $h \in \mathbb{R}$ such that $0 < |h| < 1$, we have $g(x + h) = 0$ for $x \in [-B, B]^c$. Then since $g(x) = 0$ for $x \in [-B, B]^c$, we have

$$g(x + h) - g(x) = 0 \quad \text{for } x \in [-B, B]^c.$$

This implies that

$$(4) \qquad \int_{\mathbb{R}} |g(x+h) - g(x)| \, \mu_L(dx) = \int_{[-B,B]} |g(x+h) - g(x)| \, \mu_L(dx).$$

Since g is a continuous function on $\mathbb{R}$ and g vanishes outside the compact set $[-B, B]$, $\max_{x \in [-B,B]} |g(x)| = \max_{x \in \mathbb{R}} |g(x)|$ exists as a positive real number. Let

$$M = \max_{x \in [-B,B]} |g(x)| = \max_{x \in \mathbb{R}} |g(x)|.$$

Then we have
$$|g(x+h) - g(x)| \le |g(x+h)| + |g(x)| \le 2M.$$

Continuity of g on $\mathbb{R}$ implies that $\lim_{h \to 0} g(x+h) = g(x)$ and then

$$\lim_{h \to 0} |g(x+h) - g(x)| = 0 \quad \text{for every } x \in \mathbb{R}.$$

Then by Theorem 7.16 (Bounded Convergence Theorem), we have

(5) $\lim_{h \to 0} \int_{[-B,B]} |g(x+h) - g(x)| \, \mu_L(dx) = \int_{[-B,B]} \lim_{h \to 0} |g(x+h) - g(x)| \, \mu_L(dx)$

$$= \int_{[-B,B]} 0 \, \mu_L(dx) = 0.$$

Combining (4) and (5), we obtain

(6) $$\lim_{h \to 0} \int_{\mathbb{R}} |g(x+h) - g(x)| \, \mu_L(dx) = 0.$$

From (3) and (6) we have

(7) $\limsup_{h \to 0} \int_{\mathbb{R}} |f(x+h) - f(x)| \, \mu_L(dx) \le \limsup_{h \to 0} \int_{\mathbb{R}} |g(x+h) - g(x)| \, \mu_L(dx) + 2\varepsilon$

$$= \lim_{h \to 0} \int_{\mathbb{R}} |g(x+h) - g(x)| \, \mu_L(dx) + 2\varepsilon$$

$$= 2\varepsilon.$$

Since this holds for every $\varepsilon > 0$, we have $\limsup_{h \to 0} \int_{\mathbb{R}} |f(x+h) - f(x)| \, \mu_L(dx) = 0$. This then implies that $\liminf_{h \to 0} \int_{\mathbb{R}} |f(x+h) - f(x)| \, \mu_L(dx) = 0$ and this in turn implies that $\lim_{h \to 0} \int_{\mathbb{R}} |f(x+h) - f(x)| \, \mu_L(dx) = 0$. ∎

Prob. 9.42. Let f be a μ_L-integrable extended real-valued $\mathfrak{M}_L$-measurable function on $\mathbb{R}$ in the measure space $(\mathbb{R}, \mathfrak{M}_L, \mu_L)$. Let $h \in \mathbb{R}$. Show that

$$\lim_{h \to \infty} \int_{\mathbb{R}} |f(x+h) - f(x)| \mu_L(dx) = 2 \int_{\mathbb{R}} |f(x)| \mu_L(dx)$$

and similarly

$$\lim_{h \to -\infty} \int_{\mathbb{R}} |f(x+h) - f(x)| \mu_L(dx) = 2 \int_{\mathbb{R}} |f(x)| \mu_L(dx).$$

Proof. Suppose f is μ_L-integrable on $\mathbb{R}$. Then Theorem 9.28 implies that for every $\varepsilon > 0$ there exists $M > 0$ such that

(1) $$\int_{[-M,M]^c} |f(x)| \, \mu_L(dx) < \varepsilon.$$

Our proof is based on the estimate (1) and Theorem 9.31 (Translation Invariance of the Lebesgue Integral on $\mathbb{R}$). We start with

(2) $$\int_{\mathbb{R}} |f(x+h) - f(x)| \, \mu_L(dx) \leq \int_{\mathbb{R}} |f(x+h)| \, \mu_L(dx) + \int_{\mathbb{R}} |f(x)| \, \mu_L(dx)$$

$$= 2 \int_{\mathbb{R}} |f(x)| \, \mu_L(dx) \quad \text{by Theorem 9.31.}$$

Next we seek a lower bound for the integral on the left of (2). By Theorem 9.31 and (1) we have

(3) $$\int_{[-M-h,M-h]^c} |f(x+h)| \, \mu_L(dx) = \int_{[-M,M]^c} |f(x)| \, \mu_L(dx) < \varepsilon.$$

Observe that

(4) $$h > 2M \Rightarrow [-M-h, M-h] \cap [-M, M] = \emptyset.$$

Let $h > 2M$. Then we have

(5) $$\int_{\mathbb{R}} |f(x+h) - f(x)| \, \mu_L(dx) \geq \int_{[-M-h,M-h]} |f(x+h) - f(x)| \, \mu_L(dx)$$

$$+ \int_{[-M,M]} |f(x+h) - f(x)| \, \mu_L(dx).$$

Let us find lower bounds for the two integrals on the right of (5). Now the inequality $\big||f(x+h)| - |f(x)|\big| \leq |f(x+h) - f(x)|$ implies that we have

$$1° \quad |f(x+h) - f(x)| \geq |f(x+h)| - |f(x)|$$

and

$$2° \quad |f(x+h) - f(x)| \geq |f(x)| - |f(x+h)|.$$

Then we have

(6) $$\int_{[-M-h,M-h]} |f(x+h) - f(x)| \mu_L(dx)$$

$$\geq \int_{[-M-h,M-h]} |f(x+h)| \, \mu_L(dx) - \int_{[-M-h,M-h]} |f(x)| \, \mu_L(dx) \quad \text{by } 1°$$

$$\geq \int_{[-M-h,M-h]} |f(x+h)| \, \mu_L(dx) - \varepsilon,$$

by the fact that $[-M-h, M-h] \subset [-M, M]^c$ and by (1). Similarly we have

(7)
$$\int_{[-M,M]} |f(x+h) - f(x)| \, \mu_L(dx)$$

$$\geq \int_{[-M,M]} |f(x)| \, \mu_L(dx) - \int_{[-M,M]} |f(x+h)| \, \mu_L(dx) \quad \text{by } 2°$$

$$\geq \int_{[-M,M]} |f(x)| \, \mu_L(dx) - \varepsilon,$$

by the fact that $[-M, M] \subset [-M-h, M-h]^c$ and by (3). Substituting (6) and (7) in (5), we have

(8)
$$\int_{\mathbb{R}} |f(x+h) - f(x)| \, \mu_L(dx) \geq \int_{[-M-h,M-h]} |f(x+h)| \, \mu_L(dx)$$

$$+ \int_{[-M,M]} |f(x)| \, \mu_L(dx) - 2\varepsilon.$$

Next we find lower bounds for the two integrals on the right of (8). We have

(9)
$$\int_{[-M-h,M-h]} |f(x+h)| \, \mu_L(dx)$$

$$= \int_{\mathbb{R}} |f(x+h)| \, \mu_L(dx) - \int_{[-M-h,M-h]^c} |f(x+h)| \, \mu_L(dx)$$

$$> \int_{\mathbb{R}} |f(x+h)| \, \mu_L(dx) - \varepsilon \quad \text{by (3)}$$

$$= \int_{\mathbb{R}} |f(x)| \, \mu_L(dx) - \varepsilon \quad \text{by Theorem 9.31.}$$

Similarly we have

(10)
$$\int_{[-M,M]} |f(x)| \, \mu_L(dx) = \int_{\mathbb{R}} |f(x)| \, \mu_L(dx) - \int_{[-M,M]^c} |f(x)| \, \mu_L(dx)$$

$$> \int_{\mathbb{R}} |f(x)| \, \mu_L(dx) - \varepsilon \quad \text{by (1).}$$

Substituting (9) and (10) in (8), we have

(11)
$$\int_{\mathbb{R}} |f(x+h) - f(x)| \, \mu_L(dx) > 2 \int_{\mathbb{R}} |f(x)| \, \mu_L(dx) - 4\varepsilon.$$

By (2) and (11) we have

(12) $\quad 2 \int_{\mathbb{R}} |f(x)| \, \mu_L(dx) - 4\varepsilon < \int_{\mathbb{R}} |f(x+h) - f(x)| \, \mu_L(dx) \leq 2 \int_{\mathbb{R}} |f(x)| \, \mu_L(dx).$

Thus we have shown that for every $\varepsilon > 0$ there exists $M > 0$ such that (12) holds for $h > 2M$. This proves that we have

$$\lim_{h \to \infty} \int_{\mathbb{R}} |f(x+h) - f(x)| \mu_L(dx) = 2 \int_{\mathbb{R}} |f(x)| \mu_L(dx).$$

The case $\lim_{h \to -\infty}$ is proved similarly. ∎

Prologue to Prob. 9.43. Given a measure space $(X, \mathfrak{A}, \mu)$. Let f be a complex valued function on a set $D \in \mathfrak{A}$. Let $\mathfrak{R}(f)$ and $\mathfrak{I}(f)$ be the real and imaginary parts of f respectively. Thus $f = \mathfrak{R}(f)+i\mathfrak{I}(f)$. We say that f is $\mathfrak{A}$-measurable, μ semi-integrable, or μ-integrable on D if both $\mathfrak{R}f$ and $\mathfrak{I}f$ are $\mathfrak{A}$-measurable, μ semi-integrable, or μ-integrable on D. If f is μ semi-integrable on D, then we define $\int_D f \, d\mu = \int_D \mathfrak{R}f \, d\mu + i \int_D \mathfrak{I}f \, d\mu$.

Prob. 9.43. Let f be a μ_L-integrable extended real-valued $\mathfrak{M}_L$-measurable function on $\mathbb{R}$.
(a) Show that for every $y \in \mathbb{R}$, the complex valued function $e^{ixy} f(x)$ for $x \in \mathbb{R}$ is $\mathfrak{M}_L$-measurable and μ_L-integrable on $\mathbb{R}$.
(b) The Fourier transform of f is a complex valued function $\hat{f}$ on $\mathbb{R}$ defined by setting
$$\hat{f}(y) = \int_{\mathbb{R}} e^{ixy} f(x)\mu_L(dx) \quad \text{for } y \in \mathbb{R}.$$
Show that $\hat{f}$ is bounded on $\mathbb{R}$ and in fact $\sup_{y \in \mathbb{R}} |\hat{f}(y)| \le 2 \int_{\mathbb{R}} |f(x)|\mu_L(dx)$.
(c) Show that $\hat{f}$ defined in (b) is continuous on $\mathbb{R}$.

Proof. (a) We have $e^{ixy} = \cos xy + i \sin xy$ and thus
$$e^{ixy} f(x) = \cos xy \cdot f(x) + i \sin xy \cdot f(x).$$

Now f is μ_L-integrable on $\mathbb{R}$ and $|\cos xy| \le 1$ for $x \in \mathbb{R}$. Thus $\cos xy \cdot f(x)$ is μ_L-integrable on $\mathbb{R}$ by Prob. 9.32. Similarly $\sin xy \cdot f(x)$ is μ_L-integrable on $\mathbb{R}$. This shows that $e^{ixy} f(x)$ is μ_L-integrable on $\mathbb{R}$.
(b) We have
$$\hat{f}(y) = \int_{\mathbb{R}} e^{ixy} f(x)\mu_L(dx) = \int_{\mathbb{R}} \cos xy \cdot f(x)\mu_L(dx) + i \int_{\mathbb{R}} \sin xy \cdot f(x)\mu_L(dx)$$

and then
$$|\hat{f}(y)| \le \left| \int_{\mathbb{R}} \cos xy \cdot f(x)\mu_L(dx) \right| + \left| \int_{\mathbb{R}} \sin xy \cdot f(x)\mu_L(dx) \right|$$
$$\le \int_{\mathbb{R}} |\cos xy| \cdot |f(x)|\mu_L(dx) + \int_{\mathbb{R}} |\sin xy| \cdot |f(x)|\mu_L(dx)$$
$$\le \int_{\mathbb{R}} |f(x)|\mu_L(dx) + \int_{\mathbb{R}} |f(x)|\mu_L(dx) = 2 \int_{\mathbb{R}} |f(x)|\mu_L(dx).$$

This shows that
$$\sup_{y \in \mathbb{R}} |\hat{f}(y)| \le 2 \int_{\mathbb{R}} |f(x)|\mu_L(dx).$$

(c) To prove the continuity of $\hat{f}$ on $\mathbb{R}$ we show the continuity of $\mathfrak{R}(\hat{f})$ and $\mathfrak{I}(\hat{f})$ on $\mathbb{R}$. Now we have
$$\mathfrak{R}(\hat{f})(y) = \int_{\mathbb{R}} \cos xy \cdot f(x)\mu_L(dx)$$

and then
(1) $\mathfrak{R}(\hat{f})(y + h) - \mathfrak{R}(\hat{f})(y) = \int_{\mathbb{R}} \left\{ \cos x(y + h) - \cos xy \right\} f(x)\mu_L(dx).$

Observe that continuity of the cosine function implies

(2) $$\lim_{h \to 0} \left\{ \cos x(y+h) - \cos xy \right\} = 0.$$

Observe also that

(3) $$\left| \left\{ \cos x(y+h) - \cos xy \right\} f(x) \right| \le 2|f(x)|,$$

where the dominating function $2|f|$ is a μ_L-integrable function. With (2) and (3), we can apply Theorem 9.20 (Dominated Convergence Theorem) to obtain

(4) $$\lim_{h \to 0} \int_{\mathbb{R}} \left\{ \cos x(y+h) - \cos xy \right\} f(x) \mu_L(dx) = \int_{\mathbb{R}} 0 \, \mu_L(dx) = 0.$$

Then from (1) and (4) we have

$$\lim_{h \to 0} \left\{ \Re(\hat{f})(y+h) - \Re(\hat{f})(y) \right\} = 0.$$

This proves the continuity of $\Re(\hat{f})$ at every $y \in \mathbb{R}$. Continuity of $\Im(\hat{f})$ on $\mathbb{R}$ is shown likewise. This completes the proof that $\hat{f}$ is continuous on $\mathbb{R}$. ∎

Prob. 9.44. Prove the following theorem:

Riemann-Lebesgue Theorem for the Fourier Transform. Let f be a μ_L-integrable extended real-valued $\mathfrak{M}_L$-measurable function on $\mathbb{R}$ and define the Fourier transform $\hat{f}(y) = \int_{\mathbb{R}} e^{ixy} f(x)\mu_L(dx)$ for $y \in \mathbb{R}$. Then we have $\lim_{y \to \infty} \hat{f}(y) = 0$ and $\lim_{y \to -\infty} \hat{f}(y) = 0$.

(Hint: Show first that by the translation invariance of the Lebesgue integral in $(\mathbb{R}, \mathfrak{M}_L, \mu_L)$,

$$\hat{f}(y) = -\int_{\mathbb{R}} e^{i(x+\frac{\pi}{y})y} f(x)\mu_L(dx) = -\int_{\mathbb{R}} e^{ixy} f\left(x - \frac{\pi}{y}\right)\mu_L(dx).$$

Then use this expression in $2\hat{f}$.)

Proof. Observe that

$$e^{ixy} = -e^{i(xy+\pi)} = -e^{i(x+\frac{\pi}{y})y}.$$

Then we have

$$\hat{f}(y) = \int_{\mathbb{R}} e^{ixy} f(x)\mu_L(dx) = -\int_{\mathbb{R}} e^{i(x+\frac{\pi}{y})y} f(x)\mu_L(dx)$$

$$= -\int_{\mathbb{R}} e^{ixy} f\left(x - \frac{\pi}{y}\right)\mu_L(dx) \quad \text{by Theorem 9.31.}$$

Then we have

$$2\hat{f}(y) = \int_{\mathbb{R}} e^{ixy} f(x)\mu_L(dx) - \int_{\mathbb{R}} e^{ixy} f\left(x - \frac{\pi}{y}\right)\mu_L(dx)$$

$$= \int_{\mathbb{R}} e^{ixy} \left\{ f(x) - f\left(x - \frac{\pi}{y}\right) \right\}\mu_L(dx).$$

Thus we have

$$2|\hat{f}(y)| \le \int_{\mathbb{R}} |e^{ixy}| \left| f(x) - f\left(x - \frac{\pi}{y}\right) \right|\mu_L(dx)$$

$$= \int_{\mathbb{R}} \left| f(x) - f\left(x - \frac{\pi}{y}\right) \right|\mu_L(dx).$$

Then we have by Prob. 9.41

$$\lim_{\frac{1}{y} \to 0} |\hat{f}(y)| \le \lim_{\frac{1}{y} \to 0} \frac{1}{2} \int_{\mathbb{R}} \left| f(x) - f\left(x - \frac{\pi}{y}\right) \right|\mu_L(dx) = 0.$$

Thus we have

$$\lim_{y \to \pm\infty} |\hat{f}(y)| = 0.$$

This is equivalent to

$$\lim_{y \to \pm\infty} \hat{f}(y) = 0.$$

This completes the proof. ∎

Prob. 9.45. Given a measure space $(X, \mathfrak{A}, \mu)$. Let f be an extended real-valued $\mathfrak{A}$-measurable function on a set $D \in \mathfrak{A}$. Let $(D_n : n \in \mathbb{N})$ be a sequence of $\mathfrak{A}$-measurable subsets of D such that $\lim_{n \to \infty} \mu(D_n) = 0$. (Note that $(D_n : n \in \mathbb{N})$ is not assumed to be a decreasing sequence.)

(a) Show that if f is μ-integrable on D then we have $\lim_{n \to \infty} \int_{D_n} f \, d\mu = 0$.

(b) Show if f is only μ semi-integrable on D then the conclusion in (a) does not hold.

Proof. **(a)** Assume that f is μ-integrable on D. To show that $\lim_{n \to \infty} \int_{D_n} f \, d\mu = 0$, we show that for every $\varepsilon > 0$ there exists $N \in \mathbb{N}$ such that

$$\left| \int_{D_n} f \, d\mu \right| < \varepsilon \quad \text{for } n \geq N.$$

Let $\varepsilon > 0$ be arbitrarily given. According to Theorem 9.26 (Uniform Absolute Continuity of the Integral with Respect to the Measure), μ-integrability of f on D implies that for every $\varepsilon > 0$ there exists $\delta > 0$ such that

$$\left| \int_E f \, d\mu \right| \leq \int_E |f| \, d\mu < \varepsilon \quad \text{when } E \subset D, E \in \mathfrak{A}, \text{ and } \mu(E) < \delta.$$

Now $\lim_{n \to \infty} \mu(D_n) = 0$ implies that there exists $N \in \mathbb{N}$ such that $\mu(D_n) < \delta$ for $n \geq N$. Then we have $\left| \int_{D_n} f \, d\mu \right| < \varepsilon$ for $n \geq N$.

(b) Consider the measure space $(\mathbb{R}, \mathfrak{M}_L, \mu_L)$. Let f be a real-valued $\mathfrak{M}_L$-measurable function on $[0, \infty)$ defined by $f(x) = x$ for $x \in [0, \infty)$. Then $\int_{[0,\infty)} f(x) \, \mu_L(dx) = \infty$ so that f is μ_L semi-integrable but not μ_L-integrable on $[0, \infty)$.

With $D = [0, \infty)$, let us define $D_n = \left[n, n + \frac{1}{n} \right)$ for $n \in \mathbb{N}$. Then we have $\mu_L(D_n) = \frac{1}{n}$ and $\lim_{n \to \infty} \mu_L(D_n) = \lim_{n \to \infty} \frac{1}{n} = 0$. On the other hand, we have

$$\int_{D_n} f(x) \, \mu_L(dx) = \int_n^{n + \frac{1}{n}} x \, dx = \frac{1}{2} \left[x^2 \right]_n^{n + \frac{1}{n}} = 1 + \frac{1}{2n^2}$$

and then

$$\lim_{n \to \infty} \int_{D_n} f(x) \, \mu_L(dx) = \lim_{n \to \infty} \left\{ 1 + \frac{1}{2n^2} \right\} = 1.$$

This shows a counter example for (b). ∎

Prob. 9.46. Construct a continuous real-valued function f on $[0, \infty)$ such that the improper Riemann integral $\int_0^\infty f(x)\, dx := \lim_{a \to \infty} \int_0^a f(x)\, dx$ exists but $\int_{[0,\infty)} f\, d\mu_L$ does not exist.

Proof. 1. Let $I_n = [(n-1)\pi, n\pi)$ for $n \in \mathbb{N}$. Then $\{I_n : n \in \mathbb{N}\}$ is a collection of disjoint intervals and $\bigcup_{n \in \mathbb{N}} I_n = [0, \infty)$.

Let $g(x) = \sin x$ for $x \in [0, \infty)$. Then define a step function h on $[0, \infty)$ by setting

$$h(x) = \frac{1}{n} \quad \text{for } x \in I_n \text{ for } n \in \mathbb{N}.$$

Define a function f on $[0, \infty)$ by setting

$$f(x) = h(x)g(x) \quad \text{for } x \in [0, \infty).$$

Observe that

1° $f = 0$ at the endpoints of the interval I_n for every $n \in \mathbb{N}$;

2° f is nonnegative on I_n when n is an odd integer and f is non-positive on I_n when n is an even integer;

3° f is continuous on $[0, \infty)$.

We have

(1)
$$\int_{(n-1)\pi}^{n\pi} f(x)\, dx = \int_{(n-1)\pi}^{n\pi} \frac{1}{n} \sin x\, dx = \frac{1}{n}\Big[\cos x \Big]_{n\pi}^{(n-1)\pi}$$

$$= \frac{1}{n}\big\{ \cos(n-1)\pi - \cos n\pi \big\} = \frac{1}{n}\big\{ (-1)^{n-1} - (-1)^n \big\}$$

$$= \frac{1}{n}(-1)^{n-1}\{1 - (-1)\} = (-1)^{n-1}\frac{2}{n}.$$

Then applying (1) we have

(2)
$$\lim_{a \to \infty} \int_0^a f(x)\, dx = 2 \sum_{n \in \mathbb{N}} (-1)^{n-1}\frac{1}{n} = 2\Big\{ 1 - \frac{1}{2} + \frac{1}{3} - \frac{1}{4} + \frac{1}{5} + \cdots \Big\} \in \mathbb{R}.$$

This shows that the improper Riemann integral $\int_0^\infty f(x)\, dx$ exists.

2. For our function f, the positive part f^+ and the negative part f^- are given by 2° above. Thus we have

$$\int_{[0,\infty)} f^+\, d\mu_L = 2\Big\{ 1 + \frac{1}{3} + \frac{1}{5} + \cdots \Big\} = \infty,$$

and similarly

$$\int_{[0,\infty)} f^-\, d\mu_L = 2\Big\{ \frac{1}{2} + \frac{1}{4} + \frac{1}{6} + \cdots \Big\} = \infty.$$

This shows that $\int_{[0,\infty)} f\, d\mu_L$ does not exist according to Observation 9.2. ∎

Prob. 9.47. Let f be a continuous nonnegative real-valued function on $[0, \infty)$ such that the improper Riemann integral $\int_0^\infty f(x)\,dx := \lim_{a \to \infty} \int_0^a f(x)\,dx$ exists. Show that we have the equality $\int_0^\infty f(x)\,dx = \int_{[0,\infty)} f\,d\mu_L$. (Note that nonnegativity of f is a crucial condition.)

Proof. Observe that the nonnegativity and $\mathfrak{M}_L$-measurability implied by continuity of f imply the existence of $\int_{[0,\infty)} f\,d\mu_L$ in $[0, \infty]$.

Let us assume that the improper Riemann integral $\int_0^\infty f(x)\,dx$ exists. Then we have

(1) $$\int_0^\infty f(x)\,dx := \lim_{a \to \infty} \int_0^a f(x)\,dx = \gamma \in [0, \infty].$$

The second equality in (1) implies

(2) $$\lim_{n \to \infty} \int_0^n f(x)\,dx = \gamma \in [0, \infty].$$

According to Theorem 7.27 we have

(3) $$\int_0^n f(x)\,dx = \int_{[0,n]} f\,d\mu_L.$$

Observe that $\big([0, n] : n \in \mathbb{N}\big)$ is an increasing sequence of $\mathfrak{M}_L$-measurable sets and $\lim_{n \to \infty} [0, n] = \bigcup_{n \in \mathbb{N}} [0, n] = [0, \infty)$. Let us define a sequence of functions $(f_n : n \in \mathbb{N})$ on $[0, \infty)$ by setting

(4) $$f_n(x) = \begin{cases} f(x) & \text{for } x \in [0, n], \\ 0 & \text{for } x \in (n, \infty). \end{cases}$$

Then $(f_n : n \in \mathbb{N})$ is an increasing sequence and $\lim_{n \to \infty} f_n(x) = f(x)$ for $x \in [0, \infty)$. Thus by Theorem 8.5 (Monotone Convergence Theorem) we have

$$\int_{[0,\infty)} f\,d\mu_L = \lim_{n \to \infty} \int_{[0,\infty)} f_n\,d\mu_L = \lim_{n \to \infty} \int_{[0,n]} f\,d\mu_L \quad \text{by (4)}$$

$$= \lim_{n \to \infty} \int_0^n f(x)\,dx = \gamma \quad \text{by (3) and (2)}$$

$$= \int_0^\infty f(x)\,dx \quad \text{by (1)}.$$

This completes the proof. ∎

Prob. 9.48. Prove the following:

(a) $\int_{[0,\infty)} xe^{-x^2} \mu_L(dx) = \frac{1}{2}$.

(b) $\int_{[0,\infty)} e^{-x^2} \mu_L(dx) < \infty$.

(c) $\lim_{n\to\infty} \int_{[0,\infty)} e^{-nx^2} \sin nx\, \mu_L(dx) = 0$.

Proof. (a) The function $f(x) = xe^{-x^2}$ for $x \in [0, \infty)$ is a continuous nonnegative real-valued function on $[0, \infty)$. Now according to Prob. 9.47, if the improper Riemann integral $\int_0^\infty xe^{-x^2}\,dx := \lim_{a\to\infty} \int_0^a xe^{-x^2}\,dx$ exists then $\int_0^\infty xe^{-x^2}\,dx = \int_{[0,\infty)} xe^{-x^2} \mu_L(dx)$.

Let us show that the improper Riemann integral $\int_0^\infty xe^{-x^2}\,dx := \lim_{a\to\infty} \int_0^a xe^{-x^2}\,dx$ exists. Now we have

$$\int_0^a xe^{-x^2}\,dx = \left[\frac{1}{2}e^{-x^2}\right]_a^0 = \frac{1}{2}\left\{1 - e^{-a^2}\right\}$$

and then

$$\int_0^\infty xe^{-x^2}\,dx := \lim_{a\to\infty} \int_0^a xe^{-x^2}\,dx = \lim_{a\to\infty} \frac{1}{2}\left\{1 - e^{-a^2}\right\} = \frac{1}{2}.$$

This shows that $\int_0^\infty xe^{-x^2}\,dx$ exists and is equal to $\frac{1}{2}$. Then $\int_{[0,\infty)} xe^{-x^2} \mu_L(dx) = \frac{1}{2}$ by Prob. 9.47.

(b) The function $g(x) = xe^{-x^2}$ for $x \in [0, \infty)$ is a continuous nonnegative real-valued function on $[0, \infty)$. This implies that $\int_{[0,\infty)} e^{-x^2} \mu_L(dx)$ exists in $[0, \infty]$.

It remains to show that $\int_{[0,\infty)} e^{-x^2} \mu_L(dx) < \infty$. Now we have

$$(1) \qquad \int_{[0,\infty)} e^{-x^2} \mu_L(dx) = \int_{[0,1]} e^{-x^2} \mu_L(dx) + \int_{[1,\infty)} e^{-x^2} \mu_L(dx).$$

Since $0 \le e^{-x^2} \le 1$ for $x \in [0, \infty)$, we have

$$(2) \qquad \int_{[0,1]} e^{-x^2} \mu_L(dx) \le \int_{[0,1]} 1\, \mu_L(dx) = 1.$$

For $x \in [1, \infty)$ we have $e^{-x^2} \le xe^{-x^2}$ and thus

$$(3) \qquad \int_{[1,\infty)} e^{-x^2} \mu_L(dx) \le \int_{[1,\infty)} xe^{-x^2} \mu_L(dx) \le \int_{[0,\infty)} xe^{-x^2} \mu_L(dx) = \frac{1}{2}.$$

Substituting (2) and (3) in (1), we have $\int_{[0,\infty)} e^{-x^2} \mu_L(dx) \le \frac{3}{2}$.

(c) We have $\lim_{n\to\infty} e^{-nx^2} \sin nx = 0$. We also have $|e^{-nx^2} \sin nx| \le e^{-x^2}$, which is μ_L-integrable on $[0, \infty)$ according to (b). Thus Theorem 9.20 (Dominated Convergence Theorem) is applicable and we have

$$\lim_{n\to\infty} \int_{[0,\infty)} e^{-nx^2} \sin nx\, \mu_L(dx) = \int_{[0,\infty)} \lim_{n\to\infty} e^{-nx^2} \sin nx\, \mu_L(dx) = \int_{[0,\infty)} 0\, d\mu_L = 0.$$

This completes the proof. ∎

Prob. 9.49. Given that the improper Riemann integral $\int_{-\infty}^{\infty} \frac{1}{\sqrt{\pi}} \exp\{-x^2\} dx = 1$ implies the Lebesgue integral $\int_{\mathbb{R}} \frac{1}{\sqrt{\pi}} \exp\{-x^2\} \mu_L(dx) = 1$. Prove the following statements:

(a) For $v > 0$ and $m \in \mathbb{R}$ we have
$$\int_{\mathbb{R}} \frac{1}{\sqrt{2\pi v}} \exp\left\{ - \frac{|x-m|^2}{2v} \right\} \mu_L(dx) = 1.$$

(b) Let f be a bounded real valued $\mathfrak{M}_L$-measurable function on $\mathbb{R}$. Suppose f is continuous at $m \in \mathbb{R}$. Then we have
$$\lim_{v \downarrow 0} \int_{\mathbb{R}} \frac{1}{\sqrt{2\pi v}} \exp\left\{ - \frac{|x-m|^2}{2v} \right\} f(x) \mu_L(dx) = f(m).$$

Proof. **(a)** According to Theorem 9.33 (Affine Transformation of the Lebesgue Integral in $\mathbb{R}$), for a μ_L semi-integrable extended real-valued $\mathfrak{M}_L$-measurable function f on $\mathbb{R}$ and for $\alpha \in \mathbb{R}$ such that $\alpha \neq 0$ and $\beta \in \mathbb{R}$, we have
$$\int_{\mathbb{R}} f(x) \mu_L(dx) = |\alpha| \int_{\mathbb{R}} f(\alpha x + \beta) \mu_L(dx).$$

Then for $f(x) = \frac{1}{\sqrt{\pi}} \exp\{-x^2\}$ for $x \in \mathbb{R}$ and for $\alpha x + \beta = \frac{x}{\sqrt{2v}} - \frac{m}{\sqrt{2v}}$ where $v > 0$ and $m \in \mathbb{R}$, we have
$$1 = \int_{\mathbb{R}} \frac{1}{\sqrt{\pi}} \exp\{-x^2\} \mu_L(dx) = \frac{1}{\sqrt{2v}} \int_{\mathbb{R}} \frac{1}{\sqrt{\pi}} \exp\left\{ - |\frac{x}{\sqrt{2v}} - \frac{m}{\sqrt{2v}}|^2 \right\} \mu_L(dx)$$
$$= \int_{\mathbb{R}} \frac{1}{\sqrt{2\pi v}} \exp\left\{ - \frac{|x-m|^2}{2v} \right\} \mu_L(dx).$$

This proves (a).

(b.1) Let f be a bounded real-valued $\mathfrak{M}_L$-measurable function on $\mathbb{R}$ that is continuous at $m \in \mathbb{R}$. We claim

$$(1) \qquad \lim_{v \downarrow 0} \int_{\mathbb{R}} \frac{1}{\sqrt{2\pi v}} \exp\left\{ - \frac{|x-m|^2}{2v} \right\} f(x) \mu_L(dx) = f(m).$$

For the sake or brevity, let us define a function $\varphi_{v,m}$ on $\mathbb{R}$, where $v > 0$ and $m \in \mathbb{R}$, by

$$(2) \qquad \varphi_{v,m}(x) = \frac{1}{\sqrt{2\pi v}} \exp\left\{ - \frac{|x-m|^2}{2v} \right\} \quad \text{for } x \in \mathbb{R}.$$

Then (1) is rewritten as

$$(3) \qquad \lim_{v \downarrow 0} \int_{\mathbb{R}} \varphi_{v,m}(x) f(x) \mu_L(dx) - f(m) = 0.$$

We proved in (a) that

$$(4) \qquad \int_{\mathbb{R}} \varphi_{v,m}(x) \mu_L(dx) = 1.$$

Then we have

$$f(m) = f(m) \cdot 1 = f(m) \int_{\mathbb{R}} \varphi_{v,m}(x) \mu_L(dx) = \int_{\mathbb{R}} \varphi_{v,m}(x) f(m) \mu_L(dx).$$

Thus (3) is equivalent to

(5) $$\lim_{v\downarrow 0} \int_{\mathbb{R}} \varphi_{v,m}(x)\{f(x) - f(m)\}\, \mu_L(dx) = 0.$$

Observe that (5) is equivalent to

(6) $$\lim_{v\downarrow 0} \left| \int_{\mathbb{R}} \varphi_{v,m}(x)\{f(x) - f(m)\}\, \mu_L(dx) \right| = 0.$$

By the non-negativity of $\varphi_{v,m}$ on $\mathbb{R}$, we have

$$\left| \int_{\mathbb{R}} \varphi_{v,m}(x)\{f(x) - f(m)\}\, \mu_L(dx) \right| \le \int_{\mathbb{R}} \varphi_{v,m}(x)|f(x) - f(m)|\, \mu_L(dx),$$

and then

$$\lim_{v\downarrow 0} \left| \int_{\mathbb{R}} \varphi_{v,m}(x)\{f(x) - f(m)\}\, \mu_L(dx) \right| \le \lim_{v\downarrow 0} \int_{\mathbb{R}} \varphi_{v,m}(x)|f(x) - f(m)|\, \mu_L(dx).$$

Thus if we show

(7) $$\lim_{v\downarrow 0} \int_{\mathbb{R}} \varphi_{v,m}(x)|f(x) - f(m)|\, \mu_L(dx) = 0,$$

then (6) follows. Thus it remains to prove (7).

(b.2) Let us analyze the behavior of $\varphi_{v,m}(x)$ as $v \downarrow 0$.

The function $\varphi_{v,m}$ on $\mathbb{R}$ as defined by (1) has a maximum point $m \in \mathbb{R}$ and we have $\varphi_{v,m}(m) = \frac{1}{\sqrt{2\pi v}} e^0 = \frac{1}{\sqrt{2\pi v}}$. Thus we have

(8) $$\lim_{v\downarrow 0} \varphi_{v,m}(m) = \lim_{v\downarrow 0} \frac{1}{\sqrt{2\pi v}} = \infty.$$

For $x \ne m$, we have

$$\varphi_{v,m}(x) = \frac{1}{\sqrt{2\pi v}} \exp\left\{ -\frac{|x - m|^2}{2v} \right\} = \frac{1}{\sqrt{2\pi v}} \frac{1}{\exp\left\{ \frac{|x-m|^2}{2v} \right\}}$$

$$\le \frac{1}{\sqrt{2\pi v}} \frac{1}{\frac{|x-m|^2}{2v}} = \sqrt{\frac{2v}{\pi}} \frac{1}{|x - m|^2},$$

and then

(9) $$\lim_{v\downarrow 0} \varphi_{v,m}(x) = 0 \quad \text{for } x \ne m.$$

Note that whereas $\lim_{v\downarrow 0} \varphi_{v,m}(x) = 0$ when $x \ne m$ according to (9), we do not necessarily have $\varphi_{v,m}(x) \downarrow$ as $v \downarrow 0$. We show next that at $x \ne m$, as $v \downarrow 0$, $\varphi_{v,m}(x)$ increases first and then decreases, the critical value of v depending on x. Observe first that $\varphi_{v,m}(x) \downarrow$ as $v \downarrow$ if and only if $\frac{d}{dv}[\varphi_{v,m}(x)] \geq 0$ and $\varphi_{v,m}(x) \uparrow$ as $v \downarrow$ if and only if $\frac{d}{dv}[\varphi_{v,m}(x)] \leq 0$. We have

$$
\begin{aligned}
\frac{d}{dv}[\varphi_{v,m}(x)] &= \frac{d}{dv}\left[\frac{1}{\sqrt{2\pi v}} \exp\left\{ -\frac{|x-m|^2}{2v} \right\} \right] \\
&= \frac{1}{\sqrt{2\pi v}}\left[-\frac{1}{2}v^{-\frac{3}{2}} \exp\left\{ -\frac{|x-m|^2}{2v} \right\} \right. \\
&\quad \left. + v^{-\frac{1}{2}}\exp\left\{ -\frac{|x-m|^2}{2v} \right\}\left\{ -\frac{|x-m|^2}{2} \right\}(-v^{-2}) \right] \\
&= \frac{1}{\sqrt{2\pi v}}\exp\left\{ -\frac{|x-m|^2}{2v} \right\}\left[-\frac{1}{2}v^{-\frac{3}{2}} + \frac{|x-m|^2}{2}v^{-\frac{5}{2}} \right] \\
&= \frac{1}{\sqrt{2\pi v}}\exp\left\{ -\frac{|x-m|^2}{2v} \right\}\frac{1}{2}v^{-\frac{3}{2}}\left[\frac{|x-m|^2}{2v} - 1 \right].
\end{aligned}
$$

Then we have

(10)
$$
\begin{cases}
\varphi_{v,m}(x) \downarrow \text{ as } v \downarrow 0 \Leftrightarrow \frac{d}{dv}[\varphi_{v,m}(x)] \geq 0 \Leftrightarrow \frac{|x-m|^2}{2v} - 1 \geq 0 \Leftrightarrow \frac{|x-m|^2}{2} \geq v \\
\varphi_{v,m}(x) \uparrow \text{ as } v \downarrow 0 \Leftrightarrow \frac{d}{dv}[\varphi_{v,m}(x)] \leq 0 \Leftrightarrow \frac{|x-m|^2}{2v} - 1 \leq 0 \Leftrightarrow \frac{|x-m|^2}{2} \leq v.
\end{cases}
$$

We restate (10) as

(11)
$$
\begin{cases}
\varphi_{v,m}(x) \uparrow \text{ as } v \downarrow 0 \text{ when } v \geq \frac{|x-m|^2}{2} \\
\varphi_{v,m}(x) \downarrow \text{ as } v \downarrow 0 \text{ when } v \leq \frac{|x-m|^2}{2}.
\end{cases}
$$

This shows that as $v \downarrow 0$, $\varphi_{v,m}(x)$ increases until v reaches the critical value $v = \frac{|x-m|^2}{2}$ and then $\varphi_{v,m}(x)$ decreases.

Let $\delta > 0$ be arbitrarily given and let

$$
D_\delta = \{x \in \mathbb{R} : |x-m| \geq \delta\} = (-\infty, m-\delta] \cup [m+\delta, \infty).
$$

Let $v \in \left(0, \frac{\delta^2}{2}\right]$ so that we have $\frac{\delta^2}{2v} \geq 1$. Then we have, according to (11),

(12)
$$
|x-m| \geq \delta \Rightarrow \frac{|x-m|^2}{2v} \geq \frac{\delta^2}{2v} \geq 1 \Rightarrow \frac{|x-m|^2}{2} \geq v
$$

$$
\Rightarrow \varphi_{v,m}(x) \downarrow \text{ as } v \in \left(0, \frac{\delta^2}{2}\right] \text{ and } v \downarrow 0.
$$

Let $v_\delta = \frac{\delta^2}{2}$. Then (12) implies

(13)
$$
\varphi_{v,m}(x) \downarrow \text{ as } v \in (0, v_\delta] \text{ and } v \downarrow 0 \quad \text{for } x \in D_\delta.
$$

In particular we have

(14)
$$
\varphi_{v,m}(x) \leq \varphi_{v_\delta,m}(x) \quad \text{for } v \in (0, v_\delta] \text{ for } x \in D_\delta.
$$

Let $B > 0$ be a bound of f on $\mathbb{R}$. Then (14) implies

(15) $\qquad \varphi_{v,m}(x)|f(x) - f(m)| \le \varphi_{v_\delta,m}(x)|f(x) - f(m)| \le 2B\varphi_{v_\delta,m}(x)$

$$\text{for } v \in (0, v_\delta] \text{ for } x \in D_\delta.$$

Now the dominating function $2B\varphi_{v_\delta,m}$ is μ_L-integrable on $\mathbb{R}$. Then by Theorem 9.20 (Dominated Convergence Theorem), we have

(16) $\displaystyle \lim_{v\downarrow 0} \int_{D_\delta} \varphi_{v,m}(x)|f(x) - f(m)|\,\mu_L(dx) = \int_{D_\delta} \lim_{v\downarrow 0} \varphi_{v,m}(x)|f(x) - f(m)|\,\mu_L(dx)$

$$= \int_{D_\delta} 0\,\mu_L(dx) = 0.$$

Next, continuity of f at $m \in \mathbb{R}$ implies that for every $\varepsilon > 0$ there exists $\delta > 0$ such that $|f(x) - f(m)| < \varepsilon$ for $x \in (m - \delta, m + \delta)$. Then we have

$$\lim_{v\downarrow 0} \int_{(m-\delta,m+\delta)} \varphi_{v,m}(x)|f(x) - f(m)|\,\mu_L(dx) \le \lim_{v\downarrow 0} \int_{(m-\delta,m+\delta)} \varphi_{v,m}(x)\varepsilon\,\mu_L(dx)$$

$$\le \lim_{v\downarrow 0} \varepsilon \int_{\mathbb{R}} \varphi_{v,m}(x)\,\mu_L(dx)$$

$$= \varepsilon \quad \text{by (4)}.$$

Since this holds for every $\varepsilon > 0$, we have

(17) $\qquad \displaystyle \lim_{v\downarrow 0} \int_{(m-\delta,m+\delta)} \varphi_{v,m}(x)|f(x) - f(m)|\,\mu_L(dx) = 0.$

Now we have $\mathbb{R} = D_\delta \cup (m - \delta, m + \delta)$ and $D_\delta \cap (m - \delta, m + \delta) = \emptyset$. Thus we have

$$\int_{\mathbb{R}} \varphi_{v,m}(x)|f(x) - f(m)|\,\mu_L(dx) = \int_{(m-\delta,m+\delta)} \varphi_{v,m}(x)|f(x) - f(m)|\,\mu_L(dx)$$

$$+ \int_{(m-\delta,m+\delta)} \varphi_{v,m}(x)|f(x) - f(m)|\,\mu_L(dx).$$

Letting $v \downarrow 0$ on both sides of the last equality and then quoting (16) and (17), we obtain

$$\lim_{v\downarrow 0} \int_{\mathbb{R}} \varphi_{v,m}(x)|f(x) - f(m)|\,\mu_L(dx) = 0.$$

This proves (7) and completes the proof. ∎

Prob. 9.50. Let f be an extended real-valued $\mathfrak{M}_L$-measurable function on $(0, \infty)$ defined by

$$f(x) = \frac{1}{1+x^2} \log(1 - e^{-x}) \quad \text{for } x \in (0, \infty).$$

Show that f is μ_L-integrable on $(0, \infty)$ and estimate the integral $\int_{(0,\infty)} f \, d\mu_L$.

Proof. 1. Consider the function

(1) $$f(x) = \frac{1}{1 + x^2} \log(1 - e^{-x}) \quad \text{for } x \in (0, \infty).$$

Observe that

$$\frac{1}{1+x^2} \in (0, 1) \qquad\qquad\qquad \text{for } x \in (0, \infty).$$

$$e^{-x} \in (0, 1) \qquad\qquad\qquad \text{for } x \in (0, \infty).$$

$$1 - e^{-x} \in (0, 1) \qquad\qquad\qquad \text{for } x \in (0, \infty).$$

$$\log(1 - e^{-x}) < 0 \qquad\qquad\qquad \text{for } x \in (0, \infty).$$

$$\log(1 - e^{-x}) < 0 \downarrow -\infty \quad \text{as } x \downarrow 0$$

and then

(2) $$f(x) = \frac{1}{1 + x^2} \log(1 - e^{-x}) < 0 \quad \text{for } x \in (0, \infty).$$

Then (2) implies that $-f > 0$ on $(0, \infty)$. Thus $\int_{(0,\infty)} (-f) \, d\mu_L$ exists in $[0, \infty]$ and then $\int_{(0,\infty)} f \, d\mu_L$ exists in $[-\infty, 0]$. It remains to show that f in μ_L-integrable on $(0, \infty)$, that is, it remains to show $\int_{(0,\infty)} f \, d\mu_L \in \mathbb{R}$.

Now f is μ_L-integrable if and only if $|f|$ is μ_L-integrable. Since $|f|$ is nonnegative valued, existence of the improper Riemann integral $\int_0^\infty |f(x)| \, dx$ would imply the equality $\int_0^\infty |f(x)| \, dx = \int_{(0,\infty)} |f| \, d\mu_L$. We show below that the improper Riemann integral $\int_0^\infty |f(x)| \, dx$ exists as a finite real number. First of all let us write

$$\int_0^\infty |f(x)| \, dx = \int_0^1 |f(x)| \, dx + \int_1^\infty |f(x)| \, dx.$$

2. Consider the improper Riemann integral

(3) $$\int_0^1 |f(x)| \, dx = \int_0^1 \frac{1}{1 + x^2} \left| \log(1 - e^{-x}) \right| \, dx$$

$$\leq \int_0^1 \left| \log(1 - e^{-x}) \right| \, dx$$

$$= \left| \int_0^1 \log(1 - e^{-x}) \, dx \right|$$

where the last equality is from the fact that $\log(1 - e^{-x}) < 0$ for all $x \in (0, 1)$.

Let us estimate the improper Riemann integral in the last member of (3). To start with, we have

$$
\begin{aligned}
1 - e^{-x} &= 1 - \left\{ 1 - \frac{x}{1!} + \frac{x^2}{2!} - \frac{x^3}{3!} + \frac{x^4}{4!} - \cdots \right\} \\
&= \left\{ \frac{x}{1!} - \frac{x^2}{2!} \right\} + \left\{ \frac{x^3}{3!} - \frac{x^4}{4!} \right\} + \cdots \\
&> \frac{x}{1!} - \frac{x^2}{2!} > x - x^2 = x(1 - x) \quad \text{for } x \in (0, 1],
\end{aligned}
$$

and then

$$
1 > 1 - e^{-x} > x(1 - x) \quad \text{for } x \in (0, 1].
$$

Taking the logarithm we have

$$
0 > \log(1 - e^{-x}) > \log x(1 - x) = \log x + \log(1 - x).
$$

This implies

$$
(4) \qquad 0 \geq \int_0^1 \log(1 - e^{-x})\, dx \geq \int_0^1 \log x\, dx + \int_0^1 \log(1 - x)\, dx.
$$

Now we have

$$
\begin{aligned}
(5) \qquad \int_0^1 \log x\, dx &= \lim_{\delta \downarrow 0} \int_\delta^1 \log x\, dx = \lim_{\delta \downarrow 0} \left[x \log x - x \right]_\delta^1 \\
&= \lim_{\delta \downarrow 0} \left\{ -1 - \delta \log \delta - \delta \right\} = -1 - \lim_{\delta \downarrow 0} (\delta \log \delta) \\
&= -1 - \lim_{\delta \downarrow 0} \frac{\log \delta}{\delta^{-1}} = -1 - \lim_{\delta \downarrow 0} \frac{\delta^{-1}}{-\delta^{-2}} \quad \text{by l'Hôspital's Rule} \\
&= -1.
\end{aligned}
$$

Similarly we have by a change of variable: $y = 1 - x$,

$$
(6) \qquad \int_0^1 \log(1 - x)\, dx = -\int_1^0 \log y\, dy = \int_0^1 \log y\, dy = -1.
$$

Substituting (5) and (6) in (4), we have

$$
(7) \qquad 0 \geq \int_0^1 \log(1 - e^{-x})\, dx \geq -2.
$$

Then by (3) and (7) we have

$$
(8) \qquad \int_0^1 |f(x)|\, dx \leq 2.
$$

Next consider the improper Riemann integral

(9)
$$\int_1^\infty |f(x)|\,dx = \int_1^\infty \frac{1}{1+x^2} \left| \log(1 - e^{-x}) \right| dx$$

$$\leq \left| \log(1 - e^{-1}) \right| \int_1^\infty \frac{1}{1+x^2}\,dx$$

$$= \left| \log(1 - e^{-1}) \right| \lim_{a \to \infty} \int_1^a \frac{1}{1+x^2}\,dx$$

$$= \left| \log(1 - e^{-1}) \right| \lim_{a \to \infty} \Big[\arctan x \Big]_1^a$$

$$= \left| \log(1 - e^{-1}) \right| \{ \frac{\pi}{2} - \frac{\pi}{4} \}$$

$$= \frac{\pi}{4} \left| \log(1 - e^{-1}) \right|.$$

3. By (8) and (9) we have

(10)
$$\int_0^\infty |f(x)|\,dx = \int_0^1 |f(x)|\,dx + \int_1^\infty |f(x)|\,dx$$

$$\leq 2 + \frac{\pi}{4} \left| \log(1 - e^{-1}) \right|.$$

This implies

(11)
$$\int_{(0,\infty)} |f|\,d\mu_L = \int_0^\infty |f(x)|\,dx \leq 2 + \frac{\pi}{4} \left| \log(1 - e^{-1}) \right| < \infty.$$

This shows that $|f|$ is μ_L-integrable on $(0, \infty)$ and thus f is μ_L-integrable on $(0, \infty)$, that is,

$$\int_{(0,\infty)} f\,d\mu_L \in \mathbb{R}.$$

Moreover we have the estimate:

$$\left| \int_{(0,\infty)} f\,d\mu_L \right| \leq \int_{(0,\infty)} |f|\,d\mu_L \leq 2 + \frac{\pi}{4} \left| \log(1 - e^{-1}) \right|.$$

(Observe that since $f < 0$ on $(0, \infty)$ we actually have $\left| \int_{(0,\infty)} f\,d\mu_L \right| = \int_{(0,\infty)} |f|\,d\mu_L$.)
This completes the proof. ∎

Prob. 9.51. Let f be a real valued function on $\mathbb{R}$ defined by

$$f(x) = \begin{cases} \frac{1}{\sqrt{x}} & \text{for } x \in (0, 1) \\ 0 & \text{for } x \in \mathbb{R} \setminus (0, 1). \end{cases}$$

Let $(r_n : n \in \mathbb{N})$ be an arbitrary enumeration of the rational numbers in $\mathbb{R}$. For every $n \in \mathbb{N}$ define a function g_n on $\mathbb{R}$ by setting

$$g_n(x) = f(x - r_n) \quad \text{for } x \in \mathbb{R},$$

and then define a function g on $\mathbb{R}$ by setting

$$g(x) = \sum_{n \in \mathbb{N}} 2^{-n} g_n(x) \quad \text{for } x \in \mathbb{R}.$$

(a) Show that f is μ_L-integrable on $\mathbb{R}$.
(b) Show that g is μ_L-integrable on $\mathbb{R}$.
(c) Show that g has the following properties :
1° $g < \infty$ a.e. on $\mathbb{R}$.
2° $\lim\limits_{x \downarrow r_n} g(x) = \infty$ for every $n \in \mathbb{N}$.
3° g is unbounded on any interval in $\mathbb{R}$.
4° g is discontinuous at every $x \in \mathbb{R}$.
(d) Show that $g^2 < \infty$ a.e. on $\mathbb{R}$ but $\int_I g^2 \, d\mu_L = \infty$ for every interval I in $\mathbb{R}$.

Proof. **(a)** Non-negativity of f implies that $\int_{\mathbb{R}} f \, d\mu_L$ is equal to the improper Riemann integral $\int_{\infty}^{\infty} f(x) \, dx$ provided that the latter exists. Now we have

$$\int_{\infty}^{\infty} f(x) \, dx = \int_0^1 \frac{1}{\sqrt{x}} \, dx = \lim_{\delta \downarrow 0} \int_\delta^1 \frac{1}{\sqrt{x}} \, dx = \lim_{\delta \downarrow 0} \left[2\sqrt{x} \right]_\delta^1 = 2.$$

This shows that $\int_{\mathbb{R}} f \, d\mu_L = 2$.

(b) Let us show that g is μ_L-integrable on $\mathbb{R}$. By Theorem 9.31 (Translation Invariance of the Lebesgue Integral on $\mathbb{R}$) we have

$$\int_{\mathbb{R}} g_n(x) \, \mu_L(dx) = \int_{\mathbb{R}} f(x - r_n) \, \mu_L(dx) = \int_{\mathbb{R}} f(x) \, \mu_L(dx) = 2.$$

Then applying Theorem 8.5 (Monotone Convergence Theorem) we have

$$\int_{\mathbb{R}} g(x) \, \mu_L(dx) = \int_{\mathbb{R}} \left\{ \sum_{n \in \mathbb{N}} \frac{1}{2^n} g_n(x) \right\} \mu_L(dx)$$

$$= \sum_{n \in \mathbb{N}} \frac{1}{2^n} \int_{\mathbb{R}} g_n(x) \, \mu_L(dx) = \sum_{n \in \mathbb{N}} \frac{1}{2^n} 2 = 2.$$

This proves the μ_L-integrability of g on $\mathbb{R}$.

(c) 1°. μ_L-integrability of g on $\mathbb{R}$ implies that $g < \infty$ a.e. on $\mathbb{R}$.
2°. Observe that we have $\lim\limits_{x \downarrow 0} f(x) = \infty$. Then we have

$$\lim_{x \downarrow r_n} g_n(x) = \lim_{x \downarrow r_n} f(x - r_n) = \lim_{y \downarrow 0} f(y) = \infty.$$

Then $g_n \le g$ on $\mathbb{R}$ implies that $\lim\limits_{x \downarrow r_n} g_n(x) \le \lim\limits_{x \downarrow r_n} g(x)$ and thus $\lim\limits_{x \downarrow r_n} g(x) = \infty$.

3°. Let us show that g is unbounded on any interval (α, β) in $\mathbb{R}$. Since the set of all rational numbers is dense in $\mathbb{R}$, there exists $n \in \mathbb{N}$ such that $r_n \in (\alpha, \beta)$. Then since $\lim_{x \downarrow r_n} g(x) = \infty$, g is not bounded on (α, β).

4°. To show that g is discontinuous at every $x \in \mathbb{R}$, let us assume the contrary, that is, let us assume that g is continuous at some $x_0 \in \mathbb{R}$. Then for every $\varepsilon > 0$ there exists $\delta > 0$ such that $|g(x) - g(x_0)| < \varepsilon$ for $x \in (x_0 - \delta, x_0 + \delta)$, that is,

$$g(x_0) - \varepsilon < g(x) < g(x_0) + \varepsilon \quad \text{for } x \in (x_0 - \delta, x_0 + \delta).$$

This contradicts property 3° that g is unbounded on any interval (α, β) in $\mathbb{R}$.

(d) We have $g < \infty$ a.e. on $\mathbb{R}$. This implies that $g^2 < \infty$ a.e. on $\mathbb{R}$. Let us show that $\int_I g^2 d\mu_L = \infty$ for every interval I in $\mathbb{R}$. Since $g^2 \geq 0$, the integral of g^2 increases when the domain of integration increases. Note also that every interval in $\mathbb{R}$ contains an open interval. Thus it suffices to show that $\int_J g^2 d\mu_L = \infty$ for every open interval J in $\mathbb{R}$.

Let (α, β) be an arbitrary open interval in $\mathbb{R}$. Let us show that there exists $n \in \mathbb{N}$ such that $\int_{(\alpha,\beta)} g_n^2 d\mu_L = \infty$.

Let $n \in \mathbb{N}$ be such that $r_n \in (\alpha, \beta)$ and consider the function $g_n(x) = f(x - r_n)$ for $x \in \mathbb{R}$. Since $f(x) = 0$ for $x \in \mathbb{R} \setminus (0, 1)$, $g_n(x) = f(x - r_n) = 0$ for $x \in \mathbb{R} \setminus (r_n, r_n + 1)$. Then we have

(1)
$$\int_{(\alpha,\beta)} g_n^2(x)\, \mu_L(dx) = \int_{(r_n, (r_n+1) \wedge \beta)} g_n^2(x)\, \mu_L(dx)$$

$$= \int_{(r_n, (r_n+1) \wedge \beta)} f^2(x - r_n)\, \mu_L(dx) = \int_{(r_n, (r_n+1) \wedge \beta)} \frac{1}{x - r_n}\, \mu_L(dx)$$

$$= \lim_{\delta \downarrow r_n} \int_\delta^{(r_n+1) \wedge \beta} \frac{1}{x - r_n}\, dx = \lim_{\delta \downarrow r_n} \Big[\log(x - r_n) \Big]_\delta^{(r_n+1) \wedge \beta}$$

$$= \log \big\{ (r_n + 1) \wedge \beta - r_n \big\} - \lim_{\delta \downarrow r_n} \log \big\{ \delta - r_n \big\}$$

$$= \log \big\{ (r_n + 1) \wedge \beta - r_n \big\} - \{-\infty\}.$$

Now if $(r_n + 1) \wedge \beta = r_n + 1$ then we have

$$\log \big\{ (r_n + 1) \wedge \beta - r_n \big\} = \log \big\{ r_n + 1 - r_n \big\} = \log 1 = 0.$$

On the other hand, if $(r_n + 1) \wedge \beta = \beta > r_n$ then we have

$$\log \big\{ (r_n + 1) \wedge \beta - r_n \big\} = \log \big\{ \beta - r_n \big\} \in \mathbb{R}.$$

Thus we have $\log \big\{ (r_n + 1) \wedge \beta - r_n \big\} := \gamma \in \mathbb{R}$. Substituting this in (1), we have

(2)
$$\int_{(\alpha,\beta)} g_n^2(x)\, \mu_L(dx) = \gamma - \{-\infty\} = \infty.$$

Finally, since $g_n^2 \leq g^2$ on $\mathbb{R}$, (2) implies that $\int_{(\alpha,\beta)} g^2(x)\, \mu_L(dx) = \infty$. ∎

Prob. 9.52. Let $a_{n,k} \geq 0$ for $n, k \in \mathbb{N}$. For $\alpha, \beta \in \mathbb{R}$, write $\alpha \wedge \beta$ for $\min\{\alpha, \beta\}$.

(a) Show that

(1) $$\lim_{n \to \infty} \sum_{k \in \mathbb{N}} \frac{1}{2^k}(a_{n,k} \wedge 1) = 0 \Leftrightarrow \lim_{n \to \infty} a_{n,k} = 0, \forall k \in \mathbb{N}.$$

(b) Show that

(2) $$\lim_{n \to \infty} \sum_{k \in \mathbb{N}} \frac{1}{2^k} \frac{a_{n,k}}{1 + a_{n,k}} = 0 \Leftrightarrow \lim_{n \to \infty} a_{n,k} = 0, \forall k \in \mathbb{N}.$$

Proof. 1. Let us prove (1). Since all the summands in (1) are nonnegative we have

$$\lim_{n \to \infty} \sum_{k \in \mathbb{N}} \frac{1}{2^k}(a_{n,k} \wedge 1) = 0 \Rightarrow \lim_{n \to \infty} \frac{1}{2^k}(a_{n,k} \wedge 1) = 0, \forall k \in \mathbb{N}$$

$$\Rightarrow \lim_{n \to \infty} (a_{n,k} \wedge 1) = 0, \forall k \in \mathbb{N}$$

$$\Rightarrow \lim_{n \to \infty} a_{n,k} = 0, \forall k \in \mathbb{N}.$$

Conversely suppose we have $\lim_{n \to \infty} a_{n,k} = 0$ for every $k \in \mathbb{N}$. Consider $(\mathbb{R}, \mathfrak{M}_L, \mu_L)$. Let $I = [0, 1) \in \mathfrak{M}_L$. Decompose I into countably many disjoint subintervals $\{I_k : k \in \mathbb{N}\}$ where I_k is defined by setting

$$I_k = \left[\sum_{j=1}^{k-1} \frac{1}{2^j}, \sum_{j=1}^{k} \frac{1}{2^j} \right) \quad \text{for } k \in \mathbb{N},$$

with the convention that $\sum_{j=1}^{0} \frac{1}{2^j} = 0$. Note that $\mu_L(I_k) = \frac{1}{2^k}$ for every $k \in \mathbb{N}$. For each $n \in \mathbb{N}$ let us define a function f_n on I by setting

$$f_n(x) = (a_{n,k} \wedge 1) \quad \text{for } x \in I_k \text{ for every } k \in \mathbb{N}.$$

Consider the sequence of functions $(f_n : n \in \mathbb{N})$ on I. Observe that our assumption $\lim_{n \to \infty} a_{n,k} = 0$ for every $k \in \mathbb{N}$ implies $\lim_{n \to \infty} (a_{n,k} \wedge 1) = 0$ for every $k \in \mathbb{N}$. Thus we have

$$\lim_{n \to \infty} f_n(x) = \lim_{n \to \infty} (a_{n,k} \wedge 1) = 0 \quad \text{for } x \in I_k.$$

Since $I = \bigcup_{k \in \mathbb{N}} I_k$, every $x \in I$ is in I_k for some $k \in \mathbb{N}$. Thus the last equality implies

$$\lim_{n \to \infty} f_n(x) = 0 \quad \text{for every } x \in I.$$

Now we have $|f_n| \leq 1$ on I for every $n \in \mathbb{N}$ and the constant function 1 on I is μ_L-integrable on I. Thus by Theorem 9.20 (Dominated Convergence Theorem) we have

$$\lim_{n \to \infty} \int_I f_n(x) \, \mu_L(dx) = \int_I \lim_{n \to \infty} f_n(x) \, \mu_L(dx) = \int_I 0 \, \mu_L(dx) = 0.$$

On the other hand for the step function f_n we have

$$\int_I f_n(x) \, \mu_L(dx) = \sum_{k \in \mathbb{N}} (a_{n,k} \wedge 1)\mu_L(I_k) = \sum_{k \in \mathbb{N}} (a_{n,k} \wedge 1)\frac{1}{2^k}.$$

Thus we have

$$\lim_{n\to\infty} \sum_{k\in\mathbb{N}} \frac{1}{2^k}(a_{n,k} \wedge 1) = 0.$$

This completes the proof of (1).

2. Let us prove (2). Since all the summands in (1) are nonnegative we have

(3)
$$\lim_{n\to\infty} \sum_{k\in\mathbb{N}} \frac{1}{2^k}\frac{a_{n,k}}{1+a_{n,k}} = 0 \Rightarrow \lim_{n\to\infty} \frac{1}{2^k}\frac{a_{n,k}}{1+a_{n,k}} = 0, \forall k \in \mathbb{N}$$

$$\Rightarrow \lim_{n\to\infty} \frac{a_{n,k}}{1+a_{n,k}} = 0, \forall k \in \mathbb{N}.$$

Let us show $\lim_{n\to\infty} a_{n,k} = 0$ for every $k \in \mathbb{N}$. Now if $a_{n,k} = 0$ for all but finitely many $n \in \mathbb{N}$ then we have $\lim_{n\to\infty} a_{n,k} = 0$. It remains to consider the case that $a_{n,k} \neq 0$ for infinitely many $n \in \mathbb{N}$. Existence or non-existence of $a_{n,k} = 0$ is irrelevant to having $\lim_{n\to\infty} a_{n,k} = 0$. Thus we drop every $a_{n,k} = 0$ from the sequence $(a_{n,k} : n \in \mathbb{N})$ and write $(a_{n,k} : n \in \mathbb{N})$ for the resulting sequence. Then we have $a_{n,k} > 0$ for every $n \in \mathbb{N}$. Then we have

(4)
$$\lim_{n\to\infty} \frac{a_{n,k}}{1+a_{n,k}} = 0 \Rightarrow \lim_{n\to\infty} \frac{1+a_{n,k}}{a_{n,k}} = \infty \Rightarrow \lim_{n\to\infty} \left\{\frac{1}{a_{n,k}} + 1\right\} = \infty$$

$$\Rightarrow \lim_{n\to\infty} \frac{1}{a_{n,k}} = \infty \Rightarrow \lim_{n\to\infty} a_{n,k} = 0.$$

Substituting (4) in (3), we have

$$\lim_{n\to\infty} \sum_{k\in\mathbb{N}} \frac{1}{2^k}\frac{a_{n,k}}{1+a_{n,k}} = 0 \Rightarrow \lim_{n\to\infty} a_{n,k} = 0, \forall k \in \mathbb{N}.$$

The converse is proved by the same argument as for proving (1). ∎

Prob. 9.53. Prove the following:

Theorem. (Dominated Convergence Theorem with Index Set $\mathbb{R}$) Given a measure space $(X, \mathfrak{A}, \mu)$. Let $D \in \mathfrak{A}$ and $(\alpha, \beta) \subset \mathbb{R}$. For every $t \in [\alpha, \beta]$, let f_t be an extended real-valued $\mathfrak{A}$-measurable function on D. Let $t_0 \in [\alpha, \beta]$. Suppose that

1° $\lim_{t \to t_0} f_t(x) = f_{t_0}(x)$ a.e. on D;

2° there exists a μ-integrable nonnegative extended real-valued $\mathfrak{A}$-measurable function g on D such that $|f_t(x)| \le g(x)$ for all $(x, t) \in D \times [\alpha, \beta]$.

Then f_t is μ-integrable on D for every $t \in [\alpha, \beta]$ and moreover we have

(1) $\lim_{t \to t_0} \int_D f_t(x) \, \mu(dx) = \int_D f_{t_0}(x) \, \mu(dx)$.

(Observe that for the particular case $(\alpha, \beta) = \mathbb{R} = (-\infty, \infty)$ we have $[\alpha, \beta] = [-\infty, \infty]$.)

Proof. μ-integrability of f_t follows from the fact that $|f_t| \le g$ on D and g is μ-integrable on D.

To prove (1) we quote the following statement.

Proposition. Let g be a real-valued function on $[\alpha, \beta]$ and let $t_0 \in [\alpha, \beta]$. Then

(2) $\lim_{t \to t_0} g(t) = \xi \in \mathbb{R}$

$\Leftrightarrow \lim_{n \to \infty} g(t_n) = \xi, \quad \forall \, (t_n : n \in \mathbb{N}) \subset [\alpha, \beta]$ such that $t_n \ne t_0$ and $\lim_{n \to \infty} t_n = t_0$.

Then, to prove (1), it suffices to show that for every sequence $(t_n : n \in \mathbb{N})$ in $[\alpha, \beta]$ such that $t_n \ne t_0$ and $\lim_{n \to \infty} t_n = t_0$, we have

(3) $\lim_{n \to \infty} \int_D f_{t_n}(x) \, \mu_L(dx) = \int_D f_{t_0}(x) \, \mu_L(dx)$.

But Theorem 9.20 (Dominated Convergence Theorem) implies (3). ∎

Prob. 9.54. Prove the following:

Proposition. Consider the measure space $(\mathbb{R}, \mathfrak{M}_L, \mu_L)$. Let $D \in \mathfrak{M}_L$ and $(\alpha, \beta) \subset \mathbb{R}$. For every $t \in [\alpha, \beta]$, let f_t be a real-valued $\mathfrak{M}_L$-measurable function on D; Suppose that

1° for every $x \in D$, $f_t(x)$ is a continuous function of $t \in [\alpha, \beta]$;

2° there exists a μ_L-integrable nonnegative extended real-valued $\mathfrak{M}_L$-measurable function g on D such that $|f_t(x)| \le g(x)$ for all $(x, t) \in D \times [\alpha, \beta]$.

Let us define a function I on $[\alpha, \beta]$ by setting $I(t) = \int_D f_t(x)\, \mu_L(dx)$ for $t \in [\alpha, \beta]$. Then I is real-valued and continuous on $[\alpha, \beta]$.

(Observe that for the particular case $(\alpha, \beta) = \mathbb{R} = (-\infty, \infty)$ we have $[\alpha, \beta] = [-\infty, \infty]$.)

Proof. The fact that $|f_t| \le g$ and g is μ_L-integrable on D implies that f_t is μ_L-integrable on D. Thus we have $I(t) = \int_D f_t(x)\, \mu_L(dx) \in \mathbb{R}$ for every $t \in [\alpha, \beta]$ and thus I is real-valued on $[\alpha, \beta]$.

To show that I is continuous on $[\alpha, \beta]$, we show that I is continuous at every $t_0 \in [\alpha, \beta]$. Let $t_0 \in [\alpha, \beta]$. According to 1°, for every $x \in D$, $f_t(x)$ is a continuous function of $t \in [\alpha, \beta]$. Then in particular continuity of $f_t(x)$ at t_0 implies that we have

$$\lim_{t \to t_0} f_t(x) = f_{t_0}(x) \quad \text{for every } x \in D.$$

This shows that condition 1° in Prob. 9.53 is satisfied. Then by (1) of Prob. 9.52 we have

$$\lim_{t \to t_0} \int_D f_t(x)\, \mu_L(dx) = \int_D f_{t_0}(x)\, \mu_L(dx),$$

that is, we have $\lim_{t \to t_0} I(t) = I(t_0)$. This proves the continuity of I at t_0. ∎

§10 Signed Measures

Prob. 10.1. Let $(X, \mathfrak{A}, \mu)$ be a measure space and let $A \in \mathfrak{A}$. Let f be an extended real-valued $\mathfrak{A}$-measurable function on A. Let $A_0 \subset A$ and $A_0 \in \mathfrak{A}$. Prove the following statements regarding the existence of $\int_A f \, d\mu$ and $\int_{A_0} f \, d\mu$ in $\overline{\mathbb{R}}$:

(a)	$\int_A f \, d\mu \in \overline{\mathbb{R}}$	$\Rightarrow \int_{A_0} f \, d\mu \in \overline{\mathbb{R}}$.
(b)	$\int_A f \, d\mu \in \mathbb{R}$	$\Rightarrow \int_{A_0} f \, d\mu \in \mathbb{R}$.
(c)	$\int_A f \, d\mu = \infty$	$\Rightarrow \int_{A_0} f \, d\mu \neq -\infty$.
(d)	$\int_A f \, d\mu = -\infty$	$\Rightarrow \int_{A_0} f \, d\mu \neq \infty$.
(e)	$\int_A f \, d\mu \in (-\infty, \infty]$	$\Rightarrow \int_{A_0} f \, d\mu \in (-\infty, \infty]$.
(f)	$\int_A f \, d\mu \in [-\infty, \infty)$	$\Rightarrow \int_{A_0} f \, d\mu \in [-\infty, \infty)$.

Proof. 1. Let us prove (a). Existence of $\int_A f \, d\mu$ in $\overline{\mathbb{R}}$ implies that at least one of the two integrals $\int_A f^+ \, d\mu$ and $\int_A f^- \, d\mu$ is finite. Now $A_0 \subset A$ implies

$$0 \leq \int_{A_0} f^+ \, d\mu \leq \int_A f^+ \, d\mu \quad \text{and} \quad 0 \leq \int_{A_0} f^- \, d\mu \leq \int_A f^- \, d\mu.$$

Thus at least one of the two integrals $\int_{A_0} f^+ \, d\mu$ and $\int_{A_0} f^- \, d\mu$ is finite. This implies that $\int_{A_0} f \, d\mu$ exists.

2. Let us prove (b). The assumption $\int_A f \, d\mu \in \mathbb{R}$ implies that $\int_A f^+ \, d\mu \in [0, \infty)$ and $\int_A f^- \, d\mu \in [0, \infty)$. These then imply that $\int_{A_0} f^+ \, d\mu \in [0, \infty)$ and $\int_{A_0} f^- \, d\mu \in [0, \infty)$. Then $\int_{A_0} f \, d\mu \in \mathbb{R} = \int_{A_0} f^+ \, d\mu - \int_{A_0} f^- \, d\mu \in \mathbb{R}$.

3. Let us prove (c). The assumption $\int_A f \, d\mu = \infty$ implies that $\int_A f^+ \, d\mu = \infty$ and $\int_A f^- \, d\mu \in [0, \infty)$. The latter of the two implies that $\int_{A_0} f^- \, d\mu \in [0, \infty)$. Then we have $\int_{A_0} f \, d\mu = \int_{A_0} f^+ \, d\mu - \int_{A_0} f^- \, d\mu \neq -\infty$.

4. Let us prove (d). The assumption $\int_A f \, d\mu = -\infty$ implies that $\int_A f^+ \, d\mu \in [0, \infty)$ and $\int_A f^- \, d\mu = \infty$. The former of the two implies that $\int_{A_0} f^+ \, d\mu \in [0, \infty)$. Then we have $\int_{A_0} f \, d\mu = \int_{A_0} f^+ \, d\mu - \int_{A_0} f^- \, d\mu \neq \infty$.

5. Let us prove (e). Assume that $\int_A f \, d\mu \in (-\infty, \infty]$. Then we have either $\int_A f \, d\mu \in \mathbb{R}$ or $\int_A f \, d\mu = \infty$. If $\int_A f \, d\mu \in \mathbb{R}$ then $\int_{A_0} f \, d\mu \in \mathbb{R}$ by (b). If $\int_A f \, d\mu = \infty$ then $\int_{A_0} f \, d\mu \neq -\infty$ by (c). Thus $\int_{A_0} f \, d\mu \in (-\infty, \infty]$.

6. (f) is proved by the same argument as for (e) above. This completes the proof. ∎

Prob. 10.2. Let $(X, \mathfrak{A}, \lambda)$ be a signed measure space. Let f be an extended real-valued $\mathfrak{A}$-measurable function on a set $A \in \mathfrak{A}$. Let $A_0 \in \mathfrak{A}$ and $A_0 \subset A$.
(a) Show that if $\int_A f \, d\lambda$ exists in $\overline{\mathbb{R}}$, then so does $\int_{A_0} f \, d\lambda$.
(b) Show that if $\int_A f \, d\lambda$ exists in $\mathbb{R}$, then so does $\int_{A_0} f \, d\lambda$.

Proof. 1. Let $\lambda = \lambda^+ - \lambda^-$ be the Jordan decomposition of λ. According to Definition 10.32, we define

$$\int_A f \, d\lambda = \int_A f \, d\lambda^+ - \int_A f \, d\lambda^-,$$

provided that $\int_A f \, d\lambda^+$, $\int_A f \, d\lambda^-$ and $\int_A f \, d\lambda^+ - \int_A f \, d\lambda^-$ exists in $\overline{\mathbb{R}}$.
 2. Let us prove (a). Assume that $\int_A f \, d\lambda$ exists in $\overline{\mathbb{R}}$. Then $\int_A f \, d\lambda^+$, $\int_A f \, d\lambda^-$ and $\int_A f \, d\lambda^+ - \int_A f \, d\lambda^-$ exist in $\overline{\mathbb{R}}$. Existence of $\int_A f \, d\lambda^+ - \int_A f \, d\lambda^-$ in $\overline{\mathbb{R}}$ implies that we have either

(1) $$\int_A f \, d\lambda^+ \in (-\infty, \infty] \quad \text{and} \quad \int_A f \, d\lambda^- \in [-\infty, \infty),$$

or

(2) $$\int_A f \, d\lambda^+ \in [-\infty, \infty) \quad \text{and} \quad \int_A f \, d\lambda^- \in (-\infty, \infty].$$

Consider case (1). In this case $\int_A f \, d\lambda^+ \in (-\infty, \infty]$ implies $\int_{A_0} f \, d\lambda^+ \in (-\infty, \infty]$ and $\int_A f \, d\lambda^- \in [-\infty, \infty)$ implies $\int_{A_0} f \, d\lambda^- \in [-\infty, \infty)$ according to Prob. 10.1. Thus $\int_{A_0} f \, d\lambda^+ - \int_{A_0} f \, d\lambda^-$ exists in $\overline{\mathbb{R}}$ and then $\int_{A_0} f \, d\lambda$ exists in $\overline{\mathbb{R}}$.
   For case (2) we show the existence of $\int_{A_0} f \, d\lambda$ in $\overline{\mathbb{R}}$ by the same argument as for case (1). This completes the proof of (a).
   3. Let us prove (b). If $\int_A f \, d\lambda$ exists in $\mathbb{R}$ then certainly $\int_A f \, d\lambda$ exists in $\overline{\mathbb{R}}$. Thus $\int_A f \, d\lambda$ is defined by setting

$$\int_A f \, d\lambda = \int_A f \, d\lambda^+ - \int_A f \, d\lambda^-.$$

Then $\int_A f \, d\lambda \in \mathbb{R}$ implies that $\int_A f \, d\lambda^+ \in \mathbb{R}$ and $\int_A f \, d\lambda^- \in \mathbb{R}$. Then Prob. 10.1 implies that $\int_{A_0} f \, d\lambda^+ \in \mathbb{R}$ and $\int_{A_0} f \, d\lambda^- \in \mathbb{R}$. Thus $\int_{A_0} f \, d\lambda^+ - \int_{A_0} f \, d\lambda^-$ exists in $\mathbb{R}$ and then by Definition 10.32 we have

$$\int_{A_0} f \, d\lambda = \int_{A_0} f \, d\lambda^+ - \int_{A_0} f \, d\lambda^- \in \mathbb{R}.$$

This completes the proof of (b). ∎

Prob. 10.3. Given a measure space $(X, \mathfrak{A}, \mu)$. Let f be a μ semi-integrable extended real-valued $\mathfrak{A}$-measurable function on X. Define a signed measure λ on $(X, \mathfrak{A})$ by setting $\lambda(E) = \int_E f \, d\mu$ for $E \in \mathfrak{A}$.
(a) Find a Hahn decomposition of $(X, \mathfrak{A}, \lambda)$ in terms of f.
(b) Find a Jordan decomposition of λ in terms of f.
(c) Find the total variation $|\lambda|$ of λ in terms of f.

Proof. The set function λ on $\mathfrak{A}$ defined by setting $\lambda(E) = \int_E f \, d\mu$ for $E \in \mathfrak{A}$ is a signed measure on $(X, \mathfrak{A})$ according to Proposition 10.3.

1. By Definition 10.13, a Hahn decomposition of the signed measure space $(X, \mathfrak{A}, \lambda)$ is a pair $\{A, B\}$ of $\mathfrak{A}$-measurable subsets of X such that $A \cap B = \emptyset$, $A \cup B = X$, A is a positive set and B is a negative set of $(X, \mathfrak{A}, \lambda)$. To find such sets A and B, let us define

(1) $A = \{x \in X : f(x) \geq 0\}$ and $B = \{x \in X : f(x) < 0\}$.

Then we have $A, B \in \mathfrak{A}$, $A \cap B = \emptyset$ and $A \cup B = X$. It remains to show that A is a positive set and B is a negative set in the signed measure space $(X, \mathfrak{A}, \lambda)$.

Let $A_0 \subset A$ and $A_0 \in \mathfrak{A}$. Then we have $\lambda(A_0) = \int_{A_0} f \, d\mu \geq 0$. This shows that A is a positive set in the signed measure space $(X, \mathfrak{A}, \lambda)$. Similarly for $B_0 \subset B$ and $B_0 \in \mathfrak{A}$, we have $\lambda(B_0) = \int_{B_0} f \, d\mu \leq 0$. This shows that B is a negative set in the signed measure space $(X, \mathfrak{A}, \lambda)$. This completes the proof that the pair $\{A, B\}$ as defined by (1) is a Hahn decomposition of $(X, \mathfrak{A}, \lambda)$.

2. By Definition 10.17, a Jordan decomposition of the signed measure λ is a pair $\{\lambda^+, \lambda^-\}$ of positive measures on $(X, \mathfrak{A})$ such that $\lambda^+ \perp \lambda^-$ and $\lambda = \lambda^+ - \lambda^-$. To find such measures λ^+ and λ^-, let us define two set functions λ^+ and λ^- on $\mathfrak{A}$ by setting, with the two sets A and B defined by (1),

(2) $\lambda^+(E) = \lambda(E \cap A) = \int_{E \cap A} f \, d\mu$ for $E \in \mathfrak{A}$.

(3) $\lambda^-(E) = -\lambda(E \cap B) = -\int_{E \cap B} f \, d\mu$ for $E \in \mathfrak{A}$.

It is easily verified that λ^+ and λ^- as defined above are measures on $(X, \mathfrak{A})$. Observe also that for every $E \in \mathfrak{A}$ we have

$$\lambda^+(E) - \lambda^-(E) = \int_{E \cap A} f \, d\mu + \int_{E \cap B} f \, d\mu = \int_{E \cap X} f \, d\mu = \int_E f \, d\mu = \lambda(E).$$

This shows that $\lambda = \lambda^+ - \lambda^-$.

It remains to show that $\lambda^+ \perp \lambda^-$. Recall that $A, B \in \mathfrak{A}$, $A \cap B = \emptyset$ and $A \cup B = X$. Then we have $\lambda^+(B) = \lambda(B \cap A) = \lambda(\emptyset) = 0$ and $\lambda^-(A) = -\lambda(A \cap B) = -\lambda(\emptyset) = 0$. This shows that $\lambda^+ \perp \lambda^-$. This completes the proof that the pair $\{\lambda^+, \lambda^-\}$ defined by (2) and (3) is a Jordan decomposition of λ.

3. By Definition 10.22, we have $|\lambda| = \lambda^+ + \lambda^-$. Thus for every $E \in \mathfrak{A}$ we have

$$|\lambda|(E) = \lambda^+(E) + \lambda^-(E) = \int_{E \cap A} f \, d\mu - \int_{E \cap B} f \, d\mu$$

$$= \int_{E \cap A} |f| \, d\mu + \int_{E \cap B} |f| \, d\mu = \int_E |f| \, d\mu.$$

This gives an expression of $|\lambda|$ in terms of f. ∎

Prob. 10.4. Let $(X, \mathfrak{A}, \lambda)$ be a signed measure space and let $\{\lambda^+, \lambda^-\}$ be a Jordan decomposition of λ. Let $E \in \mathfrak{A}$. Prove the following statements:
$$\lambda^+(E) = \sup\{\lambda(E_0) : E_0 \subset E, E_0 \in \mathfrak{A}\},$$
$$\lambda^-(E) = -\inf\{\lambda(E_0) : E_0 \subset E, E_0 \in \mathfrak{A}\}.$$

Proof. Let $\{A, B\}$ be a Hahn decomposition of $(X, \mathfrak{A}, \lambda)$. According to Theorem 10.21 we have

$$\lambda^+(E) = \lambda(E \cap A) \quad \text{for } E \in \mathfrak{A}.$$

$$\lambda^-(E) = -\lambda(E \cap B) \quad \text{for } E \in \mathfrak{A}.$$

Then for $E \in \mathfrak{A}$, we have

(1) $$\lambda^+(E) = \lambda(E \cap A) \leq \sup\{\lambda(E_0) : E_0 \subset E, E_0 \in \mathfrak{A}\}.$$

Conversely for $E_0 \subset E$ such that $E_0 \in \mathfrak{A}$, we have

$$\lambda(E_0) = \lambda^+(E_0) - \lambda^-(E_0) \leq \lambda^+(E_0) \leq \lambda^+(E),$$

and then

(2) $$\sup\{\lambda(E_0) : E_0 \subset E, E_0 \in \mathfrak{A}\} \leq \lambda^+(E).$$

By (1) and (2), we have

(3) $$\lambda^+(E) = \sup\{\lambda(E_0) : E_0 \subset E, E_0 \in \mathfrak{A}\}.$$

Next observe that $\lambda^- = (-\lambda)^+$. Then we have

(4) $$\lambda^-(E) = (-\lambda)^+(E) = \sup\{(-\lambda)(E_0) : E_0 \subset E, E_0 \in \mathfrak{A}\} \quad \text{by (3)}$$
$$= -\inf\{\lambda(E_0) : E_0 \subset E, E_0 \in \mathfrak{A}\}.$$

This completes the proof. ∎

Prob. 10.5. Let $(X, \mathfrak{A}, \lambda)$ be a signed measure space. Prove the following statements regarding the infinity of the signed measure λ:
(a) If there exists $E \in \mathfrak{A}$ such that $\lambda(E) = \infty$ then $\lambda(X) = \infty$.
(b) If there exists $E \in \mathfrak{A}$ such that $\lambda(E) = -\infty$ then $\lambda(X) = -\infty$.

Proof. 1. Let us prove (a). Suppose there exists $E \in \mathfrak{A}$ such that $\lambda(E) = \infty$. By Definition 10.1, a signed measure λ can assume at most one of the two infinite values ∞ and $-\infty$. This implies that $\lambda(E^c) \in (-\infty, \infty]$. Then we have $\lambda(X) = \lambda(E) + \lambda(E^c) = \infty$.
 2. (b) is proved similarly. ∎

Prob. 10.6. Given a signed measure space $(X, \mathfrak{A}, \lambda)$.
(a) Show that if $E \in \mathfrak{A}$ and $\lambda(E) > 0$, then there exists a subset E_0 of E which is a positive set for λ with $\lambda(E_0) \geq \lambda(E)$.
(a) Show that if $E \in \mathfrak{A}$ and $\lambda(E) < 0$, then there exists a subset E_0 of E which is a negative set for λ with $\lambda(E_0) \leq \lambda(E)$.

Proof. 1. Let us prove (a). Let $\{A, B\}$ be a Hahn decomposition for $(X, \mathfrak{A}, \lambda)$. Then A is a positive set and B is a negative set in $(X, \mathfrak{A}, \lambda)$. Suppose $E \in \mathfrak{A}$ and $\lambda(E) > 0$. Let $E_0 = A \cap E$. Being a $\mathfrak{A}$-measurable subset of a positive set A, E_0 is a positive set by Observation 10.10. Let us show $\lambda(E_0) \geq \lambda(E)$. Now by the finite additivity of λ on $\mathfrak{A}$ we have

(1) $\qquad \lambda(E) = \lambda(A \cap E) + \lambda(B \cap E) = \lambda(E_0) + \lambda(B \cap E)$.

Now $B \cap E$, being a $\mathfrak{A}$-measurable subset of a positive set B, is a negative set by Observation 10.10. Thus $\lambda(B \cap E) \leq 0$. This and (1) imply that $\lambda(E) \leq \lambda(E_0)$.
 2. (b) is proved by the same argument. ∎

Prob. 10.7. Let $(X, \mathfrak{A}, \lambda)$ be a signed measure space. Prove the following statements:
(a) Let A be a positive set for λ. If A^c is a negative set, then A is a maximal positive set.
(b) If A is a maximal positive set for λ, then A^c is a negative set for λ.
(c) Let A be a positive set for λ. Then A is a maximal positive set if and only if A^c is a negative set for λ.
(a′) Let B be a negative set for λ. If B^c is a positive set, then B is a maximal negative set.
(b′) If B is a maximal negative set for λ, then B^c is a positive set for λ.
(c′) Let B be a negative set for λ. Then B is a maximal negative set if and only if B^c is a positive set for λ.
(d) A set A is a maximal positive set for λ if and only if A^c is a maximal negative set for λ.
(d′) A set B is a maximal negative set for λ if and only if B^c is a maximal positive set for λ.

Proof. 1. Let us prove (a). Let A be a positive set for λ and suppose A^c is a negative set for λ. To show that A is a maximal positive set, we show that for an arbitrary positive set A' the set $A' \setminus A$ is a null set for λ according to Definition 10.12. Now $A' \setminus A = A' \cap A^c$, which is a $\mathfrak{A}$-measurable subset of the positive set A' and is thus a positive set. On the other hand, $A' \cap A^c$ is a $\mathfrak{A}$-measurable subset of the negative set A^c and is thus a negative set. Then $A' \setminus A$ is both a positive set and a negative set and is therefore a null set.

2. Let us prove (b). Let A be a maximal positive set for λ. Let us show that A^c is a negative set for λ. Suppose A^c is not a negative set. Then there exists a $\mathfrak{A}$-measurable subset E of A^c such that $\lambda(E) > 0$. Then by Prob. 10.6, there exists a subset E_0 of E which is a positive set and having $\lambda(E_0) \geq \lambda(E)$. Then since A is a maximal positive set and E_0 is a positive set, the set $E_0 \setminus A$ is a null set according to Definition 10.12. But $E_0 \subset A^c$ is disjoint from A. Thus $E_0 \setminus A = E_0$ and $\lambda(E_0 \setminus A) = \lambda(E_0) \geq \lambda(E) > 0$. This contradicts the fact that $E_0 \setminus A$ is a null set. Therefore A^c must be a negative set.

3. (c) follows from (a) and (b).

4. $(a′)$, $(b′)$ and $(c′)$ are proved in the same way as (a), (b) and (c) respectively.

5. Let us prove (d). Suppose A is a maximal positive set for λ. Then A^c is a negative set according to (b). Now A^c is a negative set and $(A^c)^c = A$ is a positive set. Then by $(a′)$, A^c is a maximal negative set.

Conversely suppose A^c is a maximal negative set. Then $(A^c)^c = A$ is a positive set according to $(b′)$. Now A is a positive set and A^c is a negative set. Then by (a), A is a maximal positive set.

6. $(d′)$ is proved in the same way as (d). ∎

Prob. 10.8. Consider the measure space $(X, \mathfrak{A}, \mu) = \big([0, 2\pi], \mathfrak{M}_L \cap [0, 2\pi], \mu_L\big)$. Define a signed measure λ on $([0, 2\pi], \mathfrak{M}_L \cap [0, 2\pi])$ by setting $\lambda(E) = \int_E \sin x \, \mu_L(dx)$ for every $E \in \mathfrak{M}_L \cap [0, 2\pi]$.

(a) Let $C = \big[\frac{4}{3}\pi, \frac{5}{3}\pi\big]$. Show that C is a negative set for λ.

(b) Show that C is not a maximal negative set for λ.

(c) Let $\varepsilon > 0$ be arbitrarily given. Find a $\mathfrak{M}_L \cap [0, 2\pi]$-measurable subset C' of C such that $\lambda(C') \geq \lambda(C)$ and $\lambda(E) > -\varepsilon$ for every $\mathfrak{M}_L \cap [0, 2\pi]$-measurable subset E of C'.

Proof. 1. Let us prove (a). Let us show first that $[\pi, 2\pi]$ is a negative set for λ. Let E be an arbitrary $\mathfrak{M}_L \cap [0, 2\pi]$-measurable subset of $[\pi, 2\pi]$. Then we have $\lambda(E) = \int_E \sin x \, \mu_L(dx) \leq 0$ since $\sin x \leq 0$ for $x \in E \subset [\pi, 2\pi]$. This shows that $[\pi, 2\pi]$ is a negative set for λ. Then C, being a $\mathfrak{M}_L \cap [0, 2\pi]$-measurable subset of the negative set $[\pi, 2\pi]$, is a negative set.

2. Let us prove (b). We are to show that C is not a maximal negative set for λ. Assume that C is a maximal negative set. Then by (b') of Prob. 10.7, C^c is a positive set for λ. This implies that every $\mathfrak{M}_L \cap [0, 2\pi]$-measurable subset of C^c is a positive set. Now we have $C^c = \big[0, \frac{4}{3}\pi\big) \cup \big(\frac{5}{3}\pi, 2\pi\big]$ and for $E := \big(\frac{5}{3}\pi, 2\pi\big]$ we have $\lambda(E) = \int_E \sin x \, \mu_L(dx) < 0$ and thus E is not a positive set. This is a contradiction. Therefore C cannot be a maximal negative set.

3. Let us prove (c). Let $\varepsilon > 0$ be arbitrarily given. Let C' be an interval contained in C and having $\mu_L(C') \leq \frac{\varepsilon}{2}$. Since $\sin x \leq 0$ for $x \in [\pi, 2\pi]$, we have

$$\lambda(C') = \int_C' \sin x \, \mu_L(dx) \geq \int_C \sin x \, \mu_L(dx) = \lambda(C).$$

Let E be an arbitrary $\mathfrak{M}_L \cap [0, 2\pi]$-measurable subset of of C'. Then we have

$$\lambda(E) = \int_E \sin x \, \mu_L(dx) \geq (-1)\mu_L(E) \quad \text{since } \sin x \geq -1$$

$$\geq (-1)\mu_L(C') \geq (-1) \cdot \frac{\varepsilon}{2} > -\varepsilon.$$

This completes the proof. ∎

Prob. 10.9. Given a signed measure space $(X, \mathfrak{A}, \lambda)$. For $E \in \mathfrak{A}$, let

$$\alpha_E = \sup \left\{ \sum_{k=1}^n |\lambda(E_k)| : \{E_1, \ldots, E_n\} \subset \mathfrak{A}, \text{ disjoint}, \bigcup_{k=1}^n E_k = E, n \in \mathbb{N} \right\},$$

and

$$\beta_E = \sup \left\{ \sum_{n \in \mathbb{N}} |\lambda(E_n)| : \{E_n : n \in \mathbb{N}\} \subset \mathfrak{A}, \text{ disjoint}, \bigcup_{n \in \mathbb{N}} E_n = E \right\}.$$

Observe that $|\lambda|(E) = \alpha_E$ according to Proposition 10.23. Show that $\alpha_E = \beta_E$.

Proof. A decomposition of E into a finite disjoint collection $\{E_1, \ldots, E_n\}$ in $\mathfrak{A}$ is a particular case of decomposing E into a countable disjoint collection $\{E_n : n \in \mathbb{N}\}$ in $\mathfrak{A}$ by treating $\{E_1, \ldots, E_n\}$ as $\{E_1, \ldots, E_n, \emptyset, \emptyset, \emptyset, \ldots\}$. This implies that $\alpha_E \leq \beta_E$. It remains to prove $\alpha_E \geq \beta_E$.

1. If there exists a countable decomposition $\{E_n : n \in \mathbb{N}\}$ such that $\sum_{n \in \mathbb{N}} |\lambda(E_n)| = \infty$ then $\beta_E = \infty$. Let $M > 0$ be arbitrarily given. Since $\sum_{n \in \mathbb{N}} |\lambda(E_n)| = \infty$, there exists $N \in \mathbb{N}$ such that $\sum_{n=1}^N |\lambda(E_n)| \geq M$. Let $E_0 = \bigcup_{n>N} E_n$. Then $\{E_1, \ldots, E_N, E_0\}$ is a finite decomposition of E and we have

$$\sum_{k=0}^N |\lambda(E_k)| = |\lambda(E_0)| + \sum_{k=1}^N |\lambda(E_k)| \geq M.$$

This implies that $\alpha_E \geq M$. Since this holds for an arbitrary $M > 0$, we have $\alpha_E = \infty = \beta_E$.

2. Consider the case that for every countable decomposition $\{E_n : n \in \mathbb{N}\}$ of E we have $\sum_{n \in \mathbb{N}} |\lambda(E_n)| < \infty$. Let $\varepsilon > 0$ be arbitrarily given. $\sum_{n \in \mathbb{N}} |\lambda(E_n)| < \infty$, there exists $N \in \mathbb{N}$ such that $\sum_{n=1}^N |\lambda(E_n)| > \sum_{n \in \mathbb{N}} |\lambda(E_n)| - \varepsilon$. Let $E_0 = \bigcup_{n>N} E_n$. Then $\{E_1, \ldots, E_N, E_0\}$ is a finite decomposition of E and we have

$$\sum_{k=0}^N |\lambda(E_k)| = |\lambda(E_0)| + \sum_{k=1}^N |\lambda(E_k)| > \sum_{n \in \mathbb{N}} |\lambda(E_n)| - \varepsilon.$$

Thus we have shown that for every countable decomposition $\{E_n : n \in \mathbb{N}\}$ of E such that $\sum_{n \in \mathbb{N}} |\lambda(E_n)| < \infty$, there exists a finite decomposition $\{E_1, \ldots, E_N, E_0\}$ such that

$$\sum_{n \in \mathbb{N}} |\lambda(E_n)| - \varepsilon < \sum_{k=0}^N |\lambda(E_k)| \leq \alpha_E,$$

and then

$$\sum_{n \in \mathbb{N}} |\lambda(E_n)| < \alpha_E + \varepsilon.$$

Then we have

$$\beta_E = \sup \left\{ \sum_{n \in \mathbb{N}} |\lambda(E_n)| : \{E_n : n \in \mathbb{N}\} \subset \mathfrak{A}, \text{ disjoint}, \bigcup_{n \in \mathbb{N}} E_n = E \right\} \leq \alpha_E + \varepsilon.$$

Since this holds for every $\varepsilon > 0$, we have $\beta_E \leq \alpha_E$. This completes the proof. ∎

Prob. 10.10. Let μ and ν be two positive measures on a measurable space $(X, \mathfrak{A})$. Suppose $\mu \perp \nu$. Show that if ρ is a positive measure on $(X, \mathfrak{A})$ and if ρ is nontrivial in the sense that $\rho(X) > 0$, then $(\mu + \rho) \perp (\nu + \rho)$ does not hold.

Proof. Suppose $(\mu+\rho) \perp (\nu+\rho)$ holds. Then there exist $C_1, C_2 \in \mathfrak{A}$ such that $C_1 \cap C_2 = \emptyset$, $C_1 \cup C_2 = X$, C_1 is a null set for $\nu + \rho$ and C_2 is a null set for $\mu + \rho$.

The fact that C_1 is a null set for $\nu + \rho$ implies that $(\nu + \rho)(C_1) = 0$. Then the fact that both ν and ρ are positive measures implies that $\nu(C_1) = 0$ and $\rho(C_1) = 0$.

Similarly the fact that C_2 is a null set for $\mu + \rho$ implies that $(\mu + \rho)(C_2) = 0$. Then the fact that both μ and ρ are positive measures implies that $\mu(C_2) = 0$ and $\rho(C_2) = 0$.

Thus we have $\rho(C_1) = 0$ and $\rho(C_2) = 0$. Then we have $\rho(X) = \rho(C_1) + \rho(C_2) = 0$. This contradicts the assumption that $\rho(X) > 0$. This shows that $(\mu + \rho) \perp (\nu + \rho)$ cannot hold. ∎

Prob. 10.11. Let μ and ν be two positive measures on a measurable space $(X, \mathfrak{A})$. Suppose for every $\varepsilon > 0$, there exists $E \in \mathfrak{A}$ such that $\mu(E) < \varepsilon$ and $\nu(E^c) < \varepsilon$. Show that $\mu \perp \nu$.

Proof. Assume that for every $\varepsilon > 0$ there exists $E \in \mathfrak{A}$ such that $\mu(E) < \varepsilon$ and $\nu(E^c) < \varepsilon$. Then for every $n \in \mathbb{N}$ there exists $E_n \in \mathfrak{A}$ such that

$$\mu(E_n) < \frac{1}{2^n} \quad \text{and} \quad \nu(E_n^c) < \frac{1}{2^n}.$$

Now we have $\sum_{n \in \mathbb{N}} \mu(E_n) \leq \sum_{n \in \mathbb{N}} \frac{1}{2^n} < \infty$ and this implies according to Theorem 6.6 (Borel-Cantelli Lemma) that

(1) $\mu\left(\limsup_{n \to \infty} E_n \right) = 0.$

Similarly we have $\sum_{n \in \mathbb{N}} \nu(E_n^c) \leq \sum_{n \in \mathbb{N}} \frac{1}{2^n} < \infty$ and this implies that

(2) $\nu\left(\limsup_{n \to \infty} E_n^c \right) = 0.$

Let $A = \limsup_{n \to \infty} E_n$ and let $B = A^c$. Then we have $\mu(A) = 0$ by (1). As for B we have

$$B = A^c = \left(\limsup_{n \to \infty} E_n \right)^c = \liminf_{n \to \infty} E_n^c \subset \limsup_{n \to \infty} E_n^c,$$

and then

$$\nu(B) \leq \nu\left(\limsup_{n \to \infty} E_n^c \right) = 0 \quad \text{by (2)},$$

so that we have $\nu(B) = 0$. Thus we have $A, B \in \mathfrak{A}$, $A \cap B = \emptyset$, $A \cup B = X$, $\mu(A) = 0$, and $\nu(B) = 0$. This shows that $\mu \perp \nu$. ∎

Prob. 10.12. Let $(X, \mathfrak{A}, \mu)$ be a σ-finite measure space and let f be a nonnegative extended real-valued $\mathfrak{A}$-measurable function on X such that $f < \infty$, μ-a.e. on X. Let λ be a measure on $(X, \mathfrak{A})$ defined by $\lambda(A) = \int_A f \, d\mu$ for $A \in \mathfrak{A}$. Show that λ is a σ-finite measure on $(X, \mathfrak{A})$, that is, there exists an increasing sequence $(G_n : n \in \mathbb{N})$ in $\mathfrak{A}$ such that $\bigcup_{n \in \mathbb{N}} G_n = X$ and $\mu(G_n) < \infty$ for every $n \in \mathbb{N}$.

Proof. Let f be a nonnegative extended real-valued $\mathfrak{A}$-measurable function on X such that $f < \infty$, μ-a.e. on X. Let f_0 be defined by

$$f_0(x) = \begin{cases} f(x) & \text{if } f(x) < \infty, \\ 0 & \text{if } f(x) = \infty. \end{cases}$$

Then f_0 is a nonnegative real-valued $\mathfrak{A}$-measurable function on X and $f = f_0$, μ-a.e. on X. Thus we have

$$(1) \qquad \lambda(A) = \int_A f \, d\mu = \int_A f_0 \, d\mu \quad \text{for } A \in \mathfrak{A}.$$

Since $(X, \mathfrak{A}, \mu)$ is a σ-finite measure space, there exists an increasing sequence $(E_n : n \in \mathbb{N})$ in $\mathfrak{A}$ such that $\bigcup_{n \in \mathbb{N}} E_n = X$ and $\mu(E_n) < \infty$ for every $n \in \mathbb{N}$. For $n \in \mathbb{N}$, let us define

$$F_n = \{x \in X : f_0(x) \in [0, n]\}.$$

Then $(F_n : n \in \mathbb{N})$ is an increasing sequence in $\mathfrak{A}$ and since $f_0 < \infty$ on X we have $\bigcup_{n \in \mathbb{N}} F_n = X$. Let $G_n = E_n \cap F_n$ for $n \in \mathbb{N}$. Then since $(E_n : n \in \mathbb{N})$ and $(F_n : n \in \mathbb{N})$ are increasing sequences in $\mathfrak{A}$, $(G_n : n \in \mathbb{N})$ is an increasing sequence in $\mathfrak{A}$. Moreover we have $\bigcup_{n \in \mathbb{N}} G_n = X$. This can be shown as follows. Let $x \in X$ be arbitrarily chosen. Since $\bigcup_{n \in \mathbb{N}} E_n = X$ and $\bigcup_{n \in \mathbb{N}} F_n = X$, there exist $n_1, n_2 \in \mathbb{N}$ such that $x \in E_{n_1}$ and $x \in F_{n_2}$. Let $n_0 = \max\{n_1, n_2\}$. Since $(E_n : n \in \mathbb{N})$ is an increasing sequence, $x \in E_{n_1}$ implies $x \in E_{n_0}$. Similarly since $(f_n : n \in \mathbb{N})$ is an increasing sequence, $x \in F_{n_2}$ implies $x \in F_{n_0}$. Thus $x \in E_{n_0} \cap F_{n_0} = G_{n_0}$. This shows that for an arbitrary $x \in X$ we have $x \in G_n$ for some $n \in \mathbb{N}$. Therefore we have $\bigcup_{n \in \mathbb{N}} G_n = X$.

Now $(G_n : n \in \mathbb{N})$ is an increasing sequence in $\mathfrak{A}$ and $\bigcup_{n \in \mathbb{N}} G_n = X$. Then $(\mathbf{1}_{G_n} : n \in \mathbb{N})$ is an increasing sequence of nonnegative real-valued $\mathfrak{A}$-measurable functions on X and $\lim_{n \to \infty} \mathbf{1}_{G_n} = \mathbf{1}_X$ on X. Then by (1) and by the Monotone Convergence Theorem, we have

$$(2) \quad \lambda(A) = \int_A f_0 \, d\mu = \int_A f_0 \mathbf{1}_X \, d\mu = \int_A f_0 \{ \lim_{n \to \infty} \mathbf{1}_{G_n} \} \, d\mu = \lim_{n \to \infty} \int_A f_0 \mathbf{1}_{G_n} \, d\mu.$$

Now let $n_0 \in \mathbb{N}$ be arbitrarily chosen. Then by (2) we have

$$\lambda(G_{n_0}) = \lim_{n \to \infty} \int_{G_{n_0}} f_0 \mathbf{1}_{G_n} \, d\mu = \lim_{n \to \infty} \int_{G_{n_0}} f_0 \mathbf{1}_{G_n} \mathbf{1}_{G_{n_0}} \, d\mu$$

$$= \int_{G_{n_0}} f_0 \mathbf{1}_{G_{n_0}} \, d\mu \leq n_0 \mu(G_{n_0}) \leq n_0 \mu(E_{n_0}) < \infty.$$

This shows that $\mu(G_n) < \infty$ for every $n \in \mathbb{N}$. ∎

Prob. 10.13. Let $(X, \mathfrak{A}, \mu)$ be a σ-finite measure space and let f be an extended real-valued $\mathfrak{A}$-measurable function on X such that $|f| < \infty$, μ-a.e. on X and f is μ semi-integrable on X. Let λ be a signed measure on $(X, \mathfrak{A})$ defined by $\lambda(A) = \int_A f \, d\mu$ for $A \in \mathfrak{A}$. Show that λ is a σ-finite signed measure on $(X, \mathfrak{A})$, that is, there exists an increasing sequence $(G_n : n \in \mathbb{N})$ in $\mathfrak{A}$ such that $\bigcup_{n \in \mathbb{N}} G_n = X$ and $\mu(G_n) \in \mathbb{R}$ for every $n \in \mathbb{N}$.

Proof. Since f in μ semi-integrable on X, that is, $\int_X f^+ \, d\mu - \int_X f^- \, d\mu$ exists in $\overline{\mathbb{R}}$, at least one of the two integrals $\int_X f^+ \, d\mu$ and $\int_X f^- \, d\mu$ is finite. Let us consider the case that

(1) $$\int_X f^- \, d\mu < \infty.$$

Now for every $A \in \mathfrak{A}$ we have

(2) $$\lambda(A) = \int_A f \, d\mu = \int_A f^+ \, d\mu - \int_A f^- \, d\mu.$$

Since f^+ is a nonnegative extended real-valued $\mathfrak{A}$-measurable function on X and $f^+ \leq |f| < \infty$, μ-a.e. on X, Prob. 10.12 implies that there exists an increasing sequence $(G_n : n \in \mathbb{N})$ in $\mathfrak{A}$ such that $\bigcup_{n \in \mathbb{N}} G_n = X$ and $\int_{G_n} f^+ \, d\mu < \infty$ for every $n \in \mathbb{N}$. On the other hand, we have $\int_{G_n} f^- \, d\mu \leq \int_X f^- \, d\mu < \infty$ by (1) for every $n \in \mathbb{N}$. Thus we have for every $n \in \mathbb{N}$

$$\lambda(G_n) = \int_{G_n} f^+ \, d\mu - \int_{G_n} f^- \, d\mu \in \mathbb{R}.$$

This completes the proof. ∎

§11 Absolute Continuity of a Measure

Prob. 11.1. Consider the Lebesgue measure space $(\mathbb{R}, \mathfrak{M}_L, \mu_L)$. Let ν be the counting measure on $\mathfrak{M}_L$, that is, ν is defined by setting $\nu(E)$ to be equal to the number of elements in $E \in \mathfrak{M}_L$ if E is a finite set and setting $\nu(E) = \infty$ if E is an infinite set.

(a) Show that $\mu_L \ll \nu$ but $\frac{d\mu_L}{d\nu}$ does not exist.

(b) Show that $\nu \ll \mu_L$ does not hold.

(c) Show that a Lebesgue decomposition of ν with respect to μ_L is impossible.

(Observe that the counting measure ν is not a σ-finite measure and this failure makes Theorem 11.10 not applicable.)

Proof. 1. Let us prove (a). We are to show that every null set for ν is a null set for μ_L. Now $\emptyset$ is the only null set for ν and we have $\mu_L(\emptyset) = 0$. This proves $\mu_L \ll \nu$.

To show that $\frac{d\mu_L}{d\nu}$ does not exist, assume the contrary. Then since μ_L is a positive measure, we have $\frac{d\mu_L}{d\nu} \geq 0$, ν a.e. on $\mathbb{R}$ by Observation 11.19. Now $\frac{d\mu_L}{d\nu}$ cannot vanish identically on $\mathbb{R}$ since that would imply $\mu_L(\mathbb{R}) = \int_{\mathbb{R}} \frac{d\mu_L}{d\nu}\, d\nu = \int_{\mathbb{R}} 0\, d\nu = 0$, a contradiction. Thus there exists $x_0 \in \mathbb{R}$ such that $\frac{d\mu_L}{d\nu}(x_0) > 0$. Then we have

$$0 = \mu_L(\{x_0\}) = \int_{\{x_0\}} \frac{d\mu_L}{d\nu}\, d\nu = \frac{d\mu_L}{d\nu}(x_0) \cdot 1 > 0.$$

This is a contradiction. This shows that $\frac{d\mu_L}{d\nu}$ cannot exist.

2. Let us prove (b). Let $x_0 \in \mathbb{R}$ be arbitrarily selected and let $E = \{x_0\}$. Then $\mu_L(E) = \mu_L(\{x_0\}) = 0$ and thus E is a null set for μ_L. Now since ν is the counting measure, we have $\nu(E) = \nu(\{x_0\}) = 1$ and thus E is not a null set for ν. This shows that $\nu \ll \mu_L$ does not hold.

3. Let us prove (c). To show that a Lebesgue decomposition of ν with respect to μ_L is impossible, let us assume the contrary. Then we have $\nu = \nu_a + \nu_s$ where $\nu_a \ll \mu_L$ and $\nu_s \perp \mu_L$.

Now $\nu_s \perp \mu_L$ implies that there exist $C_1, C_2 \in \mathfrak{M}_L$ such that $C_1 \cap C_2 = \emptyset$, $C_1 \cup C_2 = \mathbb{R}$, C_1 is a null set for μ_L and C_2 is a null set for ν_s. Let us observe that $C_2 \neq \emptyset$ for otherwise we would have $C_1 = \mathbb{R}$, that is, the null set C_1 for μ_L is equal to $\mathbb{R}$, a contradiction. Now that $C_2 \neq \emptyset$, select $x_0 \in C_2$ arbitrarily. Then $\{x_0\}$, as a subset of the null set C_2 for ν_s, is a null set for ν_s and thus we have $\nu_s(\{x_0\}) = 0$. Then we have

$$1 = \nu(\{x_0\}) = \nu_a(\{x_0\}) + \nu_s(\{x_0\}) = \nu_a(\{x_0\}).$$

On the other hand, since $\nu_a \ll \mu_L$ and $\mu_L(\{x_0\}) = 0$ we have $\nu_a(\{x_0\}) = 0$. This contradicts the last equality. This proves that a Lebesgue decomposition of ν with respect to μ_L is impossible. ∎

Prob. 11.2. Let μ and ν be two positive measures on a measurable space $(X, \mathfrak{A})$.
(a) Show that if for every $\varepsilon > 0$ there exists $\delta > 0$ such that $\nu(E) < \varepsilon$ for every $E \in \mathfrak{A}$ with $\mu(E) < \delta$, then $\nu \ll \mu$.
(b) Show that the converse of (a) does not hold by constructing a counter-example.
(c) Show that if ν is a finite positive measure, then the converse of (a) holds.

Proof. 1. Let us prove (a). Assume that for every $\varepsilon > 0$ there exists $\delta > 0$ such that $\nu(E) < \varepsilon$ for every $E \in \mathfrak{A}$ with $\mu(E) < \delta$.

Now let $A \in \mathfrak{A}$ and $\mu(A) = 0$. Let $\varepsilon > 0$ be arbitrarily given. Then since $\mu(A) = 0$, we have $\mu(A) < \delta$ and hence $\nu(A) < \varepsilon$ by the assumption. Since this holds for every $\varepsilon > 0$, we have $\nu(A) = 0$. This proves that $\nu \ll \mu$.

2. Let us construct an example to show that the converse of (a) is false. Consider the measure space $(\mathbb{R}, \mathfrak{M}_L, \mu_L)$. Let us define a function f on $\mathbb{R}$ by setting

$$f(x) = \begin{cases} \frac{1}{|x|} & \text{for } x \in \mathbb{R} \setminus \{0\}, \\ 0 & \text{for } x = 0. \end{cases}$$

Let us define a set function ν on $\mathfrak{M}_L$ by setting

$$\nu(E) = \int_E f(x)\, \mu_L(dx) = \int_E \frac{1}{|x|}\, \mu_L(dx) \quad \text{for } E \in \mathfrak{M}_L.$$

Then ν is a positive measure on $(\mathbb{R}, \mathfrak{M}_L)$ by Proposition 8.12 and we have $\nu \ll \mu_L$.

Let $\varepsilon > 0$ be arbitrarily given. Then for any $\delta > 0$ however small it may be, if we let $E = \left(-\frac{\delta}{3}, \frac{\delta}{3}\right)$ then we have $\mu_L(E) = \frac{2}{3}\delta < \delta$ but we have

$$\nu(E) = \int_{\left(-\frac{\delta}{3}, \frac{\delta}{3}\right)} \frac{1}{|x|}\, \mu_L(dx) = \infty > \varepsilon.$$

3. Let us prove (c). Assume that $\nu(X) < \infty$ and $\nu \ll \mu$. Let us show that for every $\varepsilon > 0$ there exists $\delta > 0$ such that $\nu(E) < \varepsilon$ for every $E \in \mathfrak{A}$ with $\mu(E) < \delta$. Assume the contrary, that is, for some $\varepsilon_0 > 0$, no matter how small $\delta > 0$ may be, there exists $E \in \mathfrak{A}$ such that $\mu(E) < \delta$ but $\nu(E_0) \geq \varepsilon_0$. Then for every $n \in \mathbb{N}$ there exists $E_n \in \mathfrak{A}$ such that $\mu(E_n) < \frac{1}{2^n}$ but $\nu(E_n) \geq \varepsilon_0$. For the sequence $(E_n : n \in \mathbb{N})$, we have $\sum_{n \in \mathbb{N}} \mu(E_n) \leq \sum_{n \in \mathbb{N}} \frac{1}{2^n} < \infty$. Then by Theorem 6.6 (Borel-Cantelli Lemma), we have

$$(1) \qquad\qquad\qquad \mu\Big(\limsup_{n \to \infty} E_n \Big) = 0.$$

On the other hand, according to (b) of Theorem 1.28, the condition $\nu(X) < \infty$ implies

$$(2) \qquad\qquad \nu\Big(\limsup_{n \to \infty} E_n \Big) \geq \limsup_{n \to \infty} \nu(E_n) \geq \varepsilon_0.$$

Now (1) and (2) contradict the assumption that $\nu \ll \mu$. Thus we have a contradiction. ∎

Prob. 11.3. Let μ be a positive measure and λ be a signed measure on a measurable space $(X, \mathfrak{A})$. Show that if for every $\varepsilon > 0$ there exists $\delta > 0$ such that $|\lambda(E)| < \varepsilon$ for every $E \in \mathfrak{A}$ with $\mu(E) < \delta$, then $\lambda \ll \mu$.

Proof. Assume that for every $\varepsilon > 0$ there exists $\delta > 0$ such that $|\lambda(E)| < \varepsilon$ for every $E \in \mathfrak{A}$ with $\mu(E) < \delta$.

Now let $A \in \mathfrak{A}$ and $\mu(A) = 0$. Let $\varepsilon > 0$ be arbitrarily given. Then since $\mu(A) = 0$, we have $\mu(A) < \delta$ and hence $|\lambda(A)| < \varepsilon$ by the assumption. Since this holds for every $\varepsilon > 0$, we have $\lambda(A) = 0$. This proves that $\lambda \ll \mu$. ∎

Prob. 11.4. Let $(X, \mathfrak{A})$ be a measurable space and let λ be a positive measure on $(X, \mathfrak{A})$. Let f be a λ semi-integrable nonnegative extended real-valued $\mathfrak{A}$-measurable function on X and let g be a λ semi-integrable extended real-valued $\mathfrak{A}$-measurable function on X. Define a positive measure μ on $(X, \mathfrak{A})$ by setting $\mu(E) = \int_E f \, d\lambda$ for $E \in \mathfrak{A}$ and define a signed measure ν on $(X, \mathfrak{A})$ by setting $\nu(E) = \int_E g \, d\lambda$ for $E \in \mathfrak{A}$. Let $\mathfrak{N}_\mu$ be the collection of all null sets in the positive measure space $(X, \mathfrak{A}, \mu)$ and let $\mathfrak{N}_\nu$ be the collection of all null sets in the signed measure space $(X, \mathfrak{A}, \nu)$. Let $Z_f = \{x \in X : f(x) = 0\}$ and $Z_g = \{x \in X : g(x) = 0\}$. Prove the following statements:
(a) $E \in \mathfrak{N}_\mu \Leftrightarrow \lambda(E \setminus Z_f) = 0$ and $E \in \mathfrak{N}_\nu \Leftrightarrow \lambda(E \setminus Z_g) = 0$.
(b) $\nu \ll \mu \Leftrightarrow \lambda(Z_f \setminus Z_g) = 0$.
(c) $\mu \perp \nu \Leftrightarrow \lambda\big((Z_f \cup Z_g)^c\big) = 0$.

Proof. Observe that $(X, \mathfrak{A}, \mu)$ is a positive measure space by Proposition 8.12 and $(X, \mathfrak{A}, \nu)$ is a signed measure space by Proposition 10.3.

1. Let us prove (a). Since μ is a positive measure on $(X, \mathfrak{A})$, we have

$$E \in \mathfrak{N}_\mu \Leftrightarrow \mu(E) = 0$$

$$\Leftrightarrow \int_E f \, d\lambda = 0$$

$$\Leftrightarrow f = 0, \lambda\text{- a.e. on } E$$

$$\Leftrightarrow \lambda(E \setminus Z_f) = 0.$$

On the other hand, since ν is a signed measure on $(X, \mathfrak{A})$, we have

$$E \in \mathfrak{N}_\nu \Leftrightarrow \nu(E_0) = 0 \quad \text{for every } E_0 \subset E, E_0 \in \mathfrak{A}$$

$$\Leftrightarrow \int_{E_0} g \, d\lambda = 0 \quad \text{for every } E_0 \subset E, E_0 \in \mathfrak{A}$$

$$\Leftrightarrow g = 0, \lambda\text{- a.e. on } E$$

$$\Leftrightarrow \lambda(E \setminus Z_g) = 0.$$

2. Let us prove (b). Observe that

(1) $$\nu \ll \mu \Leftrightarrow \mathfrak{N}_\mu \subset \mathfrak{N}_\nu$$

$$\Leftrightarrow \big[E \in \mathfrak{A}; E \in \mathfrak{N}_\mu \Rightarrow E \in \mathfrak{N}_\nu\big]$$

$$\Leftrightarrow \big[E \in \mathfrak{A}; \lambda(E \setminus Z_f) = 0 \Rightarrow \lambda(E \setminus Z_g) = 0\big] \quad \text{by (a)}.$$

Consider the following two statements:
$1°$ $E \in \mathfrak{A}; \lambda(E \setminus Z_f) = 0 \Rightarrow \lambda(E \setminus Z_g) = 0$.
$2°$ $\lambda(Z_f \setminus Z_g) = 0$.
Let us show that $1° \Leftrightarrow 2°$. Assume $1°$. Then since $Z_f \in \mathfrak{A}$ and $\lambda(Z_f \setminus Z_f) = \lambda(\emptyset) = 0$, we have $\lambda(Z_f \setminus Z_g) = 0$ by $1°$ and thus $2°$ holds. Conversely let us assume $2°$. Let us

observe that for any three subsets A, B and C of X we have

(2) $$A \setminus C \subset (A \setminus B) \cup (B \setminus C).$$

This set inclusion is proved by observing

$$A \setminus C = A \cap C^c = \left[(A \cap B) \cup (A \cap B^c)\right] \cap C^c$$
$$= (A \cap B \cap C^c) \cup (A \cap B^c \cap C^c)$$
$$\subset (B \cap C^c) \cup (A \cap B^c) = (B \setminus C) \cup (A \setminus B).$$

Now applying (2) we have

$$E \setminus Z_g \subset (E \setminus Z_f) \cup (Z_f \setminus Z_g)$$

and then

$$\lambda(E \setminus Z_g) \leq \lambda(E \setminus Z_f) + \lambda(Z_f \setminus Z_g).$$

Then assumption of $\lambda(E \setminus Z_f) = 0$ and $2°$ implies $\lambda(E \setminus Z_g) = 0$ so that $1°$ holds. This completes the proof of $1° \Leftrightarrow 2°$. Substituting this in (1) we have

$$\nu \ll \mu \Leftrightarrow \lambda(Z_f \setminus Z_g) = 0.$$

3. Let us prove (c).

3.1. Let us show that $\mu \perp \nu \Rightarrow \lambda\big((Z_f \cup Z_g)^c\big) = 0$. Suppose we have $\mu \perp \nu$. Then there exist $C_1, C_2 \in \mathfrak{A}$ such that $C_1 \cap C_2 = \emptyset$, $C_1 \cup C_2 = X$, $C_1 \in \mathfrak{N}_\nu$ and $C_2 \in \mathfrak{N}_\mu$. Now $C_1 \in \mathfrak{N}_\nu$ implies $\lambda(C_1 \setminus Z_g) = 0$ and $C_2 \in \mathfrak{N}_\mu$ implies $\lambda(C_2 \setminus Z_f) = 0$ according to (a). Let us show

(3) $$(Z_f \cup Z_g)^c \subset (C_1 \setminus Z_g) \cup (C_2 \setminus Z_f).$$

Let $x \in (Z_f \cup Z_g)^c$. Then $x \notin Z_f$ and $x \notin Z_g$. Since $x \in X = C_1 \cup C_2$, we have either $x \in C_1$ or $x \in C_2$. If $x \in C_1$ then $x \in C_1 \setminus Z_g$. If $x \in C_2$ then $x \in C_2 \setminus Z_f$. Thus $x \in (C_1 \setminus Z_g) \cup (C_2 \setminus Z_f)$. This proves (3). Then (3) implies

$$\lambda\big((Z_f \cup Z_g)^c\big) \leq \lambda(C_1 \setminus Z_g) + \lambda(C_2 \setminus Z_f) = 0.$$

3.2. Let us show that $\lambda\big((Z_f \cup Z_g)^c\big) = 0 \Rightarrow \mu \perp \nu$. Suppose we have $\lambda\big((Z_f \cup Z_g)^c\big) = 0$. Let $Q = (Z_f \cup Z_g)^c$. Then we have $\lambda(Q) = 0$. Let $C_2 = Z_f \in \mathfrak{A}$ and let $C_1 = (Z_g \setminus Z_f) \cup Q \in \mathfrak{A}$. Clearly we have $C_1 \cap C_2 = \emptyset$ and $C_1 \cup C_2 = X$. Now $\lambda(C_2 \setminus Z_f) = \lambda(\emptyset) = 0$. This implies that $C_2 \in \mathfrak{N}_\mu$ by (a). We have $\lambda(C_1 \setminus Z_g) = \lambda(Q) = 0$. This implies that $C_1 \in \mathfrak{N}_\nu$ by (a). This proves that $\mu \perp \nu$. $\blacksquare$

Prob. 11.5. Let $(X, \mathfrak{A})$ be a measurable space and let λ be a σ-finite positive measure on $(X, \mathfrak{A})$. Let f be a λ-integrable nonnegative extended real-valued $\mathfrak{A}$-measurable function on X with $\lambda(Z_f) = 0$ where $Z_f = \{x \in X : f(x) = 0\}$ and let g be a λ-integrable extended real-valued $\mathfrak{A}$-measurable function on X.
Define a positive measure μ on $(X, \mathfrak{A})$ by setting $\mu(E) = \int_E f \, d\lambda$ for $E \in \mathfrak{A}$ and define a signed measure ν on $(X, \mathfrak{A})$ by setting $\nu(E) = \int_E g \, d\lambda$ for $E \in \mathfrak{A}$.
Prove the following statements:
(a) μ is a σ-finite positive measure and ν is a σ-finite signed measure on $(X, \mathfrak{A})$.
(b) $\nu \ll \mu$.
(c) The Radon-Nikodym derivative $\frac{d\nu}{d\mu}$ exists and $\frac{d\nu}{d\mu} = \frac{g}{f}$, λ-a.e. on X.

Proof. 1. Let us prove (a). Now μ is a positive measure on $(X, \mathfrak{A})$ by Proposition 8.12 and ν is a signed measure on $(X, \mathfrak{A})$ by Proposition 10.3. Then the λ-integrability of f on X implies the σ-finiteness of μ and the λ-integrability of g on X implies the σ-finiteness of ν whether λ itself is σ-finite or not. Let us verify this for ν for instance. Suppose the signed measure ν is not σ-finite. Then for every countable disjoint collection $\{E_n : n \in \mathbb{N}\} \subset \mathfrak{A}$ such that $\bigcup_{n \in \mathbb{N}} E_n = X$ there exists $n_0 \in \mathbb{N}$ such that $|\nu(E_{n_0})| = \infty$. On the other hand we have $\nu(E_{n_0}) = \int_{E_{n_0}} g \, d\lambda$ and the λ-integrability of g on X implies that $\int_{E_{n_0}} g \, d\lambda \in \mathbb{R}$ and then $|\nu(E_{n_0})| = \left| \int_{E_{n_0}} g \, d\lambda \right| \in \mathbb{R}$. This is a contradiction. Thus ν must be σ-finite.
2. Let us prove (b). According to (b) of Prob. 11.4, we have

$$\nu \ll \mu \Leftrightarrow \lambda(Z_f \setminus Z_g) = 0.$$

But we have $\lambda(Z_f) = 0$ and then $\lambda(Z_f \setminus Z_g) \le \lambda(Z_f) = 0$. Thus we have $\nu \ll \mu$.
3. Let us prove (c). We have $\nu \ll \mu$ by (b). Also μ and ν are σ-finite by (a). Thus the Radon-Nikodym derivative $\frac{d\nu}{d\mu}$ exists by Theorem 11.14 (Radon-Nikodym Theorem).
The definition $\mu(E) = \int_E f \, d\lambda$ for $E \in \mathfrak{A}$ implies that $\mu \ll \lambda$. Also λ and μ are σ-finite. Thus the Radon-Nikodym derivative $\frac{d\mu}{d\lambda}$ exists and $\frac{d\mu}{d\lambda} = f$, λ-a.e. on X by Theorem 11.14. Similarly the definition $\nu(E) = \int_E g \, d\lambda$ for $E \in \mathfrak{A}$ implies that $\nu \ll \lambda$. Then the σ-finiteness of λ and ν implies the existence of the Radon-Nikodym derivative $\frac{d\nu}{d\lambda}$ and $\frac{d\nu}{d\lambda} = g$, λ-a.e. on X by Theorem 11.14.
Then by Theorem 11.23 (Chain Rule of Radon-Nikodym Derivatives) we have

$$\frac{d\nu}{d\lambda} = \frac{d\nu}{d\mu} \cdot \frac{d\mu}{d\lambda}, \quad \text{that is,} \quad g = \frac{d\nu}{d\mu} \cdot f, \quad \lambda\text{-a.e. on } X.$$

Then we have

$$\frac{d\nu}{d\mu} = \frac{g}{f}, \quad \lambda\text{-a.e. on } X.$$

This completes the proof. ∎

Prob. 11.6. Let $(X, \mathfrak{A})$ be a measurable space and let λ be a σ-finite positive measure on $(X, \mathfrak{A})$. Let f be a λ-integrable nonnegative extended real-valued $\mathfrak{A}$-measurable function on X and let g be a λ-integrable extended real-valued $\mathfrak{A}$-measurable function on X. Define a σ-finite positive measure μ on $(X, \mathfrak{A})$ by setting $\mu(E) = \int_E f \, d\lambda$ for $E \in \mathfrak{A}$ and define a σ-finite signed measure ν on $(X, \mathfrak{A})$ by setting $\nu(E) = \int_E g \, d\lambda$ for $E \in \mathfrak{A}$. According to Theorem 11.10 (Lebesgue Decomposition) there exist two signed measures ν_a and ν_s on $(X, \mathfrak{A})$ such that $\nu = \nu_a + \nu_s$, $\nu_a \ll \mu$ and $\nu_s \perp \mu$. Determine the two signed measures ν_a and ν_s in terms of the two functions f and g.

Proof. We are to find descriptions of the two signed measures ν_a and ν_s in terms of the two functions f and g. For this purpose let us introduce the notation $Z_h = \{x \in X : h(x) = 0\}$ for an arbitrary extended real-valued function h on X.

With our functions f and g given above, let us define two extended real-valued $\mathfrak{A}$-measurable functions g_1 and g_2 on X by setting

$$(1) \qquad g_1(x) = \begin{cases} 0 & \text{for } x \in Z_f, \\ g(x) & \text{for } x \in Z_f^c. \end{cases}$$

$$(2) \qquad g_2(x) = \begin{cases} g(x) & \text{for } x \in Z_f, \\ 0 & \text{for } x \in Z_f^c. \end{cases}$$

We claim that we have the following description of ν_a and ν_s

$$(3) \qquad \nu_a(E) = \int_E g_1 \, d\lambda \quad \text{for } E \in \mathfrak{A}.$$

$$(4) \qquad \nu_s(E) = \int_E g_2 \, d\lambda \quad \text{for } E \in \mathfrak{A}.$$

It remains to prove (3) and (4). For this purpose define two signed measures ν_1 and ν_2 on $(X, \mathfrak{A})$ by setting

$$(5) \qquad \nu_1(E) = \int_E g_1 \, d\lambda \quad \text{for } E \in \mathfrak{A}.$$

$$(6) \qquad \nu_2(E) = \int_E g_2 \, d\lambda \quad \text{for } E \in \mathfrak{A}.$$

It remains to show that $\nu = \nu_1 + \nu_2$, $\nu_1 \ll \mu$ and $\nu_2 \perp \mu$.

1. By (1) and (2), we have $g_1 + g_2 = g$. Then we have for every $E \in \mathfrak{A}$

$$\nu(E) = \int_E g \, d\lambda = \int_E \{g_1 + g_2\} \, d\lambda = \int_E g_1 \, d\lambda + \int_E g_2 \, d\lambda = \nu_1(E) + \nu_2(E).$$

This proves that $\nu = \nu_1 + \nu_2$.

2. According to (b) of Prob. 11.4, we have

$$\nu_1 \ll \mu \Leftrightarrow \lambda(Z_f \setminus Z_{g_1}) = 0.$$

But (1) implies that $Z_f \subset Z_{g_1}$ and then $\lambda(Z_f \setminus Z_{g_1}) = \lambda(\emptyset) = 0$. Thus we have $\nu_1 \ll \mu$.

3. According to (c) of Prob. 11.4, we have

$$\mu \perp \nu_2 \Leftrightarrow \lambda\big((Z_f \cup Z_{g_2})^c\big) = 0 \Leftrightarrow \lambda(Z_f^c \cap Z_{g_2}^c) = 0.$$

Now (2) implies that $Z_{g_2} \supset Z_f^c$ and then $Z_{g_2}^c \cap Z_f^c = \emptyset$ so that $\lambda(Z_f^c \cap Z_{g_2}^c) = \lambda(\emptyset) = 0$. Thus we have $\mu \perp \nu_2$. This completes the proof. ■

Prob. 11.7. Let μ and ν be two positive measures on a measurable space $(X, \mathfrak{A})$. Suppose the Radon-Nikodym derivative $\frac{d\nu}{d\mu}$ exists so that $\nu \ll \mu$.

(a) Show that if $\frac{d\nu}{d\mu} > 0$, μ-a.e. on X, then $\mu \ll \nu$ and thus $\mu \sim \nu$.

(b) Show that if $\frac{d\nu}{d\mu} > 0$, μ-a.e. on X and if μ and ν are σ-finite, then $\frac{d\mu}{d\nu}$ exists and moreover we have $\frac{d\mu}{d\nu} = \left(\frac{d\nu}{d\mu}\right)^{-1}$, μ- and ν-a.e. on X.

Proof. 1. Let us prove (a). Suppose $\frac{d\nu}{d\mu} > 0$, μ-a.e. on X. To show that $\mu \ll \nu$, let $E \in \mathfrak{A}$ and $\nu(E) = 0$. Then since $\frac{d\nu}{d\mu} > 0$, μ-a.e. on X and $\int_E \frac{d\nu}{d\mu} \, d\mu = \nu(E) = 0$, we have $\mu(E) = 0$ by (d) of Lemma 8.2. This shows that $\mu \ll \nu$.

2. Let us prove (b). If $\frac{d\nu}{d\mu} > 0$, μ-a.e. on X then we have $\mu \ll \nu$ according to (a). If μ and ν are σ-finite then the Radon-Nikodym derivative $\frac{d\mu}{d\nu}$ exists according to Theorem 11.14 (Radon-Nikodym Theorem). Then according to Corollary 11.26, σ-finiteness of μ and ν and existence of $\frac{d\nu}{d\mu}$ and $\frac{d\mu}{d\nu}$ imply that

$$\frac{d\nu}{d\mu} \cdot \frac{d\mu}{d\nu} = 1, \quad \mu\text{- and } \nu\text{-a.e. on } X.$$

Thus we have

$$\frac{d\mu}{d\nu} = \left(\frac{d\nu}{d\mu}\right)^{-1}, \quad \mu\text{- and } \nu\text{-a.e. on } X.$$

This completes the proof. ∎

Prob. 11.8. Given a measurable space $(X, \mathfrak{A})$. Let μ be a σ-finite positive measure on $(X, \mathfrak{A})$ and let f be a μ-integrable nonnegative extended real-valued $\mathfrak{A}$-measurable function on X such that $f \neq 0$, μ-a.e. on X. Define a set function ν on $\mathfrak{A}$ by setting $\nu(E) = \int_E f \, d\mu$ for every $E \in \mathfrak{A}$.
(a) Show that ν is a σ-finite positive measure on $(X, \mathfrak{A})$.
(b) Show that $\nu \ll \mu$ and moreover $\frac{d\nu}{d\mu}$ exists and $\frac{d\nu}{d\mu} = f$, μ-a.e. on X.
(c) Show that $\frac{d\mu}{d\nu}$ exists and $\frac{d\mu}{d\nu} = \frac{1}{f}$, μ- and ν-a.e. on X.

Proof. 1. Let us prove (a). The set function ν on $\mathfrak{A}$ is a positive measure on $(X, \mathfrak{A})$ according to Proposition 8.12. Let us show that ν is σ-finite. Suppose not. Then for any decomposition of X into a countable disjoint collection $\{E_n : n \in \mathbb{N}\} \subset \mathfrak{A}$, there exists $n_0 \in \mathbb{N}$ such that $\nu(E_{n_0}) = \infty$. Then we have $\int_{E_{n_0}} f \, d\mu = \nu(E_{n_0}) = \infty$. This contradicts the μ-integrability of f on X. Therefore ν must be σ-finite.

2. Let us prove (b). Let $E \in \mathfrak{A}$. Then $\mu(E) = 0$ implies $\nu(E) = 0$ by the definition of $\nu(E)$. This shows that $\nu \ll \mu$. Then the σ-finiteness of μ and ν imply that the Radon-Nikodym derivative $\frac{d\nu}{d\mu}$ exists. Now since $\nu(E) = \int_E f \, d\mu$ for every $E \in \mathfrak{A}$, the function f is a Radon-Nikodym derivative of ν with respect to μ according to Definition 11.1. Then by (b) (Uniqueness of the Radon-Nikodym Derivative) of Proposition 11.2, we have $\frac{d\nu}{d\mu} = f$, μ-a.e. on X.

3. Let us prove (c). By definition we have $\nu(E) = \int_E f \, d\mu$ for every $E \in \mathfrak{A}$. Since $f \geq 0$ on X and $f \neq 0$, μ-a.e. on X, $\nu(E) = 0$ implies $\mu(E) = 0$ according to (d) of Lemma 8.2. This shows that we have $\mu \ll \nu$. Then since μ and ν are σ-finite $\mu \ll \nu$ implies that $\frac{d\mu}{d\nu}$ exists according to Theorem 11.14 (Radon-Nikodym Theorem).

Now that μ and ν are σ-finite and $\frac{d\nu}{d\mu}$ and $\frac{d\mu}{d\nu}$ exist, we have according to Corollary 11.26

$$\frac{d\nu}{d\mu} \cdot \frac{d\mu}{d\nu} = 1, \quad \mu\text{- and } \nu\text{-a.e. on } X.$$

Thus we have

$$\frac{d\mu}{d\nu} = \frac{1}{\frac{d\nu}{d\mu}} = \frac{1}{f}, \quad \mu\text{- and } \nu\text{-a.e. on } X.$$

This completes the proof. ∎

Prob. 11.9. Let μ and ν be two σ-finite positive measures on a measurable space $(X, \mathfrak{A})$.
(a) Show that if we let $\lambda = \mu + \nu$ then λ is a σ-finite positive measure.
(b) Show that $\mu \ll \lambda$ and $\nu \ll \lambda$ and the Radon-Nikodym derivatives $\frac{d\mu}{d\lambda}$ and $\frac{d\nu}{d\lambda}$ exist and
moreover $\frac{d\mu}{d\lambda} \geq 0$ and $\frac{d\nu}{d\lambda} \geq 0$, λ-a.e. on X.
(c) Show that

$$\mu \perp \nu \Leftrightarrow \frac{d\mu}{d\lambda} \cdot \frac{d\nu}{d\lambda} = 0, \lambda\text{-a.e. on } X$$

Proof. 1. Let us show that λ is a σ-finite measure on $(X, \mathfrak{A})$. Now since μ is a σ-finite measure on $(X, \mathfrak{A})$ there exists an increasing sequence $(A_n : n \in \mathbb{N})$ in $\mathfrak{A}$ such that $\bigcup_{n \in \mathbb{N}} A_n = X$ and $\mu(A_n) < \infty$ for every $n \in \mathbb{N}$. Similarly since ν is a σ-finite measure on $(X, \mathfrak{A})$ there exists an increasing sequence $(B_n : n \in \mathbb{N})$ in $\mathfrak{A}$ such that $\bigcup_{n \in \mathbb{N}} B_n = X$ and $\nu(B_n) < \infty$ for every $n \in \mathbb{N}$. Let $C_n = A_n \cap B_n$ for $n \in \mathbb{N}$. Then $(C_n : n \in \mathbb{N})$ is an increasing sequence in $\mathfrak{A}$. We also have $\bigcup_{n \in \mathbb{N}} C_n = X$. (To show this let $x \in X$ be arbitrarily chosen. Then $x \in A_{n_1}$ for some $n_1 \in \mathbb{N}$ and $x \in B_{n_2}$ for some $n_2 \in \mathbb{N}$ so that $x \in A_{n_1} \cap B_{n_2}$. Let $n_0 = \max\{n_1, n_2\}$. Then $A_{n_1} \subset A_{n_0}$ and $B_{n_2} \subset B_{n_0}$ so that $x \in A_{n_1} \cap B_{n_2} \subset A_{n_0} \cap B_{n_0} = C_{n_0} \subset \bigcup_{n \in \mathbb{N}} C_n$. Thus $X \subset \bigcup_{n \in \mathbb{N}} C_n$ and hence $\bigcup_{n \in \mathbb{N}} C_n = X$.) Now for every $n \in \mathbb{N}$ we have

$$\lambda(C_n) = \mu(C_n) + \nu(C_n) \leq \mu(A_n) + \nu(B_n) < \infty.$$

This shows that λ is a σ-finite measure on $(X, \mathfrak{A})$.

2. For every $E \in \mathfrak{A}$ we have $\lambda(E) = \mu(E) + \nu(E) \geq 0$. Thus if $\lambda(E) = 0$ then $\mu(E) = 0$ and $\nu(E) = 0$. This shows that $\mu \ll \lambda$ and $\nu \ll \lambda$.

Now that $\mu \ll \lambda$ and $\nu \ll \lambda$ and μ, ν and λ are σ-finite, the Radon-Nikodym derivatives $\frac{d\mu}{d\lambda}$ and $\frac{d\nu}{d\lambda}$ exist according to Theorem 11.14 (Radon-Nikodym Theorem). Moreover we have $\frac{d\mu}{d\lambda} \geq 0$ and $\frac{d\nu}{d\lambda} \geq 0$, λ-a.e. on X according to (b) of Observation 11.19.

3. Suppose $\mu \perp \nu$. Then there exist $A, B \in \mathfrak{A}$ such that $A \cap B = \emptyset$, $A \cup B = X$, $\mu(B) = 0$, and $\nu(A) = 0$. Now $\mu(B) = \int_B \frac{d\mu}{d\lambda} d\lambda$ and $\frac{d\mu}{d\lambda} \geq 0$, λ-a.e. on X. Thus $\mu(B) = 0$ implies that $\frac{d\mu}{d\lambda} = 0$, λ-a.e. on B. Similarly we have $\nu(A) = \int_A \frac{d\nu}{d\lambda} d\lambda$ and $\frac{d\nu}{d\lambda} \geq 0$, λ-a.e. on X. Then $\nu(A) = 0$ implies that $\frac{d\nu}{d\lambda} = 0$, λ-a.e. on A. Therefore $\frac{d\mu}{d\lambda} \cdot \frac{d\nu}{d\lambda} = 0$, λ-a.e. on A and $\frac{d\mu}{d\lambda} \cdot \frac{d\nu}{d\lambda} = 0$, λ-a.e. on B. Then since $A \cup B = X$, we have $\frac{d\mu}{d\lambda} \cdot \frac{d\nu}{d\lambda} = 0$, λ-a.e. on X.

4. Conversely suppose $\frac{d\mu}{d\lambda} \cdot \frac{d\nu}{d\lambda} = 0$, λ-a.e. on X. Then there exists a null set X_0 in $(X, \mathfrak{A}, \lambda)$ such that $\frac{d\mu}{d\lambda} \cdot \frac{d\nu}{d\lambda} = 0$ on $X \setminus X_0$. Then for every $x \in X \setminus X_0$, at least one of $\frac{d\mu}{d\lambda}(x)$ and $\frac{d\nu}{d\lambda}(x)$ is equal to 0. Let

$$X_1 = \left\{ x \in X \setminus X_0 : \frac{d\mu}{d\lambda}(x) = 0 \right\} \in \mathfrak{A},$$

$$X_2 = \left\{ x \in X \setminus X_0 : \frac{d\nu}{d\lambda}(x) = 0 \right\} \in \mathfrak{A}.$$

Now $X_1 \cup X_2 = X \setminus X_0$ but X_1 and X_2 need not be disjoint. Thus let us define

$$X_3 = X_1 \cap X_2,$$

$$A = X_1 \cup X_0,$$

$$B = X_2 \setminus X_3.$$

Then we have $A, B \in \mathfrak{A}$, $A \cap B = \emptyset$ and $A \cup B = X$. Moreover we have

$$\mu(A) = \int_A \frac{d\mu}{d\lambda}\, d\lambda = \int_{X_1 \cup X_0} \frac{d\mu}{d\lambda}\, d\lambda = \int_{X_1} 0\, d\lambda = 0,$$

$$\nu(B) = \int_B \frac{d\nu}{d\lambda}\, d\lambda = \int_{X_2 \setminus X_3} \frac{d\nu}{d\lambda}\, d\lambda = \int_{X_2 \setminus X_3} 0\, d\lambda = 0.$$

This shows that $\mu \perp \nu$. ∎

Prob. 11.10. Given a measurable space $(X, \mathfrak{A})$. For $A \in \mathfrak{A}$, let us write $\mathfrak{A} \cap A$ for the σ-algebra of subsets of A consisting of subsets of A of the type $E \cap A$ where $E \in \mathfrak{A}$. Prove the following theorem related to the Lebesgue Decomposition Theorem:

Theorem. Let μ and ν be σ-finite positive measures on a measurable space $(X, \mathfrak{A})$. There exist $A, B \in \mathfrak{A}$ such that $A \cap B = \emptyset$, $A \cup B = X$, $\mu \sim \nu$ on the measurable space $(A, \mathfrak{A} \cap A)$ and $\mu \perp \nu$ on the measurable space $(B, \mathfrak{A} \cap B)$.

Proof. 1. Let $\lambda = \mu + \nu$. Then by Prob. 11.9, λ is a σ-finite positive measure, $\mu \ll \lambda$ and $\nu \ll \lambda$ and the Radon-Nikodym derivatives $\frac{d\mu}{d\lambda}$ and $\frac{d\nu}{d\lambda}$ exist and moreover $\frac{d\mu}{d\lambda} \geq 0$ and $\frac{d\nu}{d\lambda} \geq 0$, λ-a.e. on X. Let

$$A = \left\{ x \in X : \frac{d\mu}{d\lambda}(x) > 0 \text{ and } \frac{d\nu}{d\lambda}(x) > 0 \right\},$$

$$B = A^c.$$

Then we have $A, B \in \mathfrak{A}$, $A \cap B = \emptyset$ and $A \cup B = X$.

2. Consider the measurable space $(A, \mathfrak{A} \cap A)$. Let us show that $\mu \sim \nu$ on $(A, \mathfrak{A} \cap A)$. Take an arbitrary member of $\mathfrak{A} \cap A$, say $E \cap A$ where $E \in \mathfrak{A}$. Suppose $\mu(E \cap A) = 0$. Then since $\frac{d\mu}{d\lambda}$ is the Radon-Nikodym derivative of μ with respect to λ on $(X, \mathfrak{A})$, we have

$$\int_{E \cap A} \frac{d\mu}{d\lambda} \, d\lambda = \mu(E \cap A) = 0.$$

Since $\frac{d\mu}{d\lambda} > 0$ on $E \cap A$, the last equality implies that $\lambda(E \cap A) = 0$ by (d) of Lemma 8.12. Then since $\nu \ll \lambda$ on $(X, \mathfrak{A})$, the equality $\lambda(E \cap A) = 0$ implies the equality $\nu(E \cap A) = 0$. This shows that $\nu \ll \mu$ on $(A, \mathfrak{A} \cap A)$. Interchanging the roles of μ and ν in the argument above we show that $\mu \ll \nu$ on $(A, \mathfrak{A} \cap A)$. Thus we have $\mu \sim \nu$ on $(A, \mathfrak{A} \cap A)$.

3. Consider the measurable space $(B, \mathfrak{A} \cap B)$. To show that $\mu \perp \nu$ on $(B, \mathfrak{A} \cap B)$, let

$$B_0 = \left\{ x \in B : \frac{d\mu}{d\lambda}(x) = 0 \text{ and } \frac{d\nu}{d\lambda}(x) = 0 \right\},$$

$$B_1 = \left\{ x \in B : \frac{d\mu}{d\lambda}(x) > 0 \text{ and } \frac{d\nu}{d\lambda}(x) = 0 \right\},$$

$$B_2 = \left\{ x \in B : \frac{d\mu}{d\lambda}(x) = 0 \text{ and } \frac{d\nu}{d\lambda}(x) > 0 \right\}.$$

Then $\{B_0, B_1, B_2\}$ is a disjoint collection in $\mathfrak{A} \cap B$ with $B_0 \cup B_1 \cup B_2 = B$. Next let $C_1 = B_1 \cup B_0$ and $C_2 = B_2$. Then $\{C_1, C_2\}$ is a disjoint collection in $\mathfrak{A} \cap B$ with $C_1 \cup C_2 = B$. Now we have

(1) $\mu(C_2) = \displaystyle\int_{C_2} \frac{d\mu}{d\lambda} \, d\lambda = \int_{B_2} \frac{d\mu}{d\lambda} \, d\lambda = \int_{B_2} 0 \, d\lambda = 0.$

(2) $\nu(C_1) = \displaystyle\int_{C_1} \frac{d\nu}{d\lambda} \, d\lambda = \int_{B_1} \frac{d\nu}{d\lambda} \, d\lambda + \int_{B_0} \frac{d\nu}{d\lambda} \, d\lambda = \int_{B_1} 0 \, d\lambda + \int_{B_0} 0 \, d\lambda = 0.$

This shows that $\mu \perp \nu$ on $(B, \mathfrak{A} \cap B)$. ∎

Prob. 11.11. Prove the following theorem related to the Lebesgue Decomposition Theorem:
Theorem. Let μ and ν be σ-finite positive measures on a measurable space $(X, \mathfrak{A})$. There exist a nonnegative extended real-valued $\mathfrak{A}$-measurable function f on X and a set $A_0 \in \mathfrak{A}$ with $\mu(A_0) = 0$ such that $\nu(E) = \int_E f \, d\mu + \nu(E \cap A_0)$ for every $E \in \mathfrak{A}$.

Proof. By Theorem 11.10 (Lebesgue Decomposition Theorem), there exist two positive measures ν_a and ν_s on $(X, \mathfrak{A})$ such that $\nu_a \ll \mu$, $\nu_s \perp \mu$, $\nu = \nu_a + \nu_s$ and moreover $\nu_a(E) = \int_E f \, d\mu$ for every $E \in \mathfrak{A}$ where f is a nonnegative extended real-valued $\mathfrak{A}$-measurable function on X.

Since $\nu_s \perp \mu$, there exist $C_1, C_2 \in \mathfrak{A}$ such that $C_1 \cap C_2 = \emptyset$, $C_1 \cup C_2 = X$, $\mu(C_2) = 0$ and $\nu_s(C_1) = 0$. Let us select $A_0 = C_2$. Then for any $E \in \mathfrak{A}$, we have

$$\nu(E) = \nu_a(E) + \nu_s(E) = \nu_a(E) + \nu_s(E \cap C_1) + \nu_s(E \cap C_2)$$

$$= \nu_a(E) + \nu_s(E \cap C_2) \quad \text{since } \nu_s(C_1) = 0$$

$$= \nu_a(E) + \nu_a(E \cap C_2) + \nu_s(E \cap C_2) \quad \text{since } \nu_a(C_2) = 0$$

$$= \nu_a(E) + \nu(E \cap C_2)$$

$$= \int_E f \, d\mu + \nu(E \cap A_0).$$

This completes the proof. ∎

§12   Monotone Functions and Functions of Bounded Variation

Prob. 12.1. Let $f, g \in BV([a, b])$.
(a) Show that $cf \in BV([a, b])$ and $V_a^b(cf) = |c| V_a^b(f)$ for every $c \in \mathbb{R}$.
(b) Show that $f + g \in BV([a, b])$ and $V_a^b(f + g) \leq V_a^b(f) + V_a^b(g)$.
(c) Construct an example for which we have $V_a^b(f + g) \neq V_a^b(f) + V_a^b(g)$.

Proof. 1. Let us prove (a). Let $\mathcal{P} \in \mathfrak{P}_{a,b}$ be given by $\mathcal{P} = \{a = x_0 \leq \cdots \leq x_n = b\}$.

$$V_a^b(cf, \mathcal{P}) = \sum_{k=1}^{n} |cf(x_k) - cf(x_{k-1})| = |c| \sum_{k=1}^{n} |f(x_k) - f(x_{k-1})| = |c| V_a^b(f, \mathcal{P}).$$

Thus we have

$$V_a^b(cf) = \sup_{\mathcal{P} \in \mathfrak{P}_{a,b}} V_a^b(cf, \mathcal{P}) = |c| \sup_{\mathcal{P} \in \mathfrak{P}_{a,b}} V_a^b(f, \mathcal{P}) = |c| V_a^b(f).$$

Since $f \in BV([a, b])$ we have $V_a^b(f) < \infty$ and this implies that $V_a^b(cf) = |c| V_a^b(f) < \infty$. This shows that $cf \in BV([a, b])$.

2. Let us prove (b). With $\mathcal{P} \in \mathfrak{P}_{a,b}$ given by $\mathcal{P} = \{a = x_0 \leq \cdots \leq x_n = b\}$, we have

$$V_a^b(f + g, \mathcal{P}) = \sum_{k=1}^{n} |(f + g)(x_k) - (f + g)(x_{k-1})|$$

$$\leq \sum_{k=1}^{n} |f(x_k) - f(x_{k-1})| + \sum_{k=1}^{n} |g(x_k) - g(x_{k-1})|$$

$$= V_a^b(f, \mathcal{P}) + V_a^b(g, \mathcal{P}).$$

Then we have

$$V_a^b(f + g) = \sup_{\mathcal{P} \in \mathfrak{P}_{a,b}} V_a^b(f + g, \mathcal{P}) \leq \sup_{\mathcal{P} \in \mathfrak{P}_{a,b}} V_a^b(f, \mathcal{P}) + \sup_{\mathcal{P} \in \mathfrak{P}_{a,b}} V_a^b(g, \mathcal{P})$$

$$= V_a^b(f) + V_a^b(g).$$

Since $f, g \in BV([a, b])$, we have $V_a^b(f) < \infty$ and $V_a^b(g) < \infty$. This then implies that $V_a^b(f + g) \leq V_a^b(f) + V_a^b(g) < \infty$. This shows that $f + g \in BV([a, b])$.

3. To prove (c), let $f(x) = x$ for $x \in [a, b]$ and let $g(x) = -x$ for $x \in [a, b]$.
Then we have $f, g \in BV([a, b])$ with $V_a^b(f) = b - a$ and $V_a^b(g) = b - a$. Thus we have $V_a^b(f) + V_a^b(g) = 2(b - a)$.

On the other hand, $(f + g)(x) = x - x = 0$ for $x \in [a, b]$ and this implies that $f + g \in BV([a, b])$ with $V_a^b(f + g) = 0$. Thus we have $V_a^b(f + g) \neq V_a^b(f) + V_a^b(g)$. ∎

Prob. 12.2. Let $f \in BV([a, b])$. Suppose that there exists $\gamma > 0$ such that $|f| \geq \gamma$ on $[a, b]$. Show that $\frac{1}{f} \in BV([a, b])$.

Proof. Let $\mathcal{P} \in \mathfrak{P}_{a,b}$ be given by $\mathcal{P} = \{a = x_0 \leq \cdots \leq x_n = b\}$. For brevity in notations, let us write $g = \frac{1}{f}$. Then we have

$$V_a^b(g, \mathcal{P}) = \sum_{k=1}^n |g(x_k) - g(x_{k-1})|.$$

Now we have

$$|g(x_k) - g(x_{k-1})| = \left|\frac{1}{f(x_k)} - \frac{1}{f(x_{k-1})}\right| = \frac{|f(x_k) - f(x_{k-1})|}{|f(x_k) f(x_{k-1})|}$$

$$\leq \frac{1}{\gamma^2}|f(x_k) - f(x_{k-1})|.$$

Then we have

$$V_a^b(g, \mathcal{P}) \leq \frac{1}{\gamma^2}\sum_{k=1}^n |f(x_k) - f(x_{k-1})| = \frac{1}{\gamma^2}V_a^b(f, \mathcal{P}) \leq \frac{1}{\gamma^2}V_a^b(f).$$

Thus we have

$$V_a^b(g) = \sup_{\mathcal{P} \in \mathfrak{P}_{a,b}} V_a^b(g, \mathcal{P}) \leq \frac{1}{\gamma^2}V_a^b(f) < \infty.$$

This shows that $g \in BV([a, b])$. $\blacksquare$

Prob. 12.3. Let $f, g \in BV([a, b])$. Show that $fg \in BV([a, b])$ and

$$V_a^b(fg) \leq \sup_{[a,b]} |f| \cdot V_a^b(g) + \sup_{[a,b]} |g| \cdot V_a^b(f).$$

Proof. Let $\mathcal{P} \in \mathfrak{P}_{a,b}$ be given by $\mathcal{P} = \{a = x_0 \leq \cdots \leq x_n = b\}$. We have

$$(1) \qquad V_a^b(fg, \mathcal{P}) = \sum_{k=1}^{n} |(fg)(x_k) - (fg)(x_{k-1})|.$$

Now

$$(fg)(x_k) - (fg)(x_{k-1}) = f(x_k)g(x_k) - f(x_{k-1})g(x_{k-1})$$

$$= f(x_k)g(x_k) - f(x_k)g(x_{k-1}) + f(x_k)g(x_{k-1}) - f(x_{k-1})g(x_{k-1})$$

and then

$$|(fg)(x_k) - (fg)(x_{k-1})| \leq |f(x_k)||g(x_k) - g(x_{k-1})| + |g(x_{k-1})||f(x_k) - f(x_{k-1})|.$$

Since $f, g \in BV([a, b])$, f and g are bounded on $[a, b]$ according to (b) of Observation 12.13 and thus we have $\sup_{[a,b]} |f| < \infty$ and $\sup_{[a,b]} |g| < \infty$. Thus we have

$$|(fg)(x_k) - (fg)(x_{k-1})| \leq \sup_{[a,b]} |f| \cdot |g(x_k) - g(x_{k-1})| + \sup_{[a,b]} |g| \cdot |f(x_k) - f(x_{k-1})|.$$

Substituting this in (1), we have

$$V_a^b(fg, \mathcal{P}) \leq \sup_{[a,b]} |f| \cdot \sum_{k=1}^{n} |g(x_k) - g(x_{k-1})| + \sup_{[a,b]} |g| \cdot \sum_{k=1}^{n} |f(x_k) - f(x_{k-1})|$$

$$= \sup_{[a,b]} |f| \cdot V_a^b(g, \mathcal{P}) + \sup_{[a,b]} |g| \cdot V_a^b(f, \mathcal{P}).$$

Then

$$(2)$$

$$V_a^b(fg) = \sup_{\mathcal{P} \in \mathfrak{P}_{a,b}} V_a^b(fg, \mathcal{P}) \leq \sup_{[a,b]} |f| \cdot \sup_{\mathcal{P} \in \mathfrak{P}_{a,b}} V_a^b(g, \mathcal{P}) + \sup_{[a,b]} |g| \cdot \sup_{\mathcal{P} \in \mathfrak{P}_{a,b}} V_a^b(f, \mathcal{P})$$

$$= \sup_{[a,b]} |f| \cdot V_a^b(g) + \sup_{[a,b]} |g| \cdot V_a^b(f).$$

Since $f, g \in BV([a, b])$, we have $V_a^b(f) < \infty$ and $V_a^b(g) < \infty$. Then (2) implies that $V_a^b(fg) < \infty$. This shows that $fg \in BV([a, b])$. ∎

Prob. 12.4. Let f be a real-valued function on $[a, b]$. Prove the following statements:
(a) $V_a^b(|f|) \leq V_a^b(f)$.
(b) If $f \in BV([a, b])$ then $|f| \in BV([a, b])$.

Proof. 1. Let us prove (a). Let $\mathcal{P} \in \mathfrak{P}_{a,b}$ be given by $\mathcal{P} = \{a = x_0 \leq \cdots \leq x_n = b\}$. Then we have

$$V_a^b(|f|, \mathcal{P}) = \sum_{k=1}^{n} \big||f|(x_k) - |f|(x_{k-1})\big| = \sum_{k=1}^{n} \big||f(x_k)| - |f(x_{k-1})|\big|$$

$$\leq \sum_{k=1}^{n} |f(x_k) - f(x_{k-1})| = V_a^b(f, \mathcal{P}).$$

Then we have

$$V_a^b(|f|) = \sup_{\mathcal{P} \in \mathfrak{P}_{a,b}} V_a^b(|f|, \mathcal{P}) \leq \sup_{\mathcal{P} \in \mathfrak{P}_{a,b}} V_a^b(f, \mathcal{P}) = V_a^b(f).$$

2. If $f \in BV([a, b])$ then $V_a^b(f) < \infty$. Then by (a) we have $V_a^b(|f|) \leq V_a^b(f) < \infty$. Thus we have $|f| \in BV([a, b])$. ∎

Prob. 12.5. Let f be a real-valued function on $[a, b]$ satisfying a Lipschitz condition, that is, there exists a constant $M > 0$ such that $|f(x') - f(x'')| \leq M|x' - x''|$ for every $x', x'' \in [a, b]$. Show that $f \in BV([a, b])$ and $V_a^b(f) \leq M(b - a)$.

Proof. Let $\mathcal{P} \in \mathfrak{P}_{a,b}$ be given by $\mathcal{P} = \{a = x_0 \leq \cdots \leq x_n = b\}$. Then we have

$$V_a^b(f, \mathcal{P}) = \sum_{k=1}^{n} |f(x_k) - f(x_{k-1})| \leq M \sum_{k=1}^{n} |x_k - x_{k-1}| = M(b - a)$$

and then

$$V_a^b(f) = \sup_{\mathcal{P} \in \mathfrak{P}_{a,b}} V_a^b(f, \mathcal{P}) \leq M(b - a) < \infty.$$

This shows that $f \in BV([a, b])$ and $V_a^b(f) \leq M(b - a)$. ∎

Prob. 12.6. Let f be a real-valued function on $[a, b]$. Suppose f is continuous on $[a, b]$ and is differentiable on (a, b) with $|f'| \leq M$ for some constant $M > 0$. Show that $f \in BV([a, b])$ and $V_a^b(f) \leq M(b - a)$.

Proof. Let $\mathcal{P} \in \mathfrak{P}_{a,b}$ be given by $\mathcal{P} = \{a = x_0 \leq \cdots \leq x_n = b\}$. Then we have

(1) $$V_a^b(f, \mathcal{P}) = \sum_{k=1}^{n} |f(x_k) - f(x_{k-1})|.$$

According to the Mean Value Theorem we have

$$f(x_k) - f(x_{k-1}) = f'(\xi)(x_k - x_{k-1}) \quad \text{where } \xi \in (x_{k-1}, x_k)$$

and then

$$|f(x_k) - f(x_{k-1})| = |f'(\xi)||x_k - x_{k-1}| \leq M|x_k - x_{k-1}|.$$

Substituting this estimate in (1), we have

(2) $$V_a^b(f, \mathcal{P}) = \sum_{k=1}^{n} |f(x_k) - f(x_{k-1})| \leq M \sum_{k=1}^{n} |x_k - x_{k-1}| = M(b - a).$$

Then we have

$$V_a^b(f) = \sup_{\mathcal{P} \in \mathfrak{P}_{a,b}} V_a^b(f, \mathcal{P}) \leq M(b - a) < \infty.$$

This shows that $f \in BV([a, b])$ and $V_a^b(f) \leq M(b - a)$. ∎

Prob. 12.7. According to Prob. 12.6, continuity of f on $[a, b]$ and existence and boundedness of the derivative f' on (a, b) imply that f is a BV function. These conditions are sufficient but not necessary for f to be a BV function. To show this find a real-valued function f on $[a, b]$ satisfying the following conditions:
1° f is continuous on $[a, b]$.
2° f' exists on (a, b).
3° f' is not bounded on (a, b).
4° $f \in BV([a, b])$.

Proof. Let $f(x) = \sqrt{x}$ for $x \in [0, 1]$. Then f is continuous on $[0, 1]$. Moreover we have $f'(x) = \frac{1}{2} \frac{1}{\sqrt{x}}$ for $x \in (0, 1)$. Thus f' is not bounded on $(0, 1)$. In fact we have $\lim_{x \downarrow 0} f'(x) = \infty$. However f is increasing on $[0, 1]$ and this implies that $f \in BV([0, 1])$ according to (d) of Observation 12.13. ∎

Prob. 12.8. Let $f(x) = \sin x$ for $x \in [0, 2\pi]$.
(a) Show that $f \in BV([0, 2\pi])$ moreover we have $V_0^{[0,2\pi]}(f) = 4$.
(b) Find the total variation function v_f of f on $[0, 2\pi]$.

Proof. 1. Observe that $[0, 2\pi] = [0, \frac{\pi}{2}] \cup [\frac{\pi}{2}, \pi] \cup [\pi, \frac{3}{2}\pi] \cup [\frac{3}{2}\pi, 2\pi]$.
On $[0, \frac{\pi}{2}]$, f increases from 0 to 1. Thus according to (d) of Observation 12.13 we have $f \in BV([0, \frac{\pi}{2}])$ with $V_0^{\frac{\pi}{2}}(f) = 1$. Similarly we have $f \in BV([\frac{\pi}{2}, \pi])$ with $V_{\frac{\pi}{2}}^{\pi}(f) = 1$; $f \in BV([\pi, \frac{3}{2}\pi])$ with $V_{\pi}^{\frac{3}{2}\pi}(f) = 1$; and $f \in BV([\frac{3}{2}\pi, 2\pi])$ with $V_{\frac{3}{2}\pi}^{2\pi}(f) = 1$.
Then by (b) of Lemma 12.15, we have

$$\begin{cases} f \in BV([0, 2\pi]). \\ V_0^{2\pi}(f) = V_0^{\frac{\pi}{2}}(f) + V_{\frac{\pi}{2}}^{\pi}(f) + V_{\pi}^{\frac{3}{2}\pi}(f) + V_{\frac{3}{2}\pi}^{2\pi}(f) = 4. \end{cases}$$

This completes the proof of (a).
2. Observe that f increases from 0 to 1 on $[0, \frac{\pi}{2}]$, decreases from 1 to -1 on $[\frac{\pi}{2}, \frac{3}{2}\pi]$ and increases from -1 to 0 on $[\frac{3}{2}\pi, 2\pi]$. We conclude that

$$v_f(x) = \begin{cases} \sin x & \text{for } x \in [0, \frac{\pi}{2}], \\ 2 - \sin x & \text{for } x \in [\frac{\pi}{2}, \frac{3}{2}\pi], \\ 4 + \sin x & \text{for } x \in [\frac{3}{2}\pi, 2\pi]. \end{cases}$$

This answers (b). ∎

Prob. 12.9. Consider the interval $[0, 1]$ in $\mathbb{R}$. For $n \in \mathbb{N}$, let $b_n = \frac{1}{n}$ and let $a_n = \frac{1}{2}(b_n + b_{n+1})$. Let f be a real-valued function on $[0, 1]$ such that $f(0) = 0$, $f(a_n) = 0$ for every $n \in \mathbb{N}$, $f(b_n) = 1$ for every $n \in \mathbb{N}$ and f is linear on $[a_n, b_n]$ and $[b_{n+1}, a_n]$ for every $n \in \mathbb{N}$.

Thus defined, f is bounded on $[0, 1]$, continuous at every $x \in (0, 1]$ and discontinuous at $x = 0$. Prove the following statements:

(a) For every $\varepsilon_0 \in (0, 1)$, we have $f \in BV\big([\varepsilon_0, 1]\big)$ and $V^1_{\varepsilon_0}(f) \le 2(n_0 - 1)$ where $n_0 \in \mathbb{N}$ such that $\frac{1}{n_0} < \varepsilon_0$.

(b) $f \notin BV\big([0, 1]\big)$.

Proof. 1. Observe that $1 = b_1 > a_1 > b_2 > a_2 > b_3 > a_3 > \cdots$ and moreover we have

$$[0, 1] = \{0\} \cup (0, 1] = \{0\} \cup \bigcup_{n \in \mathbb{N}} [b_{n+1}, b_n]$$

and

$$[b_{n+1}, b_n] = [b_{n+1}, a_n] \cup [a_n, b_n] \quad \text{for } n \in \mathbb{N}.$$

Now f decreases from 1 to 0 on $[b_{n+1}, a_n]$ and increases from 0 to 1 on $[a_n, b_n]$. This implies according to (d) of Observation 12.13 that $f \in BV\big([b_{n+1}, a_n]\big)$ with $V^{a_n}_{b_{n+1}}(f) = 1$ and $f \in BV\big([a_n, b_n]\big)$ with $V^{b_n}_{a_n}(f) = 1$. Then by (b) of Lemma 12.15, we have

(1)
$$\begin{cases} f \in BV\big([b_{n+1}, b_n]\big). \\ V^{b_n}_{b_{n+1}}(f) = V^{a_n}_{b_{n+1}}(f) + V^{b_n}_{a_n}(f) = 2. \end{cases}$$

Observe that for $n \ge 2$, we have

$$[b_n, b_1] = [b_n, b_{n-1}] \cup [b_{n-1}, b_{n-2}] \cup \cdots \cup [b_2, b_1].$$

Then applying (b) of Lemma 12.15, we have for $n \ge 2$

(2)
$$\begin{cases} f \in BV\big([b_n, b_1]\big). \\ V^{b_1}_{b_n}(f) = V^{b_{n-1}}_{b_n}(f) + V^{b_{n-2}}_{b_{n-1}}(f) + \cdots + V^{b_1}_{b_2}(f) = 2(n - 1). \end{cases}$$

2. Let $\varepsilon_0 \in (0, 1)$ be arbitrarily given. Then since $\lim_{n \to \infty} b_n = \lim_{n \to \infty} \frac{1}{n} = 0$ there exists $n_0 \in \mathbb{N}$ such that $\frac{1}{n_0} < \varepsilon_0$. Then we have $[\varepsilon_0, 1] \subset [b_{n_0}, b_1]$. Then since $f \in BV\big([b_{n_0}, b_1]\big)$ and $V^{b_1}_{b_{n_0}}(f) = 2(n_0 - 1)$ according to (2), (a) of Lemma 12.15 implies that $f \in BV\big([\varepsilon_0, 1]\big)$ and $V^1_{\varepsilon_0}(f) \le 2(n_0 - 1)$.

3. To show that $f \notin BV\big([0, 1]\big)$, let us assume the contrary. Then since $[b_n, b_1] \subset [0, 1]$ for $n \ge 2$, we have

$$V^1_0(f) \ge V^{b_1}_{b_n}(f) = 2(n - 1).$$

Since this holds for every $n \ge 2$, we have $V^1_0(f) = \infty$, contradicting the assumption that $f \in BV\big([0, 1]\big)$. ∎

Prob. 12.10. Let f be a real-valued function on $\left[0, \frac{2}{\pi}\right]$ defined by
$$f(x) = \begin{cases} \sin \frac{1}{x} & \text{for } x \in \left(0, \frac{2}{\pi}\right], \\ 0 & \text{for } x = 0. \end{cases}$$

Prove the following statements:

(a) For every $\varepsilon_0 > 0$ we have $f \in BV\left(\left[\varepsilon_0, \frac{2}{\pi}\right]\right)$. Moreover for $\varepsilon_0 = \left(\frac{\pi}{2} + n_0 2\pi\right)^{-1} > 0$

where $n_0 \in \mathbb{N}$, we have $V_{\varepsilon_0}^{\frac{2}{\pi}}(f) = 4n_0$.

(b) $f \notin BV\left(\left[0, \frac{2}{\pi}\right]\right)$.

Proof. 1. Observe that the function $g(x) = \sin x$ for $x \in [0, \infty)$ is a continuous function assuming the maximal value 1 at $x = \frac{\pi}{2} + k2\pi$ for $k \in \mathbb{Z}_+$ and the minimal value -1 at $x = \frac{3}{2}\pi + k2\pi$ for $k \in \mathbb{Z}_+$ and g is monotone on every interval between two adjacent extremum points.

Thus our function f assumes the maximal value 1 at $x = \left(\frac{\pi}{2} + k2\pi\right)^{-1}$ for $k \in \mathbb{Z}_+$ and minimal value -1 at $x = \left(\frac{3}{2}\pi + k2\pi\right)^{-1}$ for $k \in \mathbb{Z}_+$ and f is monotone on every interval between two adjacent extremum points. Let

$$a_k = \left(\frac{\pi}{2} + k2\pi\right)^{-1} \quad \text{and} \quad b_k = \left(\frac{3}{2}\pi + k2\pi\right)^{-1} \quad \text{for } k \in \mathbb{Z}_+.$$

Then we have

(1)
$$\begin{cases} f(a_k) = 1 \text{ and } f(b_k) = -1 \quad \text{for } k \in \mathbb{Z}_+. \\ 0 < b_n < a_n < b_{n-1} < a_{n-1} < \cdots < b_1 < a_1 < b_0 < a_0 = \frac{2}{\pi}. \\ [a_k, a_{k-1}] = [a_k, b_{k-1}] \cup [b_{k-1}, a_{k-1}] \quad \text{for } k \geq 1. \end{cases}$$

On $[a_k, b_{k-1}]$, f decreases from 1 to -1. Thus by (d) of Observation 12.13, we have $f \in BV\left([a_k, b_{k-1}]\right)$ with $V_{a_k}^{b_{k-1}}(f) = 2$. Similarly f increases from -1 to 1 on $[b_{k-1}, a_{k-1}]$ and this implies that $f \in BV\left([b_{k-1}, a_{k-1}]\right)$ with $V_{b_{k-1}}^{a_{k-1}}(f) = 2$ by (d) of Observation 12.13. Then since $[a_k, a_{k-1}] = [a_k, b_{k-1}] \cup [b_{k-1}, a_{k-1}]$, (b) of Lemma 12.15 implies that we have

(2)
$$\begin{cases} f \in BV\left([a_k, a_{k-1}]\right). \\ V_{a_k}^{a_{k-1}}(f) = V_{a_k}^{b_{k-1}}(f) + V_{b_{k-1}}^{a_{k-1}}(f) = 4. \end{cases}$$

Observe that for $n \in \mathbb{N}$ we have

$$[a_n, a_0] = [a_n, a_{n-1}] \cup [a_{n-1}, a_{n-2}] \cup \cdots \cup [a_1, a_0].$$

Then according to (b) of Lemma 12.15, (1) implies that for $n \in \mathbb{N}$ we have

(3)
$$\begin{cases} f \in BV\left([a_n, a_0]\right). \\ V_{a_n}^{a_0}(f) = V_{a_n}^{a_{n-1}}(f) + V_{a_{n-1}}^{a_{n-2}}(f) + \cdots + V_{a_1}^{a_0} = 4n. \end{cases}$$

2. Let us prove (a). Let $\varepsilon_0 = a_{n_0} = \left(\frac{\pi}{2} + n_0 2\pi\right)^{-1} > 0$ where $n_0 \in \mathbb{N}$. Then by (3) we have $f \in BV\left([\varepsilon_0, \frac{2}{\pi}]\right)$ with $V_{\varepsilon_0}^{\frac{2}{\pi}}(f) = 4n_0$.

Next let $\varepsilon_0 > 0$ be arbitrarily given. Then since $a_k \downarrow 0$ as $k \to \infty$ there exists $n_0 \in \mathbb{N}$ such that $a_{n_0} < \varepsilon_0$ and then $[\varepsilon_0, \frac{2}{\pi}] \subset [a_{n_0}, \frac{2}{\pi}]$. According to (3), we have $f \in BV([a_{n_0}, \frac{2}{\pi}])$. Then by (a) of Lemma 12.15, we have $f \in BV([\varepsilon_0, \frac{2}{\pi}])$.

3. To show that $f \notin BV([0, \frac{2}{\pi}])$, assume the contrary, that is, assume that $f \in BV([0, \frac{2}{\pi}])$. Since $[a_n, a_0] \subset [0, \frac{2}{\pi}]$ for every $n \in \mathbb{N}$, we have by (a) of Lemma 12.15 and (3)

$$V_0^{\frac{2}{pi}}(f) \geq V_{a_n}^{a_0}(f) = 4n.$$

Since this holds for every $n \in \mathbb{N}$, we have $V_0^{\frac{2}{pi}}(f) = \infty$, contradicting the assumption that $f \in BV([0, \frac{2}{\pi}])$. Thus we must have $f \notin BV([0, \frac{2}{\pi}])$. ∎

Prob. 12.11. Let f be a real-valued function on $[a, b]$ such that $f \in BV([a, x])$ for every $x \in [a, b)$ and consider the total variation function $v_f(x) = V_a^x(f)$ for $x \in [a, b)$.
(a) Show that $f \in BV([a, b])$ if and only if v_f is bounded on $[a, b)$, or equivalently,

$$\lim_{x \uparrow b} v_f(x) < \infty \Leftrightarrow f \in BV([a, b]).$$

(b) Assume that we have $\lim_{x \uparrow b} v_f(x) < \infty$. Show that $\lim_{x \uparrow b} v_f(x) = V_a^b(f)$ if and only if f
    is continuous at b.

(c) Assume that we have $\lim_{x \uparrow b} v_f(x) < \infty$. Construct an example for $\lim_{x \uparrow b} v_f(x) \neq V_a^b(f)$.

Proof. 1. Let us prove (a). Assume that we have $\lim_{x \uparrow b} v_f(x) < \infty$. Since v_f is an increasing function on $[a, b)$, the last assumption implies that v_f is bounded above on $[a, b)$. Thus there exists $M > 0$ such that $v_f(x) \leq M$ for $x \in [a, b)$. Let an arbitrary $\mathcal{P} \in \mathfrak{P}_{a,b}$ be given by $\mathcal{P} = \{a = x_0 \leq \cdots \leq x_n = b\}$. Then we have

$$(1) \quad V_a^b(f, \mathcal{P}) = \sum_{k=1}^{n} |f(x_k) - f(x_{k-1})| = \sum_{k=1}^{n-1} |f(x_k) - f(x_{k-1})| + |f(b) - f(x_{n-1})|$$

$$\leq V_a^{x_{n-1}}(f) + |f(b) - f(x_{n-1})| = v_f(x_{n-1}) + |f(b) - f(x_{n-1})|$$

$$\leq M + |f(b) - f(x_{n-1})|.$$

Now we have

$$|f(b) - f(x_{n-1})| = |f(b) - f(x_{n-1}) - f(a) + f(a)|$$

$$\leq |f(b)| + |f(x_{n-1}) - f(a)| + |f(a)|$$

$$\leq |f(b)| + V_a^{x_{n-1}}(f) + |f(a)|$$

$$\leq |f(b)| + v_f(x_{n-1}) + |f(a)|$$

$$\leq |f(b)| + M + |f(a)|.$$

Substituting this estimate in (1) we have

$$(2) \qquad\qquad V_a^b(f, \mathcal{P}) \leq 2M + |f(a)| + |f(b)|.$$

Then we have

$$V_a^b(f) = \sup_{\mathcal{P} \in \mathfrak{P}_{a,b}} V_a^b(f, \mathcal{P}) \leq 2M + |f(a)| + |f(b)| < \infty.$$

This shows that $f \in BV([a, b])$.

Conversely assume that $f \in BV([a, b])$. Then according to (a) of Lemma 12.15, we have $f \in BV([a, x])$ and $V_a^x(f) \leq V_a^b(f) < \infty$ for every $x \in [a, b)$. Thus we have

$$0 \leq v_f(x) \leq V_a^b(f) < \infty \quad \text{for } x \in [a, b),$$

and then
$$\lim_{x \uparrow b} v_f(x) \leq V_a^b(f) < \infty.$$

This completes the proof of (a).

2. Let us prove (b). Observe first that $V_a^b(f) = v_f(b)$. Thus $\lim_{x \uparrow b} v_f(x) = V_a^b(f)$ if and only if v_f is continuous at b. But according to Theorem 12.22, v_f is continuous at b if and only if f is continuous at b.

3. Let f be a real-valued function on $[0, 1]$ defined by

$$f(x) = \begin{cases} 0 & \text{for } x \in [0, 1), \\ 1 & \text{for } x = 1. \end{cases}$$

Then we have $v_f(x) = 0$ for $x \in [0, 1)$ so that $\lim_{x \uparrow 1} v_f(x) = 0$. On the other hand we have $v_f(1) = 1$, that is, $V_0^1(f) = 1$. Thus $\lim_{x \uparrow 1} v_f(x) \neq V_0^1(f)$. ∎

Prob. 12.12. Let $p \in (0, \infty)$ and let f be a real-valued continuous function on $\left[0, \frac{2}{\pi}\right]$ defined by setting

$$f(x) = \begin{cases} x^p \sin \frac{1}{x} & \text{for } x \in \left(0, \frac{2}{\pi}\right], \\ 0 & \text{for } x = 0. \end{cases}$$

(a) Show that $f \notin BV\left(\left[0, \frac{2}{\pi}\right]\right)$ if $p \in (0, 1]$.

(b) Show that $f \in BV\left(\left[0, \frac{2}{\pi}\right]\right)$ if $p \in (1, \infty)$.

Proof. 1. The proof is based on the following estimates:

$$(1) \qquad \begin{cases} \sum_{k \in \mathbb{N}} \frac{1}{k^p} < \infty & \text{if } p \in (1, \infty). \\ \sum_{k \in \mathbb{N}} \frac{1}{k^p} = \infty & \text{if } p \in (0, 1]. \end{cases}$$

The function $\sin x$ for $x \in (0, \infty)$ assumes the value 1 at $x = \frac{\pi}{2} + 2k\pi$ for $k \in \mathbb{Z}_+$ and assumes the value -1 at $x = \frac{3}{2}\pi + 2k\pi$ for $k \in \mathbb{Z}_+$. Then the function $\sin \frac{1}{x}$ for $x \in (0, \infty)$ assumes the value 1 at $\frac{1}{x} = \frac{\pi}{2} + 2k\pi$, that is, $x = \left(\frac{\pi}{2} + 2k\pi\right)^{-1}$, for $k \in \mathbb{Z}_+$ and assumes the value -1 at $\frac{1}{x} = \frac{3}{2}\pi + 2k\pi$, that is, $x = \left(\frac{3}{2}\pi + 2k\pi\right)^{-1}$, for $k \in \mathbb{Z}_+$. Let us set

$$(2) \qquad \begin{cases} a_k = \left(\frac{\pi}{2} + 2k\pi\right)^{-1} = \frac{1}{1+4k} \frac{2}{\pi} & \text{for } k \in \mathbb{Z}_+. \\ b_k = \left(\frac{3}{2}\pi + 2k\pi\right)^{-1} = \frac{1}{3+4k} \frac{2}{\pi} & \text{for } k \in \mathbb{Z}_+. \end{cases}$$

Observe that we have

$$(3) \qquad f(a_k) = a_k^p \text{ and } f(b_k) = -b_k^p \quad \text{for } k \in \mathbb{Z}_+.$$

Observe also that for every $n \in \mathbb{Z}_+$ we have

$$(4) \qquad 0 < b_n < a_n < b_{n-1} < a_{n-1} < \cdots < b_1 < a_1 < b_0 < a_0 = \frac{2}{\pi}.$$

2. Let us prove (a), that is, let us prove that $f \notin BV\left(\left[0, \frac{2}{\pi}\right]\right)$ if $p \in (0, 1]$. For $n \in \mathbb{N}$, let $\mathcal{P}_n$ be a partition of $\left[0, \frac{2}{\pi}\right]$ given by

$$\mathcal{P}_n = \left\{0 < b_n < a_n < b_{n-1} < a_{n-1} < \cdots < b_1 < a_1 < b_0 < a_0 = \frac{2}{\pi}\right\}.$$

Then we have

$$(5) \qquad V_0^{\frac{2}{\pi}}(f, \mathcal{P}_n) = |f(a_0) - f(b_0)| + |f(b_0) - f(a_1)|$$

$$+ |f(a_1) - f(b_1)| + |f(b_1) - f(a_2)|$$

$$\vdots$$

$$+ |f(a_n) - f(b_n)| + |f(b_n) - f(0)|$$

$$\geq \sum_{k=0}^{n} |f(a_k) - f(b_k)| = \sum_{k=0}^{n} \left\{a_k^p + b_k^p\right\}.$$

Now we have

$$\sum_{k=0}^{n} \{a_k^p + b_k^p\} = \sum_{k=0}^{n} \left\{ \left(\frac{1}{1+4k}\frac{2}{\pi}\right)^p + \left(\frac{1}{3+4k}\frac{2}{\pi}\right)^p \right\}$$

$$\geq 2 \sum_{k=0}^{n} \left(\frac{1}{4+4k}\frac{2}{\pi}\right)^p = \frac{2^{p+1}}{(4\pi)^p} \sum_{k=0}^{n} \frac{1}{(1+k)^p}.$$

Substituting this estimate in (5), we obtain

(6) $$V_0^{\frac{2}{\pi}}(f, \mathcal{P}_n) \geq \frac{2^{p+1}}{(4\pi)^p} \sum_{k=0}^{n} \frac{1}{(1+k)^p}.$$

Then we have

$$V_0^{\frac{2}{\pi}}(f) = \sup_{\mathcal{P} \in \mathfrak{P}_{0,\frac{2}{\pi}}} V_0^{\frac{2}{\pi}}(f, \mathcal{P}) \geq \sup_{n \in \mathbb{N}} V_0^{\frac{2}{\pi}}(f, \mathcal{P}_n)$$

$$\geq \frac{2^{p+1}}{(4\pi)^p} \sup_{n \in \mathbb{N}} \left\{ \sum_{k=0}^{n} \frac{1}{(1+k)^p} \right\} \quad \text{by (6)}$$

$$= \frac{2^{p+1}}{(4\pi)^p} \sum_{k \in \mathbb{Z}_+} \frac{1}{(1+k)^p} = \infty \quad \text{by (1)}.$$

This proves that $f \notin BV\left([0, \frac{2}{\pi}]\right)$ if $p \in (0, 1]$.

3. Let us prove (b), that is, let us prove that $f \in BV\left([0, \frac{2}{\pi}]\right)$ if $p \in (1, \infty)$.

On $[b_k, a_k]$ where $k \in \mathbb{Z}_+$, f increases from $f(b_k)$ to $f(a_k)$. Thus (d) of Observation 12.13 implies

$$\begin{cases} f \in BV\left([b_k, a_k]\right). \\ V_{b_k}^{a_k}(f) = f(a_k) - f(b_k) = a_k^p + b_k^p. \end{cases}$$

Similarly on $[a_k, b_{k-1}]$ where $k \in \mathbb{N}$, f decreases from $f(a_k)$ to $f(b_{k-1})$ and thus we have

$$\begin{cases} f \in BV\left([a_k, b_{k-1}]\right). \\ V_{a_k}^{b_{k-1}}(f) = f(a_k) - f(b_{k-1}) = a_k^p + b_{k-1}^p. \end{cases}$$

Now $[b_k, a_k] \cup [a_k, b_{k-1}] = [b_k, b_{k-1}]$. Then by (b) of Lemma 12.15, we have

$$\begin{cases} f \in BV\left([b_k, b_{k-1}]\right). \\ V_{b_k}^{b_{k-1}}(f) = V_{b_k}^{a_k}(f) + V_{a_k}^{b_{k-1}}(f) = 2a_k^p + b_{k-1}^p + b_k^p. \end{cases}$$

Observe that for $n \in \mathbb{Z}_+$, we have

$$[b_n, a_0] = [b_n, b_{n-1}] \cup [b_{n-1}, b_{n-2}] \cup \cdots \cup [b_1, b_0] \cup [b_0, a_0].$$

Then (b) of Lemma 12.15 implies that $f \in BV([b_n, a_0]$ and moreover we have

$$(7) \qquad V_{b_n}^{a_0}(f) = \sum_{k=1}^{n} V_{b_k}^{b_{k-1}}(f) + V_{b_0}^{a_0}(f)$$

$$= \sum_{k=1}^{n} \{2a_k^p + b_{k-1}^p + b_k^p\} + \{a_0^p + b_0^p\}$$

$$\leq 2 \sum_{k=0}^{n} \{a_k^p + b_k^p\}.$$

Now we have

$$(8) \qquad \sum_{k=0}^{n} \{a_k^p + b_k^p\} = \sum_{k=0}^{n} \left\{ \left(\frac{1}{1+4k}\frac{2}{\pi}\right)^p + \left(\frac{1}{3+4k}\frac{2}{\pi}\right)^p \right\}$$

$$\leq 2 \sum_{k=0}^{n} \left(\frac{1}{1+4k}\frac{2}{\pi}\right)^p = 2\left(\frac{2}{\pi}\right)^p + 2 \sum_{k=1}^{n} \left(\frac{1}{4k}\frac{2}{\pi}\right)^p$$

$$= 2\left(\frac{2}{\pi}\right)^p + \frac{2}{(2\pi)^p} \sum_{k=1}^{n} \frac{1}{k^p}.$$

Substituting (8) in (7), we have

$$V_{b_n}^{a_0}(f) \leq 4\left(\frac{2}{\pi}\right)^p + \frac{4}{(2\pi)^p} \sum_{k=1}^{n} \frac{1}{k^p}.$$

Then by (1) we have

$$(9) \qquad \lim_{n \to \infty} V_{b_n}^{a_0}(f) \leq \left(\frac{2}{\pi}\right)^p + \frac{4}{(2\pi)^p} \lim_{n \to \infty} \left\{ \sum_{k=1}^{n} \frac{1}{k^p} \right\} < \infty.$$

Consider the total variation function $v_f(x) = V_0^x(f)$ for $x \in \left[0, \frac{2}{\pi}\right]$. We have

$$V_{b_n}^{a_0}(f) = v_f(a_0) - v_f(b_n).$$

Continuity of f implies continuity of v_f by Theorem 12.22. Since $\lim_{n \to \infty} b_n = 0$, we have

$$(10) \quad \lim_{n \to \infty} V_{b_n}^{a_0}(f) = v_f(a_0) - \lim_{n \to \infty} v_f(b_k) = v_f(a_0) - v_f(0) = v_f(a_0) = V_0^{a_0}(f).$$

By (9) and (10), we have $V_0^{a_0}(f) < \infty$. This proves that $f \in BV\left([0, \frac{2}{\pi}]\right)$ if $p \in (1, \infty)$. $\blacksquare$

Prob. 12.13. (A Generalization of Prob. 12.12) Let $p \in (0, \infty)$ and let f be a real-valued continuous function on $[0, b]$ where $b > 0$ defined by setting
$$f(x) = \begin{cases} x^p \sin \frac{1}{x} & \text{for } x \in (0, b], \\ 0 & \text{for } x = 0. \end{cases}$$
(a) Show that $f \notin BV([0, b])$ if $p \in (0, 1]$.
(b) Show that $f \in BV([0, b])$ if $p \in (1, \infty)$.

Proof. 1. Let us prove (a). Assume $p \in (0, 1]$.

1.1. Consider the case $b \leq \frac{2}{\pi}$, that is, $[0, b] \subset [0, \frac{2}{\pi}]$. To show that $f \notin BV([0, b])$, let us assume the contrary, that is, let us assume that $f \in BV([0, b])$. Now a simple calculation shows that f' is continuous on $[b, \frac{2}{\pi}]$ and hence bounded on $[b, \frac{2}{\pi}]$. Thus by Prop. 12.5 we have $f \in BV([b, \frac{2}{\pi}])$. Then we have $f \in BV([0, \frac{2}{\pi}])$ by (b) of Lemma 12.15. This contradicts (a) of Prob. 12.12. Therefore we must have $f \notin BV([0, b])$.

1.2. Consider the case $\frac{2}{\pi} < b$, that is, $[0, \frac{2}{\pi}] \subset [0, b]$. To show that $f \notin BV([0, b])$, let us assume the contrary. Then we have $f \in BV([0, b])$. This implies that $f \in BV([0, \frac{2}{\pi}])$ by (a) of Lemma 12.15. This contradicts (a) of Prob. 12.12. Therefore we must have $f \notin BV([0, b])$.

2. Let us prove (b). Assume $p \in (1, \infty)$.

2.1. Consider the case $b \leq \frac{2}{\pi}$, that is, $[0, b] \subset [0, \frac{2}{\pi}]$. By (b) of Prob. 12.12, we have $f \in BV([0, \frac{2}{\pi}])$. This implies that $f \in BV([0, b])$ by (a) of Lemma 12.15.

2.2. Consider the case $\frac{2}{\pi} < b$, that is, $[0, \frac{2}{\pi}] \subset [0, b]$. By (b) of Prob. 12.12, we have $f \in BV([0, \frac{2}{\pi}])$. It is easily shown that f' is continuous on $[\frac{2}{\pi}, b]$ and hence bounded on $[\frac{2}{\pi}, b]$. Then by Prob. 12.6, we have $f \in BV([\frac{2}{\pi}, b])$. Then by (b) of Lemma 12.15, we have $f \in BV([0, b])$. ∎

Prob. 12.14. Construct a sequence $(f_i : i \in \mathbb{N})$ of real-valued functions on $[0, b]$ such that
1° $f_i \in BV([0, b])$ for every $i \in \mathbb{N}$.
2° $f = \lim_{i \to \infty} f_i$ exists on $[0, b]$ but $f \notin BV([0, b])$.

Proof. For $i \in \mathbb{N}$ let us define a function f_i on $[0, b]$ by setting
$$f_i(x) = \begin{cases} x^{1+\frac{1}{i}} \sin \frac{1}{x} & \text{for } x \in (0, b], \\ 0 & \text{for } x = 0, \end{cases}$$
and define a function f on $[0, b]$ by setting
$$f(x) = \begin{cases} x \sin \frac{1}{x} & \text{for } x \in (0, b], \\ 0 & \text{for } x = 0. \end{cases}$$

According to Prob. 12.13, we have $f_i \in BV([0, b])$ for every $i \in \mathbb{N}$ and $f \notin BV([0, b])$. Now $\lim_{i \to \infty} x^{1+\frac{1}{i}} = x$ and this implies that $\lim_{i \to \infty} f_i = f$ on $[0, b]$. ∎

Prob. 12.15. Let $(f_i : i \in \mathbb{N})$ and f be real-valued functions on an interval $[a, b]$ such that $\lim_{i \to \infty} f_i(x) = f(x)$ for $x \in [a, b]$. Show that $V_a^b(f) \le \liminf_{i \to \infty} V_a^b(f_i)$.
(There are such cases that $\liminf_{i \to \infty} V_a^b(f_i) = \infty$ or even $\lim_{i \to \infty} V_a^b(f_i) = \infty$. See Prob. 12.16.
Thus $V_a^b(f) \le \liminf_{i \to \infty} V_a^b(f_i)$ does not imply $V_a^b(f) < \infty$.)

Proof. For a sequence $(c_i : i \in \mathbb{N})$ in $\mathbb{R}$ and $c \in \mathbb{R}$, we have

(1)
$$\lim_{i \to \infty} c_i = c \Rightarrow \lim_{i \to \infty} |c_i| = |c| = \left| \lim_{i \to \infty} c_i \right|.$$

(This can be proved as follows. Suppose we have $\lim_{i \to \infty} c_i = c$. Then for every $\varepsilon > 0$ there exists $N \in \mathbb{N}$ such that $|c_i - c| < \varepsilon$ for $i \ge N$. Then we have $\big||c_i| - |c|\big| \le |c_i - c| < \varepsilon$ for $i \ge N$. This shows that $\lim_{i \to \infty} |c_i| = |c| = \left| \lim_{i \to \infty} c_i \right|$.)

Let $\mathcal{P} \in \mathfrak{P}_{a,b}$ be given by $\mathcal{P} = \{a = x_0 \le \cdots \le x_n = b\}$. Then we have

(2)
$$V_a^b(f, \mathcal{P}) = \sum_{k=1}^n |f(x_k) - f(x_{k-1})| = \sum_{k=1}^n \left| \lim_{i \to \infty} f_i(x_k) - \lim_{i \to \infty} f_i(x_{k-1}) \right|$$

$$= \sum_{k=1}^n \left| \lim_{i \to \infty} \{ f_i(x_k) - f_i(x_{k-1}) \} \right|$$

$$= \sum_{k=1}^n \lim_{i \to \infty} |f_i(x_k) - f_i(x_{k-1})| \quad \text{by (1)}$$

$$= \lim_{i \to \infty} \sum_{k=1}^n |f_i(x_k) - f_i(x_{k-1})|$$

$$= \lim_{i \to \infty} V_a^b(f_i, \mathcal{P}) = \liminf_{i \to \infty} V_a^b(f_i, \mathcal{P}).$$

Now we have $V_a^b(f_i, \mathcal{P}) \le V_a^b(f_i)$ for every $i \in \mathbb{N}$ and then we have

(3)
$$\liminf_{i \to \infty} V_a^b(f_i, \mathcal{P}) \le \liminf_{i \to \infty} V_a^b(f_i).$$

Substituting (3) in (2), we have

(4)
$$V_a^b(f, \mathcal{P}) \le \liminf_{i \to \infty} V_a^b(f_i) \quad \text{for every } \mathcal{P} \in \mathfrak{P}_{a,b}.$$

Then we have
$$V_a^b(f) = \sup_{\mathcal{P} \in \mathfrak{P}_{a,b}} V_a^b(f, \mathcal{P}) \le \liminf_{i \to \infty} V_a^b(f_i).$$

This completes the proof. ∎

Prob. 12.16. For each $i \in \mathbb{N}$, let us partition $[0, 1]$ into 2^i subintervals of length $\frac{1}{2^i}$. Thus $[0, 1] = \bigcup_{k=1}^{2^i} I_k$ where $I_k = \left[\frac{k-1}{2^i}, \frac{k}{2^i}\right]$ for $k = 1, \ldots, 2^i$. Let f_i be a real-valued function on $[0, 1]$ defined by setting

$$f_i(x) = \begin{cases} 1 & \text{for } x = \frac{k}{2^i} \text{ for } k = 0, \ldots, 2^i \\ 0 & \text{elswhere on } [0, 1]. \end{cases}$$

Prove the following statements:

(a) $f_i \in BV\big([0, 1]\big)$ for every $i \in \mathbb{N}$ and $\lim_{i \to \infty} V_0^1(f_i) = \infty$.

(b) $f = \lim_{i \to \infty} f_i$ exists on $[0, 1]$ and $f \notin BV\big([0, 1]\big)$.

Proof. 1. Let us prove (a).

On every subinterval I_k, where $k = 1, \ldots, 2^i$, f_i decreases first from 1 to 0 and then increases from 0 to 1. This implies that $f_i \in BV(I_k)$ with $V_{\frac{k-1}{2^i}}^{\frac{k}{2^i}}(f_i) = 2$ according to (d) of Observation 12.13 and (b) of Lemma 12.15.

Then by (b) of Lemma 12.15 we have $f_i \in BV\big([0, 1]\big)$ and we have

$$V_0^1(f_i) = \sum_{k=1}^{2^i} V_{\frac{k-1}{2^i}}^{\frac{k}{2^i}}(f_i) = 2^i \cdot 2 = 2^{i+1}.$$

Then we have

$$\lim_{i \to \infty} V_0^1(f_i) = \lim_{i \to \infty} 2^{i+1} = \infty.$$

This completes the proof of (a).

2. Let us prove (b). For every $i \in \mathbb{N}$, let $\mathcal{B}_i = \left\{\frac{k}{2^i} : k = 0, \cdots, 2^i\right\}$ and call the members of the set $\mathcal{B}_i$ binary rational numbers of denomination i in $[0, 1]$. Observe that $(\mathcal{B}_i : i \in \mathbb{N})$ is an increasing sequence of sets, that is, $\mathcal{B}_1 \subset \mathcal{B}_2 \subset \mathcal{B}_3 \subset \cdots$. Let $\mathcal{B} = \bigcup_{i \in \mathbb{N}} \mathcal{B}_i$ and call the members of the set $\mathcal{B}$ binary rational numbers in $[0, 1]$. The definition of our function f_i is then restated as follows

$$f_i(x) = \begin{cases} 1 & \text{for } x \in \mathcal{B}_i, \\ 0 & \text{elsewhere on } [0, 1]. \end{cases}$$

Then $(f_i(x) : i \in \mathbb{N})$ is an increasing sequence for every $x \in [0, 1]$. Indeed we have

$$(0, 0, 0, \ldots) \qquad\qquad \text{for } x \notin \mathcal{B}.$$

$$(0_1, \ldots, 0_{i_0-1}, 1, 1, 1, \ldots) \quad \text{for } x \in \mathcal{B}_{i_0} \text{ where } i_0 > 1.$$

$$(1, 1, 1, \ldots) \qquad\qquad \text{for } x \in \mathcal{B}_1.$$

Now that $(f_i(x) : i \in \mathbb{N})$ is an increasing sequence of real numbers for every $x \in [0, 1]$, $f(x) = \lim_{i \to \infty} f_i(x)$ exists for every $x \in [0, 1]$. Moreover we have

$$f(x) = \begin{cases} 1 & \text{for } x \in \mathcal{B}, \\ 0 & \text{elsewhere on } [0, 1]. \end{cases}$$

Let us show that $f \notin BV([0, 1])$. For each $i \in \mathbb{N}$, let $\mathcal{P}_i$ be a partition of $[0, 1]$ whose partition points are given by

$$\begin{cases} \mathcal{B}_i = \{\frac{k}{2^i} : k = 0, \cdots, 2^i\}, \\ \text{irrational number } \xi_k \in I_k = [\frac{k-1}{2^i}, \frac{k}{2^i}] \text{for } k = 1, \ldots, 2^i. \end{cases}$$

On I_k where $k = 1, \ldots, 2^i$, we have

$$\left| f(\xi_k) - f\left(\frac{k-1}{2^i}\right) \right| + \left| f\left(\frac{k}{2^i}\right) - f(\xi_k) \right| = 1 + 1 = 2.$$

Thus we have

$$V_0^1(f, \mathcal{P}_i) = 2^i \cdot 2 = 2^{i+1}.$$

Then we have

$$V_0^1(f) = \sup_{\mathcal{P} \in \mathfrak{P}_{0,1}} V_0^1(f, \mathcal{P}) \geq \sup_{i \in \mathbb{N}} V_0^1(f, \mathcal{P}_i) = \sup_{i \in \mathbb{N}} 2^{i+1} = \infty.$$

Thus we have $V_0^1(f) = \infty$. This shows that $f \notin BV([0, 1])$. ∎

Prob. 12.17. Let $f \in BV([a, b])$ and $g \in BV([c, d])$. Assume that $f([a, b]) \subset [c, d]$ so that $g \circ f$ is defined on $[a, b]$. Construct an example such that $g \circ f \notin BV([a, b])$.

Proof. 1. Let us define two real-valued functions f and g on $[0, 1]$ by setting

$$f(x) = \begin{cases} x^2 |\sin \frac{1}{x}| & \text{for } x \in (0, 1], \\ 0 & \text{for } x = 0, \end{cases}$$

and

$$g(x) = \sqrt{x} \quad \text{for } x \in [0, 1].$$

We have $f([0, 1]) \subset [0, 1]$, which is the domain of definition of g so that $g \circ f$ is defined on $[0, 1]$. Now g is monotone on $[0, 1]$ and this implies that $g \in BV([0, 1])$ by (d) of Observation 12.13. To verify that $f \in BV([0, 1])$, let us define a real-valued function f_0 on $[0, 1]$ by setting

$$f_0(x) = \begin{cases} x^2 \sin \frac{1}{x} & \text{for } x \in (0, 1], \\ 0 & \text{for } x = 0. \end{cases}$$

Then $f_0 \in BV([0, 1])$ by (b) of Prob. 12.13. Then since $f = |f_0|$, we have $f \in BV([0, 1])$ according to Prob. 12.4.

2. Let us show that $g \circ f \notin BV([0, 1])$. For brevity let us write $h = g \circ f$. Then

$$h(x) = (g \circ f)(x) = \sqrt{f(x)} = \begin{cases} x\sqrt{|\sin \frac{1}{x}|} & \text{for } x \in (0, 1], \\ 0 & \text{for } x = 0. \end{cases}$$

Let us show that $h \notin BV([0, 1])$.

The function $|\sin x|$ for $x \in (0, \infty)$ assumes the value 0 at $x = k\pi$ for $k \in \mathbb{N}$ and assumes the value 1 at $x = \frac{\pi}{2} + k\pi$ for $k \in \mathbb{N}$. So does the function $\sqrt{|\sin x|}$ for $x \in (0, \infty)$. Then the function $\sqrt{|\sin \frac{1}{x}|}$ for $x \in (0, \infty)$ assumes the value 0 at $a_k = (k\pi)^{-1}$ for $k \in \mathbb{N}$ and assumes the value 1 at $b_k = (\frac{\pi}{2} + k\pi)^{-1}$ for $k \in \mathbb{N}$. Observe that we have

$$h(a_k) = 0 \text{ and } h(b_k) = b_k \quad \text{for } k \in \mathbb{N}.$$

For every $n \in \mathbb{N}$, we have

$$0 < b_n < a_n < b_{n-1} < a_{n-1} < \cdots < b_1 < a_1 < 1.$$

Define $b_0 = 1$ and define $a_{n+1} = 0$ for the time being. We still have $h(a_{n+1}) = h(0) = 0$. Consider a partition $\mathcal{P}_n$ of $[0, 1]$ defined by

$$\mathcal{P}_n = \{0 = a_{n+1} < b_n < a_n < b_{n-1} < a_{n-1} < \cdots < b_1 < a_1 < b_0 = 1\}.$$

We have

$$
\begin{aligned}
V_0^1(h, \mathcal{P}_n) &= \sum_{k=0}^{n} |h(a_{k+1}) - h(b_k)| + \sum_{k=1}^{n} |h(b_k) - h(a_k)| \\
&= \sum_{k=0}^{n} |h(b_k)| + \sum_{k=1}^{n} |h(b_k)| = h(1) + 2 \sum_{k=1}^{n} b_k \\
&= h(1) + 2 \sum_{k=1}^{n} \frac{1}{\frac{\pi}{2} + k\pi} \\
&\geq h(1) + 2 \sum_{k=1}^{n} \frac{1}{(k+1)\pi} = h(1) + \frac{2}{\pi} \sum_{k=1}^{n} \frac{1}{k+1}.
\end{aligned}
$$

Then we have

$$
\begin{aligned}
V_0^1(h) &= \sup_{\mathcal{P} \in \mathfrak{P}_{0,1}} V_0^1(h, \mathcal{P}) \geq \sup_{n \in \mathbb{N}} V_0^1(h, \mathcal{P}_n) \\
&\geq \sup_{n \in \mathbb{N}} \left\{ h(1) + \frac{2}{\pi} \sum_{k=1}^{n} \frac{1}{k+1} \right\} \\
&= h(1) + \frac{2}{\pi} \sup_{n \in \mathbb{N}} \left\{ \sum_{k=1}^{n} \frac{1}{k+1} \right\} \\
&= \infty.
\end{aligned}
$$

This shows that $h \notin BV([0,1])$. ∎

Prob. 12.18. Let $f \in BV([a,b])$ and $g \in BV([c,d])$. Assume that $f([a,b]) \subset [c,d]$ so that $g \circ f$ is defined on $[a,b]$. Assume further that f is an increasing function on $[a,b]$. Show that $g \circ f \in BV([a,b])$.

Proof. If $g \in BV([c,d])$ then $g = g_1 - g_2$ where g_1 and g_2 are increasing functions on $[c,d]$ according to Theorem 12.18 (Jordan Decomposition). Then we have

$$
g \circ f = (g_1 - g_2) \circ f = g_1 \circ f - g_2 \circ f.
$$

Now f is an increasing function on $[a,b]$ and g_1 is an increasing function on $[c,d]$. This implies that $g_1 \circ f$ is an increasing function on $[a,b]$. Similarly $g_2 \circ f$ is an increasing function on $[a,b]$. Then by Theorem 12.18 we have $g \circ f \in BV([a,b])$. ∎

Prob. 12.19. Let f be a real-valued continuous and BV function on $[0, 1]$. Show that

$$\lim_{n \to \infty} \sum_{i=1}^{n} \left| f\left(\tfrac{i}{n}\right) - f\left(\tfrac{i-1}{n}\right) \right|^2 = 0.$$

Proof. Continuity of f on the compact set $[0, 1]$ implies uniform continuity of f on $[0, 1]$. Thus for every $\varepsilon > 0$ there exists $\delta > 0$ such that

(1) $\qquad |f(t') - f(t'')| < \varepsilon \quad \text{for } t', t'' \in [0, 1] \text{ such that } |t' - t''| < \delta.$

Let $N \in \mathbb{N}$ be so large that $\tfrac{1}{N} < \delta$. Then for every $n \geq N$ we have $\tfrac{1}{n} < \delta$. Then for $n \geq N$ we have

(2) $\qquad \left| f\left(\tfrac{i}{n}\right) - f\left(\tfrac{i-1}{n}\right) \right| < \varepsilon \quad \text{for } i = 1, \ldots, n.$

Then we have

(3) $\qquad \displaystyle\sum_{i=1}^{n} \left| f\left(\tfrac{i}{n}\right) - f\left(\tfrac{i-1}{n}\right) \right|^2 = \sum_{i=1}^{n} \left| f\left(\tfrac{i}{n}\right) - f\left(\tfrac{i-1}{n}\right) \right| \cdot \left| f\left(\tfrac{i}{n}\right) - f\left(\tfrac{i-1}{n}\right) \right|$

$$\leq \varepsilon \sum_{i=1}^{n} \left| f\left(\tfrac{i}{n}\right) - f\left(\tfrac{i-1}{n}\right) \right| \quad \text{by (2)}$$

$$\leq \varepsilon V_0^1(f).$$

Observe that $f \in BV\big([0, 1]\big)$ implies $V_0^1(f) < \infty$. Next, (3) implies

(4) $\qquad \displaystyle\limsup_{n \to \infty} \sum_{i=1}^{n} \left| f\left(\tfrac{i}{n}\right) - f\left(\tfrac{i-1}{n}\right) \right|^2 \leq \varepsilon V_0^1(f).$

Since (4) holds for every $\varepsilon > 0$, we have

(5) $\qquad \displaystyle\limsup_{n \to \infty} \sum_{i=1}^{n} \left| f\left(\tfrac{i}{n}\right) - f\left(\tfrac{i-1}{n}\right) \right|^2 = 0.$

The nonnegativity of the sum in (5) implies that

$$0 \leq \liminf_{n \to \infty} \sum_{i=1}^{n} \left| f\left(\tfrac{i}{n}\right) - f\left(\tfrac{i-1}{n}\right) \right|^2 \leq \limsup_{n \to \infty} \sum_{i=1}^{n} \left| f\left(\tfrac{i}{n}\right) - f\left(\tfrac{i-1}{n}\right) \right|^2 = 0.$$

This shows that the limit inferior and the limit superior are both equal to 0. This implies that the limit exists and is equal to 0, that is,

$$\lim_{n \to \infty} \sum_{i=1}^{n} \left| f\left(\tfrac{i}{n}\right) - f\left(\tfrac{i-1}{n}\right) \right|^2 = 0.$$

This completes the proof. ∎

Prob. 12.20. Let f be real-valued and differentiable everywhere on (a, b). Show that the derivative f' is a Borel measurable function on (a, b).

Proof. We say that a real-valued function g on (a, b) is Borel measurable if it is a $\mathfrak{B}_{\mathbb{R}}/\mathfrak{B}_{\mathbb{R}}$-measurable mapping of (a, b) into $\mathbb{R}$.

Now if f is differentiable at $x \in (a, b)$ then f is continuous at x. Then if f is differentiable everywhere on (a, b) then f is continuous at every $x \in (a, b)$ and hence continuous on (a, b). Since every continuous function on (a, b) is a $\mathfrak{B}_{\mathbb{R}}/\mathfrak{B}_{\mathbb{R}}$-measurable mapping of (a, b) into $\mathbb{R}$ according to Theorem 4.27, f is a $\mathfrak{B}_{\mathbb{R}}/\mathfrak{B}_{\mathbb{R}}$-measurable mapping of (a, b) into $\mathbb{R}$.

Since f' exists everywhere on (a, b) we have

$$f'(x) = \lim_{n \to \infty} \frac{f(x + \frac{1}{n}) - f(x)}{\frac{1}{n}} \quad \text{for } x \in (a, b).$$

For $n \in \mathbb{N}$ define

$$g_n(x) = x + \frac{1}{n} \quad \text{for } x \in (a, b),$$

and then define

$$f_n = f \circ g_n \quad \text{that is,} \quad f_n(x) = f\big(g_n(x)\big) = f\big(x + \tfrac{1}{n}\big) \quad \text{for } x \in (a, b).$$

Then we have

$$f'(x) = \lim_{n \to \infty} n\{f_n(x) - f(x)\} \quad \text{for } x \in (a, b).$$

Now g_n is a continuous function on (a, b) and thus g_n is a $\mathfrak{B}_{\mathbb{R}}/\mathfrak{B}_{\mathbb{R}}$-measurable mapping of (a, b) into $\mathbb{R}$. Then since f is a $\mathfrak{B}_{\mathbb{R}}/\mathfrak{B}_{\mathbb{R}}$-measurable mapping, $f_n = f \circ g_n$ is a $\mathfrak{B}_{\mathbb{R}}/\mathfrak{B}_{\mathbb{R}}$-measurable mapping by Theorem 1.40 (Chain Rule of Measurable Mappings). Then $n\{f_n - f\}$ is a $\mathfrak{B}_{\mathbb{R}}/\mathfrak{B}_{\mathbb{R}}$-measurable mapping by Theorem 4.11 and Theorem 4.12. Then $\lim_{n \to \infty} n\{f_n - f\}$ is a $\mathfrak{B}_{\mathbb{R}}/\mathfrak{B}_{\mathbb{R}}$-measurable mapping by Theorem 4.22. This shows that f' is a $\mathfrak{B}_{\mathbb{R}}/\mathfrak{B}_{\mathbb{R}}$-measurable mapping. ∎

Prob. 12.21. Prove the following theorem:

(Fubini's Differentiation Theorem for Series of Increasing Functions) *Let* $(f_n : n \in \mathbb{N})$ *be a sequence of real-valued increasing functions on* $[a, b]$. *Assume that the series* $s(x) = \sum_{n \in \mathbb{N}} f_n(x)$ *converges for every* $x \in [a, b]$. *Then we have* $s'(x) = \sum_{n \in \mathbb{N}} f_n'(x)$ *for* μ_L-*a.e.* $x \in [a, b]$.

Proof. 1. Since f_n is a real-valued increasing function on $[a, b]$ for every $n \in \mathbb{N}$ and since $s(x) = \sum_{n \in \mathbb{N}} f_n(x)$ converges for every $x \in [a, b]$, s is a real-valued increasing function on $[a, b]$.

By Theorem 12.10 (Lebesgue), the derivative s' exists and is nonnegative μ_L-a.e. on $[a, b]$, and moreover s' is $\mathfrak{M}_L$-measurable and μ_L-integrable on $[a, b]$ so that s' is real-valued μ_L-a.e. on $[a, b]$. Thus there exists a null set E in the measure space $(\mathbb{R}, \mathfrak{M}_L, \mu_L)$ contained in $[a, b]$ such that $s'(x) \in [0, \infty)$ for $x \in [a, b] \setminus E$. Since f_n is real-valued and increasing on $[a, b]$, by Theorem 12.10 similarly the derivative f_n' exists and is nonnegative μ_L-a.e. on $[a, b]$, and moreover f_n' is $\mathfrak{M}_L$-measurable and μ_L-integrable on $[a, b]$ so that f_n' is real-valued μ_L-a.e. on $[a, b]$. Thus there exists a null set E_n in the measure space $(\mathbb{R}, \mathfrak{M}_L, \mu_L)$ contained in $[a, b]$ such that $f_n'(x) \in [0, \infty)$ for $x \in [a, b] \setminus E_n$.

Let $A = E \cup \left(\bigcup_{n \in \mathbb{N}} E_n \right)$, a null set in the measure space $(\mathbb{R}, \mathfrak{M}_L, \mu_L)$ contained in $[a, b]$. Then we have

$$(1) \qquad \{ f_n'(x) : n \in \mathbb{N} \} \subset [0, \infty) \text{ and } s'(x) \in [0, \infty) \quad \text{for } x \in [a, b] \setminus A.$$

Consider the sequence of partial sums $(s_n : n \in \mathbb{N})$ of the series $s = \sum_{n \in \mathbb{N}} f_n$, that is, $s_n = \sum_{k=1}^{n} f_k$ for $n \in \mathbb{N}$ and $s = \lim_{n \to \infty} s_n$. Now $\sum_{n \in \mathbb{N}} f_n' = \lim_{n \to \infty} \sum_{k=1}^{n} f_k' = \lim_{n \to \infty} s_n'$. Thus to prove the theorem it remains to show that

$$(2) \qquad s'(x) = \lim_{n \to \infty} s_n'(x) \quad \text{for } x \in [a, b] \setminus A.$$

2. Let us consider the particular case that f_n is a nonnegative real-valued increasing function on $[a, b]$ for every $n \in \mathbb{N}$. Now nonnegativity of f_n implies

$$(3) \qquad s_n(x) \uparrow s(x) \quad \text{for } x \in [a, b].$$

Let $h > 0$. Since f_{n+1} is an increasing function on $[a, b]$, we have

$$\frac{f_{n+1}(x + h) - f_{n+1}(x)}{h} \geq 0$$

and then

$$\frac{s_n(x + h) - s_n(x)}{h} \leq \frac{s_n(x + h) - s_n(x)}{h} + \frac{f_{n+1}(x + h) - f_{n+1}(x)}{h}$$

$$= \frac{s_{n+1}(x + h) - s_{n+1}(x)}{h}$$

$$\leq \frac{s(x + h) - s(x)}{h} \quad \text{by (3)}.$$

Letting $h \downarrow 0$, we have $s_n'(x) \le s_{n+1}'(x) \le s'(x) < \infty$ for $x \in [a, b] \setminus A$. Thus we have

(4) $s_n'(x) \uparrow$ and $\lim_{n \to \infty} s_n'(x) \le s'(x) < \infty$ for $x \in [a, b] \setminus A$.

It remains to show that actually we have the equality

(5) $\lim_{n \to \infty} s_n'(x) = s'(x)$ for $x \in [a, b] \setminus A$.

Since $s_n'(x) \uparrow$, to prove (5) it suffices to show that there exists a subsequence $(n_k : k \in \mathbb{N})$ of the sequence $(n : n \in \mathbb{N})$ such that

(6) $\lim_{k \to \infty} s_{n_k}'(x) = s'(x)$ for $x \in [a, b] \setminus A$.

Now we have $s_n(b) \uparrow s(b)$ and therefore $\lim_{n \to \infty} \{s(b) - s_n(b)\} = 0$. Choose n_1 so that $s(b) - s_{n_1}(b) < \frac{1}{2}$. Choose $n_2 > n_1$ so that $s(b) - s_{n_2}(b) < \frac{1}{2^2}$. Choose $n_3 > n_2$ so that $s(b) - s_{n_3}(b) < \frac{1}{2^3}$ and so on. Then we have

(7) $\sum_{k \in \mathbb{N}} \{s(b) - s_{n_k}(b)\} \le \sum_{k \in \mathbb{N}} \frac{1}{2^k} = 1 < \infty.$

Since $s - s_{n_k} = \sum_{n \ne n_k} f_n$ and this sum is an increasing function on $[a, b]$, we have

(8) $0 \le s(x) - s_{n_k}(x) \le s(b) - s_{n_k}(b).$

By (7) and (8), we have

(9) $\sum_{k \in \mathbb{N}} \{s(x) - s_{n_k}(x)\} \le \sum_{k \in \mathbb{N}} \{s(b) - s_{n_k}(b)\} < \infty.$

Now $(s - s_{n_k} : n \in \mathbb{N})$ is a sequence of nonnegative increasing functions on $[a, b]$ and the series $\sum_{k \in \mathbb{N}} \{s - s_{n_k}\}$ converges on $[a, b]$ according to (9). Thus (1) applies and we have

(10) $\sum_{k \in \mathbb{N}} \{s'(x) - s_{n_k}'(x)\} < \infty$ for $x \in [a, b] \setminus A$.

Convergence of the series in (10) implies

                         $\lim_{k \to \infty} \{s'(x) - s_{n_k}'(x)\} = 0$ for $x \in [a, b] \setminus A$.

This proves (6) and then (5).

3. Consider the general case, that is, we do not assume that f_n is nonnegative for every $n \in \mathbb{N}$. Let

                 $g_n(x) = f_n(x) - f_n(a)$ for $x \in [a, b]$.

Since f_n is a real-valued increasing function on $[a, b]$, g_n is a nonnegative real-valued increasing function on $[a, b]$. Since $s(x) = \sum_{n \in \mathbb{N}} f_n(x)$ converges for every $x \in [a, b]$,

             $t(x) := \sum_{n \in \mathbb{N}} g_n(x) = \sum_{n \in \mathbb{N}} \{f_n(x) - f_n(a)\} = s(x) - s(a)$

converges for every $x \in [a, b]$. Then by our result in **2**, we have $t'(x) = \sum_{n \in \mathbb{N}} g_n'(x)$ for μ_L-a.e. $x \in [a, b]$, that is, $s'(x) = \sum_{n \in \mathbb{N}} f_n'(x)$ for μ_L-a.e. $x \in [a, b]$. ∎

§13   Absolutely Continuous Functions

Prob. 13.1. Let f and g be real-valued absolutely continuous functions on $[a, b]$. Show that the following functions are absolutely continuous on $[a, b]$:
(a) cf where $c \in \mathbb{R}$,
(b) $f + g$,
(c) fg,
(d) $1/f$ provided that $f(x) \neq 0$ for every $x \in [a, b]$.
(e) $|f|$.

Proof. 1. Let us prove (a). Suppose f is a real-valued absolutely continuous function on $[a, b]$. Let $c \in \mathbb{R}$ and consider the function cf on $[a, b]$. Let us show that cf is absolutely continuous on $[a, b]$.

Now a constant function on $[a, b]$ is trivially absolutely continuous on $[a, b]$. If $c = 0$ then cf is a constant function on $[a, b]$ and then it is absolutely continuous on $[a, b]$. Consider the case $c \neq 0$. In this case we have $|c| > 0$. Let $\varepsilon > 0$. Since f is absolutely continuous on $[a, b]$ there exists $\delta > 0$ such that for any finite collection of non-overlapping closed intervals $\{[a_k, b_k] : k = 1, \ldots, n\}$ contained in $[a, b]$ with $\sum_{k=1}^{n} (b_k - a_k) < \delta$ we have $\sum_{k=1}^{n} |f(b_k) - f(a_k)| < \frac{\varepsilon}{|c|}$. Then we have

$$\sum_{k=1}^{n} |(cf)(b_k) - (cf)(a_k)| = |c| \sum_{k=1}^{n} |f(b_k) - f(a_k)| < |c| \frac{\varepsilon}{|c|} = \varepsilon.$$

This shows that cf is absolutely continuous on $[a, b]$.

2. Let us prove (b). Suppose f and g are real-valued absolutely continuous functions on $[a, b]$. Let us show that $f + g$ is absolutely continuous on $[a, b]$.

Let $\varepsilon > 0$. Since f is absolutely continuous on $[a, b]$ there exists $\delta' > 0$ such that for any finite collection of non-overlapping closed intervals $\{[a_k', b_k'] : k = 1, \ldots, n'\}$ contained in $[a, b]$ with $\sum_{k=1}^{n'} (b_k' - a_k') < \delta'$ we have $\sum_{k=1}^{n'} |f(b_k') - f(a_k')| < \frac{\varepsilon}{2}$.

Similarly the absolute continuity of g on $[a, b]$ implies that there exists $\delta'' > 0$ such that for any finite collection of non-overlapping closed intervals $\{[a_k'', b_k''] : k = 1, \ldots, n''\}$ contained in $[a, b]$ with $\sum_{k=1}^{n''} (b_k'' - a_k'') < \delta''$ we have $\sum_{k=1}^{n''} |f(b_k'') - f(a_k'')| < \frac{\varepsilon}{2}$.

Let $\delta = \min\{\delta', \delta''\} > 0$. Let $\{[a_k, b_k] : k = 1, \ldots, n\}$ be an arbitrary finite collection of non-overlapping closed intervals contained in $[a, b]$ with $\sum_{k=1}^{n} (b_k - a_k) < \delta$. Then we have

$$\sum_{k=1}^{n} |(f + g)(b_k) - (f + g)(a_k)| = \sum_{k=1}^{n} |\{f(b_k) + g(b_k)\} - \{f(a_k) + g(a_k)\}|$$

$$\leq \sum_{k=1}^{n} \{|f(b_k) - f(a_k)| + |g(b_k) - g(a_k)|\} = \sum_{k=1}^{n} |f(b_k) - f(a_k)| + \sum_{k=1}^{n} |g(b_k) - g(a_k)|$$

$$< \frac{\varepsilon}{2} + \frac{\varepsilon}{2} = \varepsilon.$$

This proves the absolute continuity of $f + g$ on $[a, b]$.

3. Let us prove (c). Suppose f and g are real-valued absolutely continuous functions on $[a, b]$. Let us show that fg is absolutely continuous on $[a, b]$. Now absolute continuity of f and g on $[a, b]$ implies that f and g are bounded on $[a, b]$ by Theorem 13.12 and (b) of Observation 12.13. Thus there exists $M > 0$ such that $|f| \leq M$ and $|g| \leq M$ on $[a, b]$.

Let $\varepsilon > 0$. Since f is absolutely continuous on $[a, b]$ there exists $\delta' > 0$ such that for any finite collection of non-overlapping closed intervals $\{[a'_k, b'_k] : k = 1, \ldots, n'\}$ contained in $[a, b]$ with $\sum_{k=1}^{n'}(b'_k - a'_k) < \delta'$ we have $\sum_{k=1}^{n'} |f(b'_k) - f(a'_k)| < \frac{\varepsilon}{2M}$.

Similarly the absolute continuity of g on $[a, b]$ implies that there exists $\delta'' > 0$ such that for any finite collection of non-overlapping closed intervals $\{[a''_k, b''_k] : k = 1, \ldots, n''\}$ contained in $[a, b]$ with $\sum_{k=1}^{n''}(b''_k - a''_k) < \delta''$ we have $\sum_{k=1}^{n''} |f(b''_k) - f(a''_k)| < \frac{\varepsilon}{2M}$.

Let $\delta = \min\{\delta', \delta''\} > 0$. Let $\{[a_k, b_k] : k = 1, \ldots, n\}$ be an arbitrary finite collection of non-overlapping closed intervals contained in $[a, b]$ with $\sum_{k=1}^{n}(b_k - a_k) < \delta$. Then we have

$$\sum_{k=1}^{n} |(fg)(b_k) - (fg)(a_k)| = \sum_{k=1}^{n} |f(b_k)g(b_k) - f(a_k)g(a_k)|$$

$$\leq \sum_{k=1}^{n} \left\{ |f(b_k)g(b_k) - f(b_k)g(a_k)| + |f(a_k)g(b_k) - f(a_k)g(a_k)| \right\}$$

$$= \sum_{k=1}^{n} \left\{ |g(b_k)||f(b_k) - f(a_k)| + |f(a_k)||g(b_k) - g(a_k)| \right\}$$

$$\leq M \sum_{k=1}^{n} |f(b_k) - f(a_k)| + M \sum_{k=1}^{n} |g(b_k) - g(a_k)|$$

$$< M \frac{\varepsilon}{2M} + M \frac{\varepsilon}{2M} = \varepsilon.$$

This proves the absolute continuity of fg on $[a, b]$.

4. Let us prove (d). Suppose f is a real-valued absolutely continuous function on $[a, b]$ such that $f(x) \neq 0$ for every $x \in [a, b]$. Let us show that the function $1/f$ is absolutely continuous on $[a, b]$.

Absolute continuity of f on $[a, b]$ implies continuity of f on $[a, b]$ and this then implies continuity of $|f|$ on $[a, b]$. This then implies the existence of a minimal point and a minimal value $m = \min_{[a,b]} |f|$ of $|f|$. Since $|f|(x) \neq 0$ for every $x \in [a, b]$, we have $m > 0$.

Let $\varepsilon > 0$. Since f is absolutely continuous on $[a, b]$ there exists $\delta > 0$ such that for any finite collection of non-overlapping closed intervals $\{[a_k, b_k] : k = 1, \ldots, n\}$ contained in $[a, b]$ with $\sum_{k=1}^{n}(b_k - a_k) < \delta$ we have $\sum_{k=1}^{n} |f(b_k) - f(a_k)| < m^2\varepsilon$. Then we have

$$\sum_{k=1}^{n} |(1/f)(b_k) - (1/f)(a_k)| = \sum_{k=1}^{n} \left| \frac{1}{f(b_k)} - \frac{1}{f(a_k)} \right| = \sum_{k=1}^{n} \frac{|f(b_k) - f(a_k)|}{|f(b_k)f(a_k)|}$$

$$\leq \frac{1}{m^2} \sum_{k=1}^{n} |f(b_k) - f(a_k)| < \frac{1}{m^2} m^2 \varepsilon = \varepsilon.$$

This proves the absolute continuity of $1/f$ on $[a, b]$.

5. Let us prove (e). Suppose f is a real-valued absolutely continuous function on $[a, b]$. Let us show that $|f|$ is absolutely continuous on $[a, b]$.

Let $\varepsilon > 0$. Since f is absolutely continuous on $[a, b]$ there exists $\delta > 0$ such that for any finite collection of non-overlapping closed intervals $\{[a_k, b_k] : k = 1, \ldots, n\}$ contained in $[a, b]$ with $\sum_{k=1}^{n} (b_k - a_k) < \delta$ we have $\sum_{k=1}^{n} |f(b_k) - f(a_k)| < \varepsilon$. Then we have

$$\sum_{k=1}^{n} \Big| |f|(b_k) - |f|(a_k) \Big| = \sum_{k=1}^{n} \Big| |f(b_k)| - |f(a_k)| \Big| \leq \sum_{k=1}^{n} |f(b_k) - f(a_k)| < \varepsilon.$$

This proves the absolute continuity of $|f|$ on $[a, b]$. ∎

Prob. 13.2. Let f be a real-valued function on $[a, b]$. Let $c \in (a, b)$. Show that if f is absolutely continuous on $[a, c]$ and $[c, b]$ then f is absolutely continuous on $[a, b]$.

Proof. 1. Let f be a real-valued function on $[a, b]$. Let $I = [x', x''] \subset [a, b]$. Let us introduce the notation: $\Delta f(I) = f(x'') - f(x')$. Then the definition of absolute continuity of a function can be reformulated as follows:

We say that a real-valued function f on $[a, b]$ is absolutely continuous on $[a, b]$ if for every $\varepsilon > 0$ there exists $\delta > 0$ such that for any finite collection $\{I_1, \ldots, I_n\}$ of non-overlapping closed intervals contained in $[a, b]$ with $\sum_{k=1}^{n} \ell(I_k) < \delta$ we have $\sum_{k=1}^{n} |\Delta f(I_k)| < \varepsilon$.

2. Let f be a real-valued function on $[a, b]$. Let $c \in (a, b)$. Suppose that f is absolutely continuous on $[a, c]$ and $[c, b]$. Let us show that f is absolutely continuous on $[a, b]$.

Absolute continuity of f on $[a, c]$ implies that for every $\varepsilon > 0$ there exists $\delta_1 > 0$ such that for any finite collection $\{I_{1,1}, \ldots, I_{1,n_1}\}$ of non-overlapping closed intervals contained in $[a, c]$ with $\sum_{k=1}^{n_1} \ell(I_{1,k}) < \delta_1$ we have $\sum_{k=1}^{n_1} |\Delta f(I_{1,k})| < \frac{\varepsilon}{2}$.

Similarly absolute continuity of f on $[c, b]$ implies that for every $\varepsilon > 0$ there exists $\delta_2 > 0$ such that for any finite collection $\{I_{2,1}, \ldots, I_{2,n_2}\}$ of non-overlapping closed intervals contained in $[c, b]$ with $\sum_{k=1}^{n_2} \ell(I_{2,k}) < \delta_2$ we have $\sum_{k=1}^{n_2} |\Delta f(I_{2,k})| < \frac{\varepsilon}{2}$.

Let $\delta = \min\{\delta_1, \delta_2\} > 0$. Let $\{I_1, \ldots, I_n\}$ be an arbitrary finite collection of non-overlapping closed intervals contained in $[a, b]$ with $\sum_{k=1}^{n} \ell(I_k) < \delta$. There are two possible cases to consider.

This first case is when c is not in the interior of I_k for any $k = 1, \ldots, n$. In this case each I_k is either contained in $[a, c]$ or contained in $[c, b]$. Let $\{I_{1,1}, \ldots, I_{1,n_1}\}$ be the collection of those contained in $[a, c]$ and let $\{I_{2,1}, \ldots, I_{2,n_2}\}$ be the collection of those contained in $[c, b]$. Note that we have

$$\begin{cases} \sum_{k=1}^{n_1} \ell(I_{1,k}) \leq \sum_{k=1}^{n} \ell(I_k) < \delta \leq \delta_1. \\ \sum_{k=1}^{n_2} \ell(I_{2,k}) \leq \sum_{k=1}^{n} \ell(I_k) < \delta \leq \delta_2. \end{cases}$$

Then we have

$$\sum_{k=1}^{n} |\Delta f(I_k)| = \sum_{k=1}^{n_1} |\Delta f(I_{1,k})| + \sum_{k=1}^{n_2} |\Delta f(I_{2,k})| < \frac{\varepsilon}{2} + \frac{\varepsilon}{2} = \varepsilon.$$

This shows that f is absolutely continuous on $[a, b]$.

The second case is when c is in the interior of I_{k_0} for some $k_0 \in \{1, \ldots, n\}$. Let us write $I_{k_0} = [a_{k_0}, b_{k_0}]$ and define $I'_{k_0} = [a_{k_0}, c]$ and $I''_{k_0} = [c, b_{k_0}]$. Then $I_1, \ldots, I_{k_0-1}, I'_{k_0}$ are contained in $[a, c]$ and $I''_{k_0}, I_{k_0+1}, \ldots, I_n$ are contained in $[c, b]$. Moreover we have

$$\begin{cases} \ell(I_1) + \cdots + \ell(I_{k_0-1}) + \ell(I'_{k_0}) \leq \sum_{k=1}^{n} \ell(I_k) < \delta \leq \delta_1. \\ \ell(I''_{k_0}) + \ell(I_{k_0+1}) + \cdots + \ell(I_n) \leq \sum_{k=1}^{n} \ell(I_k) < \delta \leq \delta_2. \end{cases}$$

Then

$$\sum_{k=1}^{n} |\Delta f(I_k)| \leq \left\{ \Delta f(I_1) + \cdots + \Delta f(I_{k_0-1}) + \Delta f(I'_{k_0}) \right\}$$

$$+ \left\{ \Delta f(I''_{k_0}) + \Delta f(I_{k_0+1}) + \cdots + \Delta f(I_n) \right\}$$

$$< \frac{\varepsilon}{2} + \frac{\varepsilon}{2} = \varepsilon.$$

This shows that f is absolutely continuous on $[a, b]$. ∎

Prob. 13.3. Let f be a real-valued absolutely continuous function on $[a, b]$. Let $E = \{x \in [a, b] : f'(x) = 0\}$. Show that $\mu_L(f(E)) = 0$.

Proof. Let f be a real-valued absolutely continuous function on $[a, b]$. Then according to Corollary 13.3 the derivative f' exists $(\mathfrak{M}_L, \mu_L)$-a.e. on $[a, b]$ and is $\mathfrak{M}_L$-measurable and μ_L-integrable on $[a, b]$. Let

$$E = \{x \in [a, b] : f'(x) = 0\} \in \mathfrak{M}_L.$$

Then by Theorem 13.5 we have

$$\mu_L^*\big(f(E)\big) \leq \int_E |f'| \, d\mu = \int_E 0 \, d\mu = 0.$$

Thus we have $\mu_L^*(f(E)) = 0$ which implies $\mu_L(f(E)) = 0$. ∎

Prob. 13.4. Let f be a real-valued function on $[a, b]$. Show that if the derivative $f'(x)$ exists at every $x \in [a, b]$ then f satisfies Lusin's Condition (N) on $[a, b]$, that is, for every $E \subset [a, b]$ which is a null set in the measure space $(\mathbb{R}, \mathfrak{M}_L, \mu_L)$, the set $f(E)$ is a null set in $(\mathbb{R}, \mathfrak{M}_L, \mu_L)$.

Proof. Suppose the derivative $f'(x)$ exists at every $x \in [a, b]$. Let us show that f satisfies Lusin's Condition (N) on $[a, b]$.

Let $E \subset [a, b]$ be a null set in the measure space $(\mathbb{R}, \mathfrak{M}_L, \mu_L)$. Being a null set in $(\mathbb{R}, \mathfrak{M}_L, \mu_L)$, E is a $\mathfrak{M}_L$-measurable set. Since f' exists at every $x \in [a, b]$, f' exists at every $x \in E$. Then by Theorem 13.5, we have

$$\mu_L^*\big(f(E)\big) \leq \int_E |f'| \, d\mu_L.$$

Since E is a null set in $(\mathbb{R}, \mathfrak{M}_L, \mu_L)$, we have $\int_E |f'| \, d\mu_L = 0$ and then $\mu_L^*\big(f(E)\big) = 0$. This shows that $f(E)$ is a null set in $(\mathbb{R}, \mathfrak{M}_L, \mu_L)$. ∎

Prob. 13.5. Let f be a real-valued function on $[a, b]$ such that f is differentiable (that is, f has a finite derivative f') at every point in $[a, b]$. Show that f is absolutely continuous on $[a, b]$ if and only if f is a function of bounded variation on $[a, b]$.

Proof. Differentiability of f at a point in $[a, b]$ implies continuity of f at the point. Thus differentiability of f at every point in $[a, b]$ implies continuity of f on $[a, b]$. Existence of the derivative f' at every point in $[a, b]$ implies that f satisfies Lusin's Condition (N) on $[a, b]$ according to Prob. 13.4. Thus our function f is continuous on $[a, b]$ and satisfies Lusin's Condition (N) on $[a, b]$. Then by Theorem 13.8 (Banach-Zarecki), f is absolutely continuous on $[a, b]$ if and only if f is a function of bounded variation on $[a, b]$. ∎

Prob. 13.6. (If a real-valued function f on $[a, b]$ is absolutely continuous on $[a, b]$ then f is differentiable at a.e. $x \in [a, b]$ by Theorem 13.18. On the other hand if a real-valued function f on $[a, b]$ is differentiable at every $x \in [a, b]$, f need not be absolutely continuous on $[a, b]$. Here is an example.)
Let f be a real-valued function on $\left[0, \frac{1}{\sqrt{\pi}}\right]$ defined by

$$f(x) = \begin{cases} x^2 \sin \frac{1}{x^2} & \text{for } x \in \left(0, \frac{1}{\sqrt{\pi}}\right] \\ 0 & \text{for } x = 0. \end{cases}$$

(a) Show that f is differentiable at every $x \in \left[0, \frac{1}{\sqrt{\pi}}\right]$ and thus f is continuous on $\left[0, \frac{1}{\sqrt{\pi}}\right]$.
(b) Show that f is not a function of bounded variation on $\left[0, \frac{1}{\sqrt{\pi}}\right]$ and thus f is not absolutely continuous on $\left[0, \frac{1}{\sqrt{\pi}}\right]$ according to Theorem 13.8 (Banach-Zarecki).

Proof. 1. For $x \in \left(0, \frac{1}{\sqrt{\pi}}\right]$, we have

$$f'(x) = 2x \sin \frac{1}{x^2} - \frac{2}{x} \cos \frac{1}{x^2} \in \mathbb{R}.$$

At $x = 0$, we have

$$f'(0) = \lim_{x \to 0} \frac{f(x) - f(0)}{x - 0} = \lim_{x \to 0} x \sin \frac{1}{x^2} = 0 \in \mathbb{R}.$$

This shows that f is differentiable at every $x \in \left[0, \frac{1}{\sqrt{\pi}}\right]$. This implies that f is continuous on $\left[0, \frac{1}{\sqrt{\pi}}\right]$.

2. Let us show that $f \notin BV\left(\left[0, \frac{1}{\sqrt{\pi}}\right]\right)$. The function $\sin x$ for $x \in [0, \infty)$ assumes the value 0 at $x = k\pi$ for $k \in \mathbb{Z}_+$ and assumes the values ± 1 at $x = \left(k + \frac{1}{2}\right)\pi$ for $k \in \mathbb{Z}_+$. Then the function $\sin \frac{1}{x^2}$ for $x \in (0, \infty)$ assumes the value 0 at $\frac{1}{x^2} = k\pi$, that is, $x = \frac{1}{\sqrt{k\pi}}$ for $k \in \mathbb{N}$ and assumes the values ± 1 at $\frac{1}{x^2} = \left(k + \frac{1}{2}\right)\pi$, that is, $x = \frac{1}{\sqrt{\left(k + \frac{1}{2}\right)\pi}}$ for $k \in \mathbb{N}$.
Let

$$a_k = \frac{1}{\sqrt{k\pi}} \quad \text{and} \quad b_k = \frac{1}{\sqrt{\left(k + \frac{1}{2}\right)}} \quad \text{for } k \in \mathbb{N}.$$

Then we have

$$f(a_k) = a_k^2 \sin k\pi = 0 \quad \text{and} \quad f(b_k) = b_k^2 \sin \left(k + \frac{1}{2}\right)\pi = \pm b_k^2 \quad \text{for } k \in \mathbb{N}.$$

Observe also that for every $n \in \mathbb{N}$ we have

$$0 < b_n < a_n < b_{n-1} < a_{n-1} < \cdots < b_1 < a_1 = \frac{1}{\sqrt{\pi}}.$$

For $n \in \mathbb{N}$, let $\mathcal{P}_n$ be a partition of $\left[0, \frac{1}{\sqrt{\pi}}\right]$ given by

$$\mathcal{P}_n = \left\{0 < b_n < a_n < b_{n-1} < a_{n-1} < \cdots < b_1 < a_1 = \frac{1}{\sqrt{\pi}}\right\}.$$

For the variation of f on $\left[0, \frac{1}{\sqrt{\pi}}\right]$ corresponding to the partition $\mathcal{P}_n$, we have

$$V_0^{\frac{1}{\sqrt{\pi}}}(f, \mathcal{P}_n) = |f(a_1) - f(b_1)| + |f(b_1) - f(a_2)|$$
$$+ |f(a_2) - f(b_2)| + |f(b_2) - f(a_3)|$$
$$\vdots$$
$$+ |f(a_n) - f(b_n)| + |f(b_n) - f(0)|$$
$$= 2\sum_{k=1}^{n} |f(b_k)| = 2\sum_{k=1}^{n} b_k^2 = 2\sum_{k=1}^{n} \frac{1}{\left(k + \frac{1}{2}\right)\pi}$$
$$\geq \frac{2}{\pi} \sum_{k=1}^{n} \frac{1}{k+1}.$$

Thus we have

$$\sup_{n\in\mathbb{N}} V_0^{\frac{1}{\sqrt{\pi}}}(f, \mathcal{P}_n) \geq \sup_{n\in\mathbb{N}} \left\{ \frac{2}{\pi} \sum_{k=1}^{n} \frac{1}{k+1} \right\} = \frac{2}{\pi} \sum_{k\in\mathbb{N}} \frac{1}{k+1} = \infty.$$

Then for the total variation of f on $\left[0, \frac{1}{\sqrt{\pi}}\right]$, we have

$$V_0^{\frac{1}{\sqrt{\pi}}}(f) = \sup_{\mathcal{P}\in\mathcal{P}_{0,1/\sqrt{\pi}}} V_0^{\frac{1}{\sqrt{\pi}}}(f, \mathcal{P}) \geq \sup_{n\in\mathbb{N}} V_0^{\frac{1}{\sqrt{\pi}}}(f, \mathcal{P}_n) \geq \infty.$$

Thus we have $V_0^{\frac{1}{\sqrt{\pi}}}(f) = \infty$ and $f \notin BV\left(\left[0, \frac{1}{\sqrt{\pi}}\right]\right)$. ∎

Prob. 13.7. Let f be a real-valued function on $[a, b]$. Show that f is absolutely continuous on $[a, b]$ if and only if it satisfies the following conditions:
1° f is continuous on $[a, b]$.
2° f satisfies Lusin's Condition (N) on $[a, b]$.
3° f' exists μ_L-a.e. on $[a, b]$ and f' is $\mathfrak{M}_L$-measurable and μ_L-integrable on $[a, b]$.

Proof. 1. Suppose f is absolutely continuous on $[a, b]$. Then by Corollary 13.3, f satisfies condition 3° and by Theorem 13.8 (Banach-Zarecki), f satisfies conditions 1° and 2°.

2. Conversely suppose f satisfies conditions 1°, 2° and 3°. To show that f is absolutely continuous on $[a, b]$, let us show first that for any $[c, d] \subset [a, b]$ we have

$$(1) \qquad\qquad |f(c) - f(d)| \leq \int_{[c,d]} |f'| \, d\mu_L.$$

Since f' exists μ_L-a.e. on $[a, b]$ and $[c, d] \subset [a, b]$, f' exists μ_L-a.e. on $[c, d]$. Thus there exists a null set F in $(\mathbb{R}, \mathfrak{M}_L, \mu_L)$ contained in $[c, d]$ such that f' exists on $E := [c, d] \setminus F$. Then we have $[c, d] = E \cup F$, $E \cap F = \emptyset$, $\mu_L(F) = 0$, and f' exists on E. Since f satisfies Lusin's Condition (N) on $[a, b]$, $\mu_L(F) = 0$ implies

$$(2) \qquad\qquad \mu_L^*\big(f(F)\big) = 0.$$

Let $\overline{f(c), f(d)}$ be the closed interval in the image space $\mathbb{R}$ of f with $f(c)$ and $f(d)$ as its two endpoints. Let $y \in \overline{f(c), f(d)}$. Since f is a continuous function on $[a, b]$ there exists $x \in [c, d]$ such that $y = f(x)$ by the Intermediate Value Theorem for Continuous Functions. Thus we have

$$(3) \qquad\qquad \overline{f(c), f(d)} \subset f\big([c, d]\big).$$

Then we have

$$\begin{aligned}
|f(c) - f(d)| &= \ell\big(\overline{f(c), f(d)}\big) = \mu_L\big(\overline{f(c), f(d)}\big) \\
&\leq \mu_L^*\big(f([c, d])\big) \quad \text{by (3)} \\
&= \mu_L^*\big(f(E \cup F)\big) = \mu_L^*\big(f(E) \cup f(F)\big) \\
&\leq \mu_L^*\big(f(E)\big) + \mu_L^*\big(f(F)\big) \\
&= \mu_L^*\big(f(E)\big) \quad \text{by (2)} \\
&\leq \int_E |f'| \, d\mu_L \quad \text{by Theorem 13.5} \\
&\leq \int_{[c,d]} |f'| \, d\mu_L.
\end{aligned}$$

This proves (1).

To show that f is absolutely continuous on $[a, b]$, we show that for every $\varepsilon > 0$ there exists $\delta > 0$ such that for any finite collection of non-overlapping closed intervals $\big\{[a_k, b_k] : k = 1, \ldots, n\big\}$ contained in $[a, b]$ with $\sum_{k=1}^n (b_k - a_k) < \delta$, we have $\sum_{k=1}^n |f(b_k) - f(a_k)| <$

ε. Now let $\varepsilon > 0$ be arbitrarily given. By Theorem 9.26, the μ_L-integrability of f' on $[a, b]$ implies that there exists $\delta > 0$ such that

(4) $\quad \int_A |f'| \, d\mu_L < \varepsilon \quad$ for any $A \subset [a, b]$ such that $A \in \mathfrak{M}_L$ and $\mu_L(A) < \delta$.

Let $\{[a_k, b_k] : k = 1, \ldots, n\}$ be an arbitrary finite collection of non-overlapping closed intervals contained in $[a, b]$ with $\sum_{k=1}^n (b_k - a_k) < \delta$. Then we have

$$\sum_{k=1}^n |f(b_k) - f(a_k)| \leq \sum_{k=1}^n \int_{[a_k, b_k]} |f'| \, d\mu_L = \int_{\bigcup_{k=1}^n [a_k, b_k]} |f'| \, d\mu_L < \varepsilon$$

where the first inequality is by (1) and the last inequality is by (4) and the fact that

$$\mu_L\left(\bigcup_{k=1}^n [a_k, b_k]\right) = \sum_{k=1}^n \mu_L\left([a_k, b_k]\right) = \sum_{k=1}^n (b_k - a_k) < \delta.$$

This completes the proof that f is absolutely continuous on $[a, b]$. ∎

Prob. 13.8. Let f be a real-valued function on $[a, b]$ satisfying the following conditions:
1° f is differentiable at every point in $[a, b]$.
2° $f \notin BV([a, b])$.
Show that f' is $\mathfrak{M}_L$-measurable but not μ_L-integrable on $[a, b]$.

Proof. Differentiability of f at every point in $[a, b]$ implies that f is continuous on $[a, b]$. Existence of f' at every point in $[a, b]$ implies that f satisfies Lusin's Condition (N) on $[a, b]$ according to Prob. 13.4. Existence of f' at every point in $[a, b]$ also implies that f' is $\mathfrak{M}_L$-measurable on $[a, b]$ according to Proposition 12.4.

Now if f' were μ_L-integrable on $[a, b]$ then f would be absolutely continuous on $[a, b]$ by Prob. 13.7 and then f would be a function of bounded variation on $[a, b]$ by Theorem 13.2, contradicting the condition 2°. ∎

Prob. 13.9. For a real-valued function f on an interval $J \subset \mathbb{R}$, let us define

$$\omega(f, J) = \sup_J f - \inf_J f \in [0, \infty].$$

Show that a real-valued function f on $[a, b]$ is absolutely continuous on $[a, b]$ if and only if for every $\varepsilon > 0$ there exists $\delta > 0$ such that for any finite collection of non-overlapping closed intervals $\{I_k : k = 1, \ldots, n\}$ contained in $[a, b]$ with $\sum_{k=1}^n \ell(I_k) < \delta$, we have $\sum_{k=1}^n \omega(f, I_k) < \varepsilon$.

Proof. 1. Suppose f satisfies the condition. Then for every $\varepsilon > 0$ there exists $\delta > 0$ such that for any finite collection of non-overlapping closed intervals $\{[a_k, b_k] : k = 1, \ldots, n\}$ contained in $[a, b]$ with $\sum_{k=1}^n (b_k - a_k) < \delta$, we have $\sum_{k=1}^n \omega(f, [a_k, b_k]) < \varepsilon$. Then since $|f(b_k) - f(a_k)| \le \sup_{[a_k, b_k]} f - \inf_{[a_k, b_k]} f$, we have

$$\sum_{k=1}^n |f(b_k) - f(a_k)| \le \sum_{k=1}^n \omega(f, [a_k, b_k]) < \varepsilon.$$

This shows that f is absolute continuous on $[a, b]$.

2. Conversely suppose f is absolutely continuous on $[a, b]$. Then for every $\varepsilon > 0$ there exists $\delta > 0$ such that for any finite collection of non-overlapping closed intervals $\{[a_k, b_k] : k = 1, \ldots, n\}$ contained in $[a, b]$ with $\sum_{k=1}^n (b_k - a_k) < \delta$, we have

$$(1) \qquad\qquad \sum_{k=1}^n |f(b_k) - f(a_k)| < \frac{\varepsilon}{3}.$$

Consider $[a_k, b_k]$ and $\omega(f, [a_k, b_k]) = \sup_{[a_k, b_k]} f - \inf_{[a_k, b_k]} f$. Since f is absolutely continuous on $[a, b]$, f is bounded on $[a, b]$. Thus $-\infty < \inf_{[a_k, b_k]} f \le \sup_{[a_k, b_k]} f < \infty$. Then there exists $x'_k \in [a_k, b_k]$ such that

$$\inf_{[a_k, b_k]} f + \frac{\varepsilon}{3n} > f(x'_k) \quad \text{that is,} \quad \inf_{[a_k, b_k]} f > f(x'_k) - \frac{\varepsilon}{3n}$$

and there exists $x''_k \in [a_k, b_k]$ such that

$$\sup_{[a_k, b_k]} f - \frac{\varepsilon}{3n} < f(x''_k) \quad \text{that is,} \quad \sup_{[a_k, b_k]} f < f(x''_k) + \frac{\varepsilon}{3n}.$$

Then

$$(2) \qquad\qquad \omega(f, [a_k, b_k]) = \sup_{[a_k, b_k]} f - \inf_{[a_k, b_k]} f$$

$$< \left\{ f(x''_k) + \frac{\varepsilon}{3n} \right\} - \left\{ f(x'_k) - \frac{\varepsilon}{3n} \right\}$$

$$\le |f(x''_k) - f(x'_k)| + \frac{2\varepsilon}{3n}.$$

Let $\alpha_k = \min\{x'_k, x''_k\}$ and $\beta_k = \max\{x'_k, x''_k\}$ and decompose $[a_k, b_k]$ into three non-overlapping closed intervals $[a_k, \alpha_k]$, $[\alpha_k, \beta_k]$, and $[\beta_k, b_k]$. We have

$$\ell([a_k, \alpha_k]) + \ell([\alpha_k, \beta_k]) = \ell([\beta_k, b_k]) = \ell([a_k, b_k]) = b_k - a_k$$

and

$$(3) \qquad \sum_{k=1}^{n} \left\{ (\alpha_k - a_k) + (\beta_k - \alpha_k) + (b_k - \beta_k) \right\} = \sum_{k=1}^{n} (b_k - a_k) < \delta.$$

Thus $\{ [a_k, \alpha_k], [\alpha_k, \beta_k], [\beta_k, b_k] : k = 1, \ldots, n \}$ is a finite collection of non-overlapping closed intervals contained in $[a, b]$ with total length less than δ.

By (2) we have

$$\omega\left(f, [a_k, b_k]\right) = \left| f(x''_k) - f(x'_k) \right| + \frac{2\varepsilon}{3n} = \left| f(\beta_k) - f(\alpha_k) \right| + \frac{2\varepsilon}{3n}$$

$$= \left| f(\alpha_k) - f(a_k) \right| + \left| f(\beta_k) - f(\alpha_k) \right| + \left| f(b_k) - f(\beta_k) \right| + \frac{2\varepsilon}{3n}.$$

Then

$$\sum_{k=1}^{n} \omega\left(f, [a_k, b_k]\right)$$

$$= \sum_{k=1}^{n} \left\{ \left| f(\alpha_k) - f(a_k) \right| + \left| f(\beta_k) - f(\alpha_k) \right| + \left| f(b_k) - f(\beta_k) \right| \right\} + \frac{2\varepsilon}{3}$$

$$< \frac{\varepsilon}{3} + \frac{2\varepsilon}{3} = \varepsilon$$

where the inequality is by (1) which is applicable on account of (3). ∎

Problems and Proofs in Real Analysis

Prob. 13.10. Definition. (Lipschitz Condition) *Let f be a real-valued function on an interval $I \subset \mathbb{R}$. We say that f satisfies a Lipschitz Condition with coefficient $M > 0$ on I if there exists $M > 0$ such that $|f(x') - f(x'')| \leq M|x' - x''|$ for any $x', x'' \in I$.*

Definition. (Condition (L)) *Let f be a real-valued function on an interval $I \subset \mathbb{R}$. Let us say that f satisfies Condition (L) on I if for every $\varepsilon > 0$ there exists $\delta > 0$ such that for any finite collection $\{[a_k, b_k] : k = 1, \ldots, n\}$ of possibly overlapping or even coinciding closed intervals contained in I with $\sum_{k=1}^{n}(b_k - a_k) < \delta$, we have $\sum_{k=1}^{n} |f(b_k) - f(a_k)| < \varepsilon$.*

Prove the following theorem:

Theorem. (Equivalence of Lipschitz Condition and Condition (L)) *Let f be a real-valued function on an interval $I \subset \mathbb{R}$. Then f satisfies a Lipschitz Condition on I if and only if it satisfies Condition (L) on I.*

Proof. 1. Let us show that a Lipschitz Condition on I implies Condition (L) on I. Assume that f satisfies a Lipschitz Condition with coefficient $M > 0$ on I.

Let $\varepsilon > 0$ be arbitrarily given. Select $\delta \in \left(0, \frac{\varepsilon}{M}\right)$. Let $\{[a_k, b_k] : k = 1, \ldots, n\}$ be an arbitrary finite collection of closed intervals contained in I with $\sum_{k=1}^{n}(b_k - a_k) < \delta$. Then since f satisfies a Lipschitz Condition with coefficient $M > 0$ on I, we have

$$\sum_{k=1}^{n} |f(b_k) - f(a_k)| \leq \sum_{k=1}^{n} M(b_k - a_k) = M \sum_{k=1}^{n}(b_k - a_k) = M\delta < M\frac{\varepsilon}{M} = \varepsilon.$$

This shows that f satisfies Condition (L) on I.

2. Let us show that Condition (L) on I implies a Lipschitz Condition on I. Assume that f satisfies Condition (L) on I.

Observe that if f satisfies Condition (L) on I then f is uniformly continuous on I.

Now by Condition (L), for $\varepsilon_0 = 1 > 0$ there exists $\delta_0 > 0$ such that for any finite collection of closed intervals $\{[a_k, b_k] : k = 1, \ldots, n\}$ contained in I with $\sum_{k=1}^{n}(b_k - a_k) < \delta_0$, we have $\sum_{k=1}^{n} |f(b_k) - f(a_k)| < 1$.

Let $[x, y] \subset I$ with $y - x < \delta_0$. Let $n \in \mathbb{N}$ and let $[a, b] \subset [x, y]$ such that

$$b - a = \frac{\delta_0}{2n}, \quad \text{that is,} \quad b = a + \frac{\delta_0}{2n}.$$

Let $[a_k, b_k] = [a, b]$ for $k = 1, \ldots, n$. Then we have

$$\sum_{k=1}^{n}(b_k - a_k) = n(b - a) = \frac{\delta_0}{2} < \delta_0.$$

This implies that we have $\sum_{k=1}^{n} |f(b_k) - f(a_k)| < 1$. Then since $a_k = a$ and $b_k = b$ for $k = 1, \ldots, n$ the last inequality reduces to $n|f(b) - f(a)| < 1$, that is,

$$|f(b) - f(a)| < \frac{1}{n} = \frac{2}{\delta_0}(b - a).$$

Thus we have

$$|f(b) - f(a)| < \frac{2}{\delta_0}(b - a) \quad \text{for } b = a + \frac{\delta_0}{2n}.$$

By the continuity of f on I, the last inequality implies

$$|f(b) - f(a)| < \frac{2}{\delta_0}(b - a) \quad \text{for } [a, b] \subset [x, y].$$

This shows that f satifies a Lipschitz Condition with coefficient $M = \frac{2}{\delta_0} > 0$ on any closed interval $[x, y] \subset I$ with $y - x < \delta_0$.

Let $[x, y]$ be an arbitrary closed interval contained in I with $0 < y - x < \infty$. Then there exists $m \in \mathbb{N}$ so large that $\frac{y-x}{m} < \delta_0$. Then we can select $x = x_0 < x_1 < \cdots < x_m = y$ so that $x_i - x_{i-1} < \delta_0$ for $i = 1, \ldots, m$. Then by our results above, f satisfies a Lipschitz Condition with coefficient $M = \frac{2}{\delta_0} > 0$ on $[x_{i-1}, x_i]$ for $i = 1, \ldots, m$. Thus we have

$$|f(x) - f(y)| \leq |f(x_0) - f(x_1)| + |f(x_1) - f(x_2)| + \cdots + |f(x_{m-1}) - f(x_m)|$$
$$\leq \frac{2}{\delta_0}\{|x_0 - x_1| + |x_1 - x_2| + \cdots + |x_{m-1} - x_m|\}$$
$$= \frac{2}{\delta_0}|x - y|.$$

This shows that f satisfies a Lipschitz Condition with coefficient $M = \frac{2}{\delta_0} > 0$ on I. ∎

Prob. 13.11. Prove the following theorem:

Theorem. *If a real-valued function f on $[a, b] \subset \mathbb{R}$ satisfies a Lipschitz Condition on $[a, b]$ then f is absolutely continuous on $[a, b]$.*

Proof. According to Prob. 13.10, every Lipschitz Condition is equivalent to Condition (L). A comparison of the definition of Condition (L) in Prob. 13.10 and Definition 13.1 for absolute continuity shows that Condition (L) on f on $[a, b]$ implies absolute continuity of f on $[a, b]$. This shows that if f satisfies a Lipschitz Condition on $[a, b]$ then f is absolutely continuous on $[a, b]$. ∎

Prob. 13.12. Let f be a real-valued function on $[a, b]$ satisfying a Lipschitz Condition with coefficient $M > 0$ on $[a, b]$. Show that f' exists $(\mathfrak{M}_L, \mu_L)$-a.e. on $[a, b]$ and $|f'| \leq M$.

Proof. If f is a real-valued function on $[a, b]$ satisfying a Lipschitz Condition with coefficient $M > 0$ on $[a, b]$, then f is absolutely continuous on $[a, b]$ according to Prob. 13.11. Then f is a function of bounded variation on $[a, b]$ according to Theorem 13.2. Then f' exists $(\mathfrak{M}_L, \mu_L)$-a.e. on $[a, b]$ according to Theorem 12.21. Now

$$f'(x) = \lim_{h \to \infty} \frac{f(x + h) - f(x)}{h}.$$

Then by the Lipschitz Condition with coefficient $M > 0$ we have

$$\left| \frac{f(x + h) - f(x)}{h} \right| \leq \frac{M|h|}{|h|} = M.$$

Thus we have

$$-M \leq \frac{f(x + h) - f(x)}{h} \leq M$$

and then

$$-M \leq \lim_{h \to \infty} \frac{f(x + h) - f(x)}{h} \leq M.$$

This shows that $|f'(x)| \leq M$. ∎

Prob. 13.13. Let f be a real-valued continuous function on $[a, b]$ such that f' exists on (a, b) and satisfies the condition $|f'(x)| \leq M$ for $x \in (a, b)$ with some $M > 0$. Show that f satisfies a Lipschitz Condition with coefficient $M > 0$ on $[a, b]$.

Proof. Assume that f' exists on (a, b) and satisfies the condition $|f'(x)| \leq M$ for $x \in (a, b)$ with some $M > 0$.

Let $x', x'' \in [a, b]$ and $x' < x''$. By the Mean Value Theorem, there exists $\xi \in (x', x'')$ such that

$$f(x'') - f(x') = f'(\xi)(x'' - x')$$

and then

$$\left| f(x'') - f(x') \right| = \left| f'(\xi) \right| |x'' - x'| \leq M|x'' - x'|.$$

This shows that f satisfies the Lipschitz Condition with coefficient $M > 0$ on $[a, b]$. ∎

Prob. 13.14. Let f be a real-valued function on $[a, b]$ satisfying a Lipschitz Condition with coefficient $M > 0$ on $[a, b]$. Let $E \subset [a, b]$. Show that we have $\mu_L^*(f(E)) \leq M\mu_L^*(E)$.

Proof. By Observation 3.2 we have

(1) $\qquad \mu_L^*(E) = \inf \left\{ \sum_{n\in\mathbb{N}} \ell(I_n) : (I_n : n \in \mathbb{N}) \subset \mathfrak{I}_c, \bigcup_{n\in\mathbb{N}} I_n \supset E \right\}.$

Then for every $\varepsilon > 0$ there exists $(I_n : n \in \mathbb{N}) \subset \mathfrak{I}_c$ contained in $[a, b]$ such that

(2) $\qquad \bigcup_{n\in\mathbb{N}} I_n \supset E \quad \text{and} \quad \sum_{n\in\mathbb{N}} \ell(I_n) \leq \mu_L^*(E) + \varepsilon.$

Then we have

(3) $\qquad f(E) \subset f\left(\bigcup_{n\in\mathbb{N}} I_n \right) = \bigcup_{n\in\mathbb{N}} f(I_n).$

Let $x', x'' \in I_n$. Then by the Lipschitz Condition with coefficient $M > 0$ on $[a, b]$ we have

$$|f(x') - f(x'')| \leq M|x' - x''| \leq M\ell(I_n).$$

This implies that

$$\sup_{I_n} f - \inf_{I_n} f \leq M\ell(I_n).$$

Thus $f(I_n)$ is contained in a closed interval J_n with $\ell(J_n) \leq M\ell(I_n)$. Then with (3) we have

$$f(E) \subset \bigcup_{n\in\mathbb{N}} f(I_n) \subset \bigcup_{n\in\mathbb{N}} J_n$$

and

$$\sum_{n\in\mathbb{N}} \ell(J_n) \leq \sum_{n\in\mathbb{N}} M\ell(I_n) \leq M\{\mu_L^*(E) + \varepsilon\}.$$

Thus we have

$$\mu_L^*(f(E)) \leq \sum_{n\in\mathbb{N}} \ell(J_n) \leq M\{\mu_L^*(E) + \varepsilon\}.$$

Since this holds for every $\varepsilon > 0$, we have $\mu_L^*(f(E)) \leq M\mu_L^*(E)$. ∎

Prob. 13.15. Let f be a real-valued function on $[a, b]$ satisfying a Lipschitz Condition with coefficient $M > 0$ on $[a, b]$. Show that f satisfies Lusin's Condition (N) on $[a, b]$.

Proof. Let us show that f satisfies Lusin's Condition (N) on $[a, b]$, that is, for every $E \subset [a, b]$ which is a null set in the measure space $(\mathbb{R}, \mathfrak{M}_L, \mu_L)$, the set $f(E)$ is a null set in $(\mathbb{R}, \mathfrak{M}_L, \mu_L)$.

Let $E \subset [a, b]$ be a null set in $(\mathbb{R}, \mathfrak{M}_L, \mu_L)$. Since E is a subset of $[a, b]$, we have according to Prob. 13.14 the estimate $\mu_L^*(f(E)) \leq M\mu_L^*(E)$. Then since E is a null set in $(\mathbb{R}, \mathfrak{M}_L, \mu_L)$ we have $\mu_L^*(E) = 0$. This then implies that $\mu_L^*(f(E)) = 0$, that is, $f(E)$ is a null set in $(\mathbb{R}, \mathfrak{M}_L, \mu_L)$. ■

Prob. 13.16. Find a real-valued function f on $[a, b]$ such that f is continuous on $[a, b]$, absolutely continuous on $[c, b]$ for every $c \in (a, b]$, but not absolutely continuous on $[a, b]$.

Proof. Let us define a real-valued function f on $\left[0, \frac{2}{\pi}\right]$ by setting

$$f(x) = \begin{cases} x \sin \frac{1}{x} & \text{for } x \in \left(0, \frac{2}{\pi}\right], \\ 0 & \text{for } x = 0. \end{cases}$$

1. Thus defined, f is continuous on $\left[0, \frac{2}{\pi}\right]$.

2. Let us show that f is absolutely continuous on $\left[c, \frac{2}{\pi}\right]$ for every $c \in \left(0, \frac{2}{\pi}\right]$. We have

$$f'(x) = \sin \frac{1}{x} - \frac{1}{x} \cos \frac{1}{x} \quad \text{for } x \in \left[c, \frac{2}{\pi}\right].$$

Thus f' is a continuous function on the compact set $\left[c, \frac{2}{\pi}\right]$ and this implies that f' is bounded on $\left[c, \frac{2}{\pi}\right]$. Then by Prob. 13.13, f satisfies a Lipschitz Condition on $\left[c, \frac{2}{\pi}\right]$. This implies that f is absolutely continuous on $\left[c, \frac{2}{\pi}\right]$ according to Prob. 13.11.

3. Let us show that f is not absolutely continuous on $\left[0, \frac{2}{\pi}\right]$. According to Prob. 12.12, we have $f \notin BV\left(\left[0, \frac{2}{\pi}\right]\right)$. Then according to Theorem 13.8 (Banach-Zarecki), f cannot be absolutely continuous on $\left[0, \frac{2}{\pi}\right]$. ■

Prob. 13.17. Prove the following theorem:

Theorem. *Let f be a real-valued function on $[a, b]$ such that*
1° f *is continuous on* $[a, b]$,
2° f *is absolutely continuous on* $[c, b]$ *for every* $c \in (a, b]$,
3° $f \in BV([a, c_0])$ *for some* $c_0 \in (a, b]$.
Then f is absolutely continuous on $[a, b]$.

Proof. To show that f is absolutely continuous on $[a, b]$, we show that for every $\varepsilon > 0$ there exists $\delta > 0$ such that for any finite collection of non-overlapping closed intervals $\{[a_k, b_k] : k = 1, \ldots, n\}$ contained in $[a, b]$ with $\sum_{k=1}^{n} (b_k - a_k) < \delta$ we have $\sum_{k=1}^{n} |f(b_k) - f(a_k)| < \varepsilon$.

1. Let $\varepsilon > 0$ be arbitrarily given. Now 3° and 1° imply that the total variation function $V_a^x(f)$ for $x \in [a, c_0]$ is a real-valued continuous function on $[a, c_0]$ by Theorem 12.22. Then since $\lim_{x \downarrow a} V_a^x(f) = 0$ there exists $c_1 \in (a, c_0]$ such that

$$(1) \qquad\qquad V_a^{c_1}(f) < \frac{\varepsilon}{2}.$$

2. Absolute continuity of f on $[c_1, b]$ implies that there exists $\eta > 0$ such that for any finite collection of non-overlapping closed intervals $\{[\alpha_k, \beta_k] : k = 1, \ldots, n\}$ contained in $[c_1, b]$ with $\sum_{k=1}^{n} (\beta_k - \alpha_k) < \eta$ we have

$$(2) \qquad\qquad \sum_{k=1}^{n} |f(\beta_k) - f(\alpha_k)| < \frac{\varepsilon}{2}.$$

3. Let $\delta = \min\{c_1 - a, \eta\} > 0$. Let $\{[a_k, b_k] : k = 1, \ldots, n\}$ be an arbitrary finite collection of non-overlapping closed intervals contained in $[a, b]$ with $\sum_{k=1}^{n} (b_k - a_k) < \delta$. We proceed to show that we have

$$(3) \qquad\qquad \sum_{k=1}^{n} |f(b_k) - f(a_k)| < \varepsilon.$$

Case 1. Suppose $\{[a_k, b_k] : k = 1, \ldots, n\} \subset [a, c_1]$. Let $\mathcal{P}$ be a partition of $[a, c_1]$ including $a_1, \ldots, a_n, b_1, \ldots, b_n$ as partition points. Then we have by (1).

$$\sum_{k=1}^{n} |f(b_k) - f(a_k)| \leq V_a^{c_1}(f, \mathcal{P}) \leq V_a^{c_1}(f) < \frac{\varepsilon}{2}.$$

This proves (3) for this case.

Case 2. Suppose $\{[a_k, b_k] : k = 1, \ldots, n\} \subset [c_1, b]$. Since $\sum_{k=1}^{n} (b_k - a_k) < \delta \leq \eta$, we have by (2)

$$\sum_{k=1}^{n} |f(b_k) - f(a_k)| < \frac{\varepsilon}{2}.$$

This proves (3) for this case.

Case 3. Suppose $\{[a_k, b_k] : k = 1, \ldots, k_0\} \subset [a, c_1]$ and $\{[a_k, b_k] : k = k_0 + 1, \ldots, n\} \subset [c_1, b]$. In this case we have

$$\sum_{k=1}^{n} |f(b_k) - f(a_k)| = \sum_{k=1}^{k_0} |f(b_k) - f(a_k)| + \sum_{k=k_0+1}^{n} |f(b_k) - f(a_k)|$$

$$< \frac{\varepsilon}{2} + \frac{\varepsilon}{2} = \varepsilon \quad \text{by Case 1 and Case 2.}$$

This proves (3) for this case.

Case 4. Suppose $c_1 \in (a_{k_0}, b_{k_0})$ for some $k_0 \in \{1, \ldots, n\}$. In this case we have

$$|f(b_{k_0}) - f(a_{k_0})| \leq |f(b_{k_0}) - f(c_1)| + |f(c_1) - f(a_{k_0})|$$

and then

$$\sum_{k=1}^{n} |f(b_k) - f(a_k)| \leq \left\{ \sum_{k=1}^{k_0-1} |f(b_k) - f(a_k)| + |f(c_1) - f(a_{k_0})| \right\}$$

$$+ \left\{ |f(b_{k_0}) - f(c_1)| + \sum_{k=k_0}^{n} |f(b_k) - f(a_k)| \right\}$$

$$< \frac{\varepsilon}{2} + \frac{\varepsilon}{2} = \varepsilon \quad \text{by Case 1 and Case 2.}$$

This proves (3) for this case. ∎

Prob. 13.18. Let $f(x) = x^p$ for $x \in [0, b]$ where $p \in [0, \infty)$. Show that f is absolutely continuous on $[0, b]$.

Proof. To show that f is absolutely continuous on $[0, b]$, we show that f satisfies conditions $1°$, $2°$ and $3°$ of Prob. 13.17.

Now $f(x) = x^p$ for $x \in [0, b]$ where $p \in [0, \infty)$ is continuous on $[0, b]$. Also f is increasing on $[0, b]$ and this implies that $f \in BV([0, b])$.

Let us show that f is absolutely continuous on $[c, b]$ for every $c \in (0, b]$. Now we have

$$f'(x) = px^{p-1} \quad \text{for } x \in [c, b].$$

Thus f' is a continuous function on the compact set $[c, b]$ and this implies that f' is bounded on $[c, b]$. Then by Prob. 13.13, f satisfies a Lipschitz Condition on $[c, b]$. This implies that f is absolutely continuous on $[c, b]$ according to Prob. 13.11.

Thus we have shown that f satisfies conditions $1°$, $2°$ and $3°$ of Prob. 13.17. Then by Prob. 13.17, f is absolutely continuous on $[0, b]$. ∎

Prob. 13.19. Let f be a real-valued continuous on $[0, b]$ such that
$1°$ $\quad f$ is continuous on $[a, b]$,
$2°$ $\quad f \in BV([a, c_0])$ for some $c_0 \in (a, b]$,
$3°$ $\quad f'(x)$ exists and $|f'(x)| \leq \frac{1}{x}$ for $x \in (0, b]$.
Show that f is absolutely continuous on $[0, b]$.

Proof. The proof is based on Prob. 13.17. Our condition $1°$ is condition $1°$ of Prob. 13.17 and our condition $2°$ is condition $3°$ of of Prob. 13.17. We show next that our condition $3°$ implies condition $2°$ of Prob. 13.17.

Let $c \in (0, b]$ be arbitrarily chosen and consider the interval $[c, b] \subset (0, b]$. By $3°$ we have

$$|f'(x)| \leq \frac{1}{x} \leq \frac{1}{c} \quad \text{for } x \in [c, b].$$

This implies that f satisfies a Lipschitz Condition with coefficient $\frac{1}{c}$ on $[c, b]$ according to Prob. 13.13. Then f is absolutely continuous on $[c, b]$ by Prob. 13.11. Then f is absolutely continuous on $[c, b]$ for every $c \in (0, b]$, satisfying condition $2°$ of Prob. 13.17.

Thus f satisfies conditions $1°$, $2°$ and $3°$ of Prob. 13.17. This implies that f is absolutely continuous on $[0, b]$ by Prob. 13.17. ∎

Prob. 13.20. Let f be a real-valued continuous function on $[0, b]$ where $b > 0$ defined with some $p \in (0, \infty)$ by setting

$$f(x) = \begin{cases} x^p \sin \frac{1}{x} & \text{for } x \in (0, b], \\ 0 & \text{for } x = 0. \end{cases}$$

Show that f is absolutely continuous on $[0, b]$ if and only if $p \in (1, \infty)$.

Proof. 1. According to Prob. 12.13, if $p \in (0, 1]$ then $f \notin BV([0, b])$. Then f cannot be absolutely continuous on $[0, b]$ by Theorem 13.8 (Banach-Zarecki).

2. According to Prob. 12.13, if $p \in (1, \infty)$ then $f \in BV([0, b])$. Note that f is continuous on $[0, b]$. Thus according to Prob. 13.17, to show that f is absolutely continuous on $[0, b]$ it remains to verify that f is absolutely continuous on $[c, b]$ for every $c \in (0, b]$.

Now we have

$$f'(x) = px^{p-1} \sin \frac{1}{x} - x^{p-2} \cos \frac{1}{x} \quad \text{for } x \in [c, b].$$

Thus f' is continuous on $[c, b]$ and then compactness of $[c, b]$ implies that f' is bounded on $[c, b]$. This implies that f satisfies a Lipschitz Condition on $[c, b]$ according to Prob. 13.13. Then f is absolutely continuous on $[c, b]$ according to Prob. 13.11. ∎

Prob. 13.21. Find an absolutely continuous function on $[a, b]$ that does not satisfy any Lipschitz Condition on $[a, b]$.

Proof. Consider a real-valued function f on $[0, b]$ defined by

$$f(x) = \sqrt{x} \quad \text{for } x \in [0, b].$$

According to Prob. 13.18, f is absolutely continuous on $[0, b]$.

To show that f does not satisfy any Lipschitz Condition on $[0, b]$, assume the contrary, that is, assume that f satisfies a Lipschitz Condition with coefficient $M > 0$ on $[0, b]$. This implies that f' exists $(\mathfrak{M}_L, \mu_L)$-a.e. on $[0, b]$ and $|f'| \leq M$ according to Prob. 13.12. But we have

$$f'(x) = \frac{1}{2} \frac{1}{\sqrt{x}} \quad \text{for } x \in (0, b],$$

and thus f' is unbounded as $x \downarrow 0$. This is a contradiction. Therefore f cannot satisfy any Lipschitz Condition on $[0, b]$. ∎

Prob. 13.22. (If f is a real-valued absolutely continuous function on $[a, b]$ then according to Theorem 13.18, f' exists μ_L-a.e. on $[a, b]$ and f' is $\mathfrak{M}_L$-measurable and μ_L-integrable on $[a, b]$ and f' is finite μ_L-a.e. on $[a, b]$.)

Let $E \subset [a, b]$ be an arbitrary non-empty null set in the measure space $(\mathbb{R}, \mathfrak{M}_L, \mu_L)$. Show that there exists a real-valued absolutely continuous increasing function f on $[a, b]$ such that $f'(x) = \infty$ for every $x \in E$.

Proof. According to Definition 3.1, the Lebesgue outer measure μ_L^* on $\mathbb{R}$ is defined for an arbitrary subset A of $\mathbb{R}$ by

(1) $$\mu_L^*(A) = \inf \left\{ \sum_{k \in \mathbb{N}} \ell(I_k) : (I_k : k \in \mathbb{N}) \subset \mathfrak{J}_o, \bigcup_{k \in \mathbb{N}} I_k \supset A \right\}.$$

Observe that for $(I_k : k \in \mathbb{N}) \subset \mathfrak{J}_o$, $\bigcup_{k \in \mathbb{N}} I_k$ is an open set in $\mathbb{R}$. On the other hand, every non-empty open set O in $\mathbb{R}$ is a union of countably many disjoint open intervals $(I_k : k \in \mathbb{N}) \subset \mathfrak{J}_o$ and moreover we have $\mu_L(O) = \sum_{k \in \mathbb{N}} \ell(I_k)$. Let $\mathfrak{O}$ be the collection of all open sets in $\mathbb{R}$. Then for an arbitrary $A \subset \mathbb{R}$ we have

(2) $$\mu_L^*(A) = \inf \left\{ \mu_L(O) : O \in \mathfrak{O}, O \supset A \right\}.$$

Let E be an arbitrary non-empty null set in the measure space $(\mathbb{R}, \mathfrak{M}_L, \mu_L)$. Then $\mu_L^*(E) = 0$ and (2) implies that there exists a decreasing sequence $(O_n : n \in \mathbb{N})$ of open sets in $\mathbb{R}$ such that $O_n \supset E$ and $\mu_L(O_n) < \frac{1}{2^n}$ for every $n \in \mathbb{N}$.

Let us define a sequence $(\varphi_n : n \in \mathbb{N})$ of nonnegative real-valued $\mathfrak{M}_L$-measurable functions on $\mathbb{R}$ by setting

$$\varphi_n = \sum_{k=1}^n \mathbf{1}_{O_k}.$$

At every $x \in \mathbb{R}$ we have $\varphi_n(x) \uparrow$ as $n \to \infty$. Let us define a function g on $\mathbb{R}$ by

$$g = \lim_{n \to \infty} \varphi_n.$$

Then g is a nonnegative extended real-valued $\mathfrak{M}_L$-measurable function on $\mathbb{R}$. Moreover

$$\int_{[a,b]} g \, d\mu_L = \int_{[a,b]} \lim_{n \to \infty} \varphi_n \, d\mu_L = \lim_{n \to \infty} \int_{[a,b]} \varphi_n \, d\mu_L$$

$$= \lim_{n \to \infty} \int_{[a,b]} \sum_{k=1}^n \mathbf{1}_{O_k} \, d\mu_L = \lim_{n \to \infty} \sum_{k=1}^n \int_{[a,b]} \mathbf{1}_{O_k} \, d\mu_L$$

$$= \lim_{n \to \infty} \sum_{k=1}^n \mu_L(O_k) \le \lim_{n \to \infty} \sum_{k=1}^n \frac{1}{2^k} = 1 < \infty,$$

where the second equality is by Theorem 8.5 (Monotone Convergence Theorem). This shows that g is μ_L-integrable on $[a, b]$.

With the μ_L-integrable nonnegative extended real-valued $\mathfrak{M}_L$-measurable function g on $[a, b]$, let us define

$$f(x) = \int_{[a,x]} g \, d\mu_L \quad \text{for } x \in [a, b].$$

Since g is nonnegative, f is an increasing function on $[a, b]$. As an indefinite integral of g, f is absolutely continuous on $[a, b]$ by Theorem 13.15.

Let us show that $f'(x) = \infty$ for every $x \in E$. Let $x \in E$ and let $n \in \mathbb{N}$ be arbitrarily chosen and fixed. Let $h > 0$ be so small that $[x, x + h] \subset O_n$. Then we have

(1)
$$\frac{f(x + h) - f(x)}{h} = \frac{1}{h}\left\{ \int_{[a, x+h]} g\, d\mu_L - \int_{[a, x]} g\, d\mu_L \right\}$$

$$= \frac{1}{h}\int_{(x, x+h]} g\, d\mu_L \geq \frac{1}{h}\int_{(x, x+h]} \varphi_n\, d\mu_L \quad \text{since } \varphi_n \leq g$$

$$= \frac{1}{h}\int_{(x, x+h]} \sum_{k=1}^{n} \mathbf{1}_{O_k}\, d\mu_L = \frac{1}{h}\sum_{k=1}^{n} \int_{(x, x+h]} \mathbf{1}_{O_k}\, d\mu_L.$$

Now we have $[x, x + h] \subset O_n \subset O_{n-1} \subset \cdots \subset O_1$. Thus we have $\int_{(x, x+h]} \mathbf{1}_{O_k}\, d\mu_L = h$ for $k = 1, \ldots, n$ and then

$$\sum_{k=1}^{n} \int_{(x, x+h]} \mathbf{1}_{O_k}\, d\mu_L = nh.$$

Substituting this into (1), we have

(2)
$$\frac{f(x + h) - f(x)}{h} \geq \frac{1}{h}nh = n.$$

This implies that
$$\lim_{h \downarrow 0} \frac{f(x + h) - f(x)}{h} \geq n.$$

Since this holds for every $n \in \mathbb{N}$, we have
$$\lim_{h \downarrow 0} \frac{f(x + h) - f(x)}{h} = \infty.$$

This shows that the right hand derivative of f at x is ∞. We show similarly that the left hand derivative of f at x is ∞. Thus the derivative of f at x is ∞. ∎

Prob. 13.23. (The Cantor-Lebesgue function τ on $[0, 1]$ is a real-valued continuous and increasing function with derivative $\tau' = 0$ a.e. on $[0, 1]$.)
Show that there exists a real-valued continuous and strictly increasing function f with derivative $f' = 0$ a.e. on $[0, 1]$.

Proof. 1. The Cantor-Lebesgue function τ on $[0, 1]$ is a real-valued continuous and increasing function with $\tau(0) = 0$ and $\tau(1) = 1$ but τ is not strictly increasing. (See Theorem 4.34.)

Let T be the Cantor ternary set. Then $T \subset [0, 1]$ and according to Theorem 4.33, $\mu_L(T) = 0$ and $G := [0, 1] \setminus T$ is a union of countably many disjoint open intervals with $\mu_L(G) = 1$. The Cantor-Lebesgue function τ on $[0, 1]$ is constant on each of the countably many disjoint open intervals constituting G so that $\tau' = 0$ on $G = [0, 1] \setminus T$.

Let us extended the domain of definition of τ from $[0, 1]$ to $\mathbb{R}$ by setting $\tau(x) = 0$ for $x \in (-\infty, 0)$ and $\tau(x) = 1$ for $x \in (1, \infty)$. Then τ is a real-valued continuous and increasing function on $\mathbb{R}$ and $\tau'(x) = 0$ for $x \in \mathbb{R} \setminus T$.

Let $\mathfrak{J}$ be the collection of all closed intervals with rational endpoints contained in $[0, 1]$. Then $\mathfrak{J}$ is a countable collection. Let $\mathfrak{J} = \{[a_n, b_n] : n \in \mathbb{N}\}$ be an arbitrary enumeration of the members of $\mathfrak{J}$. In this expression, a_n and b_n are rational numbers contained in $[0, 1]$.

For every $n \in \mathbb{N}$, let us define a mapping of $\mathbb{R}$ into $\mathbb{R}'$ by setting

$$(1) \qquad \varphi_n(x) = \frac{x - a_n}{b_n - a_n} \quad \text{for } x \in \mathbb{R}.$$

Then φ_n is a one-to-one mapping of $\mathbb{R}$ onto $\mathbb{R}'$ and φ_n is absolutely continuous on every finite closed interval in $\mathbb{R}$. The inverse function φ_n^{-1} is given by

$$\varphi_n^{-1}(y) = (b_n - a_n)y + a_n \quad \text{for } y \in \mathbb{R}'.$$

Thus φ_n^{-1} is absolutely continuous on every finite closed interval in $\mathbb{R}'$.

For every $n \in \mathbb{N}$, let us define a function on $\mathbb{R}$ by setting

$$(2) \qquad \psi_n(x) = \tau \circ \varphi_n(x) \quad \text{for } x \in \mathbb{R}.$$

Since φ_n and τ are real-valued continuous and increasing functions on $\mathbb{R}$, $\psi_n = \tau \circ \varphi_n$ is a real-valued continuous and increasing function on $\mathbb{R}$. Since $\tau(y) \in [0, 1]$ for every $y \in \mathbb{R}'$, we have $\psi_n(x) \in [0, 1]$ for every $x \in \mathbb{R}$.

Let us define a function f on $\mathbb{R}$ by setting

$$(3) \qquad f(x) = \sum_{n \in \mathbb{N}} \frac{1}{2^n} \psi_n(x) \quad \text{for } x \in \mathbb{R}.$$

Since $\psi_n(x) \in [0, 1]$ for every $x \in \mathbb{R}$ and every $n \in \mathbb{N}$, we have

$$f(x) = \sum_{n \in \mathbb{N}} \frac{1}{2^n} \psi_n(x) \le \sum_{n \in \mathbb{N}} \frac{1}{2^n} = 1.$$

Thus the series in (3) converges for every $x \in \mathbb{R}$ and f is defined on $\mathbb{R}$ and $f(x) \in [0, 1]$ for all $x \in \mathbb{R}$. Now for every $n \in \mathbb{N}$, ψ_n is an increasing function on $\mathbb{R}$. This implies that f is an increasing function on $\mathbb{R}$.

2. Let us show that f is continuous on $\mathbb{R}$. Now f is defined by $f = \sum_{n\in\mathbb{N}} \frac{1}{2^n}\psi_n$. Consider $\left(\sum_{k=1}^n \frac{1}{2^k}\psi_k : n \in \mathbb{N}\right)$, the sequence of partial sums of the series defining f. Since ψ_k is continuous on $\mathbb{R}$ for every $k \in \mathbb{N}$, $\sum_{k=1}^n \frac{1}{2^k}\psi_k$ is continuous on $\mathbb{R}$. Then in order to show that f is continuous on $\mathbb{R}$, it suffices to show that the sequence of partial sums converges uniformly on $\mathbb{R}$. Thus we show that the sequence of partial sums satisfies the Uniform Cauchy Criterion on $\mathbb{R}$, that is, for every $\varepsilon > 0$ there exists $N \in \mathbb{N}$ such that

$$\left| \sum_{k=1}^m \frac{1}{2^k}\psi_k(x) - \sum_{k=1}^n \frac{1}{2^k}\psi_k(x) \right| < \varepsilon \quad \text{for all } x \in \mathbb{R} \text{ when } m \geq n \geq N,$$

that is,

(4) $$\left| \sum_{k=n+1}^m \frac{1}{2^k}\psi_k(x) \right| < \varepsilon \quad \text{for all } x \in \mathbb{R} \text{ when } m \geq n \geq N.$$

Now since $\sum_{n\in\mathbb{N}} \frac{1}{2^n} = 1$, for every $\varepsilon > 0$ there exists $N \in \mathbb{N}$ such that $\sum_{k=n+1}^m \frac{1}{2^k} < \varepsilon$ for $m \geq n \geq N$. With such $N \in \mathbb{N}$, we have

$$\left| \sum_{k=n+1}^m \frac{1}{2^k}\psi_k(x) \right| \leq \sum_{k=n+1}^m \frac{1}{2^k} < \varepsilon \quad \text{for } m \geq n \geq N.$$

This proves (4) and shows that f is continuous on $\mathbb{R}$.

3. Let us show that $f' = 0$ a.e. on $\mathbb{R}$. By Prob. 12.21 (Fubini's Differentiation Theorem for Series of Increasing Functions), we have

(5) $$f' = \sum_{n\in\mathbb{N}} \frac{1}{2^n}\psi_n' \quad \text{a.e. on } \mathbb{R}.$$

Let us show first that $\psi_n' = 0$ a.e. on $\mathbb{R}$ for every $n \in \mathbb{N}$. Now $\psi_n = \tau \circ \varphi_n$. If φ_n' exists at $x_0 \in \mathbb{R}$ and τ' exists at $\varphi_n(x_0)$, then ψ_n' exists at $x_0 \in \mathbb{R}$ and $\psi_n'(x_0) = \tau'\big(\varphi_n(x_0)\big)\varphi_n'(x_0)$. Now $\tau'(y)$ exists and $\tau'(y) = 0$ for $y \in \mathbb{R}' \setminus T$. As we noted above, φ_n maps $\mathbb{R}$ one-to-one onto $\mathbb{R}'$ and φ_n^{-1} maps $\mathbb{R}'$ one-to-one onto $\mathbb{R}$ and φ_n^{-1} is absolutely continuous on every finite closed interval in $\mathbb{R}'$ and thus by Lusin's Condition (N), φ_n^{-1} maps the null set $T \subset \mathbb{R}'$ into a null set E_n in $\mathbb{R}$. Then for $x \in \mathbb{R} \setminus E_n$ we have $\varphi_n(x) \in \mathbb{R}' \setminus T$ so that

$$\psi_n'(x) = \tau'\big(\varphi_n(x)\big)\varphi_n'(x) = 0 \cdot \frac{1}{b_n - a_n} = 0.$$

This shows that for every $n \in \mathbb{N}$ there exists a null set $E_n \subset \mathbb{R}$ such that $\psi_n'(x) = 0$ for $x \in \mathbb{R} \setminus E_n$.

Let $E = \bigcup_{n\in\mathbb{N}} E_n$. Then E is a null set in $\mathbb{R}$ and $\psi_n'(x) = 0$ for all $n \in \mathbb{N}$ for $x \in \mathbb{R} \setminus E$. Then by (5) we have

$$f'(x) = \sum_{n\in\mathbb{N}} \frac{1}{2^n}\psi_n'(x) = 0 \quad \text{for } x \in \mathbb{R} \setminus E.$$

This shows that $f' = 0$ a.e. on $\mathbb{R}$.

4. Finally let us show that f is strictly increasing on $[0, 1]$. Let us show first that if α and β are rational numbers in $[0, 1]$ and $\alpha < \beta$ then $f(\alpha) < f(\beta)$. By the definition of $\mathfrak{J} = \{[a_n, b_n] : n \in \mathbb{N}\}$ we have $[\alpha, \beta] = [a_{n_0}, b_{n_0}]$ for some $n_0 \in \mathbb{N}$. Since

$$\varphi_{n_0}(x) = \frac{x - a_{n_0}}{b_{n_0} - a_{n_0}} \quad \text{for } x \in \mathbb{R},$$

we have $\varphi_{n_0}(a_{n_0}) = 0$ and $\varphi_{n_0}(b_{n_0}) = 1$. Since $\tau(0) = 0$ and $\tau(1) = 1$, we have

$$\psi_{n_0}(a_{n_0}) = \tau \circ \varphi_{n_0}(a_{n_0}) = \tau(0) = 0,$$

$$\psi_{n_0}(b_{n_0}) = \tau \circ \varphi_{n_0}(b_{n_0}) = \tau(1) = 1.$$

Thus we have $\psi_{n_0}(a_{n_0}) < \psi_{n_0}(b_{n_0})$, that is, $\psi_{n_0}(\alpha) < \psi_{n_0}(\beta)$. On the other hand for every $n \in \mathbb{N}$, ψ_n is an increasing function on $\mathbb{R}$ so that we have $\psi_n(\alpha) \le \psi_n(\beta)$. Thus we have

$$\sum_{n \in \mathbb{N}} \frac{1}{2^n} \psi_n(\alpha) < \sum_{n \in \mathbb{N}} \frac{1}{2^n} \psi_n(\beta),$$

that is, $f(\alpha) < f(\beta)$.

Now let $x_1, x_2 \in [0, 1]$ and $x_1 < x_2$. Then there exist rational numbers α and β such that $x_1 < \alpha < \beta < x_2$. Then by our result above we have $f(\alpha) < f(\beta)$. Then since f is an increasing function on $\mathbb{R}$, we have $f(x_1) \le f(\alpha) < f(\beta) \le f(x_2)$. Thus $f(x_1) < f(x_2)$. This shows that f is strictly increasing on $[0, 1]$. ∎

Prob. 13.24. Let f be a real-valued function on $[a, b]$. Suppose $f([a, b]) \subset [c, d]$. Let g be a real-valued function on $[c, d]$. Suppose f satisfies a Lipschitz Condition with coefficient $M_1 > 0$ on $[a, b]$ and g satisfies a Lipschitz Condition with coefficient $M_2 > 0$ on $[c, d]$. Show that $g \circ f$ satisfies a Lipschitz Condition with coefficient $M_2 M_1 > 0$ on $[a, b]$.

Proof. Let $x', x'' \in [a, b]$. Then we have $f(x'), f(x'') \in [c, d]$. Then we have

$$\left|(g \circ f)(x') - (g \circ f)(x'')\right| = \left|g(f(x')) - g(f(x''))\right|$$
$$\leq M_2 \left|f(x') - f(x'')\right|$$
$$\leq M_2 M_1 |x' - x''|.$$

This shows that $g \circ f$ satisfies a Lipschitz Condition with coefficient $M_2 M_1 > 0$ on $[a, b]$. ∎

Prob. 13.25. Let f be a real-valued absolutely continuous function on $[a, b]$. Suppose $f([a, b]) \subset [c, d]$. Let g be a real-valued function on $[c, d]$ satisfying a Lipschitz Condition with coefficient $M > 0$ on $[c, d]$. Show that $g \circ f$ is absolutely continuous on $[a, b]$.

Proof. To show that $g \circ f$ is an absolutely continuous function on $[a, b]$, we verify that for every $\varepsilon > 0$ there exists $\delta > 0$ such that for any finite collection of non-overlapping closed intervals $\{[a_k, b_k] : k = 1, \ldots, n\}$ contained in $[a, b]$ with $\sum_{k=1}^{n}(b_k - a_k) < \delta$, we have $\sum_{k=1}^{n} |(g \circ f)(b_k) - (g \circ f)(a_k)| < \varepsilon$.

Let $\varepsilon > 0$ be arbitrarily given. Since f is absolutely continuous on $[a, b]$ there exists $\delta > 0$ such that for any finite collection $\{[a_k, b_k] : k = 1, \ldots, n\}$ of non-overlapping closed intervals contained in $[a, b]$ with $\sum_{k=1}^{n}(b_k - a_k) < \delta$, we have

$$\sum_{k=1}^{n} \left|f(b_k) - f(a_k)\right| < \frac{\varepsilon}{M}.$$

Now $f(a_k), f(b_k) \in [c, d]$. Then since g satisfies a Lipschitz Condition with coefficient $M > 0$ on $[c, d]$, we have

$$\left|(g \circ f)(b_k) - (g \circ f)(a_k)\right| = \left|g(f(b_k)) - g(f(a_k))\right| \leq M \left|f(b_k) - f(a_k)\right|$$

and then

$$\sum_{k=1}^{n} \left|(g \circ f)(b_k) - (g \circ f)(a_k)\right| \leq M \sum_{k=1}^{n} \left|f(b_k) - f(a_k)\right| < M \frac{\varepsilon}{M} = \varepsilon.$$

This shows that $g \circ f$ is absolutely continuous on $[a, b]$. ∎

Prob. 13.26. Let f be a real-valued absolutely continuous function on $[a, b]$. Suppose $f([a, b]) \subset [c, d]$. Let g be a real-valued absolutely continuous function on $[c, d]$. Assume further that f is increasing on $[a, b]$. Show that $g \circ f$ is absolutely continuous on $[a, b]$.

Proof. To show that $g \circ f$ is an absolutely continuous function on $[a, b]$, we verify that for every $\varepsilon > 0$ there exists $\delta > 0$ such that for any finite collection of non-overlapping closed intervals $\{[a_k, b_k] : k = 1, \ldots, n\}$ contained in $[a, b]$ with $\sum_{k=1}^{n} (b_k - a_k) < \delta$, we have $\sum_{k=1}^{n} |(g \circ f)(b_k) - (g \circ f)(a_k)| < \varepsilon$.

Let $\varepsilon > 0$ be arbitrarily given. Since g is absolutely continuous on $[c, d]$ there exists $\eta > 0$ such that for any finite collection of non-overlapping closed intervals $\{[c_k, d_k] : k = 1, \ldots, n\}$ contained in $[c, d]$ with $\sum_{n=1}^{n} (d_k - c_k) < \eta$, we have

$$(1) \qquad \sum_{k=1}^{n} |g(d_k) - g(c_k)| < \varepsilon.$$

Since f is absolutely continuous on $[a, b]$, for our $\eta > 0$ there exists $\delta > 0$ such that for any finite collection $\{[a_k, b_k] : k = 1, \ldots, n\}$ of non-overlapping closed intervals contained in $[a, b]$ with $\sum_{k=1}^{n} (b_k - a_k) < \delta$, we have

$$(2) \qquad \sum_{k=1}^{n} |f(b_k) - f(a_k)| < \eta.$$

Assume that the collection $\{[a_k, b_k] : k = 1, \ldots, n\}$ is so numbered that we have

$$a \le a_1 \le b_1 \le a_2 \le b_2 \le \cdots \le a_n \le b_n \le b.$$

Then since f increases on $[a, b]$ we have

$$c \le f(a_1) \le f(b_1) \le f(a_2) \le f(b_2) \le \cdots \le f(a_n) \le f(b_n) \le d.$$

Then $\{[f(a_k), f(b_k)] : k = 1, \ldots, n\}$ is a finite collection of non-overlapping closed intervals contained in $[c, d]$ with $\sum_{k=1}^{n} |f(b_k) - f(a_k)| < \eta$. Thus by (1) we have

$$\sum_{k=1}^{n} |g(f(b_k)) - g(f(a_k))| < \varepsilon,$$

that is, we have

$$(3) \qquad \sum_{k=1}^{n} |(g \circ f)(b_k) - (g \circ f)(a_k)| < \varepsilon.$$

This shows that $g \circ f$ is absolutely continuous on $[a, b]$. ∎

Prob. 13.27. Let f and g be two real-valued functions defined on $[0, 1]$ by

$$f(x) = \sqrt{x} \quad \text{for } x \in [0, 1],$$

and

$$g(x) = \begin{cases} x^2 |\sin \frac{1}{x}| & \text{for } x \in (0, 1], \\ 0 & \text{for } x = 0. \end{cases}$$

(a) Show that f and g are absolutely continuous on $[0, 1]$.
(b) Show that $g \circ f$ is absolutely continuous on $[0, 1]$.
(c) Show that $f \circ g$ is not absolutely continuous on $[0, 1]$.
(Hint for (c)): Show that $f \circ g$ is not of bounded variation on $[0, 1]$.)

Proof. 1. Let us prove (a). Now f is absolutely continuous on $[0, 1]$ by Prob. 13.18. Regarding g let us define a function h on $[0, 1]$ by setting

$$h(x) = \begin{cases} x^2 \sin \frac{1}{x} & \text{for } x \in (0, 1], \\ 0 & \text{for } x = 0. \end{cases}$$

Then h is absolutely continuous on $[0, 1]$ by Prob. 13.20. Then by (e) of Prob. 13.1, $|h|$ is absolutely continuous on $[0, 1]$. But $|h| = g$. Thus g is absolutely continuous on $[0, 1]$.

2. Let us prove (b). Now f and g are absolutely continuous on $[0, 1]$ by (a) and f is an increasing function on $[0, 1]$. Then $g \circ f$ is absolutely continuous on $[0, 1]$ by Prob. 13.26.

3. Let us prove (c). According to Theorem 13.8 (Banach-Zarecki), every absolutely continuous function is a function of bounded variation. Therefore to show that $f \circ g$ is not absolutely continuous on $[0, 1]$ it suffices to show that $f \circ g \notin BV\big([0, 1]\big)$.

Let us show that $f \circ g \notin BV\big([0, 1]\big)$. Now we have

$$(f \circ g)(x) = \sqrt{g(x)} = \begin{cases} \sqrt{x^2 |\sin \frac{1}{x}|} = x \sqrt{|\sin \frac{1}{x}|} & \text{for } x \in (0, 1]. \\ 0 & \text{for } x = 0. \end{cases}$$

For brevity let us write

$$(1) \qquad h(x) = \begin{cases} x \sqrt{|\sin \frac{1}{x}|} & \text{for } x \in (0, 1]. \\ 0 & \text{for } x = 0. \end{cases}$$

Then it remains to show that $h \notin BV\big([0, 1]\big)$.

Consider the function $|\sin x|$ for $x \in (0, \infty)$. Now $|\sin x|$ is a nonnegative real-valued and continuous function assuming the minimum value 0 at $x = k\pi$ for $k \in \mathbb{N}$ and assuming the maximum value 1 at $x = \frac{\pi}{2} + k\pi$ for $k \in \mathbb{N}$. Then $\sqrt{|\sin x|}$ assumes the minimum value 0 at $x = k\pi$ for $k \in \mathbb{N}$ and assumes the maximum value 1 at $x = \frac{\pi}{2} + k\pi$ for $k \in \mathbb{N}$. Thus $\sqrt{|\sin \frac{1}{x}|}$ assumes the minimum value 0 at $a_k = (k\pi)^{-1}$ for $k \in \mathbb{N}$ and assumes the maximum value 1 at $b_k = \left(\frac{\pi}{2} + k\pi\right)^{-1}$ for $k \in \mathbb{N}$. Observe that we have

$$(2) \qquad h(a_k) = 0 \quad \text{and} \quad h(b_k) = b_k = \left(\frac{\pi}{2} + k\pi\right)^{-1} \quad \text{for } k \in \mathbb{N}.$$

For every $n \in \mathbb{N}$, we have

$$0 < b_n < a_n < b_{n-1} < a_{n-1} < \cdots < b_1 < a_1 < 1.$$

Consider a partition $\mathcal{P}_n$ of $[0, 1]$ defined by

$$\mathcal{P}_n = \big\{ 0 < b_n < a_n < b_{n-1} < a_{n-1} < \cdots < b_1 < a_1 < 1 \big\}.$$

We have

$$
\begin{aligned}
V_0^1(h, \mathcal{P}_n) = &\; \big|h(1) - h(a_1)\big| + \big|h(a_1) - h(b_1)\big| \\
&+ \big|h(b_1) - h(a_2)\big| + \big|h(a_2) - h(b_2)\big| \\
&+ \big|h(b_2) - h(a_3)\big| + \big|h(a_3) - h(b_3)\big| \\
&\;\;\vdots \\
&+ \big|h(b_{n-1}) - h(a_n)\big| + \big|h(a_n) - h(b_n)\big| \\
&+ \big|h(b_n) - h(0)\big|.
\end{aligned}
$$

Substituting (2) for $h(a_k)$ in the last expression we have

$$
\begin{aligned}
V_0^1(h, \mathcal{P}_n) = &\; h(1) + h(b_1) \\
&+ h(b_1) + h(b_2) \\
&+ h(b_2) + h(b_3) \\
&\;\;\vdots \\
&+ h(b_{n-1}) + h(b_n) \\
&+ h(b_n) - h(0).
\end{aligned}
$$

Thus we have

$$
V_0^1(h, \mathcal{P}_n) = h(1) + 2 \sum_{k=1}^n h(b_k) = h(1) + 2 \sum_{k=1}^n \Big(\frac{\pi}{2} + k\pi\Big)^{-1} \quad \text{by (2)}
$$

$$
= h(1) + 2 \sum_{k=1}^n \frac{1}{\frac{\pi}{2} + k\pi} \geq h(1) + \frac{2}{\pi} \sum_{k=1}^n \frac{1}{k+1}.
$$

Then we have

$$
V_0^1(h) = \sup_{\mathcal{P} \in \mathfrak{P}_{0,1}} V_0^1(h, \mathcal{P}) \geq \sup_{n \in \mathbb{N}} V_0^1(h, \mathcal{P}_n) = \sup_{n \in \mathbb{N}} \Big\{ h(1) + \frac{2}{\pi} \sum_{k=1}^n \frac{1}{k+1} \Big\} = \infty.
$$

This shows that $h \notin BV\big([0, 1]\big)$ and completes the proof. ∎

Prob. 13.28. Let f be an absolutely continuous function on $[a, b]$ and g be an absolutely continuous function on $[c, d]$. Assume that $f([a, b]) \subset [c, d]$. Show that $g \circ f$ is absolutely continuous on $[a, b]$ if and only if $g \circ f$ is of bounded variation on $[a, b]$.

Proof. 1. If $g \circ f$ is absolutely continuous on $[a, b]$, then $g \circ f \in BV\big([a, b]\big)$ according to Theorem 13.8 (Banach-Zarecki).

2. Conversely suppose $g \circ f \in BV\big([a, b]\big)$.

Now since f is absolutely continuous on $[a, b]$, f is continuous on $[a, b]$ according to Theorem 13.8. Similarly absolute continuity of g on $[c, d]$ implies continuity of g on $[c, d]$. Then $g \circ f$ is continuous on $[a, b]$.

Absolute continuity of f on $[a, b]$ implies that f satisfies Condition (N) on $[a, b]$ according to Theorem 13.8. Similarly absolute continuity of g on $[c, d]$ implies that g satisfies Condition (N) on $[c, d]$. Then it is verified immediately that $g \circ f$ satisfies Condition (N) on $[a, b]$.

Now that $g \circ f$ is continuous on $[a, b]$, satisfies Condition (N) on $[a, b]$ and $g \circ f \in BV\big([a, b]\big)$, $g \circ f$ is absolutely continuous on $[a, b]$ by Theorem 13.8. ∎

Prob. 13.29. For $p \in (1, \infty)$, let
$$f(x) = \begin{cases} x^p \sin \frac{1}{x} & \text{for } x \in (0, b], \\ 0 & \text{for } x = 0. \end{cases}$$
(Thus defined, f is absolutely continuous on $[0, b]$ by Prob. 13.20.)

(a) Show that f' exists on $[0, b]$ and f' is continuous on $(0, b]$.

(b) Show that f' is μ_L-integrable on $[0, b]$.

(c) Show that when $p \in (2, \infty)$, f' is continuous on $[0, b]$.

(d) Show that when $p \in (1, 2]$, $\lim_{x \downarrow 0} f'(x)$ does not exist.

(e) Show that when $p \in [2, \infty)$, f' is bounded on $[0, b]$.

(f) Show that when $p \in (1, 2)$, f' is not bounded on $[0, b]$.

(g) Show that when $p \in [2, \infty)$, f satisfies a Lipschitz Condition on $[0, b]$.

(h) Show that when $p \in (1, 2)$, f does not satisfy any Lipschitz Condition on $[0, b]$.

Proof. (a) We have

(1) $$f'(x) = px^{p-1} \sin \frac{1}{x} - x^{p-2} \cos \frac{1}{x} \quad \text{for } x \in (0, b].$$

This shows that f' exists on $(0, b]$ and f' is continuous on $(0, b]$. We also have

(2) $$f'(0) = \lim_{h \downarrow 0} \frac{f(h) - f(0)}{h} = \lim_{h \downarrow 0} \frac{1}{h}\left\{ h^p \sin \frac{1}{h} - 0 \right\} = \lim_{h \downarrow 0} h^{p-1} \sin \frac{1}{h} = 0$$

since $p - 1 > 0$. This shows that f' exists on $[0, b]$.

(b) By (1) we have
$$|f'(x)| = \left| px^{p-1} \sin \frac{1}{x} - x^{p-2} \cos \frac{1}{x} \right| \leq px^{p-1} + x^{p-2}.$$

Then we have
$$\int_{[0,b]} |f'(x)| \, \mu_L(dx) \leq \int_{[0,b]} \left\{ px^{p-1} + x^{p-2} \right\} \mu_L(dx)$$
$$= \left[x^p + \frac{x^{p-1}}{p-1} \right]_0^b = b^p - \frac{b^{p-1}}{p-1} < \infty.$$

This shows that f' is μ_L-integrable on $[0, b]$.

(c) We showed in (a) that f' is continuous on $(0, b]$ and $f'(0) = 0$. Assume that $p \in (2, \infty)$. Then we have $p - 2 > 0$ and $p - 1 > 1$ and then

$$\lim_{x \downarrow 0} f'(x) = \lim_{x \downarrow 0} \left\{ px^{p-1} \sin \frac{1}{x} - x^{p-2} \cos \frac{1}{x} \right\} \quad \text{by (1)}$$

$$= 0 = f'(0) \quad \text{by (2)}.$$

This proves the continuity of f' at $x = 0$. Therefore f' is continuous on $[0, b]$.

(d) Assume that $p \in (1, 2]$. Then we have $p - 1 > 0$ and $p - 2 \leq 0$. Now $f'(x)$ for $x \in (0, b]$ is given by (1). Then we have

$$\lim_{x \downarrow 0} \left\{ px^{p-1} \sin \frac{1}{x} \right\} = 0 \quad \text{but} \quad \not\exists \lim_{x \downarrow 0} \left\{ -x^{p-2} \cos \frac{1}{x} \right\}.$$

This shows that $\lim\limits_{x\downarrow 0} f'(x)$ does not exist.

(e) Assume that $p \in [2, \infty)$. Then we have $p - 1 \geq 1$ and $p - 2 \geq 0$. Then for $x \in (0, b]$

$$
(3) \qquad |f'(x)| = \left| px^{p-1} \sin\frac{1}{x} - x^{p-2} \cos\frac{1}{x} \right|
$$

$$
\leq px^{p-1} + x^{p-2} \leq pb^{p-1} + b^{p-2}.
$$

Thus f' is bounded on $(0, b]$. Then since $f'(0) = 0$, f' is bounded on $[0, b]$.

(f) Assume that $p \in (1, 2)$. Then we have $p - 1 > 0$ and $p - 2 < 0$. Then $px^{p-1} \sin\frac{1}{x}$ for $x \in (0, b]$ is bounded on $(0, b]$ but $x^{p-2} \cos\frac{1}{x}$ for $x \in (0, b]$ is not bounded on $(0, b]$. This shows that f' as given by (1) is not bounded on $(0, b]$. Thus f' is not bounded on $[0, b]$.

(g) When $p \in [2, \infty)$, f' is bounded on $[0, b]$ as we showed in (e) and indeed we have according to (3) and (2)

$$
(4) \qquad |f'(x)| \leq pb^{p-1} + b^{p-2} \quad \text{for } x \in [0, b].
$$

According to Prob. 13.13, (4) implies that f satisfies a Lipschitz Condition with coefficient $pb^{p-1} + b^{p-2} > 0$ on $[0, b]$.

(h) According to Prob. 13.12, if f satisfies a Lipschitz Condition with coefficient $M > 0$ on $[0, b]$ then f' exists $(\mathfrak{M}_L, \mu_L)$-a.e. on $[0, b]$ and $|f'| \leq M$. Now when $p \in (1, 2)$, f' is not bounded on $[0, b]$ as we showed in (f). Thus f cannot satisfy any Lipschitz Condition on $[0, b]$. ∎

Prob. 13.30. Let
$$f(x) = \begin{cases} x^2 \cos \frac{\pi}{x^2} & \text{for } x \in (0, b], \\ 0 & \text{for } x = 0. \end{cases}$$

(a) Show that f' exists on $[0, b]$.

(b) Show that f is absolutely continuous on $[a, b]$ and f' is μ_L-integrable on $[a, b]$ for every $a \in (0, b]$.

(c) Show that f is not absolutely continuous on $[0, b]$ and f' is not μ_L-integrable on $[0, b]$. (Hint: With $\alpha_n = \left(2n + \frac{1}{2}\right)^{-1/2}$ and $\beta_n = (2n)^{-1/2}$ for $n \in \mathbb{N}$, compute $\int_{[\alpha_n, \beta_n]} f' \, d\mu_L$.)

Proof. 1. We have

(1) $$f'(x) = 2x \cos \frac{\pi}{x^2} + \frac{\pi}{x} \sin \frac{\pi}{x^2} \quad \text{for } x \in (0, b],$$

and

(2) $$f'(0) = \lim_{h \downarrow 0} \frac{1}{h} \{ f(h) - f(0) \} = \lim_{h \downarrow 0} \frac{1}{h} \left\{ h^2 \cos \frac{\pi}{h^2} - 0 \right\} = 0.$$

This shows that f' exists on $[0, b]$.

2. Let $a \in (0, b]$. Then f', as given by (1), is bounded on $[a, b]$. This implies, according to Prob. 13.13, that f satisfies a Lipschitz Condition on $[a, b]$. This then implies, according to Prob. 13.11, that f is absolutely continuous on $[a, b]$. Then by Theorem 13.18 we have

$$\int_{[a,b]} f' \, d\mu_L = f(b) - f(a) = b^2 \cos \frac{\pi}{b^2} - a^2 \cos \frac{\pi}{a^2} \in \mathbb{R}.$$

This shows that f' is μ_L-integrable on $[a, b]$.

3. We showed above that f is absolutely continuous on $[a, b]$ for every $a \in (0, b]$. This implies that f is absolutely continuous on $[\alpha, \beta] \subset [a, b]$. Then by Theorem 13.18 we have

(3) $$\int_{[\alpha,\beta]} f' \, d\mu_L = f(\beta) - f(\alpha) = \beta^2 \cos \frac{\pi}{\beta^2} - \alpha^2 \cos \frac{\pi}{\alpha^2}.$$

For $n \in \mathbb{N}$, let us define

(4) $$\alpha_n = \frac{1}{\sqrt{2n + \frac{1}{2}}} < \frac{1}{\sqrt{2n}} = \beta_n.$$

Then by (3) we have

(5) $$\int_{[\alpha_n, \beta_n]} f' \, d\mu_L = \frac{1}{2n} \cos 2n\pi - \frac{1}{2n + \frac{1}{2}} \cos \left(2n + \frac{1}{2}\right)\pi$$
$$= \frac{1}{2n} \cdot 1 - \frac{1}{2n + \frac{1}{2}} \cdot 0 = \frac{1}{2n}.$$

Observe that (4) implies that $\alpha_{n+1} < \beta_{n+1} < \alpha_n < \beta_n$. Then $\{(\alpha_n, \beta_n] : n \in \mathbb{N}\}$ is a disjoint collection of intervals and $\bigcup_{n\in\mathbb{N}}(\alpha_n, \beta_n] = \left(0, \frac{1}{\sqrt{2}}\right]$. Then we have

$$\int_{\left(0, \frac{1}{\sqrt{2}}\right]} f' \, d\mu_L = \int_{\bigcup_{n\in\mathbb{N}}(\alpha_n, \beta_n]} f' \, d\mu_L = \sum_{n\in\mathbb{N}} \int_{[\alpha_n, \beta_n]} f' \, d\mu_L$$

$$= \sum_{n\in\mathbb{N}} \frac{1}{2n} = \infty \quad \text{by (5)}.$$

Thus f' is not μ_L-integrable on $\left(0, \frac{1}{\sqrt{2}}\right]$. Then f' is not μ_L-integrable on $[0, b]$. Then according to Theorem 13.18, f is not absolutely continuous on $[0, b]$. ∎

Prob. 13.31. Let $D \subset [a, b]$ and $D \in \mathfrak{M}_L$ with $\mu_L(D) > 0$. Show that for every $\lambda \in [0, 1]$, there exists $c \in [a, b]$ such that $\mu_L\left(D \cap [a, c]\right) = \lambda \cdot \mu_L(D)$.
(Hint: Consider the Lebesgue integral of $\mathbf{1}_{D\cap[a,x]}$ for $x \in [a, b]$.)

Proof. Let $D \subset [a, b]$ and $D \in \mathfrak{M}_L$. Then $\mathbf{1}_D$ is a μ_L-integrable $\mathfrak{M}_L$-measurable real-valued function on $[a, b]$ and we have $\int_{[a,b]} \mathbf{1}_D \, d\mu_L = \mu_L(D) < \infty$.

Consider an indefinite integral of the μ_L-integrable function $\mathbf{1}_D$ on $[a, b]$ defined by

$$(1) \qquad\qquad F(x) = \int_{[a,x]} \mathbf{1}_D \, d\mu_L \quad \text{for } x \in [a, b].$$

By Theorem 13.15, F is a continuous function on $[a, b]$. Moreover (1) implies that

$$(2) \qquad\qquad \begin{cases} F(a) = \int_{\{a\}} \mathbf{1}_D \, d\mu_L = 0. \\ F(b) = \int_{[a,b]} \mathbf{1}_D \, d\mu_L = \mu_L(D). \end{cases}$$

Let $\lambda \in [0, 1]$. Then we have according to (2)

$$(3) \qquad\qquad \lambda \cdot \mu_L(D) \in \left[0, \mu_L(D)\right] = \left[F(a), F(b)\right].$$

Then the Intermediate Value Theorem for Continuous Functions implies that there exists $c \in [a, b]$ such that

$$(4) \qquad\qquad F(c) = \lambda \cdot \mu_L(D).$$

On the other hand we have

$$(5) \qquad\qquad F(c) = \int_{[a,c]} \mathbf{1}_D \, d\mu_L = \int_{[a,b]} \mathbf{1}_{D\cap[a,c]} \, d\mu_L \quad \text{by (1)}$$

$$= \mu_L\left(D \cap [a, c]\right).$$

Thus by (4) and (5), we have $\mu_L\left(D \cap [a, c]\right) = \lambda \cdot \mu_L(D)$. ∎

Prob. 13.32. Let $D \subset \mathbb{R}$, $D \in \mathfrak{M}_L$, and $\mu_L(D) \in (0, \infty)$. Show that for every $\lambda \in (0, 1)$, there exists $\xi \in \mathbb{R}$ such that $\mu_L(D \cap (-\infty, \xi]) = \lambda \cdot \mu_L(D)$.

Proof. Let

(1) $$D_n = D \cap (-\infty, n] \quad \text{for } n \in \mathbb{Z}.$$

Then we have

(2) $$\cdots \subset D_{-3} \subset D_{-2} \subset D_{-1} \subset D_0 \subset D_1 \subset D_2 \subset D_3 \subset \cdots$$

We have $\lim_{n\uparrow\infty} D_n = \bigcup_{n\in\mathbb{Z}} D_n = D$ and then by (a) of Theorem 1.26 we have

(3) $$\lim_{n\uparrow\infty} \mu_L(D_n) = \mu_L\left(\lim_{n\uparrow\infty} D_n\right) = \mu_L(D).$$

On the other hand we have $\lim_{n\downarrow-\infty} D_n = \bigcap_{n\in\mathbb{Z}} D_n = \emptyset$ and for negative $n \in \mathbb{Z}$ we have $D_n \subset D_0$ and $\mu_L(D_0) \leq \mu_L(D) < \infty$. Then by (b) of Theorem 1.26 we have

(4) $$\lim_{n\downarrow-\infty} \mu_L(D_n) = \mu_L\left(\lim_{n\downarrow-\infty} D_n\right) = \mu_L(\emptyset) = 0.$$

Let $\lambda \in (0, 1)$ be arbitrarily chosen and then let $0 < \lambda_1 < \lambda < \lambda_2 < 1$. Then (3) and (4) imply that there exist $n_1, n_2 \in \mathbb{Z}, n_1 < n_2$, such that

(5) $$\mu_L(D_{n_1}) \leq \lambda_1 \cdot \mu_L(D) < \lambda \cdot \mu_L(D) < \lambda_2 \cdot \mu_L(D) \leq \mu_L(D_{n_2}).$$

Let us define $\lambda_1', \lambda_2' \in [0, 1]$ by setting

(6) $$\begin{cases} \lambda_1' = \mu_L(D_{n_1})/\mu_L(D). \\ \lambda_2' = \mu_L(D_{n_2})/\mu_L(D). \end{cases}$$

Then (5) and (6) imply

(7) $$\lambda_1' < \lambda_1 < \lambda < \lambda_2 < \lambda_2'.$$

Consider the function $\mathbf{1}_{D\cap(n_1,n_2]}$ defined on $[n_1, n_2]$. Let us define a real-valued function F on $[n_1, n_2]$ by setting

(8) $$F(x) = \int_{[n_1,x]} \mathbf{1}_{D\cap(n_1,n_2]} \, d\mu_L + \lambda_1' \cdot \mu_L(D) \quad \text{for } x \in [n_1, n_2].$$

Thus defined, F is an indefinite integral of the μ_L-integrable function $\mathbf{1}_{D\cap(n_1,n_2]}$ and therefore F is continuous on $[n_1, n_2]$ by Theorem 13.15. Moreover we have

(9) $$F(n_1) = \lambda_1' \cdot \mu_L(D)$$

and

(10) $F(n_2) = \mu_L\big(D \cap (n_1, n_2]\big) + \lambda'_1 \cdot \mu_L(D)$ by (8)

$= \mu_L\big(D \cap (-\infty, n_2]\big) - \mu_L\big(D \cap (-\infty, n_1]\big) + \lambda'_1 \cdot \mu_L(D)$

$= \mu_L(D_{n_2}) - \mu_L(D_{n_1}) + \lambda'_1 \cdot \mu_L(D)$ by (1)

$= \lambda'_2 \cdot \mu_L(D) - \lambda'_1 \cdot \mu_L(D) + \lambda'_1 \cdot \mu_L(D)$ by (6)

$= \lambda'_2 \cdot \mu_L(D).$

Thus the continuous function F on $[n_1, n_2]$ assumes the value $\lambda'_1 \cdot \mu_L(D)$ at $x = n_1$ and the value $\lambda'_2 \cdot \mu_L(D)$ at $x = n_2$. Now from (7) we have

$$\lambda \cdot \mu_L(D) \in \Big(\lambda'_1 \cdot \mu_L(D), \lambda'_2 \cdot \mu_L(D)\Big).$$

Thus by the Intermediate Value Theorem for Continuous Functions, there exists $\xi \in [n_1, n_2]$ such that $F(\xi) = \lambda \cdot \mu_L(D)$. Then we have

$$\lambda \cdot \mu_L(D) = F(\xi) = \int_{[n_1, \xi]} \mathbf{1}_{D \cap (n_1, n_2]} \, d\mu_L + \lambda'_1 \cdot \mu_L(D) \quad \text{by (8)}$$

$$= \int_{[n_1, \xi]} \mathbf{1}_{D \cap (n_1, n_2]} \, d\mu_L + \mu_L(D_{n_1}) \quad \text{by (6)}$$

$$= \int_{\mathbb{R}} \mathbf{1}_{D \cap (n_1, \xi]} \, d\mu_L + \int_{\mathbb{R}} \mathbf{1}_{D \cap (-\infty, n_1]} \, d\mu_L \quad \text{by (1)}$$

$$= \int_{\mathbb{R}} \mathbf{1}_{D \cap (-\infty, \xi]} \, d\mu_L = \mu_L\big(D \cap (-\infty, \xi]\big).$$

This completes the proof. ∎

Prob. 13.33. Definition. Let $E \subset \mathbb{R}$, $E \in \mathfrak{M}_L$, and I be a finite interval in $\mathbb{R}$. We call $(\ell(I))^{-1} \mu_L(E \cap I)$ the average density of E over the interval I. For an arbitrary $x \in \mathbb{R}$, the density of E at x is defined by

$$p_E(x) = \lim_{h \downarrow 0} \frac{1}{2h} \mu_L(E \cap (x - h, x + h)).$$

Show that

$$p_E(x) = \begin{cases} 1 & \text{for a.e. } x \in E, \\ 0 & \text{for a.e. } x \in E^c. \end{cases}$$

Proof. To find $p_E(x)$ for $x \in \mathbb{R}$, choose $a, b \in \mathbb{R}$ such that $a < x < b$. Then we have

$$
\begin{aligned}
(1) \qquad p_E(x) &= \lim_{h \downarrow 0} \frac{1}{2h} \mu_L(E \cap (x - h, x + h)) \\
&= \lim_{h \downarrow 0} \frac{1}{2h} \int_{[a,b]} \mathbf{1}_{E \cap (x-h,x+h)} \, d\mu_L \\
&= \lim_{h \downarrow 0} \frac{1}{2h} \int_{[a,b] \cap (x-h,x+h)} \mathbf{1}_E \, d\mu_L \\
&= \lim_{h \downarrow 0} \frac{1}{2h} \left\{ \int_{[a,x+h)} \mathbf{1}_E \, d\mu_L - \int_{[a,x-h]} \mathbf{1}_E \, d\mu_L \right\}.
\end{aligned}
$$

Consider an indefinite integral of the μ_L-integrable function $\mathbf{1}_E$ on $[a, b]$ given by

$$(2) \qquad\qquad F(x) = \int_{[a,x]} \mathbf{1}_E \, d\mu_L \quad \text{for } x \in [a, b].$$

According to Theorem 13.15, we have:

1°   The derivative F' exists a.e. on $[a, b]$ and $F' = \mathbf{1}_E$ a.e. on $[a, b]$.

Recall Definition 12.1 for the right hand derivative $D_r F$ and the left hand derivative $D_\ell F$ given by

$$(3) \qquad\qquad (D_r F)(x) = \lim_{h \downarrow 0} \frac{F(x + h) - F(x)}{h}.$$

$$(4) \qquad\qquad (D_\ell F)(x) = \lim_{h \uparrow 0} \frac{F(x + h) - F(x)}{h}.$$

By Definition 12.1, if $F'(x)$ exists then $F'(x) = (D_r F)(x) = (D_\ell F)(x)$. Then we have

$$(5) \qquad F'(x) = (D_r F)(x) = \lim_{h \downarrow 0} \frac{1}{h} \{ F(x + h) - F(x) \} \quad \text{by (3)}$$

$$= \lim_{h \downarrow 0} \frac{1}{h} \left\{ \int_{[a,x+h]} \mathbf{1}_E \, d\mu_L - \int_{[a,x]} \mathbf{1}_E \, d\mu_L \right\}.$$

Similarly we have

(6)
$$F'(x) = (D_\ell F)(x) = \lim_{h \uparrow 0} \frac{1}{h}\{F(x+h) - F(x)\} \quad \text{by (4)}$$

$$= \lim_{h \downarrow 0} \frac{1}{-h}\{F(x-h) - F(x)\}$$

$$= -\lim_{h \downarrow 0} \frac{1}{h}\{F(x-h) - F(x)\}$$

$$= -\lim_{h \downarrow 0} \frac{1}{h}\left\{ \int_{[a,x-h]} 1_E \, d\mu_L - \int_{[a,x]} 1_E \, d\mu_L \right\}.$$

Adding (5) and (6) we have

$$2F'(x) = \lim_{h \downarrow 0} \frac{1}{h}\left\{ \int_{[a,x+h]} 1_E \, d\mu_L - \int_{[a,x-h]} 1_E \, d\mu_L \right\}$$

and then

(7)
$$F'(x) = \lim_{h \downarrow 0} \frac{1}{2h}\left\{ \int_{[a,x+h]} 1_E \, d\mu_L - \int_{[a,x-h]} 1_E \, d\mu_L \right\}.$$

Now (7) and (1) show that $p_E(x) = F'(x)$ for any x for which $F'(x)$ exists. Then by 1° we have

$$p_E(x) = F'(x) = 1_E(x), \quad \text{a.e. on } [a, b].$$

Thus we have

$$p_E(x) = \begin{cases} 1 & \text{for a.e. } x \in E, \\ 0 & \text{for a.e. } x \in E^c. \end{cases}$$

This completes the proof. ∎

§16   The L^p Spaces

Prob. 16.1. Let f be a strictly positive real-valued $\mathfrak{M}_L$-measurable function on $[0, 1]$. Prove

$$\left\{ \int_{[0,1]} f \, d\mu \right\} \left\{ \int_{[0,1]} \frac{1}{f} \, d\mu \right\} \geq 1.$$

Proof. By Theorem 16.14 (Hölder's Inequality), we have

$$1 = \int_{[0,1]} 1 \, d\mu_L = \int_{[0,1]} \sqrt{f} \, \frac{1}{\sqrt{f}} \, d\mu_L$$

$$\leq \left\{ \int_{[0,1]} \left(\sqrt{f} \right)^2 d\mu_L \right\}^{\frac{1}{2}} \left\{ \int_{[0,1]} \left(\frac{1}{\sqrt{f}} \right)^2 d\mu_L \right\}^{\frac{1}{2}}$$

$$= \left\{ \int_{[0,1]} f \, d\mu_L \right\}^{\frac{1}{2}} \left\{ \int_{[0,1]} \frac{1}{f} \, d\mu_L \right\}^{\frac{1}{2}}.$$

Taking square on both sides, we have

$$1 \leq \left\{ \int_{[0,1]} f \, d\mu_L \right\} \left\{ \int_{[0,1]} \frac{1}{f} \, d\mu_L \right\}.$$

This completes the proof. ∎

Prob. 16.2. Let $(X, \mathfrak{A}, \mu)$ be a measure space. Let f be an extended complex-valued $\mathfrak{A}$-measurable function on X. Let $0 < p < q < \infty$. Show that

$$\|f\|_p \le \|f\|_q \{\mu(X)\}^{\frac{1}{p} - \frac{1}{q}},$$

provided that the product exists. (The product does not exist if one factor is equal to 0 and the other is equal to ∞.)

Contrast this result with Corollary 16.16 which states that for $p, q \in (1, \infty)$ such that $\frac{1}{p} + \frac{1}{q} = 1$ we have $\|f\|_1 \le \|f\|_p \{\mu(X)\}^{\frac{1}{q}}$.

Proof. Let $r = \frac{q}{p} > 1$ and let s be the conjugate of r, that is, $\frac{1}{r} + \frac{1}{s} = 1$. This implies

$$\frac{1}{s} = 1 - \frac{1}{r} = 1 - \frac{p}{q} = \frac{q - p}{q}.$$

Now we have by Theorem 16.14 (Hölder's Inequality),

$$\int_X |f|^p \, d\mu = \int_X |f|^p \cdot 1 \, d\mu \le \left\{ \int_X |f|^{pr} \, d\mu \right\}^{\frac{1}{r}} \left\{ \int_X 1^s \, d\mu \right\}^{\frac{1}{s}}$$

$$= \left\{ \int_X |f|^q \, d\mu \right\}^{\frac{p}{q}} \{\mu(X)\}^{\frac{1}{s}}$$

$$= \left\{ \int_X |f|^q \, d\mu \right\}^{\frac{p}{q}} \{\mu(X)\}^{\frac{q-p}{q}}.$$

Then we have

$$\|f\|_p = \left\{ \int_X |f|^p \, d\mu \right\}^{\frac{1}{p}} \le \left\{ \int_X |f|^q \, d\mu \right\}^{\frac{1}{q}} \{\mu(X)\}^{\frac{q-p}{pq}}$$

$$= \|f\|_q \{\mu(X)\}^{\frac{1}{p} - \frac{1}{q}}.$$

This completes the proof. ∎

Prob. 16.3. Let $(X, \mathfrak{A}, \mu)$ be a measure space. Let $\vartheta \in (0, 1)$ and let $p, q, r \geq 1$, $p, q \geq r$, be related by

$$\frac{1}{r} = \frac{\vartheta}{p} + \frac{1 - \vartheta}{q}.$$

Show that for every extended complex-valued $\mathfrak{A}$-measurable function f on X we have

$$\|f\|_r \leq \|f\|_p^{\vartheta} \|f\|_q^{1-\vartheta}.$$

Proof. Multiplying the equality $\frac{1}{r} = \frac{\vartheta}{p} + \frac{1-\vartheta}{q}$ by r we have

$$1 = \frac{r\vartheta}{p} + \frac{r(1 - \vartheta)}{q} = \frac{1}{\frac{p}{r\vartheta}} + \frac{1}{\frac{q}{r(1-\vartheta)}}.$$

Now since $p \geq r$ and $\vartheta \in (0, 1)$ we have $\frac{p}{r\vartheta} > 1$. Similarly since $q \geq r$ and $1 - \vartheta \in (0, 1)$ we have $\frac{q}{r(1-\vartheta)} > 1$. Thus $\frac{p}{r\vartheta}$ and $\frac{q}{r(1-\vartheta)}$ are conjugates. Then by Theorem 16.14 (Hölder's Inequality) we have

$$\int_X |f|^r \, d\mu = \int_X |f|^{r\vartheta} \cdot |f|^{r(1-\vartheta)} \, d\mu$$

$$\leq \left\{ \int_X \left(|f|^{r\vartheta} \right)^{\frac{p}{r\vartheta}} \, d\mu \right\}^{\frac{r\vartheta}{p}} \left\{ \int_X \left(|f|^{r(1-\vartheta)} \right)^{\frac{q}{r(1-\vartheta)}} \, d\mu \right\}^{\frac{r(1-\vartheta)}{q}}$$

$$= \left\{ \int_X |f|^p \, d\mu \right\}^{\frac{r\vartheta}{p}} \left\{ \int_X |f|^q \, d\mu \right\}^{\frac{r(1-\vartheta)}{q}}$$

$$= \|f\|_p^{r\vartheta} \|f\|_q^{r(1-\vartheta)}.$$

This implies

$$\|f\|_r = \left\{ \int_X |f|^r \, d\mu \right\}^{\frac{1}{r}} \leq \|f\|_p^{\vartheta} \|f\|_q^{1-\vartheta}.$$

This completes the proof. ∎

Prob. 16.4. Given a measure space $(X, \mathfrak{A}, \mu)$. Let f be an extended real-valued $\mathfrak{A}$-measurable function on X such that $\int_X |f|^p \, d\mu < \infty$ for some $p \in (0, \infty)$. Show that for every $\varepsilon > 0$ there exists $M > 0$ such that for the truncation of f at level M, that is, $f^{[M]} = (f \wedge M) \vee (-M)$, we have $\int_X |f - f^{[M]}|^p \, d\mu < \varepsilon$.

Proof. The assumption $\int_X |f|^p \, d\mu < \infty$ implies that $|f|^p < \infty$, a.e. on X and then $|f| < \infty$, a.e. on X. Then we have

$$\lim_{M \to \infty} f^{[M]} = f, \text{ a.e. on } X.$$

Note also that we have $|f^{[M]}| \leq |f|$ and this implies

$$|f - f^{[M]}|^p \leq \{|f| + |f^{[M]}|\}^p \leq 2^p |f|^p.$$

By the assumption $\int_X |f|^p \, d\mu < \infty$, $|f|^p$ is μ-integrable on X and then so is $2^p |f|^p$. Then by Theorem 9.20 (Dominated Convergence Theorem), we have

$$\lim_{M \to \infty} \int_X |f - f^{[M]}|^p \, d\mu = \int_X \lim_{M \to \infty} |f - f^{[M]}|^p \, d\mu = \int_X 0 \, d\mu = 0.$$

This implies that for every $\varepsilon > 0$ there exists $M > 0$ such that $\int_X |f - f^{[M]}|^p \, d\mu < \varepsilon$. ∎

An Alternate Proof. According to Theorem 9.26 (Uniform Absolute Continuity of Integral with respect to Measure), the assumption $\int_X |f|^p \, d\mu < \infty$ implies that for every $\varepsilon > 0$ there exists $\delta > 0$ such that

$$(1) \qquad \int_E |f|^p \, d\mu < \frac{\varepsilon}{2^p} \quad \text{for every } E \in \mathfrak{A} \text{ such that } \mu(E) < \delta.$$

For every $M > 0$, define

$$(2) \qquad D_M = \{x \in X : |f(x)| > M\}.$$

Then we have

$$(3) \qquad f^{[M]} = f \quad \text{on } X \setminus D_M.$$

Now we have

$$M^p \mu(D_M) \leq \int_{D_M} |f|^p \, d\mu \leq \int_X |f|^p \, d\mu < \infty$$

and then

$$\mu(D_M) \leq \frac{1}{M^p} \int_X |f|^p \, d\mu \quad \text{and} \quad \lim_{M \to \infty} \mu(D_M) = 0.$$

Let $M > 0$ be so large that $\mu(D_M) < \delta$. Then we have

$$\int_X |f - f^{[M]}|^p \, d\mu = \int_{D_M} |f - f^{[M]}|^p \, d\mu \leq \int_{D_M} \{|f| + |f^{[M]}|\}^p \, d\mu \quad \text{by (3)}$$

$$\leq \int_{D_M} \{2|f|\}^p \, d\mu = 2^p \int_{D_M} |f|^p \, d\mu < 2^p \frac{\varepsilon}{2^p} = \varepsilon \quad \text{by (1)}.$$

This completes the proof. ∎

Prob. 16.5. Let $(X, \mathfrak{A}, \mu)$ be a measure space and let $p, q \in [1, \infty]$ be conjugates, that is, $\frac{1}{p} + \frac{1}{q} = 1$. Let $(f_n : n \in \mathbb{N}) \subset L^p(X, \mathfrak{A}, \mu)$ and $f \in L^p(X, \mathfrak{A}, \mu)$ and similarly $(g_n : n \in \mathbb{N}) \subset L^q(X, \mathfrak{A}, \mu)$ and $g \in L^q(X, \mathfrak{A}, \mu)$. Show that if $\lim_{n \to \infty} \|f_n - f\|_p = 0$ and $\lim_{n \to \infty} \|g_n - g\|_q = 0$ then $\lim_{n \to \infty} \|f_n g_n - fg\|_1 = 0$.

Proof. We have
$$f_n g_n - fg = f_n g_n - f_n g + f_n g - fg$$
and then by Theorem 16.14 (Hölder's Inequality) and Theorem 16.40 (Hölder's Inequality)

$$\|f_n g_n - fg\|_1 \leq \|f_n g_n - f_n g\|_1 + \|f_n g - fg\|_1$$
$$= \|f_n(g_n - g)\|_1 + \|(f_n - f)g\|_1$$
$$\leq \|f_n\|_p \|g_n - g\|_q + \|f_n - f\|_p \|g\|_q,$$

and then

$$\lim_{n \to \infty} \|f_n g_n - fg\|_1 \leq \|f_n\|_p \lim_{n \to \infty} \|g_n - g\|_q + \|g\|_q \lim_{n \to \infty} \|f_n - f\|_p = 0.$$

This completes the proof. ∎

Prob. 16.6. Let $(X, \mathfrak{A}, \mu)$ be a measure space and let $p \in [1, \infty)$. Let $(f_n : n \in \mathbb{N})$ be a sequence in $L^p(X, \mathfrak{A}, \mu)$ and $f \in L^p(X, \mathfrak{A}, \mu)$ be such that $\lim_{n \to \infty} \|f_n - f\|_p = 0$. Let $(g_n : n \in \mathbb{N})$ be a sequence of complex-valued $\mathfrak{A}$-measurable functions on X such that $|g_n| \leq M$ for every $n \in \mathbb{N}$ and let g be complex-valued $\mathfrak{A}$-measurable function on X such that $\lim_{n \to \infty} g_n = g$ a.e. on X. Show that $\lim_{n \to \infty} \|g_n f_n - gf\|_p = 0$.

Proof. We have

$$|g_n f_n - gf| = |g_n f_n - g_n f + g_n f - gf| \leq |g_n| |f_n - f| + |f| |g_n - g|,$$

and then

$$|g_n f_n - gf|^p \leq \Big\{ |g_n| |f_n - f| + |f| |g_n - g| \Big\}^p$$

$$\leq 2^p \Big\{ |g_n|^p |f_n - f|^p + |f|^p |g_n - g|^p \Big\}^p \quad \text{by Lemma 16.7}$$

$$\leq 2^p \Big\{ M^p |f_n - f|^p + |f|^p |g_n - g|^p \Big\}^p,$$

and thus we have

(1) $$\int_X |g_n f_n - gf|^p \, d\mu \leq 2^p M^p \int_X |f_n - f|^p \, d\mu + 2^p \int_X |f|^p |g_n - g|^p \, d\mu.$$

For the first term on the right side of (1), $\lim_{n \to \infty} \|f_n - f\|_p = 0$ implies $\lim_{n \to \infty} \|f_n - f\|_p^p = 0$ and thus we have

(2) $$\lim_{n \to \infty} 2^p M^p \int_X |f_n - f|^p \, d\mu = 2^p M^p \lim_{n \to \infty} \|f_n - f\|_p^p = 0.$$

For the second term on the right side of (1), observe that $|f|^p |g_n - g|^p \leq |f|^p (2M)^p$ which is μ-integrable on X since $|f|^p$ is. Thus by Theorem 9.20 (Dominated Convergence Theorem) we have

(3) $$\lim_{n \to \infty} 2^p \int_X |f|^p |g_n - g|^p \, d\mu = 2^p \int_X \lim_{n \to \infty} |f|^p |g_n - g|^p \, d\mu = 2^p \int_X 0 \, d\mu = 0.$$

Then by (1), (2) and (3), we have

(4) $$\lim_{n \to \infty} \|g_n f - gf\|_p^p = \lim_{n \to \infty} \int_X |g_n f_n - gf|^p \, d\mu = 0.$$

Then (4) implies $\lim_{n \to \infty} \|g_n f - gf\|_p = 0$. This completes the proof. ∎

Prob. 16.7. Definition. (Hölder Condition) *Let f be a real-valued function on an interval $I \subset \mathbb{R}$. We say that f satisfies a Hölder Condition with coefficient $C > 0$ and order $p \in (0, \infty)$ on I if we have $|f(x') - f(x'')| \leq C|x' - x''|^p$ for any $x', x'' \in I$. Thus the Lipschitz Condition is a Hölder Condition of order 1.*
Prove the following theorem:
Theorem. *Let f be a real-valued function on an interval $I \subset \mathbb{R}$. Suppose f satisfies a Hölder Condition with coefficient $C > 0$ and order $p \in [1, \infty)$ on I. Then we have:*
(a) *f satisfies a Lipschitz Condition with coefficient $C > 0$ on I.*
(b) *f satisfies Condition (L) on I.*
(c) *f is absolutely continuous on every $[a, b] \subset I$.*

Proof. 1. Let us prove (a). Assume that f satisfies a Hölder Condition with coefficient $C > 0$ and order $p \in [1, \infty)$ on I, that is, we have

$$(1) \qquad |f(x') - f(x'')| \leq C|x' - x''|^p \quad \text{for any } x', x'' \in I.$$

With $p \in [1, \infty)$, define a real-valued function g on $[0, 1]$ by setting $g(x) = x^p$ for $x \in [0, 1]$. Then $p \in [1, \infty)$ implies

$$(2) \qquad g(x) = x^p \leq x \quad \text{for } x \in [0, 1].$$

Let $x, y \in I$ be such that $x < y$ and $y - x \leq 1$. Then for any $x', x'' \in [x, y]$ we have

$$(3) \qquad |f(x') - f(x'')| \leq C|x' - x''|^p \leq C|x' - x''|$$

by (1) and (2). This shows that f satisfies a Lipschitz Condition with coefficient $C > 0$ on any $[x, y] \subset I$ with $y - x \leq 1$.
Consider an arbitrary $[x, y] \subset I$. For a sufficiently large $m \in \mathbb{N}$ we have $\frac{y-x}{m} \leq 1$. Then we can select $x = x_0 < x_1 < \cdots < x_m = y$ such that $x_i - x_{i-1} \leq 1$ for $i = 1, \ldots, m$. Then for any $x', x'' \in [x, y]$ such that $x' < x''$ we have

$$x' \leq x_n < x_{n+1} < \cdots < x_{n+k} \leq x'' \quad \text{where } n \geq 0 \text{ and } n + k \leq m.$$

Then we have

$$|f(x') - f(x'')| \leq |f(x') - f(x_n)| + |f(x_n) - f(x_{n+1})| + \cdots + |f(x_{n+k}) - f(x'')|$$
$$\leq C\Big\{ |x' - x_n| + |x_n - x_{n+1}| + \cdots + |x_{n+k} - x''| \Big\} \quad \text{by (3)}$$
$$= C|x' - x''|.$$

This shows that f satisfies a Lipschitz Condition with coefficient $C > 0$ on I.
2. According to Prob. 13.10, a Lipschitz Condition and Condition (L) are equivalent. Thus (a) implies (b).
3. According to Prob. 13.11, Lipschitz Condition on I implies absolute continuity of f on $[a, b] \subset I$. Thus (a) implies (c). ∎

Prob. 16.8. Let f be a real-valued function on $[a, b] \subset \mathbb{R}$ with $\ell([a, b]) = 1$. Assume that f satisfies a Lipschitz Condition with coefficient $C > 0$ on $[a, b]$. Show that f satisfies a Hölder Condition with coefficient $C > 0$ and order $p \in (0, 1]$ on $[a, b]$.

Proof. We have

(1) $\qquad\qquad |f(x') - f(x'')| \le C|x' - x''| \quad$ for any $x', x'' \in [a, b]$.

With $p \in (0, 1]$, define a real-valued function g on $[0, 1]$ by setting $g(x) = x^p$ for $x \in [0, 1]$. Then $p \in (0, 1]$ implies

(2) $\qquad\qquad\qquad g(x) = x^p \ge x \quad$ for $x \in [0, 1]$.

Then for $x', x'' \in [a, b]$, by applying (1) and (2), we have

(3) $\qquad\qquad |f(x') - f(x'')| \le C|x' - x''| \le C|x' - x''|^p$.

This shows that f satisfies a Hölder Condition with coefficient $C > 0$ and order $p \in (0, 1]$ on $[a, b]$. ∎

Prob. 16.9. Let f be an extended real-valued $\mathfrak{M}_L$-measurable function on $[a, b]$ such that $\int_{[a,b]} |f|^p \, d\mu_L < \infty$ for some $p \in (1, \infty)$. Let $q \in (1, \infty)$ be the conjugate of p.
(a) Show that f is μ_L-integrable on $[a, b]$.
(b) Let F be an indefinite integral of f on $[a, b]$. Show that F satisfies a Hölder Condition with coefficient $\|f\|_p$ and order $\frac{1}{q}$ on $[a, b]$, that is, we have

$$|F(x') - F(x'')| \leq \|f\|_p |x' - x''|^{1/q} \quad \text{for } x', x'' \in [a, b].$$

(c) Show that for every $x \in [a, b]$, we have

$$\lim_{h \downarrow 0} \frac{F(x+h) - F(x)}{h^{1/q}} = 0.$$

Proof. 1. Let us prove (a). Let $q \in (1, \infty)$ be the conjugate of $p \in (1, \infty)$, that is, $\frac{1}{p} + \frac{1}{q} = 1$. Then by Theorem 16.14 (Hölder's Inequality), we have

$$\int_{[a,b]} |f| \, d\mu_L = \int_{[a,b]} |f| \cdot 1 \, d\mu_L \leq \left\{ \int_{[a,b]} |f|^p \, d\mu_L \right\}^{\frac{1}{p}} \left\{ \int_{[a,b]} 1^q \, d\mu_L \right\}^{\frac{1}{q}}$$

$$= \left\{ \int_{[a,b]} |f|^p \, d\mu_L \right\}^{\frac{1}{p}} (b-a)^{\frac{1}{q}} < \infty.$$

This shows that f is μ_L-integrable on $[a, b]$.

2. Let us prove (b). The indefinite integral F of f on $[a, b]$ is given by

$$F(x) = \int_{[a,x]} f \, d\mu_L + C_0 \quad \text{for } x \in [a, b].$$

Then for $x', x'' \in [a, b]$ such that $x' < x''$ we have by Theorem 16.14 (Hölder's Inequality)

$$|F(x'') - F(x')| = \left| \int_{[x',x'']} f \, d\mu_L \right| \leq \int_{[x',x'']} |f| \, d\mu_L = \int_{[x',x'']} |f| \cdot 1 \, d\mu_L$$

$$\leq \left\{ \int_{[x',x'']} |f|^p \, d\mu_L \right\}^{\frac{1}{p}} \left\{ \int_{[x',x'']} 1^q \, d\mu_L \right\}^{\frac{1}{q}}$$

$$\leq \left\{ \int_{[a,b]} |f|^p \, d\mu_L \right\}^{\frac{1}{p}} |x'' - x'|^{\frac{1}{q}}$$

$$= \|f\|_p |x'' - x'|^{\frac{1}{q}}.$$

3. Let us prove (c). We have

$$F(x+h) - F(x) = \int_{[a,x+h]} f \, d\mu_L - \int_{[a,x]} f \, d\mu_L = \int_{(x,x+h]} f \, d\mu_L.$$

Then by Theorem 16.14 (Hölder's Inequality), for $x \in [a, b]$ we have

$$|F(x+h) - F(x)| = \left| \int_{(x,x+h]} f \, d\mu_L \right| \leq \int_{(x,x+h]} |f| \cdot 1 \, d\mu_L$$

$$\leq \left\{ \int_{(x,x+h]} |f|^p \, d\mu_L \right\}^{\frac{1}{p}} \left\{ \int_{(x,x+h]} 1^q \, d\mu_L \right\}^{\frac{1}{q}}$$

$$= \left\{ \int_{(x,x+h]} |f|^p \, d\mu_L \right\}^{\frac{1}{p}} h^{\frac{1}{q}}$$

and then

(1)
$$\frac{|F(x+h) - F(x)|}{h^{\frac{1}{q}}} \leq \left\{ \int_{(x,x+h]} |f|^p \, d\mu_L \right\}^{\frac{1}{p}}.$$

We claim

(2)
$$\lim_{h \downarrow 0} \left\{ \int_{(x,x+h]} |f|^p \, d\mu_L \right\} = 0.$$

Since $|f|^p$ is μ_L-integrable on $[a, b]$, by Theorem 9.26 (Uniform Absolute Continuity of Integral) for every $\varepsilon > 0$ there exists $\delta > 0$ such that $\int_D |f|^p \, d\mu_L < \varepsilon$ for any $D \in \mathfrak{M}_L$ such that $D \subset [a, b]$ and $\mu_L(D) < \delta$. Then we have $\int_{(x,x+h]} |f|^p \, d\mu_L < \varepsilon$ for $h \in (0, \delta)$. This proves (2). Then by (1) and (2) we have

$$\lim_{h \downarrow 0} \frac{|F(x+h) - F(x)|}{h^{\frac{1}{q}}} \leq \lim_{h \downarrow 0} \left\{ \int_{(x,x+h]} |f|^p \, d\mu_L \right\}^{\frac{1}{p}} = 0,$$

which is equivalent to

$$\lim_{h \downarrow 0} \frac{F(x+h) - F(x)}{h^{\frac{1}{q}}} = 0.$$

This completes the proof. ∎

Prob. 16.10. Let f be an extended real-valued $\mathfrak{M}_L$-measurable function on $[0, 1]$ such that $\int_{[0,1]} |f|^p \, d\mu_L < \infty$ for some $p \in [1, \infty)$. Let $q \in (1, \infty]$ be the conjugate of p, that is, $\frac{1}{p} + \frac{1}{q} = 1$. Let $a \in (0, 1]$. Show that

$$\lim_{a \downarrow 0} \frac{1}{a^{1/q}} \int_{[0,a]} |f| \, d\mu_L = 0.$$

Proof. 1. For the case that $p = 1$ we have $q = \infty$ and then $1/q = 0$ and $a^{1/q} = a^0 = 1$. Then we have

$$\lim_{a \downarrow 0} \frac{1}{a^{1/q}} \int_{[0,a]} |f| \, d\mu_L = \lim_{a \downarrow 0} \left\{ 1 \cdot \int_{[0,a]} |f| \, d\mu_L \right\} = \lim_{a \downarrow 0} \int_{[0,a]} |f| \, d\mu_L = 0.$$

2. Consider the case that $p \in (1, \infty)$. In this case we also have $q \in (1, \infty)$. Then by Theorem 16.14 (Hölder's Inequality) we have

$$\int_{[0,a]} |f| \, d\mu_L = \int_{[0,a]} |f| \cdot 1 \, d\mu_L$$

$$\leq \left\{ \int_{[0,a]} |f|^p \, d\mu_L \right\}^{\frac{1}{p}} \left\{ \int_{[0,a]} 1^q \, d\mu_L \right\}^{\frac{1}{q}}$$

$$= \left\{ \int_{[0,a]} |f|^p \, d\mu_L \right\}^{\frac{1}{p}} a^{\frac{1}{q}}$$

and then

$$\frac{1}{a^{\frac{1}{q}}} \int_{[0,a]} |f| \, d\mu_L \leq \left\{ \int_{[0,a]} |f|^p \, d\mu_L \right\}^{\frac{1}{p}}.$$

Now by assumption we have $\int_{[0,1]} |f|^p \, d\mu_L < \infty$. This implies that $\lim_{a \downarrow 0} \int_{[0,a]} |f|^p \, d\mu_L = 0$ and then $\lim_{a \downarrow 0} \left\{ \int_{[0,a]} |f|^p \, d\mu_L \right\}^{1/p} = 0$. Then we have

$$\lim_{a \downarrow 0} \frac{1}{a^{\frac{1}{q}}} \int_{[0,a]} |f| \, d\mu_L \leq \lim_{a \downarrow 0} \left\{ \int_{[0,a]} |f|^p \, d\mu_L \right\}^{\frac{1}{p}} = 0.$$

This completes the proof. ∎

Prob. 16.11. Prove the following theorem:

Theorem. (Uniform Absolute Continuity of Integral Relative to Measure for a Class with Bounded Norms in L^p Space) *Let $(X, \mathfrak{A}, \mu)$ be a measure space and let $p_0 \in (0, \infty)$. Let $\{f_\alpha : \alpha \in A\} \subset L^{p_0}(X, \mathfrak{A}, \mu)$ such that $\sup_{\alpha \in A} \|f_\alpha\|_{p_0} < \infty$. Let $p \in (0, p_0)$. Then*

(a)

$$\lim_{\lambda \to \infty} \sup_{\alpha \in A} \int_{\{X : |f_\alpha|^p > \lambda\}} |f_\alpha|^p \, d\mu = 0.$$

(b) *For every $\varepsilon > 0$ there exists $\delta > 0$ such that for every $E \in \mathfrak{A}$ with $\mu(E) < \delta$ we have*

$$\int_E |f_\alpha|^p \, d\mu < \varepsilon \quad \text{for all } \alpha \in A.$$

(Hint for (a): Let $p \in (0, p_0)$. Then for $0 < \eta < \xi$ we have $\xi^p = \xi^{p-p_0} \xi^{p_0} \leq \eta^{p-p_0} \xi^{p_0}$.)

Proof. 1. Let us prove (a). Let $p \in (0, p_0)$. Then for $0 < \eta < \xi$ we have

$$\xi^p = \xi^{p-p_0} \xi^{p_0} \leq \eta^{p-p_0} \xi^{p_0}$$

and this implies

$$\int_{\{|f_\alpha|^p > \eta^p\}} |f_\alpha|^p \, d\mu \leq \int_{\{|f_\alpha|^p > \eta^p\}} \eta^{p-p_0} |f_\alpha|^p \, d\mu \leq \eta^{p-p_0} \|f_\alpha\|_{p_0}^{p_0}$$

and then

$$\sup_{\alpha \in A} \int_{\{|f_\alpha|^p > \eta^p\}} |f_\alpha|^p \, d\mu \leq \eta^{p-p_0} \sup_{\alpha \in A} \|f_\alpha\|_{p_0}^{p_0}.$$

Then since $\lim_{\eta \to \infty} \eta^{p-p_0} = 0$ and $\sup_{\alpha \in A} \|f_\alpha\|_{p_0}^{p_0} < \infty$, we have

$$\lim_{\eta \to \infty} \sup_{\alpha \in A} \int_{\{|f_\alpha|^p > \eta^p\}} |f_\alpha|^p \, d\mu \leq \lim_{\eta \to \infty} \eta^{p-p_0} \sup_{\alpha \in A} \|f_\alpha\|_{p_0}^{p_0} = 0.$$

Then writing λ for η^p we have

$$(1) \qquad \lim_{\lambda \to \infty} \sup_{\alpha \in A} \int_{\{|f_\alpha|^p > \lambda\}} |f_\alpha|^p \, d\mu = 0.$$

2. Let us prove (b). Now (1) implies that for every $\varepsilon > 0$ there exists $\lambda > 0$ such that

$$\int_{\{|f_\alpha|^p > \lambda\}} |f_\alpha|^p \, d\mu < \frac{\varepsilon}{2} \quad \text{for all } \alpha \in A.$$

Then for every $E \in \mathfrak{A}$ we have

$$\int_E |f_\alpha|^p \, d\mu = \int_{E \cap \{|f_\alpha|^p > \lambda\}} |f_\alpha|^p \, d\mu + \int_{E \cap \{|f_\alpha|^p \leq \lambda\}} |f_\alpha|^p \, d\mu < \frac{\varepsilon}{2} + \lambda \mu(E).$$

Let $\delta = \frac{\varepsilon}{2\lambda}$. Then for every $E \in \mathfrak{A}$ with $\mu(E) < \delta$ we have

$$\int_E |f_\alpha|^p \, d\mu < \frac{\varepsilon}{2} + \lambda \frac{\varepsilon}{2\lambda} = \varepsilon.$$

This completes the proof. ∎

Prob. 16.12. Prove the following theorem:

Theorem. (Bounded Convergence Theorem for Norms) *Let $(X, \mathfrak{A}, \mu)$ be a finite measure space and let $p_0 \in (0, \infty)$. Let $(f_n : n \in \mathbb{N})$ be a sequence and f be an element in $L^{p_0}(X, \mathfrak{A}, \mu)$ such that*

1° $\lim_{n \to \infty} f_n = f$, μ-a.e. on X.

2° $\|f_n\|_{p_0} \leq M$ *for all* $n \in \mathbb{N}$.

Then for every $p \in (0, p_0)$ *we have* $\lim_{n \to \infty} \|f_n - f\|_p = 0$.

(Hint: Apply Egoroff's Theorem and Prob. 16.12.)

Proof. Let us write f_0 for f. Then $f_0 \in L^{p_0}(X, \mathfrak{A}, \mu)$ and thus we have $\|f_0\|_{p_0} < \infty$. This and 2° imply that we have

(1) $$\sup_{n \in \mathbb{Z}_+} \|f_n\|_{p_0} < \infty.$$

Let $p \in (0, p_0)$. According to Prob. 16.11, condition (1) above implies that for every $\varepsilon > 0$ there exists $\delta > 0$ such that for every $E \in \mathfrak{A}$ with $\mu(E) < \delta$ we have

(2) $$\int_E |f_n|^p \, d\mu < \varepsilon \quad \text{for all } n \in \mathbb{Z}_+.$$

Now by 1° we have $\lim_{n \to \infty} f_n = f_0$, μ-a.e. on X and $(X, \mathfrak{A}, \mu)$ a finite measure space. Thus according to Theorem 6.12 (Egoroff) for every $\delta > 0$ there exists $E \in \mathfrak{A}$ with $\mu(E) < \delta$ such that $\lim_{n \to \infty} f_n = f_0$ uniformly on E^c. This implies that for every $\eta > 0$ there exists $N \in \mathbb{N}$ such that $|f_n - f_0| < \eta$ on E^c for $n \geq N$ and then

$$\int_{E^c} |f_n - f_0|^p \, d\mu < \eta^p \mu(E^c) \quad \text{for } n \geq N.$$

This shows that we have

(3) $$\lim_{n \to \infty} \int_{E^c} |f_n - f_0|^p \, d\mu = 0.$$

Then we have

$$\int_X |f_n - f_0|^p \, d\mu = \int_{E^c} |f_n - f_0|^p \, d\mu + \int_E |f_n - f_0|^p \, d\mu$$

$$\leq \int_{E^c} |f_n - f_0|^p \, d\mu + 2^p \left\{ \int_E |f_n|^p \, d\mu \int_E |f_0|^p \, d\mu \right\}$$

$$\leq \int_{E^c} |f_n - f_0|^p \, d\mu + 2^p 2\varepsilon \quad \text{by (2)},$$

where the first inequality is based on the inequality $|\alpha + \beta|^p = 2^p \{|\alpha|^p + |\beta|^p\}$ in Lemma 16.7. Then we have

(4) $$\limsup_{n \to \infty} \int_X |f_n - f_0|^p \, d\mu \leq \lim_{n \to \infty} \int_{E^c} |f_n - f_0|^p \, d\mu + 2^{p+1}\varepsilon$$

$$= 2^{p+1}\varepsilon \quad \text{by (3)}.$$

Since (4) holds for every $\varepsilon > 0$ we have

$$\limsup_{n \to \infty} \int_X |f_n - f_0|^p \, d\mu = 0.$$

This then implies

$$\liminf_{n \to \infty} \int_X |f_n - f_0|^p \, d\mu = 0$$

and then we have

$$\lim_{n \to \infty} \int_X |f_n - f_0|^p \, d\mu = 0.$$

Thus we have $\displaystyle\lim_{n \to \infty} \|f_n - f_0\|_p^p = 0$ and then $\displaystyle\lim_{n \to \infty} \|f_n - f_0\|_p = 0$. ∎

Prob. 16.13. Let $(X, \mathfrak{A}, \mu)$ be a finite measure space and let $p, q \in (1, \infty)$ be conjugates. Let $(f_n : n \in \mathbb{N})$ be a sequence and f be an element in $L^p(X, \mathfrak{A}, \mu)$ such that
1° $\lim\limits_{n\to\infty} f_n = f$, μ-a.e. on X.
2° $\|f_n\|_p \leq M$ for all $n \in \mathbb{N}$.
Show that
(a) $\lim\limits_{n\to\infty} \int_A f_n g \, d\mu = \int_A f g \, d\mu$ for every $A \in \mathfrak{A}$ for every $g \in L^q(X, \mathfrak{A}, \mu)$.
(b) $\lim\limits_{n\to\infty} \int_A f_n \, d\mu = \int_A f \, d\mu$ for every $A \in \mathfrak{A}$.
(Hint: Egoroff's Theorem)

Proof. 1. Observe first of all that by 1°, Theorem 8.13 (Fatou's Lemma) and 2° we have

$$\int_X |f|^p \, d\mu = \int_X \lim_{n\to\infty} |f_n|^p \, d\mu \leq \liminf_{n\to\infty} \int_X |f_n|^p \, d\mu \leq M^p.$$

This implies that $\|f\|_p \leq M$.

  2. Let us prove (a). Let $A \in \mathfrak{A}$. Let $g \in L^q(X, \mathfrak{A}, \mu)$. Then we have $\int_A |g|^q \, d\mu \leq \int_X |g|^q \, d\mu < \infty$. Thus $|g|^q$ is μ-integrable on A. Then by Theorem 9.26 (Uniform Absolute Continuity of the Integral with Respect to the Measure), for every $\varepsilon > 0$ there exists $\delta > 0$ such that

(1) $$\int_E |g|^q \, d\mu < \varepsilon \quad \text{for every } E \subset A \text{ such that } E \in \mathfrak{A} \text{ and } \mu(E) < \delta.$$

Since $\mu(A) \leq \mu(X) < \infty$ and 1° holds, Theorem 6.12 (Egoroff) implies that there exists $E \subset A$, $E \in \mathfrak{A}$ with $\mu(E) < \delta$ such that

(2) $$\lim_{n\to\infty} f_n = f \quad \text{uniformly on } A \setminus E.$$

Now we have

(3) $$\left| \int_A f_n g \, d\mu - \int_A f g \, d\mu \right| \leq \int_A |f_n - f| |g| \, d\mu$$
$$= \int_{A \setminus E} |f_n - f| |g| \, d\mu + \int_E |f_n - f| |g| \, d\mu.$$

  Consider the first integral in the last member of (3). Now (2) implies that for $1 > 0$ there exists $N \in \mathbb{N}$ such that $|f_n - f| \leq 1$ on $A \setminus E$ for $n \geq N$. Then we have

(4) $$|f_n - f| |g| \leq |g| \quad \text{on } A \setminus E \text{ for } n \geq N.$$

Now $g \in L^q(X, \mathfrak{A}, \mu)$. Thus by Theorem 16.14 (Hölder's Inequality) we have

$$\int_X |g| \, d\mu = \int_X |g| \cdot 1 \, d\mu \leq \left\{ \int_X |g|^p \, d\mu \right\}^{\frac{1}{p}} \left\{ \int_X 1^q \, d\mu \right\}^{\frac{1}{q}} = \|g\|_p \mu(X)^{\frac{1}{q}} < \infty.$$

This shows that $|g|$, the bounding function in (4), is μ-integrable on X and hence μ-integrable on $A \setminus E$. Then by Theorem 9.20 (Dominated Convergence Theorem), we have

(5) $$\lim_{n\to\infty} \int_{A \setminus E} |f_n - f| |g| \, d\mu = \int_{A \setminus E} \lim_{n\to\infty} |f_n - f| |g| \, d\mu = \int_{A \setminus E} 0 \cdot |g| \, d\mu = 0.$$

For the second integral in the last member of (3) we have by Theorem 16.14 (Hölder)

(6) $\displaystyle\int_E |f_n - f| |g| \, d\mu \le \left\{ \int_E |f_n - f|^p \, d\mu \right\}^{\frac{1}{p}} \left\{ \int_E |g|^q \, d\mu \right\}^{\frac{1}{q}}$

$\qquad\qquad \le \| f_n - f \|_p \varepsilon^{\frac{1}{q}} \quad \text{by (1)}$

$\qquad\qquad \le \left\{ \| f_n \|_p + \| f \|_p \right\} \varepsilon^{\frac{1}{q}} \quad \text{by Theorem 16.17 (Minkowski)}$

$\qquad\qquad \le 2M \varepsilon^{\frac{1}{q}} \quad \text{by } 2° \text{ and } \mathbf{1}.$

Then by (3), (5) and (6), we have

(7) $\displaystyle \limsup_{n \to \infty} \left| \int_A f_n g \, d\mu - \int_A f g \, d\mu \right| \le \lim_{n \to \infty} \int_{A \setminus E} |f_n - f| |g| \, d\mu + 2M \varepsilon^{\frac{1}{q}} = 2M \varepsilon^{\frac{1}{q}}.$

Since this holds for every $\varepsilon > 0$ we have

$$\limsup_{n \to \infty} \left| \int_A f_n g \, d\mu - \int_A f g \, d\mu \right| = 0.$$

This then implies

$$\liminf_{n \to \infty} \left| \int_A f_n g \, d\mu - \int_A f g \, d\mu \right| = 0$$

and then

$$\lim_{n \to \infty} \left| \int_A f_n g \, d\mu - \int_A f g \, d\mu \right| = 0.$$

This then implies

$$\lim_{n \to \infty} \left\{ \int_A f_n g \, d\mu - \int_A f g \, d\mu \right\} = 0,$$

and then

$$\lim_{n \to \infty} \int_A f_n g \, d\mu = \int_A f g \, d\mu.$$

3. Let us prove (b). Let $A \in \mathfrak{A}$. Since $\mu(X) < \infty$, we have

$$\int_X (\mathbf{1}_E)^q \, d\mu = \int_X \mathbf{1}_E \, d\mu = \int_E 1 \, d\mu = \mu(E) < \infty.$$

Thus $\mathbf{1}_A \in L^q(X, \mathfrak{A}, \mu)$. Then (b) is a particular case of (a) in which $g = \mathbf{1}_A$. ∎

Prob. 16.14. Let $(X, \mathfrak{A}, \mu)$ be a σ-finite measure space and let $p \in [1, \infty)$. Let f be an element and $(f_n : n \in \mathbb{N})$ be a sequence in $L^p(X, \mathfrak{A}, \mu)$ such that $\lim_{n \to \infty} \|f_n - f\|_p = 0$.

(a) Show that for every $\varepsilon > 0$ there exists $\delta > 0$ such that for every $E \in \mathfrak{A}$ with $\mu(E) < \delta$ we have

$$\int_E |f_n|^p \, d\mu < \varepsilon \quad \text{for all } n \in \mathbb{N}.$$

(b) Show that for every $\varepsilon > 0$ there exists $A \in \mathfrak{A}$ with $\mu(A) < \infty$ such that

$$\int_{X \setminus A} |f_n|^p \, d\mu < \varepsilon \quad \text{for all } n \in \mathbb{N}.$$

Proof. 1. Let us prove (a). We have

$$|f_n| = |f_n - f + f| \le |f_n - f| + |f|$$

and then by Lemma 16.7 we have

$$|f_n|^p \le \big||f_n - f| + |f|\big|^p \le 2^p \big\{ |f_n - f|^p + |f|^p \big\}.$$

Let $E \in \mathfrak{A}$. Then by the last inequality we have

(1)
$$\begin{aligned}
\int_E |f_n|^p \, d\mu &\le 2^p \Big\{ \int_E |f_n - f|^p \, d\mu + \int_E |f|^p \, d\mu \Big\} \\
&\le 2^p \Big\{ \int_X |f_n - f|^p \, d\mu + \int_E |f|^p \, d\mu \Big\} \\
&= 2^p \Big\{ \|f_n - f\|_p^p + \int_E |f|^p \, d\mu \Big\}.
\end{aligned}$$

Let $\varepsilon > 0$ be arbitrarily given. Since $\lim_{n \to \infty} \|f_n - f\|_p = 0$, there exists $N \in \mathbb{N}$ such that

(2)
$$\|f_n - f\|_p < \Big\{ \frac{1}{2} \frac{\varepsilon}{2^p} \Big\}^{\frac{1}{p}} \quad \text{for } n > N.$$

Since $f \in L^p(X, \mathfrak{A}, \mu)$, $|f|^p$ is μ-integrable on X. Then by Theorem 9.26 (Uniform Absolute Continuity of the Integral with Respect to the Measure), there exists $\delta_0 > 0$ such that

(3)
$$\int_E |f|^p \, d\mu < \frac{1}{2} \frac{\varepsilon}{2^p} \quad \text{for every } E \in \mathfrak{A} \text{ with } \mu(E) < \delta_0.$$

Applying (2) and (3) to (1), we have

(4)
$$\int_E |f_n|^p \, d\mu < \frac{\varepsilon}{2} + \frac{\varepsilon}{2} = \varepsilon \quad \text{for every } E \in \mathfrak{A} \text{ with } \mu(E) < \delta_0 \text{ for } n > N.$$

For $n = 1, \ldots, N$, since $f_n \in L^p(X, \mathfrak{A}, \mu)$, $|f_p|^p$ is μ-integrable on X. Then according to Theorem 9.26 there exists $\delta_n > 0$ such that

(5)
$$\int_E |f_n|^p \, d\mu < \varepsilon \quad \text{for every } E \in \mathfrak{A} \text{ with } \mu(E) < \delta_n \text{ for } n = 1, \ldots, N.$$

Let $\delta = \min\{\delta_0, \delta_1, \ldots, \delta_N\}$. Then summarizing (4) and (5) we have $\int_E |f_n|^p \, d\mu < \varepsilon$ for every $E \in \mathfrak{A}$ with $\mu(E) < \delta$ for $n \in \mathbb{N}$. This proves (a).

2. Let us prove (b). Let h be an extended complex-valued $\mathfrak{A}$-measurable function on X. Then since $(X, \mathfrak{A}, \mu)$ is a σ-finite measure space, μ-integrability of h on X implies that for every $\varepsilon > 0$ there exists $A \in \mathfrak{A}$ with $\mu(A) < \infty$ such that

$$(6) \qquad \int_{X \setminus A} |h| \, d\mu < \varepsilon.$$

Since (1) holds for every $E \in \mathfrak{A}$, it holds in particular for $X \setminus E$ and thus we have

$$(7) \qquad \int_{X \setminus E} |f_n|^p \, d\mu \leq 2^p \left\{ \|f_n - f\|_p^p + \int_{X \setminus E} |f|^p \, d\mu \right\}.$$

Let $\varepsilon > 0$ be arbitrarily given. Then since $\lim_{n \to \infty} \|f_n - f\|_p = 0$, there exists $N \in \mathbb{N}$ such that

$$(8) \qquad \|f_n - f\|_p < \left\{ \frac{1}{2} \frac{\varepsilon}{2^p} \right\}^{\frac{1}{p}} \quad \text{for } n > N.$$

Since $f \in L^p(X, \mathfrak{A}, \mu)$, $|f|^p$ is μ-integrable on X. Then according to (6) there exists $E_0 \in \mathfrak{A}$ with $\mu(E_0) < \infty$ such that

$$(9) \qquad \int_{X \setminus E_0} |f|^p \, d\mu < \frac{1}{2} \frac{\varepsilon}{2^p}.$$

Substituting (8) and (9) in (7), we have

$$(10) \qquad \int_{X \setminus E_0} |f|^p \, d\mu < \frac{\varepsilon}{2} + \frac{\varepsilon}{2} = \varepsilon \quad \text{for } n > N.$$

For $n = 1, \ldots, N$, since $|f_n|^p$ is μ-integrable on X, according to (6) there exists $E_n \in \mathfrak{A}$ with $\mu(E_n) < \infty$ such that

$$(11) \qquad \int_{X \setminus E_n} |f_n|^p \, d\mu < \varepsilon \quad \text{for } n = 1, \ldots, N.$$

Let $A = \bigcup_{n=0}^N E_n$. Then $\mu(A) < \infty$. Since $E_n \subset A$ for $n = 0, \ldots, N$ we have

$$(12) \qquad \int_{X \setminus A} |f_n|^p \, d\mu \leq \int_{X \setminus E_n} |f_n|^p \, d\mu < \varepsilon \quad \text{for } n = 0, \ldots, N.$$

By (10) and (12), we have

$$\int_{X \setminus A} |f_n|^p \, d\mu < \varepsilon \quad \text{for all } n \in \mathbb{N}.$$

This proves (b). ∎

Prob. 16.15. Prove the following theorem on the convergence of a sequence of integrals:

Theorem. *Let $(X, \mathfrak{A}, \mu)$ be a σ-finite measure space and let $p \in [1, \infty)$. Let $(f_n : n \in \mathbb{N})$ be a sequence in $L^p(X, \mathfrak{A}, \mu)$ satisfying the following conditions:*

1° *for every $\varepsilon > 0$ there exists $\delta > 0$ such that for every $E \in \mathfrak{A}$ with $\mu(E) < \delta$ we have*

$$\int_E |f_n|^p \, d\mu < \varepsilon \quad \text{for all } n \in \mathbb{N}.$$

2° *for every $\varepsilon > 0$ there exists $A \in \mathfrak{A}$ with $\mu(A) < \infty$ such that*

$$\int_{X \setminus A} |f_n|^p \, d\mu < \varepsilon \quad \text{for all } n \in \mathbb{N}.$$

Let f be an extended complex-valued $\mathfrak{A}$-measurable function on X such that $|f| < \infty$, μ-a.e. on X and $\lim_{n \to \infty} f_n = f$, μ-a.e. on X. Then $f \in L^p(X, \mathfrak{A}, \mu)$ and moreover we have $\lim_{n \to \infty} \|f_n - f\|_p = 0$ and in particular $\lim_{n \to \infty} \int_X |f_n|^p \, d\mu = \int_X |f|^p \, d\mu$.

Proof. 1. Let us show first that $f \in L^p(X, \mathfrak{A}, \mu)$. Thus we are to show that $\int_X |f|^p \, d\mu < \infty$. Let $\varepsilon > 0$ be arbitrarily given. Now according to 1°, there exists $\delta > 0$ such that

$$(1) \qquad \int_E |f_n|^p \, d\mu < \varepsilon \quad \text{for every } E \in \mathfrak{A} \text{ with } \mu(E) < \delta \text{ for all } n \in \mathbb{N}.$$

Then according to 2°, there exists $A \in \mathfrak{A}$ with $\mu(A) < \infty$ such that

$$(2) \qquad \int_{X \setminus A} |f_n|^p \, d\mu < \varepsilon \quad \text{for all } n \in \mathbb{N}.$$

Since $\lim_{n \to \infty} f_n = f$, μ-a.e. on X, we have $\lim_{n \to \infty} f_n = f$, μ-a.e. on A. Since $\mu(A) < \infty$, according to Theorem 6.12 (Egoroff) for our $\delta > 0$ there exists $B \in \mathfrak{A}$, $B \subset A$, with $\mu(B) < \delta$ such that $\lim_{n \to \infty} f_n = f$ uniformly on $A \setminus B$. Thus for every $\eta > 0$ there exists $N \in \mathbb{N}$ such that $|f_n - f| < \eta$ on $A \setminus B$ for $n \geq N$ and then

$$\int_{A \setminus B} |f_n - f|^p \, d\mu \leq \eta^p \mu(A \setminus B) \quad \text{for } n \geq N.$$

Since $\mu(A \setminus B) < \infty$, the arbitrariness of $\eta > 0$ in the last estimate implies that we have

$$(3) \qquad \lim_{n \to \infty} \int_{A \setminus B} |f_n - f|^p \, d\mu = 0.$$

Now $A \setminus B \in \mathfrak{A}$. If we let $\mathfrak{A}|_{A \setminus B} = \{E \cap (A \setminus B) : E \in \mathfrak{A}\}$ then $\mathfrak{A}|_{A \setminus B}$ is a σ-algebra of subsets of $A \setminus B$ and $(A \setminus B, \mathfrak{A}|_{A \setminus B}, \mu)$ is a measure space. Consider the Banach space $L^p(A \setminus B, \mathfrak{A}|_{A \setminus B}, \mu)$. Since $f_n \in L^p(X, \mathfrak{A}, \mu)$, we have $\int_{A \setminus B} |f_n|^p \, d\mu \leq \int_X |f_n|^p \, d\mu < \infty$. Thus $f_n \in L^p(A \setminus B, \mathfrak{A}|_{A \setminus B}, \mu)$. By (3) we have $\int_{A \setminus B} |f_n - f|^p \, d\mu < \infty$ for sufficiently large $n \in \mathbb{N}$ and thus $f_n - f \in L^p(A \setminus B, \mathfrak{A}|_{A \setminus B}, \mu)$ for sufficiently large

$n \in \mathbb{N}$. Then $f = (f - f_n) + f_n \in L^p(A \setminus B, \mathfrak{A}|_{A \setminus B}, \mu)$ since $L^p(A \setminus B, \mathfrak{A}|_{A \setminus B}, \mu)$ is a linear space. Thus we have

(4) $$\int_{A \setminus B} |f|^p \, d\mu < \infty.$$

Since $\mu(B) < \delta$, we have $\int_B |f_n|^p \, d\mu < \varepsilon$ for all $n \in N$ according to (1). Then by Theorem 8.13 (Fatou's Lemma), we have

(5) $$\int_B |f|^p \, d\mu \le \liminf_{n \to \infty} \int_B |f_n|^p \, d\mu \le \varepsilon.$$

From (2), we have by Theorem 8.13 again

(6) $$\int_{X \setminus A} |f|^p \, d\mu \le \liminf_{n \to \infty} \int_{X \setminus A} |f_n|^p \, d\mu \le \varepsilon.$$

Now $\{A \setminus B, B, X \setminus A\}$ is a disjoint collection in $\mathfrak{A}$ and the union of the three sets is equal to X. Thus by (4), (5) and (6) we have

$$\int_X |f|^p \, d\mu = \int_{A \setminus B} |f|^p \, d\mu + \int_B |f|^p \, d\mu + \int_{X \setminus A} |f|^p \, d\mu < \infty.$$

This shows that $f \in L^p(X, \mathfrak{A}, \mu)$.

2. Let us show that $\lim_{n \to \infty} \|f_n - f\|_p = 0$. For the sake of convenience in subsequent argument, let us write f_0 for f. Let $\varepsilon > 0$ be arbitrarily given. Since $f_0 \in L^p(X, \mathfrak{A}, \mu)$ as we showed above, $|f_0|^p$ is μ-integrable on X. Then by Theorem 9.26 (Uniform Absolute Continuity of the Integral with Respect to the Measure), there exists $\delta_0 > 0$ such that

(7) $$\int_E |f_0|^p \, d\mu < \varepsilon \quad \text{for every } E \in \mathfrak{A} \text{ with } \mu(E) < \delta_0.$$

With $\delta > 0$ in (1), let $\delta' = \min\{\delta, \delta_0\}$. Then combining (1) and (7) we have

(8) $$\int_E |f_n|^p \, d\mu < \varepsilon \quad \text{for every } E \in \mathfrak{A} \text{ with } \mu(E) < \delta' \text{ for all } n \in \mathbb{Z}_+.$$

Since $\int_X |f_0|^p \, d\mu < \infty$ and $(X, \mathfrak{A}, \mu)$ is a σ-finite measure space, there exists $A_0 \in \mathfrak{A}$ with $\mu(A_0) < \infty$ such that

(9) $$\int_{X \setminus A_0} |f_0|^p \, d\mu < \varepsilon.$$

With A in (2), let $A' = A \cup A_0$. Then $\mu(A') < \infty$ and by (2) and (9) we have

(10) $$\int_{X \setminus A'} |f_n|^p \, d\mu < \varepsilon \quad \text{for all } n \in \mathbb{Z}_+.$$

Now $\lim_{n \to \infty} F_n = f_0, \mu$-a.e. on X and hence $\lim_{n \to \infty} F_n = f_0, \mu$-a.e. on A'. Since $\mu(A') < \infty$, by Theorem 6.12 (Egoroff) for our $\delta' > 0$ there exists $B' \in \mathfrak{A}$, $B' \subset A'$ with $\mu(B') < \delta'$

such that $\lim_{n\to\infty} f_n = f_0$ uniformly on $A' \setminus B'$. Then by the same argument as the one employed to obtain (3) we have

(11)
$$\lim_{n\to\infty} \int_{A'\setminus B'} |f_n - f_0|^p \, d\mu = 0.$$

Since $\mu(B') < \delta'$, we have by (8)

(12)
$$\int_{B'} |f_n|^p \, d\mu < \varepsilon \quad \text{for all } n \in \mathbb{Z}_+.$$

Since $\{A' \setminus B', B', X \setminus A'\}$ is a disjoint collection in $\mathfrak{A}$ and the union of the three sets is equal to X, we have

(13)
$$\int_X |f_n - f_0|^p \, d\mu = \int_{A'\setminus B'} |f_n - f_0|^p \, d\mu + \int_{B'} |f_n - f_0|^p \, d\mu + \int_{X\setminus A'} |f_n - f_0|^p \, d\mu$$
$$\leq \int_{A'\setminus B'} |f_n - f_0|^p \, d\mu + 2^p \left\{ \int_{B'} |f_n|^p \, d\mu + \int_{B'} |f_0|^p \, d\mu \right\}$$
$$+ 2^p \left\{ \int_{X\setminus A'} |f_n|^p \, d\mu + \int_{X\setminus A'} |f_0|^p \, d\mu \right\},$$

where the inequality is by means of Lemma 16.7. Then applying (11), (12) and (10) to (13), we have

$$\limsup_{n\to\infty} \int_X |f_n - f_0|^p \, d\mu \leq 0 + 2^p 2\varepsilon + 2^p 2\varepsilon = 2^{p+2}\varepsilon.$$

Since this holds for every $\varepsilon > 0$ we have $\limsup_{n\to\infty} \int_X |f_n - f_0|^p \, d\mu = 0$. This then implies that we have $\liminf_{n\to\infty} \int_X |f_n - f_0|^p \, d\mu = 0$ and then $\lim_{n\to\infty} \int_X |f_n - f_0|^p \, d\mu = 0$, that is, $\lim_{n\to\infty} \|f_n - f_0\|_p^p = 0$. This implies that $\lim_{n\to\infty} \|f_n - f_0\|_p = 0$. Then according to Theorem 16.25, $\lim_{n\to\infty} \|f_n - f_0\|_p = 0$ implies $\lim_{n\to\infty} \|f_n\|_p = \|f_0\|_p$, that is, $\lim_{n\to\infty} \left\{ \int_X |f_n|^p \, d\mu \right\}^{\frac{1}{p}} = \left\{ \int_X |f_0|^p \, d\mu \right\}^{\frac{1}{p}}$ and this implies $\lim_{n\to\infty} \int_X |f_n|^p \, d\mu = \int_X |f_0|^p \, d\mu$. This completes the proof. ∎

Prob. 16.16. Definition. (**Equicontinuity**) *Let* $\{f_\alpha : \alpha \in A\}$ *be a collection of real-valued functions on an interval* $I \subset \mathbb{R}$. *We say that the collection* $\{f_\alpha : \alpha \in A\}$ *is equicontinuous on* I *if for every* $\varepsilon > 0$ *there exists* $\delta > 0$ *such that* $|f_\alpha(x') - f_\alpha(x'')| < \varepsilon$ *for all* $\alpha \in A$ *when* $x', x'' \in I$ *are such that* $|x' - x''| < \delta$.
Prove the following theorem:

Theorem. *Let* $\{f_\alpha : \alpha \in A\}$ *be a collection of real-valued functions on* $[a, b] \subset \mathbb{R}$ *such that* f_α *is differentiable and the derivative* f'_α *is bounded on* $[a, b]$ *and* $\sup_{\alpha \in A} \int_{[a,b]} |f'_\alpha|^2 \, d\mu_L \leq M$ *where* $M > 0$. *Then* $\{f_\alpha : \alpha \in A\}$ *is equicontinuous on* $[a, b]$.

Proof. Let $[x', x''] \subset [a, b]$. Then boundedness of f'_α on $[a, b]$ implies boundedness of f'_α on $[x', x'']$. Then according to Theorem 13.23, we have

$$(1) \qquad f_\alpha(x'') - f_\alpha(x') = \int_{[x',x'']} f'_\alpha(x) \, \mu_L(dx),$$

and then applying Theorem 16.14 (Hölder's Inequality) we have

$$(2) \quad |f_\alpha(x'') - f_\alpha(x')| \leq \int_{[x',x'']} |f'_\alpha(x)| \, \mu_L(dx) = \int_{[x',x'']} 1 \cdot |f'_\alpha(x)| \, \mu_L(dx)$$

$$\leq \left\{ \int_{[x',x'']} 1^2 \, \mu_L(dx) \right\}^{\frac{1}{2}} \left\{ \int_{[x',x'']} |f'_\alpha(x)|^2 \, \mu_L(dx) \right\}^{\frac{1}{2}}$$

$$\leq \sqrt{x'' - x'} \cdot \left\{ \int_{[a,b]} |f'_\alpha(x)|^2 \, \mu_L(dx) \right\}^{\frac{1}{2}}$$

$$\leq \sqrt{x'' - x'} \cdot M \quad \text{for all } \alpha \in A.$$

Let $\varepsilon > 0$ be arbitrarily given. Let $\delta = \frac{\varepsilon^2}{M^2}$. Then for $[x', x''] \subset [a, b]$ such that $|x' - x''| < \delta$ we have

$$(3) \qquad |f_\alpha(x'') - f_\alpha(x')| \leq M \cdot \sqrt{x'' - x'} < M\sqrt{\delta} = M \frac{\varepsilon}{M} = \varepsilon \quad \text{for all } \alpha \in A.$$

This proves the equicontinuity of $\{f_\alpha : \alpha \in A\}$ on $[a, b]$. ∎

Prob. 16.17. Given a measure space $(X, \mathfrak{A}, \mu)$. Let $D \in \mathfrak{A}$ and let $(f_n : n \in \mathbb{N})$ and f be extended real-valued $\mathfrak{A}$-measurable and μ-integrable functions on D such that

$1°\quad \lim_{n\to\infty} f_n = f$ a.e. on D,

$2°\quad \lim_{n\to\infty} \int_D |f_n|\, d\mu = \int_D |f|\, d\mu$.

Show that for every $E \in \mathfrak{A}$ such that $E \subset D$ we have

(a) $\lim_{n\to\infty} \int_E |f_n|\, d\mu = \int_E |f|\, d\mu$,

(b) $\lim_{n\to\infty} \int_E f_n\, d\mu = \int_E f\, d\mu$.

Proof. 1. Let us prove (a). Observe that $D \in \mathfrak{A}$ implies that $\mathfrak{A}|_D = \{A \cap D : A \in \mathfrak{A}\}$ is a σ-algebra of subsets of D. Consider the measure space $(D, \mathfrak{A}|_D, \mu)$ and the Banach space $L^1(D, \mathfrak{A}|_D, \mu)$. Then according to Theorem 16.28, assumption of conditions $1°$ and $2°$ implies

(1) $$\lim_{n\to\infty} \|f_n - f\|_1 = 0, \quad \text{that is,} \quad \lim_{n\to\infty} \int_D |f_n - f|\, d\mu = 0.$$

Let $E \in \mathfrak{A}$ and $E \subset D$. Consider the measure space $(E, \mathfrak{A}|_E, \mu)$ and the Banach space $L^1(E, \mathfrak{A}|_E, \mu)$. Then $E \subset D$ and (1) imply

(2) $$\lim_{n\to\infty} \int_E |f_n - f|\, d\mu \leq \lim_{n\to\infty} \int_D |f_n - f|\, d\mu = 0.$$

Thus we have $\lim_{n\to\infty} \|f_n - f\|_1 = 0$ in the Banach space $L^1(E, \mathfrak{A}|_E, \mu)$. According to Theorem 16.25 this implies

(3) $$\lim_{n\to\infty} \|f_n\|_1 = \|f\|_1, \quad \text{that is,} \quad \lim_{n\to\infty} \int_E |f_n|\, d\mu = \int_E |f|\, d\mu.$$

This proves (a).

2. To prove (b), observe that

(4) $$\left| \int_E f_n\, d\mu - \int_E f\, d\mu \right| = \left| \int_E (f_n - f)\, d\mu \right| \leq \int_E |f_n - f|\, d\mu = \|f_n - f\|_1.$$

We showed above that $\lim_{n\to\infty} \|f_n - f\|_1 = 0$ in the Banach space $L^1(E, \mathfrak{A}|_E, \mu)$. Applying this to (4) we have

$$\lim_{n\to\infty} \left| \int_E f_n\, d\mu - \int_E f\, d\mu \right| = 0.$$

This implies that

$$\lim_{n\to\infty} \int_E f_n\, d\mu = \int_E f\, d\mu.$$

This proves (b). ∎

Prob. 16.18. If condition 2° in Prob. 16.17 is replaced by

3° $\lim_{n\to\infty} \int_D f_n \, d\mu = \int_D f \, d\mu$,

then **(a)** and **(b)** in Prob. 16.17 do not hold. Construct an example to show this.

Proof. Consider the measure space $(\mathbb{R}, \mathfrak{M}_L, \mu_L)$. Let $D = (0, 1)$ and let $(f_n : n \in \mathbb{N})$ and f be real-valued $\mathfrak{M}_L$-measurable and μ_L-integrable functions on D defined by setting

$$f_n(t) = \begin{cases} n & \text{for } t \in \left(0, \frac{1}{n}\right) \\ 0 & \text{for } t \in \left[\frac{1}{n}, 1 - \frac{1}{n}\right] \\ -n & \text{for } t \in \left(1 - \frac{1}{n}, 1\right) \end{cases}$$

and

$$f(t) = 0 \quad \text{for } t \in (0, 1).$$

Then we have

$$\lim_{n\to\infty} f_n(t) = f(t) \quad \text{for } t \in D,$$

$$\lim_{n\to\infty} \int_D f_n \, d\mu_L = \lim_{n\to\infty} 0 = 0 = \int_D f \, d\mu_L,$$

$$\lim_{n\to\infty} \int_D f_n \, d\mu_L = \lim_{n\to\infty} 2 = 2 \neq 0 = \int_D |f| \, d\mu_L.$$

This verifies that condition 1° of Prob. 16.17 and condition 3° are satisfied but condition 2° of Prob. 16.17 is not satisfied.

Now let $E = \left(0, \frac{1}{2}\right) \subset D$. Then we have

$$\lim_{n\to\infty} \int_E |f_n| \, d\mu_L = \lim_{n\to\infty} 1 = 1 \neq 0 = \int_E |f| \, d\mu_L,$$

$$\lim_{n\to\infty} \int_E f_n \, d\mu_L = \lim_{n\to\infty} 1 = 1 \neq 0 = \int_E f \, d\mu_L.$$

This shows that **(a)** and **(b)** in Prob. 16.17 do not hold here. ∎

Prob. 16.19. Consider a Banach space $L^p(X, \mathfrak{A}, \mu)$ where $p \in (0, \infty)$. Let f be an element and $(f_n : n \in \mathbb{N})$ be a sequence in $L^p(X, \mathfrak{A}, \mu)$ such that

1° $\lim_{n \to \infty} f_n = f$ a.e. on X,

2° $\lim_{n \to \infty} \|f_n\|_p = \|f\|_p$.

According to Theorem 16.28, assumption of 1° and 2° implies that $\lim_{n \to \infty} \|f_n - f\|_p = 0$.
Show that this is not true for a Banach space $L^\infty(X, \mathfrak{A}, \mu)$, that is, $\lim_{n \to \infty} f_n = f$ a.e. on X,
and $\lim_{n \to \infty} \|f_n\|_\infty = \|f\|_\infty$ do not imply $\lim_{n \to \infty} \|f_n - f\|_\infty = 0$.

Proof. Consider $L^\infty(\mathbb{R}, \mathfrak{M}_L, \mu_L)$. Let f be an element and $(f_n : n \in \mathbb{N})$ be a sequence in $L^\infty(\mathbb{R}, \mathfrak{M}_L, \mu_L)$ defined by setting

$$f(x) = \begin{cases} 0 & \text{for } x \in (-\infty, 0) \\ 1 & \text{for } x \in [0, \infty) \end{cases}$$

and

$$f_n(x) = \begin{cases} 0 & \text{for } x \in (-\infty, 0] \\ 1 & \text{for } x \in [\frac{1}{n}, \infty) \\ \text{linear} & \text{on } [0, \frac{1}{n}]. \end{cases}$$

We have $\|f\|_\infty = 1 < \infty$ and $\|f_n\|_\infty = 1 < \infty$, showing that $f, f_n \in L^\infty(\mathbb{R}, \mathfrak{M}_L, \mu_L)$.
Now we have $\lim_{n \to \infty} f_n = f$ on $\mathbb{R}$ and $\lim_{n \to \infty} \|f_n\|_\infty = \lim_{n \to \infty} 1 = 1 = \|f\|_\infty$.
Let us show that $\lim_{n \to \infty} \|f_n - f\|_\infty = 0$ does not hold. Observe that

$$(f - f_n)(x) = \begin{cases} 0 & \text{for } x \in (-\infty, 0) \\ 1 & \text{for } x = 0 \\ 0 & \text{for } x \in [\frac{1}{n}, \infty) \\ \text{linear} & \text{on } [0, \frac{1}{n}]. \end{cases}$$

This shows that $\|f_n - f\|_\infty = \|f - f_n\|_\infty = 1$ for every $n \in \mathbb{N}$. Then we have

$$\lim_{n \to \infty} \|f_n - f\|_\infty = \lim_{n \to \infty} 1 = 1 \neq 0.$$

This completes the proof. ∎

Prob. 16.20. Consider a real-valued function $f(x) = e^{-x^2/2}$ for $x \in \mathbb{R}$.
Show that $f \in L^p(\mathbb{R}, \mathfrak{M}_L, \mu_L)$ for every $p \in (0, \infty]$ and moreover $\lim_{p \to \infty} \|f\|_p = \|f\|_\infty$.

Proof. 1. Our function f is essentially bounded on $\mathbb{R}$ and the infimum of its essential
bounds is equal to 1. Thus $f \in L^\infty(\mathbb{R}, \mathfrak{M}_L, \mu_L)$ and $\|f\|_\infty = 1$.
  2. Consider the case $p \in (0, \infty)$. We have

$$(1) \qquad \int_{\mathbb{R}} f(x) \, \mu_L(dx) = \int_{-\infty}^{\infty} e^{-\frac{x^2}{2}} \, dx = \sqrt{2\pi}.$$

Now we have

$$(2) \qquad \int_{\mathbb{R}} |f(x)|^p \, \mu_L(dx) = \int_{\mathbb{R}} e^{-p\frac{x^2}{2}} \, \mu_L(dx).$$

Let $u = \sqrt{p}\,x$. Then $x = p^{-1/2}u$ and $dx = p^{-1/2}\,du$. Then we have

$$\int_{\mathbb{R}} |f(x)|^p \, \mu_L(dx) = \left\{ \int_{\mathbb{R}} e^{-\frac{u^2}{2}} \, \mu_L(du) \right\} p^{-\frac{1}{2}}$$
$$= \sqrt{2\pi} \, p^{-\frac{1}{2}} \quad \text{by (1)}.$$

Thus we have

$$(3) \qquad \|f\|_p = \left\{ \int_{\mathbb{R}} |f(x)|^p \, \mu_L(dx) \right\}^{\frac{1}{p}} = \left\{ \sqrt{2\pi} \, p^{-\frac{1}{2}} \right\}^{\frac{1}{p}} = (2\pi)^{\frac{1}{2p}} p^{-\frac{1}{2p}} < \infty.$$

This shows that $f \in L^p(\mathbb{R}, \mathfrak{M}_L, \mu_L)$ for every $p \in (0, \infty)$.
  3. Let us show that $\lim_{p \to \infty} \|f\|_p = \|f\|_\infty$. We have shown above that $\|f\|_p = (2\pi)^{\frac{1}{2p}} p^{-\frac{1}{2p}}$. We have $\lim_{p \to \infty} (2\pi)^{\frac{1}{2p}} = (2\pi)^0 = 1$. Let us investigate $\lim_{p \to \infty} p^{-\frac{1}{2p}}$. We
have

$$\lim_{p \to \infty} \log p^{-\frac{1}{2p}} = \lim_{p \to \infty} \left(-\frac{1}{2p} \log p \right) = -\frac{1}{2} \lim_{p \to \infty} \frac{\log p}{p} = 0.$$

This implies that $\lim_{p \to \infty} p^{-\frac{1}{2p}} = 1$. Then we have $\lim_{p \to \infty} \|f\|_p = \lim_{p \to \infty} (2\pi)^{\frac{1}{2p}} p^{-\frac{1}{2p}} = 1 = \|f\|_\infty$. This completes the proof. ∎

Prob. 16.21. Construct a measure space $(X, \mathfrak{A}, \mu)$ and a real-valued $\mathfrak{A}$-measurable function f on X such that $f \in L^p(X, \mathfrak{A}, \mu)$ for every $p \in (0, \infty)$ but $f \notin L_\infty(X, \mathfrak{A}, \mu)$.

Proof. 1. Consider the measure space $(\mathbb{R}, \mathfrak{M}_L, \mu_L)$. Now $[0, \infty) \in \mathfrak{M}_L$ and this implies that $\mathfrak{M}_L|_{[0,\infty)} = \{E \cap [0, \infty) : E \in \mathfrak{M}_L\}$ is a σ-algebra of subsets of $[0, \infty)$ and thus we have a measure space $([0, \infty), \mathfrak{M}_L|_{[0,\infty)}, \mu_L)$.

2. Consider the measure space $([0, \infty), \mathfrak{M}_L|_{[0,\infty)}, \mu_L)$ defined above. Let g be a real-valued function on $[0, \infty)$ defined by setting

(1) $$g(x) = e^{-\frac{x^2}{2}} \quad \text{for } x \in [0, \infty).$$

We have

(2) $$\int_{[0,\infty)} g(x)\,\mu_L(dx) = \int_{[0,\infty)} e^{-\frac{x^2}{2}}\,\mu_L(dx) = \tfrac{1}{2}\sqrt{2\pi} = \sqrt{\tfrac{\pi}{2}}.$$

Let us define a set function λ on $\mathfrak{M}_L|_{[0,\infty)}$ by setting

(3) $$\lambda(E) = \int_E g(x)\,\mu_L(dx) \quad \text{for } E \in \mathfrak{M}_L|_{[0,\infty)}.$$

By Proposition 10.3, λ is a measure on the measurable space $([0, \infty), \mathfrak{M}_L|_{[0,\infty)})$ and thus we have a measure space $([0, \infty), \mathfrak{M}_L|_{[0,\infty)}, \lambda)$. By Definition 11.1, g is the Radon-Nikodym derivative of λ with respect to μ_L. Then according to Theorem 11.21, for every extended real-valued $\mathfrak{M}_L|_{[0,\infty)}$-measurable function f on $[0, \infty)$ such that $\int_{[0,\infty)} f\,d\lambda$ exists, we have

(4) $$\int_{[0,\infty)} f(x)\,\lambda(dx) = \int_{[0,\infty)} f(x)g(x)\,\mu_L(dx).$$

3. In the measure space $([0, \infty), \mathfrak{M}_L|_{[0,\infty)}, \lambda)$, let f be a real-valued function on $[0, \infty)$ defined by setting

(5) $$f(x) = x \quad \text{for } x \in [0, \infty).$$

Then f is $\mathfrak{M}_L|_{[0,\infty)}$-measurable on $[0, \infty)$ but not essentially bounded on $[0, \infty)$. Thus we have $\|f\|_\infty = \infty$ and $f \notin L_\infty([0, \infty), \mathfrak{M}_L|_{[0,\infty)}, \lambda)$.

4. Let us show that $\|f\|_p < \infty$ and therefore $f \in L^p([0, \infty), \mathfrak{M}_L|_{[0,\infty)}, \lambda)$ for $p \in (0, \infty)$. Now we have

(6) $$\|f\|_p^p = \int_{[0,\infty)} |f(x)|^p\,\lambda(dx) = \int_{[0,\infty)} x^p\,\lambda(dx)$$
$$= \int_{[0,\infty)} x^p e^{-\frac{x^2}{2}}\,\mu_L(dx) \quad \text{by (4) and (1).}$$

Let us estimate the last integral. We have $x^p < e^x$ for $x \in [0, \infty)$. Thus we have

(7) $$\int_{[0,\infty)} x^p e^{-\frac{x^2}{2}}\,\mu_L(dx) \le \int_{[0,\infty)} e^x e^{-\frac{x^2}{2}}\,\mu_L(dx)$$
$$= \int_{[0,\infty)} e^{-\frac{x^2}{2}+x}\,\mu_L(dx).$$

Now we have $-\frac{x^2}{2} + x = -\frac{1}{2}(x-1)^2 + \frac{1}{2}$ and then $e^{-\frac{x^2}{2}+x} = e^{-\frac{1}{2}(x-1)^2}e^{\frac{1}{2}}$. Then

$$
(8) \qquad \int_{[0,\infty)} e^{-\frac{x^2}{2}+x} \mu_L(dx) = e^{\frac{1}{2}} \int_{[0,\infty)} e^{-\frac{1}{2}(x-1)^2} \mu_L(dx)
$$

$$
\leq e^{\frac{1}{2}} \int_{\mathbb{R}} e^{-\frac{1}{2}(x-1)^2} \mu_L(dx)
$$

$$
= e^{\frac{1}{2}} \int_{\mathbb{R}} e^{-\frac{1}{2}x^2} \mu_L(dx)
$$

$$
= e^{\frac{1}{2}}\sqrt{2\pi} < \infty,
$$

where the second equality is by Theorem 9.31 (Translation Invariance of the Lebesgue Integral on $\mathbb{R}$. Substituting (7) and (8) in (6), we have $\|f\|_p^p < \infty$ and then $\|f\|_p < \infty$ for every $p \in (0, \infty)$. This completes the proof that $f \in L^p([0, \infty), \mathfrak{M}_L|_{[0,\infty)}, \lambda)$ for $p \in (0, \infty)$. ∎

Prob. 16.22. Let $(X, \mathfrak{A}, \mu)$ be a σ-finite measure space with $\mu(X) = \infty$.
(a) Show that there exists a disjoint sequence $(E_n : n \in \mathbb{N})$ in $\mathfrak{A}$ such that $\bigcup_{n\in\mathbb{N}} E_n = X$ and $\mu(E_n) \in [1, \infty)$ for every $n \in \mathbb{N}$.
(b) Show that there exists a real-valued $\mathfrak{A}$-measurable function f on X such that $f \notin L^1(X, \mathfrak{A}, \mu)$ and $f \in L^p(X, \mathfrak{A}, \mu)$ for all $p \in (1, \infty]$.

Proof. 1. Since $(X, \mathfrak{A}, \mu)$ is a σ-finite measure space, there exists a disjoint sequence $(F_k : k \in \mathbb{N})$ in $\mathfrak{A}$ such that $\bigcup_{k\in\mathbb{N}} F_k = X$ and $\mu(F_k) < \infty$ for every $k \in \mathbb{N}$. We have $\sum_{k\in\mathbb{N}} \mu(F_k) = \mu(\bigcup_{k\in\mathbb{N}} F_k) = \mu(X) = \infty$. Thus there exists $n_1 \in \mathbb{N}$ such that $\sum_{k=1}^{n_1} \mu(F_k) \in [1, \infty)$. Let $E_1 = \bigcup_{k=1}^{n_1} F_k$. Then $\mu(E_1) = \sum_{k=1}^{n_1} \mu(F_k) \in [1, \infty)$. Since $\sum_{k\geq n_1+1} \mu(F_k) = \infty$, there exists $n_2 \in \mathbb{N}$ such that $\sum_{k=n_1+1}^{n_2} \mu(F_k) \in [1, \infty)$. Let $E_2 = \bigcup_{k=n_1+1}^{n_2} F_k$. Then $\mu(E_2) = \sum_{k=n_1+1}^{n_2} \mu(F_k) \in [1, \infty)$. Iterating this process indefinitely, we obtain a disjoint sequence $(E_n : n \in \mathbb{N})$ in $\mathfrak{A}$ such that $\mu(E_n) \in [1, \infty)$ for every $n \in \mathbb{N}$. Moreover since $\bigcup_{k\in\mathbb{N}} F_k = X$ and since every F_k is a subset of some E_n, we have $\bigcup_{n\in\mathbb{N}} E_n = X$.

2. We have shown above the existence of a disjoint sequence $(E_n : n \in \mathbb{N})$ in $\mathfrak{A}$ such that $\bigcup_{n\in\mathbb{N}} E_n = X$ and $\mu(E_n) \in [1, \infty)$ for every $n \in \mathbb{N}$. Let us define a real-valued $\mathfrak{A}$-measurable function f on X by setting

$$(1) \qquad f(x) = \frac{1}{\mu(E_n)} \frac{1}{n} \quad \text{for } x \in E_n \text{ for } n \in \mathbb{N}.$$

Then we have

$$(2) \qquad \int_X |f| \, d\mu = \sum_{n\in\mathbb{N}} \int_{E_n} |f| \, d\mu = \sum_{n\in\mathbb{N}} \frac{1}{\mu(E_n)} \frac{1}{n} \mu(E_n) = \sum_{n\in\mathbb{N}} \frac{1}{n} = \infty.$$

Thus $f \notin L^1(X, \mathfrak{A}, \mu)$.
For $p \in (1, \infty)$, we have

$$(3) \qquad \int_X |f|^p \, d\mu = \sum_{n\in\mathbb{N}} \int_{E_n} |f|^p \, d\mu = \sum_{n\in\mathbb{N}} \frac{1}{\mu(E_n)^p} \frac{1}{n^p} \mu(E_n) = \sum_{n\in\mathbb{N}} \frac{1}{\mu(E_n)^{p-1}} \frac{1}{n^p}.$$

Since $p \in (1, \infty)$ and $p - 1 > 0$ and $\mu(E_n) \geq 1$, we have $\mu(E_n)^{p-1} \geq 1$ and then $\frac{1}{\mu(E_n)^{p-1}} \leq 1$. Substituting this in (3), we have

$$(4) \qquad \int_X |f|^p \, d\mu \leq \sum_{n\in\mathbb{N}} \frac{1}{n^p} < \infty.$$

This shows that $f \in L^p(X, \mathfrak{A}, \mu)$ for all $p \in (1, \infty)$.
Finally, since $\mu(E_n) \geq 1$ we have $\frac{1}{\mu(E_n)} \leq 1$ for every $n \in \mathbb{N}$ and then f as defined by (1) is bounded on X. Thus $\|f\|_\infty < \infty$ and then $f \in L^\infty(X, \mathfrak{A}, \mu)$. ∎

Prob. 16.23. Let $(X, \mathfrak{A}, \mu)$ be an arbitrary measure space. Let f be an extended complex-valued $\mathfrak{A}$-measurable function on X such that $|f| < \infty$, μ-a.e. on X. Suppose that $fg \in L^1(X, \mathfrak{A}, \mu)$ for every $g \in L^1(X, \mathfrak{A}, \mu)$. Show that $f \in L^\infty(X, \mathfrak{A}, \mu)$.

Proof. To show $f \in L^\infty(X, \mathfrak{A}, \mu)$, let us assume the contrary, that is, $f \notin L^\infty(X, \mathfrak{A}, \mu)$, that is, $\|f\|_\infty = \infty$. Let us define a disjoint sequence $(E_n : n \in \mathbb{Z}_+)$ in $\mathfrak{A}$ by setting

$$E_n = \left\{ x \in X : |f(x)| \in [n^3, (n+1)^3) \right\} \quad \text{for } n \in \mathbb{Z}_+.$$

Note first that since $|f| < \infty$, μ-a.e. on X there exists a null set A in $(X, \mathfrak{A}, \mu)$ such that $\bigcup_{n \in \mathbb{Z}_+} E_n = X \setminus A$. Let us show next that $\mu(E_n) > 0$ for infinitely many $n \in \mathbb{Z}_+$. Suppose $\mu(E_n) > 0$ for only finitely many $n \in \mathbb{Z}_+$. Let N be the greatest of the finitely many $n \in \mathbb{Z}_+$ with $\mu(E_n) > 0$. Then we have $|f(x)| \le (N+1)^3 < \infty$ for μ-a.e. $x \in X$ so that $\|f\|_\infty < \infty$. This contradicts the assumption that $\|f\|_\infty = \infty$. Therefore $\mu(E_n) > 0$ for infinitely many $n \in \mathbb{Z}_+$. Thus there exists a subsequence $(n_k : k \in \mathbb{N})$ of the sequence $(n : n \in \mathbb{Z}_+)$ such that $\mu(E_{n_k}) > 0$ for every $k \in \mathbb{N}$ and $\mu(E_n) = 0$ for every $n \notin (n_k : k \in \mathbb{N})$. Note that since $(n_k : k \in \mathbb{N})$ is a subsequence of $(n : n \in \mathbb{Z}_+)$ it is possible that $n_1 = 0$. In any case we have $n_k > 0$ for all $k \ge 2$. Let us define a nonnegative real-valued $\mathfrak{A}$-measurable function g on X by setting

$$g(x) = \begin{cases} \dfrac{1}{\mu(E_{n_1})} & \text{for } x \in E_{n_1}, \\[2mm] \dfrac{1}{\mu(E_{n_k})} \dfrac{1}{n_k^2} & \text{for } x \in E_{n_k}, k \ge 2, \\[2mm] 0 & \text{for } x \in X \setminus \bigcup_{k \in \mathbb{N}} E_{n_k}. \end{cases}$$

Then we have

$$\int_X g \, d\mu = \int_{\bigcup_{k \in \mathbb{N}} E_{n_k}} g \, d\mu = \sum_{k \in \mathbb{N}} \int_{E_{n_k}} g \, d\mu = 1 + \sum_{k \ge 2} \frac{1}{n_k^2} < \infty,$$

this shows that $g \in L^1(X, \mathfrak{A}, \mu)$. On the other hand, we have

$$\int_X |fg| \, d\mu = \int_{\bigcup_{k \in \mathbb{N}} E_{n_k}} |fg| \, d\mu = \sum_{k \in \mathbb{N}} \int_{E_{n_k}} |fg| \, d\mu \ge n_1^3 + \sum_{k \ge 2} n_k^3 \frac{1}{n_k^2} \ge \sum_{k \in \mathbb{N}} n_k = \infty.$$

Thus $fg \notin L^1(X, \mathfrak{A}, \mu)$. This contradicts the assumption that $fg \in L^1(X, \mathfrak{A}, \mu)$ for every $g \in L^1(X, \mathfrak{A}, \mu)$. Therefore we must have $f \in L^\infty(X, \mathfrak{A}, \mu)$. ∎

Prob. 16.24. Let $(X, \mathfrak{A}, \mu)$ be a σ-finite measure space with $\mu(X) = \infty$. Let $p, q \in [1, \infty]$ be such that $\frac{1}{p} + \frac{1}{q} = 1$. Let f be an extended complex-valued $\mathfrak{A}$-measurable function on X such that $|f| < \infty$, μ-a.e. on X. Suppose that $fg \in L^1(X, \mathfrak{A}, \mu)$ for every $g \in L^q(X, \mathfrak{A}, \mu)$. Show that $f \in L^p(X, \mathfrak{A}, \mu)$.

Proof. 1. Consider the case $p = \infty$ and $q = 1$. According to Prob. 16.23, if $fg \in L^1(X, \mathfrak{A}, \mu)$ for every $g \in L^1(X, \mathfrak{A}, \mu)$, then $f \in L^\infty(X, \mathfrak{A}, \mu)$.

2. Consider the case $p \in [1, \infty)$ and $q \in (1, \infty]$. To show that $f \in L^p(X, \mathfrak{A}, \mu)$, let us assume the contrary, that is, $f \notin L^p(X, \mathfrak{A}, \mu)$, that is, $\int_X |f|^p \, d\mu = \infty$.

Let us define a measure λ on $(X, \mathfrak{A})$ by setting

$$(1) \qquad\qquad \lambda(A) = \int_A |f|^p \, d\mu \quad \text{for } A \in \mathfrak{A}.$$

Since $|f|^p < \infty$, μ-a.e. on X and since $(X, \mathfrak{A}, \mu)$ is a σ-finite measure space, the measure λ defined by (1) is a σ-finite measure according to Prob. 10.13. Note that we also have $\lambda(X) = \int_X |f|^p \, d\mu = \infty$. Thus according to Prob. 16.22, there exists an extended real-valued $\mathfrak{A}$-measurable function h on X such that $h \notin L^1(X, \mathfrak{A}, \mu)$ and $h \in L^r(X, \mathfrak{A}, \mu)$ for all $r \in (1, \infty]$ so that in particular

$$(2) \qquad\qquad h \notin L^1(X, \mathfrak{A}, \mu) \text{ and } h \in L^q(X, \mathfrak{A}, \mu).$$

Let us define an extended real-valued $\mathfrak{A}$-measurable function g on X by setting

$$(3) \qquad\qquad g = h|f|^{p-1}.$$

Let us verify that $g \in L^q(X, \mathfrak{A}, \mu)$ and $fg \notin L^1(X, \mathfrak{A}, \mu)$. Now we have

$$(4) \qquad\qquad |g|^q = |h|^q |f|^{(p-1)q} = |h|^q |f|^p.$$

Since $\lambda(A) = \int_A |f|^p \, d\mu$ for every $A \in \mathfrak{A}$, the Radon-Nikodym derivative $\frac{d\lambda}{d\mu} = |f|^p$. Then by (4) and by Theorem 11.21 we have

$$\int_X |g|^q \, d\mu = \int_X |h|^q |f|^p \, d\mu = \int_X |h|^q \frac{d\lambda}{d\mu} \, d\mu = \int_X |h|^q \, d\lambda < \infty,$$

since $h \in L^q(X, \mathfrak{A}, \mu)$. This shows that $g \in L^q(X, \mathfrak{A}, \mu)$. On the other hand, we have

$$(5) \qquad\qquad |fg| = |f||h||f|^{p-1} = |h||f|^p$$

and then

$$\int_X |fg| \, d\mu = \int_X |h||f|^p \, d\mu = \int_X |h| \frac{d\lambda}{d\mu} \, d\mu = \int_X |h| \, d\lambda = \infty,$$

since $h \notin L^1(X, \mathfrak{A}, \mu)$. This shows that $fg \notin L^1(X, \mathfrak{A}, \mu)$. This shows that there exists $g \in L^q(X, \mathfrak{A}, \mu)$ such that $fg \notin L^1(X, \mathfrak{A}, \mu)$. This contradicts the assumption. Therefore we must have $f \in L^p(X, \mathfrak{A}, \mu)$. ∎

Prob. 16.25. (An Extension of Hölder's Inequality) Let $(X, \mathfrak{A}, \mu)$ be a measure space. For an arbitrary $n \in \mathbb{N}$, let $f_1, \ldots, f_n$ be n extended complex-valued $\mathfrak{A}$-measurable functions on X such that $|f_1|, \ldots, |f_n| < \infty$ a.e. on X. Let $p_1, \ldots, p_n \in (1, \infty)$ be such that $1/p_1 + \cdots + 1/p_n = 1$. Show that

(1) $$\| f_1 \cdots f_n \|_1 \le \| f_1 \|_{p_1} \cdots \| f_n \|_{p_n},$$

where the equality holds if and only if there exist $A_1, \ldots, A_n > 0$ such that

(2) $$A_1 |f_1|^{p_1} = \cdots = A_n |f_n|^{p_n} \quad \text{a.e. on } X. \quad .$$

Proof. We prove (1) and (2) by induction on n.

For $n = 2$, (1) and (2) hold by Theorem 16.14 (Hölder's Inequality).

Suppose (1) and (2) hold for some $n \ge 2$. Let us show that (1) and (2) hold for $n + 1$. Let $p_1, \ldots, p_{n+1} \in (1, \infty)$ be such that $1/p_1 + \cdots + 1/p_{n+1} = 1$. Let $q \in (1, \infty)$ be such that $1/p_1 + \cdots + 1/p_n = 1/q$ so that $1/q + 1/p_{n+1} = 1$. Then by Theorem 16.14 we have

(3) $$\int_X |f_1 \cdots f_{n+1}| \, d\mu \le \left\{ \int_X |f_1 \cdots f_n|^q \, d\mu \right\}^{1/q} \left\{ \int_X |f_{n+1}|^{p_{n+1}} \, d\mu \right\}^{1/p_{n+1}},$$

where the equality holds if and only if there exist $A, B > 0$ such that

(4) $$A|f_1 \cdots f_n|^q = B|f_{n+1}|^{p_{n+1}} \quad \text{a.e. on } X.$$

Now since $1/p_1 + \cdots + 1/p_n = 1/q$, we have $\frac{1}{p_1/q} + \cdots + \frac{1}{p_n/q} = 1$. Thus by our induction hypothesis we have

$$\int_X |f_1 \cdots f_n|^q \, d\mu \le \left\{ \int_X |f_1|^{q\frac{p_1}{q}} \, d\mu \right\}^{q/p_1} \cdots \left\{ \int_X |f_n|^{q\frac{p_n}{q}} \, d\mu \right\}^{q/p_n}$$
$$= \left\{ \int_X |f_1|^{p_1} \, d\mu \right\}^{q/p_1} \cdots \left\{ \int_X |f_n|^{p_n} \, d\mu \right\}^{q/p_n},$$

and then

(5) $$\left\{ \int_X |f_1 \cdots f_n|^q \, d\mu \right\}^{1/q} \le \left\{ \int_X |f_1|^{p_1} \, d\mu \right\}^{1/p_1} \cdots \left\{ \int_X |f_n|^{p_n} \, d\mu \right\}^{1/p_n},$$

where the equality holds if and only if there exist $A_1, \ldots, A_n > 0$ such that

$$A_1 |f_1|^{q\frac{p_1}{q}} = \cdots = A_n |f_n|^{q\frac{p_n}{q}} \quad \text{a.e. on } X,$$

that is,

(6) $$A_1 |f_1|^{p_1} = \cdots = A_n |f_n|^{p_n} \quad \text{a.e. on } X.$$

Substituting (5) in (3), we have (1) for $n + 1$.

Next let us prove (2) for $n + 1$. Observe first that from $1/p_1 + \cdots + 1/p_n = 1/q$ we have $q = (p_1 \cdots p_n)/(p_1 + \cdots + p_n)$. From (6) we have

$$|f_k| = \left(\frac{A_1}{A_k} |f_1|^{p_1} \right)^{\frac{1}{p_k}} = \left(A_1 |f_1|^{p_1} \right)^{\frac{1}{p_k}} A_k^{-\frac{1}{p_k}} \quad \text{for } k = 1, \ldots, n$$

and then

$$(7) \qquad A|f_1 \cdots f_n|^q = A\left\{ \prod_{k=1}^{n} \left(A_1|f_1|^{p_1}\right)^{\frac{1}{p_k}} \right\}^q \left\{ \prod_{k=1}^{n} A_k^{-\frac{q}{p_k}} \right\}$$

$$= AA_1^{\{\frac{1}{p_1}+\cdots+\frac{1}{p_n}\}q}|f_1|^{p_1\{\frac{1}{p_1}+\cdots+\frac{1}{p_n}\}q}\left\{ \prod_{k=1}^{n} A_k^{\frac{q}{p_k}} \right\}^{-1}$$

$$= AA_1|f_1|^{p_1}\left\{ \prod_{k=1}^{n} A_k^{\frac{q}{p_k}} \right\}^{-1}.$$

Combining (4) and (7), we have

$$|f_{n+1}|^{p_{n+1}} = \frac{A}{B}A_1|f_1|^{p_1}\left\{ \prod_{k=1}^{n} A_k^{\frac{q}{p_k}} \right\}^{-1} \quad \text{a.e. on } X.$$

Let us set

$$A_{n+1} := \frac{B}{A}\left\{ \prod_{k=1}^{n} A_k^{\frac{q}{p_k}} \right\}.$$

Then we have

$$A_{n+1}|f_{n+1}|^{p_{n+1}} = A_1|f_1|^{p_1} \quad \text{a.e. on } X.$$

Then by (6) we have

$$A_1|f_1|^{p_1} = \cdots = A_{n+1}|f_{n+1}|^{p_{n+1}} \quad \text{a.e. on } X.$$

This proves (2) for $n + 1$.

Thus we have shown that if (1) and (2) hold for some $n \geq 2$ then (1) and (2) hold for $n + 1$. Then since (1) and (2) hold for $n = 2$, (1) and (2) hold for every $n \geq 2$ by induction on n. ∎

§17   Relation among the L^p Spaces

Prob. 17.1. Consider the measure space $([0, 1), \mathfrak{M}_L|_{[0,1)}, \mu_L)$. Then $[0, 1) = \bigcup_{k \in \mathbb{N}} D_k$ where $D_k = \left[\sum_{i=1}^{k-1} 2^{-i}, \sum_{i=1}^{k} 2^{-i} \right)$ for $k \in \mathbb{N}$.

Let f be a function on $[0, 1)$ defined by setting $f(t) = \sqrt{2^k}$ for $t \in D_k$ for $k \in \mathbb{N}$.

Determine the values of $p \in (0, \infty)$ for which we have $f \in L^p([0, 1), \mathfrak{M}_L|_{[0,1)}, \mu_L)$.

Proof. Observe that $\bigcup_{k \in \mathbb{N}} D_k$ is a disjoint union of countably many intervals D_k and $\ell(D_k) = 1/2^k$. Let $p \in (0, \infty)$. Then we have

$$\int_{[0,1)} |f|^p \, d\mu_L = \sum_{k \in \mathbb{N}} \frac{2^{\frac{kp}{2}}}{2^k} = \sum_{k \in \mathbb{N}} \left\{ 2^{\frac{p}{2}-1} \right\}^k.$$

Now a geometric series with ratio $r \in \mathbb{R}$, $\sum_{k \in \mathbb{N}} r^k$, converges if and only if $|r| < 1$ and when $|r| < 1$ then we have $\sum_{k \in \mathbb{N}} r^k = \frac{r}{1-r}$. Thus we have

$$\int_{[0,1)} |f|^p \, d\mu_L < \infty \Leftrightarrow \left| 2^{\frac{p}{2}-1} \right| < 1.$$

Observe that

$$\left| 2^{\frac{p}{2}-1} \right| < 1 \Leftrightarrow -1 < 2^{\frac{p}{2}-1} < 1 \Leftrightarrow 2^{\frac{p}{2}-1} < 1 \Leftrightarrow 2^{1-\frac{p}{2}} > 1$$

$$\Leftrightarrow \log_2 \left\{ 2^{1-\frac{p}{2}} \right\} > \log_2 1$$

$$\Leftrightarrow 1 - \frac{p}{2} > 0$$

$$\Leftrightarrow p < 2.$$

Thus we have

$$\int_{[0,1)} |f|^p \, d\mu_L < \infty \Leftrightarrow p < 2.$$

This shows that $f \in L^p([0, 1), \mathfrak{M}_L|_{[0,1)}, \mu_L)$ if and only if $p < 2$. ∎

Prob. 17.2. Given a measure space $(X, \mathfrak{A}, \mu)$. Let $f \in L^p(X, \mathfrak{A}, \mu)$ for some $p \in (0, \infty)$ and $g \in L^\infty(X, \mathfrak{A}, \mu)$. Show that $fg \in L^p(X, \mathfrak{A}, \mu)$ and $\|fg\|_p \leq \|f\|_p \|g\|_\infty$.

Proof. Let $f \in L^p(X, \mathfrak{A}, \mu)$ for some $p \in (0, \infty)$. Then we have $\int_X |f|^p \, d\mu < \infty$.
Let $g \in L^\infty(X, \mathfrak{A}, \mu)$. Then $\|g\|_\infty < \infty$ and $|g| \leq \|g\|_\infty$ a.e. on X.
Then for $f \in L^p(X, \mathfrak{A}, \mu)$ and $g \in L^\infty(X, \mathfrak{A}, \mu)$, we have

$$\int_X |fg|^p \, d\mu = \int_X |f|^p |g|^p \, d\mu \leq \int_X |f|^p \|g\|_\infty^p \, d\mu = \|g\|_\infty^p \int_X |f|^p \, d\mu < \infty.$$

This shows that $fg \in L^p(X, \mathfrak{A}, \mu)$. Moreover we have

$$\|fg\|_p = \left\{ \int_X |fg|^p \, d\mu \right\}^{\frac{1}{p}} \leq \left\{ \|g\|_\infty^p \int_X |f|^p \, d\mu \right\}^{\frac{1}{p}} = \|g\|_\infty \|f\|_p.$$

This completes the proof. ∎

Prob. 17.3. Prove the following theorem:

Theorem. (Bounded Convergence Theorem for Norms) *Let* $(X, \mathfrak{A}, \mu)$ *be a finite measure space and let* $1 \le p < p_0$. *Let* $f \in L^{p_0}(X, \mathfrak{A}, \mu)$ *and* $(f_n : n \in \mathbb{N}) \subset L^{p_0}(X, \mathfrak{A}, \mu)$ *be such that*

1° $\lim_{n \to \infty} f_n = f$, μ-*a.e. on* X,

2° $\|f_n\|_{p_0} \le C$ *for all* $n \in \mathbb{N}$ *for some* $C > 0$.

Then we have $\lim_{n \to \infty} \|f_n - f\|_p = 0$.

Proof. 1. Since $\mu(X) < \infty$ and $1 \le p < p_0$, we have $L^{p_0}(X, \mathfrak{A}, \mu) \subset L^p(X, \mathfrak{A}, \mu)$ by Theorem 17.4. Thus we have

$$f \in L^p(X, \mathfrak{A}, \mu) \quad \text{and} \quad (f_n : n \in \mathbb{N}) \subset L^p(X, \mathfrak{A}, \mu).$$

Since $f \in L^p(X, \mathfrak{A}, \mu)$, $|f|^p$ is μ-integrable on X. Then according to Theorem 9.26 (Uniform Absolute Continuity of Integral with Respect to Measure), for every $\varepsilon > 0$ there exists $\delta > 0$ such that $\int_E |f|^p \, d\mu < \varepsilon$ whenever $E \in \mathfrak{A}$ and $\mu(E) < \delta$. In this connection let us agree to select $\delta < \varepsilon$. (This convention is applied in proving (5) below.)

Let us select $E \in \mathfrak{A}$ with $\mu(E) < \delta$ so that we have

(1) $\int_E |f|^p \, d\mu < \varepsilon.$

Since $\mu(X) < \infty$ and condition 1° holds, Theorem 6.12 (Egoroff) implies that for any $\delta > 0$ we can select $E \in \mathfrak{A}$ with $\mu(E) < \delta$ such that

(2) $\lim_{n \to \infty} f_n = f \quad \text{uniformly on } E^c.$

Our selection $E \in \mathfrak{A}$ with $\mu(E) < \delta$ then satisfies both (1) and (2).

Now (2) implies $\lim_{n \to \infty} |f_n - f|^p = 0$ uniformly on E^c. Then for every $\eta > 0$ there exists $N \in \mathbb{N}$ such that $|f_n - f|^p < \eta$ on E^c for $n \ge N$ and then $\int_{E^c} |f_n - f|^p \, d\mu \le \eta \, \mu(E^c)$ for $n \ge N$. Since this holds for every $\eta > 0$ we have

(3) $\lim_{n \to \infty} \int_{E^c} |f_n - f|^p \, d\mu = 0.$

Then we have

(4) $\displaystyle \int_X |f_n - f|^p \, d\mu = \int_{E^c} |f_n - f|^p \, d\mu + \int_E |f_n - f|^p \, d\mu$

$\displaystyle \qquad \le \int_{E^c} |f_n - f|^p \, d\mu + \int_E 2^p \{|f_n|^p + |f|^p\} \, d\mu \quad \text{by Lemma 16.7}$

$\displaystyle \qquad = \int_{E^c} |f_n - f|^p \, d\mu + 2^p \int_E |f_n|^p \, d\mu + 2^p \int_E |f|^p \, d\mu$

$\displaystyle \qquad \le \int_{E^c} |f_n - f|^p \, d\mu + 2^p \int_E |f_n|^p \, d\mu + 2^p \varepsilon \quad \text{by (1)}.$

Next let us estimate $\int_E |f_n|^p \, d\mu$. Now we have $1 \le p < p_0$ and thus $r := p_0/p > 1$. Let $q \in (1, \infty)$ be the conjugate of our $r \in (1, \infty)$. Then by Theorem 16.14 (Hölder's Inequality), we have

$$(5) \quad \int_E |f_n|^p \, d\mu = \int_X |f_n|^p \cdot 1_E \, d\mu \le \left\{ \int_X |f_n|^{pr} \, d\mu \right\}^{1/r} \left\{ \int_X 1_E^q \, d\mu \right\}^{1/q}$$

$$= \left\{ \int_X |f_n|^{p_0} \, d\mu \right\}^{p/p_0} \{\mu(E)\}^{1/q}$$

$$= \|f_n\|_{p_0}^p \varepsilon^{1/q} \quad \text{since } \mu(E) < \delta < \varepsilon$$

$$\le C^p \varepsilon^{1/q} \quad \text{by } 2^\circ.$$

Substituting (5) in (4), we have

$$(6) \quad \int_X |f_n - f|^p \, d\mu \le \int_{E^c} |f_n - f|^p \, d\mu + 2^p C^p \varepsilon^{1/q} + 2^p \varepsilon.$$

Then we have

$$\limsup_{n \to \infty} \int_X |f_n - f|^p \, d\mu \le \lim_{n \to \infty} \int_{E^c} |f_n - f|^p \, d\mu + 2^p C^p \varepsilon^{1/q} + 2^p \varepsilon$$

$$= 2^p \{C^p \varepsilon^{1/q} + \varepsilon\} \quad \text{by (3).}$$

Since this holds for every $\varepsilon > 0$, we have $\limsup_{n \to \infty} \int_X |f_n - f|^p \, d\mu = 0$. This then implies $\liminf_{n \to \infty} \int_X |f_n - f|^p \, d\mu = 0$ and consequently $\lim_{n \to \infty} \int_X |f_n - f|^p \, d\mu = 0$. Thus we have $\lim_{n \to \infty} \|f_n - f\|_p^p = 0$ and then $\lim_{n \to \infty} \|f_n - f\|_p = 0$. This completes the proof. ∎

390 Problems and Proofs in Real Analysis

Prob. 17.4. Definition. *Let* $(X, \mathfrak{A}, \mu)$ *be a measure space and let* $p_1, p_2 \in (0, \infty]$. *We define* $L^{p_1}(X, \mathfrak{A}, \mu) + L^{p_2}(X, \mathfrak{A}, \mu)$ *to be a linear space consisting of all* $f_1 + f_2$ *where* $f_1 \in L^{p_1}(X, \mathfrak{A}, \mu)$ *and* $f_2 \in L^{p_2}(X, \mathfrak{A}, \mu)$.
Prove the following theorem:
Theorem. *Let* $(X, \mathfrak{A}, \mu)$ *be a measure space and let* $0 < p_1 < p < p_2 \leq \infty$. *Then*

$$L^p(X, \mathfrak{A}, \mu) \subset \left[L^{p_1}(X, \mathfrak{A}, \mu) + L^{p_2}(X, \mathfrak{A}, \mu) \right],$$

that is, every $f \in L^p(X, \mathfrak{A}, \mu)$ *can be represented as* $f = f_1 + f_2$ *where* $f_1 \in L^{p_1}(X, \mathfrak{A}, \mu)$ *and* $f_2 \in L^{p_2}(X, \mathfrak{A}, \mu)$.

Proof. 1. Let $f \in L^p(X, \mathfrak{A}, \mu)$. Let us define

(1)
$$\begin{cases} D_1 = \{x \in X : |f(x)| > 1\} \in \mathfrak{A} \\ D_2 = \{x \in X : |f(x)| \leq 1\} \in \mathfrak{A} \end{cases}$$

and

(2)
$$\begin{cases} f_1 = f \cdot \mathbf{1}_{D_1} \\ f_2 = f \cdot \mathbf{1}_{D_2}. \end{cases}$$

Then $D_1 \cap D_2 = \emptyset$ and $D_1 \cup D_2 = X$ and thus $f = f_1 + f_2$ on X.

2. Let us show that $f_1 \in L^{p_1}(X, \mathfrak{A}, \mu)$. Now we have

$$|f_1|^{p_1} = |f \cdot \mathbf{1}_{D_1}|^{p_1} = |f|^{p_1} \cdot \mathbf{1}_{D_1} \leq |f|^p \cdot \mathbf{1}_{D_1},$$

since $|f| > 1$ on D_1 and $p_1 < p$. Then we have

$$\int_X |f_1|^{p_1} \, d\mu \leq \int_X |f|^p \cdot \mathbf{1}_{D_1} \, d\mu \leq \int_X |f|^p \, d\mu < \infty.$$

This shows that $f_1 \in L^{p_1}(X, \mathfrak{A}, \mu)$.

3. Let us show that $f_2 \in L^{p_2}(X, \mathfrak{A}, \mu)$. We consider first the case $p_2 < \infty$. Now we have

$$|f_2|^{p_2} = |f \cdot \mathbf{1}_{D_2}|^{p_2} = |f|^{p_2} \cdot \mathbf{1}_{D_2} \leq |f|^p \cdot \mathbf{1}_{D_2},$$

since $|f| \leq 1$ on D_2 and $p < p_2$. Then we have

$$\int_X |f_2|^{p_2} \, d\mu \leq \int_X |f|^p \cdot \mathbf{1}_{D_2} \, d\mu \leq \int_X |f|^p \, d\mu < \infty.$$

This shows that $f_2 \in L^{p_1}(X, \mathfrak{A}, \mu)$ for the case $p_2 < \infty$.

For the case $p_2 = \infty$, we have $|f_2| = |f \cdot \mathbf{1}_{D_2}| \leq 1$ and thus $\|f_2\|_\infty = 1 < \infty$ and then $f_2 \in L^\infty(X, \mathfrak{A}, \mu)$. This completes the proof. ∎

Prob. 17.5. Given a measure space $(X, \mathfrak{A}, \mu)$ with $\mu(X) \in (0, \infty)$. Let f be an arbitrary extended complex-valued $\mathfrak{A}$-measurable function on X. By Definition 17.1, we have $M_p(f) = \|f\|_p/\mu(X)^{1/p}$ for $p \in (0, \infty)$ and $M_\infty(f) = \|f\|_\infty$. Prove the following statements:

(a) For $p \in (0, \infty)$, we have $\lim_{p \to \infty} M_p(f) = M_\infty(f)$.

(b) For $p \in [1, \infty)$, we have $M_p(f) \uparrow M_\infty(f)$ as $p \uparrow \infty$.

(c) For $p \in (0, \infty)$, we have $\lim_{p \to \infty} \|f\|_p = \|f\|_\infty$.

(d) For $p \in [1, \infty)$, we have $\|f\|_p \uparrow \|f\|_\infty$ as $p \uparrow \infty$, provided that $\mu(X) \in (0, 1]$.

Proof. 1. Let us prove (a) and (b). Observe that (b) implies (a). Therefore it suffices to prove (b).

Let us prove (b). According to Theorem 17.4, for $p \in [1, \infty)$ we have $M_p(f) \uparrow$ as $p \uparrow \infty$. It remains to show that $M_p(f) \uparrow M_\infty(f)$ as $p \uparrow \infty$.

Consider first the case $M_\infty(f) = \infty$, that is, $\|f\|_\infty = \infty$. In this case f is not essentially bounded on X so that for every $B \geq 0$ we have $\mu\{x \in X : |f(x)| > B\} > 0$. Let $B > 1$ be arbitrarily fixed and let

$$C = \mu\{x \in X : |f(x)| > B\} > 0.$$

Now we have

$$M_p(f) = \frac{1}{\mu(X)^{\frac{1}{p}}} \|f\|_p = \frac{1}{\mu(X)^{\frac{1}{p}}} \left\{ \int_X |f|^p \, d\mu \right\}^{\frac{1}{p}}$$

and then

$$M_p^p(f) = \frac{1}{\mu(X)} \left\{ \int_X |f|^p \, d\mu \right\} \geq \frac{1}{\mu(X)} \int_{\mu\{x \in X : |f(x)| > B\}} |f|^p \, d\mu \geq \frac{1}{\mu(X)} B^p C.$$

Then we have

$$\lim_{p \to \infty} M_p^p(f) \geq \lim_{p \to \infty} \frac{1}{\mu(X)} B^p C = \infty.$$

This shows that $\lim_{p \to \infty} M_p(f) = \infty = M_\infty(f)$.

Next consider first the case $M_\infty(f) < \infty$, that is, $\|f\|_\infty < \infty$. Now $\|f\|_\infty$ is the infimum of all essential bounds of f on X. Thus when $\|f\|_\infty < \infty$ then for every $\varepsilon > 0$, $\|f\|_\infty - \varepsilon$ is not an essential bound of f on X and thus we have

$$\mu\{x \in X : |f(x)| > \|f\|_\infty - \varepsilon\} > 0.$$

For brevity let us write

$$E = \{x \in X : |f(x)| > \|f\|_\infty - \varepsilon\}.$$

Then we have

$$0 < \frac{\mu(E)}{\mu(X)} \leq 1.$$

Now we have

$$\|f\|_p^p = \int_X |f|^p \, d\mu \geq \int_E |f|^p \, d\mu \geq \{\|f\|_\infty - \varepsilon\}^p \mu(E)$$

and then

$$M_p(f) = \frac{\|f\|_p}{\mu(X)^{1/p}} \geq \left\{\|f\|_\infty - \varepsilon\right\}\left\{\frac{\mu(E)}{\mu(X)}\right\}^{\frac{1}{p}}.$$

Since $M_p(f) \uparrow$ as $p \uparrow \infty$ by Theorem 17.4, $\lim_{p\to\infty} M_p(f)$ exists. Then we have

$$\lim_{p\to\infty} M_p(f) \geq \left\{\|f\|_\infty - \varepsilon\right\} \lim_{p\to\infty}\left\{\frac{\mu(E)}{\mu(X)}\right\}^{\frac{1}{p}} = \left\{\|f\|_\infty - \varepsilon\right\}.$$

Since this holds for every $\varepsilon > 0$, we have $\lim_{p\to\infty} M_p(f) \geq \|f\|_\infty$.

On the other hand we have $M_p(f) \leq M_\infty(f) = \|f\|_\infty$ for $p \in [1, \infty)$ according to Theorem 17.4 and then $\lim_{p\to\infty} M_p(f) \leq \|f\|_\infty$. Therefore we have $\lim_{p\to\infty} M_p(f) = \|f\|_\infty$. This proves (b).

2. Let us prove (c). We have

$$\lim_{p\to\infty} \|f\|_p = \lim_{p\to\infty}\left\{\mu(X)^{\frac{1}{p}} M_p(f)\right\} = \left\{\lim_{p\to\infty} \mu(X)^{\frac{1}{p}}\right\}\left\{\lim_{p\to\infty} M_p(f)\right\}$$

$$= 1 \cdot M_\infty(f) = \|f\|_\infty \quad \text{by (a)}.$$

3. Let us prove (d). Let $p \in [1, \infty)$. By (b), $M_p(f) \uparrow M_\infty(f)$ as $p \uparrow \infty$, that is,

(1) $\|f\|_p/\mu(X)^{\frac{1}{p}} \uparrow \|f\|_\infty$ as $p \uparrow \infty$.

Now we have

(2) $\mu(X)^{\frac{1}{p}} \uparrow 1$ as $p \uparrow \infty$ when $\mu(X) \in (0, 1]$,

(3) $\mu(X)^{\frac{1}{p}} \downarrow 1$ as $p \uparrow \infty$ when $\mu(X) \in [1, \infty)$.

Assume that $\mu(X) \in (0, 1]$. Then we have

$$\|f\|_p = \left\{\|f\|_p/\mu(X)^{\frac{1}{p}}\right\}\mu(X)^{\frac{1}{p}} \uparrow \|f\|_\infty \cdot 1 \quad \text{by (1) and (2)}.$$

This proves (d). ∎

Prob. 17.6. Given a measure space $(X, \mathfrak{A}, \mu)$ with $\mu(X) \in (0, \infty)$. Let $f \in L^\infty(X, \mathfrak{A}, \mu)$ be such that $\|f\|_\infty \in (0, \infty)$ and let $\alpha_n = \int_X |f|^n \, d\mu = \|f\|_n^n$ for $n \in \mathbb{N}$. Show that we have $\lim\limits_{n \to \infty} \dfrac{\alpha_{n+1}}{\alpha_n} = \|f\|_\infty$.

Proof. For $n \in \mathbb{N}$, we have $n \in [1, \infty)$. Thus by (b) of Prob. 17.5 we have

$$(1) \qquad\qquad M_n(f) \uparrow M_\infty(f) \quad \text{as } n \to \infty.$$

Now $|f| \leq \|f\|_\infty$, a.e. on X and this implies

$$\alpha_{n+1} = \int_X |f|^{n+1} \, d\mu = \int_X |f| \, |f|^n \, d\mu \leq \|f\|_\infty \int_X |f|^n \, d\mu = \|f\|_\infty \alpha_n$$

and then

$$\frac{\alpha_{n+1}}{\alpha_n} \leq \|f\|_\infty$$

and thus

$$(2) \qquad\qquad \limsup_{n \to \infty} \frac{\alpha_{n+1}}{\alpha_n} \leq \|f\|_\infty.$$

By Definition 17.1, we have $M_n(f) = \|f\|_n / \mu(X)^{1/n}$ and $M_\infty(f) = \|f\|_\infty$. Then we have

$$\alpha_n = \|f\|_n^n = \mu(X) M_n(f)^n$$

and thus

$$(3) \qquad\qquad \frac{\alpha_{n+1}}{\alpha_n} = \frac{M_{n+1}(f)^{n+1}}{M_n(f)^n} = \left\{ \frac{M_{n+1}(f)}{M_n(f)} \right\}^n M_{n+1}(f).$$

According to (1) we have $M_n(f) \uparrow$ as $n \to \infty$. Thus we have

$$\frac{M_{n+1}(f)}{M_n(f)} \geq 1 \quad \text{and then} \quad \left\{ \frac{M_{n+1}(f)}{M_n(f)} \right\}^n \geq 1.$$

Substituting this in (3), we have

$$\frac{\alpha_{n+1}}{\alpha_n} \geq M_{n+1}(f).$$

Then we have by (1)

$$(4) \qquad\qquad \liminf_{n \to \infty} \frac{\alpha_{n+1}}{\alpha_n} \geq \lim_{n \to \infty} M_{n+1}(f) = M_\infty(f) = \|f\|_\infty.$$

By (2) and (4) we have

$$(5) \qquad\qquad \limsup_{n \to \infty} \frac{\alpha_{n+1}}{\alpha_n} \leq \|f\|_\infty \leq \liminf_{n \to \infty} \frac{\alpha_{n+1}}{\alpha_n}.$$

This shows that $\liminf\limits_{n \to \infty} \dfrac{\alpha_{n+1}}{\alpha_n} = \limsup\limits_{n \to \infty} \dfrac{\alpha_{n+1}}{\alpha_n} = \|f\|_\infty$ so that $\lim\limits_{n \to \infty} \dfrac{\alpha_{n+1}}{\alpha_n}$ exists and $\lim\limits_{n \to \infty} \dfrac{\alpha_{n+1}}{\alpha_n} = \|f\|_\infty$. ∎

Prob. 17.7. Let $(X, \mathfrak{A}, \mu)$ be a measure space with $\mu(X) \in (0, \infty)$ and let $1 \leq p_1 < p_2 \leq \infty$. Then according to Theorem 17.4 we have $L^{p_2}(X, \mathfrak{A}, \mu) \subset L^{p_1}(X, \mathfrak{A}, \mu)$ and thus the norm $\| \cdot \|_{p_1}$ on $L^{p_1}(X, \mathfrak{A}, \mu)$ serves as a norm on $L^{p_2}(X, \mathfrak{A}, \mu)$.

Show by constructing an example that $L^{p_2}(X, \mathfrak{A}, \mu)$ need not be complete with respect to the norm $\| \cdot \|_{p_1}$ and thus $L^{p_2}(X, \mathfrak{A}, \mu)$ need not be a Banach space with respect to the norm $\| \cdot \|_{p_1}$.

(Hint for an example: $L^2\big((0, 1], \mathfrak{M}_L|_{(0,1]}, \mu_L\big) \subset L^1\big((0, 1], \mathfrak{M}_L|_{(0,1]}, \mu_L\big)$.)

Proof. Consider the measure space $\big((0, 1], \mathfrak{M}_L|_{(0,1]}, \mu_L\big)$. Then we have $\mu_L\big((0, 1]\big) = 1 \in (0, \infty)$. Thus by Theorem 17.4 we have

(1) $$L^2\big((0, 1], \mathfrak{M}_L|_{(0,1]}, \mu_L\big) \subset L^1\big((0, 1], \mathfrak{M}_L|_{(0,1]}, \mu_L\big).$$

For brevity in notations let us write $L^1\big((0, 1]\big)$ and $L^2\big((0, 1]\big)$ for $L^1\big((0, 1]), \mathfrak{M}_L|_{(0,1]}, \mu_L\big)$ and $L^2\big((0, 1], \mathfrak{M}_L|_{(0,1]}, \mu_L\big)$ respectively. Then by (1) the norm $\| \cdot \|_1$ on $L^1\big((0, 1]\big)$ is also a norm on $L^2\big((0, 1]\big)$.

By definition we say that $L^2\big((0, 1]\big)$ is complete with respect to the norm $\| \cdot \|_1$ if for every Cauchy sequence $(f_n : n \in \mathbb{N}) \subset L^2\big((0, 1]\big)$ with respect to the norm $\| \cdot \|_1$ there exists $g \in L^2\big((0, 1]\big)$ such that $\lim_{n \to \infty} \| f_n - g \|_1 = 0$. Thus to show that $L^2\big((0, 1]\big)$ is not complete with respect to the norm $\| \cdot \|_1$, we show that there exists a Cauchy sequence $(f_n : n \in \mathbb{N}) \subset L^2\big((0, 1]\big)$ with respect to the norm $\| \cdot \|_1$ such that $\lim_{n \to \infty} \| f_n - g \|_1 = 0$ for some $g \in L^1\big((0, 1]\big) \setminus L^2\big((0, 1]\big)$. We construct such a Cauchy sequence $(f_n : n \in \mathbb{N}) \subset L^2\big((0, 1]\big)$ below.

Let us define

(2) $$g(x) = \frac{1}{\sqrt{x}} \quad \text{for } x \in (0, 1].$$

Then we have

$$\|g\|_1 = \int_{(0,1]} |g(x)| \, \mu_L(dx) = \int_0^1 \frac{1}{\sqrt{x}} \, dx = \big[2\sqrt{x}\big]_0^1 = 2 < \infty,$$

$$\|g\|_2 = \Big\{ \int_{(0,1]} |g(x)|^2 \, \mu_L(dx) \Big\}^{\frac{1}{2}} = \Big\{ \int_0^1 \frac{1}{x} \, dx \Big\}^{\frac{1}{2}} = \infty.$$

This shows that we have $g \in L^1\big((0, 1]\big)$ and $g \notin L^2\big((0, 1]\big)$ and thus

(3) $$g \in L^1\big((0, 1]\big) \setminus L^2\big((0, 1]\big).$$

Next let us define a sequence $(f_n : n \in \mathbb{N})$ of functions on $(0, 1]$ by setting

(4) $$f_n(x) = \mathbf{1}_{(\frac{1}{n}, 1]}(x) f(x) = \mathbf{1}_{(\frac{1}{n}, 1]}(x) \frac{1}{\sqrt{x}} \quad \text{for } x \in (0, 1].$$

Then we have

$$\|f_n\|_2 = \Big\{ \int_{(0,1]} |f_n(x)|^2 \, \mu_L(dx) \Big\}^{\frac{1}{2}} = \Big\{ \int_{(0,1]} \Big| \mathbf{1}_{(\frac{1}{n}, 1]}(x) \frac{1}{\sqrt{x}} \Big|^2 \, \mu_L(dx) \Big\}^{\frac{1}{2}}$$

$$= \Big\{ \int_{\frac{1}{n}}^1 \frac{1}{x} \, dx \Big\}^{\frac{1}{2}} = \Big\{ \big[\log x\big]_{\frac{1}{n}}^1 \Big\}^{\frac{1}{2}} < \infty.$$

This shows that we have

(5) $$\left(f_n : n \in \mathbb{N}\right) \subset L^2\big((0, 1]\big).$$

Now we have

$$\|f_n - g\|_1 = \int_{(0,1]} |f_n - g|\, d\mu_L = \int_{(0,1]} \left|1_{(\frac{1}{n},1]}(x)\frac{1}{\sqrt{x}} - \frac{1}{\sqrt{x}}\right| \mu_L(dx)$$

$$= \int_{(0,\frac{1}{n}]} \frac{1}{\sqrt{x}}\, \mu_L(dx) = \left[2\sqrt{x}\right]_0^{\frac{1}{n}} = 2\sqrt{\frac{1}{n}}$$

and then

(6) $$\lim_{n\to\infty} \|f_n - g\|_1 = \lim_{n\to\infty} 2\sqrt{\frac{1}{n}} = 0.$$

Now (6) implies that $(f_n : n \in \mathbb{N})$ is a Cauchy sequence with respect to the norm $\|\cdot\|_1$. This shows that our $(f_n : n \in \mathbb{N}) \subset L^2\big((0, 1]\big)$ is a Cauchy sequence with respect to the norm $\|\cdot\|_1$ such that $\lim_{n\to\infty} \|f_n - g\|_1 = 0$ where $g \in L^1\big((0, 1]\big) \setminus L^2\big((0, 1]\big)$. ∎

Prob. 17.8. Let $(X, \mathfrak{A}, \mu)$ be an arbitrary measure space. Let f be an extended real-valued $\mathfrak{A}$-measurable function on X such that $f \in L^1(X, \mathfrak{A}, \mu) \cap L^\infty(X, \mathfrak{A}, \mu)$. Show that $f \in L^p(X, \mathfrak{A}, \mu)$ for every $p \in [1, \infty]$.

Proof. The identically vanishing function on X is contained in both $L^1(X, \mathfrak{A}, \mu)$ and $L^\infty(X, \mathfrak{A}, \mu)$. Thus $L^1(X, \mathfrak{A}, \mu) \cap L^\infty(X, \mathfrak{A}, \mu) \neq \emptyset$.

Let $f \in L^1(X, \mathfrak{A}, \mu) \cap L^\infty(X, \mathfrak{A}, \mu)$. Then since $f \in L^1(X, \mathfrak{A}, \mu)$, we have

$$(1) \qquad\qquad \int_X |f| \, d\mu < \infty.$$

Since $f \in L^\infty(X, \mathfrak{A}, \mu)$, we have

$$(2) \qquad\qquad \|f\|_\infty < \infty \quad \text{and} \quad |f| \leq \|f\|_\infty, \text{ a.e. on } X.$$

Let

$$E = \big\{ x \in X : |f(x)| \geq 1 \big\}.$$

Then we have

$$\infty > \int_X |f| \, d\mu \geq \int_E |f| \, d\mu \geq 1 \cdot \mu(E).$$

Thus we have $\mu(E) < \infty$. On E^c we have $|f| < 1$. Then for any $p \in [1, \infty)$ we have $|f|^p \leq |f|$ on E^c. Then we have

$$\int_X |f|^p \, d\mu = \int_E |f|^p \, d\mu + \int_{E^c} |f|^p \, d\mu$$

$$\leq \|f\|_\infty^p \mu(E) + \int_{E^c} |f| \, d\mu$$

$$\leq \|f\|_\infty^p \mu(E) + \int_X |f| \, d\mu$$

$$< \infty.$$

This shows that $f \in L^p(X, \mathfrak{A}, \mu)$ for every $p \in [1, \infty)$. ∎

Prob. 17.9. Let $(X, \mathfrak{A}, \mu)$ be a measure space and let $0 < p_1 < p_2 \leq \infty$. Then whereas $L^{p_1}(X, \mathfrak{A}, \mu) \cap L^{p_2}(X, \mathfrak{A}, \mu) \neq \emptyset$, neither $L^{p_1}(X, \mathfrak{A}, \mu) \supset L^{p_2}(X, \mathfrak{A}, \mu)$ nor $L^{p_1}(X, \mathfrak{A}, \mu) \subset L^{p_2}(X, \mathfrak{A}, \mu)$ holds in general. (See Prob. 17.11.) Prove the following theorem on $L^{p_1}(X, \mathfrak{A}, \mu) \cap L^{p_2}(X, \mathfrak{A}, \mu)$:

Theorem. *Let $(X, \mathfrak{A}, \mu)$ be a measure space and let $0 < p_1 < p < p_2 \leq \infty$. Then we have*

$$\left[L^{p_1}(X, \mathfrak{A}, \mu) \cap L^{p_2}(X, \mathfrak{A}, \mu)\right] \subset L^p(X, \mathfrak{A}, \mu).$$

Indeed for every $f \in L^{p_1}(X, \mathfrak{A}, \mu) \cap L^{p_2}(X, \mathfrak{A}, \mu)$ there exists $\lambda \in (0, 1)$ such that

$$\|f\|_p \leq \|f\|_{p_1}^{\lambda} \|f\|_{p_2}^{1-\lambda} < \infty.$$

Proof. 1. To show that $L^{p_1}(X, \mathfrak{A}, \mu) \cap L^{p_2}(X, \mathfrak{A}, \mu) \subset L^p(X, \mathfrak{A}, \mu)$, we show that if $f \in L^{p_1}(X, \mathfrak{A}, \mu) \cap L^{p_2}(X, \mathfrak{A}, \mu)$ then $f \in L^p(X, \mathfrak{A}, \mu)$, that is, we show that if $\|f\|_{p_1} < \infty$ and $\|f\|_{p_2} < \infty$ then $\|f\|_p < \infty$. We show this by showing that there exist $\alpha, \beta \in \mathbb{R}$ such that

$$\|f\|_p \leq \|f\|_{p_1}^{\alpha} \|f\|_{p_2}^{\beta} < \infty.$$

Specifically we show that there exists $\lambda \in (0, 1)$ such that

(1) $$\|f\|_p \leq \|f\|_{p_1}^{\lambda} \|f\|_{p_2}^{1-\lambda} < \infty.$$

2. Let $\lambda \in \mathbb{R}$ be defined by

(2) $$\frac{1}{p_1(\lambda p)^{-1}} + \frac{1}{p_2[(1-\lambda)p]^{-1}} = 1, \quad \text{that is,} \quad \frac{\lambda p}{p_1} + \frac{(1-\lambda)p}{p_2} = 1.$$

Let us verify that $\lambda \in (0, 1)$.

For the case $p_2 = \infty$, (2) implies $\frac{\lambda p}{p_1} = 1$ and then we have $\lambda = \frac{p_1}{p} \in (0, 1)$.

For the case $p_2 < \infty$, the alternate expression in (2) implies

$$p_2\lambda + p_1(1 - \lambda) = \frac{p_1 p_2}{p},$$

$$(p_2 - p_1)\lambda = \frac{p_1 p_2}{p} - p_1 = \frac{p_1 p_2 - p_1 p}{p} = \frac{p_1(p_2 - p)}{p},$$

(3) $$\lambda = \frac{p_1}{p} \frac{p_2 - p}{p_2 - p_1} \in (0, 1).$$

This verifies that $\lambda \in (0, 1)$.

3. Let us prove (1) for the case $p_2 = \infty$. Now we have $f \in L^{p_1}(X, \mathfrak{A}, \mu)$ and $f \in L^{\infty}(X, \mathfrak{A}, \mu)$ so that we have $\|f\|_{p_1} < \infty$ and $\|f\|_{\infty} < \infty$. Now

$$|f|^p = |f|^{p_1}|f|^{p-p_1} \leq |f|^{p_1}\|f\|_{\infty}^{p-p_1} \quad \text{since } |f| \leq \|f\|_{\infty}, \text{ a.e. on } X.$$

Then we have

$$\int_X |f|^p \, d\mu \leq \left\{\int_X |f|^{p_1} \, d\mu\right\}\|f\|_{\infty}^{p-p_1} = \|f\|_{p_1}^{p_1}\|f\|_{\infty}^{p-p_1}$$

and then

$$\|f\|_p = \left\{ \int_X |f|^p \, d\mu \right\}^{1/p} \leq \|f\|_{p_1}^{p_1/p} \|f\|_\infty^{(p-p_1)/p} = \|f\|_{p_1}^\lambda \|f\|_\infty^{1-\lambda}.$$

This proves (1) for the case $p_2 = \infty$.

4. Let us prove (1) for the case $p_2 < \infty$. Let us observe that if $a, b \in \mathbb{R}$ are conjugates, that is, if a and b satisfy the condition $\frac{1}{a} + \frac{1}{b} = 1$, then $b = \frac{a}{a-1}$. Thus if $a \in (1, \infty)$ then we have $b \in (1, \infty)$ also. Now by (2), $\frac{p_1}{\lambda p}$ and $\frac{p_2}{(1-\lambda)p}$ are conjugates. We showed above that for the case $p_2 < \infty$ the equality (3) holds. Now (3) implies

(4) $$\frac{p_1}{\lambda p} = \frac{p_2 - p_1}{p_2 - p} \in (1, \infty).$$

By our observation above, (4) implies

(5) $$\frac{p_2}{(1-\lambda)p} \in (1, \infty).$$

Now that the conjugates $\frac{p_1}{\lambda p}$ and $\frac{p_2}{(1-\lambda)p}$ are in $(1, \infty)$ Theorem 16.14 (Hölder's Inequality) is applicable. Thus we have

$$\int_X |f|^p \, d\mu = \int_X |f|^{\lambda p} |f|^{(1-\lambda)p} \, d\mu$$

$$\leq \left\{ \int_X |f|^{\lambda p \frac{p_1}{\lambda p}} \, d\mu \right\}^{\frac{\lambda p}{p_1}} \left\{ \int_X |f|^{(1-\lambda)p \frac{p_2}{(1-\lambda)p}} \, d\mu \right\}^{\frac{(1-\lambda)p}{p_2}}$$

$$= \left\{ \int_X |f|^{p_1} \, d\mu \right\}^{\frac{\lambda p}{p_1}} \left\{ \int_X |f|^{p_2} \, d\mu \right\}^{\frac{(1-\lambda)p}{p_2}}$$

$$= \|f\|_{p_1}^{\lambda p} \|f\|_{p_2}^{(1-\lambda)p},$$

and then

$$\|f\|_p = \left\{ \int_X |f|^p \, d\mu \right\}^{\frac{1}{p}} \leq \|f\|_{p_1}^\lambda \|f\|_{p_2}^{1-\lambda} < \infty.$$

This proves (1) for the case $p_2 < \infty$. This completes the proof. ∎

Prob. 17.10. Consider $L^p(X, \mathfrak{A}, \mu)$ for $p \in (0, \infty]$. By Remark 16.21, Theorem 16.23 and Theorem 16.43, $\| \cdot \|_p$ is a norm on $L^p(X, \mathfrak{A}, \mu)$ if and only if $p \in [1, \infty]$. Prove the following theorem:

Theorem. *Let $(X, \mathfrak{A}, \mu)$ be an arbitrary measure space and let $1 \le p < q \le \infty$. For brevity, let us write $L^p(X)$ for $L^p(X, \mathfrak{A}, \mu)$ and $L^q(X)$ for $L^q(X, \mathfrak{A}, \mu)$. Then $L^p(X) \cap L^q(X)$ is a linear space. Let us define a function $\| \cdot \|_*$ on $L^p(X) \cap L^q(X)$ by setting*

$$\|f\|_* = \|f\|_p + \|f\|_q \quad \text{for } f \in L^p(X) \cap L^q(X).$$

Then $\| \cdot \|_*$ is a norm on $L^p(X) \cap L^q(X)$ and moreover $L^p(X) \cap L^q(X)$ is a Banach space with respect to the norm $\| \cdot \|_*$.

Proof. 1. Let us show that $L^p(X) \cap L^q(X)$ is a linear space. Let $f_1, f_2 \in L^p(X) \cap L^q(X)$ and $c_1, c_2 \in \mathbb{C}$. Then $f_1, f_2 \in L^p(X)$ and $f_1, f_2 \in L^q(X)$. Since $f_1, f_2 \in L^p(X)$ and $L^p(X)$ is a linear space, we have $c_1 f_1 + c_2 f_2 \in L^p(X)$. Similarly $f_1, f_2 \in L^q(X)$ implies that $c_1 f_1 + c_2 f_2 \in L^q(X)$. Thus $c_1 f_1 + c_2 f_2 \in L^p(X) \cap L^q(X)$. This shows that $L^p(X) \cap L^q(X)$ is a linear space.

2. To show that $\| \cdot \|_*$ is a norm on $L^p(X) \cap L^q(X)$, we are to verify the following:
1° $\|f\|_* \in [0, \infty)$ for every $f \in L^p(X) \cap L^q(X)$.
2° $\|f\|_* = 0$ if and only if $f = \mathbf{0}$.
3° $\|cf\|_* = |c| \, \|f\|_*$ for every $f \in L^p(X) \cap L^q(X)$ and $c \in \mathbb{C}$.
4° $\|f_1 + f_2\|_* \le \|f_1\|_* + \|f_2\|_*$ for $f_1, f_2 \in L^p(X) \cap L^q(X)$.
2.1. We have $\|f\|_p \in [0, \infty)$ and $\|f\|_q \in [0, \infty)$. Then $\|f\|_* = \|f\|_p + \|f\|_q \in [0, \infty)$.
2.2. To verify 2°, observe that $\|\mathbf{0}\|_* = \|\mathbf{0}\|_p + \|\mathbf{0}\|_q = 0 + 0 = 0$ and conversely

$$\big| \|f\|_* = 0 \big| \Rightarrow \big| \|f\|_p + \|f\|_q = 0 \big| \Rightarrow \big| \|f\|_p = 0 \text{ and } \|f\|_q = 0 \big| \Rightarrow \big| f = \mathbf{0} \big|.$$

2.3. To verify 3°, observe that we have

$$\|cf\|_* = \|cf\|_p + \|cf\|_q = |c| \, \|f\|_p + |c| \, \|f\|_q$$

$$= |c| \big\{ \|f\|_p + \|f\|_q \big\} = |c| \, \|f\|_*.$$

2.4. Let us verify 4°. We have

$$\|f_1 + f_2\|_* = \|f_1 + f_2\|_p + \|f_1 + f_2\|_q$$

$$\le \big\{ \|f_1\|_p + \|f_2\|_p \big\} + \big\{ \|f_1\|_q + \|f_2\|_q \big\}$$

$$= \big\{ \|f_1\|_p + \|f_1\|_q \big\} + \big\{ \|f_2\|_p + \|f_2\|_q \big\}$$

$$= \|f_1\|_* + \|f_2\|_*.$$

This completes the proof that $\| \cdot \|_*$ is a norm on $L^p(X) \cap L^q(X)$.

3. To show that the linear space $L^p(X) \cap L^q(X)$ is a Banach space with respect to the norm $\| \cdot \|_*$ we show that $L^p(X) \cap L^q(X)$ is complete with respect to the norm $\| \cdot \|_*$, that is, for every Cauchy sequence $(f_n : n \in \mathbb{N}) \subset L^p(X) \cap L^q(X)$ with respect to the norm $\| \cdot \|_*$ there exists $f \in L^p(X) \cap L^q(X)$ such that $\lim_{n \to \infty} \|f_n - f\|_* = 0$.

Let $(f_n : n \in \mathbb{N}) \subset L^p(X) \cap L^q(X)$ be a Cauchy sequence with respect to the norm $\| \cdot \|_*$. Now $\| \cdot \|_p \leq \| \cdot \|_*$ implies that our $(f_n : n \in \mathbb{N}) \subset L^p(X)$ is a Cauchy sequence with respect to the norm $\| \cdot \|_p$ on $L^p(X)$. Similarly $\| \cdot \|_q \leq \| \cdot \|_*$ implies that our $(f_n : n \in \mathbb{N}) \subset L^q(X)$ is a Cauchy sequence with respect to the norm $\| \cdot \|_q$ on $L^q(X)$. Then since $L^p(X)$ and $L^q(X)$ are Banach spaces according to Theorem 16.23 and Theorem 16.43, there exists $f \in L^p(X)$ such that $\lim_{n \to \infty} \|f_n - f\|_p = 0$ and similarly there exists $g \in L^q(X)$ such that $\lim_{n \to \infty} \|f_n - g\|_q = 0$. By Proposition 16.25, $\lim_{n \to \infty} \|f_n - f\|_p = 0$ implies that there exists a subsequence $(f_{n_k} : k \in \mathbb{N})$ such that $\lim_{k \to \infty} f_{n_k} = f$ a.e. on X. Then since $\lim_{n \to \infty} \|f_n - g\|_q = 0$ and $(f_{n_k} : k \in \mathbb{N})$ is a subsequence of $(f_n : n \in \mathbb{N})$, we have $\lim_{k \to \infty} \|f_{n_k} - g\|_q = 0$. Then by Proposition 16.25, there exists a subsequence $(f_{n_{k_\ell}} : \ell \in \mathbb{N})$ of $(f_{n_k} : k \in \mathbb{N})$ such that $\lim_{\ell \to \infty} f_{n_{k_\ell}} = g$ a.e. on X. But $\lim_{k \to \infty} f_{n_k} = f$ a.e. on X implies $\lim_{\ell \to \infty} f_{n_{k_\ell}} = f$ a.e. on X. This shows that $f = g$ a.e. on X. Thus f and g are the same element in $L^p(X) \cap L^q(X)$. Thus we have $\lim_{n \to \infty} \|f_n - f\|_p = 0$ and $\lim_{n \to \infty} \|f_n - f\|_q = 0$. Then we have

$$\lim_{n \to \infty} \|f_n - f\|_* = \lim_{n \to \infty} \left\{ \|f_n - f\|_p + \|f_n - f\|_q \right\}$$

$$= \lim_{n \to \infty} \|f_n - f\|_p + \lim_{n \to \infty} \|f_n - f\|_q = 0.$$

This completes the proof. ∎

Prob. 17.11. Given a measure space $(\mathbb{R}, \mathfrak{M}_L, \mu_L)$. Let $0 < p < q \leq \infty$. Prove the following statements by constructing counterexamples:
(a) $L^p(\mathbb{R}, \mathfrak{M}_L, \mu_L) \not\supset L^q(\mathbb{R}, \mathfrak{M}_L, \mu_L)$.
(b) $L^p(\mathbb{R}, \mathfrak{M}_L, \mu_L) \not\subset L^q(\mathbb{R}, \mathfrak{M}_L, \mu_L)$.

Proof. (Our construction is based on the well-known result in Calculus that for $r \in \mathbb{R}$ we have $\sum_{n \in \mathbb{N}} 1/n^r < \infty$ if and only if $r > 1$. The construction is different for the case $q < \infty$ and the case $q = \infty$.)
 1. Let us prove (a) for the case $q < \infty$.
 To show $L^p(\mathbb{R}, \mathfrak{M}_L, \mu_L) \not\supset L^q(\mathbb{R}, \mathfrak{M}_L, \mu_L)$, let us construct $f \in L^q(\mathbb{R}, \mathfrak{M}_L, \mu_L)$ such that $f \notin L^p(\mathbb{R}, \mathfrak{M}_L, \mu_L)$. Let us define a real-valued function f on $\mathbb{R}$ by setting

$$f = \sum_{n \in \mathbb{N}} \frac{1}{n^{1/p}} \mathbf{1}_{(n-1,n]}.$$

Then we have $\int_{\mathbb{R}} |f|^p \, d\mu_L = \sum_{n \in \mathbb{N}} \frac{1}{n} = \infty$ so that $f \notin L^p(\mathbb{R}, \mathfrak{M}_L, \mu_L)$. On the other hand we have

$$\int_{\mathbb{R}} |f|^q \, d\mu_L = \sum_{n \in \mathbb{N}} \frac{1}{n^{q/p}} = \sum_{n \in \mathbb{N}} \frac{1}{n^{q/p}} < \infty, \quad \text{since } q/p > 1.$$

This shows that $f \in L^q(\mathbb{R}, \mathfrak{M}_L, \mu_L)$.
 2. Let us prove (a) for the case $q = \infty$.
 To show $L^p(\mathbb{R}, \mathfrak{M}_L, \mu_L) \not\supset L^\infty(\mathbb{R}, \mathfrak{M}_L, \mu_L)$, let us construct $f \in L^\infty(\mathbb{R}, \mathfrak{M}_L, \mu_L)$ such that $f \notin L^p(\mathbb{R}, \mathfrak{M}_L, \mu_L)$.
 Let $f = 1$ on $\mathbb{R}$. Then f is bounded on $\mathbb{R}$ and $\|f\|_\infty = 1 < \infty$. This implies that $f \in L^\infty(\mathbb{R}, \mathfrak{M}_L, \mu_L)$. On the other hand we have $\int_{\mathbb{R}} |f|^p \, d\mu_L = \int_{\mathbb{R}} 1 \, d\mu_L = \infty$. This implies that $f \notin L^p(\mathbb{R}, \mathfrak{M}_L, \mu_L)$.
 3. Let us prove (b) for the case $q < \infty$.
 To show $L^p(\mathbb{R}, \mathfrak{M}_L, \mu_L) \not\subset L^q(\mathbb{R}, \mathfrak{M}_L, \mu_L)$, let us construct $f \in L^p(\mathbb{R}, \mathfrak{M}_L, \mu_L)$ such that $f \notin L^q(\mathbb{R}, \mathfrak{M}_L, \mu_L)$.
 Since $q > p$ we have $q/p = 1 + \delta$ where $\delta > 0$. Let $(I_n : n \in \mathbb{N})$ be a disjoint sequence of intervals with $\mu_L(I_n) = 1/n^{2+\delta}$ for $n \in \mathbb{N}$. Let f be a real-valued function on $\mathbb{R}$ defined by setting

$$f = \sum_{n \in \mathbb{N}} n^{1/p} \mathbf{1}_{I_n}.$$

Then we have

$$\int_{\mathbb{R}} |f|^p \, d\mu_L = \sum_{n \in \mathbb{N}} n \frac{1}{n^{2+\delta}} = \sum_{n \in \mathbb{N}} \frac{1}{n^{1+\delta}} < \infty, \quad \text{since } 1 + \delta > 1.$$

This shows that $f \in L^p(\mathbb{R}, \mathfrak{M}_L, \mu_L)$. On the other hand we have

$$\int_{\mathbb{R}} |f|^q \, d\mu_L = \sum_{n \in \mathbb{N}} n^{q/p} \frac{1}{n^{2+\delta}} = \sum_{n \in \mathbb{N}} \frac{n^{1+\delta}}{n^{2+\delta}} = \sum_{n \in \mathbb{N}} \frac{1}{n} = \infty.$$

This shows that $f \notin L^q(\mathbb{R}, \mathfrak{M}_L, \mu_L)$.

4. Let us prove (b) for the case $q = \infty$.

To show $L^p(\mathbb{R}, \mathfrak{M}_L, \mu_L) \not\subset L^\infty(\mathbb{R}, \mathfrak{M}_L, \mu_L)$, let us construct $f \in L^p(\mathbb{R}, \mathfrak{M}_L, \mu_L)$ such that $f \notin L^\infty(\mathbb{R}, \mathfrak{M}_L, \mu_L)$.

Let $(I_n : n \in \mathbb{N})$ be a disjoint sequence of intervals with $\mu_L(I_n) = 1/2^n n^p$ for $n \in \mathbb{N}$. Let f be a real-valued function on $\mathbb{R}$ defined by setting

$$f = \sum_{n \in \mathbb{N}} n \, \mathbf{1}_{I_n}.$$

Then for every $n \in \mathbb{N}$, we have $\{x \in \mathbb{R} : |f(x)| > n - 1\} \supset I_n\}$ and thus we have

$$\mu_L\{x \in \mathbb{R} : |f(x)| > n - 1\} \geq \mu_L(I_n) = \frac{1}{2^n n^p} > 0.$$

This shows that f is not essentially bounded on $\mathbb{R}$ and thus $\|f\|_\infty = \infty$ and $f \notin L^\infty(\mathbb{R}, \mathfrak{M}_L, \mu_L)$. On the other hand we have

$$\int_{\mathbb{R}} |f|^p \, d\mu_L = \sum_{n \in \mathbb{N}} n^p \frac{1}{2^n n^p} = \sum_{n \in \mathbb{N}} \frac{1}{2^n} = 2 < \infty.$$

This shows that $f \in L^p(\mathbb{R}, \mathfrak{M}_L, \mu_L)$. ∎

Prob. 17.12. Given a measure space $(X, \mathfrak{A}, \mu)$.
(a) Suppose $(X, \mathfrak{A}, \mu)$ has a $\mathfrak{A}$-measurable set of arbitrarily small positive measure, that is, for every $\varepsilon > 0$ there exists $E \in \mathfrak{A}$ such that $\mu(E) \in (0, \varepsilon]$. Show that there exists a disjoint sequence $(E_n : n \in \mathbb{N})$ in $\mathfrak{A}$ with $\mu(E_n) > 0$ for every $n \in \mathbb{N}$ such that $\lim_{n\to\infty} \mu(E_n) = 0$.
(b) Suppose $(X, \mathfrak{A}, \mu)$ has a $\mathfrak{A}$-measurable set of arbitrarily large finite measure, that is, for every $M > 0$ there exists $E \in \mathfrak{A}$ such that $\mu(E) \in [M, \infty)$. Show that for every $\gamma > 0$ there exists a disjoint sequence $(E_n : n \in \mathbb{N})$ in $\mathfrak{A}$ with $\mu(E_n) \in [\gamma, \infty)$ for every $n \in \mathbb{N}$. (This problem is a preparation for the next.)

Proof. 1. Let us prove (a). Given a measure space $(X, \mathfrak{A}, \mu)$. Assume that for every $\varepsilon > 0$ there exists $E \in \mathfrak{A}$ such that $\mu(E) \in (0, \varepsilon]$. Then for every $n \in \mathbb{N}$ there exists $A_n \in \mathfrak{A}$ with

$$(1) \qquad \mu(A_n) \in \left(0, \frac{1}{2^{2n-1}}\right].$$

Consider the sequence $(A_n : n \in \mathbb{N})$. We have

$$\mu(A_1) \in \left(0, \frac{1}{2}\right], \ \mu(A_2) \in \left(0, \frac{1}{2^3}\right], \ \mu(A_3) \in \left(0, \frac{1}{2^5}\right], \ \dots.$$

Let us define a sequence $(E_n : n \in \mathbb{N}) \subset \mathfrak{A}$ by setting

$$(2) \qquad E_n = A_n \setminus \bigcup_{k \geq n+1} A_k \quad \text{for } n \in \mathbb{N},$$

that is,

$$E_1 = A_1 \setminus \left(A_2 \cup A_3 \cup A_4 \cup \cdots\right),$$
$$E_2 = A_2 \setminus \left(A_3 \cup A_4 \cup A_5 \cup \cdots\right),$$
$$E_3 = A_3 \setminus \left(A_4 \cup A_5 \cup A_6 \cup \cdots\right),$$
$$\vdots$$

Let us show that the sequence $(E_n : n \in \mathbb{N})$ is a disjoint sequence. Let $m, n \in \mathbb{N}$ and $n < m$. Then by (2) we have

$$E_n = A_n \setminus \bigcup_{k \geq n+1} A_k,$$
$$E_m = A_m \setminus \bigcup_{k \geq m+1} A_k \subset A_m \subset \bigcup_{k \geq n+1} A_k.$$

This implies that $E_n \cap E_m = \emptyset$. This shows that $(E_n : n \in \mathbb{N})$ is a disjoint sequence.
Let us show that $\mu(E_n) > 0$ for every $n \in \mathbb{N}$ and $\lim_{n\to\infty} \mu(E_n) = 0$. We have

$$(3) \qquad \mu(E_n) \leq \mu(A_n) \leq \frac{1}{2^{2n-1}} \quad \text{by (2) and (1).}$$

This implies that $\lim_{n\to\infty} \mu(E_n) = 0$. It remains to show that $\mu(E_n) > 0$ for every $n \in \mathbb{N}$. Now we have

(4) $$\mu(E_n) \geq \mu(A_n) - \mu\Big(\bigcup_{k\geq n+1} A_k\Big) \geq \mu(A_n) - \sum_{k\geq n+1} \mu(A_k).$$

Observe that

$$\sum_{k\geq n+1} \mu(A_k) \leq \frac{1}{2^{2n+1}} + \frac{1}{2^{2n+3}} + \frac{1}{2^{2n+5}} + \cdots \quad \text{by (1)}$$

$$= \frac{1}{2^{2n+1}}\Big\{1 + \frac{1}{2^2} + \frac{1}{2^4} + \cdots\Big\}$$

$$= \frac{1}{2^{2n+1}}\frac{1}{1-\frac{1}{4}} = \frac{4}{3}\frac{1}{2^{2n+1}}$$

$$= \frac{1}{3}\frac{1}{2^{2n-1}} \leq \frac{1}{3}\mu(A_n).$$

Substituting this into (4), we have

(5) $$\mu(E_n) \geq \mu(A_n) - \frac{1}{3}\mu(A_n) = \frac{2}{3}\mu(A_n) > 0.$$

This shows that $\mu(E_n) > 0$ for every $n \in \mathbb{N}$.

2. Let us prove (b). Given a measure space $(X, \mathfrak{A}, \mu)$. Assume that for every $M > 0$ there exists $E \in \mathfrak{A}$ such that $\mu(E) \in [M, \infty)$. Then for every $\gamma > 0$ we can select a sequence $(A_n : n \in \mathbb{N}) \subset \mathfrak{A}$ such that

$$\mu(A_1) \in [\gamma, \infty),$$

$$\mu(A_2) \in [\mu(A_1) + \gamma, \infty),$$

$$\mu(A_3) \in [\mu(A_1) + \mu(A_2)\gamma, \infty),$$

$$\vdots$$

(6) $$\mu(A_n) \in \Big[\sum_{k=1}^{n-1} \mu(A_k) + \gamma, \infty\Big)$$

$$\vdots$$

Then let us define a sequence $(E_n : n \in \mathbb{N}) \subset \mathfrak{A}$ by setting

(7) $$E_1 = A_1 \quad \text{and} \quad E_n = A_n \setminus \bigcup_{k=1}^{n-1} A_k \quad \text{for } n \geq 2.$$

Let us show that $(E_n : n \in \mathbb{N})$ is a disjoint sequence, that is, $E_n \cap E_m = \emptyset$ when $n < m$. For the case that $n = 1$ and $m \geq 2$ we have $E_1 = A_1$ and $E_m = A_m \setminus \bigcup_{k=1}^{m-1} A_k$. Now $\bigcup_{k=1}^{m-1} A_k \supset A_1 = E_1$. Thus $E_1 \cap E_m = \emptyset$ when $m \geq 2$.

Consider the case that $2 \leq n < m$. According to (6), we have $E_n = A_n \setminus \bigcup_{k=1}^{n-1} A_k$ and $E_m = A_m \setminus \bigcup_{k=1}^{m-1} A_k$. Since $m > n$ we have $m - 1 \geq n$ and then $\bigcup_{k=1}^{m-1} A_k \supset A_n$. Thus $E_m = A_m \setminus \bigcup_{k=1}^{m-1} A_k \subset A_m \setminus A_n$. But $E_n \subset A_n$. Thus $E_m \subset A_m \setminus A_n \subset A_m \setminus E_n$. Then we have $E_m \cap E_n = \emptyset$. This completes the proof that $(E_n : n \in \mathbb{N})$ is a disjoint sequence.

It remains to show that $\mu(E_n) \in [\gamma, \infty)$ for every $n \in \mathbb{N}$. Observe first that by (7) we have $\mu(E_1) = \mu(A_1) \in [\gamma, \infty)$. For $n \geq 2$ we have

$$\mu(E_n) = \mu\left(A_n \setminus \bigcup_{k=1}^{n-1} A_k \right) \quad \text{by (7)}$$

$$\geq \mu(A_n) - \mu\left(\bigcup_{k=1}^{n-1} A_k \right)$$

$$\geq \mu(A_n) - \sum_{k=1}^{n-1} \mu(A_k)$$

$$\geq \sum_{k=1}^{n-1} \mu(A_k) + \gamma - \sum_{k=1}^{n-1} \mu(A_k) \quad \text{by (6)}$$

$$= \gamma.$$

This shows that $\mu(E_n) \in [\gamma, \infty)$ for $n \geq 2$. ∎

Prob. 17.13. Prove the following theorem:

Theorem. *Let* $(X, \mathfrak{A}, \mu)$ *be an arbitrary measure space and let* $0 < p < q \leq \infty$. *Then we have* $L^p(X, \mathfrak{A}, \mu) \not\subset L^q(X, \mathfrak{A}, \mu)$ *if and only if* $(X, \mathfrak{A}, \mu)$ *has a* $\mathfrak{A}$-*measurable set of arbitrarily small positive measure.*

Proof. 1. Suppose $(X, \mathfrak{A}, \mu)$ has a $\mathfrak{A}$-measurable set of arbitrarily small positive measure. To show that $L^p(X, \mathfrak{A}, \mu) \not\subset L^q(X, \mathfrak{A}, \mu)$ we construct a function f on X such that $f \in L^p(X, \mathfrak{A}, \mu)$ and $f \notin L^q(X, \mathfrak{A}, \mu)$. Our construction is done separately for the case that $q < \infty$ and the case that $q = \infty$.

Consider the case that $q < \infty$. According to Prob. 17.12 there exists a disjoint sequence $(E_n : n \in \mathbb{N})$ in $\mathfrak{A}$ with $\mu(E_n) > 0$ for every $n \in \mathbb{N}$ such that $\lim_{n \to \infty} \mu(E_n) = 0$. By taking a subsequence of $(E_n : n \in \mathbb{N})$ if necessary we have a disjoint sequence $(E_n : n \in \mathbb{N})$ in $\mathfrak{A}$ such that

$$(1) \qquad\qquad 0 < \mu(E_n) \leq \frac{1}{2^n} \quad \text{for every } n \in \mathbb{N}.$$

Then let

$$(2) \qquad\qquad \lambda_n = \frac{1}{2^n} \frac{1}{\mu(E_n)} \quad \text{for every } n \in \mathbb{N}.$$

Let $\gamma > 0$ be such that $\gamma^q = 2$. Define a real-valued function f on X by setting

$$(3) \qquad\qquad f = \sum_{n \in \mathbb{N}} \gamma^n \lambda_n^{\frac{1}{p}} \mathbf{1}_{E_n}.$$

Then we have

$$(4) \qquad \int_X |f|^p \, d\mu = \sum_{n \in \mathbb{N}} \gamma^{np} \lambda_n \mu(E_n) = \sum_{n \in \mathbb{N}} \gamma^{np} \frac{1}{2^n} = \sum_{n \in \mathbb{N}} \left(\frac{\gamma^p}{2} \right)^n.$$

Since $\gamma^q = 2$ and $p < q$, we have $\gamma^p < 2$ and then $\frac{\gamma^p}{2} < 1$. Thus the last member of (4) is a convergent geometric series and this shows that $\int_X |f|^p \, d\mu < \infty$ and then $f \in L^p(X, \mathfrak{A}, \mu)$. On the other hand we have

$$\int_X |f|^q \, d\mu = \sum_{n \in \mathbb{N}} \gamma^{nq} \lambda_n^{\frac{q}{p}} \mu(E_n).$$

Since $\lambda_n \geq 1$ by (2) and (1) and since $\frac{q}{p} > 1$, we have $\lambda_n^{\frac{q}{p}} \geq \lambda_n$. Substituting this in the last equality, we have

$$(5) \qquad \int_X |f|^q \, d\mu \geq \sum_{n \in \mathbb{N}} \gamma^{nq} \lambda_n \mu(E_n) = \sum_{n \in \mathbb{N}} \gamma^{nq} \frac{1}{2^n} = \sum_{n \in \mathbb{N}} \left(\frac{\gamma^q}{2} \right)^n = \sum_{n \in \mathbb{N}} 1 = \infty.$$

This shows that $\int_X |f|^q \, d\mu = \infty$ and then $f \notin L^q(X, \mathfrak{A}, \mu)$. Thus our function f is in $L^p(X, \mathfrak{A}, \mu)$ but not in $L^q(X, \mathfrak{A}, \mu)$.

Consider the case that $q = \infty$. According to Prob. 17.12 there exists a disjoint sequence $(E_n : n \in \mathbb{N})$ in $\mathfrak{A}$ with $\mu(E_n) > 0$ for every $n \in \mathbb{N}$ such that $\lim_{n\to\infty} \mu(E_n) = 0$. By taking a subsequence of $(E_n : n \in \mathbb{N})$ if necessary we have a disjoint sequence $(E_n : n \in \mathbb{N})$ in $\mathfrak{A}$ such that

(6) $$0 < \mu(E_n) \le \frac{1}{n}\frac{1}{2^n} \quad \text{for every } n \in \mathbb{N}.$$

Let us define a function f on X by setting

(7) $$f = \sum_{n\in\mathbb{N}} n^{\frac{1}{p}} 1_{E_n}.$$

Then we have

(8) $$\int_X |f|^p \, d\mu = \sum_{n\in\mathbb{N}} n \, \mu(E_n) = \sum_{n\in\mathbb{N}} n\frac{1}{n}\frac{1}{2^n} = \sum_{n\in\mathbb{N}} \frac{1}{2^n} = 2 < \infty.$$

This shows that $f \in L^p(X, \mathfrak{A}, \mu)$. On the other hand, the infimum of the essential bounds of f on X, $\|f\|_\infty = \infty$. Thus $f \notin L^\infty(X, \mathfrak{A}, \mu)$. Thus our function f is in $L^p(X, \mathfrak{A}, \mu)$ but not in $L^\infty(X, \mathfrak{A}, \mu)$.

2. Conversely suppose $L^p(X, \mathfrak{A}, \mu) \not\subset L^q(X, \mathfrak{A}, \mu)$. We are to show that $(X, \mathfrak{A}, \mu)$ has a $\mathfrak{A}$-measurable set of arbitrarily small positive measure. Thus we are to prove the following statement:

(9) $$\left[L^p(X, \mathfrak{A}, \mu) \not\subset L^q(X, \mathfrak{A}, \mu) \right]$$
$$\Rightarrow \left[\text{for every } \varepsilon > 0, \exists E \in \mathfrak{A} \text{ such that } \mu(E) \in (0, \varepsilon] \right].$$

We prove this statement by proving its contra-positive, that is,

(10) $$\left[\exists c > 0 \text{ such that if } E \in \mathfrak{A} \text{ and } \mu(E) \ne 0 \text{ then } \mu(E) \ge c \right]$$
$$\Rightarrow \left[L^p(X, \mathfrak{A}, \mu) \subset L^q(X, \mathfrak{A}, \mu) \right].$$

Let us assume that there exists $c > 0$ such that if $E \in \mathfrak{A}$ and $\mu(E) \ne 0$ then $\mu(E) \ge c$. We proceed to show that $L^p(X, \mathfrak{A}, \mu) \subset L^q(X, \mathfrak{A}, \mu)$.

Let $f \in L^p(X, \mathfrak{A}, \mu)$. Then we have $\int_X |f|^p \, d\mu < \infty$. Let us decompose X into the following $\mathfrak{A}$-measurable subsets:

$$E_0 = \left\{ x \in X : |f(x)|^p = 0 \right\} = \left\{ x \in X : f(x) = 0 \right\},$$
$$E_k = \left\{ x \in X : |f(x)|^p \in (k-1, k] \right\} \quad \text{for } k \in \mathbb{N},$$
$$E_\infty = \left\{ x \in X : |f(x)|^p = \infty \right\} = \left\{ x \in X : |f(x)| = \infty \right\}.$$

Now $\int_X |f|^p \, d\mu < \infty$ implies $\mu(E_\infty) = 0$. This in turn implies $\int_{E_\infty} |f|^p \, d\mu = 0$. We have $\int_{E_0} |f|^p \, d\mu = 0$. Next observe that $\int_{E_1} |f|^p \, d\mu \le \int_X |f|^p \, d\mu < \infty$. Regarding E_k

for $k \geq 2$, we have

(11)
$$\infty > \int_X |f|^p \, d\mu = \int_{\bigcup_{k \in \mathbb{N}} E_k} |f|^p \, d\mu = \sum_{k \in \mathbb{N}} \int_{E_k} |f|^p \, d\mu$$

$$\geq \sum_{k \in \mathbb{N}} (k-1)\mu(E_k) = \sum_{k \geq 2} (k-1)\mu(E_k).$$

The convergence of the last series $\sum_{k \geq 2} (k-1)\mu(E_k)$ implies that $\mu(E_k) < \infty$ for $k \geq 2$. It also implies that $\lim_{k \to \infty} (k-1)\mu(E_k) = 0$ and then $\lim_{k \to \infty} \mu(E_k) = 0$. Then for our constant $c > 0$ in (10) there exists $N \in \mathbb{N}$ such that $\mu(E_k) < c$ for $k \geq N$. According to our assumption, if $\mu(E_n) \neq 0$ then $\mu(E_k) \geq c$. Thus $\mu(E_k) = 0$ for $k \geq N$. Thus we have shown that we have

1° $\mu(E_\infty) = 0$.
2° $\int_{E_1} |f|^p \, d\mu < \infty$.
3° $\mu(E_k) < \infty$ for $k \geq 2$.
4° $\mu(E_k) = 0$ for all but finitely many $k \in \mathbb{N}$.

2.1. Consider the case $p < q < \infty$. Let us show that $\int_X |f|^q \, d\mu < \infty$ so that $f \in L^q(X, \mathfrak{A}, \mu)$. Let us decompose X into the following $\mathfrak{A}$-measurable subsets:

$$F_0 = \left\{ x \in X : |f(x)|^q = 0 \right\} = \left\{ x \in X : f(x) = 0 \right\},$$

$$F_k = \left\{ x \in X : |f(x)|^q \in (k-1, k] \right\} \quad \text{for } k \in \mathbb{N},$$

$$F_\infty = \left\{ x \in X : |f(x)|^q = \infty \right\} = \left\{ x \in X : |f(x)| = \infty \right\}.$$

It is obvious from the definitions that we have

5° $F_\infty = E_\infty$.
6° $F_0 = E_0$.

Observe further that

$$E_1 = \left\{ x \in X : |f(x)|^p \in (0, 1] \right\} = \left\{ x \in X : |f(x)| \in (0, 1] \right\}$$

$$= \left\{ x \in X : |f(x)|^q \in (0, 1] \right\} = F_1.$$

Thus we have

7° $F_1 = E_1$.

Now we have

(12)
$$\int_X |f|^q \, d\mu = \int_{\bigcup_{k \in \mathbb{N}} F_k} |f|^q \, d\mu = \sum_{k \in \mathbb{N}} \int_{F_k} |f|^q \, d\mu$$

$$= \int_{F_1} |f|^q \, d\mu + \sum_{k \geq 2} \int_{F_k} |f|^q \, d\mu.$$

Now $F_1 = E_1$. On F_1 we have $|f| \in (0, 1]$. Then $p < q$ implies that $|f|^q \leq |f|^p$ on F_1. Then we have by 2°

(13)
$$\int_{F_1} |f|^q \, d\mu \leq \int_{E_1} |f|^p \, d\mu < \infty.$$

It remains to show that

(14)
$$\sum_{k\geq 2}\int_{F_k}|f|^q\,d\mu < \infty.$$

According to (11), we have $\sum_{k\geq 2}(k-1)\mu(E_k) < \infty$. For $k \geq 2$, we have $k \leq 3(k-1)$ as can be verified immediately. Thus we have

(15)
$$\sum_{k\geq 2}k\,\mu(E_k) \leq \sum_{k\geq 2}3(k-1)\mu(E_k) = 3\sum_{k\geq 2}(k-1)\mu(E_k) < \infty.$$

According to 3° and 4° we have $0 < \mu(E_k) < \infty$ for only finitely many $k \geq 2$. Thus we have $2 \leq k_1 < k_2 < \cdots < k_m$ such that $\mu(E_{k_1}), \mu(E_{k_2}), \ldots, \mu(E_{k_m}) \in (0, \infty)$ and $\mu(E_k) = 0$ for any $k \neq 1, k_1, k_2, \ldots, k_m$. Let

(16)
$$B = \sum_{j=1}^{m}\mu(E_{k_j}) \in (0, \infty),$$

(17)
$$C = \min\{\mu(E_{k_j}) : j = 1, \ldots, m\} \in (0, \infty),$$

(18)
$$\gamma = \frac{B}{C} \in [1, \infty).$$

Then by (17) and (15) we have

(19)
$$\sum_{k\geq 2}k\,C \leq \sum_{k\geq 2}k\,\mu(E_k) < \infty.$$

Now $F_0 \cup \left(\bigcup_{k\in\mathbb{N}} F_k\right) \cup F_\infty = X = E_0 \cup \left(\bigcup_{k\in\mathbb{N}} E_k\right) \cup E_\infty$. This and 5°, 6° and 7° imply that $\bigcup_{k\geq 2} F_k = \bigcup_{k\geq 2} E_k$. Then for $k \geq 2$ we have $F_k \subset \bigcup_{\ell\geq 2} E_\ell$ and then

(20)
$$\mu(F_k) \leq \mu\left(\bigcup_{\ell\geq 2} E_\ell\right) \leq \sum_{\ell\geq 2}\mu(E_\ell) = \sum_{j=1}^{m}\mu(E_{k_j}).$$

Then we have

$$\sum_{k\geq 2}\int_{F_k}|f|^q\,d\mu \leq \sum_{k\geq 2}k\,\mu(F_k) \leq \sum_{k\geq 2}k\left\{\sum_{j=1}^{m}\mu(E_{k_j})\right\} \quad \text{by (20)}$$

$$= \sum_{k\geq 2}k\,\gamma\,C = \gamma\sum_{k\geq 2}k\,C < \infty \quad \text{by (18) and (19).}$$

This proves (14). Then substituting (13) and (14) into (12), we have $\int_X |f|^q\,d\mu < \infty$. This shows that $f \in L^q(X, \mathfrak{A}, \mu)$ and therefore $L^p(X, \mathfrak{A}, \mu) \subset L^q(X, \mathfrak{A}, \mu)$.

2.2. Consider the case $p < q = \infty$. Our function $f \in L^p(X, \mathfrak{A}, \mu)$ defined by (7) is essentially bounded on X since $\mu(E_k) = 0$ for all but finitely many $k \in \mathbb{N}$. Thus $f \in L^\infty(X, \mathfrak{A}, \mu)$. This shows that we have $L^p(X, \mathfrak{A}, \mu) \subset L^\infty(X, \mathfrak{A}, \mu)$. ∎

Prob. 17.14. Prove the following theorem:

Theorem. *Let $(X, \mathfrak{A}, \mu)$ be an arbitrary measure space and let $0 < p < q \leq \infty$. Then we have $L^p(X, \mathfrak{A}, \mu) \subset L^q(X, \mathfrak{A}, \mu)$ if and only if there exists $c > 0$ such that every $E \in \mathfrak{A}$ with $\mu(E) \neq 0$ has $\mu(E) \geq c$. Thus we have $L^p(X, \mathfrak{A}, \mu) \subset L^q(X, \mathfrak{A}, \mu)$ if and only if*

$$\inf \{\mu(E) : E \in \mathfrak{A} \text{ and } \mu(E) \neq 0\} > 0.$$

Proof. By Prob. 17.13, we have the following two valid statements:

(1) $\left[L^p(X, \mathfrak{A}, \mu) \not\subset L^q(X, \mathfrak{A}, \mu)\right]$

$\Rightarrow \left[(X, \mathfrak{A}, \mu) \text{ has a } \mathfrak{A}\text{-measurable set of arbitrarily small positive measure}\right]$

and

(2) $\left[(X, \mathfrak{A}, \mu) \text{ has a } \mathfrak{A}\text{-measurable set of arbitrarily small positive measure}\right]$

$\Rightarrow \left[L^p(X, \mathfrak{A}, \mu) \not\subset L^q(X, \mathfrak{A}, \mu)\right].$

Now the contra-positive statement of (1) is given by

(3) $\left[L^p(X, \mathfrak{A}, \mu) \subset L^q(X, \mathfrak{A}, \mu)\right]$

$\Rightarrow \left[\text{There exists } c > 0 \text{ such that every } E \in \mathfrak{A} \text{ with } \mu(E) \neq 0 \text{ has } \mu(E) \geq c\right].$

Similarly the contra-positive statement of (2) is given by

(4) $\left[\text{There exists } c > 0 \text{ such that every } E \in \mathfrak{A} \text{ with } \mu(E) \neq 0 \text{ has } \mu(E) \geq c\right]$

$\Rightarrow \left[L^p(X, \mathfrak{A}, \mu) \subset L^q(X, \mathfrak{A}, \mu)\right].$

Then combining (3) and (4) we have

$$\left[L^p(X, \mathfrak{A}, \mu) \subset L^q(X, \mathfrak{A}, \mu)\right]$$

$\Leftrightarrow \left[\text{There exists } c > 0 \text{ such that every } E \in \mathfrak{A} \text{ with } \mu(E) \neq 0 \text{ has } \mu(E) \geq c\right].$

This completes the proof. ∎

Prob. 17.15. Prove the following theorem:

Theorem. *Let $(X, \mathfrak{A}, \mu)$ be an arbitrary measure space and let $0 < p < q \leq \infty$. Then we have $L^p(X, \mathfrak{A}, \mu) \not\supset L^q(X, \mathfrak{A}, \mu)$ if and only if $(X, \mathfrak{A}, \mu)$ has a $\mathfrak{A}$-measurable set of arbitrarily large finite measure.*

Proof. 1. Suppose $(X, \mathfrak{A}, \mu)$ has a $\mathfrak{A}$-measurable set of arbitrarily large finite measure. Then by Prob. 17.12 there exists a disjoint sequence $(E_n : n \in \mathbb{N})$ in $\mathfrak{A}$ such that $\mu(E_n) \in [1, \infty)$ for every $n \in \mathbb{N}$. To show that $L^p(X, \mathfrak{A}, \mu) \not\supset L^q(X, \mathfrak{A}, \mu)$, we construct a function f on X such that $f \in L^q(X, \mathfrak{A}, \mu)$ and $f \notin L^p(X, \mathfrak{A}, \mu)$. This is done differently for the case $q < \infty$ and the case $q = \infty$.

    1.1. Consider the case $q < \infty$. Let us define a real-valued $\mathfrak{A}$-measurable function f on X by setting

$$(1) \qquad f = \sum_{n \in \mathbb{N}} \frac{1}{n^{1/p}} \frac{1}{\mu(E_n)^{1/p}} \mathbf{1}_{E_n}.$$

We have

$$(2) \qquad \int_X |f|^q \, d\mu = \sum_{n \in \mathbb{N}} \frac{1}{n^{q/p}} \frac{1}{\mu(E_n)^{q/p}} \mu(E_n).$$

Since $\mu(E_n) \geq 1$ and $q/p > 1$, we have $\mu(E_n)^{q/p} \geq \mu(E_n)$ and then $\frac{1}{\mu(E_n)^{q/p}} \mu(E_n) \leq 1$. Then we have

$$(3) \qquad \int_X |f|^q \, d\mu \leq \sum_{n \in \mathbb{N}} \frac{1}{n^{q/p}} < \infty.$$

This shows that $f \in L^q(X, \mathfrak{A}, \mu)$. On the other hand we have

$$(4) \qquad \int_X |f|^p \, d\mu = \sum_{n \in \mathbb{N}} \frac{1}{n} \frac{1}{\mu(E_n)} \mu(E_n) = \sum_{n \in \mathbb{N}} \frac{1}{n} = \infty.$$

This shows that $f \notin L^p(X, \mathfrak{A}, \mu)$.

    1.2. Consider the case $q = \infty$. Let us define a function f on X by setting

$$(5) \qquad f = \sum_{n \in \mathbb{N}} \mathbf{1}_{E_n}.$$

Then we have $|f| \leq 1$ on X and indeed the infimum of the essential bounds of f on X is equal to 1. Thus we have $\|f\|_\infty = 1 < \infty$ and therefore $f \in L^\infty(X, \mathfrak{A}, \mu)$. On the other hand we have

$$(6) \qquad \int_X |f|^p \, d\mu = \sum_{n \in \mathbb{N}} \left\{ \int_{E_n} 1^p \, d\mu \right\} = \sum_{n \in \mathbb{N}} \mu(E_n) \geq \sum_{n \in \mathbb{N}} 1 = \infty.$$

This shows that $f \notin L^p(X, \mathfrak{A}, \mu)$.

2. Let us prove the converse. Thus we are to prove the following statement:

(7) $\left[L^p(X, \mathfrak{A}, \mu) \not\supset L^q(X, \mathfrak{A}, \mu) \right]$

$\Rightarrow \left[(X, \mathfrak{A}, \mu) \text{ has a } \mathfrak{A}\text{-measurable set of arbitrarily large finite measure} \right]$.

We prove this statement by proving its contra-positive, that is

(8) $\left[\exists B > 0 \text{ such that } \mu(E) \le B \text{ for every } E \in \mathfrak{A} \right]$

$\Rightarrow \left[L^p(X, \mathfrak{A}, \mu) \supset L^q(X, \mathfrak{A}, \mu) \right]$.

Now if there exists $B > 0$ such that $\mu(E) \le B$ for every $E \in \mathfrak{A}$ then in particular we have $\mu(X) \le B$ and thus $\mu(X) < \infty$. According to Theorem 17.4 and Theorem 17.13, $\mu(X) < \infty$ implies that $L^p(X, \mathfrak{A}, \mu) \supset L^q(X, \mathfrak{A}, \mu)$. ∎

Prob. 17.16. Prove the following theorem:

Theorem. *Let $(X, \mathfrak{A}, \mu)$ be an arbitrary measure space and let $0 < p < q \leq \infty$. Then we have $L^p(X, \mathfrak{A}, \mu) \supset L^q(X, \mathfrak{A}, \mu)$ if and only if there exists $B > 0$ such that $\mu(E) \leq B$ for every $E \in \mathfrak{A}$. Thus we have $L^p(X, \mathfrak{A}, \mu) \supset L^q(X, \mathfrak{A}, \mu)$ if and only if $\mu(X) < \infty$.*

Proof. By Prob. 17.15, we have the following two valid statements:

(1) $\qquad \left[L^p(X, \mathfrak{A}, \mu) \not\supset L^q(X, \mathfrak{A}, \mu) \right]$

$\qquad \Rightarrow \left[(X, \mathfrak{A}, \mu) \text{ has a } \mathfrak{A}\text{-measurable set of arbitrarily large finite measure} \right]$

and

(2) $\qquad \left[(X, \mathfrak{A}, \mu) \text{ has a } \mathfrak{A}\text{-measurable set of arbitrarily large finite measure} \right]$

$\qquad \Rightarrow \left[L^p(X, \mathfrak{A}, \mu) \not\supset L^q(X, \mathfrak{A}, \mu) \right].$

Now the contra-positive statement of (1) is given by

(3) $\qquad \left[L^p(X, \mathfrak{A}, \mu) \supset L^q(X, \mathfrak{A}, \mu) \right]$

$\qquad \Rightarrow \left[\text{There exists } B > 0 \text{ such that } \mu(E) \leq B \text{ for every } E \in \mathfrak{A} \right].$

Similarly the contra-positive statement of (2) is given by

(4) $\qquad \left[\text{There exists } B > 0 \text{ such that } \mu(E) \leq B \text{ for every } E \in \mathfrak{A} \right]$

$\qquad \Rightarrow \left[L^p(X, \mathfrak{A}, \mu) \supset L^q(X, \mathfrak{A}, \mu) \right].$

Then combining (3) and (4) we have

$\qquad \left[L^p(X, \mathfrak{A}, \mu) \supset L^q(X, \mathfrak{A}, \mu) \right]$

$\qquad \Leftrightarrow \left[\text{There exists } B > 0 \text{ such that } \mu(E) \leq B \text{ for every } E \in \mathfrak{A} \right].$

This completes the proof. ∎

Prob. 17.17. Let $f \in L^p(\mathbb{R}, \mathfrak{M}_L, \mu_L)$ where $p \in [1, \infty)$. With $h \in \mathbb{R}$, let us define a function f_h on $\mathbb{R}$ by setting $f_h(x) = f(x+h)$ for $x \in \mathbb{R}$. Show that $f_h \in L^p(\mathbb{R}, \mathfrak{M}_L, \mu_L)$ for every $h \in \mathbb{R}$ and $\lim_{h \to 0} \| f - f_h \|_p = 0$.

Proof. 1. Let us show that $f_h \in L^p(\mathbb{R}, \mathfrak{M}_L, \mu_L)$ and moreover $\| f_h \|_p = \| f \|_p$. By Theorem 9.31 (Translation Invariance of the Lebesgue Integral on $\mathbb{R}$), we have

$$\int_{\mathbb{R}} |f_h(x)|^p \, \mu_L(dx) = \int_{\mathbb{R}} |f(x+h)|^p \, \mu_L(dx) = \int_{\mathbb{R}} |f(x)|^p \, \mu_L(dx) < \infty.$$

This shows that $f_h \in L^p(\mathbb{R}, \mathfrak{M}_L, \mu_L)$ and moreover $\| f_h \|_p = \| f \|_p$.

2. Let us show that $\lim_{h \to 0} \| f - f_h \|_p = 0$.

According to Theorem 17.10, $C_c(\mathbb{R})$ is dense in $L^p(\mathbb{R}, \mathfrak{M}_L, \mu_L)$ for $p \in [1, \infty)$. Thus for every $\varepsilon > 0$ there exists $g \in C_c(\mathbb{R})$ such that $\| f - g \|_p < \varepsilon$. Let us write

$$\| f - f_h \|_p \leq \| f - g \|_p + \| g - g_h \|_p + \| g_h - f_h \|_p.$$

By Theorem 9.31 (Translation Invariance), we have $\| g_h - f_h \|_p = \| g - f \|_p < \varepsilon$. Thus we have

(1) $$\| f - f_h \|_p \leq \| g - g_h \|_p + 2\varepsilon.$$

Let us estimate $\| g - g_h \|_p$. Since $g \in C_c(\mathbb{R})$, g is continuous on $\mathbb{R}$ and vanishes outside of a finite closed interval $[-M, M]$. Then for $|h| < 1$, g_h vanishes outside of $[-M-1, M+1]$. Thus we have for $|h| < 1$

$$\| g - g_h \|_p^p = \int_{\mathbb{R}} |g(x) - g_h(x)|^p \, \mu_L(dx) = \int_{[-M-1,M+1]} |g(x) - g(x+h)|^p \, \mu_L(dx).$$

Then continuity of g on the bounded closed interval $J := [-M-1, M+1]$ implies that $\max_{x \in J} |g(x)|$ exists and then we have

$$|g(x) - g(x+h)|^p \leq \{2 \max_{x \in J} |g(x)|\}^p \quad \text{for } x \in [-M-1, M+1].$$

Thus by Theorem 7.16 (Bounded Convergence Theorem), we have

(2) $$\lim_{h \to 0} \| g - g_h \|_p^p = \lim_{h \to 0} \int_{[-M-1,M+1]} |g(x) - g(x+h)|^p \, \mu_L(dx)$$

$$= \int_{[-M-1,M+1]} \lim_{h \to 0} |g(x) - g(x+h)|^p \, \mu_L(dx)$$

$$= \int_{[-M-1,M+1]} 0 \, \mu_L(dx) = 0 \quad \text{by the continuity of } g.$$

This implies $\lim_{h \to 0} \| g - g_h \|_p = 0$. Applying this to (1), we have

$$\limsup_{h \to 0} \| f - f_h \|_p \leq 2\varepsilon.$$

Since this holds for every $\varepsilon > 0$, we have $\limsup_{h \to 0} \| f - f_h \|_p = 0$. This then implies $\liminf_{h \to 0} \| f - f_h \|_p = 0$ and then $\lim_{h \to 0} \| f - f_h \|_p = 0$. ∎

Prob. 17.18. Let $f \in L^p(\mathbb{R}, \mathfrak{M}_L, \mu_L)$ and $g \in L^q(\mathbb{R}, \mathfrak{M}_L, \mu_L)$ where $p, q \in (1, \infty)$ and are conjugates. Let us define a real-valued function F on $\mathbb{R}$ by setting

$$F(h) = \int_{\mathbb{R}} f(x+h)g(x)\mu_L(dx) \quad \text{for } h \in \mathbb{R}.$$

Show that F is a bounded continuous function on $\mathbb{R}$.

Proof. 1. With $f \in L^p(\mathbb{R}, \mathfrak{M}_L, \mu_L)$ and $h \in \mathbb{R}$, let us define a function f_h on $\mathbb{R}$ by setting

(1) $$f_h(x) = f(x+h) \quad \text{for } x \in \mathbb{R}.$$

Let us show that $f_h \in L^p(\mathbb{R}, \mathfrak{M}_L, \mu_L)$ and moreover $\|f_h\|_p = \|f\|_p$. By Theorem 9.31 (Translation Invariance of the Lebesgue Integral on $\mathbb{R}$), we have

$$\int_{\mathbb{R}} |f_h(x)|^p \mu_L(dx) = \int_{\mathbb{R}} |f(x+h)|^p \mu_L(dx) = \int_{\mathbb{R}} |f(x)|^p \mu_L(dx) < \infty.$$

This shows that $f_h \in L^p(\mathbb{R}, \mathfrak{M}_L, \mu_L)$ and moreover $\|f_h\|_p = \|f\|_p$. Then we have

(2) $$|F(h)| \leq \int_{\mathbb{R}} |f(x+h)g(x)|\mu_L(dx) = \int_{\mathbb{R}} |f_h(x)g(x)|\mu_L(dx)$$

$$\leq \|f_h\|_p\|g\|_q = \|f\|_p\|g\|_q < \infty \quad \text{by Theorem 16.14 (Hölder)}.$$

This shows that F is a bounded real-valued function on $\mathbb{R}$.

    2. Let us prove the continuity of F on $\mathbb{R}$. Observe that for $h, \Delta h \in \mathbb{R}$ we have $h + \Delta h \in \mathbb{R}$ and thus by (1) we have

(3) $$f_{h+\Delta h}(x) = f(x+h+\Delta h) = (f_h)_{\Delta h}(x) \quad \text{for } x \in \mathbb{R}.$$

As we showed above for f_h, we have $f_{h+\Delta h} \in L^p(\mathbb{R}, \mathfrak{M}_L, \mu_L)$ and moreover we have $\|f_{h+\Delta h}\|_p = \|f\|_p = \|f_h\|_p$. Then we have

(4) $$\left|F(h + \Delta h) - F(h)\right| = \left|\int_{\mathbb{R}} f(x+h+\Delta h)g(x)\,d\mu_L - \int_{\mathbb{R}} f(x+h)g(x)\,d\mu_L\right|$$

$$= \left|\int_{\mathbb{R}} f_{h+\Delta h}(x)g(x)\,d\mu_L - \int_{\mathbb{R}} f_h(x)g(x)\,d\mu_L\right|$$

$$= \left|\int_{\mathbb{R}} \{f_{h+\Delta h}(x) - f_h(x)\}g(x)\,d\mu_L\right|$$

$$\leq \int_{\mathbb{R}} |f_{h+\Delta h}(x) - f_h(x)||g(x)|\,d\mu_L$$

$$\leq \|(f_{h+\Delta h} - f_h\|_p\|g\|_q \quad \text{by Theorem 16.14 (Hölder)}$$

$$= \|(f_h)_{\Delta h} - f_h\|_p\|g\|_q \quad \text{by (3)}.$$

According to Prob. 17.17, we have $\lim_{\Delta h \to 0} \|(f_h)_{\Delta h} - f_h\|_p = 0$. This implies that we have $\lim_{\Delta h \to 0} \left|F(h + \Delta h) - F(h)\right| = 0$. This proves the continuity of F at every $h \in \mathbb{R}$. ∎

Prob. 17.19. Consider $(\mathbb{R}, \mathfrak{M}_L, \mu_L)$. Let $E \in \mathfrak{M}_L$ with $\mu_L(E) < \infty$. For $h \in \mathbb{R}$, consider the translated set $E - h$. Let us define a real-valued function F on $\mathbb{R}$ by setting

$$F(h) = \mu_L\big((E - h) \cap E\big) \quad \text{for } h \in \mathbb{R}.$$

Show that F is a bounded continuous function on $\mathbb{R}$.

Proof. We have

(1) $$F(h) = \mu_L\big((E - h) \cap E\big) = \int_{\mathbb{R}} \mathbf{1}_{(E-h) \cap E}\, d\mu_L \quad \text{for } h \in \mathbb{R}.$$

Observe that

(2) $$\mathbf{1}_{(E-h) \cap E} = \mathbf{1}_{E-h} \cdot \mathbf{1}_E$$

and

(3) $$\mathbf{1}_{E-h}(x) = \mathbf{1}_E(x + h) \quad \text{for } x \in \mathbb{R}.$$

Now the function $\mathbf{1}_E$ assumes only two values 0 and 1. This implies that $\mathbf{1}_E^2 = \mathbf{1}_E$. Then

$$\int_{\mathbb{R}} \mathbf{1}_E^2\, d\mu_L = \int_{\mathbb{R}} \mathbf{1}_E\, d\mu_L = \mu_L(E) < \infty.$$

This shows that $\mathbf{1}_E \in L^2(\mathbb{R}, \mathfrak{M}_L, \mu_L)$. Then we have $\mathbf{1}_{E-h} \in L^2(\mathbb{R}, \mathfrak{M}_L, \mu_L)$ by the same argument. Observe that $p = 2$ and $q = 2$ are conjugates since $\frac{1}{2} + \frac{1}{2} = 1$. Then we have

$$\begin{aligned} F(h) &= \int_{\mathbb{R}} \mathbf{1}_{(E-h) \cap E}\, d\mu_L = \int_{\mathbb{R}} \mathbf{1}_{E-h} \cdot \mathbf{1}_E\, d\mu_L \quad \text{by (1) and (2)} \\ &= \int_{\mathbb{R}} \mathbf{1}_E(x + h) \cdot \mathbf{1}_E(x)\, d\mu_L \quad \text{by (3)}. \end{aligned}$$

Then by Prob. 17.18, F is a bounded continuous function on $\mathbb{R}$. $\blacksquare$

Prob. 17.20. Let $f \in L^p(\mathbb{R}, \mathfrak{M}_L, \mu_L)$ and $g \in L^q(\mathbb{R}, \mathfrak{M}_L, \mu_L)$ where $p, q \in (1, \infty)$ are conjugates. With $h \in \mathbb{R}$, let $f_h(x) = f(x+h)$ and $g_h(x) = g(x+h)$ for $x \in \mathbb{R}$. Show that
(a) $\lim\limits_{h \to 0} \|f_h\, g - f\, g\|_1 = 0$.
(b) $\lim\limits_{h \to 0} \|f_h\, g_h - f\, g\|_1 = 0$.

Proof. Let us prove (b). (The proof of (a) is simpler.)
    1. Let us show first that $f_h \in L^p(\mathbb{R}, \mathfrak{M}_L, \mu_L)$ and $g_h \in L^q(\mathbb{R}, \mathfrak{M}_L, \mu_L)$. By Theorem 9.31 (Translation Invariance of the Lebesgue Integral on $\mathbb{R}$), we have

$$\int_{\mathbb{R}} |f_h(x)|^p\, \mu_L(dx) = \int_{\mathbb{R}} |f(x+h)|^p\, \mu_L(dx) = \int_{\mathbb{R}} |f(x)|^p\, \mu_L(dx) < \infty.$$

This shows that $f_h \in L^p(\mathbb{R}, \mathfrak{M}_L, \mu_L)$ and moreover $\|f_h\|_p = \|f\|_p$. Similarly by Theorem 9.31 we have

$$\int_{\mathbb{R}} |g_h(x)|^q\, \mu_L(dx) = \int_{\mathbb{R}} |g(x+h)|^q\, \mu_L(dx) = \int_{\mathbb{R}} |g(x)|^q\, \mu_L(dx) < \infty.$$

This shows that $g_h \in L^q(\mathbb{R}, \mathfrak{M}_L, \mu_L)$ and moreover $\|g_h\|_q = \|g\|_q$.
    2. If $f \in L^p(\mathbb{R}, \mathfrak{M}_L, \mu_L)$ and $g \in L^q(\mathbb{R}, \mathfrak{M}_L, \mu_L)$ then $fg \in L^1(\mathbb{R}, \mathfrak{M}_L, \mu_L)$ by Theorem 16.14 (Hölder's Inequality). Similarly we have $f_h\, g,\ f\, g_h,\ f_h\, g_h \in L^1(\mathbb{R}, \mathfrak{M}_L, \mu_L)$. Then since $L^1(\mathbb{R}, \mathfrak{M}_L, \mu_L)$ is a linear space we have

$$\|f_h\, g_h - f\, g\|_1 = \|f_h\, g_h - f_h\, g + f_h\, g - f\, g\|_1 \le \|f_h\, g_h - f_h\, g\|_1 + \|f_h\, g - f\, g\|_1$$

$$\le \|f_h\|_p \|g_h - g\|_q + \|f_h - f\|_p \|g\|_q \quad \text{by Theorem 16.14 (Hölder).}$$

By Prob. 17.17, we have $\lim\limits_{h \to 0} \|g_h - g\|_q = 0$ and $\lim\limits_{h \to 0} \|f_h - f\|_p = 0$. Thus we have

$$\lim_{h \to 0} \|f_h\, g_h - f\, g\|_1 \le \lim_{h \to 0} \left\{ \|f_h\|_p \|g_h - g\|_q \right\} + \lim_{h \to 0} \left\{ \|f_h - f\|_p \|g\|_q \right\}$$

$$= \|f\|_p \cdot 0 + 0 \cdot \|g\|_q = 0.$$

This shows that $\lim\limits_{h \to 0} \|f_h\, g_h - f\, g\|_1 = 0$. ∎

Prob. 17.21. Let $(a_n : n \in \mathbb{N})$ and $(b_n : n \in \mathbb{N})$ be two sequences of real numbers. Let $p, q \in (1, \infty)$ be such that $1/p + 1/q = 1$. Show that

$$\sum_{n \in \mathbb{N}} |a_n b_n| \le \Big\{ \sum_{n \in \mathbb{N}} |a_n|^p \Big\}^{1/p} \Big\{ \sum_{n \in \mathbb{N}} |b_n|^q \Big\}^{1/q}.$$

Proof. Consider the counting measure space $(\mathbb{N}, \mathfrak{P}(\mathbb{N}), \nu)$. Let us define two real-valued functions f and g on $\mathbb{N}$ by setting

(1)
$$\begin{cases} f(n) = a_n & \text{for } n \in \mathbb{N}, \\ g(n) = b_n & \text{for } n \in \mathbb{N}. \end{cases}$$

By Hölder's Inequality (Theorem 16.14), we have

(2)
$$\int_{\mathbb{N}} |fg| \, d\nu \le \Big\{ \int_{\mathbb{N}} |f|^p \, d\nu \Big\}^{1/p} \Big\{ \int_{\mathbb{N}} |g|^q \, d\nu \Big\}^{1/q}.$$

By (1) we have

(3)
$$\int_{\mathbb{N}} |fg| \, d\nu = \sum_{n \in \mathbb{N}} |f(n) g(n)| \, \nu(\{n\}) = \sum_{n \in \mathbb{N}} |a_n b_n| \cdot 1$$

(4)
$$\int_{\mathbb{N}} |f|^p \, d\nu = \sum_{n \in \mathbb{N}} |f(n)|^p \, \nu(\{n\}) = \sum_{n \in \mathbb{N}} |a_n|^p \cdot 1$$

(5)
$$\int_{\mathbb{N}} |g|^q \, d\nu = \sum_{n \in \mathbb{N}} |g(n)|^q \, \nu(\{n\}) = \sum_{n \in \mathbb{N}} |b_n|^q \cdot 1.$$

Substituting (3), (4), and (5) into (2), we have the proof. ∎

An Alternate Proof. Consider the measure space $\big((0, \infty), \mathfrak{M}_L|_{0,\infty}), \mu_L \big)$. In $(0, \infty)$, let $I_n = (n - 1, n]$ for $n \in \mathbb{N}$. Then $(0, \infty) = \bigcup_{n \in \mathbb{N}} I_n$. Define two real-valued functions f and g on $\mathbb{N}$ by setting

(1)
$$\begin{cases} f(x) = a_n & \text{for } x \in I_n, n \in \mathbb{N}, \\ g(x) = b_n & \text{for } x \in I_n, n \in \mathbb{N}. \end{cases}$$

By Hölder's Inequality (Theorem 16.14), we have

(2)
$$\int_{(0,\infty)} |fg| \, d\mu_L \le \Big\{ \int_{(0,\infty)} |f|^p \, d\mu_L \Big\}^{1/p} \Big\{ \int_{(0,\infty)} |g|^q \, d\mu_L \Big\}^{1/q}.$$

By (1) we have

(3)
$$\int_{(0,\infty)} |fg| \, d\mu_L = \sum_{n \in \mathbb{N}} \int_{I_n} |a_n b_n| \, d\mu_L = \sum_{n \in \mathbb{N}} |a_n b_n| \cdot 1$$

(4)
$$\int_{(0,\infty)} |f|^p \, d\mu_L = \sum_{n \in \mathbb{N}} \int_{I_n} |a_n|^p \, d\mu_L = \sum_{n \in \mathbb{N}} |a_n|^p \cdot 1$$

(5)
$$\int_{(0,\infty)} |g|^q \, d\mu_L = \sum_{n \in \mathbb{N}} \int_{I_n} |b_n|^q \, d\mu_L = \sum_{n \in \mathbb{N}} |b_n|^q \cdot 1.$$

Substituting (3), (4), and (5) into (2), we have the proof. ∎

A Third Proof. Recall that for $p, q \in (1, \infty)$ such that $1/p + 1/q = 1$ and $\alpha, \beta \in \mathbb{R}$, we have the inequality

$$(1) \qquad |\alpha\beta| \leq \frac{|\alpha|^p}{p} + \frac{|\beta|^q}{q}.$$

Let us prove

$$(2) \qquad \sum_{n\in\mathbb{N}} |a_n b_n| \leq \Big\{ \sum_{n\in\mathbb{N}} |a_n|^p \Big\}^{1/p} \Big\{ \sum_{n\in\mathbb{N}} |b_n|^q \Big\}^{1/q}.$$

If $\sum_{n\in\mathbb{N}} |a_n|^p = 0$, then $a_n = 0$ for every $n \in \mathbb{N}$ so that $\sum_{n\in\mathbb{N}} |a_n b_n| = 0$ and then (2) holds trivially. Similarly if $\sum_{n\in\mathbb{N}} |b_n|^q = 0$ then (2) holds trivially.

Let us consider the case that $\sum_{n\in\mathbb{N}} |a_n b_n| > 0$ and $\sum_{n\in\mathbb{N}} |b_n|^q > 0$. For brevity let us write

$$(3) \qquad A = \Big\{ \sum_{n\in\mathbb{N}} |a_n|^p \Big\}^{1/p} \quad \text{and} \quad \Big\{ \sum_{n\in\mathbb{N}} |b_n|^q \Big\}^{1/q}.$$

For $n \in \mathbb{N}$, let us write

$$(4) \qquad \alpha_n = \frac{|a_n|}{A} \quad \text{and} \quad \beta_n = \frac{|b_n|}{B}.$$

By (1) we have for every $n \in \mathbb{N}$

$$(5) \qquad |\alpha_n \beta_n| \leq \frac{|\alpha_n|^p}{p} + \frac{|\beta_n|^q}{q}.$$

Summing over $n \in \mathbb{N}$, we have

$$\sum_{n\in\mathbb{N}} |\alpha_n \beta_n| \leq \frac{1}{p} \sum_{n\in\mathbb{N}} |\alpha_n|^p + \frac{1}{q} \sum_{n\in\mathbb{N}} |\beta_n|^q,$$

that is,

$$\sum_{n\in\mathbb{N}} \frac{|a_n b_n|}{AB} \leq \frac{1}{p} \sum_{n\in\mathbb{N}} \frac{|a_n|^p}{A^p} + \frac{1}{q} \sum_{n\in\mathbb{N}} \frac{|b_n|^q}{B^q} = \frac{1}{p} + \frac{1}{q} = 1.$$

Then we have

$$\sum_{n\in\mathbb{N}} |a_n b_n| \leq AB = \Big\{ \sum_{n\in\mathbb{N}} |a_n|^p \Big\}^{1/p} \Big\{ \sum_{n\in\mathbb{N}} |b_n|^q \Big\}^{1/q}.$$

This completes the proof. ∎

Prob. 17.22. Let $(a_n : n \in \mathbb{N})$ and $(b_n : n \in \mathbb{N})$ be two sequences of real numbers. Let $p \in (0, 1)$ and $q \in (-\infty, 0)$ be such that $1/p + 1/q = 1$. Show that

$$\sum_{n \in \mathbb{N}} |a_n b_n| \geq \left\{ \sum_{n \in \mathbb{N}} |a_n|^p \right\}^{1/p} \left\{ \sum_{n \in \mathbb{N}} |b_n|^q \right\}^{1/q}.$$

Proof. Consider the measure space $\left((0, \infty), \mathfrak{M}_L |_{0, \infty)}, \mu_L \right)$. In $(0, \infty)$, let $I_n = (n - 1, n]$ for $n \in \mathbb{N}$. Then $(0, \infty) = \bigcup_{n \in \mathbb{N}} I_n$. Define two real-valued functions f and g on $\mathbb{N}$ by setting

(1)
$$\begin{cases} f(x) = a_n & \text{for } x \in I_n, n \in \mathbb{N}, \\ g(x) = b_n & \text{for } x \in I_n, n \in \mathbb{N}. \end{cases}$$

By Hölder's Inequality (Theorem 16.54), we have

(2)
$$\int_{(0,\infty)} |fg| \, d\mu_L \geq \left\{ \int_{(0,\infty)} |f|^p \, d\mu_L \right\}^{1/p} \left\{ \int_{(0,\infty)} |g|^q \, d\mu_L \right\}^{1/q}.$$

By (1) we have

(3)
$$\int_{(0,\infty)} |fg| \, d\mu_L = \sum_{n \in \mathbb{N}} \int_{I_n} |a_n b_n| \, d\mu_L = \sum_{n \in \mathbb{N}} |a_n b_n| \cdot 1$$

(4)
$$\int_{(0,\infty)} |f|^p \, d\mu_L = \sum_{n \in \mathbb{N}} \int_{I_n} |a_n|^p \, d\mu_L = \sum_{n \in \mathbb{N}} |a_n|^p \cdot 1$$

(5)
$$\int_{(0,\infty)} |g|^q \, d\mu_L = \sum_{n \in \mathbb{N}} \int_{I_n} |b_n|^q \, d\mu_L = \sum_{n \in \mathbb{N}} |b_n|^q \cdot 1.$$

Substituting (3), (4), and (5) into (2), we have the proof. ∎

Prob. 17.23. Let $(\varepsilon_n : n \in \mathbb{N})$ be a sequence of positive numbers. Let $p \in (0, \infty)$. Prove

$$\sum_{n \in \mathbb{N}} \varepsilon_n < \infty \Rightarrow \sum_{n \in \mathbb{N}} \varepsilon_n^p < \infty.$$

Proof. 1. Consider the case $p \in [1, \infty)$. Now $\sum_{n \in \mathbb{N}} \varepsilon_n < \infty$ implies $\lim_{n \to \infty} \varepsilon_n = 0$. Thus there exists $N \in \mathbb{N}$ such that $\varepsilon_n \leq 1$ for $n \geq N$. Then since $p \in [1, \infty)$ and $\varepsilon_n \leq 1$ for $n \geq N$, we have $\varepsilon_n^p \leq \varepsilon_n$ for $n \geq N$. Then we have $\sum_{n \geq N} \varepsilon_n^p \leq \sum_{n \geq N} \varepsilon_n < \infty$ and

$$\sum_{n \in \mathbb{N}} \varepsilon_n^p = \sum_{n=1}^{N} \varepsilon_n^p + \sum_{n > N} \varepsilon_n^p < \infty.$$

2. Consider the case $p \in (0, 1)$. Then $\sqrt{p} \in (0, 1)$ also. Let $r \in \mathbb{R}$ be the conjugate of $\sqrt{p}$, that is,

(1)
$$\frac{1}{\sqrt{p}} + \frac{1}{r} = 1.$$

This implies that $r \in (-\infty, 0)$. Indeed we have

$$\frac{1}{r} = 1 - \frac{1}{\sqrt{p}} = \frac{\sqrt{p} - 1}{\sqrt{p}} \quad \text{and} \quad r = \frac{\sqrt{p}}{\sqrt{p} - 1} \in (-\infty, 0).$$

Let us write $\varepsilon_n = \sum_{n \in \mathbb{N}} \varepsilon_n^{\sqrt{p}} \varepsilon_n^{1 - \sqrt{p}}$ and let $a_n = \varepsilon_n^{\sqrt{p}}$ and $b_n = \varepsilon_n^{1 - \sqrt{p}}$ and consider two sequences of real numbers $(a_n : n \in \mathbb{N})$ and $(b_n : n \in \mathbb{N})$. According to Prob. 17.22, (1) implies

(2)
$$\sum_{n \in \mathbb{N}} \varepsilon_n = \sum_{n \in \mathbb{N}} a_n b_n \geq \left\{ \sum_{n \in \mathbb{N}} a_n^{\sqrt{p}} \right\}^{\frac{1}{\sqrt{p}}} \left\{ \sum_{n \in \mathbb{N}} b_n^r \right\}^{\frac{1}{r}}$$

$$= \left\{ \sum_{n \in \mathbb{N}} (\varepsilon_n^{\sqrt{p}})^{\sqrt{p}} \right\}^{\frac{1}{\sqrt{p}}} \left\{ \sum_{n \in \mathbb{N}} (\varepsilon_n^{1 - \sqrt{p}})^r \right\}^{\frac{1}{r}}$$

$$= \left\{ \sum_{n \in \mathbb{N}} \varepsilon_n^p \right\}^{\frac{1}{\sqrt{p}}} \left\{ \sum_{n \in \mathbb{N}} (\varepsilon_n^{1 - \sqrt{p}})^r \right\}^{\frac{1}{r}}.$$

Now we have

(3)
$$C := \left\{ \sum_{n \in \mathbb{N}} (\varepsilon_n^{1 - \sqrt{p}})^r \right\}^{\frac{1}{r}} \in (0, \infty).$$

Then dividing (2) by $C > 0$, we have

(4)
$$\left\{ \sum_{n \in \mathbb{N}} \varepsilon_n^p \right\}^{\frac{1}{\sqrt{p}}} \leq \frac{1}{C} \left\{ \sum_{n \in \mathbb{N}} \varepsilon_n \right\},$$

and then

$$\sum_{n \in \mathbb{N}} \varepsilon_n^p \leq \frac{1}{C^{\sqrt{p}}} \left\{ \sum_{n \in \mathbb{N}} \varepsilon_n \right\}^{\sqrt{p}} < \infty.$$

This completes the proof. ∎

Prob. 17.24. According to Theorem 17.21, we have $\ell^p \subset \ell^q$ for $0 < p < q \leq \infty$. Show that $\ell^p \neq \ell^q$.

Proof. To show that $\ell^p \neq \ell^q$ we find a sequence of real numbers $(a_n : n \in \mathbb{N})$ such that $(a_n : n \in \mathbb{N}) \in \ell^q$ but $(a_n : n \in \mathbb{N}) \notin \ell^p$.

1. Consider the case $p < q < \infty$. Let

$$a_n = \frac{1}{n^{1/p}} \quad \text{for } n \in \mathbb{N}.$$

Then we have

$$\sum_{n \in \mathbb{N}} |a_n|^p = \sum_{n \in \mathbb{N}} \left| \frac{1}{n^{1/p}} \right|^p = \sum_{n \in \mathbb{N}} \frac{1}{n} = \infty,$$

$$\sum_{n \in \mathbb{N}} |a_n|^q = \sum_{n \in \mathbb{N}} \left| \frac{1}{n^{1/p}} \right|^p q = \sum_{n \in \mathbb{N}} \frac{1}{n^{q/p}} < \infty \quad \text{since } q/p > 1.$$

This shows that $(a_n : n \in \mathbb{N}) \in \ell^q$ but $(a_n : n \in \mathbb{N}) \notin \ell^p$.

2. Consider the case $p < q = \infty$. Let $a > 0$ and let $a_n = a$ for every $n \in \mathbb{N}$ and consider $(a_n : n \in \mathbb{N})$. Since $|a_n| = a$ for every $n \in \mathbb{N}$, we have $(a_n : n \in \mathbb{N}) \in \ell^\infty$. On the other hand, we have $\sum_{n \in \mathbb{N}} |a_n|^p = \sum_{n \in \mathbb{N}} a^p = \infty$ and this implies that $(a_n : n \in \mathbb{N}) \notin \ell^p$. ∎

Prob. 17.25. Let $0 < p < q \leq \infty$. Show that $\ell^p \not\supset \ell^q$.

Proof. Let us prove the statement $\ell^p \not\supset \ell^q$ by contradiction argument. Thus assume that the statement $\ell^p \not\supset \ell^q$ is false. Then we have $\ell^p \supset \ell^q$. Now according to Theorem 17.21 we have $\ell^p \subset \ell^q$. Thus we have $\ell^p = \ell^q$. This contradicts Prob. 17.24. Therefore the statement $\ell^p \not\supset \ell^q$ must be valid. ∎

An Alternate Proof. The proof is based on the following result in Calculus:

$$\sum_{n \in \mathbb{N}} \frac{1}{n^p} < \infty \qquad \text{for } p \in (1, \infty]$$
$$\sum_{n \in \mathbb{N}} \frac{1}{n^p} = \infty \qquad \text{for } p \in (0, 1].$$

1. Consider the case $0 < p < q < \infty$. There exists $r > 0$ such that $p < 1/r < q$. Then we have $0 < rp < 1 < rq < \infty$. Let $a_n = 1/n^r$ for every $n \in \mathbb{N}$ and consider the sequence $(a_n : n \in \mathbb{N})$. We have

$$\sum_{n \in \mathbb{N}} |a_n|^p = \sum_{n \in \mathbb{N}} \left|\frac{1}{n^r}\right|^p = \sum_{n \in \mathbb{N}} \frac{1}{n^{rp}} = \infty \quad \text{since } rp < 1,$$

$$\sum_{n \in \mathbb{N}} |a_n|^q = \sum_{n \in \mathbb{N}} \left|\frac{1}{n^r}\right|^q = \sum_{n \in \mathbb{N}} \frac{1}{n^{rq}} < \infty \quad \text{since } rq > 1.$$

This shows that $(a_n : n \in \mathbb{N}) \in \ell^q$ but $(a_n : n \in \mathbb{N}) \notin \ell^p$. Thus $\ell^p \not\supset \ell^q$.

2. The case $0 < p < q = \infty$ is treated in the same way as in **2.** of the Proof of Prop. 17.24. ∎

§18   Bounded Linear Functionals on the L^p Spaces

Prob. 18.1. Let $(X, \mathfrak{A}, \mu)$ be an arbitrary measure space and let $p \in (1, \infty)$. Consider $(f_n : n \in \mathbb{N}) \subset L^p(X, \mathfrak{A}, \mu)$ and $f \in L^p(X, \mathfrak{A}, \mu)$. Show that the sequence $(f_n : n \in \mathbb{N})$ converges weakly to f in $L^p(X, \mathfrak{A}, \mu)$ if and only if

1° $\quad \|f_n\|_p \leq M$ for all $n \in \mathbb{N}$ for some $M \geq 0$,

2° $\quad \lim_{n \to \infty} \int_E f_n \, d\mu = \int_E f \, d\mu$ for every $E \in \mathfrak{A}$ with $\mu(E) < \infty$.

Proof. 1. Suppose $(f_n : n \in \mathbb{N})$ converges weakly to f in $L^p(X, \mathfrak{A}, \mu)$. Let $q \in (1, \infty)$ be the conjugate of $p \in (1, \infty)$. Then we have

$$\lim_{n \to \infty} \int_X f_n g \, d\mu = \int_X f g \, d\mu \quad \text{for every } g \in L^q(X, \mathfrak{A}, \mu).$$

For each $n \in \mathbb{N}$, let L_n be a functional on $L^q(X, \mathfrak{A}, \mu)$ defined by setting

$$L_n(g) = \int_X f_n g \, d\mu \quad \text{for } g \in L^q(X, \mathfrak{A}, \mu).$$

Then L_n is a bounded linear functional on $L^q(X, \mathfrak{A}, \mu)$ with $\|L_n\|_* = \|f_n\|_p$ according to Theorem 18.1. Thus $(L_n : n \in \mathbb{N})$ is a collection of bounded linear functionals on $L^q(X, \mathfrak{A}, \mu)$ and for each fixed $g \in L^q(X, \mathfrak{A}, \mu)$ we have

$$|L_n(g)| \leq \|L_n\|_* \|g\|_q \quad \text{for all } n \in \mathbb{N}.$$

Thus by Corollary 15.50 to Theorem 15.49 (Uniform Boundedness Theorem), there exists $M \geq 0$ such that $\|L_n\|_* \leq M$ for all $n \in \mathbb{N}$. Then since $\|L_n\|_* = \|f_n\|_p$ as we showed above, we have $\|f_n\|_p \leq M$ for all $n \in \mathbb{N}$. This proves 1°.

If $E \in \mathfrak{A}$ and $\mu(E) < \infty$ then $\int_X (\mathbf{1}_E)^q \, d\mu = \int_X \mathbf{1}_E \, d\mu = \mu(E) < \infty$ so that we have $\mathbf{1}_E \in L^q(X, \mathfrak{A}, \mu)$. Then since $(f_n : n \in \mathbb{N})$ converges weakly to f in $L^p(X, \mathfrak{A}, \mu)$, we have $\lim_{n \to \infty} \int_X f_n \mathbf{1}_E \, d\mu = \int_X f \mathbf{1}_E \, d\mu$, that is, $\lim_{n \to \infty} \int_E f_n \, d\mu = \int_E f \, d\mu$. This proves 2°.

2. Conversely assume 1° and 2°. Let us show that $(f_n : n \in \mathbb{N})$ converges weakly to f in $L^p(X, \mathfrak{A}, \mu)$, that is,

(1) $$\lim_{n \to \infty} \int_X f_n g \, d\mu = \int_X f g \, d\mu \quad \text{for every } g \in L^q(X, \mathfrak{A}, \mu).$$

Let h be a μ-integrable simple function on $(X, \mathfrak{A}, \mu)$, that is,

(2) $$h = \sum_{i=1}^{k} c_i \mathbf{1}_{E_i},$$

where $\{E_i : i = 1, \ldots, k\}$ is a disjoint collection in $\mathfrak{A}$ with $\mu(E_i) < \infty$ for $i = 1, \ldots, k$ and $c_i \in \mathbb{C}$ for $i = 1, \ldots, k$.

Let $S_0(X, \mathfrak{A}, \mu)$ be the collection of all μ-integrable simple functions on $(X, \mathfrak{A}, \mu)$. Then $S_0(X, \mathfrak{A}, \mu)$ is a dense subset of $L^q(X, \mathfrak{A}, \mu)$ according to Theorem 18.3.

Let $h \in S_0(X, \mathfrak{A}, \mu)$ be as given by (2). For the set E_i in the expression (2) we have $\mu(E_i) < \infty$. Thus assumption of condition $2°$ implies $\lim_{n \to \infty} \int_{E_i} f_n \, d\mu = \int_{E_i} f \, d\mu$, that is,

$$(3) \qquad \lim_{n \to \infty} \int_X f_n \mathbf{1}_{E_i} \, d\mu = \int_X f \mathbf{1}_{E_i} \, d\mu.$$

Then by the linearity of the integral over the integrand, we have

$$(4) \qquad \lim_{n \to \infty} \int_X f_n h \, d\mu = \lim_{n \to \infty} \int_X f_n \Big\{ \sum_{i=1}^{k} c_i \mathbf{1}_{E_i} \Big\} \, d\mu \quad \text{by (2)}$$

$$= \lim_{n \to \infty} \sum_{i=1}^{k} c_i \Big\{ \int_X f_n \mathbf{1}_{E_i} \Big\} \, d\mu$$

$$= \lim_{n \to \infty} \sum_{i=1}^{k} c_i \Big\{ \int_X f \mathbf{1}_{E_i} \Big\} \, d\mu \quad \text{by (3)}$$

$$= \int_X f \Big\{ \sum_{i=1}^{k} c_i \mathbf{1}_{E_i} \Big\} \, d\mu$$

$$= \int_X f h \, d\mu \quad \text{by (2)}.$$

This proves (1) for the particular case that $g = h \in S_0(X, \mathfrak{A}, \mu)$.

Consider bounded linear functionals L_n and L on $L^q(X, \mathfrak{A}, \mu)$ defined by

$$(5) \qquad \begin{cases} L_n(g) = \int_X f_n g \, d\mu & \text{for } g \in L^q(X, \mathfrak{A}, \mu). \\ L(g) = \int_X f g \, d\mu & \text{for } g \in L^q(X, \mathfrak{A}, \mu). \end{cases}$$

In terms of L_n and L, the statement (1) is transcribed as

$$(6) \qquad \lim_{n \to \infty} L_n(g) = L(g) \quad \text{for every } g \in L^q(X, \mathfrak{A}, \mu).$$

Thus it remains to prove (6). Observe first that in terms of L_n and L our result (4) above is transcribed as

$$(7) \qquad \lim_{n \to \infty} L_n(h) = L(h) \quad \text{for every } h \in S_0(X, \mathfrak{A}, \mu).$$

To prove (6), we show that for every $\varepsilon > 0$ there exists $N \in \mathbb{N}$ such that

$$(8) \qquad |L_n(g) - L(g)| < \varepsilon \quad \text{for } n \geq N.$$

Let

$$B = \sup \big\{ \|f_n\|_p, n \in \mathbb{N}, \|f\|_p \big\} \in (0, \infty).$$

Let $\varepsilon > 0$ be arbitrarily given. Since $S_0(X, \mathfrak{A}, \mu)$ is a dense subset of $L^q(X, \mathfrak{A}, \mu)$ there exists $h \in S_0(X, \mathfrak{A}, \mu)$ such that

$$\|h - g\|_q < \frac{\varepsilon}{3B}.$$

Then (7) implies that there exists $N \in \mathbb{N}$ such that

$$|L_n(h) - L(h)| < \frac{\varepsilon}{3} \quad \text{for } n \geq N.$$

Then for $n \geq N$ we have

$$
\begin{aligned}
|L_n(g) - L(g)| &\leq |L_n(g) - L_n(h)| + |L_n(h) - L(h)| + |L(h) - L(g)| \\
&\leq \|L_n\|_* \|g - h\|_q + |L_n(h) - L(h)| + \|L\|_* \|h - g\|_q \\
&= \|f_n\|_p \|g - h\|_q + |L_n(h) - L(h)| + \|f\|_p \|h - g\|_q \\
&< B\frac{\varepsilon}{3B} + \frac{\varepsilon}{3} + B\frac{\varepsilon}{3B} = \varepsilon.
\end{aligned}
$$

This proves (8) and completes the proof of (1). ∎

Prob. 18.2. Let $(X, \mathfrak{A}, \mu)$ be a σ-finite measure space. Let $p \in [1, \infty)$ and $q \in (1, \infty]$ be conjugates, that is, $1/p + 1/q = 1$. Let $g_1, g_2 \in L^q(X, \mathfrak{A}, \mu)$. Assume that we have

$$\int_X fg_1\, d\mu = \int_X fg_2\, d\mu \quad \text{for every } f \in L^p(X, \mathfrak{A}, \mu).$$

Show that $g_1 = g_2$, μ-a.e. on $(X, \mathfrak{A}, \mu)$.

Proof. Let $g_1, g_2 \in L^q(X, \mathfrak{A}, \mu)$ and define two functionals L_{g_1} and L_{g_2} on $L^p(X, \mathfrak{A}, \mu)$ by setting

$$
\begin{cases}
L_{g_1}(f) = \int_X fg_1\, d\mu & \text{for } f \in L^p(X, \mathfrak{A}, \mu). \\
L_{g_2}(f) = \int_X fg_2\, d\mu & \text{for } f \in L^p(X, \mathfrak{A}, \mu).
\end{cases}
$$

Then L_{g_1} and L_{g_2} are bounded linear functionals on $L^p(X, \mathfrak{A}, \mu)$ by Theorem 18.1. The assumption that $\int_X fg_1\, d\mu = \int_X fg_2\, d\mu$ for every $f \in L^p(X, \mathfrak{A}, \mu)$ implies that we have $L_{g_1}(f) = L_{g_2}(f)$ for every $f \in L^p(X, \mathfrak{A}, \mu)$, that is, we have $L_{g_1} = L_{g_2}$.

Let $L = L_{g_1} = L_{g_2}$. By Theorem 18.6 (Riesz Representation Theorem) there exists a unique $g \in L^q(X, \mathfrak{A}, \mu)$ such that

$$L(f) = \int_X fg\, d\mu \quad \text{for } f \in L^p(X, \mathfrak{A}, \mu).$$

This implies that $g_1 = g = g_2$, that is, $g_1 = g_2$, μ-a.e. on $(X, \mathfrak{A}, \mu)$. ∎

Prob. 18.3. Let $(X, \mathfrak{A}, \mu)$ be a σ-finite measure space. Let $p \in [1, \infty]$. Let $f_0 \in L^p(X, \mathfrak{A}, \mu)$ be such that for every bounded linear functional L on $L^p(X, \mathfrak{A}, \mu)$ we have $L(f_0) = 0$. Show that $f_0 = 0 \in L^p(X, \mathfrak{A}, \mu)$.

Proof. Since $f_0 \in L^p(X, \mathfrak{A}, \mu)$, we have $f_0 = \Re f_0 + i \Im f_0$ where $\Re f_0$ and $\Im f_0$ are extended real-valued $\mathfrak{A}$-measurable functions on X and $\int_X |f_0|^p \, d\mu < \infty$. Now we have $f_0 = 0 \in L^p(X, \mathfrak{A}, \mu)$ if and only if $\Re f_0 = 0$ and $\Im f_0 = 0$, μ-a.e. on X.

To show that $f_0 = 0 \in L^p(X, \mathfrak{A}, \mu)$, we assume the contrary. Then at least one of the two statements "$\Re f_0 = 0$, μ-a.e. on X" and "$\Im f_0 = 0$, μ-a.e. on X" is false. Suppose the statement "$\Re f_0 = 0$, μ-a.e. on X" is false. Then we have

$$\mu(E) > 0 \quad \text{where} \quad E := \{x \in X : \Re f_0(x) \neq 0\}.$$

Let

$$E^+ = \{x \in X : \Re f_0(x) > 0\},$$
$$E^- = \{x \in X : \Re f_0(x) < 0\}.$$

Then $E^+, E^- \in \mathfrak{A}$, $E^+ \cap E^- = \emptyset$, $E^+ \cup E^- = E$ so that $\mu(E^+) + \mu(E^-) = \mu(E) > 0$ and this implies that at least one of $\mu(E^+)$ and $\mu(E^-)$ is positive. Suppose $\mu(E^+) > 0$.

Since $(X, \mathfrak{A}, \mu)$ is a σ-finite measure space, there exists a countable collection $\{D_m : m \in \mathbb{N}\}$ of disjoint $\mathfrak{A}$-measurable subsets of X such that $\bigcup_{m \in \mathbb{N}} D_m = X$ and $\mu(D_m) < \infty$ for every $m \in \mathbb{N}$. Then we have $E^+ = E^+ \cap X = \bigcup_{m \in \mathbb{N}} (E^+ \cap D_m)$ and

$$\sum_{m \in \mathbb{N}} \mu(E^+ \cap D_m) = \mu(E^+) > 0.$$

This implies that there exists $m_0 \in \mathbb{N}$ such that $\mu(E^+ \cap D_{m_0}) > 0$. then we have

$$0 < \mu(E^+ \cap D_{m_0}) \leq \mu(D_{m_0}) < \infty.$$

For each $k \in \mathbb{N}$, let

$$F_k = \left\{ x \in E^+ \cap D_{m_0} : f(x) > \tfrac{1}{k} \right\}.$$

Then we have $\bigcup_{k \in \mathbb{N}} F_k = E^+ \cap D_{m_0}$ and

$$\sum_{k \in \mathbb{N}} \mu(F_k) \geq \mu \left(\bigcup_{n \in \mathbb{N}} F_k \right) = \mu(E^+ \cap D_{m_0}) > 0.$$

This implies that there exists $k_0 \in \mathbb{N}$ such that

$$0 < \mu(F_{k_0}) \leq \mu(D_{m_0}) < \infty.$$

Consider the function $\mathbf{1}_{F_{k_0}}$ on X. For our $p \in [1, \infty]$, let $q \in [1, \infty]$ be its conjugate, that is, $1/p + 1/q = 1$.

If $p \in (1, \infty)$ then $q \in (1, \infty)$. Now

$$\int_X |\mathbf{1}_{F_{k_0}}|^q \, d\mu = \int_X \mathbf{1}_{F_{k_0}} \, d\mu = \mu(F_{k_0}) < 0.$$

Thus $\mathbf{1}_{F_{k_0}} \in L^q(X, \mathfrak{A}, \mu)$.

If $p = 1$ then $q = \infty$. Since the function $\mathbf{1}_{F_{k_0}}$ is bounded on X, we have $\mathbf{1}_{F_{k_0}} \in L^\infty(X, \mathfrak{A}, \mu) = L^q(X, \mathfrak{A}, \mu)$.

If $p = \infty$ then $q = 1$. Now

$$\int_X |\mathbf{1}_{F_{k_0}}|^1 \, d\mu = \int_X \mathbf{1}_{F_{k_0}} \, d\mu = \mu(F_{k_0}) < 0.$$

Thus $\mathbf{1}_{F_{k_0}} \in L^1(X, \mathfrak{A}, \mu) = L^q(X, \mathfrak{A}, \mu)$.

We have shown then that $\mathbf{1}_{F_{k_0}} \in L^q(X, \mathfrak{A}, \mu)$. According to Theorem 18.1, for $g \in L^q(X, \mathfrak{A}, \mu)$ if we define a function on $L^p(X, \mathfrak{A}, \mu)$ by setting

$$F(f) = \int_X f g \, d\mu \quad \text{for } f \in L^p(X, \mathfrak{A}, \mu),$$

then F is a bounded linear functional on $L^p(X, \mathfrak{A}, \mu)$. Let us define

$$F(f) = \int_X f \mathbf{1}_{F_{k_0}} \, d\mu \quad \text{for } f \in L^p(X, \mathfrak{A}, \mu).$$

Then F is a bounded linear functional on $L^p(X, \mathfrak{A}, \mu)$. Then since every bounded linear functional on $L^p(X, \mathfrak{A}, \mu)$ vanishes at f_0, we have

$$\int_X f_0 \mathbf{1}_{F_{k_0}} \, d\mu = 0.$$

On the other hand, we have

$$\int_X f_0 \mathbf{1}_{F_{k_0}} \, d\mu = \int_X \Re f_0 \mathbf{1}_{F_{k_0}} \, d\mu + i \int_X \Im f_0 \mathbf{1}_{F_{k_0}} \, d\mu$$

and

$$\int_X \Re f_0 \mathbf{1}_{F_{k_0}} \, d\mu \geq \frac{1}{k_0} \mu(F_{k_0}) > 0,$$

and thus

$$\int_X f_0 \mathbf{1}_{F_{k_0}} \, d\mu \neq 0.$$

This is a contradiction. We show similarly that the assumption that the statement "$\Im f_0 = 0$, μ-a.e. on X" is false leads to a contradiction. ∎

Prob. 18.4. Let $(X, \mathfrak{A}, \mu)$ be a σ-finite measure space. Let $p, q \in [1, \infty]$ be such that $1/p + 1/q = 1$. Let f be an extended real-valued $\mathfrak{A}$-measurable function on X such that $|f| < \infty$, μ-a.e. on X. Define a sequence of functions $(f_n : n \in \mathbb{N})$ by setting

$$f(x) = \begin{cases} f(x) & \text{if } |f(x)| \le n, \\ 0 & \text{otherwise.} \end{cases}$$

Suppose that for every extended real-valued function $g \in L^q(X, \mathfrak{A}, \mu)$ we have

$$\lim_{n \to \infty} \int_X f_n g \, d\mu \in \mathbb{R}.$$

Show that $\int_X fg \, d\mu$ exists in $\mathbb{R}$ for every $g \in L^q(X, \mathfrak{A}, \mu)$ and then $f \in L^p(X, \mathfrak{A}, \mu)$.

Proof. Let $g \in L^q(X, \mathfrak{A}, \mu)$ and write $g = g^+ - g^-$. Then $g^+ \le |g|$ on X. This implies that $\int_X |g^+|^q \, d\mu \le \int_X |g|^q \, d\mu < \infty$ and thus $g^+ \in L^q(X, \mathfrak{A}, \mu)$. Similarly we have $g^- \in L^q(X, \mathfrak{A}, \mu)$.

For $x \in X$ such that $|f(x)| < \infty$, we have $\lim_{n \to \infty} f_n(x) = f(x)$ and in particular if $f(x) \ge 0$ then $f_n(x) \uparrow f(x)$ and if $f(x) \le 0$ then $f_n(x) \downarrow f(x)$. Let us decompose X into three disjoint $\mathfrak{A}$-measurable sets defined by

$$A = \{x \in X : f(x) \in [0, \infty)\},$$
$$B = \{x \in X : f(x) \in (-\infty, 0)\},$$
$$C = \{x \in X : f(x) = \pm\infty\}.$$

Then we have $f_n \uparrow f$ on A, $f_n \downarrow f$ on B, and $\mu(C) = 0$. Now since $\mathbf{1}_A, \mathbf{1}_B \in [0, 1]$ on X, we have $\mathbf{1}_A g^+ \le g^+, \mathbf{1}_A g^- \le g^-, \mathbf{1}_B g^+ \le g^+$, and $\mathbf{1}_B g^- \le g^-$. This implies then

(9) $$\mathbf{1}_A g^+, \mathbf{1}_A g^-, \mathbf{1}_B g^+, \mathbf{1}_B g^- \in L^q(X, \mathfrak{A}, \mu).$$

Then (3) and assumption (2) imply

(10) $$\lim_{n \to \infty} \int_X f_n \mathbf{1}_A g^+ \, d\mu \in \mathbb{R},$$

(11) $$\lim_{n \to \infty} \int_X f_n \mathbf{1}_B g^+ \, d\mu \in \mathbb{R},$$

(12) $$\lim_{n \to \infty} \int_X f_n \mathbf{1}_A g^- \, d\mu \in \mathbb{R},$$

(13) $$\lim_{n \to \infty} \int_X f_n \mathbf{1}_B g^- \, d\mu \in \mathbb{R}.$$

Regarding the integral in (4), we have $\int_X f_n \mathbf{1}_A g^+ \, d\mu = \int_A f_n g^+ \, d\mu$. Since $f_n \uparrow f$ on A and since $g^+ \ge 0$, we have $f_n g^+ \uparrow f g^+$ on A. Thus by Theorem 8.5 (Monotone Convergence Theorem) and by (4), we have

(14) $$\int_A fg^+ \, d\mu = \lim_{n \to \infty} \int_A f_n g^+ \, d\mu = \lim_{n \to \infty} \int_X f_n \mathbf{1}_A g^+ \, d\mu \in \mathbb{R}.$$

Similarly regarding the integral in (5), we have $\int_X f_n 1_B g^+ \, d\mu = \int_B f_n g^+ \, d\mu$. Since $f_n \downarrow f$ on B and since $g^+ \geq 0$, we have $f_n g^+ \downarrow f g^+$ on B. Then by Theorem 9.17 (Generalized Monotone Convergence Theorem) and by (5), we have

(15) $$\int_B f g^+ \, d\mu = \lim_{n \to \infty} \int_B f_n g^+ \, d\mu = \lim_{n \to \infty} \int_X f_n 1_B g^+ \, d\mu \in \mathbb{R}.$$

Then by (8) and (9), we have

(16) $$\int_X f g^+ \, d\mu = \int_A f g^+ \, d\mu + \int_B f g^+ \, d\mu \in \mathbb{R}.$$

We show similarly

(17) $$\int_X f g^- \, d\mu = \int_A f g^- \, d\mu + \int_B f g^- \, d\mu \in \mathbb{R}.$$

Then by (10) and (11), we have

$$\int_X f g \, d\mu = \int_X f \{ g^+ - g^- \} \, d\mu = \int_X f g^+ \, d\mu - \int_X f g^- \, d\mu \in \mathbb{R}.$$

This shows that $\int_X f g \, d\mu$ exists in $\mathbb{R}$ for every $g \in L^q(X, \mathfrak{A}, \mu)$. Then by Theorem 18.8, we have $f \in L^p(X, \mathfrak{A}, \mu)$. ∎

§22 Lebesgue-Stieltjes Measure Spaces

Prolog to Prob. 22.1 and Prob. 22.2. The Lebesgue-Stieltjes measure λ_g determined by a real-valued increasing function g on $\mathbb{R}$ is a σ-finite measure on a semialgebra of subsets of $\mathbb{R}$ which contains $\mathfrak{B}_\mathbb{R}$. Prob. 22.1 asserts that every finite measure μ on $(\mathbb{R}, \mathfrak{B}_\mathbb{R})$ can be extended to be a finite Lebesgue-Stieltjes measure. Prob. 22.2 gives a necessary and sufficient condition for a measure μ on $(\mathbb{R}, \mathfrak{B}_\mathbb{R})$ to be extendible to a Lebesgue-Stieltjes measure.

Prob. 22.1. Let μ be a finite measure on $(\mathbb{R}, \mathfrak{B}_\mathbb{R})$ and let $\mu(\mathbb{R}) = C < \infty$. Define a function F on $\mathbb{R}$ by setting $F(t) = \mu\big((-\infty, t]\big)$ for $t \in \mathbb{R}$.
(a) Show that F is a nonnegative real-valued increasing and right-continuous function on $\mathbb{R}$ such that $\lim_{t \to -\infty} F(t) = 0$ and $\lim_{t \to \infty} F(t) = C$.
(b) Show that for the Lebesgue-Stieltjes measure λ_F determined by the real-valued increasing function F we have $\lambda_F = \mu$ on $\mathfrak{B}_\mathbb{R}$.

Proof. **1.** By its definition F is a nonnegative increasing function on $\mathbb{R}$ and, since $\mu\big((-\infty, t]\big) \leq \mu(\mathbb{R}) < \infty$, F is real-valued on $\mathbb{R}$. Consider a decreasing sequence $\big((-\infty, -n] : n \in \mathbb{N}\big)$ in $\mathfrak{B}_\mathbb{R}$. Since $\mu(\mathbb{R}) < \infty$, we have

$$\lim_{n \to \infty} \mu\big((-\infty, -n]\big) = \mu\big(\lim_{n \to \infty} (-\infty, -n]\big)$$

$$= \mu\big(\bigcap_{n \in \mathbb{N}} (-\infty, -n]\big) = \mu(\emptyset) = 0.$$

Then we have

$$\lim_{t \to -\infty} F(t) = \lim_{n \to \infty} F(-n) = \lim_{n \to \infty} \mu\big((-\infty, -n]\big) = 0.$$

Similarly since $\lim_{n \to \infty} (-\infty, n] = \bigcup_{n \in \mathbb{N}} (-\infty, n]$ we have

$$\lim_{t \to \infty} F(t) = \lim_{n \to \infty} F(n) = \lim_{n \to \infty} \mu\big((-\infty, n]\big)$$

$$= \mu\big(\bigcup_{n \in \mathbb{N}} (-\infty, n]\big) = \mu(\mathbb{R}) = C.$$

To show that F is right-continuous at every $t_0 \in \mathbb{R}$, it suffices to show that for every strictly decreasing sequence $(t_n : n \in \mathbb{N})$ in $\mathbb{R}$ such that $t_n \downarrow t_0$ we have $F(t_0) = \lim_{n \to \infty} F(t_n)$. Now $t < t_n$ implies $(-\infty, t_0] \subset (-\infty, t_n]$ and then we have $\mu\big((-\infty, t_0]\big) \leq \mu\big((-\infty, t_n]\big)$, that is, $F(t_0) \leq F(t_n)$ for every $n \in \mathbb{N}$. Thus $F(t_0) \leq \lim_{n \to \infty} F(t_n)$. Suppose $F(t_0) \neq \lim_{n \to \infty} F(t_n)$. Then $F(t_0) > \lim_{n \to \infty} F(t_n)$. This implies that there exists $n_0 \in \mathbb{N}$ such that $F(t_0) > F(t_{n_0})$, that is, $\mu\big((-\infty, t_0]\big) > \mu\big((-\infty, t_{n_0}]\big)$, a contradiction. Therefore we must have $F(t_0) = \lim_{n \to \infty} F(t_n)$.
   2. Consider the Lebesgue-Stieltjes measure λ_F determined by the real-valued increasing function F defined on a σ-algebra $\mathfrak{M}(\lambda_F^*)$ of subsets of $\mathbb{R}$. We have $\mathfrak{B}_\mathbb{R} \subset \mathfrak{M}(\lambda_F^*)$ by Theorem 22.4.

Consider the two measures λ_F and μ on $\mathfrak{B}_{\mathbb{R}}$. For the semialgebra $\mathfrak{I}_{oc}$ of subsets of $\mathbb{R}$, we have $\mathfrak{B}_{\mathbb{R}} = \sigma(\mathfrak{I}_{oc})$. According to Theorem 21.11, if $\lambda_F = \mu$ on the semialgebra $\mathfrak{I}_{oc}$ then $\lambda_F = \mu$ on $\sigma(\mathfrak{I}_{oc}) = \mathfrak{B}_{\mathbb{R}}$. Thus to show that $\lambda_F = \mu$ on $\mathfrak{B}_{\mathbb{R}}$ it suffices to show that $\lambda_F = \mu$ on $\mathfrak{I}_{oc}$. Let $I \in \mathfrak{I}_{oc}$ be given by $I = (a, b]$. Then by Proposition 22.7, we have

$$\lambda_F\big((a, b]\big) = F(b) - F(a) = \mu\big((-\infty, b]\big) - \mu\big((-\infty, a]\big)$$
$$= \mu\big((-\infty, b] \setminus (-\infty, a]\big) = \mu\big((a, b]\big).$$

If $I \in \mathfrak{I}_{oc}$ is given by $I = (a, \infty)$ then

$$\lambda_F\big((a, \infty)\big) = \lim_{n \to \infty} \lambda_F\big((a, n]\big) = \lim_{n \to \infty} \big\{ F(n) - F(a) \big\}$$
$$= \lim_{n \to \infty} \big\{ \mu\big((-\infty, n]\big) - \mu\big((-\infty, a]\big) \big\}$$
$$= \lim_{n \to \infty} \mu\big((-\infty, n] \setminus (-\infty, a]\big)$$
$$= \lim_{n \to \infty} \mu\big((a, n]\big) = \mu\big((a, \infty)\big).$$

If $I \in \mathfrak{I}_{oc}$ is given by $I = (-\infty, b]$ then

$$\lambda_F\big((-\infty, b]\big) = \lim_{n \to \infty} \lambda_F\big((-n, b]\big) = \lim_{n \to \infty} \big\{ F(b) - F(-n) \big\}$$
$$= \lim_{n \to \infty} \big\{ \mu\big((-\infty, b]\big) - \mu\big((-\infty, -n]\big) \big\}$$
$$= \lim_{n \to \infty} \mu\big((-\infty, b] \setminus (-\infty, -n]\big)$$
$$= \lim_{n \to \infty} \mu\big((-n, b]\big) = \mu\big((-\infty, b]\big).$$

This completes the proof that $\lambda_F = \mu$ on $\mathfrak{I}_{oc}$. ∎

Prob. 22.2. Let μ be a measure on $(\mathbb{R}, \mathfrak{B}_{\mathbb{R}})$ such that $\mu\big((-k, k)\big) < \infty$ for every $k \in \mathbb{N}$. Define a function F on $\mathbb{R}$ by setting

$$F(t) = \begin{cases} \mu\big((0, t]\big) & \text{for } t > 0, \\ 0 & \text{for } t = 0, \\ -\mu\big((t, 0]\big) & \text{for } t < 0. \end{cases}$$

(a) Show that F is a real-valued increasing right-continuous function on $\mathbb{R}$.
(b) Let λ_F be the Lebesgue-Stieltjes measure determined by the real-valued increasing function F on $\mathbb{R}$. Show that $\lambda_F = \mu$ on $\mathfrak{B}_{\mathbb{R}}$.

Proof. 1. Let us show that F is real-valued on $\mathbb{R}$. Let $t \in \mathbb{R}$ be arbitrarily selected. Let $k \in \mathbb{N}$ be so large that $t \in (-k, k)$. If $t > 0$ then $F(t) = \mu\big((0, t]\big) \le \mu\big((-k, k)\big) < \infty$. If $t < 0$ then $F(t) = -\mu\big((t, 0]\big)$. Now $\mu\big((t, 0]\big) \le \mu\big((-k, k)\big) < \infty$. Thus we have $F(t) = -\mu\big((t, 0]\big) > -\infty$. For $t = 0$ we have $F(0) = 0$ by definition. Therefore we have $F(t) \in \mathbb{R}$ for every $t \in \mathbb{R}$.

Let us show that F is increasing on $\mathbb{R}$. Consider $(0, \infty)$. For $0 < t' < t''$, we have $F(t') = \mu\big((0, t']\big) \le \mu\big((0, t'']\big) = F(t'')$. This shows that F is increasing on $(0, \infty)$. Consider next $(-\infty, 0)$. For $t' < t'' < 0$, we have $(t', 0] \supset (t'', 0]$ and $\mu\big((t', 0]\big) \ge \mu\big((t'', 0]\big)$. Then $F(t') = -\mu\big((t', 0]\big) \le -\mu\big((t'', 0]\big) = F(t'')$. This shows that F is increasing on $(-\infty, 0)$. This and the fact that $F \le 0$ on $(-\infty, 0)$, $F(0) = 0$, and $F \ge 0$ on $(0, \infty)$, imply that F is increasing on $\mathbb{R}$.

2. Let us show that F is right-continuous on $\mathbb{R}$. To show this, it suffices to show that for every $t_0 \in \mathbb{R}$ and every strictly decreasing sequence $(t_n : n \in \mathbb{N})$ such that $t_n \downarrow t_0$ we have $F(t_0) = \lim_{n \to \infty} F(t_n)$.

Consider the case $t_0 > 0$. Since $t_0 < t_n$, we have $(0, t_0] \subset (0, t_n]$ and then $\mu\big((0, t_0]\big) \le \mu\big((0, t_n]\big)$, that is, $F(t_0) \le F(t_n)$ for every $n \in \mathbb{N}$ and hence $F(t_0) \le \lim_{n \to \infty} F(t_n)$. Suppose $F(t_0) \ne \lim_{n \to \infty} F(t_n)$. Then $F(t_0) > \lim_{n \to \infty} F(t_n)$ and thus there exists $n_0 \in \mathbb{N}$ such that $F(t_0) > F(t_{n_0})$, that is, $\mu\big((0, t_0]\big) > \mu\big((0, t_{n_0}]\big)$, a contradiction. Therefore we must have $F(t_0) = \lim_{n \to \infty} F(t_n)$.

Next consider the case $t_0 < 0$. Since $t_0 < t_n$, we have $(t_0, 0] \supset (t_n, 0]$ and then $\mu\big((t_0, 0]\big) \ge \mu\big((t_n, 0]\big)$ so that $F(t_0) = -\mu\big((t_0, 0]\big) \le -\mu\big((t_n, 0]\big) = F(t_n)$ for every $n \in \mathbb{N}$. Thus $F(t_0) \le \lim_{n \to \infty} F(t_n)$. Suppose $F(t_0) \ne \lim_{n \to \infty} F(t_n)$. Then $F(t_0) > \lim_{n \to \infty} F(t_n)$ and thus there exists $n_0 \in \mathbb{N}$ such that $F(t_0) > F(t_{n_0})$, that is, $-\mu\big((t_0, 0]\big) > -\mu\big((t_n, 0]\big)$ and $\mu\big((t_0, 0]\big) < \mu\big((t_n, 0]\big)$, a contradiction. Therefore we must have $F(t_0) = \lim_{n \to \infty} F(t_n)$.

Finally consider the case $t_0 = 0$. Since $0 = t_0 < t_n$, we have $F(t_n) = \mu\big((0, t_n]\big) \ge 0 = F(0)$ for every $n \in \mathbb{N}$. Thus $F(0) \le \lim_{n \to \infty} F(t_n)$. Now $\big((0, t_n] : n \in \mathbb{N}\big)$ is a decreasing sequence in $\mathfrak{B}_{\mathbb{R}}$ and for sufficiently large $k \in \mathbb{N}$ we have $(0, t_n] \subset (-k, k)$ for all $n \in \mathbb{N}$ and $\mu\big((-k, k)\big) < \infty$. Thus we have $\lim_{n \to \infty} \mu\big((0, t_n]\big) = \mu\big(\bigcup_{n \in \mathbb{N}}(0, t_n]\big) = \mu(\emptyset) = 0$. Then $\lim_{n \to \infty} F(t_n) = \lim_{n \to \infty} \mu\big((0, t_n]\big) = 0 = F(0)$. This proves the right-continuity of F at $t_0 = 0$.

3. Consider the Lebesgue-Stieltjes measure λ_F determined by the real-valued increasing function F defined on a σ-algebra $\mathfrak{M}(\lambda_F^*)$ of subsets of $\mathbb{R}$. We have $\mathfrak{B}_{\mathbb{R}} \subset \mathfrak{M}(\lambda_F^*)$ by Theorem 22.4.

Consider the two measures λ_F and μ on $\mathfrak{B}_{\mathbb{R}}$. For the semialgebra $\mathfrak{I}_{oc}$ of subsets of $\mathbb{R}$, we have $\mathfrak{B}_{\mathbb{R}} = \sigma(\mathfrak{I}_{oc})$. According to Theorem 21.11, if $\lambda_F = \mu$ on the semialgebra $\mathfrak{I}_{oc}$ then $\lambda_F = \mu$ on $\sigma(\mathfrak{I}_{oc}) = \mathfrak{B}_{\mathbb{R}}$. Thus to show that $\lambda_F = \mu$ on $\mathfrak{B}_{\mathbb{R}}$ it suffices to show that $\lambda_F = \mu$ on $\mathfrak{I}_{oc}$. Let $I \in \mathfrak{I}_{oc}$ be given by $I = (a, b]$. There are several cases to consider.
Case $0 < a < b$:

$$\lambda_F((a, b]) = F(b) - F(a) = \mu((0, b]) - \mu((0, a])$$
$$= \mu((0, b] \setminus (0, a]) = \mu((a, b]).$$

Case $a < 0 < b$:

$$\lambda_F((a, b]) = F(b) - F(a) = \mu((0, b]) - \{-\mu((a, 0])\}$$
$$= \mu((a, 0]) + \mu((0, b]) = \mu((a, 0] \cup (0, b])$$
$$= \mu((a, b]).$$

Case $a < b < 0$:

$$\lambda_F((a, b]) = F(b) - F(a) = -\mu((b, 0]) - \{-\mu((a, 0])\}$$
$$= \mu((a, 0]) - \mu((b, 0]) = \mu((a, 0] \setminus (b, 0])$$
$$= \mu((a, b]).$$

Case $0 = a < b$:

$$\lambda_F((a, b]) = F(b) - F(a) = F(b) - F(0)$$
$$= \mu((0, b]) = \mu((a, b]).$$

Case $a < b = 0$:

$$\lambda_F((a, b]) = F(b) - F(a) = F(0) - F(a)$$
$$= -\{-\mu((a, 0])\} = \mu((a, b]).$$

This verifies that $\lambda_F = \mu$ on $\mathfrak{I}_{oc}$. ∎

Comment on Prob. 22.2. If μ_g is a Lebesgue-Stieltjes measure determined by a real-valued increasing function g on $\mathbb{R}$, then $\mu_g((-k, k)) = g(k-) - g(-k+) < \infty$ for every $k \in \mathbb{N}$. Prob. 22.2 asserts that a measure μ on $(\mathbb{R}, \mathfrak{B}_{\mathbb{R}})$ can be extended to be a Lebesgue-Stieltjes measure if and only if μ satisfies the condition that $\mu((-k, k)) < \infty$ for every $k \in \mathbb{N}$. Thus in particular every finite measure μ on $(\mathbb{R}, \mathfrak{B}_{\mathbb{R}})$ can be extended to be a Lebesgue-Stieltjes measure.

Prob. 22.3. *Let F be a real-valued right-continuous increasing function on $\mathbb{R}$ such that $F = 0$ on $(-\infty, 0)$. Let λ_F be the Lebesgue-Stieltjes measure on $\mathbb{R}$ determined by F.*
(a) *Show that*

$$\text{(1)} \qquad \int_{[0,\infty)} t\,\lambda_F(dt) = \int_{[0,\infty)} \lambda_F\big((t,\infty)\big)\,\mu_L(dt).$$

(b) *Show that if in addition we have $\lim_{t\to\infty} F(t) = C < \infty$ then*

$$\text{(2)} \qquad \int_{[0,\infty)} t\,\lambda_F(dt) = \int_{[0,\infty)} \{C - F(t)\}\,\mu_L(dt).$$

Proof. 1. The Lebesgue-Stieltjes measure λ_F is defined on a σ-algebra $\mathfrak{M}(\lambda_F^*)$ of subsets of $\mathbb{R}$ and $\mathfrak{B}_{\mathbb{R}} \subset \mathfrak{M}(\lambda_F^*)$ by Theorem 22.4. By Proposition 22.7 we have for $0 \le t' < t''$

$$\text{(3)} \qquad \lambda_F\big([0, t']\big) = F(t'+) - F(0-) = F(t'),$$

$$\lambda_F\big((t', t'']\big) = F(t''+) - F(t'+) = F(t'') - F(t').$$

Consider the integral $\int_{[0,\infty)} t\,\lambda_F(dt)$. Let $\varphi(t) = t$ for $t \in [0, \infty)$. Then we have

$$\text{(4)} \qquad \int_{[0,\infty)} t\,\lambda_F(dt) = \int_{[0,\infty)} \varphi(t)\,\lambda_F(dt).$$

Consider a sequence of $\mathfrak{B}_{\mathbb{R}}$-measurable functions on $[0, \infty)$ given by $\big(\mathbf{1}_{[0,n]} : n \in \mathbb{N}\big)$. We have $\mathbf{1}_{[0,n]} \uparrow \mathbf{1}_{[0,\infty)}$ on $[0, \infty)$ as $n \to \infty$. Then by the Monotone Convergence Theorem (Theorem 8.5), we have

$$\text{(5)} \qquad \int_{[0,\infty)} \varphi(t)\,\lambda_F(dt) = \lim_{n\to\infty} \int_{[0,\infty)} \varphi(t)\mathbf{1}_{[0,n]}(t)\,\lambda_F(dt)$$

$$= \lim_{n\to\infty} \int_{[0,n]} \varphi(t)\,\lambda_F(dt).$$

To evaluate the last integral in (5), let us evaluate $\int_{[0,a]} \varphi(t)\,\lambda_F(dt)$ for an arbitrary $a > 0$. For each $k \in \mathbb{N}$, let us decompose the interval $[0, a]$ into 2^k subintervals $J_{k,\ell}$ defined by

$$J_{k,1} = \Big[0, \frac{a}{2^k}\Big] \quad \text{and} \quad J_{k,\ell} = \Big((\ell-1)\frac{a}{2^k}, \ell\frac{a}{2^k}\Big] \quad \text{for } \ell = 2, 3, \dots, 2^k.$$

Let $\big(\varphi_k : k \in \mathbb{N}\big)$ be a sequence of simple function on $[0, a]$ defined by

$$\varphi_k = \sum_{\ell=1}^{2^k} \ell\frac{a}{2^k}\mathbf{1}_{J_{k,\ell}}.$$

Note that $\varphi_k \downarrow \varphi$ on $[0, a]$ as $k \to \infty$ and $\varphi_k(t) \in [0, a]$ for $t \in [0, a]$ and $k \in \mathbb{N}$. Now $\varphi_k \le a$ on $[0, a]$ and by (3)

$$\int_{[0,a]} a\,\lambda_F(dt) = a\lambda_F\big([0, a]\big) = a\{F(a+) - F(0-)\} = aF(a) < \infty.$$

Thus by the Dominated Convergence Theorem (Theorem 9.20) we have

(6) $$\int_{[0,a]} \varphi(t)\, \lambda_F(dt) = \lim_{k\to\infty} \int_{[0,a]} \varphi_k(t)\, \lambda_F(dt).$$

Next let us evaluate $\int_{[0,a]} \varphi_k(t)\, \lambda_F(dt)$ for each $k \in \mathbb{N}$. Now since φ_k is a simple function, we have by (3)

(7) $$\int_{[0,a]} \varphi_k(t)\, \lambda_F(dt) = \sum_{\ell=1}^{2^k} \ell \frac{a}{2^k}\, \lambda_F(J_{k,\ell})$$

$$= \sum_{\ell=1}^{2^k} \ell \frac{a}{2^k}\left\{F\left(\ell \frac{a}{2^k}\right) - F\left((\ell-1)\frac{a}{2^k}\right)\right\}.$$

To simplify the sum in (7), consider a sequence of real numbers $\left(A_\ell : \ell = 0, 1, \ldots, m\right)$. By simple computation we have

(8) $$\sum_{\ell=1}^{m} \ell\{A_\ell - A_{\ell-1}\} = m A_m - \sum_{\ell=0}^{m-1} A_\ell.$$

Let

$$A_\ell = F\left(\ell \frac{a}{2^k}\right) \quad \text{for } \ell = 0, 1, \ldots, 2^k.$$

Then by (7) and (8) we have

(9) $$\int_{[0,a]} \varphi_k(t)\, \lambda_F(dt) = \frac{a}{2^k}\left\{2^k F\left(2^k \frac{a}{2^k}\right) - \sum_{\ell=0}^{2^k-1} F\left(\ell \frac{a}{2^k}\right)\right\}$$

$$= aF\left(2^k \frac{a}{2^k}\right) - \sum_{\ell=0}^{2^k-1} F\left(\ell \frac{a}{2^k}\right)\frac{a}{2^k}$$

$$= aF(a) - \sum_{\ell=0}^{2^k-1} F\left(\varphi_k\left(\ell \frac{a}{2^k}\right)\right)\frac{a}{2^k}$$

$$= \int_{[0,a]} F(a)\, \mu_L(dt) - \int_{[0,a]} F\left(\varphi_k(t)\right) \mu_L(dt).$$

Now $\varphi_k(t) \downarrow t$ as $k \to \infty$ at every $t \in [0, a]$. Then since F is right-continuous at t, we have $\lim_{k\to\infty} F\left(\varphi_k(t)\right) = F(t)$. Thus we have

(10) $$\lim_{k\to\infty} \int_{[0,a]} \varphi_k(t)\, \lambda_F(dt) = \int_{[0,a]} F(a)\, \mu_L(dt) - \lim_{k\to\infty} \int_{[0,a]} F\left(\varphi_k(t)\right) \mu_L(dt)$$

$$= \int_{[0,a]} F(a)\, \mu_L(dt) - \int_{[0,a]} F(t)\, \mu_L(dt).$$

$$= \int_{[0,a]} \{F(a) - F(t)\}\, \mu_L(dt)$$

$$= \int_{[0,a]} \lambda_F\big((t, a]\big)\, \mu_L(dt).$$

By (6) and (10) we have

(11) $$\int_{[0,a]} \varphi(t)\,\lambda_F\,(dt) = \int_{[0,a]} \lambda_F\big((t,a]\big)\,\mu_L(dt).$$

Substituting (11) into (5), we have

(12) $$\int_{[0,\infty)} \varphi(t)\,\lambda_F(dt) = \lim_{n\to\infty} \int_{[0,n]} \lambda_F\big((t,n]\big)\,\mu_L(dt)$$

$$= \lim_{n\to\infty} \int_{[0,\infty)} \mathbf{1}_{[0,n]}(t)\lambda_F\big((t,n]\big)\,\mu_L(dt).$$

Now $\big((t,n] : n \in \mathbb{N}\big)$ is an increasing sequence in $\mathfrak{B}_{\mathbb{R}}$ and $\lim_{n\to\infty}(t,n] = (t,\infty)$ and then we have $\lim_{n\to\infty}\lambda_F\big((t,n]\big) = \lambda_F\big((t,\infty)\big)$. Thus applying the Monotone Convergence Theorem (Theorem 8.5) to the right side of (12), we have

$$\int_{[0,\infty)} \varphi(t)\,\lambda_F\,(dt) = \int_{[0,\infty)} \lambda_F\big((t,\infty)\big)\,\mu_L(dt).$$

This proves (1).

2. Suppose $\lim_{t\to\infty} F(t) = C < \infty$. By Theorem 22.8 we have

$$\lambda_F(\mathbb{R}) = \lim_{t\to\infty} F(t) - \lim_{t\to-\infty} F(t) = C - 0 = C,$$

$$C - F(t) = \lambda_F(\mathbb{R}) - \lambda_F\big((-\infty,t]\big) = \lambda_F\big((t,\infty)\big).$$

Substituting the last equality in (1), we have (2). ∎

Comment on Prob. 22.3. In Prob. 22.3, F is a real-valued right-continuous increasing function on $\mathbb{R}$ such that $F = 0$ on $(-\infty,0)$ and $\lim_{t\to\infty} F(t) = C < \infty$. The last conditions ensure that for the Lebesgue-Stieltjes measure λ_F determined by F we have not only $\lambda_F(\mathbb{R}) = C < \infty$ but also $\lambda_F\big((-\infty,0)\big) = 0$ and thus $\int_{\mathbb{R}} t\,\lambda_F(dt) = \int_{[0,\infty)} t\,\lambda_F(dt)$. If we replace the condition that $F = 0$ on $(-\infty,0)$ with the less stringent condition that $\lim_{t\to-\infty} F(t) = 0$, then we still have $\lambda_F(\mathbb{R}) = C$ but we may not have $\lambda_F\big((-\infty,0)\big) = 0$ and $\int_{\mathbb{R}} t\,\lambda_F(dt)$ may not exist.

Let us construct a real-valued right-continuous increasing function F on $\mathbb{R}$ such that $\lim_{t\to-\infty} F(t) = 0$ and $\lim_{t\to\infty} F(t) = C < \infty$ for which $\int_{\mathbb{R}} t\,\lambda_F(dt)$ does not exist.

For $p > 1$ the series $\sum_{k\in\mathbb{N}} \frac{1}{k^p}$ converges. Let $S(p) = \sum_{k\in\mathbb{N}} \frac{1}{k^p} < \infty$. Let us define a function G on $[0,\infty)$ by setting

$$G(t) = \begin{cases} 0 & \text{for } t = 0, \\ \sum_{k=1}^{n} \frac{1}{k^p} & \text{for } t \in (n-1,n], n \in \mathbb{N}. \end{cases}$$

Thus defined, G is a real-valued left-continuous increasing function on $[0,\infty)$ and we have $\lim_{t\to\infty} G(t) = S(P)$. Let us extend the definition of G to $\mathbb{R}$ by setting

$$G(t) = -G(|t|) \quad \text{for } t \in (-\infty,0).$$

Then G is a real-valued increasing function on $\mathbb{R}$ and $\lim_{t \to -\infty} G(t) = -S(P)$. Note that G is left-continuous on $[0, \infty)$ and right-continuous on $(-\infty, 0)$. Let G_r be the right-continuous modification of G and define a function F on $\mathbb{R}$ by setting

$$F(t) = G_r(t) + S(p).$$

Then F is a real-valued right-continuous increasing function on $\mathbb{R}$ with $\lim_{t \to -\infty} F(t) = 0$ and $\lim_{t \to \infty} F(t) = 2S(P)$. Note that for the Lebesgue-Stieltjes measure λ_F determined by F, we have

(1) $$\lambda_F\big((n-1, n]\big) = F(n) - F(n-1) = \frac{1}{n^p},$$

(2) $$\lambda_F\big((-n, -n+1]\big) \doteq F(-n+1) - F(-n) = \frac{1}{n^p}.$$

Let our $p > 1$ be such that $p \in (1, 2)$, that is, $p - 1 < 1$. Let us show that $\int_{\mathbb{R}} t \, \lambda_F(dt)$ does not exist. Let φ be a function on $\mathbb{R}$ defined by $\varphi(t) = t$. Then to show that $\int_{\mathbb{R}} t \, \lambda_F(dt)$ does not exist, we show that $\int_{\mathbb{R}} \varphi(t) \lambda_F(dt)$ does not exist. Consider $\varphi = \varphi^+ - \varphi^-$, the decomposition of φ into its positive part and negative part. To show that $\int_{\mathbb{R}} \varphi(t) \lambda_F(dt)$ does not exist, we show that $\int_{\mathbb{R}} \varphi^+(t) \lambda_F(dt) = \infty$ and $\int_{\mathbb{R}} \varphi^-(t) \lambda_F(dt) = \infty$. Now by the symmetry of the function $|\varphi|$ with respect to $0 \in \mathbb{R}$ and by the symmetry of the measure λ_F with respect to $0 \in \mathbb{R}$, we have $\int_{\mathbb{R}} \varphi^+(t) \lambda_F(dt) = \int_{\mathbb{R}} \varphi^-(t) \lambda_F(dt)$. Thus it remains to show that $\int_{\mathbb{R}} \varphi^+(t) \lambda_F(dt) = \infty$. Now $\int_{\mathbb{R}} \varphi^+(t) \lambda_F(dt) = \int_{[0,\infty)} \varphi^+(t) \lambda_F(dt)$. Consider a function ψ on $[0, \infty)$ defined by $\psi = \sum_{n \in \mathbb{N}}(n-1)\mathbf{1}_{[n-1,n)}$. We have $0 \leq \psi \leq \varphi^+$ on $[0, \infty)$ and thus

(3) $$\int_{\mathbb{R}} \psi(t) \lambda_F(dt) \leq \int_{\mathbb{R}} \varphi^+(t) \lambda_F(dt).$$

Now we have

(4) $$\int_{\mathbb{R}} \psi(t) \lambda_F(dt) = \sum_{n \in \mathbb{N}}(n-1)\frac{1}{n^p} = \sum_{n \in \mathbb{N}}\frac{n}{n^p} - \sum_{n \in \mathbb{N}}\frac{1}{n^p}$$
$$= \sum_{n \in \mathbb{N}}\frac{1}{n^{p-1}} - S(p) = \infty$$

where the last equality is by the fact that $\sum_{n \in \mathbb{N}}\frac{1}{n^{p-1}} = \infty$ since $p - 1 < 1$. By (3) and (4) we have $\int_{\mathbb{R}} \varphi^+(t) \lambda_F(dt) = \infty$. This completes the proof that $\int_{\mathbb{R}} t \, \lambda_F(dt)$ does not exist. ∎

Prob. 22.4. Prove the following theorem:
Theorem. (Cavalieri's Principle) *Let $(X, \mathfrak{A}, \mu)$ be a measure space and let f be a non-negative real-valued $\mathfrak{A}$-measurable function on X. Then we have*

$$\int_X f \, d\mu = \int_{[0,\infty)} \mu\{x \in X : f(x) > t\} \, \mu(dt).$$

(Hint: Theorem 9.34, Prob. 22.1 and Prob. 22.3.)

Proof. 1. Since f is a real-valued $\mathfrak{A}$-measurable function on X, f is a $\mathfrak{A}/\mathfrak{B}_{\mathbb{R}}$-measurable mapping of X into $\mathbb{R}$. Let ν be the image measure of μ on $\mathfrak{B}_{\mathbb{R}}$ by the mapping f. Thus we have

$$\nu = \mu \circ f^{-1} \quad \text{on } \mathfrak{B}_{\mathbb{R}}.$$

Note that since f is nonnegative real-valued, we have $f^{-1}((-\infty, 0)) = \emptyset$ and thus $\nu((-\infty, 0)) = 0$. Thus by Theorem 9.34 (Integration by Image Measure) we have

(1) $$\int_X f \, d\mu = \int_{\mathbb{R}} t \, \nu(dt) = \int_{[0,\infty)} t \, \nu(dt).$$

2. Consider the case that $(X, \mathfrak{A}, \mu)$ is a finite measure space. In this case, $\nu = \mu \circ f^{-1}$ is a finite measure on $(\mathbb{R}, \mathfrak{B}_{\mathbb{R}})$. Then according to Prob. 22.1, if we define $F(t) = \nu((-\infty, t])$ for $t \in \mathbb{R}$ then F is a nonnegative real-valued right-continuous increasing function on $\mathbb{R}$ such that $\lim_{t \to -\infty} F(t) = 0$ and $\lim_{t \to \infty} F(t) = C = \nu(\mathbb{R})$ and moreover for the Lebesgue-Stieltjes measure λ_F determined by F we have $\nu = \lambda_F$ on $\mathfrak{B}_{\mathbb{R}}$. Thus we have

(2) $$\int_{[0,\infty)} t \, \nu(dt) = \int_{[0,\infty)} t \, \lambda_F(dt).$$

By Prob. 22.3, we have

(3) $$\int_{[0,\infty)} t \, \lambda_F(dt) = \int_{[0,\infty)} \lambda_F((t, \infty)) \, \mu_L(dt).$$

Since $\lambda_F = \nu$ on $\mathfrak{B}_{\mathbb{R}}$, we have

(4) $$\lambda_F((t, \infty)) = \nu((t, \infty)) = \mu \circ f^{-1}(t, \infty) = \mu\{x \in X : f(x) > t\}.$$

By (1), (2), (3), and (4), we have

(5) $$\int_X f \, d\mu = \int_{[0,\infty)} \mu\{x \in X : f(x) > t\} \, \mu_L(dt).$$

3. Consider the case that $(X, \mathfrak{A}, \mu)$ is a σ-finite measure space. Then there exists an increasing sequence $(A_n : n \in \mathbb{N})$ in $\mathfrak{A}$ such that $\bigcup_{n \in \mathbb{N}} A_n = X$ and $\mu(A_n) < \infty$ for every $n \in \mathbb{N}$. Then $(A_n, \mathfrak{A} \cap A_n, \mu)$, the restriction of the measure space $(X, \mathfrak{A}, \mu)$ to A_n, is a finite measure space and thus by (5) we have

$$\int_{A_n} f \, d\mu = \int_{[0,\infty)} \mu\{x \in A_n : f(x) > t\} \, \mu_L(dt),$$

that is, we have

(6) $$\int_X \mathbf{1}_{A_n} f \, d\mu = \int_{[0,\infty)} \mu\{x \in A_n : f(x) > t\} \, \mu_L(dt).$$

Then by the Monotone Convergence Theorem (Theorem 8.5), we have

(7) $$\int_X f \, d\mu = \lim_{n\to\infty} \int_X \mathbf{1}_{A_n} f \, d\mu = \lim_{n\to\infty} \int_{[0,\infty)} \mu\{x \in A_n : f(x) > t\} \, \mu_L(dt).$$

For each $n \in \mathbb{N}$, let $E_n = \{x \in A_n : f(x) > t\} \in \mathfrak{A}$. Since $(A_n : n \in \mathbb{N})$ is an increasing sequence in $\mathfrak{A}$, $(E_n : n \in \mathbb{N})$ is an increasing sequence in $\mathfrak{A}$. Then

$$\lim_{n\to\infty} E_n = \bigcup_{n\in\mathbb{N}} E_n = \bigcup_{n\in\mathbb{N}} \{x \in A_n : f(x) > t\}$$

$$= \{x \in \bigcup_{n\in\mathbb{N}} A_n : f(x) > t\}$$

$$= \{x \in X : f(x) > t\}$$

and thus we have $\lim_{n\to\infty} \mu(E_n) = \mu\left(\lim_{n\to\infty} E_n\right) = \mu\{x \in X : f(x) > t\}$. Then by the Monotone Convergence Theorem we have

(8) $$\lim_{n\to\infty} \int_{[0,\infty)} \mu\{x \in A_n : f(x) > t\} \, \mu_L(dt) = \int_{[0,\infty)} \mu\{x \in X : f(x) > t\} \, \mu_L(dt).$$

By (7) and (8) we have

(9) $$\int_X f \, d\mu = \int_{[0,\infty)} \mu\{x \in X : f(x) > t\} \, \mu_L(dt).$$

4. Consider the general case that $(X, \mathfrak{A}, \mu)$ is an arbitrary measure space. Let

(10) $$X_0 = \{x \in X : f(x) > 0\} \in \mathfrak{A}.$$

Note that since f is nonnegative on X, we have

(11) $$X_0^c = \{x \in X : f(x) = 0\} \in \mathfrak{A}.$$

Now (11) implies that $\int_{X_0^c} f \, d\mu = 0$ and hence

(12) $$\int_X f \, d\mu = \int_{X_0} f \, d\mu.$$

We also have for $t \in [0, \infty)$

$$\{x \in X : f(x) > t\} = \{x \in X_0 : f(x) > t\} \cup \{x \in X_0^c : f(x) > t\}$$

$$= \{x \in X_0 : f(x) > t\} \cup \emptyset$$

and then for $t \in [0, \infty)$

(13)
$$\mu\{x \in X : f(x) > t\} = \mu\{x \in X_0 : f(x) > t\}.$$

The set X_0 is either σ-finite with respect to μ or not σ-finite with respect to μ. We treat these two cases separately below.

4.1. Suppose X_0 is σ-finite with respect to μ. In this case we have

$$\int_X f \, d\mu = \int_{X_0} f \, d\mu \quad \text{by (12)}$$

$$= \int_{[0,\infty)} \mu\{x \in X_0 : f(x) > t\} \mu_L(dt) \quad \text{by (9)}$$

$$= \int_{[0,\infty)} \mu\{x \in X : f(x) > t\} \mu_L(dt) \quad \text{by (13)}.$$

4.2. Suppose X_0 is not σ-finite with respect to μ. Let us show that in this case we have

(14)
$$\int_X f \, d\mu = \infty,$$

and

(15)
$$\int_{[0,\infty)} \mu\{x \in X : f(x) > t\} \mu_L(dt) = \infty.$$

With (14) and (15), we have

$$\int_X f \, d\mu = \infty = \int_{[0,\infty)} \mu\{x \in X : f(x) > t\} \mu_L(dt)$$

and the proof is complete.

To prove (14), let $I_1 = [1, \infty]$ and $I_k = [\frac{1}{k}, \frac{1}{k-1})$ for $k \geq 2$. Then $\{I_k : k \in \mathbb{N}\}$ is a disjoint collection and $\bigcup_{k \in \mathbb{N}} I_k = (0, \infty]$. Let $E_k = \{x \in X_0 : f(x) \in I_l\}$ for $k \in \mathbb{N}$. Disjointness of the collection $\{I_k : k \in \mathbb{N}\}$ implies that $\{E_k : k \in \mathbb{N}\}$ is a disjoint collection in $\mathfrak{A}$. The fact that $f > 0$ on X_0 implies that $\bigcup_{k \in \mathbb{N}} E_k = X_0$. Then since X_0 is not σ-finite with respect to μ, there exists $k_0 \in \mathbb{N}$ such that $\mu(E_{k_0}) = \infty$. Then we have

$$\int_X f \, d\mu = \int_{X_0} f \, d\mu = \int_{\bigcup_{k \in \mathbb{N}} E_k} f \, d\mu = \sum_{k \in \mathbb{N}} \int_{E_k} f \, d\mu$$

$$\geq \int_{E_{k_0}} f \, d\mu \geq \frac{1}{k_0} \mu(E_{k_0}) = \infty.$$

This proves (14).

Let us prove (15). For $n \in \mathbb{N}$, let $F_n = \{x \in X_0 : f(x) > \frac{1}{n}\}$. Then $(F_n : n \in \mathbb{N})$ is an increasing sequence in $\mathfrak{A}$. Since $f > 0$ on X_0, we have $\bigcup_{n \in \mathbb{N}} F_n = X_0$. Then since X_0 is not σ-finite with respect to μ, there exists $n_0 \in \mathbb{N}$ such that $\mu(F_{n_0}) = \infty$. Thus we have

$\mu\{x \in X_0 : f(x) > \frac{1}{n_0}\} = \infty$. Then since $\mu\{x \in X_0 : f(x) > t\}$ is a decreasing function of $t \in [0, \infty)$ we have

$$\mu\{x \in X_0 : f(x) > t\} = \infty \quad \text{for } t \in \left[0, \tfrac{1}{n_0}\right].$$

Then we have

$$\int_{[0,\infty)} \mu\{x \in X : f(x) > t\}\, \mu_L(dt)$$

$$= \int_{[0,\infty)} \mu\{x \in X_0 : f(x) > t\}\, \mu_L(dt) \quad \text{by (13)}$$

$$\geq \int_{[0,\frac{1}{n_0}]} \mu\{x \in X_0 : f(x) > t\}\, \mu_L(dt)$$

$$= \infty \cdot \mu_L\left(\left[0, \tfrac{1}{n_0}\right]\right) = \infty.$$

This proves (15). ∎

§23 Product Measure Spaces

Prob. 23.1. Consider the product measure space $\big(\mathbb{R} \times \mathbb{R}, \sigma(\mathfrak{B}_{\mathbb{R}} \times \mathfrak{B}_{\mathbb{R}}), \mu_L \times \mu_L\big)$. Let $D = \big\{(x, y) \in \mathbb{R} \times \mathbb{R} : x = y\big\}$. Show that $D \in \sigma(\mathfrak{B}_{\mathbb{R}} \times \mathfrak{B}_{\mathbb{R}})$ and $(\mu_L \times \mu_L)(D) = 0$.

Proof. 1. For $n \in \mathbb{N}$, let

$$D_n = \big\{(x, y) \in [-n, n] \times [-n, n] : x = y\big\}.$$

For $k \in \mathbb{N}$ and $m \in \mathbb{Z}$, let

$$S_{k,m} = \Big[\frac{m-1}{2^k}, \frac{m}{2^k}\Big] \times \Big[\frac{m-1}{2^k}, \frac{m}{2^k}\Big] \in \mathfrak{B}_{\mathbb{R}} \times \mathfrak{B}_{\mathbb{R}}.$$

Then $S_{k,m}$ is a square in the xy-plane and has non-empty intersection with D. Indeed the union of $2 \times n \times 2^k$ such squares contains D_n. Let $E_{n,k}$ be the union of the $2 \times n \times 2^k$ squares. Then $E_{n,k} \in \sigma(\mathfrak{B}_{\mathbb{R}} \times \mathfrak{B}_{\mathbb{R}})$ and $D_n \subset E_{n,k}$. Since $(\mu_L \times \mu_L)(S_{k,m}) = (1/2^k)^2$ we have

$$(1) \qquad (\mu_L \times \mu_L)(E_{n,k}) = 2 \times n \times 2^k \cdot (\mu_L \times \mu_L)(S_{k,m}) = 2n \cdot 2^k \cdot \Big(\frac{1}{2^k}\Big)^2 = \frac{2n}{2^k}.$$

Now $(E_{n,k} : k \in \mathbb{N})$ is a decreasing sequence. If $(x, y) \in \mathbb{R} \times \mathbb{R}$ is such that $x \neq y$ then for sufficiently large $k \in \mathbb{N}$ we have $(x, y) \notin E_{n,k}$. Thus we have

$$\lim_{k \to \infty} E_{n,k} = \bigcap_{k \in \mathbb{N}} E_{n,k} = D_n.$$

This shows that $D_n \in \sigma(\mathfrak{B}_{\mathbb{R}} \times \mathfrak{B}_{\mathbb{R}})$. Since $(E_{n,k} : k \in \mathbb{N})$ is a decreasing sequence and $(\mu_L \times \mu_L)(E_{n,1}) = n < \infty$ by (1), we have

$$(\mu_L \times \mu_L)(D_n) = (\mu_L \times \mu_L)\big(\lim_{k \to \infty} E_{n,k}\big)$$

$$= \lim_{k \to \infty} (\mu_L \times \mu_L)(E_{n,k}) \quad \text{by (b) of Theorem 1.26}$$

$$= \lim_{k \to \infty} \frac{2n}{2^k} = 0 \quad \text{by (1).}$$

This shows that we have

$$(2) \qquad D_n \in \sigma(\mathfrak{B}_{\mathbb{R}} \times \mathfrak{B}_{\mathbb{R}}) \quad \text{and} \quad (\mu_L \times \mu_L)(D_n) = 0.$$

2. Observe that $(D_n : n \in \mathbb{N})$ is an increasing sequence and $\lim_{n \to \infty} D_n = \bigcup_{n \in \mathbb{N}} D_n = D$. This shows that $D \in \sigma(\mathfrak{B}_{\mathbb{R}} \times \mathfrak{B}_{\mathbb{R}})$. Then we have

$$(\mu_L \times \mu_L)(D) = (\mu_L \times \mu_L)\big(\lim_{n \to \infty} D_n\big)$$

$$= \lim_{n \to \infty} (\mu_L \times \mu_L)(D_n) \quad \text{by (a) of Theorem 1.26}$$

$$= \lim_{n \to \infty} 0 = 0 \quad \text{by (2).}$$

This completes the proof. ∎

Prob. 23.2. Given the product measure space $\big(\mathbb{R} \times \mathbb{R}, \sigma(\mathfrak{B}_\mathbb{R} \times \mathfrak{B}_\mathbb{R}), \mu_L \times \mu_L\big)$. Let f be a real-valued continuous function on $\mathbb{R}$. Consider the graph of f which is a subset of $\mathbb{R} \times \mathbb{R}$ defined by

$$G = \big\{(x, y) \in \mathbb{R} \times \mathbb{R} : y = f(x) \text{ for } x \in \mathbb{R}\big\}.$$

Let us write $\big(\mathbb{R} \times \mathbb{R}, \mathfrak{A}, \mu_L \times \mu_L\big)$ for the completion of $\big(\mathbb{R} \times \mathbb{R}, \sigma(\mathfrak{B}_\mathbb{R} \times \mathfrak{B}_\mathbb{R}), \mu_L \times \mu_L\big)$. Show that G is $\mathfrak{A}$-measurable and $(\mu_L \times \mu_L)(G) = 0$.

Proof. Let $\big(\mathbb{R} \times \mathbb{R}, \mathfrak{A}, \mu_L \times \mu_L\big)$ be the completion of $\big(\mathbb{R} \times \mathbb{R}, \sigma(\mathfrak{B}_\mathbb{R} \times \mathfrak{B}_\mathbb{R}), \mu_L \times \mu_L\big)$. According to Prob. 5.3, every subset of a null set in $\big(\mathbb{R} \times \mathbb{R}, \sigma(\mathfrak{B}_\mathbb{R} \times \mathfrak{B}_\mathbb{R}), \mu_L \times \mu_L\big)$ is $\mathfrak{A}$-measurable. Thus to show that G is $\mathfrak{A}$-measurable, we show that G is a subset of null set in $\big(\mathbb{R} \times \mathbb{R}, \sigma(\mathfrak{B}_\mathbb{R} \times \mathfrak{B}_\mathbb{R}), \mu_L \times \mu_L\big)$. We proceed as follows.

1. For definiteness of notations let us write $\mathbb{R}_x \times \mathbb{R}_y$ for $\mathbb{R} \times \mathbb{R}$. With $[a, b] \subset \mathbb{R}_x$, let

$$(1) \qquad G_{[a,b]} = \big\{(x, y) \in \mathbb{R}_x \times \mathbb{R}_y : y = f(x) \text{ for } x \in [a, b]\big\}.$$

Let us show that for every $\varepsilon > 0$ there exists a set $E \in \sigma(\mathfrak{B}_\mathbb{R} \times \mathfrak{B}_\mathbb{R})$ such that

$$(2) \qquad G_{[a,b]} \subset E \quad \text{and} \quad (\mu_L \times \mu_L)(E) < \varepsilon.$$

Since f is continuous on $\mathbb{R}_x$, it is continuous on $[a, b]$ and then it is uniformly continuous on $[a, b]$. Then for every $\varepsilon > 0$ there exists $\delta > 0$ such that

$$(3) \qquad |f(x') - f(x'')| < \frac{\varepsilon}{b - a} \quad \text{whenever } x', x'' \in [a, b] \text{ and } |x' - x''| < \delta.$$

Let $m \in \mathbb{N}$ be so large that $\frac{b-a}{m} < \delta$. Let us partition $[a, b]$ into m closed subintervals of equal length $\{I_1, \dots, I_m\}$. Then we have length $\ell(I_k) < \delta$ for $k = 1, \dots, m$. Now continuity of f on the finite closed interval I_k implies that $J_k := f(I_k)$ is a finite closed interval in $\mathbb{R}_y$. Moreover (3) implies that length $\ell(J_k) \leq \frac{\varepsilon}{b-a}$. By the definitions of I_k and J_k we have $(x, f(x)) \in I_k \times J_k$ for $x \in I_k$. This implies

$$G_{[a,b]} \subset \bigcup_{k=1}^{m} (I_k \times J_k).$$

Let $E = \bigcup_{k=1}^{m}(I_k \times J_k)$. Then we have $G_{[a,b]} \subset E$ and moreover

$$(\mu_L \times \mu_L)(E) = (\mu_L \times \mu_L)\Big(\bigcup_{k=1}^{m}(I_k \times J_k)\Big) = \sum_{k=1}^{m}(\mu_L \times \mu_L)(I_k \times J_k)$$

$$= \sum_{k=1}^{m}\mu_L(I_k)\mu_L(J_k) \leq \frac{\varepsilon}{b-a}\sum_{k=1}^{m}\ell(I_k) = \varepsilon,$$

since $\mu_L(J_k) = \ell(J_k) \leq \frac{\varepsilon}{b-1}$. This proves (2).

We showed above that for every $\varepsilon > 0$ there exists a set $E \in \sigma(\mathfrak{B}_\mathbb{R} \times \mathfrak{B}_\mathbb{R})$ such that $G_{[a,b]} \subset E$ and $(\mu_L \times \mu_L)(E) < \varepsilon$. This implies that for every $k \in \mathbb{N}$ there exists a set $E_k \in \sigma(\mathfrak{B}_\mathbb{R} \times \mathfrak{B}_\mathbb{R})$ such that

$$(4) \qquad G_{[a,b]} \subset E_k \quad \text{and} \quad (\mu_L \times \mu_L)(E_n) < \frac{1}{k}.$$

Let $F = \bigcap_{k \in \mathbb{N}} E_k$. Since $G_{[a,b]} \subset E_k$ for every $k \in \mathbb{N}$, we have $G_{[a,b]} \subset F$.

Since $E_k \in \sigma(\mathfrak{B}_{\mathbb{R}} \times \mathfrak{B}_{\mathbb{R}})$ for every $k \in \mathbb{N}$, we have $F = \bigcap_{k \in \mathbb{N}} E_k \in \sigma(\mathfrak{B}_{\mathbb{R}} \times \mathfrak{B}_{\mathbb{R}})$. We have $(\mu_L \times \mu_L)(F) = (\mu_L \times \mu_L)\big(\bigcap_{k \in \mathbb{N}} E_k\big) \leq (\mu_L \times \mu_L)(E_k) < \frac{1}{k}$ for every $k \in \mathbb{N}$ and thus $(\mu_L \times \mu_L)(F) = 0$. Thus F is a null set in $\big(\mathbb{R} \times \mathbb{R}, \sigma(\mathfrak{B}_{\mathbb{R}} \times \mathfrak{B}_{\mathbb{R}}), \mu_L \times \mu_L\big)$. Then $G_{[a,b]}$, a subset of F, is $\mathfrak{A}$-measurable and $(\mu_L \times \mu_L)(G_{[a,b]}) \leq (\mu_L \times \mu_L)(F) = 0$.

2. For $n \in \mathbb{N}$, consider $G_{[-n,n]}$, a particular case of $G_{[a,b]}$ considered above, that is,

(5)
$$G_{[-n,n]} = \big\{(x,y) \in \mathbb{R}_x \times \mathbb{R}_y : y = f(x) \text{ for } x \in [-n,n]\big\}.$$

Then $G_{[-n,n]}$ is $\mathfrak{A}$-measurable and $(\mu_L \times \mu_L)(G_{[-n,n]}) = 0$. Now $\big(G_{[-n,n]} : n \in \mathbb{N}\big)$ is an increasing sequence of $\mathfrak{A}$-measurable sets and

$$\lim_{n \to \infty} G_{[-n,n]} = \bigcup_{n \in \mathbb{N}} G_{[-n,n]} = G.$$

Then since $G_{[-n,n]}$ is $\mathfrak{A}$-measurable for every $n \in \mathbb{N}$, G is $\mathfrak{A}$-measurable. We also have

$$(\mu_L \times \mu_L)(G) = (\mu_L \times \mu_L)\big(\lim_{n \to \infty} G_{[-n,n]}\big) = \lim_{n \to \infty} (\mu_L \times \mu_L)(G_{[-n,n]})$$

$$= \lim_{n \to \infty} 0 = 0.$$

This completes the proof that G is $\mathfrak{A}$-measurable and $(\mu_L \times \mu_L)(G) = 0$. ∎

Prob. 23.3. Given the product measure space $\left(\mathbb{R} \times \mathbb{R}, \sigma(\mathfrak{B}_\mathbb{R} \times \mathfrak{B}_\mathbb{R}), \mu_L \times \mu_L\right)$. Let f be a real-valued function of bounded variation on $[a, b]$. Consider the graph of f defined by

$$G = \left\{(x, y) \in \mathbb{R} \times \mathbb{R} : y = f(x) \text{ for } x \in [a, b]\right\}.$$

Show that $G \in \sigma(\mathfrak{B}_\mathbb{R} \times \mathfrak{B}_\mathbb{R})$ and $(\mu_L \times \mu_L)(G) = 0$.

Proof. Let $\left(\mathbb{R} \times \mathbb{R}, \mathfrak{A}, \mu_L \times \mu_L\right)$ be the completion of $\left(\mathbb{R} \times \mathbb{R}, \sigma(\mathfrak{B}_\mathbb{R} \times \mathfrak{B}_\mathbb{R}), \mu_L \times \mu_L\right)$. According to Prob. 5.3, every subset of a null set in $\left(\mathbb{R} \times \mathbb{R}, \sigma(\mathfrak{B}_\mathbb{R} \times \mathfrak{B}_\mathbb{R}), \mu_L \times \mu_L\right)$ is $\mathfrak{A}$-measurable. Thus to show that G is $\mathfrak{A}$-measurable, we show that G is a subset of null set in $\left(\mathbb{R} \times \mathbb{R}, \sigma(\mathfrak{B}_\mathbb{R} \times \mathfrak{B}_\mathbb{R}), \mu_L \times \mu_L\right)$. We proceed as follows.

1. For definiteness of notations let us write $\mathbb{R}_x \times \mathbb{R}_y$ for $\mathbb{R} \times \mathbb{R}$. With $[a, b] \subset \mathbb{R}_x$, let

(1) $G_{[a,b]} = \left\{(x, y) \in \mathbb{R}_x \times \mathbb{R}_y : y = f(x) \text{ for } x \in [a, b]\right\}.$

Since f is a function of bounded variation on $[a, b]$ we can write $f = f_1 - f_2$ where $f_i, i = 1, 2$, are real-valued increasing functions on $[a, b]$. For f_i and $x \in [a, b]$, define the jump of f_i at x by setting $\Delta_i(x) = f_i(x+) - f_i(x-)$. Then since f_i is a real-valued increasing function on $[a, b]$, if f_i is continuous at $x \in [a, b]$ then $\Delta_i(x) = 0$ and if f_i is discontinuous at $x \in [a, b]$ then $\Delta_i(x) > 0$.

Let $\varepsilon > 0$. Since $f_i(b) - f_i(b) < \infty$, f_i can have only finitely many points of discontinuity with jump $\Delta_i(x) \geq \varepsilon/4(b - a)$. Let $a = a_0 < a_1 < \cdots < a_m = b$ be such that all points of discontinuity of $f_i, i = 1, 2$, with jump $\Delta_i(x) \geq \varepsilon/4(b - a)$ are included in the set $\{a_0, a_1, \ldots, a_m\}$.

Consider (a_{k-1}, a_k) where $k = 1, \ldots, m$. Now f_1 and f_2 do not have points of discontinuity with jump $\Delta_i(x) \geq \varepsilon/4(b-a)$. Thus we can select $a_{k-1} = b_0 < b_1 < \cdots < b_n = a_k$ such that

$$f_i(b_\ell) - f_i(b_{\ell-1}) < \frac{\varepsilon}{4(b - a)} \quad \text{for } \ell = 1, \ldots, n \text{ and } i = 1, 2.$$

Then for $x \in [b_{\ell-1}, b_\ell]$, we have

$$|f(x) - f(b_{\ell-1})| \leq \left\{f_1(x) - f_1(b_{\ell-1})\right\} + \left\{f_2(x) - f_2(b_{\ell-1})\right\} < \frac{\varepsilon}{2(b - a)}.$$

Let us define

$$J_\ell = \left[f(b_{\ell-1}) - \frac{\varepsilon}{2(b - a)}, f(b_{\ell-1}) + \frac{\varepsilon}{2(b - a)}\right] \subset \mathbb{R}_y.$$

Then for $x \in [b_{\ell-1}, b_\ell]$, we have $(x, f(x)) \in [b_{\ell-1}, b_\ell] \times J_\ell$. Thus we have

$$G_{[b_{\ell-1}, b_\ell]} \subset [b_{\ell-1}, b_\ell] \times J_\ell.$$

This implies then

$$G_{[a_{k-1}, a_k]} = \bigcup_{\ell=1}^{n} G_{[b_{\ell-1}, b_\ell]} \subset \bigcup_{\ell=1}^{n} [b_{\ell-1}, b_\ell] \times J_\ell.$$

Let us define

$$E_k = \bigcup_{\ell=1}^{n} [b_{\ell-1}, b_\ell] \times J_\ell \in \sigma(\mathfrak{B}_\mathbb{R} \times \mathfrak{B}_\mathbb{R}).$$

Then we have $G_{[a_{k-1},a_k]} \subset E_k$ and moreover we have

$$(\mu_L \times \mu_L)(E_k) = (\mu_L \times \mu_L)\left(\bigcup_{\ell=1}^{n}[b_{\ell-1},b_\ell] \times J_\ell\right) = \sum_{\ell=1}^{n}(\mu_L \times \mu_L)\left([b_{\ell-1},b_\ell] \times J_\ell\right)$$

$$= \sum_{\ell=1}^{n}\mu_L([b_{\ell-1},b_\ell])\mu_L(J_\ell) = \sum_{\ell=1}^{n}(b_\ell - b_{\ell-1})\frac{\varepsilon}{b-a} = (a_k - a_{k-1})\frac{\varepsilon}{b-a}.$$

Consider $G_{[a,b]} = \bigcup_{k=1}^{m} G_{[a_{k-1},a_k]}$. Let $E = \bigcup_{k=1}^{m} E_k \in \sigma(\mathfrak{B}_\mathbb{R} \times \mathfrak{B}_\mathbb{R})$. Then $G_{[a,b]} \subset E$ and moreover

$$(\mu_L \times \mu_L)(E) = \sum_{k=1}^{m}(\mu_L \times \mu_L)(E_k) = \sum_{k=1}^{m}(a_k - a_{k-1})\frac{\varepsilon}{b-a} = \varepsilon.$$

This shows that for every $\varepsilon > 0$ there exists $E \in \sigma(\mathfrak{B}_\mathbb{R} \times \mathfrak{B}_\mathbb{R})$ such that $G_{[a,b]} \subset E$ and $(\mu_L \times \mu_L)(E) < \varepsilon$. Then for every $n \in \mathbb{N}$ there exists $E_n \in \sigma(\mathfrak{B}_\mathbb{R} \times \mathfrak{B}_\mathbb{R})$ such that $G_{[a,b]} \subset E_n$ and $(\mu_L \times \mu_L)(E_n) < \frac{1}{n}$. Then we have $\bigcap_{n \in \mathbb{N}} E_n \in \sigma(\mathfrak{B}_\mathbb{R} \times \mathfrak{B}_\mathbb{R})$ and $G_{[a,b]} \subset \bigcap_{n \in \mathbb{N}} E_n$ and moreover $(\mu_L \times \mu_L)\left(\bigcap_{n \in \mathbb{N}} E_n\right) \leq (\mu_L \times \mu_L)(E_n) < \frac{1}{n}$ for every $n \in \mathbb{N}$ so that $(\mu_L \times \mu_L)\left(\bigcap_{n \in \mathbb{N}} E_n\right) = 0$. This shows that $G_{[a,b]}$ is $\mathfrak{A}$-measurable and $(\mu_L \times \mu_L)(G_{[a,b]}) \leq (\mu_L \times \mu_L)\left(\bigcap_{n \in \mathbb{N}} E_n\right) = 0$.

2. For $n \in \mathbb{N}$, consider $G_{[-n,n]}$, a particular case of $G_{[a,b]}$ considered above, that is,

$$(2) \qquad G_{[-n,n]} = \left\{(x,y) \in \mathbb{R}_x \times \mathbb{R}_y : y = f(x) \text{ for } x \in [-n,n]\right\}.$$

Then $G_{[-n,n]}$ is $\mathfrak{A}$-measurable and $(\mu_L \times \mu_L)(G_{[-n,n]}) = 0$. Now $\left(G_{[-n,n]} : n \in \mathbb{N}\right)$ is an increasing sequence of $\mathfrak{A}$-measurable sets and

$$\lim_{n \to \infty} G_{[-n,n]} = \bigcup_{n \in \mathbb{N}} G_{[-n,n]} = G.$$

Then since $G_{[-n,n]}$ is $\mathfrak{A}$-measurable for every $n \in \mathbb{N}$, G is $\mathfrak{A}$-measurable. We also have

$$(\mu_L \times \mu_L)(G) = (\mu_L \times \mu_L)\left(\lim_{n \to \infty} G_{[-n,n]}\right) = \lim_{n \to \infty}(\mu_L \times \mu_L)(G_{[-n,n]})$$

$$= \lim_{n \to \infty} 0 = 0.$$

This completes the proof that G is $\mathfrak{A}$-measurable and $(\mu_L \times \mu_L)(G) = 0$. ∎

Prob. 23.4. Let $(X, \mathfrak{A}, \mu)$ be a finite measure space and $(Y, \mathfrak{B}, \nu)$ be a non σ-finite measure space given by

$$X = Y = [0, 1],$$

$\mathfrak{A} = \mathfrak{B} = \mathfrak{B}_{[0,1]}$, the σ-algebra of the Borel sets in $[0, 1]$,

$\mu = \mu_L$ and ν is the counting measure.

Consider the product measurable space $(X \times Y, \sigma(\mathfrak{A} \times \mathfrak{B}))$ and a subset in it defined by $E = \{(x, y) \in X \times Y : x = y\}$.
(a) Show that $E \in \sigma(\mathfrak{A} \times \mathfrak{B})$.
(b) Show that

$$\int_X \left\{ \int_Y \mathbf{1}_E \, d\nu \right\} d\mu \neq \int_Y \left\{ \int_X \mathbf{1}_E \, d\mu \right\} d\nu.$$

(This shows that the σ-finiteness condition on $(X, \mathfrak{A}, \mu)$ and $(Y, \mathfrak{B}, \nu)$ in Theorem 23.17 (Tonelli's Theorem) cannot be dropped.)

Proof. 1. Let us prove (a). Let us define two functions f and g on $X \times Y$ by setting

$$f(x, y) = x \quad \text{for } (x, y) \in [0, 1] \times [0, 1],$$

$$g(x, y) = x \quad \text{for } (x, y) \in [0, 1] \times [0, 1].$$

Then f and g are $\sigma(\mathfrak{A} \times \mathfrak{B})$-measurable functions. This implies

$$E = \{(x, y) \in X \times Y : x = y\} = \{(x, y) \in X \times Y : f(x, y) = g(x, y)\} \in \sigma(\mathfrak{A} \times \mathfrak{B}).$$

2. Let us prove (b). For every $x \in X$, we have $\int_Y \mathbf{1}_E(x, y) \, \nu(dy) = \nu(\{x\}) = 1$ and then we have

$$\int_X \left\{ \int_Y \mathbf{1}_E(x, y) \, \nu(dy) \right\} \mu(dx) = \int_{[0,1]} 1 \, \mu_L(dx) = 1.$$

On the other hand, for every $y \in Y$, we have $\int_X \mathbf{1}_E(x, y) \, \mu(dx) = \mu_L(\{y\}) = 0$ and then

$$\int_Y \left\{ \int_X \mathbf{1}_E(x, y) \, \mu(dx) \right\} \nu(dy) = \int_{[0,1]} 0 \, \nu(dy) = 0.$$

This shows that the two iterated integrals are not equal. ∎

Prob. 23.5. Given two σ-finite measure spaces $(X, \mathfrak{A}, \mu)$ and $(Y, \mathfrak{B}, \nu)$ where
$$X = Y = \mathbb{Z}_+,$$
$$\mathfrak{A} = \mathfrak{B} = \mathfrak{P}(\mathbb{Z}_+), \text{ the } \sigma\text{-algebra of all subsets of } \mathbb{Z}_+,$$
$$\mu \text{ and } \nu \text{ are the counting measures.}$$
Consider the product measure space $(X \times Y, \sigma(\mathfrak{A} \times \mathfrak{B}), \mu \times \nu)$. Define a function f on $X \times Y$ by

$$f(x, y) = \begin{cases} 1 + 2^{-x} & \text{when } x = y, \\ -1 - 2^{-x} & \text{when } x = y + 1, \\ 0 & \text{otherwise.} \end{cases}$$

Show that
(a) $\int_X \{\int_Y f \, d\nu\} d\mu \neq \int_Y \{\int_X f \, d\mu\} d\nu$,
(b) $\int_{X \times Y} |f| \, d(\mu \times \nu) = \infty$,
(c) $\int_{X \times Y} f \, d(\mu \times \nu)$ does not exist.

Proof. (Draw the graph of $X \times Y = \mathbb{Z}_+ \times \mathbb{Z}_+$. Then mark the points (x, y) satisfying the condition $x = y$ and the points (x, y) satisfying the condition $x = y + 1$.)
1. Let us prove (a). We have $X = \mathbb{Z}_+$. Then we have

$$\int_Y f(x, y) \, \nu(dy) = \begin{cases} 2 & \text{when } x = 0, \\ 0 & \text{when } x \in \mathbb{N}. \end{cases}$$

Then we have

$$\int_X \left\{ \int_Y f(x, y) \, \nu(dy) \right\} \mu(dx) = 2.$$

On the other hand from $Y = \mathbb{Z}_+$ we have

$$\int_X f(x, y) \, \mu(dx) = 2^{-y} - 2^{-(y+1)} \quad \text{for } y \in \mathbb{Z}_+,$$

and then

$$\int_Y \left\{ \int_X f(x, y) \, \mu(dx) \right\} \nu(dy) = \sum_{y \in \mathbb{Z}_+} \{ 2^{-y} - 2^{-(y+1)} \}$$

$$= \left\{ \sum_{n \in \mathbb{Z}_+} 2^{-n} \right\} \left(1 - \frac{1}{2} \right) = 2 \cdot \frac{1}{2} = 1.$$

This verifies (a).
2. (b) follows from the fact that $\int_{X \times Y} |f| \, d(\mu \times \nu) \geq 1 + 1 + 1 + \cdots = \infty$.
3. To prove (c), let us write $f = f^+ - f^-$. It is easily shown that $\int_{X \times Y} f^+ \, d(\mu \times \nu) = \infty$ and $\int_{X \times Y} f^- \, d(\mu \times \nu) = \infty$. This shows that $\int_{X \times Y} f \, d(\mu \times \nu)$ does not exist. ∎

Prob. 23.6. Consider the product measure space $(X \times Y, \sigma(\mathfrak{A} \times \mathfrak{B}), \mu \times \nu)$ of two σ-finite measure spaces $(X, \mathfrak{A}, \mu)$ and $(Y, \mathfrak{B}, \nu)$. Let f be an extended real-valued $\mathfrak{A}$-measurable and μ-integrable function on X and g be an extended real-valued $\mathfrak{B}$-measurable and ν-integrable function on Y. Let

$$h(x, y) = \begin{cases} f(x)g(y) & \text{for } (x, y) \in X \times Y \text{ for which the product is defined,} \\ 0 & \text{otherwise.} \end{cases}$$

(a) Show that h is $\sigma(\mathfrak{A} \times \mathfrak{B})$-measurable on $X \times Y$.
(b) Show that h is $\mu \times \nu$ integrable on $X \times Y$ and moreover we have the equality

$$\int_{X \times Y} h \, d(\mu \times \nu) = \left\{ \int_X f \, d\mu \right\} \left\{ \int_Y g \, d\nu \right\}.$$

Proof. 1. Observe that the product $f(x)g(y)$ is undefined if and only if $f(x) = 0$ and $g(y) = \pm\infty$ or $f(x) = \pm\infty$ and $g(y) = 0$.

Since f is μ-integrable on X, the set $N_f = \{x \in X : f(x) = \pm\infty\} \in \mathfrak{A}$ is a null set in the measure space $(X, \mathfrak{A}, \mu)$. Let us modify the definition of f by setting $f(x) = 0$ for $x \in N_f$.

Similarly since g is ν-integrable on Y, the set $N_g = \{y \in Y : g(y) = \pm\infty\} \in \mathfrak{B}$ is a null set in the measure space $(Y, \mathfrak{B}, \nu)$. Let us modify the definition of g by setting $g(y) = 0$ for $y \in N_g$. Then we have

$$h(x, y) = f(x)g(y) \quad \text{for } (x, y) \in X \times Y.$$

2. Let us show that h is $\sigma(\mathfrak{A} \times \mathfrak{B})$-measurable on $X \times Y$. Since f is a $\mathfrak{A}$-measurable real-valued function on X, for every $\alpha \in \mathbb{R}$ we have $\{x \in X : f(x) \leq \alpha\} \in \mathfrak{A}$. Then for f regarded as a real-valued function on $X \times Y$, we have

$$\{(x, y) \in X \times Y : f(x) \leq \alpha\} = \{x \in X : f(x) \leq \alpha\} \times Y \in \mathfrak{A} \times \mathfrak{B} \subset \sigma(\mathfrak{A} \times \mathfrak{B}).$$

This shows that f is $\sigma(\mathfrak{A} \times \mathfrak{B})$-measurable on $X \times Y$. Similarly g is $\sigma(\mathfrak{A} \times \mathfrak{B})$-measurable on $X \times Y$. Then h as the product of f and g is $\sigma(\mathfrak{A} \times \mathfrak{B})$-measurable on $X \times Y$.

3. We prove (b) by quoting Theorem 23.19 (Fubini-Tonelli). For this purpose let us show

$$\int_X \left\{ \int_Y |h| \, d\nu \right\} d\mu = \int_X \left\{ \int_Y |f(x)| \, |g(y)| \, \nu(dy) \right\} \mu(dx)$$
$$= \int_X |f(x)| \left\{ \int_Y |g(y)| \, \nu(dy) \right\} \mu(dx)$$
$$= \left\{ \int_X |f(x)| \, \mu(dx) \right\} \left\{ \int_Y |g(y)| \, \nu(dy) \right\} < \infty.$$

By Theorem 23.19 (Fubini-Tonelli), the finiteness of the iterated integral $\int_X \left\{ \int_Y |h| \, d\nu \right\} d\mu$ implies that $h = fg$ is $\mu \times \nu$ integrable on $X \times Y$ and moreover we have

$$\int_{X \times Y} h \, d(\mu \times \nu) = \left\{ \int_X f \, d\mu \right\} \left\{ \int_Y g \, d\nu \right\}.$$

This completes the proof. ∎

Prob. 23.7. Consider the product measure space $(\mathbb{R}_x \times \mathbb{R}_y, \sigma(\mathfrak{B}_\mathbb{R} \times \mathfrak{B}_\mathbb{R}), \mu_L \times \mu_L)$. Let g_1 and g_2 be real-valued continuous functions on $[a, b] \subset \mathbb{R}_x$ such that $g_1(x) < g_2(x)$ for $x \in [a, b]$ and let

$$D_x = \{(x, y) \in \mathbb{R}_x \times \mathbb{R}_y : y \in [g_1(x), g_2(x)] \text{ for } x \in [a, b]\}.$$

Let h_1 and h_2 be real-valued continuous functions on $[c, d] \subset \mathbb{R}_y$ such that $h_1(y) < h_2(y)$ for $y \in [c, d]$ and let

$$D_y = \{(x, y) \in \mathbb{R}_x \times \mathbb{R}_y : x \in [h_1(y), h_2(y)] \text{ for } y \in [c, d]\}.$$

Let $D = D_x \cap D_y$ and let f be a real-valued continuous function on D.
(a) Show that $D \in \sigma(\mathfrak{B}_\mathbb{R} \times \mathfrak{B}_\mathbb{R})$ and moreover D is a compact set in $\mathbb{R}_x \times \mathbb{R}_y$.
(b) Show that f is $\mu_L \times \mu_L$-integrable on D and furthermore we have the equalities:

$$\int_D f \, d(\mu_L \times \mu_L) = \int_{[a,b]} \left[\int_{[g_1(x), g_2(x)]} f(x, y) \, \mu_L(dy) \right] \mu_L(dx)$$

$$= \int_{[c,d]} \left[\int_{[h_1(y), h_2(y)]} f(x, y) \, \mu_L(dx) \right] \mu_L(dy).$$

Proof. 1. Let us show that $D_x \in \sigma(\mathfrak{B}_\mathbb{R} \times \mathfrak{B}_\mathbb{R})$. For $n \in \mathbb{N}$, let

$$a = x_{n,0} < \cdots < x_{n,2^n} = b \quad \text{where} \quad x_{n,k} - x_{n,k-1} = \frac{b-a}{2^n} \quad \text{for } k = 1, \ldots, 2^n,$$

and let

$$I_{n,k} = [x_{n,k-1}, x_{n,k}]; \quad m_{n,k} = \min_{x \in I_{n,k}} g_1(x); \quad M_{n,k} = \max_{x \in I_{n,k}} g_2(x) \quad \text{for } k = 1, \ldots, 2^n.$$

Then let

(1) $$E_n = \bigcup_{k=1}^{2^n} I_{n,k} \times [m_{n,k}, M_{n,k}] \in \sigma(\mathfrak{B}_\mathbb{R} \times \mathfrak{B}_\mathbb{R}).$$

Note that $D_x \subset E_n$ for every $n \in \mathbb{N}$ and that $(E_n : n \in \mathbb{N})$ is a decreasing sequence of sets. Let $E = \bigcap_{n \in \mathbb{N}} E_n \in \sigma(\mathfrak{B}_\mathbb{R} \times \mathfrak{B}_\mathbb{R})$. Then we have $D_x \subset E$. Then to show that $D_x \in \sigma(\mathfrak{B}_\mathbb{R} \times \mathfrak{B}_\mathbb{R})$, we show that $D_x = E$. This is done as follows.

Since $D_x \subset E$, to show that $D_x = E$ we show that if $(\xi, \eta) \notin D_x$ then $(\xi, \eta) \notin E$. Suppose $(\xi, \eta) \notin D_x$. Now if $\xi \notin [a, b]$ then $(\xi, \eta) \notin E_n$ for every $n \in \mathbb{N}$ and then $(\xi, \eta) \notin E$. Then consider the case that $\xi \in [a, b]$. Now since $(\xi, \eta) \notin D_x$ and $\xi \in [a, b]$ we have either $\eta < g_1(\xi)$ or $\eta > g_2(\xi)$. Consider for instance the case that $\eta > g_2(\xi)$. Let $c = \eta - g_2(\xi) > 0$. Since g_2 is continuous on $[a, b]$, it is uniformly continuous on $[a, b]$. Thus there exists $\delta > 0$ such that $|g_2(x') - g_2(x'')| < \frac{c}{2}$ whenever $x', x'' \in [a, b]$ and $|x' - x''| < \delta$. Let $n_0 \in \mathbb{N}$ be so large that $(b - a)/2^{n_0} < \delta$. Let $k_0 = 1, \ldots, 2^{n_0}$ be such that $\xi \in I_{n_0, k_0}$. (Since the closed intervals $I_{n_0, k}$ are not disjoint and have common endpoints, such k_0 may not be unique.) Now we have $|g_2(\xi) - M_{n_0, k_0}| < \frac{c}{2}$ and then

(2) $$|\eta - M_{n_0, k_0}| \geq |\eta - g_2(\xi)| - |g_2(\xi) - M_{n_0, k_0}| \geq c - \frac{c}{2} = \frac{c}{2} > 0.$$

(The triangle inequality $|a + b| \leq |a| + |b|$ in $\mathbb{R}$ implies the inequality $\big||a| - |b|\big| \leq |a - b|$ as shown in Observation 15.6. Then for an arbitrary $c \in \mathbb{R}$ we have

$$\big||a - c| - |b - c|\big| \leq \big|(a - c) - (b - c)\big| = |a - b|$$

and then

(3) $|a - b| \geq |a - c| - |c - b|$.

The inequality (2) above is an application of the inequality (3).) Now by (2) we have

$$(\xi, \eta) \notin I_{n_0, k_0} \times \big[m_{n_0, k_0}, M_{n_0, k_0}\big] \subset E_{n_0}.$$

Then we have $(\xi, \eta) \notin \bigcap_{n \in \mathbb{N}} E_n = E$. This completes the proof that $D_x = E$ and then $D_x \in \sigma(\mathfrak{B}_{\mathbb{R}} \times \mathfrak{B}_{\mathbb{R}})$.

By similar argument we show that $D_y \in \sigma(\mathfrak{B}_{\mathbb{R}} \times \mathfrak{B}_{\mathbb{R}})$. Then since $\sigma(\mathfrak{B}_{\mathbb{R}} \times \mathfrak{B}_{\mathbb{R}})$ as a σ-algebra of subsets of $\mathbb{R}_x \times \mathbb{R}_y$ is closed under intersection, we have $D = D_x \cap D_y \in \sigma(\mathfrak{B}_{\mathbb{R}} \times \mathfrak{B}_{\mathbb{R}})$.

The set E_n defined by (1) is a bounded closed set in $\mathbb{R}_x \times \mathbb{R}_y$ and is thus a compact set in $\mathbb{R}_x \times \mathbb{R}_y$. Then $D_x = \bigcap_{n \in \mathbb{N}} E_n$ is a compact set in $\mathbb{R}_x \times \mathbb{R}_y$. Similarly D_y is a compact set in $\mathbb{R}_x \times \mathbb{R}_y$. Then $D = D_x \cap D_y$ is a compact set in $\mathbb{R}_x \times \mathbb{R}_y$.

2. We showed above that D is a compact set in $\mathbb{R}_x \times \mathbb{R}_y$ and $D \in \sigma(\mathfrak{B}_{\mathbb{R}} \times \mathfrak{B}_{\mathbb{R}})$. Let f be a real-valued continuous function on D. Then continuity of f on D and compactness of D imply that f is bounded on D. Also continuity of f on D and $\sigma(\mathfrak{B}_{\mathbb{R}} \times \mathfrak{B}_{\mathbb{R}})$-measurability of D imply that f is $\sigma(\mathfrak{B}_{\mathbb{R}} \times \mathfrak{B}_{\mathbb{R}})$-measurable on D. Observe also that since D is a bounded set in $\mathbb{R}_x \times \mathbb{R}_y$ we have $(\mu_L \times \mu_L)(D) < \infty$. Then boundedness of f on D and finiteness of $(\mu_L \times \mu_L)(D)$ imply that $\int_D |f| \, d(\mu_L \times \mu_L) < \infty$. This shows that the function f on D is $\mu_L \times \mu_L$-integrable on D. Then the function $\mathbf{1}_D \cdot f$ on $\mathbb{R}_x \times \mathbb{R}_y$ is $\mu_L \times \mu_L$-integrable on $\mathbb{R}_x \times \mathbb{R}_y$. Observe also that we have

(4) $\displaystyle \int_D f \, d(\mu_L \times \mu_L) = \int_{\mathbb{R}_x \times \mathbb{R}_y} \mathbf{1}_D \cdot f \, d(\mu_L \times \mu_L).$

According to Theorem 23.18 (Fubini), $\mu_L \times \mu_L$-integrability of the function $\mathbf{1}_D \cdot f$ on $\mathbb{R}_x \times \mathbb{R}_y$ implies both the equality

(5) $\displaystyle \int_{\mathbb{R}_x \times \mathbb{R}_y} \mathbf{1}_D \cdot f \, d(\mu_L \times \mu_L) = \int_{\mathbb{R}_x} \Big[\int_{\mathbb{R}_y} \mathbf{1}_D(x, y) f(x, y) \, \mu_L(dy) \Big] \mu_L(dx)$

$$= \int_{[a,b]} \Big[\int_{[g_1(x), g_2(x)]} f(x, y) \, \mu_L(dy) \Big] \mu_L(dx)$$

and the equality

(6) $\displaystyle \int_{\mathbb{R}_x \times \mathbb{R}_y} \mathbf{1}_D \cdot f \, d(\mu_L \times \mu_L) = \int_{\mathbb{R}_y} \Big[\int_{\mathbb{R}_x} \mathbf{1}_D(x, y) f(x, y) \, \mu_L(dx) \Big] \mu_L(dy)$

$$= \int_{[c,d]} \Big[\int_{[h_1(y), h_2(y)]} f(x, y) \, \mu_L(dx) \Big] \mu_L(dy).$$

Then combining (4) and (5), we have

(7) $$\int_D f \, d(\mu_L \times \mu_L) = \int_{[a,b]} \left[\int_{[g_1(x),g_2(x)]} f(x,y) \, \mu_L(dy) \right] \mu_L(dx),$$

and combining (4) and (6), we have

(8) $$\int_D f \, d(\mu_L \times \mu_L) = \int_{[c,d]} \left[\int_{[h_1(y),h_2(y)]} f(x,y) \, \mu_L(dx) \right] \mu_L(dy).$$

With (7) and (8), the proof of (b) is complete. ∎

Prob. 23.8. Consider the following integrals:

(a) $\quad \int_0^2 \int_x^2 y^2 \sin xy \, dy \, dx$

(b) $\quad \int_0^\pi \int_x^\pi \frac{\sin y}{y} \, dy \, dx$

(c) $\quad \int_0^1 \int_{2y}^2 \cos x^2 \, dx \, dy$

(d) $\quad \int_0^1 \int_y^1 x^2 e^{xy} \, dx \, dy$

(e) $\quad \int_0^8 \int_{\sqrt[3]{x}}^2 \frac{1}{y^4+1} \, dy \, dx$

(f) $\quad \int_0^2 \int_0^{4-x^2} \frac{xe^{2y}}{4-y} \, dy \, dx.$

Determine the domain of integration as a subset of the xy-plane and then evaluate the integral.

Proof. (a) $\int_0^2 \int_x^2 y^2 \sin xy \, dy \, dx$
The domain of integration is a rectangular triangle in the xy-plane bounded by three lines:
$x = 0$, $y = 2$ and $y = x$.

$$\int_0^2 \int_x^2 y^2 \sin xy \, dy \, dx = \int_0^2 \int_0^y y^2 \sin xy \, dx \, dy = \int_{y=0}^{y=2} y^2 \left[\frac{-1}{y} \cos xy \right]_{x=0}^{x=y} dy$$

$$= \int_{y=0}^{y=2} \{ -y \cos y^2 + y \} \, dy = \left[-\frac{1}{2} \sin y^2 + \frac{y^2}{2} \right]_0^2$$

$$= 2 - \frac{1}{2} \sin 4.$$

(b) $\int_0^\pi \int_x^\pi \frac{\sin y}{y} \, dy \, dx$
The domain of integration is a rectangular triangle in the xy-plane bounded by three lines:
$x = 0$, $y = \pi$ and $y = x$.

$$\int_0^\pi \int_x^\pi \frac{\sin y}{y} \, dy \, dx = \int_{y=0}^{y=\pi} \int_{x=0}^{x=y} \frac{\sin y}{y} \, dx \, dy = \int_{y=0}^{y=\pi} \frac{\sin y}{y} \left[x \right]_{x=0}^{x=y} dy$$

$$= \int_0^\pi \sin y \, dy = \left[\cos y \right]_\pi^0 = 2.$$

(c) $\int_0^1 \int_{2y}^2 \cos x^2 \, dx \, dy$
The domain of integration is a rectangular triangle in the xy-plane bounded by three lines:
$y = 0$, $x = 2$ and $x = 2y$.

$$\int_0^1 \int_{2y}^2 \cos x^2 \, dx \, dy = \int_{x=0}^{x=2} \left\{ \int_{y=0}^{y=\frac{x}{2}} \cos x^2 \, dy \right\} dx = \int_{x=0}^{x=2} \cos x^2 \left[y \right]_{y=0}^{y=\frac{x}{2}} dx$$

$$= \int_0^2 \frac{x}{2} \cos x^2 \, dx = \frac{1}{4} \left[\sin x^2 \right]_0^2 = \frac{1}{4} \sin 4.$$

(d) $\int_0^1 \int_y^1 x^2 e^{xy} \, dx \, dy$
The domain of integration is a rectangular triangle in the xy-plane bounded by three lines:

$y = 0, x = 1$ and $x = y$.

$$\int_0^1 \int_y^1 x^2 e^{xy} \, dx \, dy = \int_{x=0}^{x=1} \left\{ \int_{y=0}^{y=x} x^2 e^{xy} \, dy \right\} dx = \int_{x=0}^{x=1} \left[x^2 \frac{1}{x} e^{xy} \right]_{y=0}^{y=x} dx$$

$$= \int_0^1 \left\{ x e^{x^2} - x \right\} dx = \left[\frac{1}{2} e^{x^2} - \frac{x^2}{2} \right]_0^1$$

$$= \left\{ \frac{1}{2} e - \frac{1}{2} \right\} - \frac{1}{2} = \frac{e}{2} - 1.$$

(e) $\int_0^8 \int_{\sqrt[3]{x}}^2 \frac{1}{y^4+1} \, dy \, dx$

The domain of integration is an area in the xy-plane bounded by a line $x = 0$, a line $y = 2$ and a curve $y = \sqrt[3]{x}$.

$$\int_0^8 \int_{\sqrt[3]{x}}^2 \frac{1}{y^4+1} \, dy \, dx = \int_{y=0}^{y=2} \left\{ \int_{x=0}^{x=y^3} \frac{1}{y^4+1} \, dx \right\} dy = \int_{y=0}^{y=2} \frac{1}{y^4+1} \left[x \right]_{x=0}^{x=y^3} dy$$

$$= \int_0^2 \frac{y^3}{y^4+1} \, dy = \frac{1}{4} \left[\ln \left(y^4 + 1 \right) \right]_0^2$$

$$= \frac{1}{4} \left\{ \ln 17 - \ln 1 \right\} = \frac{\ln 17}{4}.$$

(f) $\int_0^2 \int_0^{4-x^2} \frac{xe^{2y}}{4-y} \, dy \, dx$

The domain of integration is an area in the xy-plane bounded by a line $x = 0$, a line $y = 0$ and a curve $y = 4 - x^2$.

$$\int_0^2 \int_0^{4-x^2} \frac{xe^{2y}}{4-y} \, dy \, dx = \int_{y=0}^{y=4} \left\{ \int_{x=0}^{x=\sqrt{4-y}} \frac{xe^{2y}}{4-y} \, dx \right\} dy = \int_{y=0}^{y=4} \frac{e^{2y}}{4-y} \left[\frac{x^2}{2} \right]_{x=0}^{x=\sqrt{4-y}} dy$$

$$= \int_{y=0}^{y=4} \frac{e^{2y}}{4-y} \frac{4-y}{2} \, dy = \int_0^4 \frac{1}{2} e^{2y} \, dy$$

$$= \frac{1}{4} \left[e^{2y} \right]_0^4 = \frac{1}{4} \{ e^8 - 1 \}. \quad \blacksquare$$

Prob. 23.9. Consider the product measure space $\left(\mathbb{R}_x \times \mathbb{R}_y, \sigma(\mathfrak{M}_L \times \mathfrak{M}_L), \mu_L \times \mu_L\right)$. Let f be a real-valued function on $\mathbb{R}_x \times \mathbb{R}_y$ defined by setting

$$f(x, y) = e^{-(x^2 + y^2)} \quad \text{for } (x, y) \in \mathbb{R}_x \times \mathbb{R}_y.$$

Show that f is $\mu_L \times \mu_L$-integrable on $\mathbb{R}_x \times \mathbb{R}_y$ and $\int_{\mathbb{R}_x \times \mathbb{R}_y} f \, d(\mu_L \times \mu_L) = \pi$.
(Hint: Apply the improper Riemann integral $\int_{-\infty}^{\infty} e^{-x^2} dx = \sqrt{\pi}$.)

Proof. The function f is continuous on $\mathbb{R}_x \times \mathbb{R}_y$ and therefore $\sigma(\mathfrak{M}_L \times \mathfrak{M}_L)$ measurable on $\mathbb{R}_x \times \mathbb{R}_y$. Then the nonnegativity of f on $\mathbb{R}_x \times \mathbb{R}_y$ implies that the integral $\int_{\mathbb{R}_x \times \mathbb{R}_y} f \, d(\mu_L \times \mu_L)$ exists in $[0, \infty]$.

According to Theorem 23.19 (Fubini-Tonelli), the condition that

$$(1) \qquad \int_{\mathbb{R}_x} \left[\int_{\mathbb{R}_y} |f| \, d\mu_L \right] d\mu_L < \infty$$

implies that f is $\mu_L \times \mu_L$-integrable on $\mathbb{R}_x \times \mathbb{R}_y$ and moreover we have the equality

$$(2) \qquad \int_{\mathbb{R}_x \times \mathbb{R}_y} f \, d(\mu_L \times \mu_L) = \int_{\mathbb{R}_x} \left[\int_{\mathbb{R}_y} f \, d\mu_L \right] d\mu_L.$$

Let us show that (1) holds. Thus we are to show that

$$\int_{\mathbb{R}_x} \left[\int_{\mathbb{R}_y} e^{-(x^2 + y^2)} \mu_L(dy) \right] \mu_L(dx) < \infty,$$

that is,

$$(3) \qquad \int_{\mathbb{R}_x} e^{-x^2} \left[\int_{\mathbb{R}_y} e^{-y^2} \mu_L(dy) \right] \mu_L(dx) < \infty.$$

Now since e^{-y^2} is nonnegative on $\mathbb{R}_y$, the integral $\int_{\mathbb{R}_y} e^{-y^2} \mu_L(dy)$ exists and moreover

$$(4) \qquad \int_{\mathbb{R}_y} e^{-y^2} \mu_L(dy) = \int_{-\infty}^{\infty} e^{-y^2} dy = \sqrt{\pi}.$$

Then by applying (4) twice, we have

$$(5) \qquad \int_{\mathbb{R}_x} e^{-x^2} \left[\int_{\mathbb{R}_y} e^{-y^2} \mu_L(dy) \right] \mu_L(dx) = \sqrt{\pi} \int_{\mathbb{R}_x} e^{-x^2} \mu_L(dx) = \pi < \infty.$$

This proves (3). Thus f is $\mu_L \times \mu_L$-integrable on $\mathbb{R}_x \times \mathbb{R}_y$ and moreover (2) holds. Thus

$$\int_{\mathbb{R}_x \times \mathbb{R}_y} e^{-(x^2 + y^2)} (\mu_L \times \mu_L)(d(x, y)) = \int_{\mathbb{R}_x} e^{-x^2} \left[\int_{\mathbb{R}_y} e^{-y^2} \mu_L(dy) \right] \mu_L(dx) = \pi \quad \text{by (5).}$$

This completes the proof. ∎

Prob. 23.10. Consider the product measure space $\left(\mathbb{R}_x \times \mathbb{R}_y, \sigma(\mathfrak{M}_L \times \mathfrak{M}_L), \mu_L \times \mu_L\right)$. Let f be a real-valued $\sigma(\mathfrak{M}_L \times \mathfrak{M}_L)$-measurable function on $\mathbb{R}_x \times \mathbb{R}_y$. Suppose further that

1° f is $\mu_L \times \mu_L$-integrable on $\mathbb{R}_x \times \mathbb{R}_y$, that is, $\int_{\mathbb{R}_x \times \mathbb{R}_y} f\, d(\mu_L \times \mu_L) \in \mathbb{R}$;

2° for every $x \in \mathbb{R}_x$, the improper Riemann integral $g(x) = \int_{-\infty}^{\infty} f(x, y)\, dy$ exists in $\mathbb{R}$ and g is $\mathfrak{M}_L$-measurable on $\mathbb{R}_x$.

3° $\int_{-\infty}^{\infty} g(x)\, dx$ exists in $\mathbb{R}$ and $\int_{-\infty}^{\infty} g(x)\, dx = \int_{\mathbb{R}_x} g(x)\, \mu_L(dx)$.

Show that we have

$$\int_{\mathbb{R}_x \times \mathbb{R}_y} f\, d(\mu_L \times \mu_L) = \int_{-\infty}^{\infty} \left\{ \int_{-\infty}^{\infty} f(x, y)\, dy \right\} dx.$$

(Thus $\int_{\mathbb{R}_x \times \mathbb{R}_y} f\, d(\mu_L \times \mu_L)$ can be evaluated as an iterated improper Riemann integral.)

Proof. According to Theorem 23.18 (Fubini), condition 1° implies that

(1) $$\int_{\mathbb{R}_x \times \mathbb{R}_y} f\, d(\mu_L \times \mu_L) = \int_{\mathbb{R}_x} \left[\int_{\mathbb{R}_y} f(x, y)\, \mu_L(dy) \right] \mu_L(dx).$$

By 2° we have

(2) $$g(x) = \int_{-\infty}^{\infty} f(x, y)\, dy \in \mathbb{R} \quad \text{for every } x \in \mathbb{R}_x.$$

Now condition 1° is equivalent to the $\mu_L \times \mu_L$-integrability of $|f|$ and thus by Theorem 23.18 (Fubini) we have

(3) $$\int_{\mathbb{R}_x} \left\{ \int_{\mathbb{R}_y} |f(x, y)|\mu_L(dy) \right\} \mu_L(dx) < \infty.$$

Then (3) implies that $\int_{\mathbb{R}_y} |f(x, y)|\, \mu_L(dy) < \infty$ for a.e. $x \in \mathbb{R}_x$. Thus there exists a null set N in the measure space $(\mathbb{R}_x, \mathfrak{M}_L, \mu_L)$ such that

(4) $$\int_{\mathbb{R}_y} |f(x, y)|\, \mu_L(dy) < \infty \quad \text{for } x \in \mathbb{R}_x \setminus N.$$

For $x \in \mathbb{R}_x \setminus N$, consider the function $1_{[-n,n]}(y) f(x, y)$ for $y \in \mathbb{R}_y$. Then we have $\lim_{n \to \infty} 1_{[-n,n]}(y) f(x, y) = f(x, y)$ and moreover $|1_{[-n,n]}(y) f(x, y)| \leq |f(x, y)|$ which is a μ_L-integrable function on $\mathbb{R}_y$. Thus by Theorem 9.20 (Dominated Convergence Theorem), we have for $x \in \mathbb{R}_x \setminus N$

(5) $$\int_{\mathbb{R}_y} f(x, y)\, \mu_L(dy) = \lim_{n \to \infty} \int_{\mathbb{R}_y} 1_{[-n,n]}(y) f(x, y)\, \mu_L(dy)$$

$$= \lim_{n \to \infty} \int_{[-n,n]} f(x, y)\, \mu_L(dy) = \lim_{n \to \infty} \int_{-n}^{n} f(x, y)\, dy$$

$$= \int_{-\infty}^{\infty} f(x, y)\, dy = g(x) \quad \text{by 2°}.$$

Thus we have

(6) $\int_{\mathbb{R}_y} f(x, y)\, \mu_L(dy) = g(x)$ for $x \in \mathbb{R}_x \setminus N$.

Then since $(\mathbb{R}_x, \mathfrak{M}_L, \mu_L)$ is a complete measure space and g is $\mathfrak{M}_L$-measurable on $\mathbb{R}_x$, (6) implies that $\int_{\mathbb{R}_y} f(x, y)\, \mu_L(dy)$ is a $\mathfrak{M}_L$-measurable function on $\mathbb{R}_x$ by (b) of Observation 4.20. Thus we have the $\mathfrak{M}_L$-measurability of $\int_{\mathbb{R}_y} f(x, y)\, \mu_L(dy)$ on $\mathbb{R}_x$ and the equality

(7) $\int_{\mathbb{R}_y} f(x, y)\, \mu_L(dy) = g(x)$ for a.e. $x \in \mathbb{R}_x$.

Substituting (7) into (1), we have

$$\int_{\mathbb{R}\times\mathbb{R}} f\, d(\mu_L \times \mu_L) = \int_{\mathbb{R}} g(x)\mu_L(dx) = \int_{-\infty}^{\infty} g(x)\, dx \quad \text{by } 3°$$
$$= \int_{-\infty}^{\infty} \left\{ \int_{-\infty}^{\infty} f(x, y)\, dy \right\} dx \quad \text{by } 2°.$$

This completes the proof.   ∎

Prob. 23.11. From the improper Riemann integral $\int_{-\infty}^{\infty} e^{-x^2} dx = \sqrt{\pi}$, we derive

(1) $\qquad \int_{[0,\infty)} e^{-\alpha x^2} \mu_L(dx) = \frac{1}{2}\sqrt{\frac{\pi}{\alpha}}$ for $\alpha \in (0, \infty)$.

Show that for every $n \in \mathbb{N}$ we have

(2) $\qquad \int_{[0,\infty)} e^{-\alpha x^2} x^{2n} \mu_L(dx) < \infty$ for $\alpha \in (0, \infty)$,

and indeed with an arbitrary $c \in (0, \alpha)$ we have

(3) $\qquad \int_{[0,\infty)} e^{-\alpha x^2} x^{2n} \mu_L(dx) \leq \frac{1}{2}\frac{n!}{c^n}\sqrt{\frac{\pi}{\alpha - c}}$.

Proof. Let us prove (3). Let $\alpha \in (0, \infty)$ be fixed and then let $c \in (0, \alpha)$ so that $\alpha - c > 0$. Now we have

$$e^{cx^2} = 1 + \frac{cx^2}{1!} + \frac{(cx^2)^2}{2!} + \cdots + \frac{(cx^2)^n}{n!} + \cdots$$

so that

$$e^{cx^2} \geq \frac{(cx^2)^n}{n!}, \quad \text{that is,} \quad x^{2n} \leq \frac{n!}{c^n}e^{cx^2}$$

and then

$$e^{-\alpha x^2} x^{2n} \leq e^{-\alpha x^2}\frac{n!}{c^n}e^{cx^2} = \frac{n!}{c^n}e^{-(\alpha - c)x^2}.$$

Then we have

$$\int_{[0,\infty)} e^{-\alpha x^2} x^{2n} \mu_L(dx) \leq \frac{n!}{c^n} \int_{[0,\infty)} e^{-(\alpha - c)x^2} \mu_L(dx)$$

$$= \frac{n!}{c^n}\frac{1}{2}\sqrt{\frac{\pi}{\alpha - c}},$$

where the last equality is by (1) which is applicable here since $\alpha - c > 0$. This proves (3). Note that (3) implies (2). This completes the proof. ∎

Prob. 23.12. From the improper Riemann integral $\int_{-\infty}^{\infty} e^{-x^2} dx = \sqrt{\pi}$, we derive

(1) $\int_{[0,\infty)} e^{-\alpha x^2} \mu_L(dx) = \frac{1}{2}\sqrt{\frac{\pi}{\alpha}}$ for $\alpha \in (0,\infty)$.

Show that for every $\alpha \in (0,\infty)$ and $n \in \mathbb{N}$ we have

(2) $\int_{[0,\infty)} e^{-\alpha x^2} x^{2n} \mu_L(dx) = 2^{-(n+1)} \cdot 1 \cdot 3 \cdot 5 \cdots (2n-1)\sqrt{\pi}\alpha^{-\frac{1}{2}-n}$,

and in particular

(3) $\int_{[0,\infty)} e^{-x^2} x^{2n} \mu_L(dx) = 2^{-(n+1)} \cdot 1 \cdot 3 \cdot 5 \cdots (2n-1)\sqrt{\pi}$.

(Hint: Apply Proposition 23.37.)

Proof. We prove (2) by induction on $n \in \mathbb{N}$. For $n = 0$, (2) reduces to (1) which is valid. Let us show that if (2) is valid for n then (2) is valid for $n + 1$. We show this by applying Proposition 23.37 by identifying our pair of variables (α, x) with the pair of variables (x, y) in Proposition 23.37.

Now assume that (2) is valid for n. We want to show that (2) is valid for $n + 1$, that is, we want to show

(4) $\int_{[0,\infty)} e^{-\alpha x^2} x^{2(n+1)} \mu_L(dx) = 2^{-(n+2)} \cdot 1 \cdot 3 \cdot 5 \cdots (2n+1)\sqrt{\pi}\alpha^{-\frac{3}{2}-n}$.

Let h be a real-valued function on $(0,\infty) \times (0,\infty)$ defined by setting

(5) $h(\alpha, x) = e^{-\alpha x^2} x^{2n}$ for $(\alpha, x) \in (0,\infty) \times (0,\infty)$.

Then we have

$0°$ $\quad \frac{\partial h}{\partial \alpha}(\alpha, x) = -e^{-\alpha x^2} x^{2(n+1)}$ for $(\alpha, x) \in (0,\infty) \times (0,\infty)$.

$1°$ $\quad h(\alpha, \cdot)$ is $\mathfrak{M}_L$-measurable on $[0,\infty)$ for every $\alpha \in (0,\infty)$.

$2°$ $\quad h(\alpha, \cdot)$ is μ_L-integrable on $[0,\infty)$ for every $\alpha \in (0,\infty)$ by (2) of Prob. 23.11.

$3°$ $\quad \left|\frac{\partial h}{\partial \alpha}(\alpha, x)\right| = e^{-\alpha x^2} x^{2(n+1)}$ which is μ_L-integrable on $[0,\infty)$ for every $\alpha \in (0,\infty)$ according to (2) of Prob. 23.11.

Thus all the conditions in Proposition 23.37 are satisfied and consequently we have according to (c) of Proposition 23.37 the equality

(6) $\int_{[0,\infty)} \frac{\partial h}{\partial \alpha}(\alpha, x) \mu_L(dx) = \frac{d}{d\alpha}\Big\{ \int_{[0,\infty)} h(\alpha, x) \mu_L(dx) \Big\}$.

Substituting $0°$ and (4) into (6), we have

(7) $-\int_{[0,\infty)} e^{-\alpha x^2} x^{2(n+1)} \mu_L(dx) = \frac{d}{d\alpha}\Big\{ \int_{[0,\infty)} e^{-x^2} x^{2n} \mu_L(dx) \Big\}$.

Now by our assumption of (2), we have

$\frac{d}{d\alpha}\Big\{ \int_{[0,\infty)} e^{-x^2} x^{2n} \mu_L(dx) \Big\} = \frac{d}{d\alpha}\Big\{ 2^{-(n+1)} \cdot 1 \cdot 3 \cdot 5 \cdots (2n-1)\sqrt{\pi}\alpha^{-\frac{1}{2}-n} \Big\}$

$\qquad\qquad\qquad = (-1)2^{-(n+2)} \cdot 1 \cdot 3 \cdot 5 \cdots (2n-1)(2n+1)\sqrt{\pi}\alpha^{-\frac{3}{2}-n}$.

Substituting this in (7), we have (4). This completes the proof. ∎

Prob. 23.13. Let $f \in L_r^1(\mathbb{R}, \mathfrak{M}_L, \mu_L)$ and let

$$\gamma := \int_{\mathbb{R}} \left\{ \int_{\mathbb{R}} f(x - y) f(y) \mu_L(dy) \right\} \mu_L(dx).$$

Show that $\gamma \geq 0$ and moreover we have $\int_{\mathbb{R}} f \, d\mu_L = \pm\sqrt{\gamma}$.

Proof. Let $f, g \in L_r^1(\mathbb{R}, \mathfrak{M}_L, \mu_L)$. According to Theorem 23.33, we have

$$\int_{\mathbb{R}} \left\{ \int_{\mathbb{R}} f(x - y) g(y) \mu_L(dy) \right\} \mu_L(dx) = \left\{ \int_{\mathbb{R}} f(x) \mu_L(dx) \right\} \left\{ \int_{\mathbb{R}} g(x) \mu_L(dx) \right\}.$$

For the particular case that $f = g$ the last equality reduces to

$$\gamma := \int_{\mathbb{R}} \left\{ \int_{\mathbb{R}} f(x - y) f(y) \mu_L(dy) \right\} \mu_L(dx) = \left\{ \int_{\mathbb{R}} f(x) \mu_L(dx) \right\}^2 \geq 0.$$

Then we have $\int_{\mathbb{R}} f \, d\mu_L = \pm\sqrt{\gamma}$. This completes the proof. ∎

Prob. 23.14. Let $f \in L^1_r(\mathbb{R}, \mathfrak{M}_L, \mu_L)$. Let $\alpha, \beta \in \mathbb{R} \setminus \{0\}$ and let
$$\gamma := \int_{\mathbb{R}} \left\{ \int_{\mathbb{R}} f(\alpha x - y) f(\beta y) \mu_L(dy) \right\} \mu_L(dx).$$
Show that $\gamma \geq 0$ and moreover we have $\int_{\mathbb{R}} f \, d\mu_L = \pm\sqrt{|\alpha\beta|\gamma}$.

Proof. Let g and h be functions on $\mathbb{R}$ defined by setting

(1) $g(x) = f(\alpha x)$ and $h(x) = f(\alpha\beta x)$ for $x \in \mathbb{R}$.

Then by (2) of Theorem 9.32 (Linear Transformation of the Lebesgue Integral on $\mathbb{R}$), we have

(2) $$\int_{\mathbb{R}} g(x) \mu_L(dx) = \frac{1}{|\alpha|} \int_{\mathbb{R}} g\left(\frac{1}{\alpha}x\right) \mu_L(dx) = \frac{1}{|\alpha|} \int_{\mathbb{R}} f(x) \mu_L(dx),$$

and similarly

(3) $$\int_{\mathbb{R}} h(x) \mu_L(dx) = \frac{1}{|\alpha\beta|} \int_{\mathbb{R}} h\left(\frac{1}{\alpha\beta}x\right) \mu_L(dx) = \frac{1}{|\alpha\beta|} \int_{\mathbb{R}} f(x) \mu_L(dx).$$

Consider the integral $\int_{\mathbb{R}} f(\alpha x - y) f(\beta y) \mu_L(dy)$ for $x \in \mathbb{R}$. Now we have
$$f(\alpha x - y) = f\left(\alpha\left(x - \frac{y}{\alpha}\right)\right) = g\left(x - \frac{y}{\alpha}\right),$$

and
$$f(\beta y) = f\left(\alpha\beta\frac{y}{\alpha}\right) = h\left(\frac{y}{\alpha}\right).$$

Then we have

(4) $$\int_{\mathbb{R}} f(\alpha x - y) f(\beta y) \mu_L(dy) = \int_{\mathbb{R}} g\left(x - \frac{y}{\alpha}\right) h\left(\frac{y}{\alpha}\right) \mu_L(dy)$$
$$= |\alpha| \int_{\mathbb{R}} g(x - y) h(y) \mu_L(dy) = |\alpha|(g * h)(x),$$

where the second equality is by (2) of Theorem 9.32. Substituting (4) in the definition of γ above, we have by (3) of Theorem 23.33

(5) $$\gamma = \int_{\mathbb{R}} |\alpha|(g * h)(x) \mu_L(dx) = |\alpha|\left[\int_{\mathbb{R}} g \, d\mu_L\right]\left[\int_{\mathbb{R}} h \, d\mu_L\right]$$
$$= |\alpha|\left[\frac{1}{|\alpha|} \int_{\mathbb{R}} f \, d\mu_L\right]\left[\frac{1}{|\alpha\beta|} \int_{\mathbb{R}} f \, d\mu_L\right] \quad \text{by (2) and (3)}$$
$$= \frac{1}{|\alpha\beta|}\left\{\int_{\mathbb{R}} f \, d\mu_L\right\}^2 \geq 0.$$

Then we have
$$\int_{\mathbb{R}} f \, d\mu_L = \pm\sqrt{|\alpha\beta|\gamma}.$$

This completes the proof. ∎

Prob. 23.15. Let $f \in L^1(\mathbb{R}, \mathfrak{M}_L, \mu_L)$. With $h > 0$ fixed, let us define a function φ_h on $\mathbb{R}$ by setting

$$\varphi_h(x) = \frac{1}{2h} \int_{[x-h,x+h]} f(t)\,\mu_L(dt) \quad \text{for } x \in \mathbb{R}.$$

(a) Show that φ_h is a real-valued uniformly continuous function on $\mathbb{R}$.
(b) Show that $\varphi_h \in L^1(\mathbb{R}, \mathfrak{M}_L, \mu_L)$ and $\|\varphi_h\|_1 \le \|f\|_1$.

Proof. 1. Let us show the uniform continuity of φ_h on $\mathbb{R}$. Now

$$\varphi_h(x) = \frac{1}{2h} \int_{[x-h,x+h]} f(t)\,\mu_L(dt)$$

$$= \frac{1}{2h} \int_{(-\infty,x+h]} f(t)\,\mu_L(dt) - \frac{1}{2h} \int_{(-\infty,x-h)} f(t)\,\mu_L(dt).$$

Let us define two real-valued functions on $\mathbb{R}$ by setting

$$F_1(x) = \int_{(-\infty,x+h]} f(t)\,\mu_L(dt) \quad \text{for } x \in \mathbb{R},$$

$$F_2(x) = \int_{(-\infty,x-h)} f(t)\,\mu_L(dt) \quad \text{for } x \in \mathbb{R}.$$

To show that φ_h is uniformly continuous on $\mathbb{R}$, we show that F_1 and F_2 are uniformly continuous on $\mathbb{R}$. To show that F_1 is uniformly continuous on $\mathbb{R}$, we show that for every $\varepsilon > 0$ there exists $\delta > 0$ such that

(1) $\qquad |F_1(x') - F_1(x'')| < \varepsilon \quad$ for any $x', x'' \in \mathbb{R}$ such that $|x' - x''| < \delta$.

Now f is μ_L-integrable on $\mathbb{R}$. Then by the absolute uniform continuity of the integral with respect to the measure, for every $\varepsilon > 0$ there exists $\delta > 0$ such that

(2) $\qquad \int_A |f|\,d\mu_L < \varepsilon \quad$ whenever $A \in \mathfrak{M}_L$ and $\mu_L(A) < \delta$.

Let $x', x'' \in \mathbb{R}$ be such that $|x' - x''| < \delta$. We may assume without loss of generality that $x' \le x''$. Then

$$|F_1(x'') - F_1(x')| = \left| \int_{(-\infty,x''+h]} f(t)\,\mu_L(dt) - \int_{(-\infty,x'+h]} f(t)\,\mu_L(dt) \right|$$

$$= \left| \int_{(x'+h,x''+h]} f(t)\,\mu_L(dt) \right|$$

$$\le \int_{(x'+h,x''+h]} |f(t)|\,\mu_L(dt)$$

$$< \varepsilon,$$

by (2) since $\mu_L\big((x'+h, x''+h]\big) = x'' - x' < \delta$. This proves the uniform continuity of F_1 on $\mathbb{R}$. The uniform continuity of F_2 on $\mathbb{R}$ is proved likewise.

2. To show that $\varphi_h \in L^1\big(\mathbb{R}, \mathfrak{M}_L, \mu_L\big)$, we show that $\|\varphi_h\|_1 < \infty$. Now we have

$$(3) \qquad \|\varphi_h\|_1 = \int_{\mathbb{R}} \big|\varphi_h(x)\big|\,\mu_L(dx) = \int_{\mathbb{R}} \frac{1}{2h}\Big|\int_{[x-h.x+h]} f(t)\,\mu_L(dt)\Big|\,\mu_L(dx)$$

$$\leq \frac{1}{2h}\int_{\mathbb{R}}\Big\{\int_{\mathbb{R}} \mathbf{1}_{[x-h.x+h]}(t)|f(t)|\,\mu_L(dt)\Big\}\,\mu_L(dx)$$

$$= \frac{1}{2h}\int_{\mathbb{R}}\Big\{\int_{\mathbb{R}} g(t,x)|f(t)|\,\mu_L(dt)\Big\}\,\mu_L(dx),$$

where g is a real-valued function on $\mathbb{R} \times \mathbb{R}$ defined by

$$(4) \qquad g(t,x) = \mathbf{1}_{[x-h.x+h]}(t) \quad \text{for } (t,x) \in \mathbb{R} \times \mathbb{R}.$$

This function is the characteristic function of a set $E \subset \mathbb{R} \times \mathbb{R}$ which is a strip in the tx-plane bounded by two parallel lines $x = t-h$ and $x = t+h$. Now E is a $\sigma(\mathfrak{M}_L \times \mathfrak{M}_L)$-measurable set in the measurable space $\big(\mathbb{R} \times \mathbb{R}, \sigma(\mathfrak{M}_L \times \mathfrak{M}_L)\big)$ and thus its characteristic function g is a $\sigma(\mathfrak{M}_L \times \mathfrak{M}_L)$-measurable function on $\big(\mathbb{R} \times \mathbb{R}, \sigma(\mathfrak{M}_L \times \mathfrak{M}_L)\big)$. Then since f is a $\mathfrak{M}_L$-measurable function on the measurable space $(\mathbb{R}, \mathfrak{M}_L)$, the function

$$g(t,x)|f(t)| \quad \text{for } (t,x) \in \mathbb{R} \times \mathbb{R}$$

is a $\sigma(\mathfrak{M}_L \times \mathfrak{M}_L)$-measurable function on $\big(\mathbb{R} \times \mathbb{R}, \sigma(\mathfrak{M}_L \times \mathfrak{M}_L)\big)$. Since $(\mathbb{R}, \mathfrak{M}_L, \mu_L)$ is a σ-finite measure space, Tonelli's Theorem is applicable and thus we have

$$(5) \qquad \int_{\mathbb{R}}\Big\{\int_{\mathbb{R}} g(t,x)|f(t)|\,\mu_L(dt)\Big\}\,\mu_L(dx) = \int_{\mathbb{R}}|f(t)|\Big\{\int_{\mathbb{R}} g(t,x)\,\mu_L(dx)\Big\}\,\mu_L(dt).$$

Observe that for fixed $x \in \mathbb{R}$, $g(\cdot, x)$ is given by

$$g(t,x) = \begin{cases} 1 & \text{for } t \in [x-h, x+h], \\ 0 & \text{for } t \in [x-h, x+h]^c. \end{cases}$$

Now

$$t \in [x-h, x+h] \Leftrightarrow x-h \leq t \leq x+h$$

$$\Leftrightarrow x \leq t+h \text{ and } t-h \leq x$$

$$\Leftrightarrow x \in [t-h, t+h].$$

Thus for fixed $t \in \mathbb{R}$, $g(t, \cdot)$ is given by

$$(6) \qquad g(t,x) = \begin{cases} 1 & \text{for } x \in [t-h, t+h], \\ 0 & \text{for } x \in [t-h, t+h]^c. \end{cases}$$

Using this in the integral on the right of (5), we have

$$\int_{\mathbb{R}}|f(t)|\Big\{\int_{\mathbb{R}} g(t,x)\,\mu_L(dx)\Big\}\,\mu_L(dt) = \int_{\mathbb{R}}|f(t)|\Big\{\int_{\mathbb{R}} \mathbf{1}_{[t-h,t+h]}(x)\,\mu_L(dx)\Big\}\,\mu_L(dt)$$

$$= 2h\int_{\mathbb{R}}|f(t)|\,\mu_L(dt) = 2h\|f\|_1.$$

Substituting this in (5) and then in (3), we have $\|\varphi_h\|_1 \leq \|f\|_1 < \infty$. ∎

Prob. 23.16. Let $(X, \mathfrak{A}, \mu)$ be a measure space. Let $D \in \mathfrak{A}$ and let f be a nonnegative real-valued $\mathfrak{A}$-measurable function on D. Consider the measure space $(\mathbb{R}, \mathfrak{B}_{\mathbb{R}}, \mu_{L})$ and the product measure space $(X \times \mathbb{R}, \sigma(\mathfrak{A} \times \mathfrak{B}_{\mathbb{R}}), \mu \times \mu_{L})$. Let E be a subset of $X \times \mathbb{R}$ such that

$$\begin{cases} E(x, \cdot) = \emptyset & \text{for } x \in D^c, \\ E(x, \cdot) = [0, f(x)) & \text{for } x \in D. \end{cases}$$

Show that $E \in \sigma(\mathfrak{A} \times \mathfrak{B}_{\mathbb{R}})$ and moreover $(\mu \times \mu_{L})(E) = \int_D f(x)\,\mu(dx)$.

Proof. Since f is a nonnegative real-valued $\mathfrak{A}$-measurable function on D, according to Lemma 8.6 there exists an increasing sequence of nonnegative simple functions $(\varphi_n : n \in \mathbb{N})$ on $(X, \mathfrak{A})$ such that

(1) $$\varphi_n \uparrow f \quad \text{on } D \quad \text{and} \quad \lim_{n \to \infty} \int_D \varphi_n\,d\mu = \int_D f\,d\mu.$$

For each $n \in \mathbb{N}$, let φ_n be represented as

(2) $$\varphi_n = \sum_{i=1}^{k_n} a_{n,i} \mathbf{1}_{D_{n,i}}$$

where $\{D_{n,i} : i = 1, \ldots, k_n\}$ is a disjoint collection in $\mathfrak{A}$ with $\bigcup_{i=1}^{k_n} D_{n,i} = D$ and $a_{n,i} \geq 0$ for $i = 1, \ldots, k_n$. Next for each $n \in \mathbb{N}$ consider the subset of $X \times \mathbb{R}$ defined by

(3) $$E_n = \bigcup_{i=1}^{k_n} D_{n,i} \times [0, a_{n,i}).$$

Now $D_{n,i} \in \mathfrak{A}$ and $[0, a_{n,i}) \in \mathfrak{B}_{\mathbb{R}}$ so that $D_{n,i} \times [0, a_{n,i}) \in \mathfrak{A} \times \mathfrak{B}_{\mathbb{R}} \subset \sigma(\mathfrak{A} \times \mathfrak{B}_{\mathbb{R}})$ for $i = 1, \ldots, k_n$ and hence $E_n \in \sigma(\mathfrak{A} \times \mathfrak{B}_{\mathbb{R}})$ and then

(4) $$(\mu \times \mu_{L})(E_n) = (\mu \times \mu_{L})(E_n)\left(\bigcup_{i=1}^{k_n} D_{n,i} \times [0, a_{n,i})\right)$$

$$= \sum_{i=1}^{k_n} (\mu \times \mu_{L})(D_{n,i} \times [0, a_{n,i})) = \sum_{i=1}^{k_n} \mu(D_{n,i})\mu_{L}([0, a_{n,i}))$$

$$= \sum_{i=1}^{k_n} a_{n,i}\mu(D_{n,i}) = \int_D \varphi_n\,d\mu.$$

Now $\varphi_n \uparrow$ implies $E_n \uparrow$ and $\varphi_n \uparrow f$ implies $E_n \uparrow E$, that is, $E = \lim_{n \to \infty} E_n \in \sigma(\mathfrak{A} \times \mathfrak{B}_{\mathbb{R}})$. Then $E_n \uparrow E$ implies

$$(\mu \times \mu_{L})(E) = \lim_{n \to \infty} (\mu \times \mu_{L})(E_n) = \lim_{n \to \infty} \int_D \varphi_n\,d\mu = \int_D f\,d\mu,$$

by (4) and (1). This completes the proof. ∎

§24 Lebesgue Measure Space on the Euclidean Space

Prob. 24.1. Let O be an open set in $\mathbb{R}^n$.
(a) Show that if $O \neq \emptyset$, then $\mu_L^n(O) > 0$.
(b) Show that if O is a bounded set, that is, it is contained in a ball with finite radius, then $\mu_L^n(O) < \infty$.

Proof. 1. In $\mathbb{R}^n$, every non-empty open set is a union of open balls. Thus if O is a non-empty open set in $\mathbb{R}^n$ then O contains an open ball $B_r(x_0)$ where $x_0 \in \mathbb{R}^n$ and $r > 0$. Now an open ball with radius $r > 0$ in $\mathbb{R}^n$ contains an open cube C with edge $2\frac{r}{\sqrt{n}}$. Thus we have

$$\mu_L^n(O) \geq \mu_L^n(C) = \left(2\frac{r}{\sqrt{n}}\right)^n > 0.$$

2. If O is a bounded open set in $\mathbb{R}^n$ then there exists an open ball $B_r(x_0)$ where $x_0 \in \mathbb{R}^n$ and $r > 0$ such that $O \subset B_r(x_0)$. Now $B_r(x_0)$ is contained in an open cube C with edge $2r$. Thus we have

$$\mu_L^n(O) \leq \mu_L^n(C) = (2r)^n < \infty.$$

This completes the proof. ∎

Prob. 24.2. Consider the Lebesgue measure space $(\mathbb{R}^{n+1}, \mathfrak{M}_L^{n+1}, \mu_L^{n+1})$ where $n \in \mathbb{N}$. Let us identify $\mathbb{R}^n$ with the subset of $\mathbb{R}^{n+1}$ given by $\mathbb{R}^n \times \{0\}$. Show that $\mu_L^{n+1}(\mathbb{R}^n) = 0$.

Proof. For $m \in \mathbb{N}$ and $k \in \mathbb{N}$, let

(1) $$R_{m,k} = \left[-m, m \right]_1 \times \cdots \times \left[-m, m \right]_n \times \left[-\frac{1}{k}, \frac{1}{k} \right] \subset \mathbb{R}^{n+1}.$$

Then for each $m \in \mathbb{N}$, $(R_{m,k} : k \in \mathbb{N})$ is a decreasing sequence and we set

(2) $$R_m := \lim_{k \to \infty} R_{m,k} = \bigcap_{k \in \mathbb{N}} R_{m,k} = \left[-m, m \right]_1 \times \cdots \times \left[-m, m \right]_n \times \{0\}.$$

Then $(R_m : m \in \mathbb{N})$ is an increasing sequence and we have

(3) $$\lim_{m \to \infty} R_m = \bigcup_{m \in \mathbb{N}} R_m = \mathbb{R}^n \times \{0\}.$$

Since $(R_{m,k} : k \in \mathbb{N})$ is a decreasing sequence, it is a sequence contained in $R_{m,1}$. Observe that $\mu_L^{n+1}(R_{m,k}) = (2m)^n \frac{2}{k}$ and in particular $\mu_L^{n+1}(R_{m,1}) = (2m)^n 2 < \infty$. Then by (b) of Theorem 1.26 we have

(4) $$\mu_L^{n+1}(R_m) = \mu_L^{n+1}\left(\lim_{k \to \infty} R_{m,k} \right) = \lim_{k \to \infty} \mu_L^{n+1}(R_{m,k}) = \lim_{k \to \infty} (2m)^n \frac{2}{k} = 0.$$

Then since $(R_m : m \in \mathbb{N})$ is an increasing sequence and (3) holds, we have by (a) of Theorem 1.26

$$\mu_L^{n+1}(\mathbb{R}^n \times \{0\}) = \mu_L^{n+1}\left(\lim_{m \to \infty} R_m \right) = \lim_{m \to \infty} \mu_L^{n+1}(R_m) = \lim_{m \to \infty} 0 = 0 \quad \text{by (4)}.$$

This completes the proof. ∎

Prob. 24.3. Prove the following theorem:

Theorem. *Consider the measure space* $(\mathbb{R}^n, \mathfrak{B}_{\mathbb{R}^n}, \mu_L^n)$. *Let ν be a measure on the measurable space* $(\mathbb{R}^n, \mathfrak{B}_{\mathbb{R}^n})$ *such that*
1° *ν is translation invariant on $\mathfrak{B}_{\mathbb{R}^n}$.*
2° *$\nu(O) > 0$ for every non-empty open set O in $\mathbb{R}^n$.*
3° *$\nu(B) < \infty$ for every bounded set $B \in \mathfrak{B}_{\mathbb{R}^n}$.*
Then there exists a constant $c \in (0, \infty)$ such that $\nu = c\,\mu_L^n$ on $\mathfrak{B}_{\mathbb{R}^n}$ and indeed c is given by $c = \nu\big((0, 1]_1 \times \cdots \times (0, 1]_n\big)$.

Proof. 1. Let us observe that if there exists $c \in (0, \infty)$ such that $\nu = c\,\mu_L^n$ on $\mathfrak{B}_{\mathbb{R}^n}$ then in particular for $(0, 1]_1 \times \cdots \times (0, 1]_n \in \mathfrak{B}_{\mathbb{R}^n}$ we have

$$\nu\big((0, 1]_1 \times \cdots \times (0, 1]_n\big) = c\,\mu_L^n\big((0, 1]_1 \times \cdots \times (0, 1]_n\big) = c \cdot 1$$

so that we have

$$c = \nu\big((0, 1]_1 \times \cdots \times (0, 1]_n\big).$$

2. (We prove our theorem by applying Theorem 21.11. We begin by verifying that the conditions in Theorem 21.11 are satisfied here.)

Observe that $\mathfrak{I}_{oc}^n$ as defined in Definition 24.2 is a semialgebra of subsets of $\mathbb{R}^n$ by Proposition 24.3 and moreover $\mathfrak{B}_{\mathbb{R}^n} = \sigma(\mathfrak{I}_{oc}^n)$ by Proposition 24.4.

A cube in $\mathbb{R}^n$ of the type $(0, 1]_1 \times \cdots \times (0, 1]_n$ is a member of $\mathfrak{I}_{oc}^n$ and moreover $\mathbb{R}^n$ is the union of countably many cubes of this type. We also have $\mu_L^n\big((0, 1]_1 \times \cdots \times (0, 1]_n\big) = 1 < \infty$. This shows that the measure μ_L^n is σ-finite on the semialgebra $\mathfrak{I}_{oc}^n$. By 3° we have $\nu\big((0, 1]_1 \times \cdots \times (0, 1]_n\big) < \infty$. Thus the measure ν is σ-finite on the semialgebra $\mathfrak{I}_{oc}^n$.

3. Let us show that

(1) $$\nu = c\,\mu_L^n \quad \text{on } \mathfrak{I}_{oc}^n \text{ where } c = \nu\big((0, 1]_1 \times \cdots \times (0, 1]_n\big).$$

For $k_1, \ldots, k_n \in \mathbb{N}$, consider a box $R = (0, \frac{1}{k_1}] \times \cdots \times (0, \frac{1}{k_n}]$. The box $(0, 1]_1 \times \cdots \times (0, 1]_n$ is the disjoint union of $k_1 \cdots k_n$ boxes each of which is a translate of R. By the translation invariance of ν each one of the $k_1 \cdots k_n$ boxes has the same ν measure as R. Thus we have $\nu\big((0, 1]_1 \times \cdots \times (0, 1]_n\big) = k_1 \cdots k_n \, \nu(R)$ and then

(2) $$\nu(R) = \frac{1}{k_1 \cdots k_n} \nu\big((0, 1]_1 \times \cdots \times (0, 1]_n\big).$$

Let $q_1, \ldots, q_n$ be positive rational numbers and consider a box $Q = (0, q_1] \times \cdots \times (0, q_n]$. Now $q_1 = \frac{m_1}{k_1}, \ldots, q_n = \frac{m_n}{k_n}$ where $k_1, \ldots, k_n, m_1, \ldots, m_n \in \mathbb{N}$. Then Q is the disjoint union of $m_1 \cdots m_n$ boxes each of which is a translate of the box $R = (0, \frac{1}{k_1}] \times \cdots \times (0, \frac{1}{k_n}]$. Thus we have

(3) $$\nu(Q) = \nu\big((0, q_1] \times \cdots \times (0, q_n]\big) = m_1 \cdots m_n \, \nu\big((0, \frac{1}{k_1}] \times \cdots \times (0, \frac{1}{k_n}]\big)$$

$$= m_1 \cdots m_n \, \nu(R) = \frac{m_1 \cdots m_n}{k_1 \cdots k_n} \, \nu\big((0, 1]_1 \times \cdots \times (0, 1]_n\big) \quad \text{by (2)}$$

$$= q_1 \cdots q_n \, \nu\big((0, 1]_1 \times \cdots \times (0, 1]_n\big).$$

Let $\alpha_1, \ldots, \alpha_n$ be positive real numbers. For each $i = 1, \ldots, n$, let $(q_{i,j} : j \in \mathbb{N})$ be a decreasing sequence of positive rational numbers such that $q_{i,j} \downarrow \alpha_i$. Then we have a decreasing sequence of boxes $\big((0, q_{1,j}] \times \cdots \times (0, q_{n,j}] : j \in \mathbb{N}\big)$ and

$$\lim_{j \to \infty} (0, q_{1,j}] \times \cdots \times (0, q_{n,j}] = \bigcap_{j \in \mathbb{N}} (0, q_{1,j}] \times \cdots \times (0, q_{n,j}]$$

$$= (0, \alpha_1] \times \cdots \times (0, \alpha_n].$$

Then by (b) of Theorem 1.26 we have

$$(4) \quad \nu\big((0, \alpha_1] \times \cdots \times (0, \alpha_n]\big) = \lim_{j \to \infty} \nu\big((0, q_{1,j}] \times \cdots \times (0, q_{n,j}]\big)$$

$$= \lim_{n \to \infty} q_{1,j} \cdots q_{n,j} \, \nu\big((0, 1]_1 \times \cdots \times (0, 1]_n\big) \quad \text{by (3)}$$

$$= \alpha_1 \cdots \alpha_n \, \nu\big((0, 1]_1 \times \cdots \times (0, 1]_n\big)$$

$$= \mu_L^n\big((0, \alpha_1] \times \cdots \times (0, \alpha_n]\big) \, \nu\big((0, 1]_1 \times \cdots \times (0, 1]_n\big).$$

Finally let $R \in \mathfrak{I}_{oc}^n$ be given by $R = (a_1, b_1] \times \cdots \times (a_n, b_n]$. Then by the translation invariance of ν on $\mathfrak{B}_{\mathbb{R}^n}$ we have

$$\nu(R) = \nu\big((a_1, b_1] \times \cdots \times (a_n, b_n]\big) = \nu\big((0, b_1 - a_1] \times \cdots \times (0, b_n - a_n]\big)$$

$$= \mu_L^n\big((0, b_1 - a_1] \times \cdots \times (0, b_n - a_n]\big) \, \nu\big((0, 1]_1 \times \cdots \times (0, 1]_n\big)$$

$$= \mu_L^n\big((a_1, b_1] \times \cdots \times (a_n, b_n]\big) \, \nu\big((0, 1]_1 \times \cdots \times (0, 1]_n\big)$$

$$= \mu_L^n(R) \, \nu\big((0, 1]_1 \times \cdots \times (0, 1]_n\big),$$

where the second equality is by the translation invariance of ν on $\mathfrak{B}_{\mathbb{R}^n}$ and the fourth equality is by the translation invariance of μ_L^n on $\mathfrak{B}_{\mathbb{R}^n}$. This proves (1).

4. We have shown that ν and $c \, \mu_L^n$ are two measures on the σ-algebra $\mathfrak{B}_{\mathbb{R}^n} = \sigma(\mathfrak{I}_{oc}^n)$ such that $\nu = c \, \mu_L^n$ on the semialgebra $\mathfrak{I}_{oc}^n$ and moreover ν and $c \, \mu_L^n$ are σ-finite on $\mathfrak{I}_{oc}^n$. Then by Theorem 21.11, we have $\nu = c \, \mu_L^n$ on $\mathfrak{B}_{\mathbb{R}^n}$. ∎

Prob. 24.4. Let ν be a measure on the measurable space $(\mathbb{R}^n, \mathfrak{B}_{\mathbb{R}^n})$. Suppose ν is translation invariant, $\nu(O) > 0$ for every nonempty open set O in $\mathbb{R}^n$ and $\nu(K) < \infty$ for every compact set K in $\mathbb{R}^n$. Show that there exists a constant $c \in (0, \infty)$ such that $\nu = c \, \mu_L^n$ on $\mathfrak{B}_{\mathbb{R}^n}$, and indeed $c = \nu\big((0, 1]_1 \times \cdots \times (0, 1]_n\big)$.

(Comment. Since a compact set K in $\mathbb{R}^n$ is a bounded closed set, the condition in Prob. 24.4 that $\nu(K) < \infty$ for every compact set K in $\mathbb{R}^n$ is less stringent than the condition in Prob. 24.3 that $\nu(B) < \infty$ for every bounded set $B \in \mathfrak{B}_{\mathbb{R}^n}$. Thus Prob. 24.4 is an extension of Prob. 24.3.)

Proof. We prove Prob. 24.4 by modifying the proof of Prob. 24.3. Specifically we verify that after replacing the condition in Prob. 24.3 that $\nu(B) < \infty$ for every bounded set $B \in \mathfrak{B}_{\mathbb{R}^n}$ with the condition in Prob. 24.4 that $\nu(K) < \infty$ for every compact set K in $\mathbb{R}^n$, we still have the σ-finiteness of ν on $\mathfrak{I}_{oc}^n$. This is shown as follows.

Consider a cube in $\mathbb{R}^n$ of the type $C = (0, 1]_1 \times \cdots \times (0, 1]_n$. Such a cube is a member of $\mathfrak{I}_{oc}^n$ and moreover $\mathbb{R}^n$ is the union of countably many such cubes. Consider a cube in $\mathbb{R}^n$ of the type $K = [0, 1]_1 \times \cdots \times [0, 1]_n$. The set K is a bounded closed set in $\mathbb{R}^n$ and hence it is a compact set in $\mathbb{R}^n$. Then we have $\nu(K) < \infty$ by our assumption on ν. Then since $C \subset K$ we have $\nu(C) \leq \nu(K) < \infty$. This implies that ν is σ-finite on $\mathfrak{I}_{oc}^n$. $\blacksquare$

Prob. 24.5. We define a curve C in $\mathbb{R}^n$ as a mapping $C = (\varphi_1, \ldots, \varphi_n)$ of an interval $[a, b] \subset \mathbb{R}$ into $\mathbb{R}^n$ where φ_i is a real-valued function on $[a, b]$ for $i = 1, \ldots, n$. The graph of the curve C is the image of $[a, b]$ by the mapping C, that is, $\Gamma = C([a, b]) = (\varphi_1, \ldots, \varphi_n)([a, b])$.

Assume that φ_i satisfies a Lipschitz condition $|\varphi_i(t') - \varphi_i(t'')| \le \alpha_i |t' - t''|$ for $t', t'' \in [a, b]$ where $\alpha_i > 0$ for $i = 1, \ldots, n$.

(a) Show that $\Gamma \in \mathfrak{B}_{\mathbb{R}^n}$.
(b) Show that when $n = 1$ we have $\mu_L(\Gamma) = \max_{[a,b]} \varphi - \min_{[a,b]} \varphi$.
(c) Show that when $n \ge 2$ we have $\mu_L^n(\Gamma) = 0$.

Proof. 1. We assume that φ_i satisfies a Lipschitz condition on $[a, b]$, that is, we assume that there exists $\alpha_i > 0$ such that

$$(1) \qquad |\varphi_i(t') - \varphi_i(t'')| \le \alpha_i |t' - t''| \quad \text{for } t', t'' \in [a, b].$$

Now condition (1) implies that φ_i is continuous on $[a, b]$. This then implies that $C = (\varphi_1, \ldots, \varphi_n)$ is a continuous mapping of $[a, b]$ into $\mathbb{R}^n$. Now $[a, b]$ is a compact set in $\mathbb{R}$ and the image by a continuous mapping of a compact set is a compact set. Thus $\Gamma = C([a, b])$ is a compact set in $\mathbb{R}^n$ and then $\Gamma \in \mathfrak{B}_{\mathbb{R}^n}$.

2. When $n = 1$, we have a real-valued continuous function φ on $[a, b]$, a finite closed interval in $\mathbb{R}$. This implies that $\Gamma = C([a, b]) = \varphi([a, b])$ is a finite closed interval in $\mathbb{R}$ given by $[\min_{[a,b]} \varphi, \max_{[a,b]} \varphi]$. Then we have $\mu_L(\Gamma) = \max_{[a,b]} \varphi - \min_{[a,b]} \varphi$.

3. Consider the case $n \ge 2$. Let us write $\mathbb{R}^n = \mathbb{R}_1 \times \cdots \times \mathbb{R}_n$ for definiteness of notation. According to Proposition 24.6 we have

$$(2) \qquad (\mu_L^n)^*(\Gamma) = \inf \left\{ \sum_{k \in \mathbb{N}} v(R_k) : (R_k : k \in \mathbb{N}) \subset \mathfrak{I}_c^n, \bigcup_{k \in \mathbb{N}} R_k \supset \Gamma \right\}.$$

To show that $\mu_L^n(\Gamma) = 0$, by Lemma 2.6 it suffices to show that $(\mu_L^n)^*(\Gamma) = 0$. This is done as follows.

For $m \in \mathbb{N}$, let

$$(3) \qquad a = t_0 < t_1 < \cdots < t_{2^m} = b \quad \text{where} \quad t_j - t_{j-1} = \frac{b-a}{2^m} \quad \text{for } j = 1, \cdots, 2^m.$$

For each $i = 1, \ldots, n$, we have

$$(4) \qquad |\varphi_i(t_{j-1}) - \varphi_i(t_j)| \le \alpha_i |t_{j-1} - t_j| = \alpha_i \frac{b-a}{2^m} \quad \text{by (1) and (3).}$$

This implies that the section of the graph of φ_i in $\mathbb{R}_i$ corresponding to the subinterval $[t_{j-1}, t_j]$ of the interval $[a, b]$ is contained in the interval $[\varphi_i(t_j), \varphi_i(t_j) + \alpha_i \frac{b-a}{2^m}] \subset \mathbb{R}_i$.

Then the section of the graph $\Gamma = C([a, b]) = (\varphi_1, \ldots, \varphi_n)([a, b])$ corresponding to the subinterval $[t_{j-1}, t_j]$ is contained in the box

$$R_{m,j} = \left[\varphi_1(t_j), \varphi_1(t_j) + \alpha_1 \frac{b-a}{2^m}\right] \times \cdots \times \left[\varphi_n(t_j), \varphi_n(t_j) + \alpha_n \frac{b-a}{2^m}\right] \quad \text{for } j = 1, \ldots, 2^m.$$

Then we have

$$(5) \qquad \Gamma \subset \bigcup_{j=1}^{2^m} R_{m,j}.$$

Note also that $\left(\bigcup_{j=1}^{2^m} R_{m,j} : m \in \mathbb{N} \right)$ is a decreasing sequence of sets. Then observe that $v(R_{m,j}) = \alpha_1 \cdots \alpha_n \left(\frac{b-a}{2^m} \right)^n$ and thus we have

$$
(6) \qquad v\left(\bigcup_{j=1}^{2^m} R_{m,j} \right) = 2^m \alpha_1 \cdots \alpha_n \left(\frac{b-a}{2^m} \right)^n = \left(\frac{1}{2^m} \right)^{n-1} \alpha_1 \cdots \alpha_n (b-a)^n.
$$

Now $n \geq 2$ implies that $n - 1 \geq 1$. Thus we have

$$
(7) \qquad \lim_{m \to \infty} v\left(\bigcup_{j=1}^{2^m} R_{m,j} \right) = \lim_{m \to \infty} \left(\frac{1}{2^m} \right)^{n-1} \alpha_1 \cdots \alpha_n (b-a)^n = 0.
$$

By (2) we have

$$
(8) \qquad \left(\mu_L^n \right)^* (\Gamma) \leq v\left(\bigcup_{j=1}^{2^m} R_{m,j} \right) \quad \text{for every } m \in \mathbb{N}.
$$

Then we have

$$
\left(\mu_L^n \right)^* (\Gamma) \leq \lim_{m \to \infty} v\left(\bigcup_{j=1}^{2^m} R_{m,j} \right) = 0 \quad \text{by (7)}.
$$

This completes the proof. ∎

§25 Differentiation on the Euclidean Space

Prob. 25.1. Let f be an extended real-valued $\mathfrak{M}_L^n$-measurable function on $\mathbb{R}^n$. Show that $f \in \mathcal{L}_{loc}^1\big(\mathbb{R}^n, \mathfrak{M}_L^n, \mu_L^n\big)$ if and only if for every $x \in \mathbb{R}^n$ there exists $r_x > 0$ such that f is μ_L^n-integrable on $B(x, r_x)$.

Proof. 1. Suppose $f \in \mathcal{L}_{loc}^1\big(\mathbb{R}^n, \mathfrak{M}_L^n, \mu_L^n\big)$. Then f is μ_L^n-integrable on every bounded set $E \in \mathfrak{M}_L^n$. Now for an arbitrary $x \in \mathbb{R}^n$ and $r > 0$ we have $B(x, r) \in \mathfrak{M}_L^n$ and $B(x, r)$ is a bounded set in $\mathbb{R}^n$. Thus f is μ_L^n-integrable on $B(x, r)$.

2. Conversely suppose for every $x \in \mathbb{R}^n$ there exists $r_x > 0$ such that f is μ_L^n-integrable on $B(x, r_x)$. To show that $f \in \mathcal{L}_{loc}^1\big(\mathbb{R}^n, \mathfrak{M}_L^n, \mu_L^n\big)$ we show that f is μ_L^n-integrable on every bounded set $E \in \mathfrak{M}_L^n$. Let E be a bounded set in $\mathbb{R}^n$ and $E \in \mathfrak{M}_L^n$. Then its closure $\overline{E}$ is a bounded closed set in $\mathbb{R}^n$ and hence it is a compact set in $\mathbb{R}^n$. Now for each $x \in \overline{E}$ consider $B(x, r_x)$. Then $\{B(x, r_x) : x \in \overline{E}\}$ is an open cover of the compact set $\overline{E}$. Thus there exists a finite subcover of $\overline{E}$, that is, there exist $x_1, \ldots, x_k \in \overline{E}$ such that

$$\overline{E} \subset \bigcup_{i=1}^{k} B(x_i, r_{x_i}).$$

Then we have

$$\int_{\overline{E}} |f| \, d\mu_L^n \leq \int_{\bigcup_{i=1}^{k} B(x_i, r_{x_i})} |f| \, d\mu_L^n \leq \sum_{i=1}^{k} \int_{B(x_i, r_{x_i})} |f| \, d\mu_L^n < \infty.$$

This shows the μ_L^n-integrability of $|f|$ on $\overline{E}$. Then since $E \subset \overline{E}$, we have the μ_L^n-integrability of $|f|$ on E. This is equivalent to the μ_L^n-integrability of f on E. ∎

Prob. 25.2. (a) Show that if f is a real-valued $\mathfrak{M}_L^n$-measurable function on $\mathbb{R}^n$ and f is bounded on every bounded subset of $\mathbb{R}^n$, then $f \in \mathcal{L}_{loc}^1\big(\mathbb{R}^n, \mathfrak{M}_L^n, \mu_L^n\big)$.
(b) Show that if f is a continuous real-valued function on $\mathbb{R}^n$, then $f \in \mathcal{L}_{loc}^1\big(\mathbb{R}^n, \mathfrak{M}_L^n, \mu_L^n\big)$.

Proof. **1.** Let us prove (a). To show that $f \in \mathcal{L}_{loc}^1\big(\mathbb{R}^n, \mathfrak{M}_L^n, \mu_L^n\big)$, we show that f is μ_L^n-integrable on every bounded set $E \in \mathfrak{M}_L^n$. Let E be an arbitrary bounded set in $\mathfrak{M}_L^n$. Now by assumption f is bounded on E. Thus there exists $M \geq 0$ such that $|f| \leq M$ on E. Also since E is a bounded set in $\mathbb{R}^n$ we have $\mu_L^n(E) < \infty$. Then we have

$$\left| \int_E f \, d\mu_L^n \right| \leq \int_E |f| \, d\mu_L^n \leq M \mu_L^n(E) < \infty.$$

Thus f is μ_L^n-integrable on E. This shows that $f \in \mathcal{L}_{loc}^1\big(\mathbb{R}^n, \mathfrak{M}_L^n, \mu_L^n\big)$.
2. Let us prove (b). If f is a continuous real-valued function on $\mathbb{R}^n$ then f is $\mathfrak{B}_{\mathbb{R}^n}$-measurable and consequently $\mathfrak{M}_L^n$-measurable on $\mathbb{R}^n$.
To show that $f \in \mathcal{L}_{loc}^1\big(\mathbb{R}^n, \mathfrak{M}_L^n, \mu_L^n\big)$, we show that f is μ_L^n-integrable on every bounded set $E \in \mathfrak{M}_L^n$. Let E be an arbitrary bounded set in $\mathfrak{M}_L^n$. Then the closure $\overline{E}$ is a bounded closed set in $\mathbb{R}^n$, that is, $\overline{E}$ is a compact set in $\mathbb{R}^n$. Then the continuity of f on $\mathbb{R}^n$ implies that $f(\overline{E})$ is a compact set and hence a bounded set in $\mathbb{R}$. Thus f is bounded on $\overline{E}$ and certainly bounded on E. Thus there exists $M \geq 0$ such that $|f| \leq M$ on E. Also since E is a bounded set in $\mathbb{R}^n$ we have $\mu_L^n(E) < \infty$. Then we have

$$\left| \int_E f \, d\mu_L^n \right| \leq \int_E |f| \, d\mu_L^n \leq M \mu_L^n(E) < \infty.$$

Thus f is μ_L^n-integrable on E. This shows that $f \in \mathcal{L}_{loc}^1\big(\mathbb{R}^n, \mathfrak{M}_L^n, \mu_L^n\big)$. ∎

Prob. 25.3. Definition. *Let* $\{h_r : r \in (0, \infty)\}$ *be a collection of real-valued functions on* $\mathbb{R}^n$. *Consider* $\{h_r(x) : r \in (0, \infty)\}$ *for* $x \in \mathbb{R}^n$.
(a) *We say that* $\lim_{r\to 0} h_r(x) = \xi$ *if there exists* $\xi \in \mathbb{R}$ *such that for every* $\varepsilon > 0$ *there exists* $\delta > 0$ *such that* $|h_r(x) - \xi| < \varepsilon$ *for* $r \in (0, \delta)$.
(b) *We say that* $\lim_{r\to 0} h_r(x) = \infty$ *if for every* $M > 0$ *there exists* $\delta > 0$ *such that* $h_r(x) \geq M$ *for* $r \in (0, \delta)$.
(c) *We say that* $\lim_{r\to 0} h_r(x) = -\infty$ *if for every* $M > 0$ *there exists* $\delta > 0$ *such that* $h_r(x) \leq -M$ *for* $r \in (0, \delta)$.
(d) *We define*

$$\limsup_{r\to 0} h_r(x) = \lim_{\delta\to 0}\Big\{ \sup_{r\in(0,\delta)} h_r(x)\Big\} \quad and \quad \liminf_{r\to 0} h_r(x) = \lim_{\delta\to 0}\Big\{ \inf_{r\in(0,\delta)} h_r(x)\Big\}.$$

Observation. Observe that $\sup_{r\in(0,\delta)} h_r(x)$ decreases as $\delta \to 0$ so that $\lim_{\delta\to 0}\Big\{ \sup_{r\in(0,\delta)} h_r(x)\Big\}$ always exists in $\overline{\mathbb{R}}$. Similarly $\inf_{r\in(0,\delta)} h_r(x)$ increases as $\delta \to 0$ so that $\lim_{\delta\to 0}\Big\{ \inf_{r\in(0,\delta)} h_r(x)\Big\}$ always exists in $\overline{\mathbb{R}}$. Therefore we have $-\infty \leq \liminf_{r\to 0} h_r(x) \leq \limsup_{r\to 0} h_r(x) \leq \infty$.

Theorem. *Let* $\{h_r : r \in (0, \infty)\}$ *be a collection of real-valued functions on* $\mathbb{R}^n$. *Consider* $\{h_r(x) : r \in (0, \infty)\}$ *for* $x \in \mathbb{R}^n$.
(a) $\lim_{r\to 0} h_r(x) = \xi \in \mathbb{R}$ *if and only if* $\liminf_{r\to 0} h_r(x) = \limsup_{r\to 0} h_r(x) = \xi$.
(b) $\lim_{r\to 0} h_r(x) = \infty$ *if and only if* $\liminf_{r\to 0} h_r(x) = \limsup_{r\to 0} h_r(x) = \infty$.
(c) $\lim_{r\to 0} h_r(x) = -\infty$ *if and only if* $\liminf_{r\to 0} h_r(x) = \limsup_{r\to 0} h_r(x) = -\infty$.
Prove the theorem above.

Proof. 1. Let us prove (a).
1.1. Suppose $\liminf_{r\to 0} h_r(x) = \limsup_{r\to 0} h_r(x) = \xi \in \mathbb{R}$. To show that $\lim_{r\to 0} h_r(x) = \xi$, we show that for every $\varepsilon > 0$ there exists $\delta_0 > 0$ such that

$$(1) \qquad\qquad |h_r(x) - \xi| < \varepsilon \quad \text{for } r \in (0, \delta_0).$$

Now by our assumption we have $\liminf_{r\to 0} h_r(x) = \xi$, that is, $\lim_{\delta\to 0}\Big\{ \inf_{r\in(0,\delta)} h_r(x)\Big\} = \xi$. This implies that for every $\varepsilon > 0$ there exists $\delta_1 > 0$ such that

$$\Big| \inf_{r\in(0,\delta)} h_r(x) - \xi\Big| < \varepsilon \quad \text{for } \delta \in (0, \delta_1).$$

This implies

$$\Big| \inf_{r\in(0,\delta_1)} h_r(x) - \xi\Big| < \varepsilon.$$

Similarly our assumption of $\limsup_{r\to 0} h_r(x) = \xi$ implies that there exists $\delta_2 > 0$ such that

$$\Big| \sup_{r\in(0,\delta_2)} h_r(x) - \xi\Big| < \varepsilon.$$

Let $\delta_0 = \min\{\delta_1, \delta_2\}$. Then we have

(2) $\qquad \left| \inf_{r \in (0, \delta_0)} h_r(x) - \xi \right| < \varepsilon \quad$ and $\quad \left| \sup_{r \in (0, \delta_0)} h_r(x) - \xi \right| < \varepsilon.$

For $r \in (0, \delta_0)$, we have

(3) $\qquad -\varepsilon < \inf_{r \in (0, \delta_0)} h_r(x) - \xi \le h_r(x) - \xi \le \sup_{r \in (0, \delta_0)} h_r(x) - \xi < \varepsilon \quad$ for $r \in (0, \delta_0)$.

This proves (1).

1.2. Conversely suppose $\lim_{r \to 0} h_r(x) = \xi \in \mathbb{R}$. Then for every $\varepsilon > 0$ there exists $\delta > 0$ such that

$$-\varepsilon < h_r(x) - \xi < \varepsilon \quad \text{for } r \in (0, \delta).$$

Then we have

$$-\varepsilon \le \inf_{r \in (0, \delta)} h_r(x) - \xi \le h_r(x) - \xi \le \sup_{r \in (0, \delta)} h_r(x) - \xi \le \varepsilon.$$

Since $\inf_{r \in (0, \delta)} h_r(x)$ increases and $\sup_{r \in (0, \delta)} h_r(x)$ decreases as $\delta \to 0$, we have

$$-\varepsilon \le \lim_{\delta \to 0} \left\{ \inf_{r \in (0, \delta)} h_r(x) \right\} - \xi \le h_r(x) - \xi \le \lim_{\delta \to 0} \left\{ \sup_{r \in (0, \delta)} h_r(x) \right\} - \xi \le \varepsilon.$$

Since this holds for every $\varepsilon > 0$, we have

$$\lim_{\delta \to 0} \left\{ \inf_{r \in (0, \delta)} h_r(x) \right\} - \xi = \lim_{\delta \to 0} \left\{ \sup_{r \in (0, \delta)} h_r(x) \right\} - \xi = 0,$$

that is,

$$\lim_{\delta \to 0} \left\{ \inf_{r \in (0, \delta)} h_r(x) \right\} = \lim_{\delta \to 0} \left\{ \sup_{r \in (0, \delta)} h_r(x) \right\} = \xi.$$

Thus we have

$$\liminf_{r \to 0} h_r(x) = \limsup_{r \to 0} h_r(x) = \xi.$$

2. Let us prove (b).

2.1. Suppose we have $\liminf_{r \to 0} h_r(x) = \limsup_{r \to 0} h_r(x) = \infty$. Then $\lim_{\delta \to 0} \left\{ \inf_{r \in (0, \delta)} h_r(x) \right\} = \liminf_{r \to 0} h_r(x) = \infty$. This implies that for every $M > 0$ there exists $\eta > 0$ such that

$$\inf_{r \in (0, \delta)} h_r(x) \ge M \quad \text{for } \delta \in (0, \eta),$$

and then

$$h_r(x) \ge M \quad \text{for } r \in (0, \eta).$$

This shows that $\lim_{r \to 0} h_r(x) = \infty$.

2.2. Conversely suppose $\lim_{r\to 0} h_r(x) = \infty$. Then for every $M > 0$ there exists $\delta > 0$ such that $h_r(x) \geq M$ for $r \in (0, \delta)$ and then

$$\inf_{r\in(0,\delta)} h_r(x) \geq M.$$

Since $\inf_{r\in(0,\delta)} h_r(x)$ increases as $\delta \to 0$, we have

$$\lim_{\delta\to 0} \left\{ \inf_{r\in(0,\delta)} h_r(x) \right\} \geq M.$$

Since this holds for every $M > 0$, we have $\lim_{\delta\to 0} \left\{ \inf_{r\in(0,\delta)} h_r(x) \right\} = \infty$, that is, we have $\liminf_{r\to 0} h_r(x) = \infty$. Then since $\liminf_{r\to 0} h_r(x) \leq \limsup_{r\to 0} h_r(x)$, we have $\limsup_{r\to 0} h_r(x) = \infty$ also.

3. (c) is proved by the same argument as for (b). ∎

Prob. 25.4. Let f be a real-valued function on $\mathbb{R}$ defined by $f(x) = 0$ if $x \in \mathbb{R}$ is rational and $f(x) = 1$ if $x \in \mathbb{R}$ is irrational.
(a) Show that $f \in \mathcal{L}^1_{loc}(\mathbb{R}, \mathfrak{M}_L, \mu_L)$.
(b) Find Af.
(c) Show that $Af = f$ a.e. on $(\mathbb{R}, \mathfrak{M}_L, \mu_L)$.

Proof. **(a)** Our function f is bounded on $\mathbb{R}$. Thus $f \in \mathcal{L}^1_{loc}(\mathbb{R}, \mathfrak{M}_L, \mu_L)$ by (a) of Prob. 25.4.
   (b) The set of rational numbers in $\mathbb{R}$ is a null set in the measure space $(\mathbb{R}, \mathfrak{M}_L, \mu_L)$. Thus $f = \mathbf{1}_{\mathbb{R}}$ a.e. on $\mathbb{R}$. Then we have for every $r > 0$ and $x \in \mathbb{R}$

$$\begin{aligned}(A_r f)(x) &= \frac{1}{\mu_L(B(x,r))} \int_{B(x,r)} f(y)\, \mu_L)(dy) \\ &= \frac{1}{\mu_L(B(x,r))} \int_{B(x,r)} \mathbf{1}_{\mathbb{R}}(y)\, \mu_L)(dy) \\ &= \frac{1}{2r} \cdot 1 \cdot 2r = 1,\end{aligned}$$

and then

$$(Af)(x) = \lim_{r\to 0} (A_r f)(x) = 1 \quad \text{for every } x \in \mathbb{R}.$$

This shows that $Af = 1$ on $\mathbb{R}$.
   (c) We showed above that $f = \mathbf{1}_{\mathbb{R}}$ a.e. on $(\mathbb{R}, \mathfrak{M}_L, \mu_L)$ and $Af = 1$ on $(\mathbb{R}, \mathfrak{M}_L, \mu_L)$. Thus $Af = f$ a.e. on $(\mathbb{R}, \mathfrak{M}_L, \mu_L)$. ∎

Prob. 25.5. (a) Show that if $f \in \mathcal{L}^1\big(\mathbb{R}^n, \mathfrak{M}_L^n, \mu_L^n\big)$, then for each $r \in (0, \infty)$ the function $A_r f$ is uniformly continuous on $\mathbb{R}^n$.
(b) Show that if $f \in \mathcal{L}_{loc}^1\big(\mathbb{R}^n, \mathfrak{M}_L^n, \mu_L^n\big)$, then for each $r \in (0, \infty)$ the function $A_r f$ is uniformly continuous on every bounded subset of $\mathbb{R}^n$.

Proof. 1. Let us prove (a). Let $f \in \mathcal{L}^1\big(\mathbb{R}^n, \mathfrak{M}_L^n, \mu_L^n\big)$. Let $r \in (0, \infty)$ be fixed. To show that the function $A_r f$ on $\mathbb{R}^n$ is uniformly continuous on $\mathbb{R}^n$, we show that for every $\varepsilon > 0$ there exists $\delta > 0$ such that

(1) $\big|(A_r f)(x') - (A_r f)(x'')\big| \le \varepsilon$ for $x', x'' \in \mathbb{R}^n$ such that $|x' - x''| < \delta$.

Observe that we have

$$(A_r f)(x') - (A_r f)(x'') = \frac{1}{M}\Big\{ \int_{B(x',r)} f(y)\, \mu_L(dy) - \int_{B(x'',r)} f(y)\, \mu_L(dy)\Big\},$$

where $M = \mu_L^n\big(B(x', r)\big) = \mu_L^n\big(B(x'', r)\big) > 0$. Then we have

$$(2) \quad \big|(A_r f)(x') - (A_r f)(x'')\big| = \frac{1}{M}\Big| \int_{B(x',r)} f(y)\, \mu_L(dy) - \int_{B(x'',r)} f(y)\, \mu_L(dy)\Big|$$

$$= \frac{1}{M}\Big| \int_{\mathbb{R}^n} \mathbf{1}_{B(x',r)}(y) f(y)\, \mu_L(dy) - \int_{\mathbb{R}^n} \mathbf{1}_{B(x'',r)}(y) f(y)\, \mu_L(dy)\Big|$$

$$\le \frac{1}{M} \int_{\mathbb{R}^n} \big|\mathbf{1}_{B(x',r)}(y) f(y) - \mathbf{1}_{B(x'',r)}(y)\big| |f(y)|\, \mu_L(dy)$$

$$\le \frac{1}{M} \int_{\mathbb{R}^n} \big|\mathbf{1}_{B(x',r)\setminus B(x'',r)}(y) + \mathbf{1}_{B(x'',r)\setminus B(x',r)}(y)\big| |f(y)|\, \mu_L(dy)$$

$$= \frac{1}{M}\Big\{ \int_{B(x',r)\setminus B(x'',r)} |f(y)|\, \mu_L(dy) + \int_{B(x'',r)\setminus B(x',r)} |f(y)|\, \mu_L(dy)\Big\}.$$

Since $f \in \mathcal{L}^1\big(\mathbb{R}^n, \mathfrak{M}_L^n, \mu_L^n\big)$, according to Theorem 9.26 (Uniform Absolute Continuity of the Integral with Respect to the Measure), for every $\varepsilon > 0$ there exists $\eta > 0$ such that

(3) $\displaystyle\int_E |f(y)|\, \mu_L^n(dy) < \varepsilon$ for any $E \in \mathfrak{M}_L^n$ such that $\mu_L^n(E) < \eta$.

Let us estimate $\mu_L^n\big(B(x', r) \setminus B(x'', r)\big)$ and $\mu_L^n\big(B(x'', r) \setminus B(x', r)\big)$. Now we have

$$\mu_L^n\big(B(x', r) \setminus B(x'', r)\big) = \int_{\mathbb{R}^n} \mathbf{1}_{B(x',r)\setminus B(x'',r)}(y)\, \mu_L^n(dy)$$

$$= \int_{\mathbb{R}^n} \big\{\mathbf{1}_{B(x',r)}(y) - \mathbf{1}_{B(x'',r)}(y)\big\} \vee 0\, \mu_L^n(dy).$$

Then we have

$$(4) \quad \lim_{x' \to x''} \mu_L^n\big(B(x', r) \setminus B(x'', r)\big) = \lim_{x' \to x''} \int_{\mathbb{R}^n} \big\{\mathbf{1}_{B(x',r)}(y) - \mathbf{1}_{B(x'',r)}(y)\big\} \vee 0\, \mu_L^n(dy)$$

$$= \int_{\mathbb{R}^n} \lim_{x' \to x''} \big\{\mathbf{1}_{B(x',r)}(y) - \mathbf{1}_{B(x'',r)}(y)\big\} \vee 0\, \mu_L^n(dy)$$

$$= \int_{\mathbb{R}^n} 0\, \mu_L^n(dy) = 0,$$

where the third equality is implied by the equality that $\lim_{x' \to x''} \mathbf{1}_{B(x',r)} = \mathbf{1}_{B(x'',r)}$, a.e. on $\mathbb{R}^n$ according to Observation 25.4. Now the equality (4) implies that for an arbitrary $\eta > 0$ there exists $\delta' > 0$ such that

$$\mu_L^n\big(B(x',r) \setminus B(x'',r)\big) < \eta \quad \text{if } x', x'' \in \mathbb{R}^n \text{ such that } |x' - x''| < \delta'.$$

Similarly for an arbitrary $\eta > 0$ there exists $\delta'' > 0$ such that

$$\mu_L^n\big(B(x'',r) \setminus B(x',r)\big) < \eta \quad \text{if } x', x'' \in \mathbb{R}^n \text{ such that } |x' - x''| < \delta''.$$

Let $\delta = \min\{\delta', \delta''\} > 0$. Then for $x', x'' \in \mathbb{R}^n$ such that $|x' - x''| < \delta$, we have

$$(5) \qquad \mu_L^n\big(B(x',r) \setminus B(x'',r)\big) < \eta \quad \text{and} \quad \mu_L^n\big(B(x'',r) \setminus B(x',r)\big) < \eta.$$

This implies according to (3) that

$$(6) \qquad \int_{B(x',r) \setminus B(x'',r)} |f(y)| \, \mu_L(dy) < \varepsilon \quad \text{and} \quad \int_{B(x'',r) \setminus B(x',r)} |f(y)| \, \mu_L(dy) < \varepsilon.$$

Substituting (6) in (2), we have

$$(7) \qquad \big|(A_r f)(x') - (A_r f)(x'')\big| \leq \frac{1}{M}\{\varepsilon + \varepsilon\} = \frac{2}{M}\varepsilon.$$

This proves (1).

2. Let us prove (b). Let $f \in \mathcal{L}_{loc}^1\big(\mathbb{R}^n, \mathfrak{M}_L^n, \mu_L^n\big)$. Let $r \in (0, \infty)$ be fixed. Let us show that the function $A_r f$ on $\mathbb{R}^n$ is uniformly continuous on every bounded subset of $\mathbb{R}^n$.

Let E be a bounded subset of $\mathbb{R}^n$. Then there exists $R > 0$ such that $E \subset B(0, R)$. Let g be a function on $\mathbb{R}^n$ defined by

$$g = \mathbf{1}_{B(0,R)}\, f \quad \text{on } \mathbb{R}^n.$$

Then since $f \in \mathcal{L}_{loc}^1\big(\mathbb{R}^n, \mathfrak{M}_L^n, \mu_L^n\big)$, we have

$$\int_{\mathbb{R}^n} |g| \, d\mu_L^n = \int_{\mathbb{R}^n} \mathbf{1}_{B(0,R)} |f| \, d\mu_L^n < \infty.$$

This shows that $g \in \mathcal{L}^1\big(\mathbb{R}^n, \mathfrak{M}_L^n, \mu_L^n\big)$. Then by (a), $A_r g$ is uniformly continuous on $\mathbb{R}^n$ and in particular uniformly continuous on $B(0, R) \supset E$. Now on $B(0, R)$ we have $g = f$. Thus $A_r f$ is uniformly continuous on $B(0, R) \supset E$. ∎

Prob. 25.6. For $f \in \mathcal{L}^1_{loc}\big(\mathbb{R}^n, \mathfrak{M}^n_L, \mu^n_L\big)$, the Hardy-Littlewood maximal function Mf of f is a nonnegative extended real-valued function on $\mathbb{R}^n$ defined by setting for $x \in \mathbb{R}^n$

$$(Mf)(x) = \sup_{r \in (0,\infty)} \frac{1}{\mu^n_L(B(x,r))} \int_{B(x,r)} |f|\, d\mu^n_L.$$

For $x \in \mathbb{R}^n$, let $\mathfrak{C}_x$ be the collection of all open balls in $\mathbb{R}^n$ that contain x. Let us define a nonnegative extended real-valued function M^*f on $\mathbb{R}^n$ by setting for $x \in \mathbb{R}^n$

$$(M^*f)(x) = \sup_{C \in \mathfrak{C}_x} \frac{1}{\mu^n_L(C)} \int_C |f|\, d\mu^n_L.$$

Show that $(Mf)(x) \leq (M^*f)(x) \leq 2^n (Mf)(x)$ for $x \in \mathbb{R}^n$.

Proof. 1. Observe that $\{B(x,r) : r \in (0,\infty)\} \subset \mathfrak{C}_x$ for each fixed $x \in \mathbb{R}^n$. This implies

$$\sup_{r \in (0,\infty)} \frac{1}{\mu^n_L(B(x,r))} \int_{B(x,r)} |f|\, d\mu^n_L \leq \sup_{C \in \mathfrak{C}_x} \frac{1}{\mu^n_L(C)} \int_C |f|\, d\mu^n_L,$$

that is, $(Mf)(x) \leq (M^*f)(x)$ for every $x \in \mathbb{R}^n$.

2. Now $\mathfrak{C}_x$ is the collection of all open balls in $\mathbb{R}^n$ that contain x. For $r \in (0,\infty)$, let $(\mathfrak{C}_x)_r$ be the collection of all open balls in $\mathbb{R}^n$ that contain x and have radius r. Then we have $\mathfrak{C}_x = \bigcup_{r \in (0,\infty)} (\mathfrak{C}_x)_r$. Thus we have

$$(1) \quad (M^*f)(x) = \sup_{C \in \mathfrak{C}_x} \frac{1}{\mu^n_L(C)} \int_C |f|\, d\mu^n_L = \sup_{r \in (0,\infty)} \left\{ \sup_{C \in (\mathfrak{C}_x)_r} \frac{1}{\mu^n_L(C)} \int_C |f|\, d\mu^n_L \right\}.$$

Now if $C \in (\mathfrak{C}_x)_r$, then $C \subset B(x,2r)$. Thus we have

$$(2) \quad \int_C |f|\, d\mu^n_L \leq \int_{B(x,2r)} |f|\, d\mu^n_L \quad \text{for } C \in (\mathfrak{C}_x)_r.$$

Also if $C \in (\mathfrak{C}_x)_r$, then C is a translate of $B(x,r)$. Similarly $B(x,2r)$ is a translate of $B(0,2r)$. Thus by Theorem 24.27 (Translation Invariance of $\big(\mathbb{R}^n, \mathfrak{M}^n_L, \mu^n_L\big)$) and Corollary 24.33 (Positive Homogeneity of $\big(\mathbb{R}^n, \mathfrak{M}^n_L, \mu^n_L\big)$), we have

$$(3) \quad \mu^n_L\big(B(x,2r)\big) = \mu^n_L\big(B(0,2r)\big) = \mu^n_L\big(2B(0,r)\big) = 2^n \mu^n_L\big(B(0,r)\big)$$

$$= 2^n \mu^n_L\big(B(x,r)\big) = 2^n \mu^n_L(C).$$

Then by (2) and (3), we have

$$\frac{1}{\mu^n_L(C)} \int_C |f|\, d\mu^n_L \leq \frac{1}{\mu^n_L(C)} \int_{B(x,2r)} |f|\, d\mu^n_L = \frac{2^n}{\mu^n_L(B(x,2r))} \int_{B(x,2r)} |f|\, d\mu^n_L,$$

and then

$$(4) \quad \sup_{C \in \mathfrak{C}_x} \frac{1}{\mu^n_L(C)} \int_C |f|\, d\mu^n_L \leq \frac{2^n}{\mu^n_L(B(x,2r))} \int_{B(x,2r)} |f|\, d\mu^n_L.$$

Then by (1) and (4), we have

$$
\begin{aligned}
(M^* f)(x) &= \sup_{r \in (0,\infty)} \left\{ \sup_{C \in (\mathfrak{C}_x)_r} \frac{1}{\mu_L^n(C)} \int_C |f| \, d\mu_L^n \right\} \\
&\le 2^n \sup_{r \in (0,\infty)} \frac{1}{\mu_L^n(B(x, 2r))} \int_{B(x,2r)} |f| \, d\mu_L^n \\
&= 2^n \sup_{r \in (0,\infty)} \frac{1}{\mu_L^n(B(x, r))} \int_{B(x,r)} |f| \, d\mu_L^n \\
&= 2^n (Mf)(x).
\end{aligned}
$$

This completes the proof that $(M^* f)(x) \le 2^n (Mf)(x)$ for $x \in \mathbb{R}^n$. ∎

Prob. 25.7. Consider $(\mathbb{R}^n, \mathfrak{M}_L^n, \mu_L^n)$. Let $E \in \mathfrak{M}_L^n$. Show directly from the definition of the symmetric density $\delta_S(E, \cdot)$ of E that $\delta_S(E, x)$ exists for a.e. $x \in \mathbb{R}^n$ and furthermore

$$\delta_S(E, x) = \lim_{r \to 0} \frac{\mu_L^n\left(E \cap B(x,r)\right)}{\mu_L^n\left(B(x,r)\right)} = \begin{cases} 1 & \text{for a.e. } x \in E, \\ 0 & \text{for a.e. } x \in E^c. \end{cases}$$

Proof. It suffices to show that for every $N \in \mathbb{N}$, we have

(1) $$\lim_{r \to 0} \frac{\mu_L^n\left(E \cap B(x,r)\right)}{\mu_L^n\left(B(x,r)\right)} = \begin{cases} 1 & \text{for a.e. } x \in E \cap B(0, N), \\ 0 & \text{for a.e. } x \in E^c \cap B(0, N). \end{cases}$$

Now we have

$$\frac{\mu_L^n\left(E \cap B(x,r)\right)}{\mu_L^n\left(B(x,r)\right)} = \frac{\int_{\mathbb{R}^n} \mathbf{1}_E(y)\, \mathbf{1}_{B(x,r)}(y)\, \mu_L^n(dy)}{\mu_L^n\left(B(x,r)\right)} = \frac{\int_{B(x,r)} \mathbf{1}_E(y)\, \mu_L^n(dy)}{\mu_L^n\left(B(x,r)\right)}.$$

Then for $x \in B(0, N)$ and $r \in (0, 1)$, we have

$$\int_{B(x,r)} \mathbf{1}_E(y)\, \mu_L^n(dy) = \int_{B(x,r)} \mathbf{1}_E(y)\, \mathbf{1}_{B(0,N+1)}(y)\, \mu_L^n(dy).$$

Thus we have

(2) $$\frac{\mu_L^n\left(E \cap B(x,r)\right)}{\mu_L^n\left(B(x,r)\right)} = \frac{\int_{B(x,r)} \mathbf{1}_E(y)\, \mathbf{1}_{B(0,N+1)}(y)\, \mu_L^n(dy)}{\mu_L^n\left(B(x,r)\right)}$$

$$\text{for } x \in B(0, N) \text{ and } r \in (0, 1).$$

Now $\mathbf{1}_E \cdot \mathbf{1}_{B(0,N+1)} \in \mathcal{L}^1(\mathbb{R}^n) \subset \mathcal{L}_{loc}^1(\mathbb{R}^n)$ so that by Theorem 25.12, we have

$$\lim_{r \to 0} \frac{\mu_L^n\left(E \cap B(x,r)\right)}{\mu_L^n\left(B(x,r)\right)} = \lim_{r \to 0} \frac{\int_{B(x,r)} \mathbf{1}_E(y)\, \mathbf{1}_{B(0,N+1)}(y)\, \mu_L^n(dy)}{\mu_L^n\left(B(x,r)\right)}$$

$$= \mathbf{1}_E(x)\, \mathbf{1}_{B(0,N+1)}(x) \quad \text{for a.e. } x \in \mathbb{R}^n.$$

$$= \mathbf{1}_E(x) \quad \text{for a.e. } x \in B(0, N)$$

$$= \begin{cases} 1 & \text{for a.e. } x \in E \cap B(0, N) \\ 0 & \text{for a.e. } x \in E^c \cap B(0, N). \end{cases}$$

This proves (1) and completes the proof. ∎

Prob. 25.8. Consider $(\mathbb{R}^n, \mathfrak{M}_L^n, \mu_L^n)$. Let $E \in \mathfrak{M}_L^n$. Show that the symmetric density $\delta_S(E, x)$ exists for every $x \in \Lambda(1_E)$ and furthermore

$$\delta_S(E, x) = \lim_{r \to 0} \frac{\mu_L^n\big(E \cap B(x,r)\big)}{\mu_L^n\big(B(x,r)\big)} = \begin{cases} 1 & \text{for every } x \in E \cap \Lambda(1_E), \\ 0 & \text{for every } x \in E^c \cap \Lambda(1_E). \end{cases}$$

Proof. (This problem is a particular case of Theorem 25.30. We present an alternate proof.)
1. By Definition 25.13, we have

$$(1) \qquad x \in \Lambda(1_E) \Leftrightarrow \lim_{r \to 0} \frac{1}{\mu_L^n\big(B(x,r)\big)} \int_{B(x,r)} \big|1_E(y) - 1_E(x)\big| \, \mu_L^n(dy) = 0.$$

According to Definition 25.15, we have

$$(2) \qquad \delta_S(E, x) = \lim_{r \to 0} \frac{\mu_L^n\big(E \cap B(x,r)\big)}{\mu_L^n\big(B(x,r)\big)} = \lim_{r \to 0} \frac{\int_{B(x,r)} 1_E(y) \, \mu_L^n(dy)}{\mu_L^n\big(B(x,r)\big)},$$

provided the limit exists.
2. Let $x \in \Lambda(1_E)$. Then by (1) we have

$$\lim_{r \to 0} \frac{1}{\mu_L^n\big(B(x,r)\big)} \int_{B(x,r)} \big|1_E(y) - 1_E(x)\big| \, \mu_L^n(dy) = 0$$

and this implies

$$(3) \qquad \lim_{r \to 0} \frac{1}{\mu_L^n\big(B(x,r)\big)} \int_{B(x,r)} \big\{1_E(y) - 1_E(x)\big\} \, \mu_L^n(dy) = 0.$$

Case (i). For $x \in E \cap \Lambda(1_E)$, we have $1_E(x) = 1$. Then (3) reduces to

$$\lim_{r \to 0} \frac{1}{\mu_L^n\big(B(x,r)\big)} \int_{B(x,r)} \big\{1_E(y) - 1\big\} \, \mu_L^n(dy) = 0,$$

and then

$$\lim_{r \to 0} \frac{1}{\mu_L^n\big(B(x,r)\big)} \left\{ \int_{B(x,r)} 1_E(y) \, \mu_L^n(dy) - \mu_L^n\big(B(x,r)\big) \right\} = 0,$$

and thus

$$\lim_{r \to 0} \frac{1}{\mu_L^n\big(B(x,r)\big)} \int_{B(x,r)} 1_E(y) \, \mu_L^n(dy) = 1.$$

Substituting this in (2), we obtain $\delta_S(E, x) = 1$.
Case (ii). For $x \in E^c \cap \Lambda(1_E)$, we have $1_E(x) = 0$. Then (3) reduces to

$$\lim_{r \to 0} \frac{1}{\mu_L^n\big(B(x,r)\big)} \int_{B(x,r)} 1_E(y) \, \mu_L^n(dy) = 0.$$

Substituting this in (2), we obtain $\delta_S(E, x) = 0$. This completes the proof. ∎

Prob. 25.9. Let $E = \left\{(\xi_1, \xi_2) \in \mathbb{R}^2 : \xi_1 \geq 0, \xi_2 \geq 0\right\} \subset \mathbb{R}^2$, where (ξ_1, ξ_2) is an orthogonal coordinate system in $\mathbb{R}^2$. Show that $\Lambda(\mathbf{1}_E) = (\partial E)^c$.

Proof. By (b) of Proposition 25.29, we have $\partial E \supset \Lambda^c(\mathbf{1}_E)$. It remains to show

$$\partial E \subset \Lambda^c(\mathbf{1}_E).$$

By Definition 25.13, we have

$$x \in \Lambda(\mathbf{1}_E) \Leftrightarrow \lim_{r \to 0} \frac{1}{\mu_L^n(B(x,r))} \int_{B(x,r)} \left| \mathbf{1}_E(y) - \mathbf{1}_E(x) \right| \mu_L^n(dy) = 0.$$

Now ∂E is the union of the positive ξ_1-axis and the positive ξ_2-axis. Let $x \in \partial E$. Since $\partial E \subset E$, we have $\mathbf{1}_E(x) = 1$. There are three separate possible cases: (i) $x = (a, 0)$ where $a > 0$; (ii) $x = (0, b)$ where $b > 0$; and (iii) $x = (0, 0)$.

  Case (i). We have $x = (a, 0)$ where $a > 0$. Let $r > 0$ be so small that $a - r > 0$. Then

$$\int_{B(x,r)} \left| \mathbf{1}_E(y) - \mathbf{1}_E(x) \right| \mu_L^n(dy) = \int_{B(x,r)} \left\{ 1 - \mathbf{1}_E(y) \right\} \mu_L^n(dy)$$

$$= \mu_L^n(B(x,r)) - \frac{1}{2}\mu_L^n(B(x,r)) = \frac{1}{2}\mu_L^n(B(x,r))$$

and thus

$$\lim_{r \to 0} \frac{1}{\mu_L^n(B(x,r))} \int_{B(x,r)} \left| \mathbf{1}_E(y) - \mathbf{1}_E(x) \right| \mu_L^n(dy) = \lim_{r \to 0} \frac{1}{2} = \frac{1}{2} \neq 0.$$

This shows that $x \notin \Lambda(\mathbf{1}_E)$ and hence $x \in \Lambda^c(\mathbf{1}_E)$.

  Case (ii). We have $x = (0, b)$ where $b > 0$. Let $r > 0$ be so small that $b - r > 0$. By the same argument as in Case (i) we have

$$\int_{B(x,r)} \left| \mathbf{1}_E(y) - \mathbf{1}_E(x) \right| \mu_L^n(dy) = \frac{1}{2}\mu_L^n(B(x,r)).$$

Then

$$\lim_{r \to 0} \frac{1}{\mu_L^n(B(x,r))} \int_{B(x,r)} \left| \mathbf{1}_E(y) - \mathbf{1}_E(x) \right| \mu_L^n(dy) = \lim_{r \to 0} \frac{1}{2} = \frac{1}{2} \neq 0.$$

This shows that $x \notin \Lambda(\mathbf{1}_E)$ and hence $x \in \Lambda^c(\mathbf{1}_E)$.

  Case (iii). We have $x = (0, 0)$. In this case we have

$$\int_{B(x,r)} \left| \mathbf{1}_E(y) - \mathbf{1}_E(x) \right| \mu_L^n(dy) = \int_{B(x,r)} \left\{ 1 - \mathbf{1}_E(y) \right\} \mu_L^n(dy)$$

$$= \mu_L^n(B(x,r)) - \frac{1}{4}\mu_L^n(B(x,r)) = \frac{3}{4}\mu_L^n(B(x,r)).$$

Then we have

$$\lim_{r \to 0} \frac{1}{\mu_L^n(B(x,r))} \int_{B(x,r)} \left| \mathbf{1}_E(y) - \mathbf{1}_E(x) \right| \mu_L^n(dy) = \lim_{r \to 0} \frac{3}{4} = \frac{3}{4} \neq 0.$$

This shows that $x \notin \Lambda(\mathbf{1}_E)$ and hence $x \in \Lambda^c(\mathbf{1}_E)$. This completes the proof that $\partial E \subset \Lambda^c(\mathbf{1}_E)$. ∎

Prob. 25.10.

Prolog. Consider $(\mathbb{R}^n, \mathfrak{M}^n_L, \mu^n_L)$. For $x \in \mathbb{R}^n$ we write $\mathcal{E} = \{E_r(x) : r \in (0, \infty)\}$ for a collection in $\mathfrak{B}_{\mathbb{R}^n}$ that shrinks nicely to x. Note that $\mathcal{S} = \{B(x, r) : r \in (0, \infty)\}$ is a particular case of $\mathcal{E}$. Let $E \in \mathfrak{M}^n_L$. By Definition 25.25 we have

$$\delta_{\mathcal{E}}(E, x) = \lim_{r \to 0} \frac{\mu^n_L\big(E \cap E_r(x)\big)}{\mu^n_L\big(E_r(x)\big)} \text{ and in particular } \delta_{\mathcal{S}}(E, x) = \lim_{r \to 0} \frac{\mu^n_L\big(E \cap B(x,r)\big)}{\mu^n_L\big(B(x,r)\big)},$$

provided that the limit exists.

By Theorem 25.28, if $\delta_{\mathcal{E}}(E, x)$ exists for every $\mathcal{E}$ then $\delta_{\mathcal{E}}(E, x)$ is independent of $\mathcal{E}$.

By Theorem 25.30, if $x \in \Lambda(1_E)$ then $\delta_{\mathcal{E}}(E, x)$ exists for every $\mathcal{E}$ and hence $\delta_{\mathcal{E}}(E, x)$ is independent of $\mathcal{E}$. In particular, if $x \in \Lambda(1_E)$ then $\delta_{\mathcal{S}}(E, x)$ exists.

Problem. Consider $(\mathbb{R}^2, \mathfrak{M}^2_L, \mu^2_L)$. Let $E = \big\{(\xi_1, \xi_2) \in \mathbb{R}^2 : \xi_1 \geq 0, \xi_2 \geq 0\big\} \subset \mathbb{R}^2$, where (ξ_1, ξ_2) is an orthogonal coordinate system in $\mathbb{R}^2$. By Prob. 25.9, ∂E is the union of the positive ξ_1-axis and the positive ξ_2-axis in the plane $\mathbb{R}^2$ and moreover $\Lambda(1_E) = (\partial E)^c$.
(a) Let $x_1 = (1, 0) \notin \Lambda(1_E)$.
(a.1) Show that $\delta_{\mathcal{S}}(E, x_1) = \frac{1}{2}$. (Thus $\mathcal{S}$ is an example of $\mathcal{E}$ such that $\delta_{\mathcal{E}}(E, x_1) = \frac{1}{2}$.)
(a.2) Construct a collection $\mathcal{E}$ such that $\delta_{\mathcal{E}}(E, x_1) = 1$.
(a.3) Construct a collection $\mathcal{E}$ such that $\delta_{\mathcal{E}}(E, x_1) = 0$.
(a.4) Construct a collection $\mathcal{E}$ such that $\delta_{\mathcal{E}}(E, x_1)$ does not exist.
(b) Let $x_0 = (0, 0) \notin \Lambda(1_E)$. Show that $\delta_{\mathcal{S}}(E, x_0) = \frac{1}{4}$.

Proof. For **(a)**, let $r > 0$ be so small that $r < 1$.
(a.1) Now $E \cap B(x_1, r)$ is the upper half of $B(x_1, r)$. Thus we have

$$\delta_{\mathcal{S}}(E, x_1) = \lim_{r \to 0} \frac{\mu^n_L\big(E \cap B(x_1, r)\big)}{\mu^n_L\big(B(x_1, r)\big)} = \lim_{r \to 0} \frac{1}{2} = \frac{1}{2}.$$

(a.2) Let $\mathcal{E} = \{E_r(x_1) : r \in (0, \infty)\}$ where $E_r(x_1)$ is the open upper half of $B(x_1, r)$. Then $E \cap E_r(x_1) = E_r(x_1)$ and thus

$$\delta_{\mathcal{E}}(E, x_1) = \lim_{r \to 0} \frac{\mu^n_L\big(E \cap E_r(x_1)\big)}{\mu^n_L\big(E_r(x_1)\big)} = \lim_{r \to 0} \frac{\mu^n_L\big(E_r(x_1)\big)}{\mu^n_L\big(E_r(x_1)\big)} = \lim_{r \to 0} 1 = 1.$$

(a.3) Let $\mathcal{E} = \{E_r(x_1) : r \in (0, \infty)\}$ where $E_r(x_1)$ is the open lower half of $B(x_1, r)$. Then $E \cap E_r(x_1) = \emptyset$ and thus

$$\delta_{\mathcal{E}}(E, x_1) = \lim_{r \to 0} \frac{\mu^n_L\big(E \cap E_r(x_1)\big)}{\mu^n_L\big(E_r(x_1)\big)} = \lim_{r \to 0} \frac{\mu^n_L\big(\emptyset\big)}{\mu^n_L\big(E_r(x_1)\big)} = \lim_{r \to 0} 0 = 0.$$

(a.4) Let $\mathcal{E} = \{E_r(x_1) : r \in (0, \infty)\}$ where $E_r(x_1)$ is the open upper half of $B(x_1, r)$ when r is rational and $E_r(x_1)$ is the open lower half of $B(x_1, r)$ when r is irrational. Then $E \cap E_r(x_1) = E_r(x_1)$ when r is rational and $E \cap E_r(x_1) = \emptyset$ when r is irrational. Thus we have

$$\frac{\mu^n_L\big(E \cap E_r(x_1)\big)}{\mu^n_L\big(E_r(x_1)\big)} = \begin{cases} 1 & \text{when } r \text{ is rational} \\ 0 & \text{when } r \text{ is irrational.} \end{cases}$$

Then $\delta_{\mathcal{E}}(E, x_1) = \lim\limits_{r \to 0} \dfrac{\mu_L^n\left(E \cap E_r(x_1)\right)}{\mu_L^n\left(E_r(x_1)\right)}$ does not exist.

(b) If $x_0 = (0, 0)$ then $E \cap B(x_0, r)$ is the first quadrant of $B(x_0, r)$. Then we have

$$\delta_{\mathcal{S}}(E, x_0) = \lim\limits_{r \to 0} \frac{\mu_L^n\left(E \cap B(x_0, r)\right)}{\mu_L^n\left(B(x_0, r)\right)} = \lim\limits_{r \to 0} \frac{1}{4} = \frac{1}{4}.$$

This completes the proof. ∎

Prob. 25.11. Let $\alpha \in (0, 1)$ and let $\vartheta \in (0, 2\pi)$ be such that $\frac{\vartheta}{2\pi} = \alpha$. Let E be a sector of the plane $\mathbb{R}^2$ sustaining an angle ϑ at $0 \in \mathbb{R}^2$. Find $\Lambda(1_E)$. Determine $\delta_{\mathcal{S}}(E, \cdot)$ and show that $\delta_{\mathcal{S}}(E, x)$ exists for every $x \in \mathbb{R}^2$.

Proof. 1. Let $\mathbb{R}^2$ be represented by $\mathbb{R}^2 = \{(\xi_1, \xi_2) : \xi_1 \in \mathbb{R}, \xi_2 \in \mathbb{R}\}$ where (ξ_1, ξ_2) is an orthogonal coordinate system in $\mathbb{R}^2$. Then $\partial E = L_1 \cup L_2$ where L_1 is the positive ξ_1-axis and L_2 is the half-line emanating from $(0, 0)$ and making an angle ϑ with L_1. By the same argument as in the proof of Prob. 25.9 we show that $\Lambda(1_E) = (\partial E)^c$.

    2. By Prob. 25.8, we have

$$(1) \qquad \delta_{\mathcal{S}}(E, x) = \lim_{r \to 0} \frac{\mu_L^n\big(E \cap B(x, r)\big)}{\mu_L^n\big(B(x, r)\big)} = \begin{cases} 1 & \text{for every } x \in E \cap \Lambda(1_E), \\ 0 & \text{for every } x \in E^c \cap \Lambda(1_E). \end{cases}$$

We still have to determine $\delta_{\mathcal{S}}(E, x)$ for $x \notin \Lambda(1_E)$, that is, $x \in \Lambda^c(1_E) = \partial E$.

    Let $x \in \partial E = L_1 \cup L_2$. There are three separate cases to consider.

    Case (i). Let $x_0 = (0, 0) \in \partial E$. Then we have

$$(2) \qquad \delta_{\mathcal{S}}(E, x_0) = \lim_{r \to 0} \frac{\mu_L^n\big(E \cap B(x_0, r)\big)}{\mu_L^n\big(B(x_0, r)\big)} = \lim_{r \to 0} \frac{\alpha \mu_L^n\big(B(x_0, r)\big)}{\mu_L^n\big(B(x_0, r)\big)} = \lim_{r \to 0} \alpha = \alpha.$$

    Case (ii). Let $x_1 = (a, 0) \in L_1 \subset \partial E$ where $a > 0$. Let $r > 0$ be so small that $E \cap B(x_1, r)$ is a half of $B(x_1, r)$. Then we have

$$(3) \qquad \delta_{\mathcal{S}}(E, x_1) = \lim_{r \to 0} \frac{\mu_L^n\big(E \cap B(x_1, r)\big)}{\mu_L^n\big(B(x_1, r)\big)} = \lim_{r \to 0} \frac{\frac{1}{2}\mu_L^n\big(B(x_1, r)\big)}{\mu_L^n\big(B(x_1, r)\big)} = \lim_{r \to 0} \frac{1}{2} = \frac{1}{2}.$$

    Case (iii). Let $x_2 = (a, a \sin \vartheta) \in L_2 \subset \partial E$ where $a > 0$. Let $r > 0$ be so small that $E \cap B(x_2, r)$ is a half of $B(x_2, r)$. Then we have

$$(4) \qquad \delta_{\mathcal{S}}(E, x_2) = \lim_{r \to 0} \frac{\mu_L^n\big(E \cap B(x_2, r)\big)}{\mu_L^n\big(B(x_2, r)\big)} = \lim_{r \to 0} \frac{\frac{1}{2}\mu_L^n\big(B(x_2, r)\big)}{\mu_L^n\big(B(x_2, r)\big)} = \lim_{r \to 0} \frac{1}{2} = \frac{1}{2}.$$

    This shows that $\delta_{\mathcal{S}}(E, x)$ exists for every $x \in \mathbb{R}^2$ and moreover $\delta_{\mathcal{S}}(E, x)$ is given by (1) for $x \in \Lambda(1_E)$ and $\delta_{\mathcal{S}}(E, x)$ is given by (2), (3) and (4) for $x \in \partial E = \Lambda^c(1_E)$. ∎

Prob. 25.12. Consider $(\mathbb{R}^n, \mathfrak{M}_L^n, \mu_L^n)$. Let $E = B(0, a) = \{x \in \mathbb{R}^n : |x| < a\}$ where $a > 0$. Find $\Lambda(1_E)$. Determine $\delta_S(E, \cdot)$ and show that $\delta_S(E, x)$ exists for every $x \in \mathbb{R}^n$.

Proof. 1. Our set E is an open ball with center at the origin and radius $a > 0$. Then $\partial E = \{x \in \mathbb{R}^n : |x| = a\}$, a spherical surface with center at the origin and radius $a > 0$. Now we have

(1) $(\partial E)^c = \{x \in \mathbb{R}^n : |x| < a\} \cup \{x \in \mathbb{R}^n : |x| > a\} := O_1 \cup O_2.$

By Definition 25.13, we have

(2) $x \in \Lambda(1_E) \Leftrightarrow \lim_{r \to 0} \dfrac{1}{\mu_L^n(B(x,r))} \int_{B(x,r)} \left|1_E(y) - 1_E(x)\right| \mu_L^n(dy) = 0.$

Let us show that

(3) $\Lambda(1_E) = (\partial E)^c, \quad \text{that is,} \quad \partial E = \Lambda^c(1_E).$

  2. By (b) of Proposition 25.29, we have $\partial E \supset \Lambda^c(1_E)$. It remains to show $\partial E \subset \Lambda^c(1_E)$. Let $x \in \partial E$. Then $|x| = a$ and thus we have

(4) $\displaystyle\int_{B(x,r)} \left|1_E(y) - 1_E(x)\right| \mu_L^n(dy) = \int_{B(x,r)} \left\{1 - 1_E(y)\right\} \mu_L^n(dy)$

$$= \mu_L^n\big(B(x,r) \cap B(0,a)\big).$$

Observe that

(5) $\displaystyle\lim_{r \to 0} \frac{\mu_L^n\big(B(x,r) \cap B(0,a)\big)}{\mu_L^n\big(B(x,r)\big)} = \frac{1}{2}.$

Then

(6) $\displaystyle\lim_{r \to 0} \frac{1}{\mu_L^n(B(x,r))} \int_{B(x,r)} \left|1_E(y) - 1_E(x)\right| \mu_L^n(dy) = \lim_{r \to 0} \frac{\mu_L^n\big(B(x,r) \cap B(0,a)\big)}{\mu_L^n\big(B(x,r)\big)}$

$$= \frac{1}{2} \neq 0,$$

so that $x \in \Lambda^c(1_E)$. Thus $\partial E \subset \Lambda^c(1_E)$. This completes the proof of (3).

  3. According to Prob. 25.8, the symmetric density $\delta_S(E, x)$ exists for every $x \in \Lambda(1_E)$ and moreover we have

(7) $\delta_S(E, x) = \begin{cases} 1 & \text{for every } x \in E \cap \Lambda(1_E), \\ 0 & \text{for every } x \in E^c \cap \Lambda(1_E). \end{cases}$

  4. To show that the symmetric density $\delta_S(E, x)$ exists for every $x \in \mathbb{R}^n$, we show that $\delta_S(E, x)$ exists for every $x \in \Lambda^c(1_E) = \partial E$. Let $x \in \partial E$. Then we have $|x| = a$ and

$$\delta_S(E, x) = \lim_{r \to 0} \frac{\mu_L^n\big(E \cap B(x,r)\big)}{\mu_L^n\big(B(x,r)\big)} = \lim_{r \to 0} \frac{\mu_L^n\big(B(0,a) \cap B(x,r)\big)}{\mu_L^n\big(B(x,r)\big)} = \frac{1}{2} \quad \text{by (5)}.$$

This completes the proof. ∎

Prob. 25.13. Let E be the set of all rational numbers in $\mathbb{R}$.
(a) Show that $\Lambda(1_E) = E^c$.
(b) Show that $\delta_S(E, x)$ exists for every $x \in \mathbb{R}$ and $\delta_S(E, x) = 0$.

Proof. (a) Let us show that $\Lambda(1_E) = E^c$. By Definition 25.1, we have

$$x \in \Lambda(1_E) \Leftrightarrow \lim_{r \to 0} \frac{1}{\mu_L(B(x,r))} \int_{B(x,r)} \left| 1_E(y) - 1_E(x) \right| \mu_L(dy) = 0.$$

(a.1) Let $x \in E$. Then $1_E(x) = 1$. Now if $y \in E$ then $1_E(y) = 1$ and if $y \in E^c$ then $1_E(y) = 0$. Then since $\mu_L(E) = 0$, $1_E(y) = 0$ for a.e. $y \in \mathbb{R}$. Thus for $x \in E$ we have

$$\lim_{r \to 0} \frac{1}{\mu_L(B(x,r))} \int_{B(x,r)} \left| 1_E(y) - 1_E(x) \right| \mu_L(dy)$$

$$= \lim_{r \to 0} \frac{1}{\mu_L(B(x,r))} \int_{B(x,r)} \left| 0 - 1 \right| \mu_L(dy)$$

$$= \lim_{r \to 0} \frac{\mu_L(B(x,r))}{\mu_L(B(x,r))} = 1.$$

This shows that if $x \in E$ then $x \notin \Lambda(1_E)$.
(a.2) Let $x \in E^c$. Then $1_E(x) = 0$. Now if $y \in E$ then $1_E(y) = 1$ and if $y \in E^c$ then $1_E(y) = 0$. Then since $\mu_L(E) = 0$, $1_E(y) = 0$ for a.e. $y \in \mathbb{R}$. Thus for $x \in E$ we have

$$\lim_{r \to 0} \frac{1}{\mu_L(B(x,r))} \int_{B(x,r)} \left| 1_E(y) - 1_E(x) \right| \mu_L(dy)$$

$$= \lim_{r \to 0} \frac{1}{\mu_L(B(x,r))} \int_{B(x,r)} \left| 0 - 0 \right| \mu_L(dy)$$

$$= \lim_{r \to 0} \frac{0}{\mu_L(B(x,r))} = 0.$$

This shows that if $x \in E^c$ then $x \in \Lambda(1_E)$.
(a.3) We showed above that if $x \in E$ then $x \notin \Lambda(1_E)$ and if $x \in E^c$ then $x \in \Lambda(1_E)$. Then since $\mathbb{R} = E \cup E^c$ and $E \cap E^c = \emptyset$ we have $\Lambda(1_E) = E^c$.
(b) By Definition 25.25, we have for every $x \in \mathbb{R}$, recalling that E is a null set in $(\mathbb{R}, \mathfrak{M}_L, \mu_L)$,

$$\delta_S(E, x) = \lim_{r \to 0} \frac{\mu_L(E \cap B(x,r))}{\mu_L(B(x,r))} = \lim_{r \to 0} \frac{0}{\mu_L(B(x,r))} = 0.$$

This shows that $\delta_S(E, x)$ exists for every $x \in \mathbb{R}$ and $\delta_S(E, x) = 0$. ∎

Prob. 25.14. Let E be the set of all irrational numbers in $\mathbb{R}$.
(a) Show that $\Lambda(\mathbf{1}_E) = E$.
(b) Show that $\delta_S(E, x)$ exists for every $x \in \mathbb{R}$ and $\delta_S(E, x) = 1$.

Proof. (a) Let us show that $\Lambda(\mathbf{1}_E) = E$. By Definition 25.1, we have

$$x \in \Lambda(\mathbf{1}_E) \Leftrightarrow \lim_{r \to 0} \frac{1}{\mu_L(B(x, r))} \int_{B(x,r)} |\mathbf{1}_E(y) - \mathbf{1}_E(x)| \, \mu_L(dy) = 0.$$

(a.1) Let $x \in E$. Then $\mathbf{1}_E(x) = 1$. Now if $y \in E$ then $\mathbf{1}_E(y) = 1$ and if $y \in E^c$ then $\mathbf{1}_E(y) = 0$. Then since $\mu_L(E^c) = 0$, $\mathbf{1}_E(y) = 1$ for a.e. $y \in \mathbb{R}$. Thus for $x \in E$ we have

$$\lim_{r \to 0} \frac{1}{\mu_L(B(x, r))} \int_{B(x,r)} |\mathbf{1}_E(y) - \mathbf{1}_E(x)| \, \mu_L(dy)$$
$$= \lim_{r \to 0} \frac{1}{\mu_L(B(x, r))} \int_{B(x,r)} |1 - 1| \, \mu_L(dy)$$
$$= \lim_{r \to 0} \frac{0}{\mu_L(B(x, r))} = 0.$$

This shows that if $x \in E$ then $x \in \Lambda(\mathbf{1}_E)$.
(a.2) Let $x \in E^c$. Then $\mathbf{1}_E(x) = 0$. Now if $y \in E$ then $\mathbf{1}_E(y) = 1$ and if $y \in E^c$ then $\mathbf{1}_E(y) = 0$. Then since $\mu_L(E) = 0$, $\mathbf{1}_E(y) = 1$ for a.e. $y \in \mathbb{R}$. Thus for $x \in E$ we have

$$\lim_{r \to 0} \frac{1}{\mu_L(B(x, r))} \int_{B(x,r)} |\mathbf{1}_E(y) - \mathbf{1}_E(x)| \, \mu_L(dy)$$
$$= \lim_{r \to 0} \frac{1}{\mu_L(B(x, r))} \int_{B(x,r)} |1 - 0| \, \mu_L(dy)$$
$$= \lim_{r \to 0} \frac{\mu_L(B(x, r))}{\mu_L(B(x, r))} = 1.$$

This shows that if $x \in E^c$ then $x \notin \Lambda(\mathbf{1}_E)$.
(a.3) We showed above that if $x \in E$ then $x \in \Lambda(\mathbf{1}_E)$ and if $x \in E^c$ then $x \notin \Lambda(\mathbf{1}_E)$. Then since $\mathbb{R} = E \cup E^c$ and $E \cap E^c = \emptyset$ we have $\Lambda(\mathbf{1}_E) = E$.
(b) By Definition 25.25, we have for every $x \in \mathbb{R}$,

$$\delta_S(E, x) = \lim_{r \to 0} \frac{\mu_L(E \cap B(x, r))}{\mu_L(B(x, r))},$$

provided the limit exists. Now $E \cap B(x, r) = \cap B(x, r) \setminus E^c$ and E^c, being the set of all rational numbers, is a null set in $(\mathbb{R}, \mathfrak{M}_L, \mu_L)$. Thus $\mu_L(E \cap B(x, r)) = \mu_L(B(x, r))$ and then

$$\delta_S(E, x) = \lim_{r \to 0} \frac{\mu_L(E \cap B(x, r))}{\mu_L(B(x, r))} == \lim_{r \to 0} \frac{\mu_L(B(x, r))}{\mu_L(B(x, r))} = 1.$$

This shows that $\delta_S(E, x)$ exists for every $x \in \mathbb{R}$ and $\delta_S(E, x) = 1$. ∎

Prob. 25.15. According to Proposition 25.29, for every $E \in \mathfrak{M}_L^n$ we have $\partial E \supset \Lambda^c(\mathbf{1}_E)$. Construct an example such that $\partial E \setminus \Lambda^c(\mathbf{1}_E) \neq \emptyset$.

Proof. Let Q be the set of all rational numbers and P be the set of all irrational numbers in $\mathbb{R}$. Then $Q, P \in \mathfrak{M}_L$ and moreover $Q \cap P = \emptyset$, $Q \cup P = \mathbb{R}$ so that $Q^c = P$ and $P^c = Q$. We also have $\partial Q = \mathbb{R}$ and $\partial P = \mathbb{R}$.

Example 1. Consider Q. By Prob. 25.13 we have $\Lambda(\mathbf{1}_Q) = Q^c = P$ and then we have $\Lambda^c(\mathbf{1}_Q) = P^c = Q$. Thus $\partial Q \setminus \Lambda^c(\mathbf{1}_Q) = \mathbb{R} \setminus Q = P \neq \emptyset$.

Example 2. Consider P. By Prob. 25.14 we have $\Lambda(\mathbf{1}_P) = P$ and then we have $\Lambda^c(\mathbf{1}_P) = P^c = Q$. Thus $\partial P \setminus \Lambda^c(\mathbf{1}_P) = \mathbb{R} \setminus Q = P \neq \emptyset$. ∎

Printed in the United States
By Bookmasters